Spintronics Handbook: Spin Transport and Magnetism, Second Edition

Nanoscale Spintronics and Applications—Volume Three

Spintronics Handbook: Spin Transport and Magnetism, Second Edition

Nanoscale Spintronics and Applications—Volume Three

Edited by
Evgeny Y. Tsymbal and Igor Žutić

CRC Press
Taylor & Francis Group
Boca Raton London New York

CRC Press is an imprint of the
Taylor & Francis Group, an **informa** business

CRC Press
Taylor & Francis Group
6000 Broken Sound Parkway NW, Suite 300
Boca Raton, FL 33487-2742

First issued in paperback 2021

ISBN 13: 978-0-367-77956-6 (pbk)
ISBN 13: 978-1-4987-6970-9 (hbk)

Library of Congress Cataloging-in-Publication Data

Names: Tsymbal, E. Y. (Evgeny Y.), editor. | Zutic, Igor, editor.
Title: Nanoscale spintronics and applications / edited by Evgeny Y. Tsymbal, Igor Zutic.
Description: Second edition. | Boca Raton : Taylor & Francis, CRC Press, 2018. | Series: Spintronics handbook ; volume 3 | Includes bibliographical references and index.
Identifiers: LCCN 2018022106 (print) | LCCN 2018048580 (ebook) | ISBN 9781498769716 (eBook General) | ISBN 9780429805264 (eBook Adobe Reader) | ISBN 9780429805257 (eBook ePub) | ISBN 9780429805240 (eBook Mobipocket) | ISBN 9781498769709 (hardback : alk. paper)
Subjects: LCSH: Spintronics.
Classification: LCC TK7874.887 (ebook) | LCC TK7874.887 .N38 2018 (print) | DDC 621.3--dc23
LC record available at https://lccn.loc.gov/2018022106

**Visit the Taylor & Francis Web site at
http://www.taylorandfrancis.com**

**and the CRC Press Web site at
http://www.crcpress.com**

Cover illustration courtesy of Markus Lindemann, Nils C. Gerhardt, and Carsten Brenner.

The e-book of this title contains full colour figures and can be purchased here: http://www.crcpress.com/9780429441189. The figures can also be found under the 'Additional Resources' tab.

Contents

SECTION VI—Spin Transport and Magnetism at the Nanoscale

SECTION VII—Applications

Foreword

S pintronics is a field of research in which novel properties of materials, especially atomically engineered magnetic multilayers, are the result of the manipulation of currents of spin-polarized electrons. Spintronics, in its most recent incarnation, is a field of research that is almost 30 years old. To date, its most significant technological impact has been in the development of a new generation of ultra-sensitive magnetic recording read heads that have powered magnetic disk drives since late 1997. These magnetoresistive read heads, which use spin-valves based on spin-dependent scattering at magnetic/non-magnetic interfaces and, since 2007, magnetic tunnel junctions (MTJs) based on spin-dependent tunneling across ultra-thin insulating layers, have a common thin film structure. These structures involve "spin engineering" to eliminate the influence of long-range magneto-dipole fields via the use of synthetic or artificial antiferromagnets, which are formed from thin magnetic layers coupled antiferromagnetically via the use of atomically thin layers of ruthenium. These structures involve the discoveries of spin-dependent tunneling in 1975, giant magnetoresistance at low temperatures in Fe/Cr in 1988, oscillatory interlayer coupling in 1989, the synthetic antiferromagnet in 1990, giant magnetoresistance at room temperature in Co/Cu and related multilayers in 1991, and the origin of giant magnetoresistance as being a result of predominant interface scattering in 1991–1993. Together, these discoveries led to the spin-valve recording read head that was introduced by IBM in 1997 and led, within a few years, to a 1,000-fold increase in the storage capacity of magnetic disk drives. This rapid pace of improvement has stalled over the past years as the difficulty of stabilizing tiny magnetic bits against thermal fluctuations whilst at the same time being able to generate large enough magnetic fields to write them, has proved intractable. The possibility of creating novel spintronic magnetic memory-storage devices to rival magnetic disk drives in capacity and to vastly exceed them in performance has emerged in the form of Racetrack Memory. This concept and the physics underlying it are discussed in this book, together with a more conventional spintronic memory, magnetic random access memory (MRAM). MRAM is based on MTJ magnetic memory bits, each one accessed in a two-dimensional cross point array via a transistor. The fundamental concept of MRAM was proposed in 1995 using local fields to write the MTJ elements. This basic concept was proven in 1999 with the subsequent demonstration of large-scale, fully integrated 64 Mbit memory chips in the following decade. Writing these same elements using spin angular momentum from sufficiently large spin-polarized currents passed through the tunnel junction elements emerged in the 1990s and is now key to the development of massive-scale MRAM chips. The second edition of this book discusses these emerging spintronic technologies as well as other breakthroughs and key advances, both fundamental and applied, in the field of spintronics.

Beyond MRAM and Racetrack Memory, this book elucidates other nascent opportunities in spintronics that do not rely directly on magneto-resistive effects, such as fault-tolerant quantum computing, non-Boolean spin-wave logic, and lasers that are enhanced by spin-polarized carriers. It is interesting that spintronics is a field of research that continues to surprise even though the fundamental property of spin was realized nearly a century ago, and the basic concept of spin-dependent scattering in magnetic materials was introduced by Neville Mott just shortly after the notion of "spin" was conceived.

Since the first edition of this book, spintronics has so much evolved that a new name of "spin-orbitronics" has been coined to describe these new discoveries and developments. In the first edition of this book, spin-orbit coupling was regarded rather negatively as a property that leads to mixing between spin-channels and the loss of spin angular momentum from spin currents to the lattice, thereby limiting the persistence of these same spin currents, both temporally and spatially. In this edition, several physical phenomena derived from spin-orbit coupling are shown to be key to the development of several new technologies, such as, in particular, the current-induced motion of a series of magnetic domain walls that underlies Racetrack Memory. This relies especially on the generation of pure spin currents via the spin Hall effect (SHE). The magnitude of the SHE was thought for some time to be very small in conventional metals, but over the past few years, this has rather been shown to be incorrect. Significant and useful SHEs have been discovered in a number of heavy materials where spin-orbit coupling is large. These spin currents can be used to help move domain walls or to help switch the magnetization direction of nanoscale magnets. Whether they can be usefully used for MRAM, however, is still a matter of debate.

Another very interesting development since the first edition of this book is the explosive increase in our understanding and knowledge of topological insulators and their cousins including, most recently, Weyl semi-metals. The number of such materials has increased astronomically and, indeed, it is now understood that a significant fraction of all extant materials are "topological". What this means, in some cases, is that the spin of the carriers is locked to their momentum leading, for example, to the Quantum Spin Hall Effect.

The very concept of these materials is derived from band inversion, which is often due to strong spin-orbit coupling. From a spintronics perspective, the novel properties of these materials can lead to intrinsic spin currents and spin accumulations that are topologically "protected" to a greater or lesser degree. The concept of topological protection is itself evolving.

Distinct from electronic topological effects are topological spin textures such as skyrmions and anti-skyrmions. The latter were only experimentally found 2 years ago. These spin textures are nano-sized magnetic objects that are related to magnetic bubbles, which are also found in magnetic materials with perpendicular magnetic anisotropy but which have boundaries or walls that are innately chiral. The chirality is determined by a vector magnetic exchange – a Dzyaloshinskii–Moriya interaction (DMI) – that is often derived from spin-orbit coupling. The DMI favors orthogonal alignment of neighboring magnetic moments in contrast to conventional ferromagnetic or antiferromagnetic exchange interactions that favor collinear magnetic arrangements. Skyrmion and anti-skyrmion spin textures have very interesting properties that could also be useful for Racetrack Memories. Typically, skyrmions and anti-skyrmions evolve from helical or conical spin textures. The magnetic phase of such systems can have complex dependences on temperature, magnetic field, and strain. Some chiral antiferromagnetic spin textures have interesting properties such as an anomalous Hall effect (AHE), which is derived from their topological chiral spin texture in the absence of any net magnetization. In practice, however, a small unbalanced moment is needed to set the material in a magnetic state with domains of the same chirality in order to evidence the AHE. On the other hand, these same chiral textures can display an intrinsic spin Hall effect whose sign is independent of the chirality of the spin texture.

The DMI interaction can also result from interfaces particularly between heavy metals and magnetic layers. Such interfacial DMIs can give rise to chiral domain walls as well as magnetic bubbles with chiral domain walls – somewhat akin to skyrmions. The tunability of the interfacial DMI via materials engineering makes it of special interest.

Thus, since the first edition of this book, chiral spin phenomena, namely chiral spin textures and domain walls, and the spin Hall effect itself, which is innately chiral, have emerged as some of the most interesting developments in spintronics. The impact of these effects was largely unanticipated. It is not too strong to say that we are now in the age of "chiraltronics"!

Another topic that has considerably advanced since the first edition of this book is the field of what is often now termed spin caloritronics, namely the use of temperature gradients to create spin currents and the use of thermal excitations of magnetic systems, i.e. magnons, for magnonic devices. Indeed, magnons carry spin angular momentum and can propagate over long distances. Perhaps here it is worth mentioning the extraordinarily long propagation distances of spin currents via magnons in antiferromagnetic systems that have recently been realized.

Recently discovered atomically thin ferromagnets reveal how the presence of spin-orbit coupling overcomes the exclusion of two-dimensional ferromagnetism expected from the Mermin-Wagner theorem. These two-dimensional materials, which are similar to graphene in that they can readily be exfoliated from bulk samples, provide a rich platform to study magnetic proximity effects and transform a rapidly growing class of van der Waals materials. Through studies of magnetic materials, it is possible to reveal their peculiar quantum manifestations. Topological insulators can become magnetic by doping with $3d$ transition metals; the quantum anomalous Hall effect has been discovered in such materials, and heterostructures that consist of magnetic and non-magnetic topological insulators have been used to demonstrate current-induced control of magnetism.

Spintronics remains a vibrant research field that spans many disciplines ranging from materials science and chemistry to physics and engineering. Based on the rich developments and discoveries over the past thirty years, one can anticipate a bountiful future.

Stuart Parkin
Director at the Max Planck Institute of
Microstructure Physics in Halle (Saale),
Germany
and
Alexander von Humboldt Professor
Martin-Luther-Universität Halle-Wittenberg,
Germany
Max Planck Institute of Microstructure Physics,
Weinberg 2,
Halle (Saale), Germany

<u>Preface</u>

The second edition of this book continues the path from the foundations of spin transport and magnetism to their potential device applications, usually referred to as spintronics. Spintronics has already left its mark on several emerging technologies, e.g., in magnetic random access memories (MRAMs), where the fundamental properties of magnetic tunnel junctions are key for device performance. Further, many intricate fundamental phenomena featured in the first edition have since evolved from an academic curiosity into the potential basis for future spintronic devices. Often, as in the case of spin Hall effects, spin-orbit torques, and electrically-controlled magnetism, the research has migrated from the initial low-temperature discovery in semiconductors to technologically more suitable room temperature manifestations in metallic systems. This path from exotic behavior to possible application continues to the present day and is reflected in the modified title of the book, which now explicitly highlights "spintronics," as its overarching scope. Exotic topics of today, for example, pertaining to topological properties, such as skyrmions, topological insulators, or even elusive Majorana fermions, may become suitable platforms for the spintronics of tomorrow. Impressive progress has been seen in the last decade in the field of spin caloritronics, which has evolved from a curious prediction 30 years ago to a vibrant field of research.

Since the first edition, there has been a significant evolution in material systems displaying spin-dependent phenomena, making it difficult to cover even the key developments in a single volume. The initially featured chapter on graphene spintronics is now complemented by a chapter on the spin-dependent properties of a broad range of two-dimensional materials that can form a myriad of heterostructures coupled by weak van der Waals forces and support superconductivity or ferromagnetism even in a single atomic layer. Exciting developments have also been seen in the field of complex oxide heterostuctures, where the non-trivial properties are driven by the interplay between the electronic, spin, and structural degrees of freedom. A particular example is the magnetism emerging in two-dimensional electron gases at oxide interfaces composed of otherwise non-magnetic constituents. The updated structure of a significantly expanded book reflects various materials developments and it is now thematically divided into three volumes, each based on broadly defined metallic and semiconductor systems or their nanoscale and applied aspects.

Spintronics becomes more and more attractive as a viable platform for propelling semiconducting technology beyond its current limits. Various schemes have been proposed to enhance the functionalities of the existing technologies based on the spin degree of freedom. Among them is the voltage control of magnetism, exploiting the non-volatile performance of ferromagnet-based devices in conjunction with their low-power operation. Another approach is utilizing spin currents carried by magnons to transport and process information. Magnon spintronics involves interesting fundamental physics and offers novel spin wave-based computing technologies and logic circuits. Optical control of magnetism is another approach, which has attracted a lot of attention due to the recent discovery of the all-optical switching of magnetization and its realization at the nanoscale. Chapters on these subjects are included in the new edition of the book.

Nearly nine decades after the discovery of superconducting proximity effects by Ragnar Holm and Walther Meissner, several new chapters now explore how a given material can be transformed through proximity effects whereby it acquires the properties of its neighbors, for example, becoming superconducting, magnetic, topologically non-trivial, or with an enhanced spin-orbit coupling. Such proximity effects not only complement the conventional methods of designing materials by doping or functionalization but can also overcome their various limitations and enable yet more unexplored spintronic applications.

We are grateful both to the authors who set aside their many priorities and contributed new chapters, which have significantly expanded the scope of this book, as well as to those who patiently provided valuable updates to their original chapters and kept this edition even more timely. The completion of the second edition was again greatly facilitated by Verona Skomski, who tirelessly collected authors' contributions and assisted their preparation for the submission to the publisher. We acknowledge the support of NSF-DMR, NSF-MRSEC, NSF-ECCS, SRC, DOE-BES, US ONR which, through the support of our research and involvement in spintronics, has also enabled our editorial work. We are thankful to our families for their support, patience, and understanding during extended periods of time when we remained focused on the completion of this edition.

Evgeny Y. Tsymbal
Department of Physics and Astronomy,
Nebraska Center for Materials and Nanoscience, University of Nebraska,
Lincoln, Nebraska 68588, USA

Igor Žutić
Department of Physics, University at Buffalo,
State University of New York, Buffalo, New York 14260, USA

About the Editors

Evgeny Y. Tsymbal is a George Holmes University Distinguished Professor at the Department of Physics and Astronomy of the University of Nebraska-Lincoln (UNL), and Director of the UNL's Materials Research Science and Engineering Center (MRSEC). He joined UNL in 2002 as an Associate Professor, was promoted to a Full Professor with Tenure in 2005 and named a Charles Bessey Professor of Physics in 2009 and George Holmes University Distinguished Professor in 2013. Prior to his appointment at UNL, he was a research scientist at University of Oxford, United Kingdom, a research fellow of the Alexander von Humboldt Foundation at the Research Center-Jülich, Germany, and a research scientist at the Russian Research Center "Kurchatov Institute," Moscow. Evgeny Tsymbal's research is focused on computational materials science aiming at the understanding of fundamental properties of advanced ferromagnetic and ferroelectric nanostructures and materials relevant to nanoelectronics and spintronics. He has published over 230 papers, review articles, and book chapters and presented over 180 invited presentations in the areas of spin transport, magnetoresistive phenomena, nanoscale magnetism, complex oxide heterostructures, interface magnetoelectric phenomena, and ferroelectric tunnel junctions. Evgeny Tsymbal is a fellow of the American Physical Society, a fellow of the Institute of Physics, UK, and a recipient of the UNL's College of Arts & Sciences Outstanding Research and Creativity Award (ORCA). His research has been supported by the National Science Foundation, Semiconductor Research Corporation, Office of Naval Research, Department of Energy, Seagate Technology, and the W. M. Keck Foundation.

Igor Žutić received his Ph.D. in theoretical physics at the University of Minnesota, after undergraduate studies at the University of Zagreb, Croatia. He was a postdoc at the University of Maryland and the Naval Research Lab. In 2005 he joined the State University of New York at Buffalo as an Assistant Professor of Physics and got promoted to an Associate Professor in 2009 and to a Full Professor in 2013. He proposed and chaired Spintronics 2001: International Conference on Novel Aspects of Spin-Polarized Transport and Spin Dynamics, at Washington DC. Work with his collaborators spans a range of topics from high-temperature superconductors, Majorana fermions, proximity effects, van der Waals materials, and unconventional magnetism, to the prediction and experimental realization of spin-based devices that are not limited to magnetoresistance. He has published over 100 refereed articles and given over 150 invited presentations on spin transport, magnetism, spintronics, and superconductivity. Igor Žutić is a recipient of the 2006 National Science Foundation CAREER Award, the 2019 State University of New York Chancellor's Award for Excellence in Scholarship and Creative Activities, the 2005 National Research Council/American Society for Engineering Education Postdoctoral Research Award, and the National Research Council Fellowship (2003–2005). His research is supported by the National Science Foundation, the Office of Naval Research, the Department of Energy, Office of Basic Energy Sciences, the Defense Advanced Research Project Agency, and the Airforce Office of Scientific Research. He is a fellow of the American Physical Society.

Contributors

Ramón Aguado
Instituto de Ciencia de
    Materiales de Madrid
Madrid, Spain

Johan Åkerman
Department of Physics
University of Gothenburg
Göteborg, Sweden

Matthias Bode
Lehrstuhl für Experimentelle Physik II
Universität Würzburg
Würzburg, Germany

Hanan Dery
Department of Electrical &
    Computer Engineering
University of Rochester
Rochester, New York

Bernard Doudin
Institut de Physique et de
    Chimie des Matériaux de
    Strasbourg (IPCMS)
Strasbourg, France

William Gallagher
IBM T.J. Watson Research Center
Yorktown Heights, New York

Richard S. Gaster
venBio
San Francisco, California

Nils C. Gerhardt
Photonics and Terahertz
    Technology
Ruhr-University Bochum
Bochum, Germany

Christian Gøthgen
Department of Physics
University at Buffalo
State University of New York
Buffalo, New York

Drew A. Hall
Department of Electrical and
    Computer Engineering
University of California
San Diego, California

Wei Han
International Center for Quantum
    Materials
Peking University
Beijing, China

Masamitsu Hayashi
National Institute for Materials
    Science
Tsukuba, Japan

Hiroshi Idzuchi
Advanced Institute for
    Material Research
Tohoku University
Sendai, Japan

Xin Jiang
IBM Almaden Research Center
San Jose, California

Csaba Józsa
Physics of Nanodevices
Zernike Institute for
    Advanced Materials
University of Groningen
Groningen, the Netherlands

Paulo E. Faria Junior
Institute for Theoretical Physics
University of Regensburg
Regensburg, Germany

Ronald Kawakami
Department of Physics
The Ohio State University
Columbus, Ohio

N.T. Kemp
Department of Physics and
    Mathematics
University of Hull
Hull, United Kingdom

Alexander Khitun
University of California
Riverside, California

Takashi Kimura
Department of Physics
Kyushu University
Fukuoka, Japan

Ilya Krivorotov
Department of Physics & Astronomy
University of California
Irvine, California

Jeongsu Lee
Institute for Theoretical Physics
University of Regensburg
Regensburg, Germany

Jung-Rok Lee
Department of Materials Science
    and Engineering
Stanford University
Stanford, California
and
Ewha Womans University
Seodaemun-gu, Seoul, South Korea

Sadamichi Maekawa
RIKEN Center for Emergent
    Matter Science (CEMS)
Wako, Japan
and
Kavli Institute for Theoretical
    Sciences (KITS)
University of Chinese Academy of
    Sciences (UCAS)
Beijing, P. R. China

Rai Moriya
Institute of Industrial Science
University of Tokyo
Tokyo, Japan

Boris E. Nadgorny
Department of Physics &
    Astronomy
Wayne State University
Detroit, Michigan

Yasuhiro Niimi
Department of Physics
Osaka University
Toyonaka, Osaka, Japan

Yoshichika Otani
The Institute for Solid State
    Physics
The University of Tokyo
Chiba, Japan
and
RIKEN ASI Quantum Nano-Scale
    Magnetism Laboratory
Saitama, Japan

Stuart Parkin
Max Planck Institute for
    Microstructure Physics
and
Martin-Luther-Universität
    Halle-Wittenberg
Halle (Saale), Germany

Rafael Ramos
Advanced Institute for Materials
    Research
Tohoku University
Sendai, Japan

Joaquín Fernández Rossier
International Iberian
    Nanotechnology Laboratory
Braga, Portugal

Eiji Saitoh
Department of Applied Physics
The University of Tokyo
Tokyo, Japan
and
Advanced Institute for Materials
    Research and Center for
    Spintronics Research
    Network
Tohoku University
Sendai, Japan

Stefano Sanvito
Centre for Research on
    Adaptive Nanostructures and
    Nanodevices
School of Physics
Trinity College Dublin
Dublin, Ireland

Guilherme M. Sipahi
Instituto de Física de São Carlos
Universidade de São Paulo
São Paulo, Brazil

Saburo Takahashi
Advanced Institute for Materials
    Research
Tohoku University
Sendai, Japan

Luc Thomas
IBM Almaden Research Center
San Jose, California

Bart J. van Wees
Physics of Nanodevices
Zernike Institute for Advanced
    Materials
University of Groningen
Groningen, the Netherlands

Elena Y. Vedmedenko
Department of Physics
University of Hamburg
Hamburg, Germany

Shan X. Wang
Departments of Materials Science
    and Electrical Engineering
Stanford University
Stanford, California

Roland Wiesendanger
Department of Physics
University of Hamburg
Hamburg, Germany

Gang Xiao
Department of Physics
Brown University
Providence, Rhode Island

Gaofeng Xu
Department of Physics
University at Buffalo
State University of New York
Buffalo, New York

Igor Žutić
Department of Physics
University at Buffalo
State University of New York
Buffalo, New York

Section VI
Spin Transport and Magnetism at the Nanoscale

Spin-Polarized Scanning Tunneling Microscopy

Matthias Bode

Fueled by the ever increasing data density in magnetic storage technology and the need for a better understanding of the physical properties of magnetic nanostructures there exists a strong demand for high-resolution magnetically sensitive microscopy techniques. The technique with the highest available resolution is spin-polarized scanning tunneling microscopy (SP-STM), which combines the atomic resolution capability of conventional STMs with spin sensitivity by making use of the tunneling magneto-resistance (TMR) effect between a magnetic tip and the magnetic sample surface. Beyond the investigation of ferromagnetic surfaces, thin films, and epitaxial nanostructures with unforeseen precision, it also allows the achievement of a long-standing dream, i.e., the real space imaging of atomic spins in antiferromagnetic surfaces. Furthermore, SP-STM allowed for the discovery and investigation of complex spin structures, like frustrated triangular structures and helical spin spirals, which were hypothesized but could not be imaged directly before.

1.1 INTRODUCTION

In Section III, Volume 1, of this book the spin transport in magnetic tunnel junctions, mostly between ferromagnetic layers separated by an insulating barrier, is discussed. In these planar devices the tunneling current is not localized but is—if the insulating barrier is sufficiently smooth—equally distributed over the area of the entire tunnel junction. Inspired by the extremely short decay length of electron tunneling across materials with a sufficiently high barrier, such as oxidic insulators, Binnig and Rohrer invented the scanning tunneling microscope in the early 1980s. In this technique a sharp tip is brought close to a sample, resulting in a local tunneling current across the vacuum barrier between the tip's foremost atom and the sample surface. With its atomic resolution imaging capability, scanning tunneling microscopy and numerous derived spectroscopy modes soon revolutionized surface science. These spectroscopy modes allow for the direct correlation between structural and a variety of electronic sample properties, such as the local work function [1–3], the analysis of image states [1, 4], and the analysis of Shockley-like [5, 6] or exchange-split surface states [7–9].

In close analogy to the development of spin-polarized electron tunneling in planar junctions, where the normal-metal electrodes are replaced by ferromagnetic electrodes to achieve spin sensitivity, it was an obvious approach to apply the same principle to STM: When scanning a magnetic surface with a magnetic tip the TMR effect should give rise to a characteristic variation of the tunneling current, the differential conductance, or other parameters which can then be analyzed to map the sample's spin structure down to the atomic level [10]. Although the principle appears to be straightforward, several obstacles had to be overcome: What is a suitable tip material? What is the magnetization direction at the apex of the tip? Can the stray field or exchange force of the tip manipulate the pristine domain structure of the sample? Is it possible to achieve atomic spin resolution? Can antiferromagnetic tips be used for imaging?

As so often in science the answers to these questions, which were largely unknown 10 years ago, seem to be quite obvious in hindsight. During the past two decades spin-polarized STM has been developed into a mature magnetically sensitive imaging tool. For the first time it opened the door for spin-sensitive imaging with a spatial resolution well below 10 nm. This enabled several breakthroughs, among them the imaging of magnetic hysteresis at the nanoscale [11], the discovery of ultra-narrow domain walls [12], the direct observation of the single-domain limit in nanomagnets [13], and the analysis of the internal spin structure of magnetic vortex cores [14]. In fact, atomic spin resolution is nowadays routinely achieved, allowing SP-STM the unrivaled investigation of antiferromagnetic surfaces [15–17]. This chapter will give a short overview of the existing capabilities of spin-polarized STM and how it is applied for mapping surface magnetic domains and spin structures.

1.2 THE CONTRAST MECHANISM AND MODES OF OPERATION

As already mentioned above the contrast mechanism of spin-polarized STM is based on the TMR effect. A thorough theoretical description was developed by Wortmann et al. [18] by generalizing the Tersoff-Hamann model originally developed for conventional, i.e., spin-averaged STM [19, 20]. This was followed by a theoretical approach by Heinze [21]; the independent-orbital model is applied to approximate the surface electronic structure. In this approach the computational demand of ab-initio calculations is significantly reduced by considering the arrangement of the magnetic surface moments only, while at the same time ignoring the accurate surface electronic structure. Especially for thin films consisting of a single element, the independent-orbital method produced results which are in excellent agreement with more demanding theoretical data and experiments performed on periodic collinear as well as non-collinear magnetic spin structures.

In both methods, ab-initio calculations and the independent-orbital model, the tunneling current I at any given tip position $\vec{r}_T$ and bias voltage U is separated into a spin-averaged and a spin-dependent contribution, I_0 and I_{SP}, respectively:

$$I(\vec{r}_T, U, \theta) = I_0(\vec{r}_T, U) + I_{SP}(\vec{r}_T, U, \theta). \qquad (1.1)$$

Here, the magnetization vectors of tip and sample, $\vec{m}_T$ and $\vec{m}_S$, include the angle θ. Using Bardeen's tunneling formalism the tunneling current can be written by the following expression:

$$I(\vec{r}_T, U, \theta) = \frac{4\pi^3 C^2 \hbar^3 e}{\kappa^2 m^2} \left[n_T \tilde{n}_S(\vec{r}_T, U) + \vec{m}_T \tilde{\vec{m}}_S(\vec{r}_T, U) \right]. \qquad (1.2)$$

Where n_T is the non–spin-polarized local density of states (LDOS) at the tip apex, $\tilde{n}_S$ is the energy-integrated LDOS of the sample, and $\vec{m}_T$ and $\tilde{\vec{m}}_S$ are the corresponding vectors of the (energy-integrated) spin-polarized (or magnetic) LDOS:

$$\tilde{\vec{m}}_S\left(\vec{r}_T, U\right) = \int\limits_{E=E_F}^{E=eU} \vec{m}_S\left(\vec{r}_T, E\right) dE, \tag{1.3}$$

with

$$\vec{m}_S = \sum \delta\left(E_\mu - E\right) \psi_\mu^{S\dagger}\left(\vec{r}_T\right) \sigma \psi_\mu^S\left(\vec{r}_T\right). \tag{1.4}$$

ψ_μ^S denotes the spinor of the sample wave function

$$\psi_\mu^S = \begin{pmatrix} \psi_{\mu\uparrow}^S \\ \psi_{\mu\downarrow}^S \end{pmatrix}, \tag{1.5}$$

and σ is Pauli's spin matrix. The interpretation of Equation 1.2 is relatively simple and very similar to Slonczewski's model [22] for planar junctions: Namely, the spin-dependent contribution to the tunneling current, I_{SP}, scales with the projection of $\tilde{\vec{m}}_S$ onto $\vec{m}_T$, or, in other words, with the cosine of the angle included between the magnetization directions of the two electrodes, $\cos\theta$.

More importantly, Wortmann and co-workers [18] gave a detailed theoretical description of the three modes of operation which are usually applied in spin-polarized scanning tunneling microscopy. First to mention is the constant-current mode where a feedback circuit acts on the z-component of the piezoelectric actuator in order to stabilize the tunneling current at a predetermined value during the lateral scanning process. Because of the very short decay length of electron tunneling across a vacuum barrier (typically of the order of Å), however, even a TMR effect of 100% will result in a relatively small variation of the tip height only, the size of which was reported to amount to a few picometers (1 pm = 10^{-12} m) only [23, 24]. Therefore, as will be shown in more detail below in Section 1.5, the constant-current mode is only suited to study atomic scale spin structures on extremely flat surfaces. This restriction is caused by the issue of separating these tiny magnetically induced height variations from structural or electronic inhomogeneities on sample surfaces that exhibit step edges or even larger topographic features. Since a single atom's diameter already exceeds 200 pm, the safe identification of the much smaller TMR-induced magnetic modulation is practically impossible.

The separation of topographic, electronic, and magnetic information can be conveniently achieved by adding a small modulation U_{mod} to the bias voltage U and detecting the resulting current modulation by lock-in technique [25]. Usual modulation amplitudes amount to millivolts (mV) to a few tens of mV. For theoretical assessment of this so-called differential conductance, or dI/dU, mode the simplifying assumption of a tip with a featureless electronic structure has to be made. While a spin-averaged measurement of the differential conductivity, $dI/dU_0(\vec{r}_T, U) \propto n_S$, gives access to the sample's non–spin-polarized local density of sample states n_S [19, 26], the application of magnetic tips leads to a spin-polarized contribution to the differential conductance, which can be written as [18]:

$$\frac{dI}{dU}(\vec{r}_T, U) = \frac{dI}{dU}(\vec{r}_T, U)_0 + \frac{dI}{dU}(\vec{r}_T, U)_{SP}$$

$$\propto n_T n_S(\vec{r}_T, E_F + eU) + \vec{m}_T \vec{m}_S(\vec{r}_T, E_F + eU). \tag{1.6}$$

The dI/dU mode detects the spin polarization within a narrow energy interval ΔE around $E_F + eU$. Although this mode has been applied to surfaces which exhibit spin-polarized surface states in proof-of-principle experiments [24, 25], it is important to stress that no particular surface electronic structure is required. As long as the dI/dU signal can be measured simultaneously with the topographic information, the imbalance of spin-up and spin-down states, which is intrinsic to any magnetic material, allows for an efficient separation between topographic, electronic, and magnetic information. It is important to note, however, that the spin-polarization of the magnetic LDOS, $\vec{m}_S$, not only changes in size but may also change its sign if different energy intervals ΔE are compared. For example, the surface (and also the tip) may exhibit a positive spin-polarization in one energy interval but a negative spin-polarization in another [25, 27]. Consequently, a high dI/dU signal does not imply that the magnetization directions of both electrodes are parallel, but rather that the magnetic LDOS in both electrodes have the same sign. As we will see in Chapter 2 the differential conductance mode allows for a rather straightforward imaging of magnetic domains at high spatial resolution on surfaces with homogeneous electronic properties. On surfaces with an inhomogeneous electronic structure, however, the separation of spin-averaged electronic from magnetic effects is more challenging. Only by comparing images which were obtained at voltages where the spin asymmetry, as defined by

$$A = \frac{dI/dU_{\uparrow\downarrow} - dI/dU_{\uparrow\uparrow}}{dI/dU_{\uparrow\downarrow} + dI/dU_{\uparrow\uparrow}}, \tag{1.7}$$

vanishes and becomes maximal the magnetic domain structure can be retrieved [28].

An alternative method for unambiguously separating magnetic from other contributions is based on an approach originally proposed by Johnson and Clarke [29], which was later refined and made useful for the imaging of magnetic domains by Wulfhekel and Kirschner [30]. In this experimental setup the tip magnetization $\vec{m}_T$ is periodically switched back and forth by a small coil wrapped around the tip shaft. With the modulation frequency exceeding the cut-off frequency of the feedback loop, the magnetic contribution to the tunneling current is no longer leveled out by the feedback circuit. Differentiation of Equation 1.2 with respect to $\vec{m}_T$ leads to [18]:

$$\frac{dI}{d\vec{m}_T}(\vec{r}_T) \propto \vec{m}_S(U), \tag{1.8}$$

i.e., the resulting modulation of the tunneling current in this so-called local TMR mode is proportional to the local magnetization of the sample at the energy corresponding to the particular bias voltage U. In the experiment the TMR signal is detected by lock-in technique.

In fact, the voltage dependence of spin-polarized STM is distinctly different from planar junctions. While the TMR effect of planar junctions generally decreases continuously with increasing bias voltage, pronounced band structure effects were observed in the STM geometry [31, 32]. A particular showcase is the Co(0001) surface which at the center of the Brillouin zone is dominated by two spin-split minority bands resulting in a virtually energy-independent polarization as shown in Figure 1.1a. Only for small tip-sample distances states with a significant $\vec{k}_\parallel$ but lower polarization may contribute to the tunneling current which leads to a dip in the TMR signal at ≈ 200 meV (Figure 1.1b). These STM-based experiments with a vacuum barrier lead to the conclusion that the so-called zero-bias anomaly in planar tunnel junctions with insulator barriers can be attributed to defect scattering in the barrier, rather than to magnon creation or spin excitations at the interfaces [32].

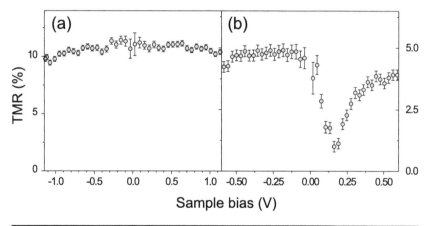

FIGURE 1.1 Bias voltage-dependent tunnel magneto-resistance (TMR) of a clean Co(0001) surface. The data were obtained at room temperature with a bulk magnetic tip stabilized at (a) 1 V, 1 nA and (b) 100 mV, 1 nA. In contrast to experiments with planar tunnel junctions, where the TMR effect monotonously decreases with increasing bias voltage, spin-polarized tunneling across a vacuum barrier reveals pronounced band structure effects. The constant polarization is a result of the particular surface band structure of Co(0001) which is dominated by two spin-split minority bands at the center of the Brillouin zone resulting in a virtually energy-independent polarization. (After Ding, H.F. et al., *Phys. Rev. Lett.* 90, 116603, 2003. With permission.)

Even though the TMR mode effectively separates structural and spin-averaged electronic contributions on one side from magnetic effects on the other side, data interpretation of measurements taken on inhomogeneous surfaces still remains challenging. As mentioned above this is caused by the fact that, in general, the spin-polarization of the magnetic LDOS $\tilde{m}s$ is different for different materials and varies or even changes sign in different energy ranges.

1.3 TIP MATERIALS

Already the initial proposal by Pierce [10] included two basic concepts for obtaining spin sensitivity, namely the use of magnetic and of GaAs tips. While magnetic tips exhibit an intrinsic spin polarization at the Fermi level, it was planned to generate a spin-polarized density of states (DOS) in GaAs tips by illumination with circularly polarized light. The physical principle is based on the direct band-gap properties of GaAs together with the 0.34 eV spin-orbit splitting of the p-like valance band. The latter leads to a fourfold degenerate $p_{3/2}$-level and a twofold degenerate $p_{1/2}$-level. While light with an energy slightly above the energy gap $E_g = 1.52$ eV of p-GaAs can excite electron from $p_{3/2}$-level into the conduction band the effective band gap for $p_{1/2}$-level states is too large. For circularly polarized light the optical selection rule mandates that $\Delta m_j = m_f - m_i = \pm 1$, where $m_{f,i}$ is the angular momentum of the final and the initial state. Therefore, for $\sigma^+ -$light the relative transmission probability into $m_j = -1/2$ states is three times higher than into $m_j = +1/2$ states (and vice versa for σ^--light). As a result, the theoretical limit of the electron spin polarization in photoemission is $P_{GaAs} = \pm 50\%$ [33]. One particularly compelling advantage of GaAs tips would be the complete absence of magnetic material in the tip which might potentially interact with the sample's magnetization. Even though a similar concept is widely used for the generation of spin-polarized electrons from planar GaAs wafers [33], experiments with GaAs tips on magnetic surfaces turned out to be plagued by competing spurious effects, such as the Faraday effect, which could never be reliably eliminated [34–38].

In contrast, magnetic tips were successfully used by different research groups on a large variety of sample systems in the three modes of operation described in the previous section. Two types of magnetic tips were used: Magnetic thin-film tips with a static magnetization in the dI/dU mode [25] and amorphous low-remanence tips in the local TMR mode [30]. In order to obtain tips with either the out-of-plane or the in-plane magnetization direction in the TMR mode, tip shapes were engineered by making use of the shape dependence of the magneto-static anisotropy. While a tapered magnetic wire is out-of-plane sensitive as it is preferentially magnetized along the wire axis to reduce its magneto-static field, the preparation of in-plane sensitive bulk tip requires a more sophisticated approach. Schlickum et al. [39] obtained in-plane magnetic contrast by electrochemically etching an approximately 1 mm diameter ring from a thin foil. Similar to the straight wire, by periodically switching the magnetization of the ring with a coil wrapped around it, the spin-dependent tunneling current between the ring and a spin-polarized sample can be measured. Since shape anisotropy is by far the dominant anisotropy contribution, the magnetization direction of TMR tips can be reliably predicted. In spite of the above-mentioned advantages of the local TMR mode, such as the clear separation of magnetic contributions to the tunneling current, and some compelling results [39], this mode is not widely applied anymore. Possible reasons are the relatively large amount of magnetic material which is brought in very close vicinity to the tip and which—in spite of attempts to reduce the stray magnetic field by using low-remanence material—might modify the sample's domain structure and issues with magneto-striction.

Another approach for measurements performed in the dI/dU mode is based on ferromagnetic thin-film tips which are grown *in situ* under ultra-high vacuum conditions. These thin-film tips are usually prepared by deposition of magnetic material onto an electrochemically etched W tip. Upon etching, the tip is introduced into the vacuum system via a load lock. Early experiments indicated that ferromagnetic films evaporated onto untreated tips are mechanically unstable and can easily be lost when approaching towards the sample surface or while scanning [40]. It was found that the film stability can significantly be improved by briefly heating the tip by electron bombardment prior to deposition to about 2200 K. The high-temperature flash removes contaminations the presence of which seems to weaken the interfacial sticking and also removes potential remains of previous magnetic coatings. However, it also melts the tip apex.

Ferromagnetic thin-film tips have several advantages over conventional bulk magnetic tips. Firstly, the amount of magnetic material is drastically reduced which helps minimize the accidental modification of the sample's magnetic domain structure by the tip's stray field. Secondly, if refractory metals are used as a tip material the non-magnetic core of the tip can be recycled many times. As evidenced by the scanning electron microscopy images shown in Figure 1.2 the melting and subsequent cooling typically results in relatively blunt tips with a typical diameter of about 1 μm [41]. Past results show that the magnetization direction of such tips is no longer governed by the shape anisotropy associated with the tapered shape of the tip, but rather by material-specific parameters, such as the surface and interface

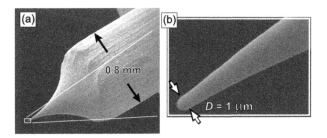

FIGURE 1.2   Scanning electron microscopy images of a flashed W tip at (a) medium and (b) higher magnification revealing a tip diameter of about 1 μm.

anisotropies of the entire system, i.e., the tip material and the thin film. For example, Fe-coated tips with a film thickness of about 10 atomic layers (AL) are usually in-plane sensitive. In contrast, Gd, GdFe alloy, and thin Cr film tips (25–45 AL) are mostly out-of-plane sensitive. These results are largely consistent with the respective easy magnetization directions of films on flat W(110) substrates [42, 43]. Since, however, the shape anisotropy may also play a role, the final magnetization direction of a particular thin-film tip is less predictable than that of the engineered TMR tips mentioned above and canted magnetization directions are frequently observed.

As compared to the TMR mode mentioned above, it may be considered the main advantage of the dI/dU mode that antiferromagnetic coatings and bulk tips can be used. By studying the magnetic coercivity and saturation of thin iron films on W(110) Kubetzka et al. [44] could show that even thin-film ferromagnetic tips unavoidably distort the film's intrinsic domain and domain wall structure. Such problems can safely be avoided by the use of antiferromagnetic tips, as, e.g., prepared by the evaporation of Cr onto the tip apex. Since antiferromagnets are compensated they exhibit no stray field. Consequently, antiferromagnetic bulk tips can be utilized as well [45, 46], thereby avoiding the rather expensive and experimentally demanding *in situ* tip cleaning and deposition procedures discussed above.

In agreement with Slonczewski's result for planar magnetic tunnel junctions [22], Equation 1.6 shows that the magnetic contrast in dI/dU maps is proportional to the projection of $\vec{m}_s$ onto $\vec{m}_t$. Since the sample system we will focus on in the next section is an hcp(0001) surface, the following discussion will focus on a system with three easy axes, i.e., six equivalent magnetic orientations. This opens up three principal cases of the relative orientation between the tip and the sample magnetization. For example, it may occur that the tip magnetization $\vec{m}_t$ lies along the sample's easy (Figure 1.3a) or hard axis (Figure 1.3b). In these cases the angle between $\vec{m}_t$ and $\vec{m}_s$ of the domain with the largest dI/dU signal is $\beta = 0°$ or $\beta = 30°$, respectively. As symbolized by hatched red lines this leads to a reduced number of contrast levels since in each case magnetization directions can be found which

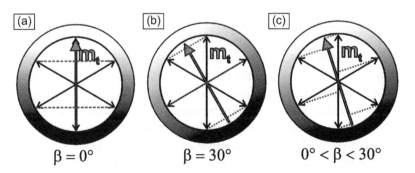

$$\beta = 0° \qquad \beta = 30° \qquad 0° < \beta < 30°$$

FIGURE 1.3   Schematic representation of the possible orientations of the tip magnetization $\vec{m}_t$ (thick arrow) with respect to the six possible easy axes magnetization directions $\vec{m}_s$ of Dy (thin arrows): (a) $\beta = 0°$, (b) $\beta = 30°$, and (c) $0° < \beta < 30°$, resulting in four, three, and six contrast levels, respectively. β denotes the azimuthal angle between $\vec{m}_t$ and $\vec{m}_s$ of the domain with the largest dI/dU signal.

are mirror-symmetric with respect to $\vec{m}_t$ and therefore result in the same dI/dU signal. Namely, we expect four contrast levels if $\beta = 0°$ and only three contrast levels if $\beta = 30°$. As shown in Figure 1.3c the intermediate case with $0° < \beta < 30°$ results in six different contrasts, such that all possible domains may be distinguished from another.

1.4 IMAGING MAGNETIC SURFACES

In the following the imaging of different kinds of magnetic surfaces will be discussed, ranging from the domain structure of a ferromagnet, the imaging of so-called topological antiferromagnetism in a layered antiferromagnet, to atomic scale spin structures of antiferromagnetically coupled nearest-neighbor atoms and complex spin spirals. In the first few examples the contrast mechanism will be revisited in detail and the particular strength and limitations of different modes of operation will be discussed. At the end of this section we will see that these enhanced magnetic imaging capabilities lead to new discoveries, such as the observation of complex spin structures in ultra-thin films.

1.4.1 LAYERED ANTIFERROMAGNETS

Since the very beginning of spin-polarized scanning tunneling microscopy the Cr(001) surface has been a favorite test sample. This choice is motivated by the fact that its topography is closely linked to the surface magnetic structure. Although Cr is generally known to be an antiferromagnet, theoretical investigations performed in the 1980s showed that the topmost layer of a Cr(001) surface is ferromagnetically ordered with an enhanced magnetic moment [47, 48]. Yet, Blügel et al. recognized that the antiferromagnetic state of Cr prevails such that the magnetization direction found on adjacent atomically flat Cr(001) planes has to alternate at monoatomic step edges. To illustrate the intimate correlation between the surface topography of Cr(001) and its magnetic structure Blügel et al. coined the term "topological antiferromagnetism" [48].

Obviously, the surfaces of topological antiferromagnets such as Cr(001) have a vanishing magnetic moment on the macroscopic scale because the moments of adjacent Cr(001) terraces cancel. As long as the surface roughness gives rise to thickness fluctuations of ≥ 1 atomic layer (AL) experimental techniques, which average over areas that are large compared to the terrace width, show no magnetic signal. Only in some few examples Cr(001) films could be grown with sufficiently smooth interfaces such that oscillations of the TMR effect could successfully be observed by spatially averaging measurements of wedge-shaped magnetic tunnel junctions [49, 50] (see also Section III, Volume 1, of this book). Accordingly, most magnetically sensitive experimental techniques are not feasible for the investigation of topological antiferromagnetic surfaces. For the development of spin-polarized scanning tunneling microscopy and spectroscopy, however, the nanoscale one-to-one correlation between topography and magnetism of Cr(001) represents an advantage as it simplifies the safe identification of magnetic contrasts.

In a pioneering experiment Wiesendanger and co-workers [23] compared line sections measured on Cr(001) with non-magnetic W and ferromagnetic CrO_2. These early experiments, which were qualitatively confirmed by Kleiber et al. (cf. Figure 2c in Ref. [24]), revealed that the apparent step height depends on the tip material. While a step height of 0.14 nm, which is consistent with the lattice constant of Cr, was found with a W tip, alternating apparent step heights were observed with a magnetic CrO_2 tip. This result was explained by the contribution of spin-polarized tunneling which—due to the constant-current mode of operation—leads to magnetization-direction dependent tip-sample distance.

As mentioned in the previous section, SP-STM operated in the constant-current mode suffers from the fact that all contributions to the signal, topography, electronic, and magnetic properties are inseparably mingled into one data channel, i.e., the vertical position of the tip (z-signal). While this mode allows for high-resolution studies of homogeneous and extremely flat surfaces, such as the topological antiferromagnet Cr(001) or for atomic resolution studies of single-element antiferromagnets (to be discussed in 1.5), it is unsuitable on strongly corrugated or inhomogeneous surfaces.

This shortcoming can be overcome by spin-polarized scanning tunneling spectroscopy. The origin of the contrast can be understood from the Cr(001) tunneling spectra shown in Figure 1.4. The spectra show a peak just at the Fermi level E_F which is known to be characteristic for Cr(001) [51] even though it is still under discussion whether the origin of this peak is a surface state [51–53] or an orbital Kondo resonance [54]. As long as non-magnetic probe tips are employed the spectra of all terraces, irrespective of their magnetization direction, are identical (not shown here). The spectra of Figure 1.4, however, were taken with an Fe-coated probe tip. The magneto-resistance effects between tip and sample lead to striking quantitative differences between spectra taken on adjacent terraces. While the peak position remains essentially identical the peak intensity varies by up to 10%. This behavior is

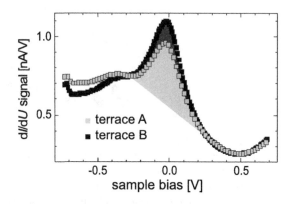

FIGURE 1.4 Tunneling spectra of Cr(001) as measured at room temperature with a Fe-coated magnetic tip on adjacent terraces. Due to the tunneling magneto-resistance effect (TMR) different peak heights can be observed.

typical for spin-polarized tunneling with the STM: The spectrum of a given surface remains qualitatively unchanged with very similar peak positions for the two domains (as the spin-averaged electronic properties of the surface are identical), but quantitative variations of the differential conductance are observed.

While it is very time consuming to take a complete tunneling spectrum at every pixel of an image, the intensity variation of the differential conductance dI/dU can be used for mapping the sample magnetic structure at certain bias voltages. For the particular tip used in the experimental run of Figure 1.4 suitable bias voltages, which can be expected to result in high magnetic contrast, would be close to the peak position, i.e., around 0 V, or at -0.6 ± 0.1 V. It shall be mentioned, however, that because of uncontrollable structural differences of the tip apex's atomic arrangement the optimal bias voltage may differ from tip to tip and has to be figured out in each individual experiment.

Since the signal is obtained with lock-in technique after adding a small bias modulation (typically a few mV) with a frequency well beyond the response frequency of the feedback loop (typically several kHz), dI/dU map can be recorded simultaneously with the topographic image. Furthermore, this technique practically separates the topography from the magnetic signal into two virtually independent data channels. As an example, Figure 1.5 shows such a data set taken with an in-plane sensitive Fe-coated probe tip at room temperature. The so-called topography image shown in Figure 1.5a represents the tip's z-position as obtained in the constant-current mode. Due to the presence of several edge and screw dislocations the sample's terrace and step structure is rather intricate and it would be essentially impossible to

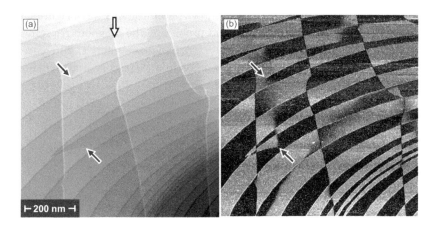

FIGURE 1.5   (a) Topography and (b) magnetic dI/dU map of a Cr(001) surface as measured with an Fe-coated probe tip at 300 K. (After Kleiber, M. et al., *Phys. Rev. Lett.* 85, 4606, 2000. With permission.) Screw (black arrows) and edge dislocations (one example marked by a white arrow) can be recognized in (a). The tunneling parameters are $I = 3$ nA and $U = -0.7$ V. Similar to other monoatomic step edges, the magnetic contrast changes across the edge dislocation due to the layered antiferromagnetic structure of Cr(001). Typically, a domain wall is found between two screw dislocation, e.g., between the two black arrows in (b).

unravel the surface magnetic domain structure from these topographic data alone. The differential conductance dI/dU map shown in Figure 1.5b, however, immediately reveals the numerous irregularities of this complex surface magnetic structure. Generally, edge dislocations as the one marked by a white arrow in Figure 1.5a do not disturb the topological antiferromagnetic order and—similar to any other monatomic step edge—lead to a reversal of the surface magnetization [55]. Screw dislocations, in contrast, appear as semi-infinite step edges which lead to a frustration of the surface magnetic order. Since they result in an additional step edge on one side of the sample only, an undistorted topological antiferromagnetic order cannot be maintained. Instead, a domain wall is generated which extends over the surface until it is annihilated at another screw dislocation [24, 55], as indicated by two black arrows in Figure 1.5b.

This example shows that surfaces with a layered or compensated magnetic structure (so-called d-type antiferromagnets) can effectively be imaged with SP-STM by making use of the differential conductance dI/dU mode. In spite of a certain cross-talk between the topography and the dI/dU signal these simultaneously measured data channels allow for an effective separation of structural and magnetic surface properties on scales ranging from the micro- down to nanometer lateral scales.

1.4.2 FERROMAGNETS

The imaging of the domain structure of ferromagnetic surfaces has been the focus of scientific interest for a long time [56]. While the magnetic exchange length of the elementary 3d transition metals Fe, Co, and Ni is sufficiently wide to be imaged by traditional magnetically sensitive microscopy techniques such as the Bitter technique, Lorentz microscopy, or magnetic force microscopy, many technologically relevant materials like rare earth metals and alloys consisting of 3d and 4f metals exhibit an exchange length well below the respective spatial resolution of the above-mentioned microscopy techniques.

One particularly illustrative example of the imaging of the domain structure of ferromagnetic surfaces by spin-polarized STM is a study on dysprosium (Dy) films by Berbil-Bautista et al. [57]. Quite contrary to most other experiments, these spin-resolved STM data were not obtained with tips which were epitaxially coated with magnetic material, but rather by dipping the tip into the Dy surface, which reliably resulted in a Dy cluster at the tip apex.

Figure 1.6a shows the topography of a 90 atomic layer (AL) thick Dy(0001) film grown on W(110). The Dy(0001) surface exhibits six equivalent magnetization directions. Therefore, according to Equation 1.6, up to six differential conductance levels can be expected if a SP-STM experiment is performed with a magnetic tip. In fact, the dI/dU map of Figure 1.6b which was acquired simultaneously with the topographic image of Figure 1.6a shows six well-separated levels of the dI/dU signal. This is particularly obvious in the histogram of Figure 1.6b, plotted in Figure 1.6c. Because of the

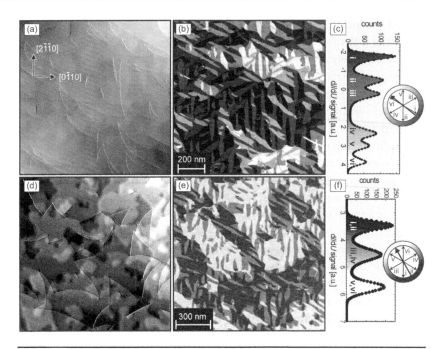

FIGURE 1.6   Examples for spin-polarized STM experiments on a ferromagnetic surface: (a) Topography and (b) spin-resolved dI/dU map of a 90 AL dysprosium film on W(110) imaged at T = 25 K (tunneling parameters: U = –1.0 V, I = 30 nA). (c) The histogram of the dI/dU map shown in (b) reveals six contrast levels. Panels (d)–(f) show corresponding data obtained in a different experimental run with a different probe tip on 45 AL Dy/W(110). Note that only three contrast levels can be recognized in the histogram of the dI/dU map because the tip magnetization direction is oriented along the sample's hard axis.

TMR effect the surface areas which are magnetized almost parallel with respect to the tip magnetization gave the highest intensity of the dI/dU signal, peak i, while those which are almost antiparallel to the tip exhibit the lowest, vi. The other magnetization directions give rise to dI/dU signals in between these two extremes (ii-v). By a variational approach this even allows for a precise determination of the tip magnetization direction, as shown in Figure 1.6c.

However, these six contrast levels can only be expected if the tip magnetization sufficiently deviates from both the easy (β = 0°) and the hard magnetic axis (β = 30°) of a surface with six equivalent magnetization directions. This situation was schematically represented in Figure 1.3c where the six magnetization directions (thin arrows) exhibit distinguishable projections onto the tip magnetization (thick arrow). In contrast, if the tip magnetization is close to the sample's easy or hard magnetic axis, as represented in Figure 1.3a, b, some projections fall onto the same vector and can no longer be distinguished, resulting in four or three signal levels, respectively. One such example is shown in Figure 1.6d–f, where the tip was accidentally magnetized along a hard magnetization axis such that the six magnetization directions of Dy(0001) lead to three separate signals only.

As a first example the high spatial resolution capability of SP-STM shall be demonstrated by an analysis of the internal spin structure of domain walls observed on Dy films. The domain structure of a 90 AL Dy film on W(110) is shown in Figure 1.7a. A histogram of the dI/dU signal intensity in Figure 1.7b again reveals six contrast levels, i.e., a situation similar to Figure 1.3a. Figure 1.7c–f show line profiles of the magnetic dI/dU signal across (c,d) two 60° domain walls, (e) a 120°, and (f) a 180° domain wall, all of which are correspondingly labeled and color-coded in Figure 1.7a, respectively. The horizontal lines in the domain wall profiles (c)–(f) represent the peak positions in the histogram of Figure 1.7b. While the dI/dU signal changes monotonously for the two 60° domain walls, a non-monotonous profile is observed in the cases of the 120° domain wall and the 180° domain wall.

The width of the domain walls amounts to between 2 and 3 nm only. It is important to note that the high spatial resolution of SP-STM not only allows for resolving the domain structure of the Dy film, but also shows the internal spin structure even of domain walls which are far below the limit of resolution of other magnetically sensitive imaging techniques. In fact, as we will see in the following section, atomic spin resolution can be achieved on appropriate surfaces.

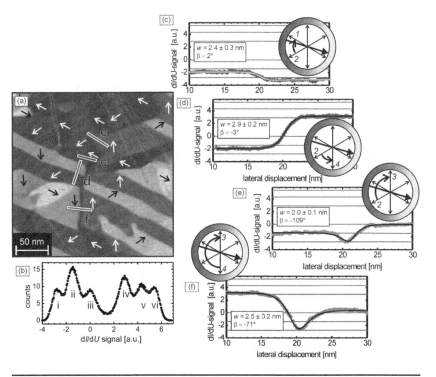

FIGURE 1.7 Domain wall profile analysis of a 90 AL Dy/W(110) film. (After Berbil-Bautista, L. et al., *Phys. Rev. B* 76, 064411, 2007. With permission.) (a) Spin-resolved dI/dU map of 90 AL Dy/W(110) showing the magnetic domain structure (T=59 K, U=−0.6 V, I=30 nA). (b) The histogram of the data shown in (a) reveals 6 peaks. The directions of $\vec{m}_s$ for the domains is indicated by arrows in (a). Line profiles and schematic representation of the path of the magnetization for (c, d) two 60° domain walls, (e) a 120°, and (f) a 180° domain wall. The respective domain walls are correspondingly labeled and color-coded in (a). The profiles are only a few nanometers wide.

1.5 IMAGING OF ANTIFERROMAGNETIC SURFACES

As already discussed in the introduction of this chapter, the imaging of anti-ferromagnetic materials represents a particular challenge to conventional magnetic microscopy methods, such as photo-electron emission microscopy (PEEM) or Lorentz microscopy, since the sample's magnetic moments cancel if averaged beyond the atomic scale. While spatially averaging polarized neutron diffraction allowed for a quite thorough understanding of the processes in bulk samples, much less is known about antiferromagnetic thin films or even nanoparticles. Although methods like Mössbauer spectroscopy and neutron diffraction were adapted to large ensembles of presumably identical nanoparticles (for a review see Ref. [58]), the analysis of the properties of single particles remained out of reach.

The well-established topographic atomic resolution capability of STM suggests that the goal of atomic spin resolution can also be achieved, but this task requires an appropriate and electrically conducting test sample. The smallest conceivable magnetic structure is a magnetically ordered surface where the orientations of magnetic moments alternate between adjacent atomic sites, i.e., a so-called g-type antiferromagnet.

1.5.1 PROOF OF PRINCIPLE

Many materials which are known to be g-type antiferromagnets in the bulk are alloys with a complex stoichiometry and cannot be stabilized at the surface. A structurally more simple sample system which can be grown epitaxially on a refractory metal is the Fe monolayer on W(001). Early spin-resolved photoemission and magneto-optical measurements showed the absence of magnetic remanence [59, 60]. Indeed, density-functional calculations performed in the local-density approximation (LDA) scheme revealed an antiferromagnetic ground state [61] which could also be confirmed using the improved generalized-gradient approximation (GGA) scheme [16].

The topography of 1.1 atomic layers (AL) of Fe epitaxially grown on W(001) is shown in Figure 1.8a. In spite of the substantial misfit between Fe and W, which amounts to approximately 10% of the lattice constants, the film grows pseudomorphic, i.e., with Fe atoms crystallographically continuing the W lattice by occupying fourfold-coordinated hollow sites of the bcc(110) surface. While the first Fe layer is completely filled, only few double-layer islands with lateral dimensions of less than 10 nm can be recognized. Indeed, atomic scale images taken with a magnetic tip reveal the atomic scale spin structure; the magnetic checkerboard pattern of the antiferromagnetic Fe monolayer can be seen in the higher resolution images shown in Figure 1.8b, c which were taken in the area marked by a hatched box approximately in the center of Figure 1.8a.

Figure 1.8b, c were simultaneously measured and show results obtained in the two magnetic imaging modes discussed above, i.e., the constant-current image and the dI/dU mode, respectively. Around two adsorbates

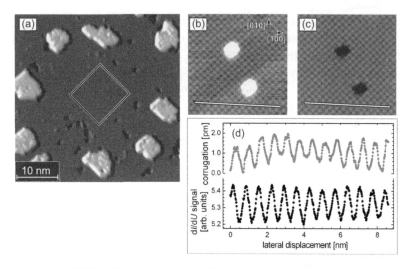

FIGURE 1.8   (a) Topography of 1.1 AL Fe on W(001). (b) The spin-resolved constant-current image and (c) a spin-resolved dI/dU map were measured simultaneously in the box of (a) and show the antiferromagnetic checkerboard pattern of the Fe monolayer on W(001) with alternating spins pointing up and down. Measurement temperature was 13 ± 1 K. Note that the two adsorbates appear as protrusions in the topography but exhibit a lower dI/dU signal strength. Line sections taken along the lines in (b) and (c) are shown in (d).

which appear as protrusions, the constant-current image clearly shows a dark and bright checkerboard pattern: due to the spin-polarized contribution to the tunneling current the tip is retracted from the sample surface whenever the integrated tip's magnetic LDOS is parallel to the sample atom and approached above atoms the magnetization direction of which is antiparallel to the tip. The line section in Figure 1.8d shows a corrugation of about 2 pm. Qualitatively the same pattern can be seen in the dI/dU map of Figure 1.8c whereby the signal-to-noise ratio is significantly improved as compared to the constant-current image of Figure 1.8b because of the frequency- and phase-sensitive detection obtained by means of a lock-in amplifier. Comparison of the dI/dU line section in Figure 1.8e with Figure 1.8d reveals that it is anti-corrugated with respect to the corresponding constant-current data which might be caused by different signs of $\tilde{\vec{m}}_S$ and $\vec{m}_S$.

It is important to note that the observation of a checkerboard pattern in an STM experiment performed with a magnetic tip is no unambiguous proof of surface antiferromagnetism. Only if no similar superstructure is observed with a non-magnetic tip a potential structural surface reconstruction as it might, for example, be caused by adsorbates can be excluded with sufficient certainty. For a definite proof it is required to reverse the magnetization direction of the ferromagnetic probe tip by an external field. If an Fe-coated tip is used, a field of about 2 T is typically sufficient to overcome the dominant magneto-static anisotropy energy. A vertical field of that magnitude safely forces the magnetization, which is parallel to the sample surface in the absence of external fields, into the out-of-plane direction. Since the exchange coupling in antiferromagnets is usually much stronger than the Zeeman energy an external field of a few Tesla leaves the sample's magnetic structure essentially unaffected and only changes the vertical component of the tip magnetization.

The described reversal of the tip magnetization and its influence on the magnetic contrast is demonstrated in the example in Figure 1.9. The middle row of Figure 1.9 shows a series of images which were taken within the same 4×4 nm^2 area. The images are aligned at the position of a native adsorbate in the center. Between successive measurements, the field direction was changed between +2.5 T and –2.5 T. Since the TMR effect depends on the relative orientation of tip and sample magnetization, the observed superstructure has to invert if the pattern is caused by spin-dependent tunneling. In contrast, the superstructure would remain unchanged if it were caused by structural effects. Careful inspection of the data indeed confirms the expected corrugation reversal. An easier evaluation is achieved by calculating sum and difference images of data taken at opposite magnetic fields. The top and bottom rows of Figure 1.9 display such images, respectively. As expected for SP-STM, the sum images show a vanishing contrast on clean areas (top). More importantly, the contrast is enhanced in the difference images in the bottom row. It is quite remarkable that the magnetic superstructure is visible even at the position of the adsorbate which indicates

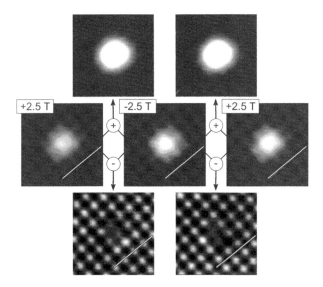

FIGURE 1.9   (middle row) Constant-current images of a 4×4 nm^2 region of an Fe monolayer on W(001) around a native adsorbate measured with an Fe-coated tip ($I = 30$ nA, $U = -40$ mV, $T = 13 \pm 1$ K). Opposite external fields were applied to force the tip magnetization out of the in-plane easy axis into perpendicular "up" and "down" directions. Since the exchange coupling within the Fe monolayer is much stronger than the Zeeman energy, the external field leaves the sample's magnetic structure unaffected. By changing the magnetization direction of the tip any magnetic contrast is inverted but topographic and spin-averaged electronic contributions remain unchanged. Consequently, the sum (top row) and the difference (bottom row) of two images recorded with opposite tip magnetization directions allow the separation of non-magnetic from magnetic contributions, respectively.

that the antiferromagnetic state of the Fe monolayer is largely unaffected by the adsorbate and that the spin-polarized LDOS of the Fe film protrudes through the adsorbate into the vacuum.

1.5.2 ANTIFERROMAGNETIC DOMAIN WALLS

The ability to image compensated spin structures with atomic resolution opens up the possibility of directly observing domain walls in antiferromagnetic materials which could only be imaged indirectly previously [62]. However, domain walls in antiferromagnets are rare and represent a metastable state. While ferromagnets reduce their total magnetic energy by domain formation, for example by avoiding magnetic charges in the flux closure configuration, domain formation in antiferromagnets is generally associated with a higher total energy state as their spin structure is already compensated and the additional exchange energy of domain walls cannot be compensated.

Accordingly, the antiferromagnetic domain wall imaged in Figure 1.10a on the surface of an antiferromagnetic monolayer of Fe on W(001) is very short [17]. Typically the length of domain walls amounts to a few nm only and they are mostly trapped at structural imperfections. The antiferromagnetic domain walls appear as an area with a faded magnetic contrast in the spin sensitive image of Figure 1.10a. At the position of the domain wall the magnetic period shifts by one atomic row, thereby separating the two domains in the lower and upper part of the image. The domain wall is very narrow as indicated by the fact that the maximum contrast is fully regained only a few atomic rows away from the domain wall center. A quantitative

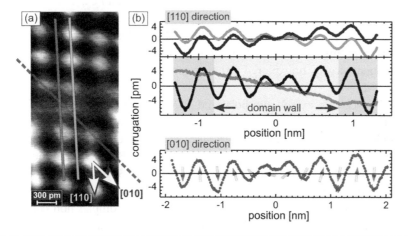

FIGURE 1.10   (a) Detailed view of a $\langle 110 \rangle$-oriented domain wall in an out-of-plane antiferromagnetic Fe monolayer on W(001) measured at $T = 13 \pm 1$ K. (b) The line sections drawn along the correspondingly colored lines in (a) along the [110] (upper panel) and the [010] direction (lower panel). The middle panel shows the sum (black) and the difference (gray) of the line profiles shown in the upper panel which were measured on adjacent atomic [110] rows. The data reveal that the wall is about 1.6 nm wide and that its out-of-plane component exhibits mirror symmetry. (After Bode, M. et al., *Nature Mater.* 5, 477, 2006 .With permission.)

understanding can be achieved by careful analysis of line sections taken along low-index crystallographic directions. The upper panel of Figure 1.10b shows data taken on adjacent atomic rows along the [110] direction, i.e., perpendicular to the wall. The middle panel shows the sum and the difference of these lines in black and gray, respectively. The difference (gray) shows an almost constant signal of opposite sign at the left and right rim of the line section, i.e., far away from the domain wall center. These regions (domains) are connected by a constant slope that extends over approximately 1.6 nm which corresponds to a domain wall width of 6–8 atomic rows. The sum (black) reveals that the average out-of-plane component of these atomic rows is mirror-symmetric. This mirror-symmetric appearance, which is also found in the line profile taken along the [010] direction shown in the lower panel of Figure 1.10b, indicates that the domain wall center is located between two atomic rows [17].

1.5.3 COMPLEX NON-COLLINEAR SPIN STRUCTURES

Structure-induced magnetic frustration [63], the competition of nearest- and next-nearest neighbor exchange interactions, or the so-called Dzyaloshinskii-Moriya interaction [64–66] may lead to complex two- or even three-dimensional spin structures. For bulk samples a variety of experimental techniques, the most prominent of which is polarized neutron scattering, have been used to study ground state magnetic properties and phase transitions. The adaption of those techniques to single thin films or even nanostructures is not practically achievable yet. On one hand this is caused by the extraordinarily small scattering cross section of neutrons with atoms. Also, the flux of neutron sources could not be increased to the same extent as synchrotron X-ray sources which increased their brilliance by many orders of magnitude over the past decades. As a result ordered complex atomic scale spin structures in these films and nanostructures remained largely unexplored experimentally prior to the invention of SP-STM.

To a certain extent spin-polarized STM can fill this gap. While being limited to conducting, relatively clean, and preferentially single-crystalline sample surfaces its atomic spin resolution capabilities should essentially allow for the imaging of arbitrary spin structures. Based on the independent orbital approximation S. Heinze predicted SP-STM images of helical and geometrically frustrated spin structures of thin films and suggested that SP-STM might even be suited for imaging "more complex non-collinear magnetic structures such as domain walls in antiferromagnets, spin-spiral states, spin glasses, or disordered states" [21].

Important experiments have confirmed these predictions. In the following, two key achievements will be presented, i.e., the successful imaging of the Néel state and the discovery of spin-orbit driven chiral spin structures on surfaces. The Néel state originates from the geometrical frustration of the magnetic coupling which cannot be matched with the crystal lattice.

A theoretically well-known example is an antiferromagnet on a triangular or hexagonal lattice. While the antiferromagnetic coupling can be maintained in the dimer and linear trimer, it cannot be realized in a compact trimer or larger clusters. Theoretically a 120° arrangement of nearest-neighbor spins was predicted, called the Néel state. One possible configuration is represented in Figure 1.11a. It consists of three sublattices (red, green, and blue) which are rotated by 120°. The independent orbital approximation by Heinze [21] allows for the simulation of spin-polarized STM images expected for two high-symmetry tip magnetization directions (Figure 1.11b, c).

Although the Néel state has been discussed for decades its direct observation was not possible until recently. Following a proposal by Wortmann et al. [18], Gao et al. performed SP-STM experiments on Mn monolayer films on Ag(111) [67]. Figure 1.11e shows the result of a constant-current image measured with a Cr-coated probe tip. As can be seen in the inset, Figure 1.11d, this high-resolution image was taken on a coalesced Mn island. While the left part exhibits a stacking fault with respect to the substrate, i.e., hcp stacking, the right part is regularly fcc-stacked. According to Gao et al. [67] this surprising result can only be explained by 30° rotation of the Mn moments as shown by arrows in the left and right of the image. It should be noted that the tip magnetization is not known. The smooth structural change in the dark transition region seems to be accompanied by a change of magnetic rotation.

Non-collinear spin structures may also result from relativistic effects. Namely, the Dzyaloshinskii-Moriya interaction, a spin-orbit–induced chiral contribution to the exchange, can result in very complex spin structures such

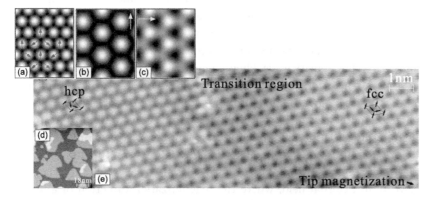

FIGURE 1.11   (a) Néel state consisting of three sublattices which are rotated by 120°. (b, c) Simulated spin-polarized STM images for two high-symmetry tip magnetization directions. (After Heinze, S., *Appl. Phys. A* 85, 407, 2006. With permission.) (d) Topography of submonolayer Mn islands are connected together. (e) Atomic scale constant-current SP-STM image taken in the rectangular area in (d). The tunneling parameters are $U = 20$ mV and $I = 20$ nA. The right part is the hcp region, and the left part is the fcc region. In the middle, the image appears darker which is considered as the smooth structural transition region from hcp to fcc. With the assumed tip magnetization orientation (bottom right), one possible configuration of the 120° Néel structure is indicated by arrows. (After Gao, C.L. et al., *Phys. Rev. Lett.* 101, 267205, 2008. With permission.)

as Skyrmion lattices [68] and two- or three-dimensional spin spirals [69–71]. However, the Dzyaloshinskii-Moriya interaction is irrelevant in most bulk materials as it cancels in inversion-symmetric lattices, such as bcc-, fcc-, or hcp-crystals. Only in some rare cases where the bulk inversion symmetry is broken, such as MnSi [72] or $Fe_{0.5}Co_{0.5}Si$ [73], the Dzyaloshinskii-Moriya interaction leads to chiral spin structures.

The situation is completely different, however, on surfaces, clusters, and in magnetic nanoparticles. Here the inversion symmetry is broken at interfaces or surfaces which separate the magnetic object from the environment, i.e., the substrate and/or the vacuum. Similar to collinear antiferromagnetic films and nanoparticles these chiral spin structures are compensated and cannot be detected by spatially averaging detection methods like neutron diffraction.

Only the invention of spin-polarized scanning tunneling microscopy allowed for the observation of ferromagnetic [70] and antiferromagnetic spin spirals [69] in numerous thin film systems [74–77]. For example, Figure 1.12a shows a constant current STM image of a Mn monolayer measured with an Fe-coated probe tip without an externally applied magnetic field [69]. A location with a few adsorbates has intentionally been chosen to find exactly the same position in later experiments. Dark and bright stripes can be recognized which correspond to rows of parallel spins oriented along the [001] direction of the sample. These stripes alone would be consistent with a row-wise antiferromagnet [15]. However, the amplitude of the stripes is not constant but varies periodically. This becomes particularly obvious in the experimental line section displayed in the bottom right panel of Figure 1.12a which shows an apparent periodicity of about 6 nm. A more detailed analysis revealed, however, that the pattern of maxima and minima is shifted by π such that a full period extends over 12 nm.

This pattern may be caused by a spin spiral but would also be consistent with a spin density wave. Phenomenologically, the difference between these two spin structures becomes obvious if we compare the size and orientation of their local magnetic moments. While the orientation of the magnetic moment changes but the size remains constant in a spin spiral, the opposite situation can be found in a spin density wave material; here the orientation is constant but the size oscillates. With this in mind it is relatively straightforward to distinguish these two cases by spin-polarized STM in an external field, where the magnetization direction of a ferromagnetic tip can be changed by the Zeeman energy. In a material with a spin density wave the minimum has to stay at the same position since it is caused by a vanishing sample's magnetic moment. In contrast, the minimum magnetic contrast in the case of a spin spiral is caused by an orthogonal orientation between tip and local sample magnetization. If the tip magnetization is rotated into the formerly orthogonal direction, the minimum of the magnetic contrast is expected to shift to a new position which is now orthogonal to the tip. In fact, this shift can clearly be observed in Figure 1.12b, c. The upper right panels of Figure 1.12a–c show

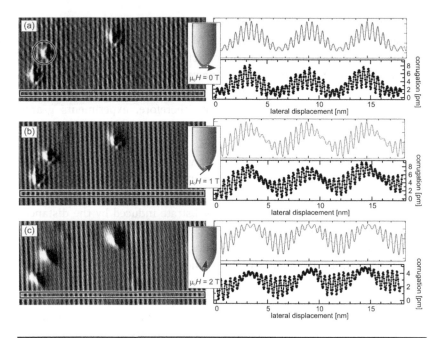

FIGURE 1.12 Magnetically sensitive constant-current images of the Mn mono-layer on W(110) (left panels), corresponding experimental (bottom right) and theoretical line sections (bottom left panels, courtesy of Stefan Heinze). (Tunnelling parameters: $I=2$ nA, $U=30$ mV, $T=13\pm1$ K.) The data show the situation at external fields of 0 T (a), 1 T (b) and 2 T (c). (After Bode, M. et al., *Nature* 447, 190, 2007. With permission.) As sketched in the insets, the external field rotates the tip magnetization from in-plane (a) to out-of-plane (c), shifting the position of maximum spin contrast. This proves that the Mn layer does not exhibit a spin-density wave but rather a cycloidal spin spiral.

simulated spin-polarized STM line sections obtained by Heinze with the independent orbital approximation [21]. An excellent agreement between experiment and simulation can be recognized.

1.6 MAGNETISM OF ATOMIC-SCALE ASSEMBLIES

Scanning tunneling microscopy is not only able to resolve surfaces on the atomic scale but also allows for the controlled assembly of nanostructures from single atoms [78]. It was a rather obvious idea to investigate the spin structure of self-organized or engineered assemblies. Indeed, important experiments showed that STM-based techniques are able to determine the *g*-value [79] and the anisotropy barrier [80] of single magnetic atoms in inelastic tunneling experiments, so-called spin-flip spectroscopy. Furthermore, detailed insight into the magnetic coupling of engineered assemblies was obtained [81], and spin-polarized STM even allowed for the measurement of magnetization curves on single atoms deposited on a nonmagnetic substrate [82].

Although no hysteresis was observed in Ref. [82], these data clearly demonstrated the ability of spin-polarized STM to detect spin signals even from the most diluted amount of magnetic material we can imagine, consisting of a few atoms only. Obviously, one expects a very small total anisotropy energy which prevents single isolated atoms or small clusters from spontaneously reversing their magnetization direction. Therefore, experiments geared towards the observation of atomic scale magnets need to be performed at very low temperatures, typically below 4 K.

Pioneering experiments in this context were performed by Khajetoorians, Wiebe, and Wiesendanger [83]. One particularly nice example, which helped unravel even the finest details of the indirect coupling behavior of magnetic atoms on a non-magnetic substrate induced by the distance-dependent Ruderman-Kittel-Kasuya-Yosida (RKKY) interaction, is shown in Figure 1.13. Linear chains of Fe atoms were assembled with the STM tip on a Cu(111) substrate. The chain length was modified between three and seven Fe atoms, out of which chains with three, four, and five atoms are presented in Figure 1.13a–c, respectively. Hereby, the Fe–Fe distance was chosen such that the RKKY coupling is antiferromagnetic, resulting in an antiparallel spin orientation of adjacent Fe atoms. The experiments were performed in a perpendicular magnetic field which is slightly higher than the critical magnetic field strength at which Zeeman energy overcomes the effective nearest-neighbor exchange interaction J_{NN}. As a result, within an Ising-like scenario

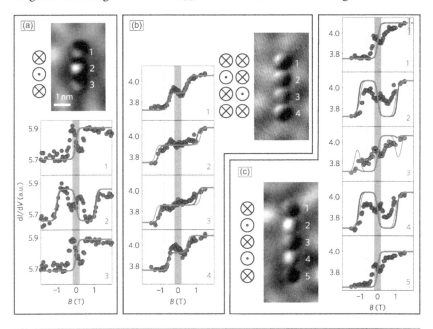

FIGURE 1.13 Spin-resolved differential conductance dI/dU maps (taken at $B \approx -0.6$ T) and magnetization curves measured on each particular atom of linear chains assembled from (a) three, (b) four, and (c) five Fe atoms, respectively. Calculated spin structures are represented schematically by symbols $\otimes$ and is $\odot$. Gray and colored lines represent theoretical magnetization curves (see text for details). (After Khajetoorians, A.A. et al., *Nature Phys.* 8, 497, 2012. With permission.)

spin structures were expected which are schematically sketched by symbols $\otimes$ and $\odot$, representing atoms that are magnetized down and up, respectively.

As evidenced by the marked alternating dark–bright contrast observed in SP-STM experiments, the single uncompensated magnetic moment caused by the ends of spin chains consisting of an odd number of atom leads to a dedicated lowest energy ground state, where the uncompensated moment is oriented parallel to the external field ($\otimes$ for the negative field used here). In contrast, the chain consisting of four Fe atoms (as well as other even Fe chains; cf. Figure 2 in Ref. [83]) display a much weaker contrast. According to the theoretical analysis performed in Ref. [83] this is due to the presence of a mixed state, which is the superposition of the states $|\otimes,\odot,\otimes,\otimes\rangle$ and $|\otimes,\otimes,\odot,\otimes\rangle$.

By performing field-dependent experiments site-specific hysteresis curves can be obtained from each Fe atom within the chain. Although the data are in very good agreement with density functional theory calculations using the full-potential Korringa–Kohn–Rostoker (KKR) Green function method [84] for large fields, significant deviations from the predicted behavior are observed around $B = 0$ T (see areas shaded gray in Figure 1.13). Here, the slope of the magnetization curves predicted within the Ising model is opposite to the measured data for odd chains and some oscillations observed on even chains are not captured by the theoretical analysis. It has been found that these differences are caused by tiny site-dependent variations of the exchange interactions and contributions beyond a simple nearest-neighbor exchange [83]. Although these variations only amount to about 10% of the total value, at small fields they affect the resulting spin structure at a qualitative level. The authors of Ref. [83] were able to qualitatively reproduce the low-field anomalies by considering an additional magnetic field, which is oriented antiparallel to the external field and saturates at $B \approx \pm 0.5$ T. However, the detailed origin of this field, which could in principle be of dipolar character, remained unclear.

1.7 TIME-RESOLVED EXPERIMENTS

Following some initial studies of superparamagnetic nanoislands [85–87] which were performed with moderate time resolution in the ms range by statically positioning the magnetic probe tip over one particular nanoisland and continuously measuring the dI/dU signal, considerable effort was undertaken towards experiments that aim for an understanding of dynamic properties of magnetic nanostructures on much shorter time scales. Pioneering all-electrical pump-probe experiments were performed by Loth et al. [88, 89] on the electron spin relaxation times in single atomic-scale nanostructures. For the detection scheme of these experiments, which is illustrated in Figure 1.14a, it is important to initially set the spin quantization axis of the magnetic tip and the sample spin by the Zeemann energy induced by a strong external magnetic field. Since thermal fluctuations affect the spin alignment of the initial state, lower measurement temperatures lead to less stringent field requirements. However, practically available superconducting

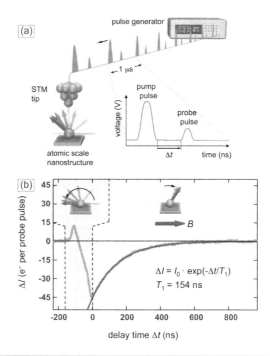

FIGURE 1.14   (a) All-electrical pump-probe detection scheme for measurements of the spin relaxation time in an SP-STM experiment. (b) Variation of the tunneling current ΔI Ê plotted as a function of the pump-probe delay time as obtained from a stroboscopic measurement on a single Fe/Cu-dimer (light green). The decay of the experimental data was fit with an exponential function (dark green), revealing a relaxation time T_1Ê $= 154$ ns. (After Loth, S. et al., *Science* 329, 1628, 2010. With permission. [Loth, S. et al., *Physik in unserer Zeit* 42, 168, 2011.])

magnets are typically limited to $B \leq 16$, thereby requiring measurement temperatures well below the boiling point of liquid helium.

This spin-aligned initial state is excited by means of an electric pump pulse with a voltage amplitude which exceeds the excitation energy of the sample spin system, followed by a probe pulse with a variable delay (see Figure 1.14a). To make sure the second pulse only probes the resulting spin configuration without further exciting the sample system the amplitude of the probe pulse must be significantly lower than the spin excitation threshold. Due to the magneto-resistance effect the tunneling current between the tip and the sample during the probe pulse depends on the momentary relative orientation of tip and sample magnetization. To determine the spin relaxation time of the excited state this stroboscopic measurement scheme is repeated with various pump-probe delays. The result of a typical measurement performed on a single nanostructure consisting of a Fe/Cu dimer is presented in Figure 1.14b [88]. At large negative delay times the probe pulse precedes the pump pulse and therefore detects the undistorted system, as indicated by a vanishing variation of the tunneling current ($\Delta I = 0$). This interval is followed by the blue-shaded region where pump and probe pulse overlap. Only after this regime the relaxation of the sample spin can be clearly recognized as an exponential recovery of the SP-STM signal. Fitting the experimental data yields a spin relaxation time of 154 ns.

1.8 OUTLOOK

The application of spin-polarized scanning tunneling microscopy and spectroscopy has contributed considerably to our modern understanding of static domain and domain wall structures in ferromagnetic, but in particular in antiferromagnetic surfaces, thin films, and nanoparticles. Beyond that, during the past 10 years the spin-polarized STM toolbox has been significantly extended and now even enables us to obtain magnetization curves and spin relaxation times on the single-atomic level. Recent progress suggests that single atom ferromagnets with spin bistability at zero field are—at least at very low temperature—achievable [90], although there remains a lot to do to transfer these findings into real-life applications.

REFERENCES

1. G. Binnig, K. H. Frank, H. Fuchs et al., Tunneling spectroscopy and inverse photoemission: Image and field states, *Phys. Rev. Lett.* **55**, 991 (1985).
2. M. Bode, R. Pascal, and R. Wiesendanger, Distance-dependent STM-study of the W(110)/Cr(15×3) surface, *Z. Phys. B* **101**, 103 (1996).
3. L. Olesen, M. Brandbyge, M. R. Sørensen et al., Apparent barrier height in scanning tunneling microscopy revisited, *Phys. Rev. Lett.* **76**, 1485 (1996).
4. T. Jung, Y. W. Mo, and F. J. Himpsel, Identification of metals in scanning tunneling microscopy via image states, *Phys. Rev. Lett.* **74**, 1641 (1995).
5. Y. Hasegawa and Ph. Avouris, Direct observation of standing wave formation at surface steps using scanning tunneling spectroscopy, *Phys. Rev. Lett.* **71**, 1071 (1993).
6. M. F. Crommie, C. P. Lutz, and D. M. Eigler, Imaging standing waves in a two-dimensional electron gas, *Nature* **363**, 524 (1993).
7. M. Bode, M. Getzlaff, A. Kubetzka, R. Pascal, O. Pietzsch, and R. Wiesendanger, Temperature-dependent exchange splitting of a surface state on a local-moment magnet: Tb(0001), *Phys. Rev. Lett.* **83**, 3017 (1999).
8. A. Bauer, A. Mühlig, D. Wegner, and G. Kaindl, Lifetime of surface states on (0001) surfaces of lanthanide metals, *Phys. Rev. B* **65**, 075421 (2002).
9. D. Wegner, A. Bauer, and G. Kaindl, Magnon-broadening of exchange-split surface states on lanthanide metals, *Phys. Rev. B* **73**, 165415 (2006).
10. D. T. Pierce, Spin-polarized electron microscopy, *Physica Scripta* **38**, 291 (1988).
11. O. Pietzsch, A. Kubetzka, M. Bode, and R. Wiesendanger, Observation of magnetic hysteresis at the nano-scale by spin polarized scanning tunneling spectroscopy, *Science* **292**, 2053 (2001).
12. M. Pratzer, H. J. Elmers, M. Bode, O. Pietzsch, A. Kubetzka, and R. Wiesendanger, Atomic-scale magnetic domain walls in quasi-one-dimensional Fe nanostripes, *Phys. Rev. Lett.* **87**, 127201 (2001).
13. A. Yamasaki, W. Wulfhekel, R. Hertel, S. Suga, and J. Kirschner, Direct observation of the single-domain limit of Fe nano-magnets by spin-polarized STM, *Phys. Rev. Lett.* **91**, 127201 (2003).
14. A. Wachowiak, J. Wiebe, M. Bode, O. Pietzsch, M. Morgenstern, and R. Wiesendanger, Direct observation of internal spin structure of magnetic vortex cores, *Science* **298**, 577 (2002).

15. S. Heinze, M. Bode, A. Kubetzka et al., Real-space imaging of two-dimensional antiferromagnetism on the atomic scale, *Science* **288**, 1805 (2000).
16. A. Kubetzka, P. Ferriani, M. Bode et al., Revealing antiferromagnetic order of the Fe monolayer on w(001): Spin-polarized scanning tunneling microscopy and first-principles calculations, *Phys. Rev. Lett.* **94**, 087204 (2005).
17. M. Bode, E. Y. Vedmedenko, K. von Bergmann et al., Atomic spin structure of antiferromagnetic domain walls, *Nat. Mater.* **5**, 477 (2006).
18. D. Wortmann, S. Heinze, Ph. Kurz, G. Bihlmayer, and S. Blügel, Resolving complex atomic-scale spin structures by spin-polarized scanning tunneling microscopy, *Phys. Rev. Lett.* **86**, 4132 (2001).
19. J. Tersoff and D. R. Hamann, Theory and application for the scanning tunneling microscope, *Phys. Rev. Lett.* **50**, 1998 (1983).
20. J. Tersoff and D. R. Hamann, Theory of the STM, *Phys. Rev. B* **31**, 805 (1985).
21. S. Heinze, Simulation of spin-polarized scanning tunneling microscopy images of nanoscale non-collinear magnetic structures, *Appl. Phys. A* **85**, 407 (2006).
22. J. C. Slonczewski, Conductance and exchange coupling of two ferromagnets separated by a tunneling barrier, *Phys. Rev. B* **39**, 6995 (1989).
23. R. Wiesendanger, H. J. Güntherodt, G. Güntherodt, R. J. Gambino, and R. Ruf, Observation of vacuum tunneling of spin-polarized electrons with the STM, *Phys. Rev. Lett.* **65**, 247 (1990).
24. M. Kleiber, M. Bode, R. Ravlić, and R. Wiesendanger, Topology-induced spin frustrations at the Cr(001) surface studied by spin-polarized scanning tunneling spectroscopy, *Phys. Rev. Lett.* **85**, 4606 (2000).
25. M. Bode, M. Getzlaff, and R. Wiesendanger, Spin-polarized vacuum tunneling into the exchange-split surface state of Gd(0001), *Phys. Rev. Lett.* **81**, 4256 (1998).
26. A. Selloni, P. Carnevali, E. Tosatti, and C. D. Chen, Voltage-dependent scanning-tunneling microscopy of a crystal surface: Graphite, *Phys. Rev. B* **31**, 2602 (1985).
27. R. Wiesendanger, M. Bode, and M. Getzlaff, Vacuum-tunneling magnetoresistance: The role of spin-polarized surface states, *Appl. Phys. Lett.* **75**, 124 (1999).
28. O. Pietzsch, A. Kubetzka, M. Bode, and R. Wiesendanger, Spin-polarized scanning tunneling spectroscopy of nanoscale cobalt islands on Cu(111), *Phys. Rev. Lett.* **92**, 057202 (2004).
29. M. Johnson and J. Clarke, Spin-polarized scanning tunneling microscope: Concept, design, and preliminary results from a prototype operated in air, *J. Appl. Phys.* **67**, 6141 (1990).
30. W. Wulfhekel and J. Kirschner, Spin-polarized scanning tunneling microscopy on ferromagnets, *Appl. Phys. Lett.* **75**, 1944 (1999).
31. M. Bode, M. Getzlaff, and R. Wiesendanger, Quantitative aspects of spin-polarized scanning tunneling spectroscopy of Gd(0001), *J. Vac. Sci. Technol. A* **17**, 2228 (1999).
32. H. F. Ding, W. Wulfhekel, J. Henk, P. Bruno, and J. Kirschner, Absence of zero-bias anomaly in spin-polarized vacuum tunneling in Co(0001), *Phys. Rev. Lett.* **90**, 116603 (2003).
33. D. T. Pierce and F. Meier, Photoemission of spin-polarized electrons from GaAs, *Phys. Rev. B* **13**, 5484 (1976).
34. R. Jansen, M. C. M. M. van der Wielen, M. W. J. Prins, D. L. Abraham, and H. van Kempen, Progress toward spin-sensitive scanning tunneling microscopy using optical orientation in GaAs, *J. Vac. Sci. Technol. B* **12**, 2133 (1994).
35. M. W. J. Prins, M. C. M. M. van der Wielen, R. Jansen, D. L. Abraham, and H. van Kempen, Photoamperic probes in scanning tunneling microscopy, *Appl. Phys. Lett.* **64**, 1207 (1994).
36. M. W. J. Prins, H. van Kempen, H. van Leuken, R. A. de Groot, W. van Roy, and J. de Boeck, Spin-dependent transport in metal/semiconductor tunnel junctions, *J. Phys.: Cond. Matter* **7**, 9447 (1995).

37. R. Jansen, M. W. J. Prins, and H. van Kempen, Theory of spin-polarized transport in photoexcited semiconductor/ferromagnet tunnel junctions, *Phys. Rev. B* **57**, 4033 (1998).

38. V. P. LaBella, D. W. Bullock, Z. Ding et al., Spatially resolved spin-injection probability for GaAs, *Science* **292**, 1518 (2001).

39. U. Schlickum, W. Wulfhekel, and J. Kirschner, Spin-polarized scanning tunneling microscope for imaging the in-plane magnetization, *Appl. Phys. Lett.* **83**, 2016 (2003).

40. M. Bode, R. Pascal, and R. Wiesendanger, Scanning tunneling spectroscopy of Fe/W(110) using iron covered probe tips, *J. Vac. Sci. Technol. A* **15**, 1285 (1997).

41. M. Bode, O. Pietzsch, A. Kubetzka, and R. Wiesendanger, Imaging magnetic nanostructures by spin-polarized scanning tunneling spectroscopy, *J. Electr. Spectr. Rel. Phen.* **114–116**, 1055 (2001).

42. H. J. Elmers and U. Gradmann, Magnetic anisotropies in Fe(110) films on W(110), *Appl. Phys. A* **51**, 255 (1990).

43. G. André, A. Aspelmeier, B. Schulz, M. Farle, and K. Baberschke, Temperature-dependence of surface and volume anisotropy in Gd/W(110), *Surf. Sci.* **326**, 275 (1995).

44. A. Kubetzka, M. Bode, O. Pietzsch, and R. Wiesendanger, Spin-polarized scanning tunneling microscopy with antiferromagnetic probe tips, *Phys. Rev. Lett.* **88**, 057201 (2002).

45. A. Li Bassi, C. S. Casari, D. Cattaneo et al., Bulk Cr tips for scanning tunneling microscopy and spin-polarized scanning tunneling microscopy, *Appl. Phys. Lett.* **91**, 173120 (2007).

46. A. Schlenhoff, S. Krause, G. Herzog, and R. Wiesendanger, Bulk Cr tips with full spatial magnetic sensitivity for spin-polarized scanning tunneling microscopy, *Appl. Phys. Lett.* **97**, 083104 (2010).

47. C. L. Fu and A. J. Freeman, Surface ferromagnetism of Cr(001), *Phys. Rev. B* **33**, 1755 (1986).

48. S. Blügel, D. Pescia, and P. H. Dederichs, Ferromagnetism versus antiferromagnetism of the Cr(001) surface, *Phys. Rev. B* **39**, 1392 (1989).

49. T. Nagahama, S. Yuasa, E. Tamura, and Y. Suzuki, Spin-dependent tunneling in magnetic tunnel junctions with a layered antiferromagnetic Cr(001) spacer: Role of band structure and interface scattering, *Phys. Rev. Lett.* **95**, 086602 (2005).

50. R. Matsumoto, A. Fukushima, K. Yakushiji et al., Spin-dependent tunneling in epitaxial Fe/Cr/MgO/Fe magnetic tunnel junctions with an ultrathin Cr(001) spacer layer, *Phys. Rev. B* **79**, 174436 (2009).

51. J. A. Stroscio, D. T. Pierce, A. Davies, R. J. Celotta, and M. Weinert, Tunneling spectroscopy of bcc(001) surface states, *Phys. Rev. Lett.* **75**, 2960 (1995).

52. T. Hänke, M. Bode, S. Krause, L. Berbil-Bautista, and R. Wiesendanger, Temperature-dependent scanning tunneling spectroscopy of Cr(001): Orbital Kondo resonance versus surface state, *Phys. Rev. B* **72**, 085453 (2005).

53. M. Budke, T. Allmers, M. Donath, and M. Bode, Surface state vs orbital Kondo resonance at Cr(001): Arguments for a surface state interpretation, *Phys. Rev. B* **77**, 233409 (2008).

54. O. Y. Kolesnychenko, R. de Kort, M. I. Katsnelson, A. I. Lichtenstein, and H. van Kempen, Real-space imaging of an orbital Kondo resonance on the Cr(001) surface, *Nature* **415**, 507 (2002).

55. R. Ravlić, M. Bode, A. Kubetzka, and R. Wiesendanger, Correlation of dislocation and domain structure of Cr(001) investigated by spin-polarized scanning tunneling microscopy, *Phys. Rev. B* **67**, 174411 (2003).

56. A. Hubert and R. Schäfer, *Magnetic Domains*, Springer (1998).

57. L. Berbil-Bautista, S. Krause, M. Bode, and R. Wiesendanger, Spin-polarized scanning tunneling microscopy and spectroscopy of ferromagnetic Dy(0001)/W(110) films, *Phys. Rev. B* **76**, 064411 (2007).

58. S. Mørup, D. E. Madsen, C. Frandsen, C. R. H. Bahl, and M. F. Hansen, Experimental and theoretical studies of nanoparticles of antiferromagnetic materials, *J. Phys. Condens. Matter* **19**, 213202 (2007).

59. G. A. Mulhollan, R. L. Fink, J. L. Erskine, and G. K. Walters, Local spin correlations in ultrathin FeW(100) films, *Phys. Rev. B* **43**, 13645 (1991).

60. J. Chen and J. L. Erskine, Surface step-induced magnetic anisotropy in thin epitaxial Fe films on W(001), *Phys. Rev. Lett.* **68**, 1212 (1992).

61. R. Wu and A. J. Freeman, Magnetic properties of Fe overlayers on W(001) and the effects of oxygen adsorption, *Phys. Rev. B* **45**, 7532 (1992).

62. N. B. Weber, H. Ohldag, H. Gomonaj, and F. U. Hillebrecht, Magnetostrictive domain walls in antiferromagnetic NiO, *Phys. Rev. Lett.* **91**, 237205 (2003).

63. U. Bhaumik and I. Bose, Collinear Néel-type ordering in partially frustrated lattices, *Phys. Rev. B* **58**, 73 (1998).

64. I. E. Dzialoshinskii, Thermodynamic theory of "weak" ferromagnetism in antiferromagnetic substances, *Sov. Phys. JETP* **5**, 1259 (1957).

65. T. Moriya, Anisotropic superexchange interaction and weak ferromagnetism, *Phys. Rev.* **120**, 91 (1960).

66. I. E. Dzyaloshinski, Theory of helicoidal structures in antiferromagnets. iii, *Sov. Phys. JETP* **20**, 665 (1965).

67. C. L. Gao, W. Wulfhekel, and J. Kirschner, Revealing the 120° antiferromagnetic Néel structure in real space: One monolayer Mn on Ag(111), *Phys. Rev. Lett.* **101**, 267205 (2008).

68. U. K. Rössler, A. N. Bogdanov, and C. Pfleiderer, Spontaneous skyrmion ground states in magnetic metals, *Nature* **442**, 797 (2006).

69. M. Bode, M. Heide, K. von Bergmann et al., Chiral magnetic order at surfaces driven by inversion asymmetry, *Nature* **447**, 190 (2007).

70. P. Ferriani, K. von Bergmann, E. Y. Vedmedenko et al., Atomic-scale spin spiral with a unique rotational sense: Mn monolayer on W(001), *Phys. Rev. Lett.* **101**, 027201 (2008).

71. M. Heide, G. Bihlmayer, and S. Blügel, Dzyaloshinskii-Moriya interaction accounting for the orientation of magnetic domains in ultrathin films: Fe/W(110), *Phys. Rev. B* **78**, 140403 (2008).

72. S. Mühlbauer, B. Binz, F. Jonietz et al., Skyrmion lattice in a chiral magnet, *Science* **323**, 915 (2009).

73. M. Uchida, Y. Onose, Y. Matsui, and Y. Tokura, Real-space observation of helical spin order, *Science* **311**, 359 (2006).

74. B. Santos, J. M. Puerta, J. I. Cerda et al., Structure and magnetism of ultra-thin chromium layers on W(110), *New J. Phys.* **10**, 013005 (2008).

75. P. Sessi, N. P. Guisinger, J. R. Guest, and M. Bode, Temperature and size dependence of antiferromagnetism in Mn nanostructures, *Phys. Rev. Lett.* **103**, 167201 (2009).

76. S. Heinze, K. von Bergmann, M. Menzel et al., Spontaneous atomic-scale magnetic skyrmion lattice in two dimensions, *Nat. Phys.* **7**, 713 (2011).

77. N. Romming, C. Hanneken, M. Menzel et al., Writing and deleting single magnetic skyrmions, *Science* **341**(6146), 636 (2013).

78. D. M. Eigler and E. K. Schweizer, Positioning single atoms with a scanning tunnelling microscope, *Nature* **344**, 524 (1990).

79. A. J. Heinrich, J. A. Gupta, C. P. Lutz, and D. M. Eigler, Single-atom spin-flip spectroscopy, *Science* **306**, 466 (2004).

80. C. F. Hirjibehedin, C.-Y. Lin, A. F. Otte et al., Large magnetic anisotropy of a single atomic spin embedded in a surface molecular network, *Science* **317**, 1199 (2007).

81. C. F. Hirjibehedin, C. P. Lutz, and A. J. Heinrich, Spin coupling in engineered atomic structures, *Science* **312**, 1021 (2006).

82. F. Meier, L. Zhou, J. Wiebe, and R. Wiesendanger, Revealing magnetic interactions from single-atom magnetization curves, *Science* **320**, 82 (2008).

83. A. A. Khajetoorians, J. Wiebe, B. Chilian, S. Lounis, S. Blügel, and R. Wiesendanger, Atom-by-atom engineering and magnetometry of tailored nanomagnets, *Nat. Phys.* **8**, 497 (2012).

84. N. Papanikolaou, R. Zeller, and P. H. Dederichs, Conceptual improvements of the KKR method, *J. Phys. Condens. Matter* **14**, 2799 (2002).

85. M. Bode, O. Pietzsch, A. Kubetzka, R. Wiesendanger, Shape-dependent thermal switching behavior of superparamagnetic nanoislands, *Phys. Rev. Lett.* **92**, 067201 (2004).

86. S. Krause, L. Berbil-Bautista, G. Herzog, M. Bode, and R. Wiesendanger, Current-induced magnetization switching with a spin-polarized scanning tunneling microscope, *Science* **317**, 1537 (2007).

87. S. Krause, G. Herzog, T. Stapelfeldt et al., Magnetization reversal of nanoscale islands: How size and shape affect the arrhenius prefactor, *Phys. Rev. Lett.* **103**, 127202 (2009).

88. S. Loth, M. Etzkorn, C. P. Lutz, D. M. Eigler, and A. J. Heinrich, Measurement of fast electron spin relaxation times with atomic resolution, *Science* **329**, 1628 (2010).

89. S. Loth, M. Morgenstern, and W. Wulfhekel, Jenseits des gleichgewichts, *Physik in unserer Zeit* **42**, 168 (2011).

90. I. G. Rau, S. Baumann, S. Rusponi et al., Reaching the magnetic anisotropy limit of a 3d metal atom, *Science* **344**, 988 (2014).

2

Point Contact Andreev Reflection Spectroscopy

Boris E. Nadgorny

CONTENTS

2.1 INTRODUCTION

It is generally accepted that spin polarization plays a major role in *spintronics* [1, 2]. Indeed, one cannot adequately discuss spin-dependent transport without addressing the question of spin imbalance in a given system. The topics included in this chapter are chosen by the association one might have with the words *spin polarization*. When we think about it, the first thing that comes to mind is probably the asymmetry in the density of states (DOS) in magnetic materials. However, spin polarization may also be related to different transport, magnetic, and optical properties of charge carriers (see Chapters 2 and 6, Volume 2), spin injection into paramagnetic materials (see Chapter 5, Volume 1, Chapter 3, Volume 2, and Chapter 4, Volume 3), transmission coefficients through interfaces and tunnel barriers (see Chapters 4, 11, and 12 in Volume 1), and different electronic properties of surfaces and the bulk. This indicates that the concept of spin polarization is not self-evident and must be defined [3]. These definitions are naturally linked to spin measurement techniques, as we will discuss later.

The efficiency of a ubiquitous spintronics device can be dramatically improved as the spin polarization approaches 100%. For example, in

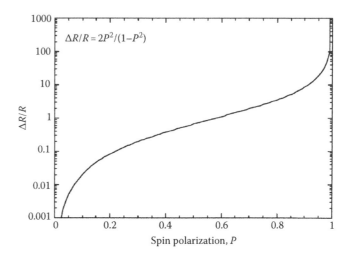

FIGURE 2.1 The efficiency of a TMR device $\Delta R/R$ as a function of spin polarization $P = P_T$ according to Julliere's model (see Section 2.2).

tunneling magnetoresistance (TMR) devices, the efficiency $\Delta R/R \to \infty$ as the spin polarization P approaches 100%, based on the qualitative Julliere's model [4]; see Figure 2.1. Obviously, this conclusion corresponds to an idealized case. For any real devices, interface scattering, spin–orbit interaction, nonzero temperatures, and other spin decoherence effects would reduce the device efficiency to finite values. Full spin polarization can be nominally realized in the extreme case of a *half-metal*, simultaneously exhibiting metallic and insulating properties at a microscopic level [5]. In half-metals, first introduced by de Groot et al. [6], only one of the energy bands (majority) crosses the Fermi level, E_F, whereas another (minority) band has a gap in the DOS. Consequently, at $T = 0$, a half-metal would only be conducting in one spin channel, whereas the other spin channel would be insulating (hence *half*-metal), so that all the electrical current is carried by the charges with the same spin. However, to achieve a high degree of transport spin polarization, half-metallicity in this narrow sense is not required. For example, if the charge carriers from the minority spin band are localized, the current will be due almost entirely to the majority carriers, resulting in the so-called *transport half-metallicity* [7, 8]. Similarly, recent dramatic advances in interface engineering, such as the use of MgO tunnel barriers [9, 10], allow one to use spin-filtering effects in magnetic tunnel junctions to obtain several hundred percent of TMR, which, to some extent, has alleviated the need for intrinsic half-metallicity.

The bulk of the chapter will be devoted to the discussion of *point contact Andreev reflection* (PCAR) spectroscopy, one of the commonly used techniques to measure spin polarization [11, 12]. This technique is based on the so-called *Andreev reflection* (AR) process [13, 14], discussed in Section 2.4, which determines the transport properties of a clean (transparent) normal-metal (N)–superconductor (S) interface, and on the observation that AR between a half-metal and a superconductor is forbidden for the quasiparticles with the energies below the superconducting gap [15]. The classical paper of Blonder et al. [16] gives the detailed description of AR in the case of an arbitrary interface. A different perspective on AR, based on the scattering theory formalism, may be found in the review by Beenakker [17]. These two approaches were used for the case of a ferromagnet (F) by Upadhyay et al. [12] and by Mazin et al. [18], among others, to create a working formalism for the analysis of the conductance data across an F/S interface, allowing one to determine the values of spin polarization of the ferromagnet. AR is intimately related to the proximity effect; the interplay between these two phenomena is discussed in the reviews by Pannetier and Courtois [19] and Buzdin [20]. In recent years, AR has also become a useful tool for studying unconventional superconductivity, such as d-wave, triplet superconductivity, multiband superconductivity, and heavy fermions. We will not address these topics here; the reviews by Kashiwaya and Tanaka [21], Mackenzie and Maeno [22], and Deutscher [23] are devoted to some of these questions; the overview of AR in graphene is given by Beenakker [24]. This chapter also provides some background material for the discussion of superconducting proximity effects in Chapter 16, Volume 1, and Majorana fermions in Chapter 14, Volume 2.

2.2 DEFINITIONS OF SPIN POLARIZATION

The problem of measuring spin polarization is twofold. While it is often assumed that the definition of spin polarization P is self-evident, similarly to magnetization, for example, which is uniquely determined by the difference between the algebraic sum of spin-up and spin-down electronic magnetic moments, this is not the case. First, the definition of P may be different for different measurement techniques and may be sensitive to experimental details. Second, while it may be straightforward to define the degree of spin polarization in the bulk of a uniform ferromagnet, the problem of spin injection, for example, require the knowledge of varying spin polarization in complex nonuniform systems, which may include various types of interfaces. As the Fermi surfaces of typical metallic ferromagnets are quite complicated, the degree of spin polarization of the injected spins may depend strongly on the crystallographic orientation and the nature of the carrier propagation across the contact [25, 26]. For example, a tunnel junction may result in different values of spin polarization compared with a ballistic point contact. Various issues important for defining the concept of spin polarization were addressed by Mazin [3]. Let us start with the most intuitive case of spin imbalance: magnetization. Magnetization spin polarization can be defined as the imbalance in the electronic DOS integrated over all occupied spin states $N_\uparrow(E)$, $N_\downarrow(E)$:

$$P_M = \frac{\int N_\uparrow(E)dE - \int N_\downarrow(E)dE}{\int N_\uparrow(E)dE + \int N_\downarrow(E)dE} = \frac{n_\uparrow - n_\downarrow}{n_\uparrow + n_\downarrow}. \tag{2.1}$$

This definition can also be applied to the case of spin injection into semiconductors, where the measured photoluminescence comes from carrier recombination [27]. The most common definition of spin polarization, P_N, is

$$P_N = \frac{N_\uparrow(E_F) - N_\downarrow(E_F)}{N_\uparrow(E_F) + N_\downarrow(E_F)} = \frac{N_\uparrow - N_\downarrow}{N_\uparrow + N_\downarrow}, \tag{2.2}$$

which is defined in terms of spin-up (down) Fermi-level DOS, $N_{\uparrow(\downarrow)}(E_F)$ (we are omitting E_F here and later). $N_{\uparrow(\downarrow)}$ is obtained by integrating over the Fermi surface S_F, where $v_{\uparrow(\downarrow)}$ are spin-up (spin-down) Fermi velocities.

$$N_{\uparrow(\downarrow)} = \int \frac{dS_F}{\mathbf{v}_{k\uparrow(\downarrow)}(2\pi)^3}. \tag{2.3}$$

Notice that P_N may differ from P_M not only in magnitude but even in sign. This definition is useful for a comparison with spin-resolved photoemission spectroscopy, which nominally measures P_N.* However, in the measurements related to transport phenomena, different definitions are more relevant. For example, AR spectroscopy in the ballistic regime measures the so-called *current-carrying* DOS $\langle Nv_x \rangle$. In a ballistic, or Sharvin, point contact of an area A, an electron going through the contact gains an energy eV [28]. As the fraction of electrons with a given wave vector k reaching the contact per unit time is Nv_x, the current through the contact is

$$I_{\uparrow(\downarrow)} = e^2 VA \langle Nv_x \rangle = e^2 VA \int \frac{v_x dS_F}{v(2\pi)^3} = e^2 VA S_{x\uparrow(\downarrow)}, \tag{2.4}$$

where S_x is the projection of the Fermi surface onto the contact plane (see Sections 2.3 and 2.5 for more detail). Thus, we can define the spin polarization of the current propagating through a ballistic or Sharvin contact as P_{Nv}:

$$P_{NV} = \frac{\langle N_\uparrow v_{x\uparrow} \rangle - \langle N_\downarrow v_{x\downarrow} \rangle}{\langle N_\uparrow v_{x\uparrow} \rangle + \langle N_\downarrow v_{x\downarrow} \rangle}. \tag{2.5}$$

Let us now consider the current in a bulk ferromagnet, assuming that it is governed by the classical Boltzmann transport equation. In this case, the current density $J_{\uparrow(\downarrow)}$ is as follows:

$$J_{\uparrow(\downarrow)} \sim \int \frac{v^2 dS_F}{\mathbf{v}_{k\uparrow(\downarrow)}(2\pi)^3} \tau_{k\uparrow(\downarrow)} = \int \frac{\mathbf{v_k} dS_F}{(2\pi)^3} \tau_{k\uparrow(\downarrow)}. \tag{2.6}$$

In general, the scattering time $\tau_{k\uparrow(\downarrow)}$ is both k and spin dependent, and the current densities need to be evaluated from Equation 2.6. By neglecting the scattering anisotropy and assuming the same relaxation time τ for spin-up and spin-down carriers, we can simplify Equation 2.6 to

---

* Strictly speaking, this is only true in the constant matrix approximation, i.e., under the assumption that the transition probabilities of photoelectrons are spin independent.

$J_{\uparrow(\downarrow)} \sim \left\langle N_{\uparrow(\downarrow)}v^2 \right\rangle \tau_{\uparrow(\downarrow)} \sim \left\langle N_{\uparrow(\downarrow)}v^2 \right\rangle \tau$, so that the spin polarization of the current can then be determined by $P_{Nv^2}{}^*$ as

$$P_{Nv^2} = \frac{J_\uparrow - J_\downarrow}{J_\uparrow + J_\downarrow} = \frac{\left\langle N_\uparrow v_\uparrow^2 \right\rangle - \left\langle N_\downarrow v_\downarrow^2 \right\rangle}{\left\langle N_\uparrow v_\uparrow^2 \right\rangle + \left\langle N_\downarrow v_\downarrow^2 \right\rangle}. \tag{2.7}$$

Equation 2.7 also applies in the case of the diffusive point contact.[†]

We may now ask: what is the spin polarization plotted in Figure 2.1? Contrary to some of the claims in the literature, it is not the spin polarization of the DOS, P_N. Julliere's derivation is based on Tedrow and Meservey's tunneling experiments [30], in which the current in a superconductor/oxide/ferromagnet (S/I/F) junction is measured as a function of voltage V. Assuming that the respective tunneling matrix elements $|M_{\uparrow\downarrow}(E)|$ are independent of energy for $V \ll E_F$, one can define the tunneling spin polarization, P_T, as

$$P_T = \frac{I_\uparrow - I_\uparrow}{I_\uparrow + I_\uparrow} = \frac{N_\uparrow \, |M_\downarrow(E)|^2 - N_\downarrow \, |M_\downarrow(E)|^2}{N_\uparrow \, |M_\downarrow(E)|^2 + N_\downarrow \, |M_\downarrow(E)|^2}. \tag{2.8}$$

Thus, in the original Julliere formula, P is defined not by $N_{\uparrow\downarrow}$ but by $N_{\uparrow\downarrow}$ multiplied by $|M_{\uparrow\downarrow}(E)|^2$. In general, the two tunneling matrix elements ($|M_\downarrow(E)|$) and ($|M_\uparrow(E)|$) are not the same, as the minority and majority carriers come from different electronic sub-bands and thus may have different masses and tunneling probabilities (e.g., Chazalviel and Yafet [31] and Hertz and Aoi [32]). This result was further emphasized in relation to the spin polarization measurements of $La_{1-x}Sr_xMnO_3$ (LSMO) [33, 34] and in the reviews by Butler [35] and Velev et al. [36]. There is a similarity with the giant magnetoresistance (GMR) effect, where the interface resistance is strongly spin dependent (for more detail see Chapter 4, Volume 1), which implies that we must separate the intrinsic spin polarization of a magnetic material from the effects of specific tunnel barriers [37–44]. A more detailed discussion of the Julliere formula is given by Belashchenko and Tsymbal in Chapter 13 and by LeClair and Moodera in Chapter 11, Volume 1.

For the tunneling current in a low transparency specular junction (defined by a δ-function), the spin polarization is reduced to P_{Nv^2} [45] and thus is also determined by Equation 2.7. Importantly, only for true half-metals does $P_N = P_{Nv} = P_{Nv^2} = 100\%$; for all other ferromagnetic materials, P_N, P_{Nv}, and P_{Nv^2} may be very different [3, 7, 46].

---

[*] This is a fairly rudimentary approximation, as even in the absence of magnetic impurities, the scattering time for different spin sub-bands is dependent on the type of wave functions (e.g., s- and d-electrons) and the DOS of the respective spin sub-bands. However, it has been fairly successfully used in a number of band structure calculations; for example, in the giant magnetoresistance (GMR) calculations by Oguchi [29].

[†] In contrast to the electric current, the spin current generally cannot be described by the continuity equation. Thus, the notion of current spin polarization can only be introduced in the limit of small spin–orbit interaction.

2.2.1 Spin Polarization Measurement Techniques

A number of techniques commonly used for the spin polarization measurements are *positron spin spectroscopy, spin-resolved photoemission spectroscopy, spin-polarized tunneling (Meservey–Tedrow technique),* and *PCAR spectroscopy.*[*]

2.2.1.1 Positron Spin Spectroscopy

Positron spin spectroscopy [49] is based on the process of positron–electron annihilation, which is dependent on the electron density at the annihilation site [50]. By analyzing the emitted γ rays, the information on the electron momentum distribution and the electron density can be extracted, and Fermi surfaces can be reconstructed. Positrons emitted from an isotope of Na-22 are partially spin polarized [51]. Switching the magnetization direction of a magnetic material allows one to obtain a spin-resolved electronic structure, as positrons annihilate preferentially with electrons of opposite spin directions. Spin-resolved positron spectroscopy allows one to analyze the electronic structure of magnetic materials [49, 52, 53], enabling the measurements of P_N in a wide temperature range. While positron spin spectroscopy can be quite effective, it is difficult to apply this technique to thin films, and its impact on the field has been rather limited due, in part, to the relative complexity of the measurements, requiring a dedicated experimental setup.

2.2.1.2 Spin-Resolved Photoemission Spectroscopy

Spin-resolved photoemission spectroscopy [54–56] is also, in principle, capable of measuring P_N (in the constant matrix approximation). In this technique, monochromatic photons are produced by a light source, with the whole setup evacuated to ultrahigh vacuum. The kinetic energy of the photoelectrons is analyzed by electrostatic analyzers as a function of the emission angles, the electron spin orientation, the photon energy, and the polarization [57]. The spin components of the beam are detected by a Mott detector, which contains a high Z target (typically Au). While the energy resolution of this technique was initially inferior to the spin polarized tunneling and PCAR techniques it has recently been significantly improved. This method is quite surface sensitive, as it measures the spin of the electrons emitted from a region close to the surface of a ferromagnet on the order of 5–20 Å [58], which corresponds approximately to the mean free path of excited electrons at tens of electron volts above the Fermi energy. Thus, while photoemission can be successfully used to study surfaces of magnetic materials, to obtain reliable values of the bulk spin polarization, samples must be prepared in situ. Detailed spin-polarized photoemission studies were performed for various types of magnetic materials, such as Fe [59–61], as well as Fe/MgO interface [62], Co [63], Ni [64], Gd [65], Fe_3O_4 [66], LSMO [67], and CrO_2 [68], among others. *Inverse photoemission* can also be applied to the studies of magnetic materials [69].

---

[*] A novel quantitative technique, measuring P_{Nv^2}, the current-induced spin wave Doppler shift, has been introduced [47]. Qualitative measurements, designed to distinguish between half-metals and other ferromagnets based on femtosecond spin excitation, have also been realized [48].

2.2.1.3 Spin-Polarized Tunneling (Meservey–Tedrow) Technique

While the previous two techniques can work at room temperature, spin-polarized tunneling and PCAR spectroscopy are inherently low-temperature techniques, as in both cases a superconductor is used. Spin-polarized tunneling in a planar junction geometry has high (sub-millielectron volt) energy resolution. Meservey and Tedrow [30, 70, 71] pioneered this technique by making ferromagnet (F)/superconductor (S) tunnel junctions and Zeeman-splitting the superconducting single-particle excitation spectrum in very thin Al films by applying a magnetic field parallel to the plane of the junction. The resulting (dI/dV) spectrum, corrected for spin–orbit and orbital depairing effects using the Maki equations [33, 72], can be used to detect P_T. As it is directly related to TMR [4], spin-polarized tunneling is the technique of choice for the study of TMR devices [73]. The drawback of the technique is the constraint of fabricating a layered device with a thin ferromagnet film on top of an oxide layer. Generally, the values of tunneling spin polarization, P_T for the same ferromagnet are sensitive to the quality of the interface [74], the degree of disorder within it [44, 75], the shape and the composition of the tunnel barrier [76, 77], and the type of bonding [78]. The Meservey–Tedrow technique is described in detail in Chapters 11 and 14, Volume 1.

2.2.1.4 Point Contact Andreev Reflection Spectroscopy

PCAR spectroscopy,* introduced by Soulen et al. [11] and Upadhyay et al. [12], is probably the most flexible technique, as it places no special constraints on the types of samples that can be studied; a large number of metals, semiconductors, and magnetic oxides have been investigated (see Table 2.2). The technique is based on Andreev reflection (AR) [13, 14], which facilitates a direct conversion of the quasiparticle current in a normal metal into a supercurrent in a superconductor, and the notion that AR is not allowed for a half-metal [15]. In contrast to spin-polarized tunneling, the PCAR technique preferentially requires a clean/no-barrier interface between a ferromagnet and a superconductor. The technique often uses the so-called *point contact*, or *Sharvin geometry*, which may be realized by using either an adjustable tip [11] or a nanoconstriction [12]. Planar AR can also be used if the transport properties of an interface are dominated by AR [79].

While the PCAR technique allows the determination of spin polarization separated from the properties of a tunnel/oxide barrier, it lacks the ability to determine *the sign* of spin polarization (in contrast to the Meservey-Tedrow and some other techniques), which is obviously a drawback. In principle, to determine the sign, one may be able to make use of the asymmetry of conductance curves resulting from the Zeeman-split Andreev reflection in high magnetic fields [80], but these measurements are complicated by spin–orbit coupling and the presence of vortices in the superconductor.

---

* While this term is often applied more broadly, e.g., to the point contact measurements of unconventional superconductors, we will limit ourselves to the spin polarization measurements.

2.3 TRANSPORT PROPERTIES OF A POINT CONTACT

2.3.1 SHARVIN RESISTANCE

The concept of the Sharvin resistance [28] was introduced by Yu. V. Sharvin, who suggested studying the Fermi surfaces of metals using the electron focusing effect in a uniform magnetic field H_z, which is similar to the principle of a beta spectrograph. Consider a group of electrons starting from the same origin $r = 0$ with the same wave vector in the direction of the field, k_z; the other two components of the wave vector may be arbitrary. As all of them have the same orbital period, they will go around an integer number of orbits in the same amount of time. Therefore, in real space, they will return to the same point at the xy plane ($x = 0$, $y = 0$), whereas z will be determined by the total number of periods, the magnetic field, and the average velocity v_z. If a significant number of these electrons (with the same k_z) were focused at the same point, a sizable additional current would flow. This effect was demonstrated by Sharvin and Fisher [81], who fabricated two very fine (point) contacts on the opposite sides of an Sn single crystal with a large mean free path, so that the electrons accelerated by the applied voltage V at one contact were focused at the other contact, which resulted in periodic oscillations of the current when the focusing conditions were fulfilled. Later, a number of studies of focusing effects were performed in the transverse geometry, particularly in a two dimentional electron gas (2DEG) [82–84]. These effects were successfully used to observe AR with two separate point contacts [85] and a single contact [86].

2.3.2 BALLISTIC (KNUDSEN) REGIME

The main requirement for the focusing effect is that the electronic elastic mean free path l should be much larger than both the radius a and the width of the contact w, so that $l \gg a,w$. This problem is analogous to the problem of a dilute gas flow through a small hole in the so-called *Knudsen regime* [87]; see Figure 2.2. The degree of "ballisticity" is determined by the *Knudsen number* $\kappa = l/a$, with $\kappa \gg 1$ defining the pure ballistic regime. Sharvin argued that in a linear regime (i.e., when eV is much smaller than E_F), as the electrons cross the area of the contact in two opposite directions, they would

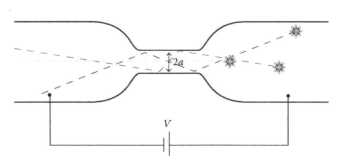

FIGURE 2.2   Ballistic (Knudsen) regime with the mean free path $l \gg a, w$.

acquire (lose) additional velocity $\delta v = eV/p_F$, where V is the voltage applied across the junction, and p_F is the Fermi momentum. The current corresponding to such a flow is $I \sim enA\delta v$, where A is the area of the contact, and n is the electron density. The additional resistance is then $R_{Sh} = V/I \sim p_F/e^2 nA = p_F/\pi e^2 na^2$. Using the Drude expression for the resistivity, $\rho = p_F/ne^2 l$ and thus, $R_{Sh} = \rho l/\pi a^2$. More accurately, by integrating over all angles, the numerical factor of 4/3 could be obtained, similarly to the kinetic gas theory, where the number of molecules hitting a unit area of a wall per second is $\frac{1}{4}nv$ rather than $\frac{1}{3}nv$, which one might assume without integration. The Sharvin resistance then becomes

$$R_{Sh} = \frac{4\rho l}{3\pi a^2} = \frac{4\rho}{3\pi a}\kappa. \tag{2.9}$$

Importantly, as in the ballistic regime, there is practically no scattering in the area of the contact, the Sharvin resistance of metals is largely independent of their material properties, as $\rho \sim 1/l$. Indeed, in the three-dimensional (3-D) geometry, the Sharvin resistance can be expressed via a number of transverse quantum modes that can propagate through the constriction $\int Ad^2k/(2\pi)^2 = k_F^2 A/4\pi$. Thus, the Sharvin resistance depends only on the properties of the Fermi surface and the area of the contact:

$$R_{Sh} = \frac{4\rho l}{3\pi a^2} = \frac{h}{2e^2}\left(\frac{k_F^2 A}{4\pi}\right)^{-1}, \tag{2.10}$$

where $2e^2/h$ is the conductance quantum for two spin channels. While the result is given here for a spherical Fermi surface with a quadratic dispersion law, it can be easily generalized for an arbitrary Fermi surface; see Section 5.2.1. Note that the semiclassical approach works well in 3-D (but not in two dimensions [2-D]), as the electron de Broglie's wave length $\lambda_F \approx 0.5$ nm is much smaller than a.

2.3.3 DIFFUSIVE (MAXWELL) REGIME

In the opposite limit, $l \gg a$, the diffusive transport regime with the so-called *Maxwell resistance* is realized. The problem can be solved exactly in oblate spheroidal coordinates [88]. The solution on both sides of the diaphragm (see Figure 2.3) is

$$\Phi = \pm V\left[1 - \frac{2}{\pi}\arctan(\xi)^{-1}\right], \tag{2.11}$$

where ξ is defined through cylindrical coordinates r and z, $(r/a)^2 = (1+\xi^2)(1-z^2/\xi^2 a^2)$, where a is the radius of a circular orifice. Thus, the Maxwell resistance is inversely proportional to the diameter of the constriction $2a$:

$$R_M = \frac{\rho}{2a}. \tag{2.12}$$

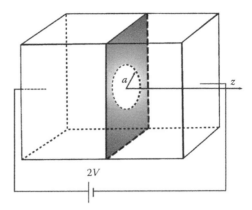

FIGURE 2.3   The Maxwell case ($l \gg a$). Electronic transport through the aperture of radius a in an insulating diaphragm separating two conducting half-spaces with the mean free path l.

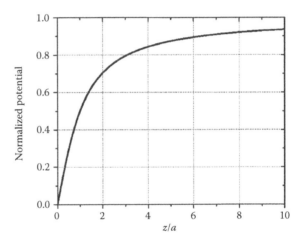

FIGURE 2.4   Voltage profile along the z axis of Figure 2.3. Note that ~75% of the total voltage drop takes place within a distance of $2a$ from the center of the orifice.

Most of the voltage drop takes place within the distance of $2a$ from the orifice (see Figure 2.4).

2.3.4 INTERMEDIATE REGIME

For the contact resistance R_C in the intermediate regime, the resistance is given by Wexler's formula [89]

$$R_C = \frac{4\rho l}{3\pi a^2} + \Gamma(\kappa)\frac{\rho}{2a} = R_{Sh}\left[1 + \frac{3\pi}{8}\Gamma(\kappa)\frac{a}{l}\right] = R_{Sh}\left[1 + \frac{3\pi}{8}\frac{\Gamma(\kappa)}{\kappa}\right], \quad (2.13)$$

where:

$\kappa = l/a$ is the Knudsen number

$\Gamma(\kappa)$ is the geometric factor on the order of unity, which varies from 0.7 when $\kappa \to \infty$ (the pure ballistic case) to 1 when $\kappa = 0$ (the pure Maxwell case)

In the intermediate case (diffusive regime), the elastic mean free path l becomes smaller than a; however, the characteristic diffusion length $\Lambda = \sqrt{l l_{in}} \geq a$, where l_{in} is an inelastic mean free path, which is strongly energy dependent. The larger κ is, the sharper the voltage drop across the contact becomes. In both cases, however, the energy is dissipated within the same length $\Lambda = \sqrt{l l_{in}}$, beyond the region of a potential drop [90].*

The exact solution in the same (linearized) approximation as used by Wexler was later found by Nikolic and Allen [91]. As the authors noted, Wexler's result is fairly accurate (within a few percent) compared with the exact solution. In the intermediate (quasi-ballistic) regime when $a \sim l$, it is expected that the contributions from both the Sharvin and the Maxwell resistances are approximately equal. Rather surprisingly, the naive approximation, which comes as the sum of the two limits,

$$R_C = R_{Sh}\left(1 + \frac{3\pi}{8}\frac{a}{l}\right) = R_{Sh}\left(1 + \frac{3\pi}{8}\frac{1}{\kappa}\right), \qquad (2.14)$$

produces fairly accurate results, with the largest difference of only about 10% (near $\kappa = 1$) compared with the exact solution. Therefore, for all practical purposes, such as the estimates of the contact size, Equation 2.14 can be used.

In the case of materials with different but *isotropic* electrical properties on both sides of the contact, the total contact resistance R_Σ can generally be expressed as an average of the two resistances, R_{C_1} and R_{C_2} [89]:

$$R_\Sigma = \frac{1}{2}R_{C_1} + \frac{1}{2}R_{C_2}. \qquad (2.15)$$

This equality has some interesting implications in the case of AR, where the resistance of the superconducting side is nominally zero. In this case (at $T = 0$), the junction resistance $R_\Sigma = \frac{1}{2}R_N$ below the gap is one half of the resistance of the normal metal. Note that this is only correct in the case of a purely ballistic transport across the junction [92]. In the ideal diffusive case (disordered contact), the resistance is $R_\Sigma = R_N$ [93–95].

2.3.5 THERMAL CASE

In the case of $\Lambda = \sqrt{l l_{in}} \ll a$, when both elastic and inelastic mean free paths become smaller than the contact size, one can define the *thermal regime* [96]. In this case, the scattering in the area of the contact is quite strong, and since most of the voltage drop also occurs there, this is where most of the energy is dissipated. This can, in principle, lead to a significant increase of the temperature in the area of the contact. Since l_{in} is strongly energy

---

* In both the diffusive and ballistic cases, the energy relaxation is nonlocal, i.e., the Joule heat through a constriction cannot be expressed through the local electric field, only through the applied voltage as GV^2.

dependent, the maximum temperature in the area of the contact T_{max} can be much larger than the bath temperature T_{bath} [97]

$$T_{max}^2 = T_{bath}^2 + \frac{V^2}{4L}, \tag{2.16}$$

where L is the Lorenz number. Thus, a contact may be in the ballistic or diffusive regime at low voltages, where one can still observe a phonon spectrum, and fall into the thermal regime only at much higher voltages, as was demonstrated in the case of Ni–Ni point contacts, where the Curie temperature of Ni ($\approx 630\,\mathrm{K}$) has presumably been reached in the thermal regime [98]. Unless the contact resistance is very small (less than 1 Ω), one is unlikely to reach the thermal regime at several millivolts, which is typically required for the PCAR measurements with conventional superconductors.

2.3.6 POINT CONTACT SPECTROSCOPY

Since the Sharvin resistance in the first approximation is independent of l, it should be largely independent of energy, except for a small (on the order of eV/E_F) variation of the DOS with the applied bias. Therefore, an ideal Sharvin contact (with $\kappa = \infty$) in this approximation is expected to be ohmic. However, this implies no scattering at all, and therefore, is unphysical. One can show [99] that for any realistic case of finite κ, the second term in Equation 2.13 leads to a strong nonlinearity of R_C. The degree of this nonlinearity, d^2V/d^2I, is dependent on the inelastic scattering rate at different energies, which, as was first noted by Yanson, is directly related to electron–phonon interaction [100]. This pioneering work enabled the study of electron–phonon as well as other quasiparticle spectra, starting the area of *point contact spectroscopy*. For further details of this technique, we refer to several extensive reviews [96, 97, 99].

There is a subtle relationship between the Sharvin and Maxwell parts of the resistance in Equation 2.13. On the one hand, the energy-dependent features in the point contact resistance are the result of scattering with quasiparticles of different types and therefore, are only possible in the presence of the diffusive part of the resistance. On the other hand, the more "ballistic" the contact is ($\kappa \to \infty$), the better is the energy selectivity, and thus the better these features are resolved. As the contact becomes more diffusive, the energy of the electron is no longer determined by the applied voltage, and eventually, the d^2V/d^2I curve washes out. Therefore, the presence of the phonon spectra is often used as a proof that the transport through a contact is ballistic or quasi-ballistic [12, 86, 102, 103].

2.3.7 CONTACT SIZE DETERMINATION

While the presence of the phonon spectra can be used to confirm that one is operating in the ballistic regime, it does not help in obtaining a quantitative estimate of the contact size. The latter can be roughly estimated by solving the quadratic Equation 2.13. However, one then has to assume the same bulk resistivity in the area of the contact, which may not be the case due to various defects. Therefore, it is preferable to determine the contact size by measuring the temperature dependence of both the residual resistivity and the contact resistance and by making use of the fact that the Sharvin resistance is independent of temperature. Following Akimenko et al. [104], we can differentiate Equation 2.13 to obtain

$$\frac{\partial R_C}{\partial T} = \frac{\partial}{\partial T}\left[\frac{4\rho l}{3\pi a^2} + \Gamma(\kappa)\frac{\rho}{2a}\right] = \frac{\Gamma(\kappa)}{2a}\frac{\partial\rho}{\partial T}. \tag{2.17}$$

We assume the function $0.7 \leq \Gamma(\kappa) \leq 1$ to be 1 to find the diameter of the contact [103, 104]:

$$d = 2a = \frac{\partial\rho/\partial T}{\partial R_C/\partial T}. \tag{2.18}$$

For the ballistic point contact of Cu and Au, for example, the following rough estimate can be used: $d = 30/\sqrt{R(\Omega)}$ (nm). For a typical resistivity of a contact ranging between 1 and 100 Ω, the contact diameter varies from 30 to 3 nm [104]; more detail can be found in Naidyuk and Yanson [105].*

2.4 ANDREEV REFLECTION: UNPOLARIZED CASE

Andreev reflection, named after Alexander F. Andreev, was introduced in 1964 [13] to explain the puzzling effects of anomalously high thermal resistance in the intermediate state of type-I superconductors, which greatly exceeded the thermal resistance in the superconducting state [106]. Moreover, in contrast to the superconducting state, the thermal resistance of the intermediate state was practically independent of the electron mean free path, even in the temperature range where the heat transfer was dominated by electrons [107]. This implied that the interface played an important role, somehow preventing electrons from crossing the N/S boundary. However, a mechanism that might be responsible for this effect was not understood.

Indeed, on arriving at an N/S interface, an electron seems to have only two obvious choices: either it goes across the interface into the superconductor or it gets reflected back to the normal metal. A single electron with energy below the superconducting gap Δ is not allowed to propagate in the superconductor as a quasiparticle (at $T = 0$ K), so the first option is not possible. On the other hand, if it simply gets reflected, this would imply that the net current through such a contact is zero. This is rather counterintuitive, as one would expect that if a part of the contact becomes superconducting, its resistance will go down compared with the all-normal case (see Section 2.3.4). Moreover, if it is going to be reflected by the usual scattering process, a large barrier might be needed to reverse the Fermi momentum of the electron. Indeed, in the case of a clean interface, the only energy scale present in

---

* Note that these estimates give the upper limit of the contact size, as they are based on the assumption that a single contact is established.

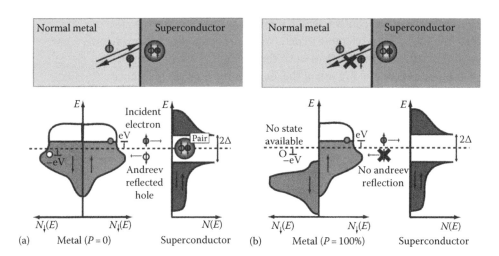

FIGURE 2.5   Schematic of quasiparticle to supercurrent conversion via Andreev reflection. (a) Andreev reflection in a normal (unpolarized) metal is unhindered, as both spin up and spin down electrons are present; (b) Andreev reflection in a half-metal is suppressed, as there are no opposite spin states for the hole to get reflected to; therefore, supercurrent conversion is blocked. (After Soulen, Jr. R.J. et al., *Science* 282, 85, 1998. With permission.)

this problem is related to the superconducting gap, which implies [108] that the change of momentum is on the order of

$$\delta p_x \sim -F_x \delta t \sim \frac{\Delta}{\xi} \frac{\xi}{v_F} \sim \frac{\Delta}{v_F} \sim p_F \left(\frac{\Delta}{\varepsilon_F} \right) \ll p_F, \tag{2.19}$$

where ξ is the coherence length. This is a classical case of an infinitely heavy ball hitting an impenetrable wall [92]. So, then, which answer will correctly describe the behavior of the electron in this case? Neither! It gets reflected—but in a highly unusual manner: its charge becomes opposite to the charge of the incoming electron, and all three components of its momentum change sign; that is, it is the *hole* that gets reflected along the same trajectory (*retro-reflected*) (see Figure 2.5a). The reflected hole also has the opposite spin.

This has a number of important consequences. (1) As the hole moving in the opposite direction is equivalent to the electron moving in the same direction, the conduction across the contact doubles (compared with the all-normal case; see Section 2.3.4). (2) As the original electron is accompanied by the electron with the opposite spin, the two of them constitute a Cooper pair that propagates inside the superconductor; thus, the quasiparticle current gets converted into the supercurrent. (3) This process is elastic, as the quasiparticle energy is the same for the incident electron and the reflected hole. Momentum is also conserved to the first order of Δ/E_F (usually a good approximation for conventional superconductors).

The thermal resistance of the intermediate state, calculated by Andreev using this mechanism, turned out to be in excellent agreement with the results of Zavaritskii [13]: the thermal resistance did not depend on the mean free path and increased exponentially at lower temperatures as the number of quasiparticles was decreasing proportionally. In other words, Andreev-reflected electrons, converted into Cooper pairs, did not carry any heat; only quasi-particles did. This mechanism has broad implications in solid-state physics, including the proximity and quantum coherence effects in N/S systems [19, 20, 109–111] and the symmetry of the order parameter [23, 112–114], as well as in other areas of physics (e.g., Jacobsen [115] and Sadzikowski and Tachibana [116]). The discussion of various proximity effects in ferromagnet/superconducting junctions is given by Eschrig in Chapter 16, Volume 1.

A different problem, rooted in the same phenomenon, was addressed by de Gennes and Saint James [117] and Saint James [14]. They considered a proximity effect of a superconductor with a normal metal and discovered that if the thickness of a normal film is small, a bound state that is an admixture of electrons and holes is created, and the DOS in the normal metal is increased by a factor of two.* These states, which are often called *de Gennes/Saint James bound states*, have been observed experimentally; see, for example, Giazotto et al. [118].

While there have been a number of indirect experiments indicating that AR is likely to play a major role in tunneling [119, 120], the first unambiguous evidence for this effect was obtained by Krylov and Sharvin [121], who used the so-called *Gantmakher (radio frequency) size effect*, in which AR manifested itself by the appearance of additional lines corresponding to the trajectories of retro-reflected quasiparticles. In addition to these types of measurements [122], the most direct demonstration of AR came from using the focusing effect in the transverse geometry [85].

Let us discuss qualitatively how AR enables the process of quasiparticle to supercurrent conversion. Adopting the description of Pippard et al. [123] and Schmidt [124], we will first consider a *clean* interface† and assume that the absolute value of the energy of an electron ε_k is less than the superconducting gap $\Delta(T)$. The order parameter then rises from 0 inside the normal region to Δ inside the superconducting region within the coherence length ξ; see Figure 2.6. As the electron with a wave vector k moves toward the interface encountering a gradually increasing gap, it becomes a quasiparticle with the energy E_k and the dispersion law

$$E_k = [\Delta^2 + (\varepsilon_k - \varepsilon_F)^2]^{1/2} = [\Delta^2 + \hbar^2 v_F^2 (k - k_F)^2]^{1/2}, \qquad (2.20)$$

---

* The author wishes to thank Prof. Guy Deutscher for drawing his attention to the largely unknown *Journal de Physique* paper of Saint James [14], which has never been translated into English. While the paper does not address the transport between normal and superconducting parts per se, it correctly identifies the origin of the DOS enhancement in the normal layer near an N/S interface, which is due to the transformation of an electron into a Cooper pair at the S interface, with a hole reflecting back to the N interface and, after another reflection from N to S, "annihilating" with the Cooper pair, thus leading to doubling of the DOS.

† We will define a clean or ideal interface as one with (1) no oxide barrier and (2) both normal and superconducting layers composed of the same material, so that no interface velocity mismatch between a superconductor and a normal metal $\gamma = v_{Fn}/v_{Fs}$ is present [16]. In practice, this can be achieved by using either a superconductor in the intermediate state or a variable thickness structure with different critical temperatures T_{c1} and T_{c2}, working at $T_{c1} < T < T_{c2}$.

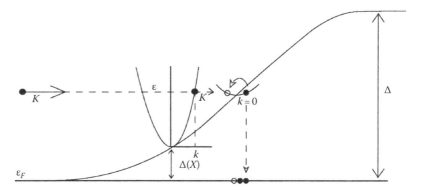

FIGURE 2.6 Quasiparticle conversion into Cooper pairs. The quasiparticle charge is gradually being transferred to the condensate. At a turning point, where the local value of the gap is equal to the quasiparticle energy, the charge of $+e$ is transferred.

which has two solutions for the momenta $k_\pm = k_F \pm [(E_k^2 - \Delta^2)]^{1/2}/\hbar v_F$ (where $+$ corresponds to the electron-like and $-$ to the hole-like particles). Accordingly, quasiparticles in a superconductor can be assigned a fractional charge [125–127].

$$q_k = \frac{\varepsilon_k}{E_k} = \frac{\varepsilon_k}{(\Delta^2 + \varepsilon_k^2)^{1/2}}. \tag{2.21}$$

As can be seen from Equation 2.21, the charge of a quasiparticle varies from the asymptotic value of -1 (pure holes) to $+1$ (pure electrons) (in this notation, the electronic charge is $+e$). As the quasiparticle moves toward the interface, its charge decreases until it reaches 0, when $k = k_F$. At this point, the group velocity of the quasiparticle becomes zero; as it crosses to another branch of the spectrum, it becomes negatively charged with a negative group velocity; that is, it becomes the Andreev-reflected hole. Importantly, in this process, the charge is continuously transferred to the condensate, and in the end, the total charge of $e - (-e) = 2e$ will have been transferred to a Cooper pair. This is equivalent to the incoming electron pairing up with another electron with the opposite spin and wave vector,* while the hole gets reflected back from the interface; see Figure 2.6.

This result can be quantitatively derived for a step-like interface assumed in the BTK model $\Delta(x) = \begin{cases} \Delta, & x > 0 \\ 0, & x < 0 \end{cases}$ [16]. As the quasiparticle current gradually decays, $J_Q = 2ev_F\exp(-x/\Lambda)$ (where $\Lambda = \hbar v_F/2\Delta \approx \xi$ is on the order of the coherence length), the supercurrent is rising at the same

---

* In the simplest model of a plane wave in the ballistic regime, quasiparticle momentum is a good quantum number for labeling the states. In the first approximation in (Δ/E_F), the momenta of reflected and transmitted particles can be considered equal to k. While the probability of the Andreev process is reduced at grazing angles, this effect is relatively small and can usually be neglected.

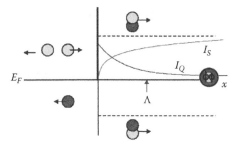

FIGURE 2.7   Quasiparticle to supercurrent conversion at a step-like interface assumed in the BTK model [20]. As the quasiparticle current I_Q decays with the characteristic length Λ, the supercurrent I_S is rising at the same rate, so that the charge conservation is satisfied.

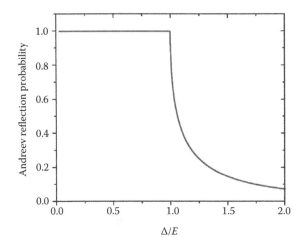

FIGURE 2.8   The probability of Andreev reflection decreases rapidly as the energy exceeds Δ. However, Andreev reflection above the superconducting gap is still possible.

rate, $J_S = 2ev_F[1 - \exp(-x/\Lambda)]$, so that the charge conservation is satisfied; see Figure 2.7. For a clean interface, the probability of AR is

$$A(E) = \begin{cases} 1 & E < \Delta \\ \dfrac{v_0^2}{u_0^2} = \dfrac{|E| - (E^2 - \Delta^2)^{1/2}}{|E| + (E^2 - \Delta^2)^{1/2}} & E > \Delta \end{cases}. \tag{2.22}$$

Note that even above the superconducting gap, AR is still possible within a narrow range of energies (which is typical for a quantum problem with sharp-edged barriers), although its probability goes down quite quickly at higher energies; see Figure 2.8. Importantly, in contrast to the tunneling case, where the transmission probability is proportional to the DOS $N_s(E)$, the transmission probability for a clean interface is $A(E) + 1$, independent of $N_s(E)$ [16]. This property of *spectral weight nonconservation* distinguishes AR from conventional tunneling. For tunnel barriers (strong scattering), the probability of AR is rapidly approaching zero (see Section 2.4.1), so that the spectral weight is almost fully conserved.

2.4.1 ARBITRARY INTERFACE WITH SCATTERING POTENTIAL (BALLISTIC LIMIT)

While the concept of a clean or ideal interface is helpful for a qualitative understanding of AR, strictly speaking, it can only be implemented when both N and S parts are identical. However, as most real interfaces are constructed of two different materials, the quasiparticle transport is not limited to the AR process, as normal reflections are also present due to (1) a scattering potential barrier at the interface (from surface oxides, impurities, etc.) and (2) different electronic structure of the N and S components, which may result in a normal reflection even in the absence of a barrier. Let us first consider a point contact geometry in the *ballistic limit* (the diffusive limit will be discussed in Section 2.5.3.4).

The effect of a tunnel barrier modeled by a δ-function potential was first described by Demers and Griffin [128] and Griffin and Demers [129]. As the strength of the surface barrier increases, the behavior of the contact gradually changes from metallic to tunneling; both can be described by a single phenomenological dimensionless parameter $Z = H/\hbar v_F$, which is related to the strength of the δ-function potential $V(x) = H\delta(x)$ introduced in a seminal paper by Blonder, Klapwijk, and Tinkham (BTK) [16], who expanded the treatment of Demers and Griffin [128] to finite voltages. The BTK model provides a solution of the Bogoliubov–de Gennes equations in a one-dimensional (1-D) case* while at the same time assuming that the order parameter and the potential drop across the interface are the step functions. Note that such boundary conditions can only be fulfilled in a 3-D case, where both the order parameter [130] and the electrical potential [131] have been shown to change on a scale comparable to a.† While, strictly speaking, the problem should be solved self-consistently to determine $\Delta(x)$ and the voltage distribution across the interface, the BTK approximation has been working surprisingly well. The parameter Z has a simple meaning: in the normal state (which can be achieved either by suppressing the superconductivity by a magnetic field or at high bias voltages), the resistance of the junction is $R_N = R_{Sh}(1 + Z^2)$. The normalized conductance in the case of a nonzero Z is shown in Figure 2.9. The total current [16] across an N/S interface is

$$I_{NS} = \left(\frac{1}{eR_{Sh}}\right)\int [1 + A(E) - B(E)][f(E - eV) - f(E)]dE, \qquad (2.23)$$

---

* The main advantage of solving 1-D rather than 3-D equations is the computational simplicity. The only 3-D case that can be exactly mapped onto a 1-D problem is a hyperbolic neck problem.

† Additionally, it is assumed that the quasiparticle distribution function in the area of the contact is largely unchanged, the assumption also only justified in 3-D.

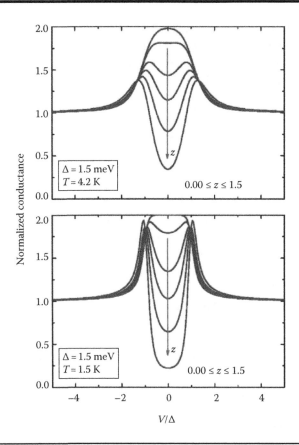

FIGURE 2.9 Normalized conductance for various values of Z (from $Z = 0$ to $Z = 1.5$) for unpolarized case, plotted using the model of Mazin et al. [18]; $\Delta = 1.5$ meV; top $T = 4.2$ K; bottom $T = 1.5$ K. Note that the conductance at zero bias is twice the normal conductance only for $Z = 0$.

where $A(E)$ and $B(E)$ are the energy-dependent probabilities of Andreev and normal reflection, respectively and $f(E)$ is the Fermi function.

As can be seen from Equation 2.23, normal reflections reduce the conductance through the N/S interface, as expected. For $Z = 0$, $B = 0$, and the probability of AR below the gap $A = 1$ (at $T = 0$ K), so that $R_{NN} = R_{Sh} = 2R_{NS}$. The conductance below the gap is thus doubled, see Equation 2.15. Thus, in this idealized case of a perfectly transparent clean interface, the NS interface conductance is completely determined by AR.

2.4.2 IMPEDANCE MISMATCH

A δ-function potential used in the BTK model implies the continuity of the derivatives of the wave functions at the interface. In an idealized case, when the superconducting and normal parts are identical, the matching conditions do not affect the transport coefficients. However, when the two banks are not the same and thus have different electronic spectra, this results in the so-called *impedance mismatch* [132], sometimes also called the *Fermi velocity/Fermi wave vector* mismatch [133–137]. Such a mismatch may lead to the normal reflection from the interface, even in the absence of a barrier, thus modifying both the reflection and transmission coefficients. In the BTK approach, which adopts a free electron model, this effect can be accounted for by introducing the effective Fermi velocity $v = \sqrt{v_1 v_2}$, where v_1 and v_2 are the Fermi velocities of the two banks. Then, Z is replaced by Z_{eff}:

$$Z_{\mathrm{eff}} = \left[Z^2 + \frac{(1-r)^2}{4r} \right]^{1/2}, \tag{2.24}$$

where r is the ratio of the two velocities $r = v_1/v_2$. Note that the transmission coefficients, which depend on Z_{eff}, will remain the same if the two banks are interchanged, in agreement with the principle of reciprocity in quantum mechanics. Then, for $Z = 0$ (no "real" barrier), the minimum value of Z_{eff} will be determined by the Fermi velocity ratio r, which implies that the value of zero bias conductance will be reduced compared with the ideal value, twice the normal conductance.

While the concept of the Fermi velocity mismatch seems physically transparent, and the minimum values of Z_{eff} are in reasonable agreement with experiments in all-metal [132] as well as metal/semiconductor [138, 139] contacts, this argument breaks down dramatically when one considers the case of heavy fermions, for example, in which, based on their large r (up to $\simeq 100$), one might expect $Z \approx 5$. This would have made efficient AR practically impossible, which contradicts many experiments in which Z is typically about an order of magnitude smaller (e.g., Naidyuk and Yanson [105] and Goll [114]). The main reason for this discrepancy is that the simple picture of the Fermi velocity mismatch can only be applied to a very restrictive case of the free electron model, with the electronic wave functions represented by plane waves. In a general case, when the electronic wave functions are *Bloch waves*, the *interface velocities* must be introduced to satisfy the matching conditions [140]. The concept of interface velocity is in line with the work of Deutscher and Nozieres [141], who argued that the Fermi velocity is not the one to be matched in the AR process.

2.5 ANDREEV REFLECTION: SPIN-POLARIZED CASE

2.5.1 IDEAL CASE: INTRODUCTION

Let us first consider qualitatively an idealized case of a clean interface. The important condition for the AR process described in Section 2.4 is that for every electronic k state, there is an equivalent hole state with the opposite spin. For a nonmagnetic metal, this is always the case, as it has two *identical overlapping* Fermi surfaces. However, for a ferromagnet, this is no longer true, as the Fermi surfaces for majority and minority electrons can be very different; so an electron from a majority sheet of the Fermi surface may not be able to find a hole state from a minority sheet to get reflected to. Therefore, in the extreme case of a half-metal ($P = 100\%$), where only one type of spin is available at the Fermi level, AR would be completely suppressed [15]; see Figure 2.5b. What will happen in the intermediate case, with the values of spin polarization $0 < P < 1$? It can be understood qualitatively if one considers the current across the interface as the sum of spin-polarized $[I_p = (I_\uparrow - I_\downarrow)]$ and unpolarized $[I_p = I - (I_\uparrow - I_\downarrow) = 2I_\downarrow]$ parts [11]; see Figure 2.10:

$$I = I_\uparrow + I_\downarrow = (I_p + I_u) = (I_\uparrow - I_\downarrow) + 2I_\downarrow$$

$$= \frac{(I_\uparrow - I_\downarrow)}{I_\uparrow + I_\downarrow} I + \frac{2I_\downarrow}{I_\uparrow + I_\downarrow} I = PI + (1 - P)I, \tag{2.25}$$

where $P = (I_\uparrow - I_\downarrow)/(I_\uparrow + I_\downarrow)$. The spin-polarized fraction of the current PI will not be able to Andreev-reflect, and therefore, will not contribute to the overall conductance, whereas the unpolarized fraction $(1 - P)I$ will reflect just as in a normal metal, with each electron contributing twice the normal conductance. For the two limiting cases $P = 0$ (an unpolarized metal) and $P = 1$ (a half-metal), the normalized conductance at zero bias (which is 2 and 0, respectively) is recovered; see Figure 2.11. Let us now take the value of $P = 25\%$, for example. In this case, 25% of the current (spin-polarized fraction) going through the F/S contact will not contribute to the conductance at all (AR is suppressed), whereas for the remaining 75% of the current, the conductance will double, and thus, the overall normalized conductance will be 1.5. By the same token, if the conductance is known, one should be able to solve the *inverse problem*, that is, to determine the spin polarization of a material based on the transport properties (conductance) of its contact with a superconductor. This observation becomes the foundation of the PCAR technique [11, 12], which we will describe later.

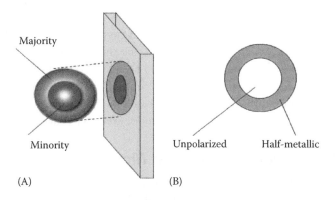

(A)   (B)

FIGURE 2.10   Current decomposition in the case of spherical Fermi surfaces for majority and minority electrons and a single band superconductor; (a) the two spin subbands projected onto the plane of a contact; (b) the number of majority and majority conductance channels (CC) in the overlapping area (white) is the same; this fraction of the current is unpolarized; the shaded area only has majority channels and thus is half-metallic (100% spin polarized) (see Section 2.5.3.1).

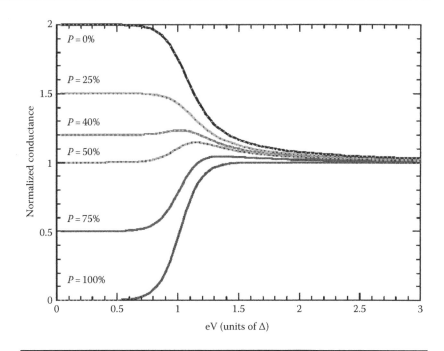

FIGURE 2.11   Normalized conductance curves for different values of spin polarization and a clean interface at $T = \Delta/10$. Every curve is uniquely determined by the value of spin polarization, P. (After Nadgorny, B. et al., *Appl. Phys. Lett.* 80, 3973, 2002. With permission.)

Note that these simple arguments are only applicable in the case of a clean interface with $Z = 0$, when there is no mixing of normal and AR (i.e., the probability of AR is 1 at $V = 0$). For an arbitrary interface with nonzero Z, there is no longer a direct relationship between the conductance at zero bias and P, and one must consider the whole conductance curve $G(V)$. Judging simply by the amplitude of the conductance curve would be misleading, as demonstrated in Figure 2.12, where the same conductance at zero bias is realized for the spin-polarized case with $P = 50\%$ and $Z = 0$ and the unpolarized case with $P = 0$, $Z = 0.55$ (both contacts are assumed to be in the ballistic regime). However, as long as one can describe the conductance curve with a unique set of P and Z values, one should be able to extract the spin polarization values; see, for example, Upadhyay et al. [12] and Mazin et al. [18]. A more complicated diffusive case, first treated by Artemenko et al. [93] for $P = 0$, has also been generalized to account for arbitrary values of P [18].

2.5.2 Comparison of Different Models

Establishing a reliable technique for probing spin polarization based on AR requires a quantitative theory. At present, there are several approaches to formulate these theories, which can be broadly divided into the Landauer–Büttiker formalism (e.g., Landauer [143]) and the Hamiltonian formalism Bardeen [144], describing tunneling in a superconductor (see, e.g., Cuevas et al. [145]). As we have mentioned in Section 2.4.1, the BTK model assumes 3-D physics, which allows one to use equilibrium distribution functions as

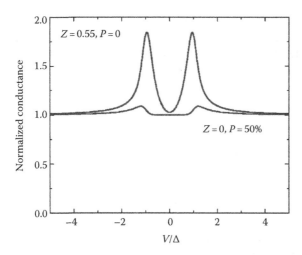

FIGURE 2.12 While the amplitudes of the two conductance curves at zero bias are the same, they correspond to different values of $P = 0$ (top curve) and $P = 50\%$ (bottom curve). Thus, to obtain the value of P for an arbitrary Z, the entire conductance curve must be analyzed.

well as sharp-edged boundary conditions, and yet solve the 1-D equations. Solving the exact problem of current transfer from a metal with an arbitrary band structure into a superconductor across a magnetically active interface self-consistently may be prohibitively complicated, even with the use of Green's function techniques, which are generally quite effective in describing transport through a point contact [113, 146].

A comparison of some of the models used in practice to extract the values of spin polarization from experimental data ([12, 18, 147] is given in a review [102]). The 3-D model used by Upadhyay et al. [12] may be able to take into account an apparent reduction of the AR probability for a fairly clean interface with small Z values as the angle of incidence becomes larger due to the kinematic restrictions related to the interface mismatch [133–135, 148].* On the other hand, to make the equations manageable in 3-D, a free electron model and a Stoner model for a ferromagnet are assumed [12, 102], which may be somewhat restrictive in view of the fact that the boundary conditions should be determined for the Bloch waves rather than for the plane waves associated with the free electrons [140, 150].

The 1-D model of Mazin et al. [18] does not make any explicit assumptions concerning the band structure of the ferromagnet but assumes the same Z for majority and minority electrons.† By introducing an evanescent wave in the minority band, Mazin et al. [18] removed the artificial conjecture

---

* In practical terms, most of the quasiparticles are approaching the interface at relatively small angles, and the difference is even less pronounced if the assumption of a clean interface is removed (the larger Z is, the more "one dimensional" the transport becomes). Overall, this effect is relatively small: a factor of 2 of the interface mismatch corresponds to about 10% reduction in transmission probability [149].

† This is a serious simplification. A more rigorous but less practical approach calls for four scattering parameters: $Z_{\uparrow\uparrow}$, $Z_{\downarrow\downarrow}$, $Z_{\uparrow\downarrow}$, and $Z_{\downarrow\uparrow}$.

of Strijkers et al. [147] that AR amplitude must be set to zero in the case of a half-metal.* This approach resulted in a more accurate determination of P by about 2%–4% [102, 151, 152]. In Section 2.5.3, we will outline the derivation of Mazin et al. [18].

2.5.3 ANDREEV REFLECTION THEORY FOR 1-D MODEL

A quantitative theory of AR should take into account (1) a different number of conductance channels (CC) open for AR for different spins, as was first noted by de Jong and Beenakker [15]; (2) finite interface resistance for both channels, as treated by Blonder et al. [16], for "nonmagnetic" channels, that is, the CC that exist in both spin directions; (3) band structure effects, such as arbitrary Fermi surfaces and arbitrary dispersion law, considered by Mazin [3] in the ballistic limit; (4) the effect of an evanescent Andreev hole on quasiparticle current in half-metallic CC, mentioned by Kashiwaya et al. [153] and Žutić and Valls [134]; and (5) AR for a ferromagnet in the diffusive case, addressed in part by Golubov [154], Jedema et al. [155], Fal'ko et al. [156], and Belzig et al. [157].

2.5.3.1 Ballistic Contact

We will first consider the ballistic contact using the assumptions of the BTK model as described in Section 2.4.2.† The conductance of such a contact is [3, 150]

$$G = \frac{e^2}{\hbar} \frac{1}{2} \langle N \mid v_x \mid \rangle A, \qquad (2.26)$$

where:

  A is the contact area
  N is the volume density of electronic states at the Fermi level
  v is the Fermi velocity, and brackets denote Fermi surface averaging

Integrating over the states with $v_{ki\sigma,x} > 0$ and summing over the band index, i, and the electron spin, σ, we obtain

$$\frac{1}{2} \langle N \mid v_x \mid \rangle = \frac{1}{(2\pi)^3} \sum_{i\sigma} \int \frac{dS_F}{\mid v_{ki\sigma} \mid} v_{ki\sigma,x}. \qquad (2.27)$$

It is instructive to derive Equation 2.27, starting with the Landauer formula for the conductance of a single electron [17], $G_0 = e^2/h$. The overall conductance is equal to G_0 times the number of CC, N_{cc}, which is defined as the number of electrons that can pass through the contact. If the translational

---

* Recently, Grein et al. [146] introduced a more general microscopic model emphasizing the effects of a barrier shape and spin mixing caused by magnetically active interfaces, which included the model of Mazin et al. [18] as a limiting case.

† In fact, the assumption of a step function order parameter is more realistic for the F/S interface, as the proximity effect there is generally less pronounced compared with the N/S case [158].

symmetry in the interface plane is preserved, then the quasi-momentum in this plane, $\mathbf{k}_\parallel$ is conserved, and N_{cc} is given by the total area of the contact times the density of the 2-D quasi-momentum. The latter is $S_x/(2\pi)^2$, where S_x is the area of the projection of the bulk Fermi surface onto the contact plane [159] (see Figure 29.10). Thus,

$$G = \frac{e^2}{h} \frac{S_x A}{(2\pi)^2} \equiv \frac{e^2}{\hbar} \frac{1}{2} \langle N \mid v_x \mid \rangle A. \tag{2.28}$$

Note that the product, $\langle N|v_x|\rangle$, which is often called the *current-carrying DOS* [160], is very different from the DOS at the Fermi level:

$$N(E_F) = \frac{1}{(2\pi)^3} \sum_{i\sigma} \int \frac{dS_F}{|v_{\mathbf{k}i\sigma}|}. \tag{2.29}$$

2.5.3.2 Diffusive Contact

In the opposite limit, the contact size is much *larger* than the mean free path. The conductance is then given by the bulk conductivity

$$\sigma = \left(\frac{e^2}{\hbar}\right)\langle Nv_x^2\rangle\tau, \tag{2.30}$$

where τ is the relaxation time. Ohm's law requires that the conductance $G = \sigma A/L$, where L is the length of the disordered region. This can be also reproduced within the random matrix approach of Beenakker [17], taking into account that now each CC ($\mathbf{k}_\parallel$) state has a finite probability for an electron to get through the disordered region, $0 \leq T \leq 1$, and

$$G = \frac{e^2}{h} \sum_\kappa T_\kappa = \frac{e^2}{h} \int_\lambda^\infty \frac{d\zeta P(\zeta)}{\cosh^2(L/\zeta)}, \tag{2.31}$$

where $\kappa \equiv \{\mathbf{k}_\parallel, i, \sigma\}$. T_κ is defined in terms of the probability distribution, $P(\zeta)$, of the localization lengths, ζ. The cutoff λ should be on the order of the mean free path l; in fact, $\lambda = 2l$ (the factor of 2 accounts for two possible directions of the electron velocity). Ohm's law requires that the conductance $G \sim 1/L$; thus, the behavior of $P(\zeta)$ at large ζ must be $const/\zeta^2$. Normalization gives $const = \lambda N_{cc}$. Substituting that in Equation 2.31, we obtain

$$G = \frac{e^2}{h} \int_\lambda^\infty \frac{\lambda N_{cc}d\zeta}{\zeta^2 \cosh^2(L/\zeta)} \approx \frac{e^2 \lambda N_{cc}}{hL}$$

$$= \frac{e^2}{\hbar} \frac{A\lambda}{\Omega L} \sum_{i\sigma} \int \frac{dS_F}{|v_\kappa|} v_{\kappa,x}, \tag{2.32}$$

where Ω is the unit cell volume.

In the constant τ approximation, used in Equation 2.30, the average mean free path $l = \sum_{i\sigma} \int \frac{dS_F}{|v_{\mathbf{k}}|} v_{\mathbf{k},x}^2 \tau \Big/ \sum_{i\sigma} \int \frac{dS_F}{|v_{\mathbf{k}}|} v_{\mathbf{k},x}$; thus, $\lambda_{\mathbf{k}} = 2v_{\mathbf{k},x}\tau$. Therefore,

$$\langle G \rangle_L = \frac{e^2}{\hbar} \frac{A}{\Omega L} \sum_{i\sigma} \int \frac{2dS_F}{|v_{\mathbf{k}}|} v_{\mathbf{k},x}^2 \tau = \frac{e^2}{\hbar} \langle Nv_x^2 \rangle \frac{A}{L} = \sigma \frac{A}{L}.$$

Thus, in the diffusive limit, the conductance is determined by $\langle Nv_x^2 \rangle$, as expected.

2.5.3.3 Random Matrix Approach: Ballistic Case

Let us reproduce the BTK results using the random matrix approach. The probabilities of four processes must be considered: (1) ordinary reflection, defined as the process where $\mathbf{k}_{\parallel}$ is conserved, while the group velocity in the direction perpendicular to the interface changes sign; (2) Andreev reflection, when $\mathbf{k}_{\parallel}$ changes to $-\mathbf{k}_{\parallel}$; and (3–4) transmission into the superconductor with or without branch crossing [16]. The conductance then is

$$\langle G \rangle_{NS} = \frac{e^2}{h} \sum_{\mathbf{k}} T_S(\mathbf{k}) = \frac{e^2}{h} \sum_{\mathbf{k}} (1 + A_{\mathbf{k}} - B_{\mathbf{k}}), \tag{2.33}$$

where A and B are the probabilities of the normal and AR, respectively. "Andreev transparency," that is, the probability of the Andreev process, can be expressed in terms of the normal transparency T_N of the interface [17]. For zero bias,

$$T_S = \frac{2T_N^2(\mathbf{k})(1+\beta^2)}{\beta^2 T_N^2 + [1+r_N^2]^2} = \frac{2T_N^2(1+\beta^2)}{\beta^2 T_N^2 + [2-T_N]^2}, \tag{2.34}$$

where:

$T_N(\mathbf{k})$ is the normal state transparency

$r_{\mathbf{k}}^2 = 1 - T_N(\mathbf{k})$ is the corresponding normal state reflectance

$\beta = V/\sqrt{|\Delta^2 - V^2|}$ is the coherence factor

A similar formula can be derived for $V > \Delta$. Equation 2.34 is equivalent to the BTK formulas. For a specular barrier, and neglecting the Fermi velocity mismatch at the interface, $T_N(\mathbf{k}) = 1/[1 + Z^2]$.

2.5.3.4 Random Matrix Approach: Diffusive Case

Let us apply the same approach to the diffusive AR contact, which can be viewed as a contact between the normal and the superconducting leads, separated, in addition to the interface, by a diffusive region. The size of the region is larger than the electronic mean free path [154]. In the zero-temperature and zero-bias limit, Equation 2.34 becomes

$$\langle G \rangle_{NS} = \frac{e^2}{h} \sum_{\mathbf{k}} T_A = \frac{e^2}{h} \sum_{\mathbf{k}} \frac{2\tilde{T}_{\mathbf{k}}^2}{(2 - \tilde{T}_{\mathbf{k}})^2}, \tag{2.35}$$

where now the normal state transmittance for the conductance channel κ is given by the sequential conductor's formula

$$\tilde{T}^{-1} - 1 = (T_N^{-1} - 1) + (t^{-1} - 1), \qquad (2.36)$$

where t is the transmittance of the diffusive region and T_N is the barrier transparency.

Using Equation 2.31 for the probability distributions, we obtain

$$\langle G_{NS} \rangle_L = \frac{e^2}{h} \sum_{\kappa} \frac{2}{(2/T_N - 2 + 2/t_\kappa - 1)^2}$$

$$= \frac{e^2}{h} \frac{\lambda N_{cc}}{L} \int_0^\infty \frac{dy}{[2(1 - T_N)/T_N + \cosh y]^2} \qquad (2.37)$$

$$= \frac{e^2}{h} \frac{\lambda N_{cc}}{L} \frac{w \cosh w - \sinh w}{\sinh^3 w},$$

where $\cosh w = 2(1 - T_N)/T_N$. For the clean (no-barrier) interface, $T_N = 1$, $w = i\pi/2$, and this expression is reduced to Equation 2.32, thus reproducing the well-known result [17, 93] that the resistance at zero bias of a diffusive Andreev contact with no additional interface barrier is the same in the superconducting and normal states.

In the following, we will derive a set of formulas for arbitrary bias, temperature, and interface resistance for both ballistic and diffusive regimes, generalizing the BTK formulas for the half-metallic CC and in the diffusive limit. These formulas are summarized in Table 2.1. $F(s)$ is defined in Section 2.5.3.5.

TABLE 2.1
Bias Dependence of the Total Interface Current in Different Regimes

	$E < \Delta$	$E > \Delta$
Ballistic nonmagnetic (BNM)	$\dfrac{2(1 + \beta^2)}{\beta^2 + (1 + 2Z^2)^2}$	$\dfrac{2\beta}{1 + \beta + 2Z^2}$
Ballistic half-metallic (BHM)	0	$\dfrac{4\beta}{(1 + \beta)^2 + 4Z^2}$
Diffusive nonmagnetic (DNM)	$\dfrac{1 + \beta^2}{2\beta} \mathrm{Im}[F(-i\beta) - F(i\beta)]$	$\beta F(\beta)$
Diffusive half-metallic (DHM)	0	$\beta F\left[(1 + \beta)^2/2 - 1\right]$

2.5.3.5 Half-Metallic Channels

By definition, half-metallic CC correspond to the $\kappa_\parallel$ allowed in only one spin direction. We will consider an incoming plane wave and the transmitted plane wave (with and without branch crossing), assuming, for simplicity, the same wave vector for all the states:

$$\psi_{in} = \begin{pmatrix} 1 \\ 0 \end{pmatrix} e^{ikx}; \quad \psi_{tr} = c \begin{pmatrix} u \\ v \end{pmatrix} e^{ikx} + d \begin{pmatrix} v \\ u \end{pmatrix} e^{-ikx}.$$

Here, u and v have the standard BTK meaning; for example, $u^2 = 1 - v^2 = (1 + \beta)/2\beta$ at $V > \Delta$. Unlike the case of Blonder et al. [16], though, here the reflected state is a combination of a plane wave and an evanescent wave:

$$\psi_{refl} = a \begin{pmatrix} 0 \\ 1 \end{pmatrix} e^{\kappa x} + b \begin{pmatrix} 1 \\ 0 \end{pmatrix} e^{-ikx}. \tag{2.38}$$

The total current is

$$\frac{G_{HS}}{G_0} = \frac{4\beta[1 + (K - 2Z)^2]}{4(\beta^2 - 1)Z(K - Z) + [1 + (K - 2Z)^2][(1 + \beta)^2 + 4Z^2]}, \tag{2.39}$$

at $V > \Delta$, where $K = \kappa/k$, and zero otherwise.

As $V \to \Delta$, $G_{HS}/G_0 \to 0$, and $G_{HS}/G_0 \to G_N/G_0 = 1/(1 + Z^2)$ as $V \to \infty$. This maximum of G_{HS}/G_0 at intermediate biases is due to the fact that although the Andreev-reflected hole does not propagate and does not carry any current, the Andreev process is, however, allowed at $V > \Delta$, thus enhancing the transparency of the barrier. This effect does not exist, though, for $Z = 0$ or for $K \to \infty$. To simplify the equations, we used $K \to \infty$ in the formulas given in Table 2.1. Note that neglecting the first term in Equation 2.38 for the normal current renormalization at $V > \Delta$ used in Srijkers et al. [147] leads to a different result: instead of $4\beta/[(1 + \beta)^2 + 4Z^2]$, it gives $[1 + \beta(1 + 2Z^2)]/[(1 + \beta)(1 + 2Z^2) + 2Z^4]$, which is discontinuous at $V = \Delta$.

2.5.3.6 Generalized Diffusive Case

Let us generalize the BTK formulas for the diffusive limit. For the nonmagnetic CC, the calculation follows Equations 2.34 and 2.36. For zero temperature and a sub-gap bias voltage $V < \Delta(T)$,

$$\langle G \rangle_{NS} = \frac{e^2}{h} \sum_{\kappa_\parallel, i} \frac{4\tilde{T}_N^2(\kappa)(1 + \beta^2)}{\beta^2 \tilde{T}_N^2(\kappa) + [2 - \tilde{T}_N(\kappa)]^2}, \tag{2.40}$$

and

$$\tilde{T}_N^{-1} = T_N^{-1} + t^{-1} - 1 = Z^2 + t^{-1}, \tag{2.41}$$

with the distribution (2.31) for t. Using this, we obtain

$$\langle G_{NS} \rangle_L = \frac{e^2}{h} \frac{\lambda N_{cc}}{L} \int_0^\infty \frac{(1 + \beta^2) dy}{\beta^2 + (2Z^2 + \cosh y)^2}. \tag{2.42}$$

Factor N_{cc} now stands for the number of CC allowed in both spin channels; λ is given by the average mean free path for the channels, and thus, the total conductance is given by $\langle Nv_x^2 \rangle$, averaged over these channels. For $Z = 0$, this results in

$$\langle \sigma_{NS} \rangle = \frac{e^2 \tau}{\Omega} \langle Nv^2 \rangle_{\downarrow\uparrow} \frac{\Delta}{V} \log \left| \frac{V + \Delta}{V - \Delta} \right|, \tag{2.43}$$

which diverges logarithmically at $V = \Delta$. For an arbitrary Z, the conductance is

$$\langle \sigma_{NS} \rangle = \frac{e^2 \tau}{\Omega} \langle Nv^2 \rangle_{\downarrow\uparrow} \frac{1 + \beta^2}{2\beta} \mathrm{Im}[F(-i\beta) - F(i\beta)],$$

where

$$F(s) = \cosh^{-1}(2Z^2 + s)/\sqrt{(2Z^2 + s)^2 - 1}.$$

Similarly, for $V > \Delta$,

$$\langle G_{NS} \rangle_L = \frac{e^2}{h} \frac{\lambda N_{cc}}{L} \int_0^\infty \frac{2\beta dy}{\beta + (2Z^2 + \cosh y)}, \tag{2.44}$$

$$\langle \sigma_{NS} \rangle = \frac{e^2 \tau}{\Omega} \langle Nv^2 \rangle_{\downarrow\uparrow} \beta F(\beta). \tag{2.45}$$

At $Z = 0$, this equation is reduced to*

$$\langle \sigma_{NS} \rangle = \frac{e^2 \tau}{\Omega} \langle Nv^2 \rangle_{\downarrow\uparrow} \frac{V}{\Delta} \log \left| \frac{V + \Delta}{V - \Delta} \right|. \tag{2.46}$$

At $V \gg \Delta$, we obtain

$$\langle \sigma_N \rangle = \frac{e^2 \tau}{\Omega} \langle Nv^2 \rangle_{\downarrow\uparrow} \frac{\cosh^{-1}(2Z^2 + 1)}{Z\sqrt{Z^2 + 1}}, \tag{2.47}$$

which should be used to normalize the conductance curve.

Finally, for the "half-metallic" CC, there is no conductance at $V < \Delta$. For $V > \Delta$,

$$\langle G_{HS} \rangle_L = \frac{e^2}{h} \frac{\lambda N_{cc}}{L} \int_0^\infty \frac{2\beta dy}{(\beta + 1)^2 + 2(2Z^2 - 1 + \cosh y)}$$

$$\langle \sigma_{HS} \rangle = \frac{e^2 \tau}{\Omega} \langle Nv^2 \rangle_{\downarrow} \beta F \left[\frac{(\beta + 1)^2}{2} - 1 \right],$$

---

* Equations 2.43 and 2.46 coincide with those obtained in Artemenko et al. [93] by a different technique.

where the arrow in the subscript shows that these channels are allowed only in one spin sub-band, as shown in Figure 2.10. Note that in the limit of $V \gg \Delta$,

$$\langle \sigma_{HS} \rangle = \frac{e^2 \tau}{\Omega} \langle N v^2 \rangle_{\downarrow} \frac{\cosh^{-1}(2Z^2 + 1)}{2Z \sqrt{Z^2 + 1}}, \tag{2.48}$$

which is exactly one half of the conductance of the nonmagnetic case.

The formulas corresponding to the generalized BTK equations for the finite spin polarization in both ballistic and diffusive limits are summarized in Table 2.1. The finite temperatures can be taken into account in the same way as in the BTK paper.

2.6 PCAR EXPERIMENTAL TECHNIQUE

2.6.1 EXPERIMENTAL GEOMETRY

To study the properties of an F/S interface, it is necessary to establish a stable contact between a ferromagnet and a superconductor. While planar junctions have been successfully fabricated for the spin polarization measurements [161–163], the point contact geometry is most commonly used. Among point contacts, there are two basic types: adjustable mechanical point contacts [11, 132, 164, 165] and nanoconstrictions fabricated in thin silicon membranes [12, 102, 166].

2.6.1.1 Point Contacts

In the most common geometry, a tip made out of a superconducting wire, which was either mechanically [7, 12] or electrochemically polished [132, 167, 168] or cut with a sharp blade [169], is employed; see Figure 2.13. In principle, reversing the role of the two materials should not affect the results, as was demonstrated, for example, for Fe and Ta [16]. While one has to be mindful of not exceeding the depairing current density limit, it is more convenient to use a superconducting tip in contact with a ferromagnetic base (the sample under study), as this allows one to study ferromagnetic films as well as bulk samples; additionally, the effect of a magnetic field on a superconductor [170] can be minimized. The properties of a contact depend largely on the design of the approach mechanism, which may vary from a simple mechanical adjustment controlled by a micrometer [11] to more versatile systems, employing a combination of differential screws, micrometers, and piezo-driven actuators [103, 171–173]. The important part of any system is the ability to maintain a stable contact, preferably within a broad temperature range. Good temperature stability is important for taking variable temperature measurements as well as for a more accurate determination of the contact size [103, 104]; see Section 2.3.7. One should also be able to control the force applied to the tip, as the sample properties may be pressure sensitive. While, compared with the nanoconstriction geometry, the tip geometry is more versatile, it has a larger uncertainty in terms of the actual contact size.*

---

* Another original technique was employed by Gonnelli et al. [174] in the measurements of MgB_2 superconductors. In this technique, a small amount of colloidal silver is placed on top of MgB_2, providing a stable point contact.

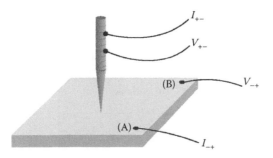

FIGURE 2.13 Schematic of a typical adjustable point contact, with two current and two voltage leads. In the most commonly used arrangement, the point is the superconductor, whereas the base is the ferromagnet (leads (A) and (B)).

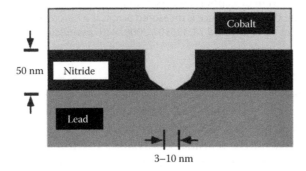

FIGURE 2.14 Schematic of a typical ferromagnet-superconductor nanocontact (nanoconstriction) device. (After Upadhyay, S.K. et al., *Phys. Rev. Lett.* 81, 3247, 1998. With permission.)

An alternative technique was used by Upadhyay et al. [12]. A silicon nitride membrane containing a tapered nanohole with a minimum diameter of 3–10 nm [175] was fabricated by chemical vapor deposition of ~50 nm thick silicon nitride; see Figure 2.14. A ferromagnet and a superconductor can then be deposited onto the opposite sides of the membrane. The important advantage of this technique is that the geometry of the contact is well defined. In addition, all the fabrication steps are typically done in situ, so that the possibility of contamination is reduced, and thus, Z is usually fairly small [102]. While it may be hard to completely exclude the accumulation of defects in the area of the contact, the presence of phonon spectra, which is routinely observed for these types of contacts, confirms that the transport is in the ballistic ($l > a$) or a quasi-ballistic ($l \approx a$) regime. Unfortunately, the fabrication of some of the most interesting new materials often requires lattice matching to a specific substrate, which limits the practical application of this method.* Ideally, the combination of an adjustable point contact with in situ sample preparation and subsequent spin polarization measurements is likely to be the most effective.

---

* Other nanofabrication techniques include nanoindentation using atomic force microscopy (AFM) [176] and focused ion beam (FIB) milling [177].

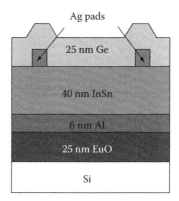

FIGURE 2.15 Example of a planar contact geometry used in the study of EuO_{1-x}. Top to bottom: 25 nm Ge cap/Ag contacts/400 nm InSn/6 nm Al/25 nm EuO_{1-x}/bulk $n+$ Si substrate. The conductance is measured across the heterostructure. (After Panguluri, R.P. et al., *Phys. Rev. B* 78, 125307, 2008. With permission.)

2.6.1.2 Planar Geometry

As an alternative to a point contact, a planar geometry—in which AR was first observed—can also be successfully employed for the spin polarization measurements [161–163]. An example of this geometry is shown in Figure 2.15. One of the potential disadvantages of this type of contact is the current nonuniformity—the same problem that is often encountered in tunnel junctions [178] and in current-perpendicular to plane (CPP) GMR devices [179], see Chapter 4, Volume 1. In addition, the contact properties are very sensitive to surface roughness and other surface defects. As the only way to change the contact properties in this case is to make a new sample, this method is rather labor intensive.

2.6.2 ELECTRICAL MEASUREMENTS

In most cases, the point contact measurements are done in the temperature range of 0.3–4 K using conventional superconductors, such as Nb, Pb, Ga, Ta, Sn, In, and Al, so that a typical maximum bias voltage $\sim 10\Delta$ is approximately 10 mV. With a typical contact resistance of 10 Ω, this results in a maximum Joule heating of about 10 μW. Thus, in most cases, the standard continuous *lock-in detection* technique can be used to measure the differential contact resistance dV/dI, especially if the measurements are done in a liquid helium bath. On the other hand, for point contact spectroscopy of high-T_c superconductors with considerably larger gaps or for contact resistances lower than 1 Ω, pulsed current techniques can be implemented to avoid heating effects, especially at lower temperatures (e.g., Rourke et al. [180]).

2.6.3 DATA ANALYSIS

2.6.3.1 General Approach

While the actual dI/dV measurements are fairly straightforward, the analysis of the data is often nontrivial, as there are more fitting parameters compared with the unpolarized case. In the standard models (e.g., Upadhyay et al. [12]; Mazin et al. [18]), there are at least two fitting parameters, P and Z; the gap Δ is typically set to the value from the Bardeen–Cooper–Schrieffer (BCS) theory for the respective superconductor. The fitting procedure involves normalizing the conductance curve dI/dV with respect to $(dI/dV)_N$, which is defined either as an asymptotic value at high bias voltages ($V \approx 10\Delta$ is usually sufficient) or as a value obtained upon the application of a critical magnetic field. A standard data analysis routine, which includes an optimization procedure minimizing the least square function, is then applied.* Additionally, a change in the value of the superconducting gap usually results in a proportional change in P [152]. It is desirable, therefore, to evaluate the gap independently by measuring the critical temperature of the superconducting transition and the conductance spectra at different temperatures. If, in addition to the superconducting gap, the values of quasiparticle lifetimes are varied as well, the number of adjustable parameters increases to four. For four parameters, the iteration procedure may lead to degenerate solutions, although a recipe to minimize this effect has been proposed [169]. Thus, while a typical accuracy of statistically averaged measurements of P is on the order of 2%–3%, a systematic error associated with the data analysis can significantly exceed this value. This problem is particularly pronounced for small values of $P \simeq 10\%$, especially when additional DOS broadening is present [102, 169].

2.6.3.2 Inelastic Scattering

The typical weak coupling BTK models described in Section 2.5 assume effectively infinite quasiparticle energy relaxation time $\tau_E \gg \Delta / \hbar$. However, different inelastic scattering mechanisms often contribute to the reduction of quasiparticle lifetime and thus broaden the BCS DOS. To account for this broadening, the Dynes parameter Γ can be introduced: $\Gamma = \hbar / \tau$ [181]. Accordingly, this will modify the quasiparticle DOS $N(E,\Gamma)$ [182, 183]:

$$N(E,\Gamma) = \mathrm{Re}\left[\frac{E + i\Gamma}{\sqrt{(E + i\Gamma)^2 - \Delta^2}} \right], \quad (2.49)$$

which will then have to be included in the model [102]. Alternatively, one might be able to account for broadening by using the original BCS quasiparticle DOS $N(E)$ but with a higher effective temperature $T_{\mathrm{eff}} = \Gamma/\sqrt{2}$ [138]. In some cases, inelastic scattering may also be accompanied by a reduction in the superconducting gap [138, 169]. For example, Chalsani et al. [102] intentionally inserted a Pt layer between a ferromagnet and a superconductor, confirming that a modification of the standard model is necessary to match the experimental data when strong spin scattering is introduced.

2.6.3.3 Possible Spin Polarization Dependence on Z

Another aspect of the procedure of evaluating the values of spin polarization involves plotting P measured for different point contacts as a function of Z

---

* It is important to keep in mind that, as in other problems of this type, this function may have multiple local minima.

and extrapolating the function $P(Z)$ to $Z = 0$. As we have discussed, there may be a difference in the barrier transparency for majority and minority carriers, as well as spin-flip processes at the interface. Kant et al. [184] proposed that P could be written as

$$P = P_0 \exp\left[-2\alpha\Psi Z^2\right], \tag{2.50}$$

where:

    P_0 is the intrinsic value of the spin polarization
    α is defined as the spin-flip scattering probability
    Ψ is the ratio of the forward and backward scattering probabilities

In this model, the factor of Z^2 is derived from multiple scattering within the interface region. Z^2 is then proportional to the number of collisions and thus, to the ratio of the contact diameter to the electron mean free path. Another possible interpretation of the spin polarization suppression in the case of finite barrier transparency is spin-flip scattering by defects at the interface, which in turn is proportional to the number of scattering events. This immediately leads to Equation 2.50.* It is obvious that for contacts with large Z and strong spin-flip scattering, P should eventually go to zero,† so the exponential function may have some validity. On the other hand, there is no statistically significant advantage in using Equation 2.50 compared with a quadratic or even a linear dependence [152].

2.6.3.4 Spreading Resistance

A specific technical problem, typically encountered in the measurements of thin films, is related to the so-called *spreading resistance* R_s ([152]; see Figure 2.13). If R_s, which corresponds to the voltage drop across the sample, is small compared with the differential resistance of the point contact, $R_c = dV/dI$ (i.e., well below $1\,\Omega$—which is usually the case for bulk samples or highly conductive films), it can safely be neglected. However, if this resistance is not insignificant, the measured $(dV/dI)_\text{meas}$ will include some contribution from the spreading resistance R_s:

$$\left(\frac{dV}{dI}\right)_\text{meas} = \left(\frac{dV}{dI}\right)_\text{cont} + R_s. \tag{2.51}$$

If this is the case, the experimentally measured conductance (which is inversely proportional to the differential resistance) does not represent the true conductance curve of the point contact, $(dI/dV)_\text{cont}$. Specifically, the apparent position of the zero bias conductance will change, and the

---

* It should be noted [152] that this argument is only applicable for the diffusive but not the ballistic case, as stated by Kant et al. [184].
† This may be difficult to confirm experimentally, as the error in the spin polarization values also tends to increase at larger Z. In practical terms, however, it is always preferable to use low Z contacts.

coherence peak will shift from $\simeq\Delta$ to higher voltages.* If the effect of spreading resistance is ignored, the extracted values of P may be seriously compromised [152].

If R_s were known, it would be easy to take R_s into account to reconstruct the actual conductance of the contact, since the current through the contact is also known. Unfortunately, it is often difficult to measure the value of R_s directly, although in some cases, it can be estimated. If the approximate value of R_s is known, the iteration procedure can be performed as a part of the fitting routine. This procedure can be set up in two different ways: either one subtracts the contribution from R_s to obtain a "virgin" conductance curve of the contact (which is different from raw experimental data and is then renormalized), or one can fit the original curve taking into account the contribution from R_s, in which case Δ will differ from the actual superconducting gap.

Alternatively, instead of analyzing individual dI/dV curves, one might analyze a temperature-dependent $R(T)$ function. This approach should allow one to separate the contribution from R_s (which is approximately temperature independent) from a strongly temperature-dependent part related to Andreev reflection [162, 185]; see Figure 2.16.

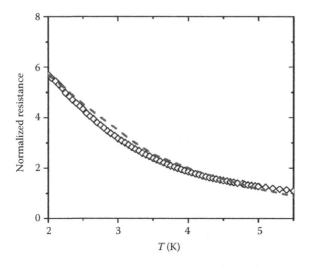

FIGURE 2.16 Numerical fit of the temperature-dependent resistivity curve of the structure shown in Figure 2.15. Dashed line is the fitting curve; open squares are the experimental data. The sharp resistivity increase below the critical temperature of InSn (~5.8 K) is due to the predominant role of Andreev reflection at the superconductor/europium oxide interface, with EuO_{1-x} nearly 90% spin polarized. (After Panguluri, R.P. et al., *Phys. Rev. B* 78, 125307, 2008. With permission.)

---

* The origin of these distortions is understandable: Andreev reflection (or the lack thereof in the case of half-metals) results in a strongly voltage-dependent conductance curve. If R_s (which is approximately ohmic) is comparable to R_c, it will result in a "flattening" of the overall conductance curve. In the extreme case when R_s is much larger than R_c, very little voltage dependence will be observed (unless $P = 100\%$ with nominally infinite resistance at zero bias).

2.6.3.5 Comparison with *ab initio* Calculations

Let us briefly discuss *ab initio* calculations of the point contacts' transparency and compare them with the experimental data. The calculations of Xia et al. [186], for example, were found to be in good agreement with the experimental results of Upadhyay et al. [12] for the nonmagnetic Cu/Pb contacts but disagreed sharply for the magnetic Ni/Pb and Co/Pb contacts. One possible reason for this disagreement is that the δ-function approximation may be insufficient here, as it does not take into account different wave function symmetries at the interface. The other explanation is related to the problem of lattice matching across the interface, which may lead to reduced magnetization of the interface layer [186]. The latter explanation seems to be consistent with the calculations of Taddei et al. [187], who had to use an enhanced magnetic moment at the interface to achieve good agreement with the experimental results [102].

2.7 SPIN POLARIZATION MEASUREMENTS

Since its inception in 1998, the use of the PCAR technique (including the AR measurements in a planar geometry) has increased dramatically. In addition to conventional transition metal ferromagnets, such as Fe, Co, Ni, and their alloys; many Heusler alloys; magnetic oxides, such as CrO_2, LSMO, Sr_2FeMoO_6, EuO, and EuS; dilute magnetic oxides, such as In_2O_3, ZnO; semiconductors, such as (Ga, Mn)As; and a large number of other materials have been investigated by this technique. In the following, we will briefly describe some of these results in detail; the summary of the measurements currently available in the literature is given in Table 2.2.

2.7.1 TRANSITION METAL FERROMAGNETS

In addition to numerous practical applications, 3d transition metal ferromagnets and their alloys have been used as a model system to develop reliable band structure calculation techniques of different flavors. However, many aspects of this seemingly simple system are still not very well understood. For example, it is difficult to reconcile the itinerant character of these materials with the positive sign of spin polarization measured by the spin-polarized tunneling technique. While it is quite reasonable to assume that this is due to larger tunneling matrix elements for *s*-electrons compared with *d*-electrons, which compensate for a higher minority DOS at the Fermi level, one can only marginally divide electrons in 3d metals into *s* and *d* types. Thus, it may be more appropriate to discuss these effects in terms of different bands with different Fermi velocities. The PCAR measurements of various Ni_xFe_{1-x} alloys, including permalloy [165], indicate that while the magnetic moment is dependent on the composition, the spin polarization values are almost composition independent; $P_{Nv} \simeq 45\%-50\%$. These results are in surprisingly good agreement with the values of $P_T \simeq 50\%$ measured by

TABLE 2.2
Results of the Spin Polarization Measurements with the PCAR Technique

Materials	Reference P (%)
Co	42 ± 2[11], 37 ± 2[12], 45 ± 2[147], 47 ± 2[184]
Fe	45 ± 2[11], 49.6 ± 1[166], 43 ± 3[147], 46 ± 3[184]
Ni	46.5 ± 1[11], 28[185], 37 ± 1[147]
CrO_2	90 ± 3.6[11], 81[188], 96 ± 4[189], 92[79], 95[176], 96 ± 1[188], 98.5[190], 96 ± 2[184, 191]
$Ni_{0.8}Fe_{0.2}$	37 ± 5[11], 35[192]
Ni_xFe_{1-x}	44 ± 3[165]
Gd	45 ± 4[184], 52[193]
Dy	50[193]
$La_{0.7}Sr_{0.3}MnO_3$	78 ± 4[11], 60–90[7], 78 ± 2[194]
$La_{0.6}Sr_{0.4}MnO_3$	83 ± 2[194]
EuO	90[162], 90[195]
EuS	80[163]
EuB_6	56[196]
$Co_{1-x}Fe_xS_2$	85[197–199], 47–61[200]
CoS_2	56[201], 64[202]
(Ga, Mn)As	85[195], 76–90[203], 83 ± 17[138]
(In, Mn)Sb	52 ± 3[139], 61 ± 1[204]
(Ga, Mn)Sb	57 ± 5[205]
Sr_2FeMoO_6	70[173], 63[206]
$SrLaVMoO_6$	50[207]
In_2O_3, $Cr:In_2O_3$	50 ± 5[208]
$Co_{1.95}Nd_{0.05}MnSi$	59[209]
$SrRuO_3$	60 ± 2[210], 51 ± 2[211], 50[212], 60[230]
$Ni_{0.76}Al_{0.24}$	50[214]
$Co_{75}Fe_{25}$	58 ± 3[215]
Fe_4N	59[216]
$Co_xFe_{80-x}B$	65[217]
FePt	42[218]
Mn_4FeGe_3	42[219]
$Co_2Cr_xFe_{1-x}Si$	56–64[220]
$Co_2FeAl_xSE_{1-x}$	57 ± 1–60 ± 5[221]
$Co_2Cr_{1-x}Fe_xAl$	54–62[222]
$Co_2MnAl_{1-x}Sn_x$	63[223]
$Co_{1.9}Fe_{0.1}CrGa$	67 ± 3[224]
Co_2CrGa	61 ± 2[225]
Co_2MnGe	35–58[225], 55–60[226]
Co_2MnSi	52–54[225], 50–55[227], 54[228], 56[229], 20[176]
Co_2FeSi	49 ± 2[227]
Co_2FeGa	59[230]
NiMnSb	58 ± 2.3[11], 44[231], 45[229]
NiFeSb	52[232], 52[233]
Ni_2MnIn	28–34[192]

(Continued)

TABLE 2.2 (CONTINUED)
Results of the Spin Polarization Measurements with the PCAR Technique

Materials	Reference P (%)
Mn_5Ge_3	42 ± 5 [234]
Fe_3Si	45 ± 5 [235]
MnAs	44–49 [236]
MnBi	63 ± 3 [237]
Mn_2Ga	40 [238]
Mn_3Ga	58 ± 3 []
Mn_3Ge	46 ± 2 [239]
$Mn_2Ru_{0.48}Ga$	54 ± 3 [240]
Mn_2FeGa	51 ± 2 [241]
$p\text{-}In_{0.96}Mn_{0.04}As$	72.5 [242]
NiMnSb	45 [243]
$La_{0.65}Ca_{0.35}MnO_3$	80 [244]
CoFeCrAl	67 ± 2 [245]
CoFeMnGe	70 ± 1 [246]
FeN	52 [247]
Co_3FeN	62 [248]
$Co_{1-x}Gdx$	55 [249]
$Fe_{1-x}V_x$	52 ± 3 [250]
$Ru_{2x}Fe_xCrSi$	53 [251]
$CeMnNi_4$	66 [252]
Th	48 ± 8 [253]
$In_{0.95}Cr_{0.05}N$	50 ± 2 [254]
Ho	42 [255]
$In_{0.95}Sn_{0.05}O_3$	58 [256]

the Meservey–Tedrow technique [257].* Similar results for permalloy have been obtained by the PCAR technique using the point contact [192] and nanoconstriction geometries [166] and in some of the spin injection experiments (e.g., Godfrey and Johnson [259]). The Doppler spin wave technique results in $P_{Nv^2} \simeq 50\%$ at room temperature [47] and $P_{Nv^2} \simeq 75\%$ at 4 K [260], in good agreement with the band structure calculations in the diffusive limit [165].

2.7.2 HALF-METALS

At present, there are many theoretical predictions of half-metallicity, far exceeding the experimental evidence for the existence of half-metals. This is understandable, as band structure calculations are normally performed

---

* Interestingly, initially, P of Ni was measured only at 8% [70]; then the value increased to 13% [258] and further to 23% [30], until the most recent result of 50% was obtained with the use of high vacuum film deposition, indicating the sensitivity of Ni and Ni-based alloys to fabrication conditions.

at zero temperature for perfectly ordered crystals, whereas the experiments are done at finite temperatures and with imperfect materials. This raises an important question of the stability of half-metallic behavior with respect to substitutions and magnetic disorder induced by finite temperatures. For example, just 5% of substitutional defects in NiMnSb may be sufficient to reduce its spin polarization dramatically and to close the minority gap [261]; similar effects have been predicted in PtMnSb [262] and in Sr_2FeMoO_6 [173]. Importantly for spintronics applications, the temperature scale for the decay of half-metallicity is defined by a relationship $T \ll T_C$ rather than $T \ll \Delta$ (e.g., Dowben and Skomski [263]; Ležaić et al. [264]; Skomski and Dowben [265]. In addition, there are other mechanisms that may effectively mix the spins and suppress half-metallicity, such as spin–orbit coupling [266, 267] and Stoner excitations (spin flips). Conversely, Andreev reflection spectroscopy of highly spin-polarized ferromagnets can be used to probe spin–orbit interaction at the F/S interface, which is also strongly dependent on the direction of magnetization in the ferromagnet; see Högl et al. [268].

Another issue is a possible difference between surface and bulk electronic structure. Does half-metallicity persist all the way to the surface, where the translational symmetry is broken? The calculations in one of the layered manganites, $LaSr_2Mn_2O_7$, show that it is a half-metal in the bulk—but not at the surface [269]. On the other hand, NiMnSb is believed not to be a half-metal at the surface, but it may preserve its half-metallicity at the interface with CdS [270]. These basic questions are closely related to the measurement techniques, many of which are surface sensitive. Different aspects of half-metals are reviewed, for example, by Irkhin and Katsnelson [271], Coey et al. [272], Ziese [273], Eschrig and Pickett [274], and Katsnelson et al. [275].

2.7.2.1 CrO_2

CrO_2 is seemingly the only material in which the predicted half-metallicity [276] has been confirmed experimentally by several different techniques. While the use of this material in practical applications is hindered by the presence of a thin ($\simeq 1.5$ nm) Cr_2O_3 surface layer, which is an antiferromagnetic insulator, the unambiguous experimental demonstration of half-metallicity is important as a matter of principle. The early PCAR results ($P \simeq 80\%$–90% [11, 188], as well as the following studies on higher-quality single crystalline films (with $P > 95\%$) [164, 189], agree well with the values of spin polarization obtained by other techniques, such as photoemission [68], the Meservey–Tedrow technique, and the AR measurements in the planar geometry [79]. In the latter work, while the probability of AR (and thus, the accuracy of the technique) is reduced rapidly as the interfacial barrier strength increases, the Meservey–Tedrow technique becomes available once the quasiparticle tunneling starts to dominate the conduction across the interface. The combined results from these measurements are shown in Figure 2.17.

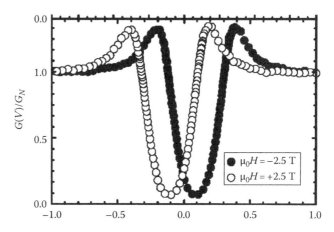

FIGURE 2.17 Spin polarization measurements of CrO_2 in a planar geometry. The conductance curve is gradually shifting on the application of a magnetic field in the plane of the junction. No features due to the possible presence of minority spins are observed in fields as high as 2.5 T. (After Parker, J.S. et al., *Phys. Rev. Lett.* 88, 196601, 2002. With permission.)

2.7.3 Other Potential Half-Metals: $Fe_xCo_{1-x}S_2$, EuO, EuS, and $Sr_xFe_{1-x}MoO_6$

2.7.3.1 $Fe_xCo_{1-x}S_2$

$Fe_xCo_{1-x}S_2$ is expected to be less sensitive to crystallographic disorder and other defects and has been predicted to be a half-metal in a wide concentration range, $0.25 < x < 0.9$, with the magnetic moment per Co of $1\mu_B$ [277]. Experimentally, the spin polarization seems to be quite sensitive to the composition [197], varying from $P = 56\%$ for pure CoS_2 [278] to the maximum value of approximately 85% at $x = 0.15$, even though the magnetic moment per Co was found to be practically constant ($1\mu_B$ within the experimental range $0.07 < x < 0.3$) [199]. While the Curie temperature of this system is probably not high enough for it to be used in applications, it provides an insight into the design of new half-metallic ferromagnets.

2.7.3.2 EuO/EuS

One of the few simple magnetic oxides, europium oxide, is also a viable candidate for half-metallicity. Recently, $\simeq 90\%$ spin polarization was measured in La-doped La_xEuO_{1-x} grown on $YAlO_3$ substrate [195], as well as in undoped EuO_{1-x} grown directly on conductively matched Si substrate with no buffer layer [162]. The latter result can potentially be used for spin injection into Si [279, 280]. The PCAR measurements are consistent with spin-resolved x-ray absorption spectroscopy, showing near-complete spin polarization [281] as well as near-complete spin-filtering effect [282] (see Chapter 14, Volume 1, Moodera and Santos). A fairly high spin polarization ($P \simeq 80\%$) has also been measured in a planar geometry for a similar material, EuS [163].

2.7.3.3 SFMO

Sr_2FeMoO_6 (SFMO) has been predicted to be 100% spin polarized in a perfectly ordered structure [283]. It is extremely difficult to fabricate perfect SFMO samples, as SFMO has a propensity for B-site defect formation, which, in turn, alters the half-metallic character of this compound [284]. For example, the band structure calculations for SFMO with 12.5% of anti-site defects results in $P_{Nv} \simeq 53\%$ [173].* These calculations are consistent with the PCAR measurements in single crystals with variable degree of disorder, resulting in $P \simeq 60\%$–70%, higher values of P corresponding to smaller degrees of disorder [173]. These values are in agreement with the earlier measurements in SFMO thin films [285, 286]. While these results indicate that there is a correlation between the degree of disorder and P, it is difficult to directly interpolate the data to a perfectly ordered system, so the question of half-metallicity of SFMO remains unresolved.

2.7.4 TRANSPORT HALF-METALS

Transport half-metals were predicted [8] and experimentally observed [7] in optimally doped $La_{0.7}Sr_{0.3}MnO_3$ (LSMO), where the minority carriers were found to be approaching the metal–insulator transition limit. The name refers to materials that may not be true half-metals in terms of their DOS but, because of the vastly different transport properties of their majority and minority carriers, carry almost 100% spin-polarized current [48, 287].

2.7.4.1 LSMO

The case of LSMO has been puzzling. On the one hand, the band structure calculations predicted about 35% DOS spin polarization for LSMO [288] and higher values for P_{Nv} and P_{Nv^2}, in good agreement with the Tedrow–Meservey measurements by Worledge and Geballe [33] and the TMR results [289], with $P = 72\%$ and $P \simeq 80\%$, respectively. On the other hand, the spin-resolved photoemission measurements of the DOS [67] gave the value of $95\% \pm 5\%$, suggesting that LSMO is a conventional half-metal.† Systematic measurements of LSMO epitaxial films and bulk single crystals by the PCAR technique [7] resulted in a broad range of P for different samples from 60% to 90%. While a correlation was found between P and the residual resistivity of the samples, ρ, surprisingly, it was the highly resistive samples that had higher P, the opposite of what one might expect for a true half-metal (see Figure 2.18). The estimates of the contact size and the mean free path indicated that the low-resistivity samples were in the ballistic regime, while the highest-resistivity samples were in the diffusive regime. Using the values of the densities of states ($N_\uparrow(E_F) = 0.58$ states/eV Mn, $N_\downarrow(E_F) = 0.27$ states/eV Mn) and the Fermi velocities ($v_{F\uparrow} = 7.4 \times 10^5$ ms^{-1}, $v_{F\downarrow} = 2.2 \times 10^5$ ms^{-1}), [288],

---

* Interestingly, the calculated spin polarization in the diffusive limit for 12.5% disorder, $P_{Nv^2} = 93.3\%$, is much higher, resembling the behavior of LSMO, as discussed later.

† For a true half-metal, all spin polarization measurement techniques should give the same value: $P_N = P_{Nv} = P_{Nv^2} = 100\%$.

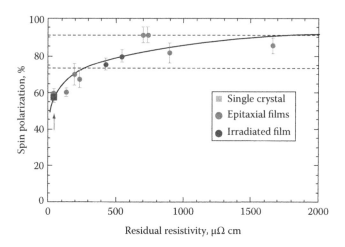

FIGURE 2.18   Spin polarization as a function of the residual resistivity of $La_{0.7}Sr_{0.3}MnO_3$ at 1.5 K. Dashed lines correspond to $P_{Nv} = 74\%$ and $P_{Nv^2} = 92\%$. Solid line is a guide to the eye. (After Nadgorny, B. et al., *Phys. Rev. B* 61, R3788, 2000. With permission.)

$P_{Nv} = 74\%$ and $P_{Nv^2} = 92\%$ were obtained, in fairly good agreement with the experimental data.* The value of P_{Nv^2}, which also corresponds to the spin polarization of the current in bulk LSMO, implies that about 96% of the current is carried by the majority electrons, as the concept of a transport half-metal implies. A record high TMR ratio found in LSMO-based structures [290] corresponds to $P_T \simeq 95\%$, in close agreement with P_{Nv^2} measured by the PCAR technique. While the consensus has yet to be reached [34, 275], the most recent theoretical [291] and experimental results [292] indicate the presence of minority spin states in LSMO.†

2.7.4.2 Geometric Spin Filtering

We can use the case of LSMO to draw an analogy between a spin-polarized current and a molecular flow. Consider a *toy model* (see Figure 2.19), where we have particles in one half of a vessel in equilibrium with another half. There are eight lighter particles (majority spins) and four heavier particles (minority spins). Suppose that the lighter particles are twice as fast as the heavier particles. The DOS "spin polarization" is $P_N = (N_1 - N_2)/(N_1 + N_2) = 33\%$. If the flow through the orifice is ballistic (*Knudsen regime*), the equilibrium is determined by the flux Nv, and the current "spin polarization" is then higher: $P_{Nv} = (N_1v_1 - N_2v_2)/(N_1v_1 + N_2v_2) = (4 - 1)/(4 + 1) = 60\%$.

---

* One might argue, though, that it should still be possible to establish a diffusive contact even in the case of low-resistivity samples (with a long mean free path) by simply increasing the size of the contact. However, in this case, instead of a single large contact, a number of smaller contacts connected in parallel will typically be established [169]. At the same time, the extracted values of the spin polarization are practically independent of the transport regime [152].

† The results from angle resolved photoemission spectroscopy (ARPES) indicate the presence of about 4% minority spin states in LSMO [292].

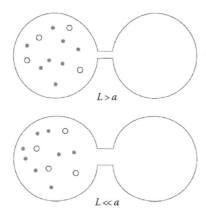

FIGURE 2.19   Geometric spin filtering. "Spin polarization" can be changed from P_{Nv} to P_{Nv^2} solely by adjusting the size of the aperture, thus changing the flow type from ballistic (Knudsen) (top) to diffusive (Maxwell) (bottom).

If the flow is diffusive (*Maxwell regime*), the equilibrium is determined by the pressure, which is proportional to the temperature and, therefore, to the square of the velocity. The spin polarization in the diffusive limit becomes higher still: $P_{Nv^2} = \left(N_1 v_1^2 - N_2 v_2^2\right)/\left(N_1 v_1^2 + N_2 v_2^2\right) = 7/9 \approx 80\%$.

Hence, one can convert the transport from ballistic to diffusive and thus change the degree of spin polarization by changing the contact geometry alone (see Figure 2.19), without changing any materials properties of the system [34].* This effect can be called *geometric spin filtering* by analogy with the spin-filtering effect in F/I/S (Meservey–Tedrow) or F/I/F (TMR) structures, where one can selectively suppress or enhance the transparency of the barrier for the majority and minority electrons [9, 10]. A systematic decrease of P as a function of the contact size a in the nanoconstriction geometry has been reported [166]. The authors attributed the observed effect to an increased spin–orbit scattering as the contact size increases. However, the results can also be explained by a gradual transition from the ballistic ($a \approx 7\,\text{nm}$) to the diffusive ($a \approx 25\,\text{nm}$) regime, as described earlier. In this case, the spin polarization of Fe would change by roughly 10% [165], in reasonable agreement with the reported results.†

2.7.5 DILUTE MAGNETIC SEMICONDUCTORS AND MAGNETIC OXIDES

2.7.5.1 (Ga, Mn) As and Other Magnetic Semiconductors

Until recently, most of the transport studies of a semiconductor/superconductor interface, starting with the pioneering work on Nb/InGaAs junctions [294], have been performed in a 2-D geometry [295]. To extend the PCAR technique

---

* The author wishes to acknowledge his discussions with J. Byers at the early stage of this work. Similar behavior has been predicted by *ab initio* calculations in cobalt nanowires [293].

† We would like to note that while in the case of LSMO, P_{Nv^2} is higher than P_{Nv}, the opposite may be true as well; in fact the two values may even have different signs [212].

to measure the spin polarization of ferromagnetic semiconductors (FmS), several conditions must be met. First, the ability to conduct the measurements at low temperatures implies the absence of strong activation behavior with the resistivity not exceeding $\cong 10\,\mathrm{m\Omega}$ cm, in which case the values of the interface mismatch [296] are typically reasonable as well. Second, the junction transparency, which is often limited by native Schottky barriers present at most Sm/S interfaces, must be high enough to ensure that AR is the dominant process below the gap. The use of high doping levels, producing carrier concentrations on the order of 10^{20}–10^{21} cm^{-3}, helps to alleviate these problems, dramatically reducing the width of the Schottky barrier [297]. This strategy has been successfully applied to a number of studies, in which FmS were measured together with matching nonmagnetic semiconductors with similar doping concentrations [138, 139]. Fortuitously, the increase of the Curie temperature in (Ga,Mn)As is associated with higher doping levels [298] at which the PCAR technique becomes available, with $P \cong 85\%$ measured in (Ga,Mn)As thin films with the use of the planar [161] and point contact geometries [138]. These measurements also indicate the increasing importance of inelastic broadening as the semiconductor band gap increases/the mobility decreases. While practically no broadening was observed in the narrow gap semiconductor (In,Mn)Sb [139, 204], it was present in (Ga,Mn)Sb [205] and even more prominently in (Ga,Mn)As [138], similarly to the effects reported for Co–Pt–Pb in the nanoconstriction geometry [102]. While the conductance broadening observed in Ga/(Ga,Mn)As was attributed primarily to the distribution of the critical temperatures in amorphous Ga used as a superconducting contact [161], additional inelastic broadening effects are also possible.

2.7.5.2 Dilute Magnetic Semiconducting Oxides

High Curie temperatures make dilute magnetic semiconducting oxides (DMSO) very attractive for applications. While the question of the nature of magnetism in these materials is still unresolved (see, e.g., the discussion by Coey [299] and Chapter 10, Volume 2), it is natural to assume that at least in conducting oxides, it is mediated by delocalized carriers [300–302]. If the magnetism is carrier mediated, the spin current is spin polarized, and it should be possible to use some of these materials for room temperature spin injection. Let us briefly discuss two materials systems: Cr-doped and undoped In_2O_3, in which carriers are created by oxygen vacancies [208], and Mn-doped and undoped ZnO, in which carriers are introduced by co-doping with Al [303]. It has been shown that the electrical and magnetic properties of $Cr:In_2O_3$ films are sensitive to the oxygen vacancy defect concentration [304]. More recent measurements of Cr-doped and undoped In_2O_3 films indicate that both are ferromagnetic with very similar magnetization, and both are almost equally spin polarized with $P \simeq 50\%$ [208], which implies that Cr is not essential for the development of magnetism in In_2O_3 with high ($\simeq 10^{21}$ cm^{-3}) concentration of oxygen vacancies.* This can be explained in

---

* Unfortunately, it is difficult to fabricate an In_2O_3 sample, which would be well conducting and nonmagnetic, to use as a standard in the PCAR measurements.

the spirit of the Zener model [302], where the development of ferromagnetism is intimately related to the interaction of delocalized charge carriers and vacancies in In_2O_3. On the other hand, Al-doped ZnO with and without Mn [303] allows direct comparison between two highly conductive materials, of which only Mn-doped is ferromagnetic (with $P \simeq 55\%$), while the nonmagnetic films without Mn is unpolarized. Based on the fairly high measured values of spin polarization in both of these systems, one can argue that ferromagnetism, at least in some DMSO, is carrier mediated.*

2.8 APPLICATION OF PCAR TO SPIN DIFFUSION LENGTH MEASUREMENTS

Spin injection and spin accumulation are of fundamental importance for spintronic devices, as discussed in Chapters 5 and 15, Volume 1, Chapters 5, 7–9, 16, 17, Volume 3. The problem of spin injection from a ferromagnet into a normal metal was first considered by Aronov [25]. Johnson and Sisbee performed the first measurements of spin relaxation and spin diffusion in aluminum using the lateral nonlocal geometry [305], see Chapter 6, Volume 1. Subsequently, a more convenient geometry for F/N/F structures has been used by Jedema et al. [306] Spin diffusion length in metals can also be determined by analyzing the thickness dependence of the current-perpendicular to plane (CPP) GMR effect [307], as well as by spin accumulation measurements via the Kerr effect, as was demonstrated by Crooker et al. [155, 302, 308].

Intuitively, the PCAR technique should be applicable to this problem as well. The simplest way to determine spin diffusion length using PCAR is to measure the spin polarization of spin current injected from a ferromagnet after it passes through a normal metal of variable thickness, as was done for Pt [309] and Au [168] see Figure 2.20a. Importantly, the implicit assumption often used for spin diffusion measurements [309]—that spin current in the normal metal decays exponentially with distance—can only be justified if the thickness of the normal layer is orders of magnitude larger than the spin diffusion length. In the most general case, however, the problem must be solved self-consistently. The presence of the spin-selective (superconductive, in the case of AR) interface results in spin accumulation due to the backflow effect at the interface. Subsequently, the spin current decay deviates from an exponential dependence, resulting in significantly modified spin profile, as well as the values of spin diffusion length based on the PCAR measurements [168]. The spin diffusion length in gold determined by this approach [168] was found to be comparable to the values of spin diffusion length obtained by other techniques [310]. We note that the backflow effect is intimately related to specific boundary conditions and thus should be relevant for other spin diffusion measurements using heterojunctions.

---

* While both measurements are done at liquid helium temperatures, the fact that both magnetization and resistivity in both systems are only weakly dependent on temperature makes it plausible (although not conclusive) that the room temperature magnetism is also carrier mediated.

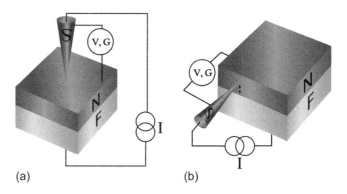

(a)   (b)

FIGURE 2.20 Spin diffusion measurements with the PCAR technique. (a) Experimental geometry using spin injection into a normal metal film of varied thickness with the superconducting tip approaching the normal layer from the top, see [309, 168]. (b) A possible alternative geometry using an STM-type nanoprobe to enable multiple measurements at different distances from the NF interface of the same sample. (After Faiz et al., cond-mat/190106756, 2019. With permission.)

One of the potential advantages of this technique, which is yet to be realized, is the ability to perform measurements at the nanoscale, using a scanning tunneling microscopy (STM) tip or other nanoprobes. The tip can then be scanned sideways across the normal layer, making several point contacts as it moves along: see Figure 2.20b. Such a geometry can potentially provide better efficiency and reproducibility, as multiple contacts can be subsequently used for the same sample and can complement the other spin diffusion measurement techniques described in this volume [305, 306, 308] (Chapter 5, Volume 1, and Chapter 6, Volume 2).

2.9 CONCLUSIONS AND FUTURE DIRECTIONS

In summary, in the two decades since its inception in 1998, PCAR spectroscopy has become one of the most useful tools in determining the values of spin polarization. While the technique employs some very basic, perhaps oversimplified, theoretical models, they seem to work well beyond their nominal applicability range to provide the practitioners of the technique with the consistent recipe to study a wide variety of magnetic materials; see Table 2.2. The alternative in terms of theoretical development is to further advance a fully self-consistent treatment. While this approach may succeed in general, it is unlikely to completely eliminate the need for phenomenological parameters, such as mixing angles, for example, which are material dependent [146]. Moreover, considering various conceptual and technical problems, such as spin–orbit interaction, proximity and inverse proximity effects, spreading resistance, and a number of other pitfalls in the PCAR technique, it may be difficult to substantially improve the accuracy of the technique beyond the current limit of 3%–5%, even if the perfect theory is developed. As the PCAR technique is inherently a low-temperature technique, it may be hard to use it fully for the measurements of high Curie temperature ferromagnets, which may have to be done at room temperatures. At the same time, it is instrumental in eliminating the effects of Stoner and other temperature-driven excitations on the ground state of half-metals, for which PCAR spectroscopy may be a technique of choice. It is tempting to apply the PCAR technique to the measurement of new classes of materials, such as topological insulators [80,311], as well as Weyl semimetals [312].

ACKNOWLEDGMENTS

I am grateful to V. Antropov, K. D. Belashchenko, N. Birge, R. A. Buhrman, A. Buzdin, G. Duetscher, A. Golubov, Yu. Lyanda-Geller, I. I. Mazin, J. S. Moodera, R. P. Panguluri, A. Petukhov, E. I. Rashba, D. Singh, B. Sinkovic, E. Tsymbal, J. Wei, S. Wolf, and I. Žutić for discussions and many useful suggestions, and to D. Dessau for sharing their experimental results prior to publication. I also wish to thank R. Panguluri and M. Faiz for their help in preparing some of the reference material and figures, and Grigori Youkov for providing most of the illustrations for this chapter.

Our work has been funded in part by the Defense Advanced Research Projects Agency (DARPA), and the National Science Foundation, and the Office of Naval Research.

REFERENCES

1. S. A. Wolf, D. D. Awschalom, R. A. Buhrman et al., Spintronics: A spin-based electronics vision for the future, *Science* **294**, 1488 (2001).
2. I. Žutić, J. Fabian, and S. Das Sarma, Spintronics: Fundamentals and applications, *Rev. Mod. Phys.* **76**, 323 (2004).
3. I. I. Mazin, How to define and calculate the degree of spin polarization in ferromagnets, *Phys. Rev. Lett.* **83**, 1427 (1999).
4. M. Julliere, Tunneling between ferromagnetic films, *Phys. Lett.* **54A**, 225 (1975).
5. W. E. Pickett and J. S. Moodera, Half metallic magnets, *Phys. Today* **54**, 39 (2001).
6. R. A. de Groot, F. M. Mueller, P. G. van Engen, and K. H. J. Buschow, New class of materials: Half-metallic ferromagnets, *Phys. Rev. Lett.* **50**, 2024 (1983).
7. B. Nadgorny, I. I. Mazin, M. Osofsky et al., Origin of high transport spin polarization in $La_{0.7}Sr_{0.3}MnO_3$: Direct evidence for minority spin states, *Phys. Rev. B* **63**, 184433 (2001).
8. W. E. Pickett and D. J. Singh, Chemical disorder and charge transport in ferromagnetic manganites, *Phys. Rev. B* **55**, R8642 (1997).
9. S. S. P. Parkin, C. Kaiser, A. Panchula, P. M. Rice, and B. Hughes, Giant tunneling magnetoresistance at room temperature with MgO (100) tunnel barriers, *Nat. Mater.* **3**, 862 (2004).
10. S. Yuasa, T. Nagahama, A. Fukushima, Y. Suzuki, and K. Ando, Giant room-temperature magnetoresistance in single-crystal Fe/MgO/Fe magnetic tunnel junctions, *Nat. Mater.* **3**, 868 (2004).
11. R. J. Soulen Jr., J. M. Byers, M. S. Osofsky et al., Measuring the spin polarization of a metal with a superconducting point contact, *Science* **282**, 85 (1998).
12. S. K. Upadhyay, A. Palanisami, R. N. Louie, and R. A. Buhrman, Probing ferromagnets with Andreev reflection, *Phys. Rev. Lett.* **81**, 3247 (1998).
13. A. F. Andreev, The thermal conductivity of the intermediate state in superconductors, *Zh. Eksp. Teor. Fiz.* **46**, 1823 (1964) [*Sov. Phys. JETP* **19**, 1228 (1964)].

14. D. Saint-James, Elementary excitations in the vicinity of a normal metal–superconducting metal contact, *J. Phys. (Paris)* **25**, 899 (1964).

15. M. J. M. de Jong and C. W. J. Beenakker, Andreev reflection in ferromagnet-superconductor junctions, *Phys. Rev. Lett.* **74**, 1657 (1995).

16. G. E. Blonder, M. Tinkham, and T. M. Klapwijk, Transition from metallic to tunneling regimes in superconducting microconstrictions: Excess current, charge imbalance, and supercurrent conversion, *Phys. Rev. B* **25**, 4515 (1982).

17. C. W. J. Beenakker, Random-matrix theory of quantum transport, *Rev. Mod. Phys.* **69**, 731 (1997).

18. I. I. Mazin, A. A. Golubov, and B. Nadgorny, Probing spin polarization with Andreev reflection: A theoretical basis, *J. Appl. Phys.* **89**, 7576 (2001).

19. B. Pannetier and H. Courtois, Andreev reflection and proximity effect, *J. Low Temp. Phys.* **118**, 599 (2000).

20. A. I. Buzdin, Proximity effects in superconductor-ferromagnet heterostructures, *Rev. Mod. Phys.* **77**, 935 (2005).

21. S. Kashiwaya and Y. Tanaka, Tunnelling effects on surface bound states in unconventional superconductors, *Rep. Prog. Phys.* **63**, 1641 (2000).

22. A. P. Mackenzie and Y. Maeno, The superconductivity of Sr_2RuO_4 and the physics of spin-triplet pairing, *Rev. Mod. Phys.* **75**, 657 (2003).

23. G. Deutscher, Andreev-Saint James reflections: A probe of cuprate superconductors, *Rev. Mod. Phys.* **77**, 109 (2005).

24. C. W. J. Beenakker, Colloquium: Andreev reflection and Klein tunneling in graphene, *Rev. Mod. Phys.* **80**, 1337 (2008).

25. A. G. Aronov, Spin injection in metals and polarization of nuclei, *Pis'ma Zh. Eksp. Teor. Fiz.* **24**, 37 (1976) [*JETP Lett.* **24**, 32 (1976)].

26. A. G. Aronov and G. E. Pikus, Spin injection into semiconductors, *Fiz. Tekh. Poluprovodn.* **10**, 1177 (1976).

27. B. T. Jonker, A. T. Hanbicki, D. T. Pierce, and M. D. Stiles, Spin nomenclature for semiconductors and magnetic metals, *J. Magn. Magn. Mater.* **277**, 24 (2004).

28. Yu. V. Sharvin, On the possible method for studying Fermi surfaces, *Zh. Eksp. Teor. Fiz.* **48**, 984 (1965) [*Sov. Phys. JETP* **21**, 655 (1965)].

29. T. Oguchi, Magnetoresistance in magnetic multilayers: A role of Fermi velocity, *Mater. Sci. Eng.* **31**, 111 (1995).

30. R. Meservey and P. M. Tedrow, Spin-polarized electron tunneling, *Phys. Rep.* **238**, 173 (1994).

31. J.-N. Chazalviel and Y. Yafet, Theory of the spin polarization of field-emitted electrons from nickel, *Phys. Rev. B* **15**, 1062 (1977).

32. J. A. Hertz and K. Aoi, Spin dependent tunneling from transition metal ferromagnets, *Phys. Rev. B* **8**, 3552 (1973).

33. D. C. Worlege and T. H. Geballe, Maki analysis of spin-polarized tunneling in an oxide ferromagnet, *Phys. Rev. B* **62**, 447 (2000).

34. B. Nadgorny, The case against half-metallicity in $La_{0.7}Sr_{0.3}MnO_3$, *J. Phys.: Condens. Matter* **19**, 315209 (2007).

35. W. H. Butler, Tunneling magnetoresistance from a symmetry filtering effect, *Sci. Technol. Adv. Mater.* **9**, 014106 (2008).

36. J. P. Velev, P. A. Dowben, E. Y. Tsymbal, S. J. Jenkins, and A. N. Caruso, Interface effects in spin-polarized metal/insulator layered structures, *Surf. Sci. Rep.* **63**, 400 (2008).

37. J. S. Moodera, J. Nowak, L. R. Kinder et al., Quantum well states in spin-dependent tunnel structures, *Phys. Rev. Lett.* **83**, 3029 (1998).

38. A. M. Bratkovsky, Assisted tunneling in ferromagnetic junctions and half-metallic oxides, *Appl. Phys. Lett.* **72**, 2334 (1998).

39. A. M. Bratkovsky, Spintronic effects in metallic, semiconductor, metal–oxide and metal–semiconductor heterostructures, *Rep. Prog. Phys.* **71**, 026502 (2008).

40. W. H. Butler, X. G. Zhang, T. C. Schulthess, and J. M. MacLaren, Spin-dependent tunneling conductance of Fe|MgO|Fe sandwiches, *Phys. Rev. B* **63**, 054416 (2001).

41. J. M. MacLaren, X.-G. Zhang, W. H. Butler, and X. Wang, Layer KKR approach to Bloch-wave transmission and reflection: Application to spin-dependent tunneling, *Phys. Rev. B* **59**, 5470 (1999).

42. A. G. Petukhov, J. Niggemann, V. N. Smelyanskiy, and V. V. Osipov, 100% spin accumulation in non-half-metallic ferromagnet–semiconductor junctions, *J. Phys.: Condens. Matter* **19**, 315205 (2007).

43. J. Mathon and A. Umerski, Theory of tunneling magnetoresistance of an epitaxial Fe/MgO/Fe(001) junction, *Phys. Rev. B* **63**, 220403 (2001).

44. E. Y. Tsymbal and D. G. Pettifor, Spin-polarized electron tunneling across a disordered insulator, *Phys. Rev. B* **58**, 432 (1998).

45. I. I. Mazin, A. A. Golubov, and A. Zaikin, "Chain Scenario" for Josephson tunneling with π shift in $YBa_2Cu_3O_7$, *Phys. Rev. Lett.* **75**, 2574 (1995).

46. C. Kaiser, S. van Dijken, S. H. Yang, H. Yang, and S. S. P. Parkin, Role of tunneling matrix elements in determining the magnitude of the tunneling spin polarization of 3d transition metal ferromagnetic alloys, *Phys. Rev. Lett.* **94**, 247203 (2005).

47. V. Vlaminck and M. Bailleul, Current-induced spin-wave Doppler shift, *Science* **322**, 410 (2008).

48. G. M. Müller, J. Walowski, M. Djordjevic et al., Spin polarization in half-metals probed by femtosecond spin excitation, *Nat. Mater.* **8**, 56 (2009).

49. K. E. H. M. Hanssen, P. E. Mijnarends, L. P. L. M. Rabou, and K. H. J. Buschow, Positron-annihilation study of the half-metallic ferromagnet NiMnSb: Experiment, *Phys. Rev. B* **42**, 1533 (1990).

50. S. Berko and A. P. Mills, Spin distribution studies in ferromagnetic metals by polarized positron annihilation experiments, *J. Phys. (Paris) Colloq.* **32**, C1 (1971).

51. P. Zwart, L. P. L. M. Rabou, G. J. Langedijk et al., in *Positron Annihilation*, P. C. Jain, R. M. Singru, and K. P. Gopinathan (Eds.), World Scientific, Singapore, p. 297, 1985.

52. S. Berko, Momentum density and Fermi surface measurements in metals by positron annihilation, in *Positron Solid State Physics*, W. Brandt and A. Dupasquier (Eds.), North-Holland, Amsterdam, the Netherlands, p. 64, 1983.

53. K. E. H. M. Hanssen and P. E. Mijnarends, Positron-annihilation study of the half-metallic ferromagnet NiMnSb: Theory, *Phys. Rev. B* **34**, 5009 (1986).

54. S. F. Alvarado, W. Eib, H. C. Siegmann, and J. P. Remeika, Photoelectron spin polarization testing of ionic structure of 3d levels in ferrites, *Phys. Rev. Lett.* **35**, 860 (1975).

55. P. D. Johnson, Spin polarized photoemission, *Rep. Prog. Phys.* **60**, 1217 (1997).

56. R. Feder, *Polarized Electrons in Surface Physics*, World Scientific, Singapore, 1985.

57. S. Hüfner, *Photoelectron Spectroscopy*, Springer, Berlin, 1995.

58. P. D. Johnson, Photoemission and Ferromagnetism, in *Core Level Spectroscopies for Magnetic Phenomena*, vol. 345, P. S. Bagus (Ed.), NATO ASI Series B: Physics, Plenum Press, New York, pp. 21–39, 1995.

59. B. T. Jonker, K.-H. Walker, E. Kisker, G. A. Prinz, and C. Carbone, Spin-polarized photoemission study of epitaxial Fe(001) films on Ag(001), *Phys. Rev. Lett.* **57**, 142 (1986).

60. E. Kisker, K. Schröder, M. Campagna, and W. Gudat, Temperature dependence of the exchange splitting of Fe by spin-resolved photoemission spectroscopy with synchrotron radiation, *Phys. Rev. Lett.* **52**, 2285 (1984).

61. D. P. Pappas, K.-P. Kämper, and H. Hopster, Reversible transition between perpendicular and in-plane magnetization in ultrathin films, *Phys. Rev. Lett.* **64**, 3179 (1990).

62. L. Plucinski, Y. Zhao, B. Sinkovic, and E. Vescovo, MgO + Fe(100) interface: A study of the electronic structure, *Phys. Rev. B* **75**, 214411 (2007).

63. W. Clemens, E. Vescovo, T. Kachel, C. Carbone, and W. Eberhardt, Spin-resolved photoemission study of the reaction of O_2 with fcc Co(100), *Phys. Rev. B* **46**, 4198 (1992).

64. B. Sinkovich, L. H. Tjeng, N. B. Brookes et al., Local electronic and magnetic structure of Ni below and above T_C: A spin-resolved circularly polarized resonant photoemission study, *Phys. Rev. Lett.* **79**, 3510 (2001).

65. H. Tang, D. Weller, T. G. Walker et al., Magnetic reconstruction of the Gd (0001) surface, *Phys. Rev. Lett.* **71**, 444 (1993).

66. Y. S. Dedkov, U. Rüdiger, and G. Güntherodt, Evidence for the half-metallic ferromagnetic state of Fe_3O_4 by spin-resolved photoelectron spectroscopy, *Phys. Rev. B* **65**, 064417 (2002).

67. J.-H. Park, E. Vescovo, H.-J. Kim, C. Kwon, R. Ramesh, and T. Venkatesan, Direct evidence for a half-metallic ferromagnet, *Nature* **392**, 794 (1998).

68. K. P. Kämper, W. Schmitt, G. Güntherodt, R. J. Gambino, and R. Ruf, CrO_2—A new half-metallic ferromagnet, *Phys. Rev. Lett.* **59**, 2788 (1987).

69. F. J. Himpsel, Exchange splitting of epitaxial fcc Fe/Cu(100) vs bcc Fe/Ag(100), *Phys. Rev. Lett.* **67**, 2363 (1991).

70. P. M. Tedrow and R. Meservey, Spin-dependent tunneling into ferromagnetic nickel, *Phys. Rev. Lett.* **26**, 192 (1971).

71. P. M. Tedrow and R. Meservey, Spin polarization of electrons tunneling from films of Fe, Co, Ni, and Gd, *Phys. Rev. B* **7**, 318 (1973).

72. K. Maki, The behavior of superconducting thin films in the presence of magnetic fields and currents, *Prog. Theor. Phys.* **32**, 29 (1964).

73. J. S. Moodera, L. R. Kinder, T. M. Wong, and R. Meservey, Large magnetoresistance at room temperature in ferromagnetic thin film tunnel junctions, *Phys. Rev. Lett.* **74**, 3273 (1995).

74. R. Meservey, P. M. Tedrow, V. R. Kalvey, and D. Paraskevopoulos, Studies of ferromagnetic metals by electron spin polarized tunneling, *J. Appl. Phys.* **50**, 1935 (1979).

75. H. Itoh, S. Shibata, T. Kumazaki, J. Inoue, and S. Maekawa, Effects of interfacial randomness on tunnel conductance and magnetoresistance in ferromagnetic tunnel junctions, *J. Phys. Soc. Jpn.* **68**, 1632 (1999).

76. S. Zhang and P. M. Levy, Models for magnetoresistance in tunnel junctions, *Eur. Phys.* **B10**, 599 (1999).

77. C. Kaiser, A. F. Panchula, and S. S. P. Parkin, Finite tunneling spin polarization at the compensation point of rare-earth-metal–transition-metal alloys, *Phys. Rev. Lett.* **95**, 047202 (2005).

78. I. I. Oleinik, E. Yu. Tsymbal, and D. G. Pettifor, Structural and electronic properties of $Co/Al_2O_3/Co$ magnetic tunnel junction from first principles, *Phys. Rev. B* **62**, 3952 (2000).

79. J. S. Parker, S. M. Watts, P. G. Ivanov, and P. Xiong, Spin polarization of CrO_2 at and across an artificial barrier, *Phys. Rev. Lett.* **88**, 196601 (2002).

80. K. Borisov, C.-Z. Chang, J. S. Moodera, and P. Stamenov, High Fermi-level spin polarization in the $(B_{1-x}Sb_x)_2Te_3$ family of topological insulators: A point contact Andreev reflection study, *Phys. Rev. B* **94**, 094415 (2017).

81. Yu. V. Sharvin and L. M. Fisher, Observation of focused electron beams in a metal, *Pis'ma Zh. Eksp. Teor. Fiz.* **1**, 54 (1965) [*JETP Lett.* **1**, 152 (1965)].

82. C. W. J. Beenakker, H. van Houten, and B. J. van Wees, Coherent electron focusing, in *Festkörperprobleme (Advances in Solid State Physics)*, vol. 29, U. Rössler (Ed.), Pergamon/Vieweg, Braunschweig, Germany, 1989.

83. V. S. Tsoi, Focusing of electrons in a metal by a transverse magnetic field, *Pis'ma Zh. Eksp. Teor. Fiz.* **19**, 114 (1974) [*JETP Lett.* **19**, 70 (1974)].

84. V. S. Tsoi, J. Bass, and P. Wyder, Studying conduction-electron/interface interactions using transverse electron focusing, *Rev. Mod. Phys.* **71**, 1641 (1999).

85. S. I. Bozhko, V. S. Tsoi, and S. E. Yakovlev, Observation of Andreev reflection with the use of transverse electron focusing effect, *Pis'ma Zh. Eksp. Teor. Fiz.* **36**, 123 (1982) [*JETP Lett.* **36**, 152 (1982)].

86. P. A. M. Benistant, A. P. van Gelder, H. van Kempen, and P. Wyder, Angular relation and energy dependence of Andreev reflection, *Phys. Rev. B* **32**, 3351 (1985).

87. M. Knudsen, *Kinetic Theory of Gases*, Methuen, London, 1934.

88. J. C. Maxwell, *A Treatise on Electricity and Magnetism*, Dover Press, New York, 1891.

89. G. Wexler, Size effect and non-local Boltzmann transport equation in orifice and disk geometry, *Proc. Phys. Soc. Lond.* **89**, 927 (1966).

90. M. Rokni and Y. Levinson, Joule heat in point contacts, *Phys. Rev. B* **52**, 1882 (1995).

91. B. Nikolić and P. Allen, Electron transport through a circular constriction, *Phys. Rev. B* **60**, 3963 (1999).

92. C. W. J. Beenakker, Why does a metal-superconductor junction have a resistance? 1999, cond-mat/9909293.

93. S. V. Artemenko, A. F. Volkov, and A. V. Zaitsev, Excess current in microbridges, *Solid State Commun.* **30**, 771 (1979).

94. C. W. J. Beenakker, Quantum transport in semiconductor-superconductor microjunctions, *Phys. Rev. B* **46**, 12841 (1992).

95. Y. V. Nazarov and T. H. Stoof, Diffusive conductors as Andreev interferometers, *Phys. Rev. Lett.* **76**, 823 (1996).

96. A. M. Duif, A. G. M. Jansen, and P. Wyder, Point-contact spectroscopy, *J. Phys.: Condens. Matter* **1**, 3157 (1989).

97. R. Holm, *Electric Contacts: Theory and Application*, 4th edn. Springer-Verlag, New York, 1967.

98. A. A. Lysykh, I. K. Yanson, O. I. Shklyarevski, and Yu. G. Naydyuk, Point-contact spectroscopy of electron-phonon interaction in alloys, *Solid State Commun.* **35**, 987 (1980).

99. A. G. M. Jansen, A. P. van Gelder, and P. Wyder, Point-contact spectroscopy in metals, *J. Phys. C: Solid State Phys.* **13**, 6073 (1980).

100. I. K. Yanson, Nonlinear effects in the electric conductivity of point junctions and electron-phonon interaction in normal metals, *Zh. Éksp. Teor. Fiz.* **66**, 1035 (1974) [*Sov. Phys. JETP* **39**, 506 (1974)].

101. Yu. G. Naidyuk and I. K. Yanson, *Point-Contact Spectroscopy*, Springer Series in Solid-State Sciences, vol. 145, Springer, Berlin, Germany, 2004.

102. P. Chalsani, S. K. Upadhyay, O. Ozatay, and R. A. Buhrman, Andreev reflection measurements of spin polarization, *Phys. Rev. B* **75**, 094417 (2007).

103. R. P. Panguluri, G. Tsoi, B. Nadgorny, S. H. Chun, and N. Samarth, Point contact spin spectroscopy of MnAs ferromagnetic films, *Phys. Rev. B* **68**, 201307 (2003).

104. A. I. Akimenko, A. B. Verkin, N. M. Ponomarenko, and I. K. Yanson, *Fiz. Nizk. Temp.* **4**, 1267 (1982) [*Sov. J. Low Temp. Phys.* **4**, 1267 (1982)].

105. Yu. G. Naidyuk and I. K. Yanson, Point-contact spectroscopy of heavy-fermion systems, *J. Phys.: Condens. Matter* **10**, 8905 (1998).

106. K. Mendelssohn and J. L. Olsen, Anomalous heat flow in superconductors, *Proc. Phys. Soc. (Lond.)* **A63**, 2 (1950); *Phys. Rev.* **80**, 859 (1950).

107. N. V. Zavaritskii, Thermal conductivity of superconductors in the intermediate state, *JETP* **38**, 1673 (1960) [*Sov. Phys. JETP* **11**, 1207 (1960)].

108. A. A. Abrikosov, *Fundamentals of the Theory of Metals*, North-Holland, Amsterdam, the Netherlands, 1988.

109. J. M. Byers and M. E. Flatté, Probing spatial correlations with nanoscale two-contact tunneling, *Phys. Rev. Lett.* **74**, 306 (1995).

110. P. Cadden-Zimansky, J. Wei, and V. Chandrasekhar, Cooper-pair-mediated coherence between two normal metals, *Nat. Phys.* **5**, 393 (2009).

111. G. Deutscher and P. G. de Gennes, Proximity effects, in *Superconductivity*, R. D. Parks (Ed.), Dekker, New York, pp. 1005–1034, 1969.

112. C. Bruder, Andreev scattering in anisotropic superconductors, *Phys. Rev. B* **41**, 4017 (1990).

113. M. Eschrig, Distribution functions in nonequilibrium theory of superconductivity and Andreev spectroscopy in unconventional superconductors, *Phys. Rev. B* **61**, 9061 (2000).

114. G. Goll, *Unconventional Superconductors: Experimental Investigation of Order Parameter Theory*, Springer Tracts in Modern Physics, vol. 214, Springer, Berlin, 2006.

115. T. Jacobson, On the origin of the outgoing black hole modes, *Phys. Rev. D* **53**, 7082 (1996).

116. M. Sadzikowski and M. Tachibana, Andreev reflection at the quark-gluon-plasma-color-flavor-locking interface, *Phys. Rev. D* **66**, 045024 (2002).

117. P. G. de Gennes and D. Saint James, Elementary excitations in the vicinity of a normal metal-superconducting metal contact, *Phys. Lett.* **4**, 151 (1963).

118. F. Giazotto, P. Pingue, F. Beltram et al., Resonant transport in Nb/GaAs AlGaAs heterostructures: Realization of the de Gennes–Saint-James model, *Phys. Rev. Lett.* **87**, 216808 (2001).

119. J. M. Rowell and W. L. McMillan, Electron interference in a normal metal induced by superconducting contracts, *Phys. Rev. Lett.* **16**, 453 (1966).

120. W. J. Tomasch, Geometrical resonance in the tunneling characteristics of superconducting Pb, *Phys. Rev. Lett.* **15**, 672 (1965).

121. I. P. Krylov and Yu. V. Sharvin, Observation of "Andreev" reflection of electrons from the boundary between a normal and superconducting phase, using the radio-frequency size effect, *Pis'ma Zh. Eksp. Teor. Fis.* **12**, 102 (1970) [*JETP Lett.* **12**, 71 (1970)].

122. S. V. Gudenko and I. P. Krylov, Radio-frequency size effect and diffusion of impurities, *Zh. Eksp. Teor. Fiz.* **86**, 2304 (1984); [*Sov. Phys. JETP* **86**, 2304 (1984)].

123. A. B. Pippard, J. G. Shepherd, and D. A. Tindall, Resistance of superconducting-normal interfaces, *Proc. R. Soc. Lond. A* **324**, 17 (1971).

124. V. V. Schmidt, *The Physics of Superconductors: Introduction to Fundamentals and Applications*, P. Muller and A. V. Ustinov (Eds.), Springer, Berlin, Germany, 1997.

125. M. Tinkham and J. Clarke, Theory of pair-quasiparticle potential difference in nonequilibrium superconductors, *Phys. Rev. Lett.* **28**, 1366 (1972).

126. C. J. Pethick and H. Smith, Relaxation and collective motion in superconductors: A two-fluid description, *Ann. Phys.* **119**, 133 (1979).

127. K. K. Likharev, Superconducting weak links, *Rev. Mod. Phys.* **51**, 101 (1979).

128. J. Demers and A. Griffin, Scattering and tunneling of electronic excitations in the intermediate state of superconductors, *Can. J. Phys.* **49**, 285 (1971).

129. A. Griffin and J. Demers, Tunneling in the normal-metal-insulator-superconductor geometry using the Bogoliubov equations of motion, *Phys. Rev. B* **4**, 2202 (1971).

130. K. K. Likharev and L. A. Yakobson, *Zh. Eksp. Teor. Fiz.* **68**, 1150 (1976) [*Sov. Phys.-JETP* **41**, 570 (1976)].

131. I. O. Kulik, A. N. Omelyanchuk, and R. I. Shekhter, *Sov. J. Low Temp. Phys.* **3**, 740 (1977).

132. G. E. Blonder and M. Tinkham, Metallic to tunneling transition in Cu-Nb point contacts, *Phys. Rev. B* **27**, 112 (1983).

133. N. A. Mortensen, K. Flensberg, and A. P. Jauho, Angle dependence of Andreev scattering at semiconductor–superconductor interfaces, *Phys. Rev. B* **59**, 10176 (1999).

134. I. Žutić and O. T. Valls, Spin-polarized tunneling in ferromagnet/unconventional superconductor junctions, *Phys. Rev. B* **60**, 6320 (1999).

135. I. Žutić and O. T. Valls, Tunneling spectroscopy for ferromagnet/superconductor junctions, *Phys. Rev. B* **61**, 1555 (2000).

136. K. Halterman and O. T. Valls, Proximity effects at ferromagnet-superconductor interfaces, *Phys. Rev. B* **65**, 014509 (2002).

137. J. Linder and A. Sudbø, Dirac fermions and conductance oscillations in s- and d-wave superconductor-graphene junctions, *Phys. Rev. Lett.* **99**, 147001 (2007).

138. R. P. Panguluri, K. C. Ku, T. Wojtowicz et al., Andreev reflection and pair-breaking effects at the superconductor/magnetic semiconductor interface, *Phys. Rev. B* **72**, 054510 (2005).

139. R. P. Panguluri, B. Nadgorny, T. Wojtowicz, W. L. Lim, X. Liu, and J. K. Furdyna, Measurement of spin polarization by Andreev reflection in ferromagnetic $In_{1-x}Mn_xSb$ epilayers, *Appl. Phys. Lett.* **84**, 4947 (2004).

140. A. A. Golubov, A. Brinkman, Y. Tanaka, I. I. Mazin, and O. V. Dolgov, Andreev spectra and subgap bound states in multiband superconductors, *Phys. Rev. Lett.* **103**, 077003 (2009).

141. G. Deutscher and P. Nozières, Cancellation of quasiparticle mass enhancement in the conductance of point contacts, *Phys. Rev. B* **50**, 13557 (1994).

142. B. Nadgorny and I. I. Mazin, High efficiency nonvolatile ferromagnet/superconductor switch, *Appl. Phys. Lett.* **80**, 3973 (2002).

143. R. Landauer, Electrical resistance of disordered one-dimensional lattices, *Philos. Mag.* **21**, 863 (1970).

144. J. Bardeen, Tunnelling from a many-particle point of view, *Phys. Rev. Lett.* **6**, 57 (1961).

145. J. C. Cuevas, A. Martín-Rodero, A. L. Yeyati, Hamiltonian approach to the transport properties of superconducting quantum point contacts, *Phys. Rev. B* **54**, 7366 (1996).

146. R. Grein, T. Löfwander, G. Metalidis, and M. Eschrig, Theory of superconductor-ferromagnet point-contact spectra: The case of strong spin polarization, *Phys. Rev. B* **81**, 094508 (2010).

147. G. J. Strijkers, Y. Ji, F. Y. Yang, C. L. Chien, and J. M. Byers, Andreev reflections at metal/superconductor point contacts: Measurement and analysis, *Phys. Rev. B* **63**, 104510 (2001).

148. B. P. Vodopyanov and L. R. Tagirov, Andreev conductance of a ferromagnet-superconductor point contact, *Pis'ma Zh. Eksp. Teor. Fiz.* **77**, 126 (2003) [*JETP Lett.* **77**, 153 (2003)]

149. I. I. Mazin, Tunneling of Bloch electrons through vacuum barrier, *Europhys. Lett.* **55**, 404 (2001).

150. K. M. Schep, P. M. Kelly, and G. E. W. Bauer, Ballistic transport and electronic structure, *Phys. Rev. B* **57**, 8907 (1998).

151. Y. Ji, G. J. Strijkers, F. Y. Yang, and C. L. Chien, Comparison of two models for spin polarization measurements by Andreev reflection, *Phys. Rev. B* **64**, 224425 (2001).

152. G. T. Woods, R. J. Soulen Jr., I. Mazin et al., Analysis of point-contact Andreev reflection spectra in spin polarization measurements, *Phys. Rev. B* **70**, 054416 (2004).

153. S. Kashiwaya, Y. Tanaka, N. Yoshida, and M. R. Beasley, Spin current in ferromagnet-insulator-superconductor junctions, *Phys. Rev. B* **60**, 3572 (1999).

154. A. A. Golubov, Interface resistance in ferromagnet/superconductor junctions, *Physica C* **327**, 46 (1999).

155. F. J. Jedema, B. J. van Wees, B. H. Hoving, A. T. Filip, and T. M. Klapwijk, Spin-accumulation-induced resistance in mesoscopic ferromagnet-superconductor junctions, *Phys. Rev. B* **60**, 16549 (1999).

156. V. I. Fal'ko, C. J. Lambert, and A. F. Volkov, Andreev reflections and magneto-resistance in ferromagnet–superconductor mesoscopic structures, *Zh. Eksp. Teor. Fiz. Pisma Red.* **69**, 497 (1999) [*JETP Lett.* **69**, 532 (1999)].

157. W. Belzig, A. Brataas, Yu. V. Nazarov, and G. E. W. Bauer, Spin accumulation and Andreev reflection in a mesoscopic ferromagnetic wire, *Phys. Rev. B* **62**, 9726 (2000).

158. J. Aarts, J. M. E. Geers, E. Brück, A. A. Golubov, and R. Coehoorn, Interface transparency of superconductor/ferromagnetic multilayers, *Phys. Rev. B* **56**, 2779 (1997).

159. W. A. Harrison, Tunneling from an independent-particle point of view, *Phys. Rev.* **123**, 85 (1961).

160. S. K. Yip, Energy-resolved supercurrent between two superconductors, *Phys. Rev. B* **58**, 5803 (1998).

161. J. G. Braden, J. S. Parker, P. Xiong, S. H. Chun, and N. Samarth, Direct measurement of the spin polarization of the magnetic semiconductor (Ga,Mn)As, *Phys. Rev. Lett.* **91**, 056602 (2003).

162. R. P. Panguluri, T. S. Santos, E. Negusse et al., Half-metallicity in europium oxide conductively matched with silicon, *Phys. Rev. B* **78**, 125307 (2008).

163. C. Ren, J. Trbovic, R. L. Kallaher et al., Measurement of the spin polarization of the magnetic semiconductor EuS with zero-field and Zeeman-split Andreev reflection spectroscopy, *Phys. Rev. B* **75**, 205208 (2007).

164. A. Anguelouch, A. Gupta, G. Xiao et al., Near-complete spin polarization in atomically-smooth chromium-dioxide epitaxial films prepared using a CVD liquid precursor, *Phys. Rev. B* **64**, 180408 (2001).

165. B. Nadgorny, R. J. Soulen Jr., M. S. Osofsky et al., Transport spin polarization of Ni_xFe_{1-x}: Electronic kinematics and band structure, *Phys. Rev. B* **61**, R3788 (2000).

166. M. Stokmaier, G. Goll, D. Weissenberger, C. Sürgers, and H. V. Löhneysen, Size dependence of current spin polarization through superconductor/ferromagnet nanocontacts, *Phys. Rev. Lett.* **101**, 147005 (2008).

167. W. K. Park and L. H. Greene, Andreev reflection and order parameter symmetry in heavy-fermion superconductors: The case of $CeCoIn_5$, *J. Phys.: Condens. Matter* **21**, 103203 (2009).

168. M. Faiz, R. P. Panguluri, B. Nadgorny et al., Backflow effect on spin diffusion near ferromagnet-superconductor interface, arXiv: 1901.06756.

169. Y. Bugoslavsky, Y. Miyoshi, S. K. Clowes et al., Possibilities and limitations of point-contact spectroscopy for measurements of spin polarization, *Phys. Rev. B* **71**, 104523 (2005).

170. Y. Miyoshi, Y. Bugoslavsky, and L. F. Cohen, Andreev reflection spectroscopy of niobium point contacts in a magnetic field, *Phys. Rev. B* **72**, 012502 (2005).

171. L. Ozyuzer, J. F. Zasadzinski, and K. E. Gray, Point contact tunnelling apparatus with temperature and magnetic field control, *Cryogenics* **38**, 911 (1998).

172. K. A. Yates, L. F. Cohen, Z. A. Ren et al., Point contact Andreev reflection spectroscopy of $NdFeAsO_{0.85}$, *Supercond. Sci. Technol.* **21**, 092003 (2008).

173. R. P. Panguluri, S. Xu, Y. Moritomo, I. V. Solovyev, and B. Nadgorny, Disorder effects in half-metallic Sr_2FeMoO_6 single crystals, *Appl. Phys. Lett.* **94**, 012501 (2009).

174. R. S. Gonnelli, D. Daghero, G. A. Ummarino et al., Direct evidence for two-band superconductivity in MgB_2 single crystals from directional point-contact spectroscopy in magnetic fields, *Phys. Rev. Lett.* **89**, 247004 (2002).

175. K. S. Ralls, R. A. Buhrman, and R. C. Tiberio, Fabrication of thin-film metal nanobridges, *Appl. Phys. Lett.* **55**, 2459 (1989).

176. E. Clifford and J. M. D. Coey, Point contact Andreev reflection by nanoindentation of polymethyl methacrylate, *Appl. Phys. Lett.* **89**, 092506 (2006).

177. F. Magnus, K. A. Yates, S. K. Clowes et al., Interface properties of Pb/InAs planar structures for Andreev spectroscopy, *Appl. Phys. Lett.* **92**, 012501 (2008).

178. R. J. M. van de Veerdonk, J. Nowak, R. Meservey, J. S. Moodera, and W. J. M. de Jonge, Current distribution effects in magnetoresistive tunnel junctions, *Appl. Phys. Lett.* **71**, 2839 (1997).

179. J. Bass and W. P. Pratt Jr., Current-perpendicular (CPP) magnetoresistance in magnetic metallic multilayers, *J. Magn. Magn. Mater.* **200**, 274 (1999).

180. P. M. Rourke, M. A. Tanatar, C. S. Turel, J. Berdeklis, C. Petrovic, and J. Y. T. Wei, Spectroscopic evidence for multiple order parameter components in the heavy fermion superconductor CeCoIn$_5$, *Phys. Rev. Lett.* **94**, 107005 (2005).

181. R. C. Dynes, V. Narayanamurti, and J. P. Garno, Direct measurement of quasiparticle-lifetime broadening in a strong-coupled superconductor, *Phys. Rev. Lett.* **41**, 1509 (1978).

182. A. Plecenik, M. Grajcar, S. Benacka, P. Seidel, and A. Pfuch, Finite-quasiparticle-lifetime effects in the differential conductance of Bi$_2$Sr$_2$CaCu$_2$Oy/Au junctions, *Phys. Rev. B* **49**, 10016 (1994).

183. H. Srikanth and A. K. Raychaudhuri, Modeling tunneling data of normal metal-oxide superconductor point contact junctions, *Physica C* **190**, 229 (1992).

184. C. H. Kant, O. Kurnosikov, A. T. Filip, P. LeClair, H. J. M. Swagten, and W. J. M. de Jonge, Origin of spin-polarization decay in point-contact Andreev reflection, *Phys. Rev. B* **66**, 212403 (2002).

185. J. Aumentado and V. Chandrasekhar, Mesoscopic ferromagnet-superconductor junctions and the proximity effect, *Phys. Rev. B* **64**, 054505 (2001).

186. K. Xia, P. J. Kelly, G. E. W. Bauer, and I. Turek, Spin-dependent transparency of ferromagnet/superconductor interfaces, *Phys. Rev. Lett.* **89**, 166603 (2002).

187. F. Taddei, S. Sanvito, and C. J. Lambert, Spin-polarized transport in F/S nanojunctions, *J. Low Temp. Phys.* **124**, 305 (2001).

188. W. J. DeSisto, P. R. Broussard, T. F. Ambrose, B. E. Nadgorny, and M. S. Osofsky, Highly spin-polarized chromium dioxide thin films prepared by chemical vapor deposition from chromyl chloride, *Appl. Phys. Lett.* **76**, 3789 (2000).

189. Y. Ji, G. J. Strijkers, F. Y. Yang et al., Determination of the spin polarization of half-metallic CrO$_2$ by point contact Andreev reflection, *Phys. Rev. Lett.* **86**, 5585 (2001).

190. K. A. Yates, W. R. Branford, F. Magnus, Y. Miyoshi, B. Morris, and L. F. Cohen, The spin polarization of CrO$_2$ revisited, *Appl. Phys. Lett.* **91**, 172504 (2007).

191. C. S. Turel, I. J. Guilaran, P. Xiong, and J. Y. T. Wei, Andreev nanoprobe of half-metallic CrO$_2$ films using superconducting cuprate tips, *Appl. Phys. Lett.* **99**, 192508 (2011).

192. L. Bocklage, J. M. Scholtyssek, U. Merkt, and G. Meier, Spin polarization of Ni$_2$MnIn and Ni$_{80}$Fe$_{20}$ determined by point-contact spectroscopy, *J. Appl. Phys.* **101**, 09J512 (2007).

193. J. M. Valentine and C. L. Chien, Determination of spin polarization of Gd and Dy by point-contact Andreev reflection, *J. Appl. Phys.* **99**, 08P902 (2006).

194. Y. Ji, C. L. Chien, Y. Tomioka, and Y. Tokura, Measurement of spin polarization of single crystals of La$_{0.7}$Sr$_{0.3}$MnO$_3$ and La$_{0.6}$Sr$_{0.4}$MnO$_3$, *Phys. Rev. B* **66**, 012410 (2002).

195. A. Schmehl, V. Vaithyanathan, A. Herrnberger et al., Epitaxial integration of the highly spin-polarized ferromagnetic semiconductor EuO with silicon and GaN, *Nat. Mater.* **6**, 882 (2007).

196. X. H. Zhang, S. von Molnár, Z. Fisk, and P. Xiong, Spin-dependent electronic states of the ferromagnetic semimetal EuB$_6$, *Phys. Rev. Lett.* **100**, 167001 (2008).

197. C. Leighton, M. Manno, A. Cady et al., Composition controlled spin polarization in Co$_{1-x}$Fe$_x$S$_2$ alloys, *J. Phys. Condens. Matter* **19**, 315219 (2007).

198. L. Wang, T. Y. Chen, C. L. Chien et al., Composition controlled spin polarization in Co$_{1-x}$Fe$_x$S$_2$: Electronics, magnetic, and thermodynamic properties, *Phys. Rev. B* **73**, 144402 (2006).

199. L. Wang, K. Umemoto, R. M. Wentzcovitch et al., Co$_{1-x}$Fe$_x$S$_2$: A tunable source of highly spin-polarized electrons, *Phys. Rev. Lett.* **94**, 056602 (2005).

200. S. F. Cheng, G. T. Woods, K. Bussmann et al., Growth and magnetic properties of single crystal $Fe_{1-x}Co_xS_2$ (x=0.35–1), *J. Appl. Phys.* **93**, 6847 (2003).

201. L. Wang, T. Y. Chen, and C. Leighton, Spin-dependent band structure effects and measurement of the spin polarization in the candidate half-metal CoS_2, *Phys. Rev. B* **69**, 094412 (2004).

202. L. Wang, T. Y. Chen, C. L. Chien, and C. Leighton, Sulfur stoichiometry effects in highly spin polarized CoS_2 single crystals, *Appl. Phys. Lett.* **88**, 232509 (2006).

203. T. W. Chiang, Y. H. Chiu, S. Y. Huang et al., Spectra broadening of point-contact Andreev reflection measurement on GaMnAs, *J. Appl. Phys.* **105**, 07C507 (2009).

204. A. Geresdi, A. Halbritter, M. Csontos et al., Nanoscale spin polarization in the dilute semiconductor (In,Mn)Sb, *Phys. Rev. B* **77**, 233304 (2008).

205. R. P. Panguluri, B. Nadgorny, T. Wojtowicz, X. Liu, and J. K. Furdyna, Inelastic scattering and spin polarization in dilute magnetic semiconductor (Ga,Mn)Sb, *Appl. Phys. Lett.* **91**, 252502 (2007).

206. W. R. Branford, S. K. Clowes, Y. V. Bugoslavsky et al., Effect of chemical substitution on the electronic properties of highly aligned thin films of $Sr_{2-x}A_xFeMoO_6$ (A = Ca,Ba,La; x = 0,0.1), *J. Appl. Phys.* **94**, 4714 (2003).

207. H. Gotoh, Y. Takeda, H. Asano, J. Zhong, A. Rajanikanth, and K. Hono, Antiferromagnetism and spin polarization in double perovskite $SrLaVMoO_6$, *Appl. Phys. Exp.* **2**, 013001 (2009).

208. R. P. Panguluri, P. Kharel, C. Sudakar et al., Ferromagnetism and spin-polarized charge carriers in In_2O_3 thin films, *Phys. Rev. B* **79**, 165208 (2009).

209. A. Rajanikanth, Y. K. Takahashi, and K. Hono, Suppression of magnon excitations in Co2MnSi Heusler alloy by Nd doping, *J. Appl. Phys.* **105**, 063916 (2009).

210. G. T. Woods, J. Sanders, S. Kolesnik et al., Measurement of the transport spin polarization of doped strontium ruthenates using point contact Andreev reflection, *J. Appl. Phys.* **104**, 083701 (2008).

211. P. Raychaudhuri, A. P. Mackenzie, J. W. Reiner, and M. R. Beasley, Transport spin polarization in $SrRuO_3$ measured through point-contact Andreev reflection, *Phys. Rev. B* **67**, 020411 (2003).

212. B. Nadgorny, M. S. Osofsky, D. J. Singh, G. T. Woods, and R. J. Soulen Jr., Measurements of spin polarization of epitaxial $SrRuO_3$ thin films, *Appl. Phys. Lett.* **82**, 427 (2003).

213. J. Sanders, G. T. Woods, P. Poddar, H. Srikanth, B. Dabrowski, and S. Kolesnik, Spin polarization measurements on polycrystalline strontium ruthenates using point-contact Andreev reflection, *J. Appl. Phys.* **97**, 10C912 (2005).

214. S. Mukhopadhyay, S. K. Dhar, and P. Raychaudhuri, Study of spin fluctuations in $Ni_{3\pm x}Al_{1\pm x}$ using point contact Andreev reflection, *Appl. Phys.* **93**, 102502 (2008).

215. S. V. Karthik, T. M. Nakatani, A. Rajanikanth, Y. K. Takahashi, and K. Hono, Spin polarization of Co-Fe alloys estimated by point contact Andreev reflection and tunneling magnetoresistance, *J. Appl. Phys.* **105**, 07C916 (2009).

216. A. Narahara, K. Ito, T. Suemasu, Y. K. Takahashi, A. Ranajikanth, and K. Hono, Spin polarization of Fe_4N thin films determined by point-contact Andreev reflection, *Appl. Phys. Lett.* **94**, 202502 (2009).

217. S. X. Huang, T. Y. Chen, and C. L. Chien, Spin polarization of amorphous CoFeB determined by point-contact Andreev reflection, *Appl. Phys. Lett.* **92**, 242509 (2008).

218. K. M. Seemann, V. Baltz, M. MacKenzie, J. N. Chapman, B. J. Hickey, and C. H. Marrows, Diffusive and ballistic current spin polarization in magnetron-sputtered L1(0) ordered epitaxial FePt, *Phys. Rev. B* **76**, 174435 (2007).

219. T. Y. Chen, C. L. Chien, and C. Petrovic, Enhanced Curie temperature and spin polarization in Mn_4FeGe_3, *Appl. Phys. Lett.* **91**, 142505 (2007).

220. S. V. Karthik, A. Rajanikanth, T. M. Nakatani et al., Effect of Cr substitution for Fe on the spin polarization of $Co_2Cr_xFe_{1-x}Si$ Heusler alloys, *J. Appl. Phys.* **102**, 043903 (2007).

221. T. M. Nakatani, A. Rajanikanth, Z. Gercsi, Y. K. Takahashi, K. Inomata, and K. Hono, Structure, magnetic property, and spin polarization of $Co_2FeAl_xSi_{1-x}$ Heusler alloys, *J. Appl. Phys.* **102**, 033916 (2007).

222. S. V. Karthik, A. Rajanikanth, Y. K. Takahashi, T. Okhubo, and K. Hono, Spin polarization of quaternary $Co_2Cr_{1-x}Fe_xAl$ Heusler alloys, *Appl. Phys. Lett.* **89**, 052505 (2006).

223. A. Rajanikanth, D. Kande, Y. K. Takahashi, and K. Hono, High spin polarization in a two phase quaternary Heusler alloy $Co_2MnAl_{1-x}Sn_x$, *J. Appl. Phys.* **101**, 09J508 (2007).

224. T. M. Nakatani, Z. Gercsi, A. Rajanikanth, Y. K. Takahashi, and K. Hono, The effect of iron addition on the spin polarization and magnetic properties of Co_2CrGa Heusler alloy, *J. Phys. D Appl. Phys.* **41**, 225002 (2008).

225. A. Rajanikanth, Y. K. Takahashi, and K. Hono, Spin polarization of Co_2MnGe and Co_2MnSi thin films with A2 and L2(1) structures, *J. Appl. Phys.* **101**, 023901 (2007).

226. S. F. Cheng, B. Nadgorny, K. Bussmann et al., Growth and magnetic properties of single crystal Co_2Mn_X(X = Si,Ge) Heusler alloys, *IEEE Trans. Magn.* **37**, 2176 (2001).

227. Z. Gercsi, A. Rajanikanth, Y. K. Takahashi et al., Spin polarization of Co_2FeSi full-Heusler alloy and tunneling magnetoresistance of its magnetic tunneling junctions, *Appl. Phys. Lett.* **89**, 082512 (2006).

228. L. J. Singh, Z. H. Barber, Y. Miyoshi, Y. Bugoslavsky, W. R. Branford, and L. F. Cohen, Structural, magnetic, and transport properties of thin films of the Heusler alloy Co_2MnSi, *Appl. Phys. Lett.* **84**, 2367 (2004).

229. L. Ritchie, G. Xiao, Y. Ji et al., Magnetic structural, and transport properties of the Heusler alloys Co_2MnSi and NiMnSb, *Phys. Rev. B* **68**, 104430 (2003).

230. M. Zhang, E. Brück, F. R. de Boer, Z. Li, and G. Wu, The magnetic and transport properties of the Co_2FeGa Heusler alloy, *J. Phys. D Appl. Phys.* **37**, 2049 (2004).

231. S. K. Clowes, Y. Miyoshi, Y. Bugoslavsky et al., Spin polarization of the transport current at the free surface of bulk NiMnSb, *Phys. Rev. B* **69**, 214425 (2004).

232. M. Zhang, Z. Liu, H. Hu et al., A new semi-Heusler ferromagnet NiFeSb: Electronic structure, magnetism and transport properties, *Solid State Commun.* **128**, 107 (2003).

233. Z. Z. Li, H. J. Tao, H. H. Wen et al., Point contact Andreev reflection measurement of spin polarization of ferromagnetic alloy NiFeSb, *Chin. Phys. Lett.* **19**, 1181 (2002).

234. R. P. Panguluri, C. Zeng, H. H. Weitering, J. M. Sullivan, S. C. Erwin, and B. Nadgorny, Spin polarization and electronic structure of ferromagnetic Mn_5Ge_3 epilayers, *Phys. Status Solidi B* **242**, R67 (2005).

235. A. Ionescu, C. A. F. Vaz, T. Trypiniotis et al., Structural, magnetic, electronic, and spin transport properties of epitaxial Fe_3Si/GaAs(001), *Phys. Rev. B* **71**, 094401 (2005).

236. R. P. Panguluri, G. Tsoi, B. Nadgorny, S. H. Chun, N. Samarth, and I. I. Mazin, Point contact spin spectroscopy of ferromagnetic MnAs epitaxial films, *Phys. Rev. B* **68**, 201307(R) (2003).

237. P. Kharel, P. Thapa, P. Lukashev et al., Transport spin polarization of high Curie temperature MnBi films, *Phys. Rev. B* **83**, 024415 (2011).

238. H. Kurt, K. Rode, M. Venkatesan, P. Stamenov, and J. M. D. Coey, $Mn_{3-x}Ga$ (0≤x≤1): Multifunctional thin film materials for spintronics and magnetic recording, *Phys. Status Solidi B* **248**, 2338 (2011).

239. H. Kurt, N. Baadji, K. Rode et al., Magnetic and electronic properties of D022-Mn_3Ge (001) films, *Appl. Phys. Lett.* **101**, 132410 (2012).

240. H. Kurt, K. Rode, P. Stamenov et al., Cubic Mn_2Ga thin films: Crossing the spin gap with ruthenium, *Phys. Rev. Lett.* **112,** 027201 (2014).

241. D. Betto, Y. C. Lau, K. Borisov et al., Structure, site-specific magnetism, and magnetotransport properties of epitaxial D022-structure Mn_2Fe_xGa thin films, *Phys. Rev. B* **96**, 024408 (2017).

242. T. Akazaki, T. Yokoyama, Y. Tanaka, H. Munekata, and H. Takayanagi, Evaluation of spin polarization in p-$In_{0.96}Mn_{0.04}As$ using Andreev reflection spectroscopy including inverse proximity effect, *Phys. Rev. B* **83**, 155212 (2011).

243. W. R. Branford, S. B. Roy, S. K. Clowes et al., Spin polarization and anomalous Hall effect in NiMnSb films, *J. Magn. Magn. Mater.* **272–276**, Supplement, E1399 (2004). Proceedings of the International Conference on Magnetism (ICM 2003).

244. A. I. D'yachenko, V. A. D'yachenko, V. Yu. Tarenkov, and V. N. Krivoruchko, Spin polarization of charge carriers and Andreev reflection in (LaCa)MnO/ superconductor point contacts, *Phys. Solid State* **48**, 432 (2006).

245. L. Bainsla, A. I. Mallick, A. A. Coelho et al., High spin polarization and spin splitting in equiatomic quaternary CoFeCrAl Heusler alloy, *J. Magn. Magn. Mater.* **394**, 82 (2015).

246. L. Bainsla, K. G. Suresh, A. K. Nigam et al., High spin polarization in CoFeMnGe equiatomic quaternary Heusler alloy, *J. Appl. Phys.* **116**, 203902 (2014).

247. N. Ji, M. S. Osofsky, V. Lauter et al., Perpendicular magnetic anisotropy and high spin-polarization ratio in epitaxial FeN thin films, *Phys. Rev. B* **84**, 245310 (2011).

248. S. Kawai, H. Ando, H. Sakakibara et al., Magnetic and transport properties of antiperovskite nitride Co_3FeN films, *IEEE Magnetics Conference (INTERMAG)*, pages 1–1, May 2015.

249. A. D. Naylor, G. Burnell, and B. J. Hickey, Transport spin polarization of the rare-earth transition-metal alloy$Co_{1-x}Gd_x$, *Phys. Rev. B* **85**, 064410 (2012).

250. M. S. Osofsky, L. Cheng, W. E. Bailey, K. Bussmann, and D. Parker, Measurement of the transport spin polarization of FeV using point-contact Andreev reflection, *Appl. Phys. Lett.* **102**, 212412 (2013).

251. I. Shigeta, O. Murayama, T. Hisamatsu et al., Spin polarization of Fe-rich ferromagnetic compounds in $Ru_{2-x}Fe_xCrSi$ Heusler alloys, *J. Phys. Chem. Solids* **72**, 604 (2011). Spectroscopies in Novel Superconductors SNS 2010.

252. S. Singh, G. Sheet, P. Raychaudhuri, and S. K. Dhar, $CeMnNi_4$: A soft ferromagnet with a high degree of transport spin polarization, *Appl. Phys. Lett.* **88**, 022506 (2006).

253. P. Stamenov and J. M. D. Coey, Fermi level spin polarization of polycrystalline thulium by point contact Andreev reflection spectroscopy, *J. Appl. Phys.* **109**, 07C713 (2011).

254. P. Thapa, G. Lawes, B. Nadgorny et al., Ferromagnetism and spin polarization in indium nitride, indium oxynitride, and Cr substituted indium oxynitride films, *Appl. Surf. Sci.* **295**, 189 (2014).

255. I. T. M. Usman, K. A. Yates, J. D. Moore et al., Evidence for spin mixing in holmium thin film and crystal samples, *Phys. Rev. B* **83**, 144518 (2011).

256. B. Xia, Y. Wu, H. W. Ho et al., A possible origin of room temperature ferromagnetism in Indium – Tin oxide thin film: Surface spin polarization and ferromagnetism, *Phys. B: Condens. Matter* **406**, 3166 (2011).

257. R. J. M. Veerdonk, J. S. Moodera, and W. J. M. de Jonge, *Conference of Digest of the 15th International Colloquium on Magnetic Films and Surfaces*, Queensland, August 4–8, 1997.

258. R. Meservey, P. M. Tedrow, V. R. Kalvey, and D. Paraskevopoulos, Studies of ferromagnetic metals by electron spin polarized tunneling, *J. Appl. Phys.* **50**, 1935 (1979).

259. R. Godfrey and M. Johnson, Spin injection in mesoscopic silver wires: Experimental test of resistance mismatch, *Phys. Rev. Lett.* **96**, 136601 (2006).

260. M. Zhu, C. L. Dennis, and R. D. McMichael, Temperature dependence of magnetization drift velocity and current polarization in $Ni_{80}Fe_{20}$ by spin-wave Doppler measurements, *Phys. Rev. B* **81**, 140407(R) (2010).

261. D. Orgassa, H. Fujiwara, T. C. Schulthess, and W. H. Butler, First-principles calculation of the effect of atomic disorder on the electronic structure of the half-metallic ferromagnet NiMnSb, *Phys. Rev. B* **60**, 13237 (1999).

262. H. Ebert and G. Schütz, Theoretical and experimental study of the electronic structure of PtMnSb, *J. Appl. Phys.* **69**, 4627 (1991).

263. P. A. Dowben and R. Skomski, Are half-metallic ferromagnets half metals? *J. Appl. Phys.* **95**, 7453 (2004).

264. M. Ležaić, Ph. Mavropoulos, J. Enkovaara, G. Bihlmayer, and S. Blügel, Thermal collapse of spin polarization in half-metallic ferromagnets, *Phys. Rev. Lett.* **97**, 026404 (2006).

265. R. Skomski and P. A. Dowben, The finite-temperature densities of states for half-metallic ferromagnets, *Europhys. Lett.* **58**, 544 (2002).

266. H. Eschrig and W. E. Pickett, Density functional theory of magnetic systems revisited, *Solid State Commun.* **118**, 123 (2001).

267. P. Mavropoulos, K. Sato, R. Zeller, P. H. Dederichs, V. Popescu, and H. Ebert, Effect of the spin–orbit interaction on the band gap of half metals, *Phys. Rev. B* **69**, 054424 (2004).

268. P. Högl, A. Matos-Abiague, I. Žutić, and J. Fabian, Magnetoanisotropic andreev reflection in ferromagnet-superconductor junctions, *Phys. Rev. Lett.* **115**, 116601 (2015).

269. P. K. Boer and R. A. de Groot, Electronic structure of the layered manganite $LaSr_2Mn_2O_7$, *Phys. Rev. B* **60**, 10758 (1999).

270. G. A. Wijs and R. A. de Groot, Towards 100% spin-polarized charge-injection: The half-metallic NiMnSb/CdS interface, *Phys. Rev. B* **64**, 020402 (2001).

271. V. Yu. Irkhin and M. I. Katsnelson, Half-metallic ferromagnets, *Usp. Fiz. Nauk* **164**, 705 (1994) [*Phys. Usp.* **37**, 659 (1994)].

272. J. M. D. Coey, M. Viret, and S. von Molnár, Mixed-valence manganites, *Adv. Phys.* **48**, 167 (1999).

273. M. Ziese, Extrinsic magnetotransport phenomena in ferromagnetic oxides, *Rep. Prog. Phys.* **65**, 143 (2002).

274. H. Eschrig and W. E. Pickett, Half metals: From formal theory to real material issues, *J. Phys.: Condens. Matter* **19**, 315203 (2007).

275. M. I. Katsnelson, V. Yu. Irkhin, L. Chioncel, A. I. Lichtenstein, and R. A. de Groot, Half-metallic ferromagnets: From band structure to many-body effects, *Rev. Mod. Phys.* **80**, 315 (2008).

276. K. Schwarz, CrO_2 predicted as a half-metallic ferromagnet, *J. Phys. F: Met. Phys.* **16**, L211 (1986).

277. I. I. Mazin, Robust half metalicity in $Fe_xCo_{1-x}S_2$, *Appl. Phys. Lett.* **77**, 3000 (2000).

278. L. Wang, T. Y. Chen, and C. Leighton, Spin-dependent band structure effects and measurement of the spin polarization in the candidate half-metal CoS_2, *Phys. Rev. B* **69**, 094412 (2004).

279. I. Appelbaum, B. Huang, and D. Monsma, Electronic measurement and control of spin transport in silicon, *Nature* **447**, 295 (2007).

280. B. T. Jonker, G. Kioseoglou, A. T. Hanbicki, C. H. Li, and P. E. Thompson, Electrical spin-injection into silicon from a ferromagnetic metal/tunnel barrier contact, *Nat. Phys.* **3**, **542** (2007).

281. P. G. Steeneken, L. H. Tjeng, I. Elfimov et al., Exchange splitting and charge carrier spin polarization in EuO, *Phys. Rev. Lett.* **88**, 047201 (2002).

282. T. S. Santos, J. S. Moodera, K. V. Raman et al., Determining exchange splitting in a magnetic semiconductor by spin-filter tunneling, *Phys. Rev. Lett.* **101**, 147201 (2008).

283. K. I. Kobayashi, T. Kimura, H. Sawada, K. Terakura, and Y. Tokura, Room-temperature magnetoresistance in an oxide material with an ordered double-perovskite structure, *Nature* **395**, 357 (1998).

284. I. V. Solovyev, Electronic structure and stability of the ferrimagnetic ordering in double perovskites, *Phys. Rev. B* **65**, 144446 (2002).

285. N. Auth, G. Jakob, T. Block, and C. Felser, Spin polarization of magnetoresistive materials by point contact spectroscopy, *Phys. Rev. B* **68**, 024403 (2003).

286. Y. Bugoslavsky, Y. Miyoshi, G. K. Perkins et al., Electron diffusivities in MgB_2 from point contact spectroscopy, *Phys. Rev. B* **72**, 224506 (2005a).

287. J. M. D. Coey and S. Sanvito, Magnetic semiconductors and half-metals, *J. Phys. D: Appl. Phys.* **37**, 988 (2004).

288. W. E. Pickett and D. J. Singh, Electronic structure and half-metallic transport in the $La_{1-x}Ca_xMnO_3$ system, *Phys. Rev. B* **53**, 1146 (1996).

289. J. Z. Sun, L. Krusin-Elbaum, P. R. Duncombe, A. Gupta, and R. B. Laibowitz, Temperature dependent, non-ohmic magnetoresistance in doped perovskite manganate trilayer junctions, *Appl. Phys. Lett.* **70**, 1769 (1997).

290. M. Bowen, M. Bibes, A. Barthélémy et al., Nearly total spin polarization in $La_{2/3}Sr_{1/3}MnO_3$ from tunneling experiments, *Appl. Phys. Lett.* **82**, 233 (2003).

291. D. I. Golosov, New correlated model of colossal magnetoresistive manganese oxides, *Phys. Rev. Lett.* **104**, 207207 (2010).

292. Z. Sun, Q. Wang, J. F. Douglas et al., Minority-spin t_{2g} states and the degree of spin polarization in ferromagnetic metallic $La_{2-2x}Sr_{1+2x}Mn_2O_7$ (x=0.38), *Sci. Rep.* **3**, 3167 (2013).

293. B. Hope and A. Horsfield, Contrasting spin-polarization regimes in Co nanowires studied by density functional theory, *Phys. Rev. B* **77**, 094442 (2008).

294. A. Kastalsky, A. W. Kleinsasser, L. H. Greene, R. Bhat, F. P. Milliken, and J. P. Harbison, Observation of pair currents in superconductor-semiconductor contacts, *Phys. Rev. Lett.* **67**, 3026 (1991).

295. B. J. van Wees, P. de Vries, P. Magnée, and T. M. Klapwijk, Excess conductance of superconductor-semiconductor interfaces due to phase conjugation between electrons and holes, *Phys. Rev. Lett.* **69**, 510 (1992).

296. I. Žutić and S. Das Sarma, Spin-polarized transport and Andreev reflection in semiconductor/superconductor hybrid structures, *Phys. Rev. B* **60**, R16322 (1999).

297. P. A. Barnes and A. Y. Cho, Nonalloyed Ohmic contacts to n-GaAs by molecular beam epitaxy, *Appl. Phys. Lett.* **33**, 651 (1978).

298. A. H. MacDonald, P. Schiffer, and N. Samarth, Ferromagnetic semiconductors: Moving beyond (Ga,Mn)As, *Nat. Mater.* **4**, 195 (2005).

299. J. M. D. Coey, Dilute magnetic oxides, *Curr. Opin. Solid State Mater. Sci.* **10**, 83 (2006).

300. M. J. Calderon and S. Das Sarma, Theory of carrier mediated ferromagnetism in dilute magnetic oxides, *Ann. Phys.* **322**, 2618 (2007).

301. T. Dietl, Origin of ferromagnetic response in diluted magnetic semiconductors and oxides, *J. Phys.: Condens. Matter* **19**, 165204 (2007).

302. A. G. Petukhov, I. Žutić, and S. C. Erwin, Thermodynamics of carrier-mediated magnetism in semiconductors, *Phys. Rev. Lett.* **99**, 257202 (2007).

303. K. A. Yates, A. J. Behan, J. R. Neal et al., Spin-polarized transport current in n-type codoped ZnO thin films measured by Andreev spectroscopy, *Phys. Rev. B* **80**, 245207 (2009).

304. J. Philip, A. Punnoose, B. I. Kim et al., Carrier-controlled ferromagnetism in transparent oxide semiconductors, *Nat. Mater.* **5**, 298 (2006).

305. M. Johnson and R. H. Silsbee, Interfacial charge-spin coupling: Injection and detection of spin magnetization in metals, *Phys. Rev. Lett.* **55**, 1790 (1985).

306. F. J. Jedema, A. T. Filip, and B. J. van Wees, Electrical spin injection and accumulation at room temperature in an all-metal mesoscopic spin valve, *Nature* **410**, 345 (2001).

307. W. P. Pratt Jr., S.-F. Lee, J. M. Slaughter, R. Loloee, P. A. Schroeder, and J. Bass, Perpendicular giant magnetoresistances of Ag/Co multilayers, *Phys. Rev. Lett.* **66**, 3060 (1991).

308. S. A. Crooker, M. Furis, X. Lou et al., Imaging spin transport in lateral ferromagnet/semiconductor structures, *Science* **309**, 2191 (2005).

309. A. Geresdi, A. Halbritter, F. Tanczikó, and G. Mihály, Direct measurement of the spin diffusion length by Andreev spectroscopy, *Appl. Phys. Lett.* **98**, 212507 (2011).

310. J. Bass and W. P. Pratt Jr., Spin-diffusion lengths in metals and alloys, and spin-flipping at metal/metal interfaces: An experimentalist's critical review, *J. Phys.: Condens. Matter* **19**, 183201 (2007).

311. C. R. Granstrom, I. Fridman, H.-C. Lei, C. Petrovic, and J. Y. T. Wei, Point-contact Andreev reflection spectroscopy on Bi_2Se_3 single crystals, *Int. J. Mod. Phys. B* **30**, 1642002 (2016).

312. L. Aggarwal, S. Gayen, S. Das, R. Kumar, V. Suess, C. Felser, C. Shekhar, and G. Sheet, Mesoscopic superconductivity and high spin polarization coexisting at metallic point contacts on Weyl semimetal TaAs, *Nature Comm.* **8**, 13974 (2017).

3

Ballistic Spin
Transport

B. Doudin and N.T. Kemp

This chapter is dedicated to the study of ballistic transport in magnetic nanostructures. Ballistic means that the charges carrying the electrical current go through the sample without experiencing scattering. This condition requires that sample dimensions are smaller than the carriers mean free path. This concept is therefore in immediate contrast with the diffusive transport model, extensively used to explain giant magnetoresistance (GMR) properties. The spin dependent resistivities of the ferromagnetic layers and the interfaces largely determine the GMR properties; however, for a diminishing sample size this is not solely the case. An example of transport occurring essentially without scattering is the tunnel regime, where the conduction of the system is ensured by evanescent waves through a dielectric barrier. The related tunnel magnetoresistance (TMR) is described in detail in Chapters 11–13, Volume 1, but let us recall that it relies on spin-dependent density of states in the ferromagnetic layers (more precisely at the interface), as well as the transmission properties of the barrier.

Ballistic transport is a topic of increasing importance, owing to technical advances in top-down miniaturization capabilities. Another fundamental reason is the potential relevance for spintronics devices. The Datta and Das transistor device [1] relies on ballistic transport of the carrier, with a spin orientation modulated by external gating. In the first section, we will review the basic concepts related to ballistic transport properties, emphasizing the key characteristic lengths and dimensionalities. We will restrict our discussion to the simplest mathematical expressions and the most simplifying hypotheses. More comprehensive textbooks on mesoscopic electronics and ballistic transport in nanostructures can be found elsewhere [2, 3]. We will emphasize the differences between semiconductors and metallic nanostructures. Most experiments on ballistic transport were performed on semiconducting materials, taking advantage of better control of their properties, and larger characteristic lengths. Experimental examples involving semiconductors are therefore more convincing and more detailed, and will be used as illustrations. We will then introduce spin-dependent ballistic transport, approaching its importance in the field of spin electronics. In the final section, experimental results on metallic structures will be presented.

The reader should be aware that this research topic is rather controversial, especially for metallic systems, where strong size reduction down to a single atom is needed. When studying spin-dependent properties, the application of an external magnetic field also complicates the interpretation of the results. We have tried to provide a fair summary of the results to date, essentially emphasizing the experiments and interpretations that have mostly reached a consensus in the community.

The emerging picture is that spin-orbit coupling (SOC) is the key physical phenomena that can be invoked to explain most of the experimental results. This chapter is therefore complementing Chapters 9, Volume 1, and 7–8, Volume 2, adding its contribution to the broad field of "spin orbitronics."

3.1 BASIC CONCEPTS

When reducing the size of a sample, one can question how geometrically restricting the flow of carriers can affect the resistance of the sample. The best pedagogical case, having the advantage of leading to analytical solutions, is the so-called Maxwell resistance. We consider a constriction of the shape of a hyperbolic volume (Figure 3.1), for which analytical solution of the Poisson's equation provides an expression for the potential, while the current is obtained by integrating Ohm's law locally over the constriction volume. The asymptotic case of an infinite curvature of the constriction, corresponding to an orifice of radius a separating two metallic half spaces, results in a conductance given by:

$$G_M \equiv \frac{1}{R_M} = 2a\sigma. \tag{3.1}$$

As a numerical example, if we consider an aperture of 10^{-6} m in Au ($\sigma = 4 \times 10^7$ Sm^{-1}), we obtain a resistance of the order of 10 mΩ.

3.1.1 THE SHARVIN MODEL

When the constriction dimensions are reduced below the length of the electron mean free path, a diffusive model for conduction cannot apply. The flow of electrons between the left and right side of the constriction is therefore the ballistic regime. At low temperatures, the mean free path of metallic samples of reasonable purity and crystallinity is exceeding a few hundred nm, and several microns for semiconductors. (Note: However, the mean free path of thin metallic films or multilayers relevant for GMR or TMR device

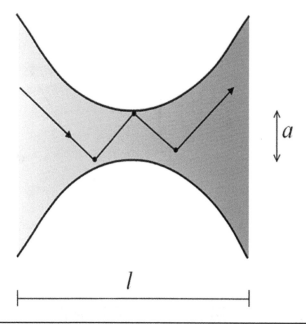

FIGURE 3.1   Sketch of a conductor with a constriction of diameter a, larger than the electronic mean free path l.

applications does not typically exceed a few tens of nm.) When an electric potential is imposed between the two sides of the constriction, the restriction to the charge flow imposed by the geometry results in an effective electrical resistance. Sharvin introduced the concept of transmission through a constriction in a material (Figure 3.2) [4], by applying perfect gas concepts to the flow of particles between two reservoirs having different total free energies. Under the hypothesis of perfect Fermi gas behavior at low temperatures and small applied voltage bias (linear approximation), the constriction along the x direction will limit the current density j by:

$$j(x) = e\langle v_x \rangle \rho(\varepsilon_F) \frac{eV}{2}, \tag{3.2}$$

where $\langle v_x \rangle$ is the average velocity along the x flow direction of the particles, corresponding to the direction of change of free energies. Assuming the usual expressions for $\langle v(x) \rangle$ and $\rho(E_F)$ in terms of the free electron gas Fermi wavevector, k_F, the Sharvin conductance is therefore of the form:

$$G_S \equiv \frac{1}{R_S} = \frac{2e^2}{h} \left(k_F \frac{a}{2} \right)^2. \tag{3.3}$$

This equation reveals that locally, Ohm's law does not apply, and that the conductance of a constriction does not depend on the resistivity of a material, but is related to the product of its density of states with the group velocity. This number is expected to be much less sensitive to impurities and temperature than the intrinsic resistivity of a material. If we take $k_F = 10^{10}$ m^{-1}, a 10 nm aperture will correspond to a resistance of a few Ω.

Experimental results on metallic systems using the so-called point contact geometry are well documented [5]. The Sharvin conductance can

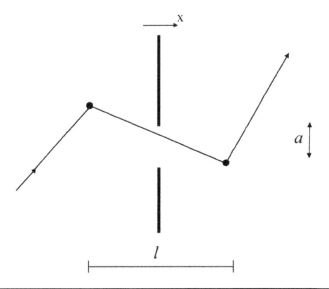

FIGURE 3.2   Sketch of a conductor with a constriction of diameter a, smaller than the mean free path l. Ohm's law does not apply, resulting in a Sharvin's conductance expressed by Equation 3.3.

be obtained by carefully pressing a sharp needle over a metallic surface. The signature of ballistic conductance is obtained by non-linearities in the current-voltage (I-V) curves, related to electron-phonon inelastic coupling. Anomalies in electrical properties are even more striking for superconductors, where the current blocking due to the superconducting bandgap is suppressed in the ballistic regime, leading to the so-called Andreev reflection. This is of particular interest for magnetic systems, as it allows the difference between $\langle v_{\uparrow x} \rangle \rho_{\uparrow}(E_F)$ and $\langle v_{\downarrow x} \rangle \rho_{\downarrow}(E_F)$, related to majority- and minority-spin electrons respectively, to be quantified. Extensive details about this topic can be found in Chapter 2, Volume 3. Such polarization of the ballistic current should, however, not be identified with the diffusive current spin polarization explaining GMR properties or the tunneling transmission spin polarization justifying the occurrence of TMR.

3.1.2 THE LANDAUER MODEL

Reducing further the size of a constriction requires a model taking into account the wave nature of the carriers of the electric current. If the sample's dimensions become comparable to, or smaller than the wavelength of the carriers, the effects of quantum confinement play a role. A simple illustration is obtained if we consider that transport between the left and right side of a one-dimensional conductor (Figure 3.3). This dimensionality restriction corresponds to stating that the width of the wire connecting the two electrodes allows only one Bloch wave function to exist in the (k_y, k_z) space, implying a diameter on the order of the Fermi wavelength λ_F. As a rule of thumb, the wavelength of conduction electrons is on the order of the distance separating

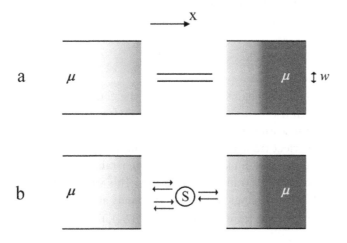

FIGURE 3.3 (a) Sketch of a conductor of width w, smaller than the conduction electrons wavelength, λ_F. In the Laudauer picture, two left and right electrons reservoirs connect the quantum conductor to the electrical circuits, with defined chemical potential μ_R and μ_L. (b) The conductor can be considered as a scattering center S, with transmission modes (arrows) connecting it to the reservoirs. The conductance relies on the number of modes and their transmission (reflection) properties. This particular example shows two modes connecting the scattering center to the left and one mode to the right.

two conduction electrons' donor sites (eventually holes in semiconductor cases). For metals, of typically one conduction electron per atom, it implies that λ_F is on the order of the atomic radius, approximately 0.3 nm. For semiconductors of high purity, this length is significantly larger, on the order of 10 nm.

Since, in one dimension, $\rho(E_F)$ is proportional to $1/\langle v(x) \rangle$, $\rho(E_F)\langle v(x) \rangle$ of Equation 3.2 remarkably simplifies to $4/h$, and the resulting conductance is of the form

$$G_0 = \frac{2e^2}{h}. \tag{3.4}$$

The factor 2 comes from the assumed spin degeneracy, and the conductance G_0 is the so-called quantum of conductance, which is $\dfrac{1}{G_0} = \dfrac{h}{2e^2} \approx 12.9 \text{ k}\Omega$.

This simple outcome can be generalized by introducing the Laudauer-Buttiker formalism, initially developed for modeling scattering at the mesoscale (Figure 3.3) [6]. In so-called two-terminal devices, the current flows between two infinite reservoirs, defined by their electrochemical potentials. They are separated from a central scattering region by two idealized semi-infinite leads, treated as an ensemble of one-dimensional conductors obeying Equation 3.4. These conductors index the possible Bloch waves travelling in the leads. The interface between the leads and reservoirs are considered perfect, so that scattering only occurs between leads and the central region. A matrix formalism is used to express the linear relationship between transmission factors of waves coming from the left and right electrodes. Diagonalization of this matrix provides a series of transmission eigenvalues, T_i, corresponding to the so-called eigenchannels. At zero bias and zero temperature bias, the conductance reduces to

$$G = \frac{2e^2}{h} \sum_i T_i. \tag{3.5}$$

At small bias (and negligible temperatures) the conductance of the system relates to the transmission probabilities, T_i, between the lead eigenchannels and the central scattering source, taking values between 0 (perfect reflection) and 1 (perfect transmission) at the Fermi level. For the case of a one-dimensional conductor, Equation 3.4 can be generalized to take into account the number of allowed transmission modes. This value $N(E_F)$ corresponds to the number of crossings of the 1-D subbands at the Fermi level. When considering an average transmission factor $\bar{T}(E_F)$, Equation 3.5 reduces to:

$$G = \frac{2e^2}{h} \bar{T}(E_F) N(E_F). \tag{3.6}$$

In the case of perfect transmission factors, $T_i = 1$, Equation 3.6 shows that the conductance is quantized in units of G_0. Experimental confirmations of such conductance quantization were reproducibly observed in both semiconductors and metallic systems. When realizing ballistic conductors of perfect transmission factors in the ballistic quantum regime, step-wise variation

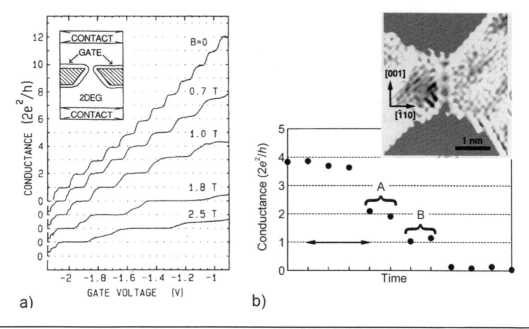

FIGURE 3.4   Experimental data showing quantization of the conductance for perfect ballistic point contacts. (a) Gated 2DEG semiconductor structure, where electrostatic gating depletes continuously the constriction between the source and drain. (After Wees, B.J. et al., *Phys. Rev. B* 38, 3625, 1988; van Wees, B.J. et al., *Phys. Rev. Lett.* 60, 848, 1988. With permission). (b) Metallic stretched Au linear constriction obtained by stretching Au under a transmission microscope, revealing the atomic nature of the contact. (After Ohnishi, H. et al., *Nature* 395, 780, 1998. With permission.)

of G is found. Figure 3.4 shows two examples of experimental realization of ballistic quantum conductors, obtained by electrostatically gating a two-dimensional electron gas (2DEG), or stretching a metallic constriction, both showing so-called *quantization of the conductance*. The experimental results on metallic systems also confirm the necessary extreme smallness of the samples.

One should, however, note that unambiguous experimental confirmation of the ballistic quantum nature of the electronic transport can only result from a four-points measurement method, where two voltage probes should be inserted between the source and drain connections. Such an experimental tour de force has been realized for a cleaved 2-DEG, using electrostatic gating to define four terminals on the resulting 1-D conductor. Under the condition that the voltage probes are noninvasive (technically corresponding to voltage lines of impedance much larger than $1/G_0$), the measured voltage drop is zero when the transmission factor reaches one, confirming that no scattering event has affected the potential energy of the particles. In other words, the observed resistance of a perfect ballistic conductor, reaching several kΩ, should be interpreted as a contact resistance, shared between the interfaces of the 1-D conductor and the source and drain leads.

The observed two-terminals conductance in multiples of the quantum value is an idealized case, corresponding to the asymptotic transmission factors of 1 and 0. In a more general case, the conductance of a small ballistic sample requires the explicit calculation of the normal modes, and their transmission factors T_i. Detailed description of the theoretical tools and their numerical outcome are beyond the scope of this chapter, and the reader can consult more dedicated texts [2, 3, 10]. Two idealized geometries can be considered as the basis for discussing the number of conduction channels, their symmetry properties and their expected transmission factors. Figure 3.5 provides a picture particularly suited to the case of metallic systems, with sample size on the order of the atomic radius. The simplest case is the pseudo-infinite 1-D wire. The derivation of the allowed modes results from solid-state physics concepts applied to a 1-D case, and the usual theoretical tools for calculating the propagating Bloch waves can be applied. The number of transmitting modes equals the number of crossings of the 1-D dispersion curves with the Fermi level. A smooth transition from a single atom width to a width reaching the mean free path value ensures that the highest transmission factor value is obtained, i.e., $T_i = 1$, if a mode is allowed to propagate in the wire. This model is useful for pedagogical purposes and for understanding which conductions channels can propagate. Calculation of the conductance is straightforward, as it corresponds to the number of allowed propagating modes. This image can be used for semiconductor systems, where the purity, size control, and fabrication techniques can result in well-defined 1-D systems in the tens of nanometers range. The other limiting case is the single atom connection, separating two semi-infinite pyramids, which is expected to match better the experiments on metals. Transmitting channels are related to wavefunctions of the contact atom and the number of propagating modes can generally be matched to the number of valence

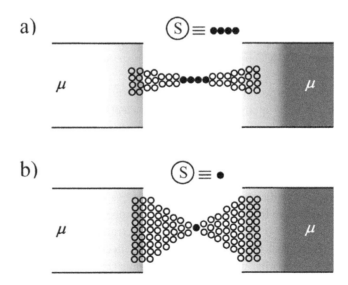

FIGURE 3.5   Sketch of the two idealized simplest pictures of atomicsized quantum contact for metallic systems. (a) Corresponds to an idealized infinite wire, with a smooth continuous transition to the two reservoirs. (b) Corresponds to a single atom connection, with two pyramidal interconnects to the infinite reservoirs.

electrons of the single atom contact. The two pyramid-shaped electrodes broaden the density of states and create an electronic cloud mismatch at the interface, resulting therefore in transmission factors deviating significantly from 1. Calculations of the transmission factors are however much less intuitive, and depend significantly on the geometrical arrangements of the pyramid atoms. One can, however, guess that atomic wavefunctions, being less localized (in particular s electrons), suffer less from the mismatch at the interface, and exhibit therefore higher transmission factors. Such a model has been particularly successful in explaining experimental results in mechanical break junctions.

The interest of keeping in mind the two types of model is to realize the importance of the dispersion curves of the electronic structures and recall that perfect transmission or reflection are only limiting cases. One should also mention that these two models are not so much different. Following the argument of Glazman et al. [11], the adiabatic condition will allow perfect transmission and no mixing of the conduction channels when the cross section of the constriction varies slowly compared to λ_F. When contemplating Figure 3.5a, b, the difference between the two geometries is mostly related to checking if the adiabatic condition is fulfilled or not. For metal systems, a contact made of two well-aligned atoms could already be considered as a reasonable approximation of an infinite wire! This might explain (among other reasons) the lack of reproducibility when performing single atom contact measurements, and the necessity of ensuring statistical information on the experimental results.

3.2 SPIN TRANSPORT IN THE BALLISTIC REGIME

Until now, the spin of the conduction particles has not been considered, except as a doubling parameter for calculating the conductance of a system. The factor 2 in Equation 3.4 explicitly expresses the spin degeneracy. In this section we will review several ways to lift this degeneracy, opening the possibility to create a spin polarized flow of particles, and to expect new magnetoresistance effects. The first part of this section describes experimental realizations on semiconductors systems, opening the possibility to create spin filters and spin detectors using point contacts in a 2DEG.

3.2.1 ZEEMAN SPLITTING

Since the discovery of the quantum conductance, the use of a large magnetic field has been shown to be a usable tool for realizing point contacts in 2DEG exhibiting quantized steps of multiples of e^2/h. The Zeeman splitting energy of the form:

$$\Delta E_Z = g\mu B, \qquad (3.7)$$

has typical order of magnitude of 0.1 meV for 2DEG systems (using a Landé factor $g = 0.4$, typical for GaAs and fields ~ few Teslas), requiring therefore experiments at temperatures below 1 K to allow conduction values of the order of $\frac{1}{2}G_0$, providing therefore a spin filtered source of carriers. Even if larger Landé factors can be obtained on other systems, Zeeman splitting remains, however, severely limited to low temperature studies.

3.2.2 SPIN ORBIT SPLITTING

Spin polarization of quantum point contacts in 2DEG systems can be further exploited in experiments involving point contact emitters and collectors, where the cyclotron ballistic motion of charge carriers is controlled through the use of an external perpendicular magnetic induction field. A large literature on such electron focusing experiments can be found [12]. Figure 3.6 shows the design of the experiment. Current flows through the detector when the distance L between the two point contacts matches a multiple of the cyclotron radius $r_c = \dfrac{\hbar k_F}{eB}$ where $k_F = \sqrt{2mE_F}/\hbar$ for free-like carriers, of effective mass m and charge e.

The emitter and detector can be gated to allow a controlled number of channels to be transmitted by the two point contacts. Potok et al. [15] showed that an additional Zeeman splitting can spin-polarize the emitted charges (by gating the emitter to e^2/h conductance), resulting in a modification of the B-dependence of the current passing through the detector. Figure 3.6 summarizes the experimental outcome, showing that the current through the detector can be significantly spin polarized. It shows that the

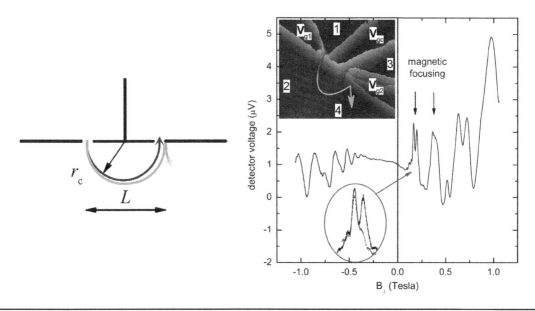

FIGURE 3.6 Spin-dependent electronic focusing experiment in GaAs 2DEG, with point contact emitters and detectors defined by top local oxidation. On the top left, the principle of the experiment is provided, with a scanning electron microscopy (SEM) picture of the corresponding device (5 × 5 mm). When varying the applied field, focusing is observed for positive field values. By applying a Zeeman splitting field of 3.5 T, the two first magnetic focusing peaks appearing for an injector of e^2/h conductance become a single peak when the injector is gated to e^2/h conductance value. Only one single spin direction (related to the black trajectory) is allowed to pass the detector point contact. (After Rokhinson, L. et al., *J. Phys.: Condens. Matter* 20, 164212, 2008. With permission.)

focusing condition, i.e., the radius of the cyclotron trajectory, is spin dependent, providing a direct illustration of SOC effects.

The coupling between orbital and spin momentum of charges results
from their relativistic interactions. For a single particle Hamiltonian in a
crystal potential, $V(\vec{r})$ one can express:

$$H = \frac{p^2}{2m} + V(\vec{r}) + \frac{\hbar}{4m^2c^2}\left[\nabla V(\vec{r}) \times \vec{p}\right] \cdot \sigma, \tag{3.8}$$

where the particle of mass m has a momentum $\vec{p}$ and a Pauli spin operator σ.

For the case of a 2DEG with a bulk origin of the symmetry-breaking
$\nabla V(\vec{r})$, one can simplify the Hamiltonian to:

$$H = \frac{1}{2m}\left(p_x + \gamma\sigma_x\right)^2 + \frac{1}{2m}\left(p_y - \gamma\sigma_y\right)^2, \tag{3.9}$$

where γ is the Dresselhaus spin orbit parameter. The resulting splitting of the
dispersion curves is illustrated in Figure 3.7, for a free-carrier-like dispersion.
A semi-classical treatment of the calculation of the cyclotron motion of the
particle shows that the radius now depends on the projection of spin along
the applied field (z-axis). One prefers to express the value of the applied field
fulfilling the focalization condition for the two spin directions:

$$B_\perp^\pm = 2\left(p_f \mp \gamma\right)/eL, \tag{3.10}$$

where p_f relates to the momentum corresponding to focalization without
SOC.

The SOC can result in splitting energies ranging from several meV to
a few tens of meV in GaAs, and is therefore significantly larger than the
Zeeman splitting. The observed variations in the detection currents relate to
the product of the injector and detector spin selectivities, reaching several

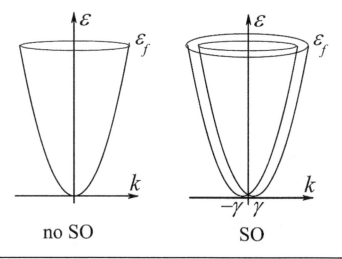

FIGURE 3.7   Dispersion curve of a free-like carrier, with the splitting due to the
spin–orbit coupling of parameter γ. The two cyclotron trajectories of Figure 3.6
correspond to the top right of the figure.

tens of percents. These devices are therefore of fundamental interest for manipulating spin populations at low temperatures, with the perspectives of performing fundamental studies. The initial experiments showed how SOC can be of use for separating the spin currents, in contrast to its detrimental effect as a source of decoherence in the spin populations. One should also emphasize that these devices allow spin electronic studies without involving magnetic elements.

3.2.3 EXCHANGE SPLITTING

When considering magnetic systems, in particular transition metal materials, one can take advantage of the huge exchange energy to lift the spin degeneracy. This splitting energy, reaching eV values, effectively splits the d-type bands of transition metals giving rise to their ferromagnetic properties. This simplest argument would therefore imply that quantized conductance in multiples of e^2/h could be found in metallic nanocontacts made from transition metal systems. In such cases, Equation 3.5 is inappropriate and must be generalized for magnetic systems. In a material where magnetization defines a spin quantization axis, we can decompose the transmission factors for majority ($\uparrow$) and minority ($\downarrow$) spin directions.

$$G = \frac{e^2}{h} \sum_i \left(T_{\uparrow i} + T_{\downarrow i} \right). \tag{3.11}$$

If we simplify the system to a single channel with left and right transmission, it becomes possible that the exchange splitting limits the band crossing at the Fermi energy to one type of spin population only. In such a case, conductance quantization in multiples of $\frac{1}{2}G_0$ is expected. Interestingly, the ballistic conductor would play the role of a spin filter, allowing only one spin population to be transmitted, creating a perfect current spin polarization. As a result, one can expect no current transmission when the two sides of the conductor are magnetized with opposite magnetization directions (antiparallel case), and a transmission of $\frac{1}{2}G_0$ when the magnetization of the system is uniform (parallel case). The ballistic conductor would play the role of a perfect (atomic size) domain wall in the antiparallel configuration. The simplified argument given here needs more complete calculations, that take into account the coherent transport through the constriction and modeling of the system in order to extract the spin-dependent transmission factors. Two illustrations are provided in Figure 3.8, both reaching a similar conclusion of strong spin filtering effects at conductance values on the order of $\frac{1}{2}G_0$ and odd multiples of $\frac{1}{2}G_0$ that are allowed in the ferromagnetic configuration only.

Occurrence of conductance quantized in units of $\frac{1}{2}G_0$ is controversial. Experiments performed under Transmission Electron Microscopy (TEM) observation, similar to Figure 3.4, showed half-integer conductance

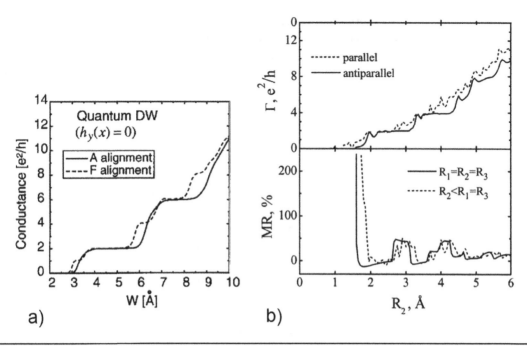

FIGURE 3.8 Free electron calculations of conductance of a ferromagnetic constriction in the quantum ballistic regime. (a) With calculated conductance as a function of the width of transition metal 1D conductor, with the conductance in the antiparallel magnetization configuration drawn in full line and conductance in the parallel configuration in dashed. (b) Same type of calculation, involving a model of a central constriction resistor R_2, separated from the electrodes by wider R_1 and R_3 resistors. (After Imamura, H. et al., *Phys. Rev. Lett.* 84, 1003, 2000. With permission.) (b) (After Zhuravlev, M.Y. et al., *Appl. Phys. Lett.* 83, 3534, 2003. With permission.)

occurring in Co [15]. However, repetitive measurements performed on mechanical break junctions (MBJ), when stretching back and forth a suspended metallic bridge lithographically defined on a flexible substrate, did not show preferred occurrence of quantized values of the conductance for transition metals contacts [18]. Many divergent experimental outcomes can be found in the literature [2, 19, 20].

Based on the intuitive arguments of the previous section, it is not surprising that magnetic junctions of high transmission single channel conductance cannot be obtained frequently. For example, a Fe 1-D wire, without imperfections, with a diameter smoothly increasing when connected to the diffusive banks, would have 7 band crossings at the Fermi level, exhibiting therefore conductance that should be $7e^2/h$. Furthermore, the s bands are expected to exhibit limited spin splitting, which always prevents full spin polarization. When one gets away from the ideal 1-D geometry, smaller transmission factors are expected, and one can expect lower measured conductance, corresponding to frequently observed values, but not guaranteeing the absence of s-type transmission. The multiple combinations of several partly conducting channels might explain the lack of preferred occurrence at simple multiples of e^2/h.

Another fundamental difficulty in making atomic spin filters is related to magnetic properties. The possibility of realizing atomic-sized domain walls remains conceptually unclear. Even though magnetic energy calculations can show that such sharp domain walls can represent special cases of stability in magnetic systems of reduced dimensions, the intrinsic magnetic properties of the atoms of the structure, as well as their exchange interaction ensuring ferromagnetism, can be modified severely for atomic-sized structures. This necessitates more steps and hypotheses in the calculations, and provides more reasons to deviate from the asymptotic behavior of infinite magnetoresistance [19].

Experimental investigations of magnetoresistance properties also exhibited a very large spectrum of results, with lack of reproducibility. A critical issue in performing measurements of atomic size structures under applied magnetic field is the intrinsic troublesome mechanical stability of the samples. Magnetostatic interactions, magnetoelastic properties of the bulk ferromagnetic electrodes, and magnetization-dependent strains can easily modify the distance separating the two large diffusive electrodes by several atomic distances, modifying therefore drastically the nanocontact and its conductance. There is now a significant consensus in the community that huge magnetoresistance values due to atomic domain wall structures are unlikely to happen, and that the published data revealing very large magnetoresistance values can be explained using models involving mechanical modifications of the nanocontacts under varying applied magnetic field [21].

However, several groups reported magnetoresistance ratios of a few tens of percents, within the expected range when considering that one or a few transmitting channels can have their transmission factors turned on and off when modifying the mutual magnetic alignment of the two connecting electrodes. More interestingly, these results all showed large anisotropic magnetoresistance (AMR) effects, which is a finding that can be related to the previously discussed SOC.

3.3 ANISOTROPIC MAGNETORESISTANCE IN THE QUANTUM LIMIT

3.3.1 AMR IN DIFFUSIVE TRANSPORT

Magnetoresistance measurements by W. Thomson showed in 1859 that the resistance of a ferromagnetic material depends on the mutual orientation of the applied magnetic field $\vec{H}$ and current $\vec{j}$ directions. This so-called anisotropic magnetoresistance (AMR) relates therefore to the experimental indication that $\rho_{//}(\vec{H} \, / \, / \vec{j})$ is, in most cases, larger than $\rho_{\perp}(\vec{H} \perp \vec{j})$, and follows the relation (in a cubic symmetry)

$$\rho(\theta) - \rho_{\perp} = (\rho_{//} - \rho_{\perp})\cos^2\theta, \qquad (3.12)$$

where θ is the angle between the current and the magnetization. This difference does not exceed a few percent at room temperature for transition metals. The mechanism by which the microscopic internal field associated with $\vec{M}$ couples to the current density in ferromagnets is the SOC interaction. It was only one century after Thomson's discovery that a scattering model, derived from Mott's two current model of conduction in transition metals, was refined to explain how SOC can lead to Equation 3.12 [22]. When the orbital motion of the d electrons is coupled to their spin and the crystal lattice, the new d electron wavefunctions are no longer eigenfunctions of S_z since H_{SO} mixes states of opposite spins. Hence, this limits the independence of majority and minority conduction channels of the Mott's model, and is an importance source of spin mixing, detrimental to GMR properties. This mixing of states is anisotropic and results in resistivity anisotropy. Potter's analytical expression for the spin-dependent relaxation rates showed that sd scattering of minority spins (i.e., the one opposite the magnetization) is responsible for the measured AMR (with $\rho_{//}>\rho_{\perp}$). It was illustrating that the sign of the AMR depends directly on the band structure of the ferromagnetic material.

The SOC in bulk metallic systems is a rather small energetic perturbation, of typically 0.1 meV magnitude. Despite its limited magnitude, SOC is involved in explaining anisotropic magnetic properties of materials, including magneto-crystalline anisotropy, magnetostriction and AMR. For magnetic systems of reduced dimensionalities and reaching atomic dimensions, one expects a significant enhancement of the SOC coupling, justifying the modification of magnetic properties and their anisotropy for ultra-thin magnetic films, islands and wires involving a few atoms only. The AMR properties should therefore be completely rethought when the electric transport approaches the ballistic regime and when the SOC magnitude increases.

3.3.2 Initial Experimental Results of AMR in Nanocontacts

Several groups reported magnetoresistance measurements varying significantly when the angle between magnetization and current was imposed. Figure 3.9 presents three different results obtained on three different types of samples. Results reported by Viret et al. [23] were obtained on mechanical break junctions, those reported by Keane et al. [24] were measured on nanocontacts made by electromigration breaking slowly a patterned nanoconstriction. Both experiments were performed at low temperatures. Data obtained by Yang et al. [25] used electroplated samples, where a gap, typically 100 nm wide, is closed by reducing metal ions from an electrolytic bath. This data was obtained in situ at room temperature, under variable reduction or oxidation (dissolution) electrochemical conditions of the junctions immersed in an ionic medium. The fact that such AMR-like properties were obtained on samples made by quite different methods, in different media and at different temperatures, strongly suggests that AMR on nanocontacts can be reproducibly observed with amplitudes much larger than those found on bulk samples. Therefore, there should be fundamental reasons to exacerbate AMR properties when the size of the sample is reduced.

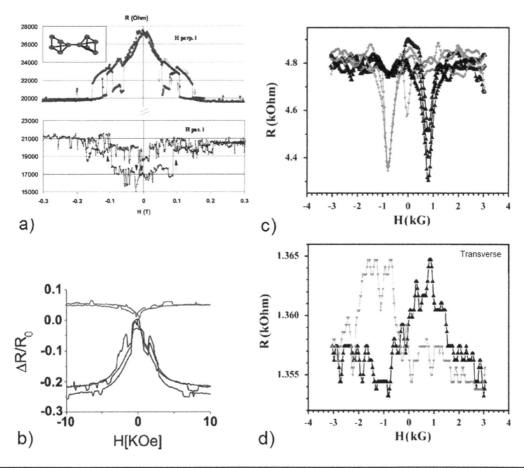

FIGURE 3.9   Angle-dependent magnetoresistance curves on Ni nanocontacts. (a) MBJ measured at 10 K. (After Viret, M. et al., *Phys. Rev. B* 66, 220401, 2002. With permission.) (b) Electromigrated Ni junction at 4.2 K. (After Keane, Z.K. et al., *Appl. Phys. Lett.* 88, 062514, 2006. With permission.) (c, d) Electroplated Ni junctions at 300 K. (After Yang, C.S. et al., *J. Magn. Magn. Mater.* 286, 186, 2005. With permission.)

3.3.3 AMR in the Ballistic Regime of Transport (BAMR)

From the previous considerations and findings, the picture of AMR of bulk materials should be completely revisited if the ballistic contribution to the sample resistance becomes dominant. A good pedagogical model can be inferred from the idealized 1-D chain of atoms with perfect transmission factors and can be used to demonstrate the concept behind AMR extended to Landauer formalism [26]. Ab-initio calculations using the pseudopotential plane wave method and symmetry arguments have shown that the weakly dispersive bands (δ bands) arising from the coupling of nearest neighbors $3d_{xy}$ and $3d_{x2-y2}$ atomic orbitals are split by the spin-orbit interaction when the magnetization is parallel to the wire, and nearly degenerate when perpendicular (Figure 3.10). The energy lifting corresponds to the SOC constant reach a magnitude of 0.1 eV, orders of

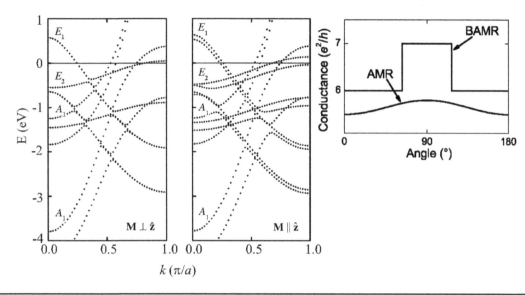

FIGURE 3.10 Calculations of the dispersion curve of a 1D Ni wire in a tight binding model, showing the splitting of the bands when magnetization and current directions are parallel. Right: comparison of angular behavior showing how AMR and BAMR can differ in magnitude and shape. (After Velev, J. et al., *Phys. Rev. Lett.* 94, 127203, 2005. With permission.)

magnitude larger than the bulk SOC energy. Because the Fermi energy lies close to the edge of the δ bands, one of the bands becomes expelled from the Fermi level when split by the SOC energy in the relevant geometry (*M* parallel to the wire). Hence, the integer number of conducting channels can be modified when the angle between magnetization and current is varied. The conductance is expected to change by e^2/h in this case because for an infinite atomic wire the conductance is simply e^2/h per band crossing the Fermi level. For the specific example of Figure 3.10 (δ bands of a Ni monatomic wire), the conductance changes from 6 e^2/h to 7 e^2/h, which is comparable in magnitude to experiments, and much larger than bulk AMR. This so-called ballistic anisotropic magnetoresistance (BAMR) is also expected to exhibit sharp angular dependence that makes the experimental signature of BAMR significantly different from bulk AMR (Figure 3.10).

It is clear that a 1-D perfect wire is a highly idealized model that must be considered as a pedagogical intuitive guideline. In fact, when trying to understand transport in magnetic contacts, one needs to model the atomic arrangement that constitutes a narrow neck and the corresponding electronic structure. A realistic modeling of the experiments must determine possible structural arrangements when making atomic contacts, compute the electronic and magnetic structures of the systems, and deduce the conductance by estimating the transmission factors of the eigenmodes of the junction. More details can be found in dedicated papers, where more information on the hypothesis for the calculations can be found [2, 19, 27]. Even though the simplicity of the infinite 1-D model can be criticized, the main conclusions of Velev et al. should be kept for more sophisticated modeling. The SOC energy is increased by the atomic size of the junction, and results in angular changes of the conduction of shape and amplitude very different from bulk AMR properties.

3.3.4 EXPERIMENTAL FINDINGS OF BAMR

Measuring AMR is obtained by keeping the sample magnetically saturated, and varying the angle the magnetization makes with the current. For applied magnetic fields above 1 T, the magnetic configuration of the samples is expected to be near saturation. Such studies therefore eliminate the problem of positioning an atomic-sized domain wall, of importance for GMR-type studies, and investigate an unambiguous magnetic configuration. Figure 3.11a shows results on Fe contacts for MBJ measured at low temperatures, and Figure 3.11b shows Co electrodeposited contacts at room temperature. Both data exhibit the expected deviations for a $\cos^2\theta$ bulk AMR variation of the resistance, with amplitude unattainable for bulk materials.

Data on Co also indicates that the sign of the effect can also be opposite. A qualitative analysis based on the arguments of Velev et al. pointed out that the dispersion curves related to the wavefunctions of a given nanocontact can vary significantly from one sample to the other, resulting in an angular variation of the number of band crossings that can be drastically modified from sample to sample. These findings also confirm that the sign of the BAMR has no fundamental reason to be positive, in contrast to AMR, where negative variations can only be observed in a few alloys.

Even though a set of experimental findings from different research teams appear to reach a consensus, one should mention another set of experiments, performed on MBJ, that revealed a voltage bias dependent AMR of nanocontacts at low temperatures (Figure 3.12, [28]). They interpreted their results with quantum fluctuations arguments, and pointed out that disorder in the nanocontact region can significantly enhance this phenomenon. The remarkable sharpness of the conductance angular transitions found on Co samples (Figure 3.11) is also reminiscent of fluctuations between two (or multiple) voltage levels found in electrical telegraph noise. It is known that such noise can reach large amplitudes in very small systems. For nanocontacts, the explanation of such two-level fluctuation is based on a model of atomic rearrangements at the constriction, possibly including magnetic fluctuations when considering magnetic systems. Such occurrence can mimic BAMR data, as recently shown by Shi and Ralph on Ni samples [29]. This discussion also recalls that steps in the conductance of metallic junctions are indicative of stable atomic configurations at the nanocontacts, and that the mechanical stability remains a critical issue, especially for magnetic systems. More detailed information on the interplay between mechanical and electronic properties of junctions can be found in the review of Agrait et al. [2]. Following the same line of reasoning, Häfner et al. put forward calculations showing that impurities can present enhanced AMR properties mimicking those of pure contacts. Recent insight into this problem was gained by

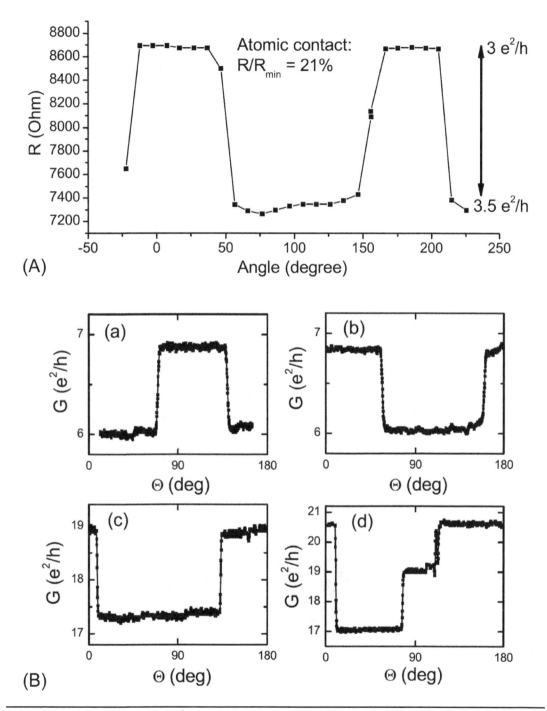

FIGURE 3.11 Angular variation of a magnetic nanocontact resistance when changing the angle between saturated magnetic state and current directions. (A) Data on Fe contacts. (After Viret, M. et al., *Eur. Phys. J. B* 51, 1, 2006, With permission.) (B) Data on Co contacts, showing step-like AMR of large amplitude, of behavior not reproducible from sample to sample. (After Sokolov, A. et al., *Nat. Nanotechnol.* 2, 171, 2007. With permission.)

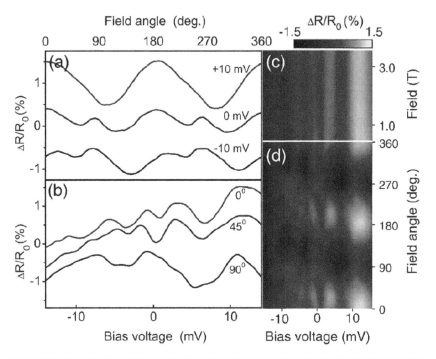

FIGURE 3.12 Variations of $R = dV/dI$ at 4.2 K in a Permalloy sample with average zero-bias resistance of 2.6 kΩ. (a) R versus field angle at different bias voltages ($B = 0.8$ T). (b) Dependence of R on V at different fixed angles of magnetic field ($B = 0.8$ T). The curves in (a) and (b) are offset vertically. (c) R as a function of V and magnetic field strength, with field directed along the x-axis. R does not have significant dependence on the magnitude of B. (d) R as a function of V and for $B = 0.8$ T. (After Bolotin, K.I. et al., *Phys. Rev. Lett.* 97, 127202 1, 2006. With permission.)

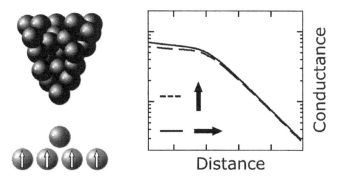

FIGURE 3.13 Tip-substrate conductance as a function of distance. The deviation from an exponential behavior indicates the onset of ballistic transport, occurring though an adatom of a magnetized substrate. The arrow indicates the orientation of the magnetization, on substrates regions related to perpendicular magnetization (full line) and inside a domain wall (dotted line). The difference between the two curves illustrates the amplitude of the anisotropy in conductance. (After Schöneberg, J. et al., *Nanolett.* 16, 1450, 2016. With permission.)

using scanning tunneling microscopy (STM), ideally suited to differentiate between impurities and single atoms, and providing experimental data of the transition between tunneling and ballistic conduction regimes [32]. Over a magnetically structured substrate, a metallic tip in contact mode with an adatom provides access to the contrast in conductivity values between transverse and longitudinal magnetic orientations, that show a value in the 10% range for pure Co adatoms (Figure 3.13).

3.4 CONCLUSIONS

We have shown that ballistic transport can be a usable way to filter spin currents, detect preferentially one spin population, and reveal new magnetoresistance applications at the nanoscale. For metallic systems, the experiments accumulated over the past few years indicated that the mechanical stability of the samples is a key issue and limitation for applications. This chapter also emphasized how SOC is the key physical phenomenon beyond all these new properties. This coupling makes the ballistic trajectories of the two spin populations of carriers different in space, which is a property that can directly be used for semiconducting systems. This coupling also explains why large anisotropies in the resistance properties can be found. This interest complements the other aspect of SOC, namely the limitation of spin coherence. This spin mixing effect is usually considered as negative, limiting the discrimination of the two spin populations in time or space. We think that both perspectives can be complementary, allowing therefore better and new understanding of quantum spin systems, as well as design of novel quantum spin devices.

REFERENCES

1. S. Datta and B. Das, Electronic analog of the electro-optic modulator, *Appl. Phys. Lett.* **56**, 665 (1990).
2. N. Agrait, A. L. Yeyati, and J. M. van Ruitenbeek, Quantum properties of atomic-sized conductors, *Phys. Rep.* **377**, 81 (2003).
3. Y. Imry, *Introduction to Mesoscopic Physics*, Oxford University Press, Oxford, 1997.
4. Y. V. Sharvin, A possible method for studying Fermi surfaces, *Sov. Phys. JETP* **21**, 655 (1965).
5. A. G. M. Jansen, A. P. van Gelder, and P. Wyder, Point-contact spectroscopy in metals, *J. Phys. C: Solid St. Phys.* **13**, 6073 (1980).
6. R. Landauer, Spatial variation of currents and fields due to localized scatterers in metallic conduction, *IBM J. Res. Dev.* **1**, 223 (1957).
7. B. J. van Wees, L. P. Kouwenhoven, H. van Houten et al., Quantized conductance of magnetoelectric subbands in ballistic point contacts, *Phys. Rev. B.* **38**, 3625 (1988).
8. B. J. van Wees, H. van Houten, C. W. J. Beenakker et al., Quantized conductance of point contacts in a two-dimensional electron gas, *Phys. Rev. Lett.* **60**, 848 (1988).
9. H. Ohnishi, Y. Kondo, and K. Takayanagi, Quantized conductance through individual rows of suspended gold atoms, *Nature* **395**, 780 (1998).
10. S. Datta, *Electronic Transport in Mesoscopic Systems*, Cambridge University Press, Cambridge, 1995.

11. L. I. Glazman, G. B. Lesovik, D. E. Khmelnitskii, and R. I. Shekhter, Reflectionless quantum transport and fundamental ballistic-resistance steps in microscopic constrictions, *JETP Lett.* **48**, 238 (1988).

12. V. S. Tsoi, J. Bass, and P. Wyder, Studying conduction-electron/interface interactions using transverse electron focusing, *Rev. Mod. Phys.* **71**, 1641 (1999).

13. R. M. Potok, J. A. Folk, C. M. Marcus, and V. Umansky, Detecting spin-polarized currents in ballistic nanostructures, *Phys. Rev. Lett.* **89**, 266602 (2002).

14. L. Rokhinson, L. N. Pfeiffer, and K. W. West, Detection of spin polarization in quantum point contacts, *J. Phys. Condens. Matter* **20**, 164212 (2008).

15. V. Rodrigues, J. Bettini, P. C. Silva, and D. Ugarte, Evidence for spontaneous spin-polarized transport in magnetic nanowires, *Phys. Rev. Lett.* **91**, 096801 (2003).

16. H. Imamura, N. Kobayashi, S. Takahashi, and S. Maekawa, Conductance quantization and magnetoresistance in magnetic point contacts, *Phys. Rev. Lett.* **84**, 1003 (2000).

17. M. Y. Zhuravlev, E. Y. Tsymbal, S. S. Jaswal, A. V. Vedyayev, and B. Dieny, Spinblockade in ferromagnetic nanocontacts, *Appl. Phys. Lett.* **83**, 3534 (2003).

18. C. Untiedt, D. M. T. Dekker, D. Djukic, and J. M. van Ruitenbeek, Absence of magnetically induced fractional quantization in atomic contacts, *Phys. Rev. B* **69**, 081401 (2004).

19. J. D. Burton and E. Y. Tsymbal, Magnetoresistive phenomena in nanoscale-magnetic contacts, in *Oxford Handbook of Nanoscience and Technology*, A. V. Narlikar and Y. Y. Fu (Eds.), vol. 1, Oxford University Press, Oxford, pp. 677–718, 2010.

20. B. Doudin and M. Viret, Ballistic magnetoresistance? *J. Phys.: Condens. Matter* **20**, 083201 (2008).

21. J. W. F. Egelhoff, L. Gan, H. Ettedgui et al., Artifacts in ballistic magnetoresistance measurements (invited), *J. Appl. Phys.* **95**, 7554 (2004).

22. R. I. Potter, Magnetoresistance anisotropy in ferromagnetic NiCu alloys, *Phys. Rev.* **B10**, 4626 (1974).

23. M. Viret, S. Berger, M. Gabureac et al., Magnetoresistance through a single nickel atom, *Phys. Rev. B* **66**, 220401 (2002).

24. Z. K. Keane, L. H. Yu, and D. Natelson, Magnetoresistance of atomic-scale electromigrated nickel nanocontacts, *Appl. Phys. Lett.* **88**, 062514 (2006).

25. C. S. Yang, C. Zhang, J. Redepenning, and B. Doudin, Anisotropy magnetoresistance of quantum ballistic nickel nanocontacts, *J. Magn. Magn. Mater.* **286**, 186 (2005).

26. J. Velev, R. F. Sabirianov, S. S. Jaswal, and E. Y. Tsymbal, Ballistic anisotropic magnetoresistance, *Phys. Rev. Lett.* **94**, 127203 (2005).

27. M. Häfner, J. K. Viljas, and J. C. Cuevas, Theory of anisotropic magnetoresistance in atomic-sized ferromagnetic metal contacts, *Phys. Rev. B* **79**, 140410(R) (2009).

28. K. I. Bolotin, F. Kuemmeth, and D. C. Ralph, Anisotropic magnetoresistance and anisotropic tunneling magnetoresistance due to quantum interference in ferromagnetic metal break junctions, *Phys. Rev. Lett.* **97**, 127202 (2006).

29. S. F. Shi and D. C. Ralph, Atomic motion in ferromagnetic break junctions, *Nat. Nano.* **2**, 522 (2007).

30. M. Viret, M. Gabureac, F. Ott et al., Giant anisotropic magneto-resistance in ferromagnetic atomic contacts, *Eur. Phys. J. B* **51**, 1 (2006).

31. A. Sokolov, C. Zhang, E. Y. Tsymbal, J. Redepenning, and B. Doudin, Quantized magnetoresistance in atomic-size contacts, *Nat. Nanotechnol.* **2**, 171 (2007).

32. J. Schöneberg, F. Otte, N. Néel et al., Ballistic anisotropic magnetoresistance of single-atom contacts, *Nanolett.* **16**, 1450 (2016).

Graphene Spintronics

Csaba Józsa and Bart J. van Wees

4.1 INTRODUCTION: OVERVIEW ON CARBON SPINTRONICS

This chapter provides an excellent introduction to graphene in the context of its significance to spintronics. The authors carefully describe early milestones and challenges in graphene spintronics. Subsequent developments related to the understanding of spin relaxation in graphene are summarized in Chapter 1, Volume 2, of this book, while recent advances in graphene spintronics can be viewed as a part of a rapidly growing field of two-dimensional van der Waals materials and their related heterostructures, presented in Chapter 5, Volume 3, of this book.

Carbon-based electronics can be separated into two categories: The first category is the "plastic electronics," which is usually based on layers of oligomers or polymers, and carrier transport is via hopping between the localized electronic states situated on adjacent molecules. The second category is based on sp^2 conjugated pure carbon, such as fullerenes ("buckyballs"), nanotubes, graphite and, last but not least, graphene. The electronic states are in principle extended, but depending on the geometry (quantum) confinement can yield 0D, 1D, 2D, or 3D systems.

In recent years, the attention for spintronics in organic systems has grown considerably (see a review on this subject in Ref. [1]). Due to the low atomic number of carbon ($Z = 6$), the spin–orbit interaction in organic molecules is usually very weak, leading to long spin relaxation times. In sp^2 conjugated carbon systems, however, carriers move with a high Fermi velocity of around 10^6 m/s, and it is not a priori clear that the simple argument from above can be used. The actual spin relaxation times and spin relaxation lengths are expected to depend on the structure (in particular, dimensionality) and extrinsic factors such as impurity potentials, defects, and edges.

Starting with 3D graphite, although spin injection and spin transport measurements on (bulk) graphite are lacking, electron spin resonance (ESR) has been used to obtain information on the g factor and spin relaxation times of the conduction electrons. From the narrow linewidths of only a few Gauss, spin relaxation times of 10–20 ns have been obtained [2, 3].

Despite the achieved successes in observing spin transport in 1D carbon nanotubes over lengths of several micrometers [1], the situation remains unclear about the ultimate values of the spin relaxation times and spin relaxation lengths which can be reached in those systems. This is connected to the difficulty in making good quality ferromagnetic contacts and excluding spurious magnetoresistance effects, which can mask or mimic the required spin-valve behavior.

The technology to study 2D graphene as an isolated layer only became available in 2005 [4]. This triggered an avalanche of experimental and theoretical studies into (mostly) the electronic properties of graphene and graphene devices. Concerning spintronics, an obvious question is how the spin-related properties of 2D graphene compare to those of 3D graphite and 1D nanotubes. This chapter will discuss in detail the technological, experimental, and conceptual aspects of spin transport and spin dynamics in graphene field-effect transistors with ferromagnetic injector and detector contacts. The results will be linked to the available theory, and future perspectives will be discussed.

4.2 GRAPHENE ELECTRONICS

Nature gives us elemental carbon in form of crystalline structures that are very different in many of their physical properties. The 3D variations of these (diamond, graphite) have been well known for a long time; 1D (nanotubes) and 0D (fullerenes) structures were discovered only some 20 years ago. One aspect that all carbon allotropes share is scientists' continuing interest in the electronic properties they display, both for fundamental research and for novel applications.

The theoretically most studied carbon allotrope is probably the 2D crystal called *graphene*, the building block of graphite and of nanotubes and fullerenes to some extent. Its hexagonal planar atomic structure (see Figure 4.1a) is a textbook example of a 2D crystal. It is simple enough to allow for band structure calculations and to be used as starting point for the more complex graphite (3D stacked graphene sheets) and carbon nanotubes (graphene sheets rolled up). In fact, its band structure was calculated already in 1947 [5] and can be described by the approximate dispersion relation in the k-space

$$E(k_x, k_y) = \pm t \sqrt{1 + 4\cos\left(\frac{\sqrt{3}}{2}k_x a\right)\cos\left(\frac{1}{2}k_y a\right) + 4\cos^2\left(\frac{1}{2}k_y a\right)}, \quad (4.1)$$

where:

    t represents the hopping integral
    a is the lattice constant

As shown in Figure 4.1, the two atoms in the unit cell are predicted to yield two band crossings per Brillouin zone in the points marked as K and K'. The

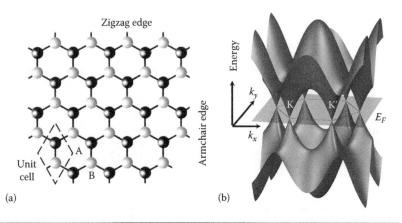

FIGURE 4.1 Graphene structure. (a) Hexagonal lattice composed of A and B sublattices. (b) Band diagram in the k-space with the valleys K, K' and the Fermi level (image by the authors).

valence and conduction bands resulting from the bands of the two equivalent sublattices A and B do not overlap (as in metals) but only touch in the charge neutrality points, resulting in a zero-gap semiconductor with a *linear dispersion* around K and K′ and two conical "valleys" where charge carriers can sit. The Fermi level (horizontal plane in Figure 4.1b) is, in the charge neutral case, situated exactly between the conduction and valence bands, that is, the Fermi surface consists only of the K, K′ points. On one hand the two valleys result in a new type of state for the electrons/holes in the graphene lattice, leading to a so-called valley degree of freedom. On the other hand, the linear dispersion resembles the relativistic Dirac fermions, for example, photons; charge carriers (electrons or holes) are moving with a (Fermi) velocity v_F independent of their energy and direction, obeying the Dirac equation $\hat{H}\begin{pmatrix}\Psi_A\\\Psi_B\end{pmatrix} = E\begin{pmatrix}\Psi_A\\\Psi_B\end{pmatrix}$ where the Hamiltonian $\hat{H} = v_F\begin{pmatrix}0 & \hat{p}_x + i\hat{p}_y\\\hat{p}_x - i\hat{p}_y & 0\end{pmatrix} = v_F\vec{\sigma}\cdot\vec{p}$ acts on the wave functions $\Psi_{A,B}$ of the sites A and B, with $\vec{\sigma}$ the Pauli matrix and $\vec{p}$ the momentum. The relativistic behavior manifests in the high Fermi velocity ($v_F \approx 10^6$ m/s) and in the effective cyclotron (transverse) mass of the particles vanishing at the Dirac neutrality point. Meanwhile, the longitudinal effective mass is $\pm\infty$, that is, the carriers cannot be accelerated or decelerated by applying a (e.g., electric) force.

In 2005, a simple but revolutionary technique was developed by Novoselov et al. at Manchester University to produce (by mechanical peeling of bulk graphite) single-layer graphene (SLG) flakes on Si/SiO$_2$ substrate on a micrometer-length scale, accessible experimentally [4].

An optical image of such single- and multilayer graphene flakes on SiO$_2$ (300 nm) substrate is shown in Figure 4.2a, accompanied by a scanning electron microscope image of the same sample contacted by metallic electrodes (Figure 4.2b). The first optically visible SLG flake (with the help of interference through a microscope) reported by the Manchester group started an avalanche of electrical transport experiments in this unique material. The existence of the valley-like dispersion relation was observed, and a minimum

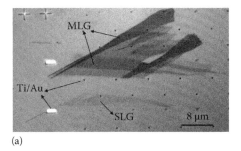

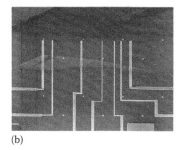

(a)

(b)

FIGURE 4.2 Single- and multilayer graphene (SLG, MLG) flakes on a Si/SiO$_2$ substrate fabricated at the University of Groningen. Optical microscope (a) shows the flakes and an array of Ti/Au markers. Same SLG flake contacted by cobalt electrodes (b), scanning electron microscope (SEM) image.

in the conductance of the graphene was measured (see Ref. [4] for single layer, respectively somewhat earlier, Ref. [6] for few layer graphene). This is illustrated in Figure 4.3 in a simple resistivity measurement on graphene using metallic contacts; the charge neutrality point is separating the metallic electron and hole conduction regimes.

The Fermi level can be raised or lowered by electrostatic gating to charge the graphene with electrons or holes at a carrier density n, creating a field-effect transistor type device. Charge carriers proved to have very high mobility μ above 10^4 cm^2/V s at room temperature (an order of magnitude larger than in Si). This value, extracted from measurements of the conductivity σ with the formula $\sigma = en\mu$, was recently further increased to 2×10^5 cm^2/V s by suspending the graphene [7–9] thus removing scattering caused by the substrate and its charged impurities [10]. Using the semiclassical relation between mobility and scattering mean free path $\sigma = en\mu = 2e^2 l \sqrt{\pi n}/h$ [11], this yields a mean free path $l \approx 50$ nm for non-suspended samples and in the range of a micrometer for the suspended ones. Further, a quantized Hall effect was detected at 4.2 K and at room temperature; its anomalous nature (due to the existence of a zero-energy Landau level) is a direct proof of the Dirac fermionic nature of the charge carriers [12]. The absence of weak localization effect was also reported [13] and the existence of percolating electron–hole puddles in the charge neutrality regime was proposed that suppress the localization and limit the conductivity. The puddles were recently measured to yield a finite local carrier density of a few 10^{11} cm^{-2} using scanning single electron transistor technique [14] and scanning electron spectroscopy [15, 16]. Bipolar supercurrent transport was also realized through graphene via the Josephson effect [17] using superconducting contacts. Relativity combined with quantum electronics of the charge carriers in an easy-to-produce 2D conductor makes the graphene a unique material to do fundamental research on. It presents richer physics than 1D (semi)

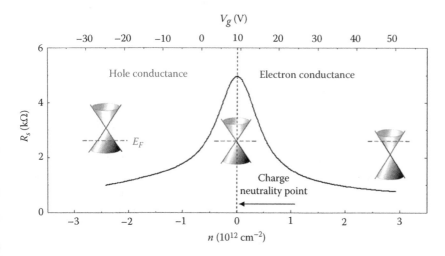

FIGURE 4.3 Sheet resistivity (square resistance) of a graphene flake measured against backgate voltage V_g (top scale). The induced charge carrier density n is shown on the bottom scale.

conductors, for example, carbon nanotubes, but does this without the band complexities of a 3D system.

The fascination with graphene does not stop, however, at condensed matter physics. For instance, the movement of the Dirac fermions in curved graphene sheets seems to conveniently resemble the behavior of relativistic quantum particles in curved spaces, offering a realistic and unprecedented experimental environment for cosmologists [18].

From the applications point of view, the convenient position of the Fermi level, the selectivity of the carrier nature (electrons versus holes), very high room temperature mobilities, electric current densities above 10^8 A/cm^2 [8], and a conductance tunable by electrostatic gating are favorable features not to be met in metallic or semiconductor thin layers, and give us a hint on future electronic devices. Carbon replacing silicon in the micro- and nano-electronics industry and single atomic thin devices is one of the (optimistic) predictions of what could be achieved in the next 50 years [19]. Additionally, the valve-type device proposed by Rycerz et al. [20] based on the selectively occupied two-valley states in graphene could lead to a new type of electronics called "valleytronics" by the authors. The very robust structure at elevated temperatures and the resistance of carbon to corrosive environments are also an important advantage. The resistivity of graphene being highly sensitive to adsorbed molecules (via charge transfer or dipole interactions), the possibility of gas- and single molecule sensing applications is being explored as well [21, 22]. The reversible hydrogenation of graphene (transforming the material into *graphane*, a 2D insulator, see, e.g., Ref. [23]) or attaching other atoms and molecules is being considered as a way to create new quasi-2D crystals with designed electronic properties.

The bottleneck in using SLG on an industrial scale is, however, the mass fabrication of high (electronic) quality flakes. The "scotch tape method" (mechanical exfoliation of bulk graphite) has enough yield for doing fundamental research on, yet it is obviously inadequate for anything more. In the last years, an immense progress is to be seen from worldwide chemistry and surface science research labs, different ways are explored for graphene production on insulating, metallic, and arbitrary surfaces. One direction that shows promising results is to chemically exfoliate graphite by oxidizing it, and then disperse the obtained graphene oxide sheets in solvents that can be used to coat by Langmuir–Schaefer [24] or Langmuir–Blodgett [25–27] technique large surfaces with single or multilayer flakes. This is followed by chemical reduction to graphene and the advantage is the control over the thickness, the coverage ratio, and the choice of substrates; however, the electronic quality of the obtained layers (low carrier mobilities, low conductance) still needs improvement. Other researchers look into the direction of epitaxial graphene growth on silicon carbide via silicon sublimation [28, 29] and chemical vapor deposition on metallic single- and polycrystalline surfaces (e.g., copper or nickel) with good lattice matching followed by chemical removal of the substrate and transfer to arbitrary surfaces [30–32]. Note that small-scale graphene growth on metals and on SiC has a long history; for a review of the early work, see Ref. [33].

4.3 ELECTRON SPIN POLARIZATION IN SINGLE-LAYER GRAPHENE

We will discuss here the effects related to spin transport in graphene. For further background, we refer to Chapters 5, Volume 1, and Chapter 11, Volume 3, on metallic spintronics and the non-local measurement technique (this will be adopted for the particular case of graphene spin valves as we will explain it later). Relevant also are Chapters 1–3, Volume 2, that treat the spin injection and relaxation processes in(to) semiconductors, as well as Chapter 6, Volume 2, on non-local effects in semiconductors. Finally, Chapter 5, Volume 2, on organic and Chapter 11, Volume 3, on molecular spintronics are recommended for understanding spintronics in graphitic devices.

Assuming the existence of a localized spin accumulation (spin imbalance) $\vec{n}_s$, proportional to the induced magnetization, in a non-spin-polarized sea of charge carriers in a graphene flake at position $x,y,z = 0$ and at time $t = 0$, created by one or another experimental method (e.g., electrical injection from a ferromagnetic contact), we can describe its spatiotemporal evolution in the diffusive limit applying basic concepts of spin transport [34]. First of all, the graphene will be considered 2D. A spin diffusion characterized by the diffusion coefficient D_s will lead to the spread of spin imbalance in the entire flake; this coefficient can be considered isotropic in the plane. Further, spin–orbit interactions will limit the lifetime of the spin imbalance (and by this, also the length scale on which the imbalance can diffuse); the scattering of spins on the local spin–orbit fields they experience is characterized by the spin relaxation time (or spin flip time) τ that includes both the spin lifetime and the dephasing time commonly denoted as $T1$ and $T2$, respectively. The characteristic length scale over which the spin imbalance can still be observed is the spin diffusion length (or spin flip length) $\lambda = \sqrt{D_s \tau}$ and it is isotropic in the (x,y) plane for an ideal graphene crystal. The presence of an external magnetic field $\vec{B}$ is accounted for by considering spin precession due to the interaction between the spins diffusing in the graphene plane and the perpendicular component of the magnetic field B_z, an interaction quantified by the gyromagnetic ratio $\gamma = g\mu_B/\hbar$ where the Landé factor is considered to be close to its free-electron value $g \approx 2$ [35]. Finally, an in-plane electric field $\vec{E}$ will lead to drift of the charge carriers of mobility μ in the graphene crystal with a velocity $\vec{v}_D = \mu\vec{E}$ parallel (holes) or antiparallel (electrons) with the field direction. This effect could not be observed in metallic spin transport, and it will add an asymmetric term to the diffusive motion of the spins as we will discuss it in detail in Section 4.5.4. (Note that this consideration is valid only for the high-density conduction regime. At the Dirac neutrality point, the mobility and, therefore, the drift velocity are ill-defined.)

To summarize the above, we can write the Bloch equation for spin accumulation completed with the drift term under an electric field

$$\frac{\partial \vec{n}_s}{\partial t} D\nabla^2 \vec{n}_s - \frac{\vec{n}_s}{\tau} + \left(\gamma\vec{B} \times \vec{n}_s\right) + \vec{v}_D \nabla \vec{n}_s. \tag{4.2}$$

Here, $\vec{n}_s = (n_s, n_y, n_z)$ describes the spin accumulation (spin imbalance) in three directions in terms of densities for each spin. In experiments, the spin electrochemical potential $\vec{\mu}_S$ is what can be measured using, for example, ferromagnetic electrodes (see later in Section 4.5), and the two are correlated via the density of states at the Fermi energy obeying $\vec{\mu}_S = V(E_F)\vec{n}_s$.

Practically, what we will need further is to calculate the spin imbalance at a certain distance L from the injection point along direction x (experimentally, this will be the injector–detector distance), that is, to find the solution to the 1D Bloch equation in the presence or absence of magnetic and electric fields. In the simplest case of spin diffusion without external fields and a constant rate of spin injection (yielding a time-independent Bloch equation), the solution is

$$n_s(x) = A\exp\left(\frac{x}{\lambda}\right) + B\exp\left(\frac{-x}{\lambda}\right), \tag{4.3}$$

a symmetric exponential decay of the spin imbalance on both sides of the injection point. This is not specific to graphene and it was measured (in metallic spin valves) and explained in Johnson and Silsbee [36, 37] and Jedema et al. [38]. In this chapter, we will only address the method to detect the spin injection and transport in graphene and model the results by the specific solutions of Equation 4.2 in multiterminal non-local geometries using ferromagnetic electrodes. The results will allow extracting parameters related both to the injection/detection of spins (the spin polarizations that play a role) and the spin transport in the graphene (the diffusion coefficient and the spin flip time and length). This will allow understanding the mechanisms behind the spin relaxation in graphene and give us a hint whether graphene-based spintronics is a realistic concept in modern nanotechnology.

4.4 BUILDING GRAPHENE-BASED MULTITERMINAL SPIN VALVES

Spin transport in graphene was measured at several research labs [39–47] with slightly different approaches and results. One aspect that is common to all experiments reported so far is the electrical injection/detection scheme based on ferromagnetic electrodes. The first step is to obtain graphene flakes, single- or multilayers, on an insulating substrate. This is followed by lithography steps, metal deposition, and liftoff to create the electrodes on a scale of a micrometer or smaller, a well-known technology in nanofabrication. A scanning electron microscope (SEM) image of a lateral SLG spin-valve device is shown in Figure 4.4a.

4.4.1 PREPARING THE SUBSTRATE

Prior to graphene production, a substrate (typically a strongly doped Si wafer covered with 300–500 nm thermal oxide) is prepared. The oxide on the back of the wafer (if any) is removed and a thick metallic layer, for example, 100 nm

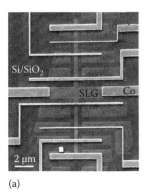

(a)

(b) Without Al_2O_3 With Al_2O_3

FIGURE 4.4 SEM images: (a) Cross-shaped graphene device etched with oxygen plasma from a large polygonal flake and contacted by Co electrodes. The etch channels are visible. (b) Comparison of graphene and SiO_2 surfaces bare (left) and with the aluminum oxide cover layer (right). The two images were made using the same SEM parameters on the same sample in two different areas. (Images by the authors.)

Ti/Au is deposited that can be used as electrode to the Si. The Si serves this way as a backgate for electrostatic charging of the graphene with holes or electrons. Further, an array of 30–50 nm thick Ti/Au markers is deposited on the silicon oxide by means of lithography and liftoff (see bright square in Figure 4.4a; the method is explained in the next paragraphs) that will help locate the graphene flakes later on. These steps result in light contamination of the oxide surface, therefore a cleaning can be performed by reactive etching in oxygen plasma [48], annealing in argon/hydrogen gas flow at 200°C–400°C [49], ultrasonication, and/or wet chemical cleaning.

4.4.2 GRAPHENE PRODUCTION IN THE LAB

In terms of electrical transport measurements, the most studied SLG is that obtained by the exfoliation technique. A simple scotch tape is used to peel off layers from a bulk graphite source. This can be natural graphite, highly oriented pyrolytic graphite (HOPG) or Kish graphite. The slightly different electronic properties of the graphene flakes from different sources are most probably related to the contaminations present in the bulk graphite and the crystalline quality of the graphene flakes that form the bulk. There is no consensus yet whether spin lifetimes and diffusion depend on the type of graphite used. Once a clean, fresh graphite surface is obtained, this is pressed against the prepared substrate. Upon removal of the scotch tape (by peeling off or dissolving its glue in an organic solvent), thin graphite flakes are left behind on the substrate, of lateral dimensions of micrometers up to a millimeter (see Figure 4.2a for an example). These are weakly attached to the oxide surface by van der Waals interaction, the strength of the interaction depending mainly on the cleanness and hydrophobicity of the surface. The thickness of the flakes (i.e., the number of graphene layers stacked) varies from 0.3–0.5 nm to μm, therefore a rigorous inspection is needed to select the single-layer samples. Two important steps are involved here. First, the number of single to few layer flakes is relatively high compared to thicker graphite pieces. The reason behind this is most probably due to the interplay of adhesion and cohesion forces [50]. Nevertheless, the flakes are only weakly attached to the substrates; detaching of large flakes (of above $100\,\mu m^2$ area) when dipped into solvents with high surface tension is a common problem and leads to loss of samples during further device fabrication steps [51]. The second surprising fact is that a single layer of carbon atoms can be made visible under an optical microscope. For certain light wavelengths, an interference occurs between the graphene and the opaque Si substrate, provided that the SiO_2 layer is of the right thickness. The effect decreases the reflection of light from the graphene/SiO_2/Si surface giving a relatively high contrast with the regions where no graphene is present. A detailed quantitative explanation for a 285 nm thick oxide is given by Blake et al. [52] and, for a range of oxide thicknesses by Tombros [53]. The preselecting can thus be done by selecting the least visible graphene flakes on the surface. The technique can be quantified by recording 8 bit grayscale digital optical images of the surface that contains flakes of different reflectivity, measuring the brightness value on the flakes Lg as well as on the substrate in the immediate vicinity of the flakes Ls, and calculating a contrast value $c = (Ls - Lg)/Ls$. Groups of discrete contrast levels can be identified this way starting from a few percent and increasing with increased thickness of the flakes. Care must be taken that the images are recorded with the same parameters (microscope aperture and magnification, exposure, wavelength of light and contrast). Following this preselecting, typical flakes from the groups are inspected by Raman spectroscopy to determine the number of layers based on the shift and width of the 2D and G peaks [54], allowing finally for a calibration of the optical microscope selection. Another method to show that a flake is SLG is to measure the half-integer quantum Hall effect that indicates the presence of Dirac fermions in the 2D lattice; however, this is cumbersome to do for all samples. Additionally, atomic force microscope (AFM) can be used to scan the surface of the graphene flakes; this serves the purpose of accurate localization of the flake and verification of its integrity and cleanness. AFM is also used to measure the thickness of the flake determining this way the number of layers, a method much debated since adsorbed contamination (e.g., water molecules) on or under the graphene flake can give erroneous thickness values for single layers [55]. Ideally, an SLG measures 0.3–0.5 nm thick.

The shape and lateral dimensions of the graphene flake can be controlled if necessary by plasma etching down to a 50 nm scale (see in Refs. [48, 56–58]) or etching on a sub-10 nm scale with thermally activated nickel nanoparticles [59]. Recently, controlled rupture of graphene involving heating–cooling cycles was also demonstrated [60] as a method to obtain graphene quantum dots of approximately 10 nm diameter. For spintronics experiments, the RF plasma etching method was employed, using argon, oxygen, or a mixture of the two gases. At 10 microbar base pressure, a flow of approximately 10 sccm of gas is ignited to form a plasma that etches graphene with a typical rate of around 0.1 nm/s. To protect parts of the graphene flake, an electron beam lithography (EBL) step is involved as follows. A thin (~100 nm) polymer resist layer is spun on the Si wafer with the graphene on it [e.g., poly(methyl methacrylate), PMMA, dissolved in *n*-amyl-acetate] and baked to form a continuous protective layer. The regions to be etched are first exposed by focused electron beam and the damaged polymer is removed using methyl isobutyl ketone solvent ("development" step). This opens a gap in the protective polymer layer allowing for the plasma to remove the parts of the graphene with an accuracy of about 50 nm. Finally, the rest of the polymer is removed by a hot acetone bath. In Figure 4.4a, an SLG etched with oxygen plasma into a cross shape is visible; the original shape of the flake can be seen from the leftover pieces (gray) on the dark Si/SiO_2 substrate.

4.4.3 Fabrication of Electrical Contacts

Next, a thin aluminum oxide (Al_2O_3) layer is created that covers the graphene flake completely or partially. The need for this will be explained in Section 4.5.2. This is done by thermal evaporation of 0.6 nm thick aluminum in an ultrahigh vacuum system with the substrate being liquid nitrogen cooled. The aluminum is then oxidized for 30–60 min in an oxygen atmosphere of about 100 mbar. The resulting oxide layer shows a roughness of 0.5–0.7 nm on the graphene surface, determined by AFM; the granular nature can be seen upon inspection with scanning electron microscope (SEM). In Figure 4.4b, two high magnification SEM images are shown, made with identical parameters on the same sample. The image on the left shows a region where no Al was deposited (due to shadowing in the evaporation system); the one on the right in comparison, illustrates the presence of the aluminum oxide both on the multilayer graphene (MLG) and on the silicon oxide.

The sequence of graphene etching and creating the aluminum oxide layer can be interchanged; this seems to decrease the chance of the graphene flake detaching from the substrate in further processing steps.

Finally, the ferromagnetic contacts are created using another step of EBL and liftoff. This process is similar to the etching described above, except the development step is followed by thermal or electron beam evaporation of a 25–40 nm thick Co layer onto the sample. The areas where the polymer is removed by development will be the contact areas where the Co grows directly on the graphene/Al_2O_3. The rest of the metallic layer is removed in acetone together with the unexposed polymer ("liftoff" step). The resolution of the technique allows creating contacts of lateral sizes below 50 nm with similar positioning accuracy. Optical lithography can also be used to speed up the fabrication process; however, its resolution is limited to a (few) hundred nanometers by the wavelength of light used and the precision of the optical mask.

4.5 SPIN TRANSPORT EXPERIMENTS IN LATERAL GRAPHENE SPIN VALVES

The simplest lateral device geometry to detect spin transport electrically is a two-terminal spin valve where the graphene is contacted by two ferromagnetic electrodes [34]. The electrodes are fabricated to have different coercive fields (e.g., by shape anisotropy that varies with the width) and therefore their magnetization can be oriented parallel or antiparallel with an external in-plane magnetic field. Such devices were built and studied by Hill et al. [39], using transparent permalloy contacts on the graphene. Magnetoresistances of a few hundred ohms were observed over an injector–detector distance of 200 nm, yet no clear distinction could be made between parallel and antiparallel configurations and the presence of spurious effects such as anisotropic magnetoresistance or Hall effect cannot be excluded.

Separating the charge current path from the voltage detection circuit and relying solely on the diffusion of spins through the device will circumvent the above problems. This was successfully performed in a four-terminal non-local geometry on metallic spin valves [36, 38], semiconductors [61], carbon nanotubes [62], and also on graphene [40, 44]. The non-local spin injection–detection circuit and mechanism is explained with the help of Figure 4.5.

A spin-polarized AC charge current I_{AC} is sent from the ferromagnetic electrode into the graphene layer through contact F3 and extracted on the far right edge of the device through F4, creating a spin accumulation P in the graphene region around F3. The spin accumulation diffuses through the graphene on both sides of the injector F3, not being tied to the charge current path. Placing a pair of ferromagnetic voltage probes F1, F2 on the left-hand side where F2 is close to the injector F3 (at a distance $L < \lambda$, the spin diffusion length), this will be sensitive to an accumulation of spins parallel to its magnetic orientation. Since F1 is further, it will detect a lower spin accumulation if any, resulting in a voltage difference $\Delta V = V_+ - V_-$ measured between F2

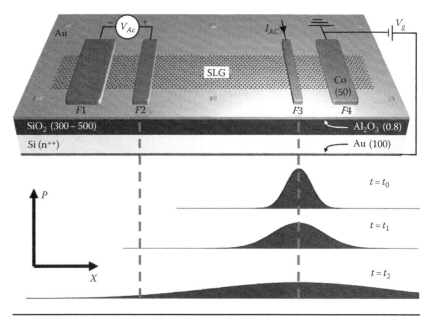

FIGURE 4.5 Non-local four-terminal spin injection geometry. P represents the injected spin polarization. See text for details.

and F1 without any charge current passing through this region. Switching the magnetization direction of the injector F3 or detector F2 by sweeping a magnetic field oriented parallel/antiparallel with the electrodes' long axes (axis y in our geometry) will change the measured voltage from positive (parallel magnetizations) to negative (antiparallel magnetizations) values. A non-local resistance can be defined as $R_{non-local} = (V_+ - V_-)/I_{AC}$. Additionally, if the outer electrodes F1, F4 are also ferromagnetic and within the distance of a few spin diffusion lengths, they will also contribute to the measured voltage by injecting or detecting spin polarization. The idea is illustrated in Figure 4.6 using the spin electrochemical potential $\mu_S = \mu_\uparrow - \mu_\downarrow$; the potentials are drawn separately for the spin-up ($\mu_\uparrow$) and spin-down ($\mu_\downarrow$) channels against the x-axis, normalized to the average.

In Figure 4.6a, we see the starting situation with all ferromagnetic electrodes oriented parallel, in the "up" direction. (A spin-valve measurement corresponding to the situation we will describe here is presented in Figure 4.7b; however, the first two steps—the switching of electrodes F4 and F1—are indiscernible in the measurement.) At moment t_0, the electrode F3 injects a spin-polarized current with the majority spin oriented "up," raising therefore the electrochemical potential for the spin-up channel (solid line) and creating a spin-up accumulation of electrons in the graphene (with the corresponding deficit of spin-down electrons) that decays exponentially along x. In the same time, electrode F4 extracts spin-up carriers, that is, creates a spin-down accumulation, rising μ for the spin-down channel (dotted line) and reducing the overall spin-up accumulation in the graphene. At distances $L \gg \lambda$ from the injectors, the spin-up and spin-down electrochemical potentials are overlapping (spin imbalance is decayed, $\mu_S \approx 0$). The detectors F2 and F1 are both sensitive to the spin-up channel (see dots) but

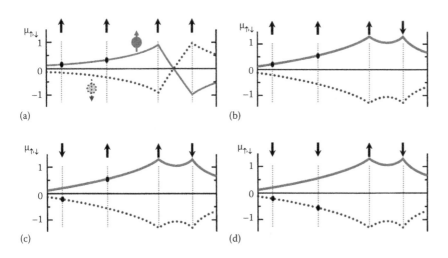

FIGURE 4.6 Electrochemical potential plotted against the x-axis for the spin-up (solid line) and spin-down channel (dashed line) in a four-terminal spin-valve device. The vertical arrows represent the magnetic orientation of the four ferromagnetic electrodes. These are switched one by one through steps (a–d).

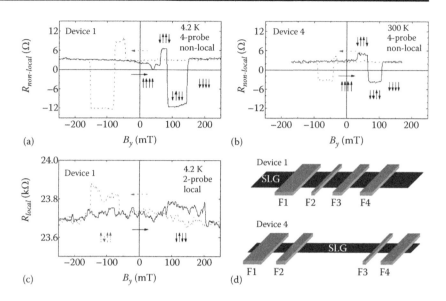

FIGURE 4.7 Spin-valve measurements on SLG in the four-terminal non-local geometry at low temperature (a) and room temperature (b) and in the two-terminal local geometry at low temperature (c). The magnetic field is swept as indicated by the horizontal arrows; the vertical arrows indicate the parallel/antiparallel combination of the electrodes' magnetizations. On (d) a sketch of the devices is drawn.

they probe this at different distances from the injectors, therefore a positive voltage difference ΔV and a positive non-local resistance will be measured between them.

Switching the magnetization of the injector F4 (Figure 4.6b) changes the spin imbalance in the device: F4 will extract now spin-down carriers, meaning the spin-up accumulation generated by F3 will be enhanced. As a result, the $\mu_\uparrow$ curve will be slightly lifted up and the difference detected by F2 and F1 will increase. Switching the magnetization of the outer detector F1 will result in the situation depicted in Figure 4.6c: F1 will probe the spin-down channel, therefore the voltage difference and corresponding non-local resistance will further increase. The next step is reorienting the inner voltage probe F2's magnetization to "down" (panel d). This will set the inner injector/detector magnetizations antiparallel. Both detectors are sensitive to the spin-down channel now, yielding $\Delta V < 0$ (the non-local resistance drops to negative values). The last step is to switch the magnetization of F3 "down"; this gives an all-parallel situation equivalent to the starting configuration with $\mu_\uparrow$ and $\mu_\downarrow$ interchanged (not shown). Therefore, the measured signal returns to the (positive) starting value.

The magnetic switching order of the four contacts is not-necessarily the one that we discussed here. In such experiments, it is governed by the width of the ferromagnetic electrodes as introduced earlier for the two-terminal case. The steps in the non-local resistance value, however, can always be traced back to the switching of individual electrodes through simple schematic representations as in Figure 4.6.

4.5.1 INJECTION/DETECTION OF SPIN-POLARIZED CURRENTS

The first set of non-local spin-valve measurements done on graphene-based spin valves at low and room temperatures was presented in 2007 by Tombros et al. and reproduced since by several groups. The examples we show here are measured in the metallic conduction regime (carrier densities in the 10^{12} cm^{-2} range); however, it was shown [40, 44, 48, 58] that the spin valves behave very similarly around the Dirac neutrality point with only a decrease of the signal amplitude. The behavior of spin injection and transport parameters in a broad density range will be discussed in Section 4.6. In Figure 4.7a, we show a spin-valve measurement at 4.2 K on a 1.4 μm wide graphene flake with uniform electrode spacing of 330 nm. A bipolar behavior and three resistance steps can be clearly identified corresponding to magnetic switches of three electrodes. The magnetic field is swept in the direction indicated by the horizontal arrows; the switching order of the electrodes is shown by the four vertical arrows for each resistance value. Non-local differential resistances up to 20 Ω were detected using AC currents of 100 nA–5 μA. Compared to the local two-terminal measurements on the same sample (Figure 4.7c) the non-local signal is considerably smaller; however, the signal-to-noise ratio is much better due to the absence of a large background resistance (the resistance of the contacts and the graphene flake itself are always included in two-terminal measurements).

A similar non-local spin-valve measurement is shown in Figure 4.7b (same scale as a); this data is obtained, however, at room temperature on a different device with 3 μm spacing between F2 and F3. (The devices are sensitive to temperature changes and therefore it proves to be very difficult to measure the same device in a broad temperature range.) The shape of the $R_{\text{non-local}}$ versus B_y curve looks similar; again, three resistance steps are clearly distinguishable. Compared to the low-temperature measurements, we see a slight increase in the noise and a reduced non-local signal due also to the larger injector–detector spacing. The observation of the resistance steps for a device where the electrode spacing is in the 3–5 μm range indicates a spin diffusion length of several micrometers. At room temperature, this value is very high, almost an order of magnitude above metallic spin valves, and carries the potential of spintronic applications where it is crucial to conserve spin information over micrometers (see Volume 3 of this book and especially Chapter 17: "Spin Logic Devices" by H. Dery).

A 3D model for the non-local resistivity measurements was given by Jedema et al. [63] for metallic spin valves. We modify this description for use in 2D on our graphene spin-valve devices and obtain the formula

$$R_{\text{non-local}} = \frac{P^2 \lambda}{2W\sigma} \exp\left(-\frac{L}{\lambda}\right), \tag{4.4}$$

where:

- W is the width of the graphene
- σ is its conductivity ($\sigma = 1/R_S$, to be determined from resistivity measurements similar to Figure 4.3)
- L is the distance between the central injector–detector electrodes
- P is the spin polarization of the contacts

The exponential dependence on L originates from the exponential decay of the spin imbalance as we move away from the injection point (Equation 4.3). The non-local resistance is also called *spin signal*; the spin signal difference between the parallel- and antiparallel geometries is the *spin-valve signal* and, ideally this is twice as large since the spin-valve measurement should be symmetric around zero. One way to determine the spin diffusion length λ is to verify the length dependence of the non-local signal. Several spin-valve measurements are needed on a set of samples with increasing injector–detector distances L while the rest of the parameters in Equation 4.4 are kept the same. This was attempted [40, 46, 48] yet it is very difficult to have the polarization P unchanged from sample to sample, even from contact to contact. The results indicate a spin diffusion length between 1 and 2 μm. Measuring the graphene's conductance σ separately, we can calculate a contact spin polarization $P \approx 10\%$ (both for injector and detector, therefore it appears squared in Equation 4.4). For a precise determination of the spin diffusion length and other spin transport parameters, one can use a more convenient technique based on the Hanle precession of the injected spins; this is related to the diffusive spin transport in graphene and thus it will be described later in Section 4.5.3.

The value of P is related to the nature of the ferromagnetic electrodes (the intrinsic spin polarization of the material) as well as to the nature of the contact between the electrode and graphene, and it is a measure of how efficiently one can inject a spin polarization into the graphene. In the following paragraphs, we will present a method to manipulate the spin injection efficiency into graphene.

4.5.2 EFFICIENCY OF SPIN INJECTION

In the diffusive transport regime, direct electrical spin injection from a ferromagnetic metal into high-impedance materials like graphene is hindered by the conductance mismatch problem. The concept of conductivity mismatch is discussed in Schmidt et al. [64], Rashba [65], and Fert and Jaffrès [66]. In simple terms, the spin accumulation created at a clean ferromagnet/graphene interface favors flowing back into the low-impedance metal injector instead of diffusing through the high-impedance graphene, therefore the efficiency of the injection is strongly reduced in such experiments [39, 41]. A 100% spin polarized injector would eliminate this problem; however, the intrinsic spin polarization of ferromagnetic metals is typically below 50% as measured in tunneling experiments by Tedrow and Meservey [67]. Another solution to improve efficiency is to reduce the contact area to ultimately a point contact [68]; this was implemented by Han et al. yielding clean, measurable non-local spin signals in the milliohm range and a polarization value $P \approx 1\%$ [46]. Yet another solution is the use of a tunnel barrier between the ferromagnet and graphene, increasing the spin-dependent interface resistance and by this combating the conductivity mismatch [65, 66]. Such a technique was successfully employed for spin injection into *n*-doped GaAs by means of a spin-sensitive Schottky tunnel barrier [61]. The role of the barrier in the graphene spin-valve devices we present here is taken by the 0.8 nm thick aluminum oxide layer that was mentioned earlier in Section 4.4.3. The high contact resistance limits the back-diffusion of spins into the injectors and allows the spin accumulation to pass underneath the opaque contacts; however, the barrier should be transparent enough to result in measurable (<1 MΩ) resistances. As we have calculated, this gives contact spin polarizations around 10%. Experimentally, it was found that contact resistances from 50 to 100 kΩ result in the highest efficiencies of spin injection, up to 18% [48, 69]. Recently successful experiments were reported [105 on SCG, respectively 106 and 107 on BLG] by using oxides other than Al_2O_3 (e.g., MgO, TiO_2-seeded MgO) that yield even higher spin polarizations at the interface reaching 26%–30% and, consequently, record levels of non-local spin-valve signals up to 130Ω.

A model for spin injection/detection in non-local ferromagnet–barrier–graphene–barrier–ferromagnet structures is constructed [48] that accounts for the effect of the conductivity mismatch in case of semitransparent contacts. It is shown that the magnitude of the spin accumulation created in graphene is drastically reduced if the contact resistance is smaller than the graphene resistance calculated over one spin relaxation length; the formula for the non-local resistance (4.4) is modified to

$$R_{\text{non-local}} = \pm \frac{2P^2\lambda}{W\sigma} \frac{(R/\lambda)^2 \exp(-L/\lambda)}{(1+2R/\lambda)^2 - \exp(-2L/\lambda)}, \qquad (4.5)$$

with the parameter $R = \sigma R_C W$ representing the spin relaxation due to the finite contact resistances R_C of the electrodes. Additionally to the prediction of a reduced non-local spin-valve signal, the model also shows how its exponential dependence on L from formula (4.4) becomes a $1/L$ type dependence for transparent contacts ($R < 10^{-6}$ m) and injector–detector distances $L \leq \lambda$.

The 0.8 nm thick oxide layer we employ is not a uniform tunnel barrier but an insulating layer with some relatively transparent regions on the nanometer scale. This is indicated, besides the AFM and SEM studies, both by the spread in the contact resistances within one sample, as well as by the I–V characteristics that do not show a tunneling behavior [48]. However, the model yielding Equation 4.5 is valid both for tunnel barriers and for pinholes in the oxide.

The interface injection/detection efficiency can be manipulated in a finalized spintronic device by electrical bias on the contacts. A study of the interplay of mechanisms behind electrical spin *detection* in lateral Fe/GaAs structures (again, using Schottky tunnel barriers) with biased *detector* contacts was recently reported [70]. The authors demonstrated the bias dependence of the tunneling spin polarization (interface effect) and of the spin transport in the GaAs before the spins reach the detection point (bulk effect). Simultaneously, a set of non-local experiments were done on graphene spin valves by Józsa et al. (with the 0.8 nm Al_2O_3 interface layer) where the *injector* electrodes were DC biased showing a change in the room temperature spin *injection* efficiency from 18% to 31% [69]. This resulted in non-local spin-valve signals up to 85 Ω. When a reverse DC bias was applied, the same device showed zero or even negative differential values for R_{SV} (the spin-valve signal reversed). A selection of these measurements is plotted in Figure 4.8a, for a variety of current biases between −5 and +5 μA. The bipolar non-local spin-valve effect is visible on all curves, and the resistance steps are clear enough to identify the switching of the two inner ferromagnetic electrodes F2, F4 from parallel to antiparallel orientation and back.

The spin-valve signals (parallel minus antiparallel) measured by the voltage probes F1, F2 are plotted in Figure 4.8c against the applied DC bias. A steep rising curve is formed, with saturation starting both for positive and negative high biases; however, further increase of the current bias was not possible due to danger of permanent damage on the graphene/oxide/cobalt interface.

The physics behind the measured biasing effect can be understood considering the relatively transparent injection regions in the oxide layer. In graphene directly around these injection points, a strong current spreading is present, on a characteristic length scale **L** (see Figure 4.8b). Here, the local electric field is large, and the carriers will undergo a strong local drift. The carriers drifting away from the injection point (for a positive DC bias and p-type graphene) reduce the backflow of the spins and thus facilitate a

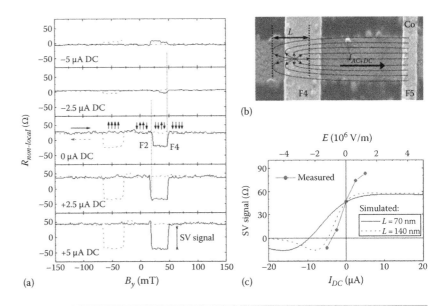

FIGURE 4.8 Spin-valve measurements with DC-biased current injector electrodes F4, F5 (a), the bias current being indicated for each measurement. The contacts are F1, F2, F4, F5; for these measurements F3 is not in the circuit. Schematic drawing of the local drift effect (b) using an SEM image of a sample. The change in the spin-valve signal is plotted on (c) against the DC bias values (bottom scale), together with simulated curves for two different values of L against the electric field (top scale). See text for details.

constant injection of high spin polarization through the interface. The upper limit of the effect is a measurement of the intrinsic spin polarization P possible to inject from the ferromagnetic electrodes, when conductivity mismatch is eliminated; the signal is reduced of course by the further diffusion and relaxation toward the contacts. On the other hand, an opposite electric field polarity (or carrier type) will result in a carrier drift toward the injection point keeping the local spin polarization high, reducing the efficiency of injection. A strong negative bias enhances the conductivity mismatch problem to a point when the drift starts to dominate the spin transport. In AC measurements, this yields a *negative differential resistance* (lower spin injection for stronger electric fields), and thus the reversed spin-valve signals in Figure 4.8 (top left). Finally, above a threshold electric field value, the strong drift effect is expected to prevent any spin imbalance to reach the detectors reducing the spin-valve signal to zero.

The complicated 2D current spreading around the injection point is modeled in Józsa et al. [69] by an approximation with a 1D drift-diffusion of the spins on the drift length scale **L** along axis x. The spin current density flowing through the region due to diffusion and drift can be written as

$$j_s(x) = -D\frac{dn_s(x)}{dx} + v_D n_s(x), \tag{4.6}$$

using the same diffusion and drift parameters as in Equation 4.2. The coupling at the edges of the drift region yields boundary conditions that on one hand involve the injection parameters such as contact resistance and spin polarization of the current source; on the other hand, assure the spin current continuity into the diffusion region, toward the detector electrodes. The model predicts a spin-valve signal on the detectors as plotted in Figure 4.8 right, solid line (drift region with **L** = 70 nm as parameter) and dashed line (**L** = 140 nm). The horizontal axis of the graph (the electric field, top scale) is connected to the measurement horizontal scale (the DC bias current, bottom scale) by the linear relationship $E = I_{DC}R_S/w$ where the graphene resistivity R_S is known but the total width of the injection regions w is not. The vertical scales are the same. Comparing quantitatively the simulations and the measurements, we conclude that electric fields in the $\pm 10^6$ V/m range must be present in the drift region, yielding a drift velocity up to 0.25×10^6 m/s. The width of the injection regions $w < 5$ nm is in this case consistent with the values based on the structure of the oxide layer.

The local drift diffusion in the graphene region around the injection points is similar to the spin drift-diffusion phenomenon on a larger scale, observed in experiments where an electric field is present in the whole graphene layer. This will be discussed in Section 4.5.4.

4.5.3 HANLE SPIN PRECESSION

In the simple four-terminal non-local geometry, injection of the spin imbalance is followed by diffusion and relaxation through the graphene layer. The spin diffusion coefficient D_s and the spin flip time τ are linked via the spin diffusion length $\lambda = \sqrt{D_s\tau}$ and they cannot be determined individually from simple spin-valve experiments (see Equations 4.4 and 4.5). An experiment that allows us to do so is the Hanle spin precession experiment [71]; additionally, it also represents a solid proof for spin injection and transport in lateral geometries [61]. The Hanle spin precession is simply the Larmor precession of the injected spin imbalance in a lateral spin valve, under the influence of a (small) perpendicular external magnetic field B_z. An illustration is given in Figure 4.9 insets. For magnetic fields that are well below the demagnetization field of the injector/detector electrodes F2, F3(left inset) the spin injection is still predominantly in-plane and the spin imbalance, while diffusing through the graphene plane, undergoes a precession and dephasing that is governed by the cross-product term in the Bloch Equation 4.2. The detector signal will, therefore, depend on the starting situation (parallel or antiparallel injector–detector electrode magnetizations), the strength of B_z, the injection efficiency (see previous section), the spin dephasing time τ, the spin diffusion coefficient D_s, and the injector–detector distance L. As the magnetic field is increased, the precession angle achieved over the distance L will increase and the signal will present an oscillatory behavior that is dampened due to dephasing and spin relaxation. Further increasing the magnetic field will cant the magnetizations of the electrodes out of the sample plane and, finally, align them with the z-axis. In this situation (see right inset in Figure 4.9), the injected spin polarization is out of plane along z; therefore no precession will occur and the detector signal will saturate at a positive value that depends on the *out-of-plane* spin relaxation time $\tau_\perp$.

Two room temperature Hanle precession measurements (non-local resistance versus orthogonal magnetic field) are plotted in Figure 4.9 in a graphene spin valve with $L = 3.1$ μm and carrier density $n \approx 5 \times 10^{11}$ cm^{-2} (holes),

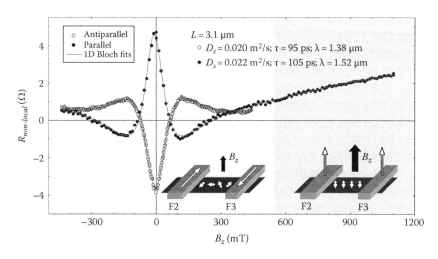

FIGURE 4.9 Hanle precession measurements for low and high magnetic fields perpendicular to the sample plane for $L = 3.1$ μm injector–detector distance. The geometry is indicated in the insets. The parallel and antiparallel configuration measurements are fitted with the Bloch Equation 4.2 to extract the spin diffusion coefficient D_s, relaxation time τ, and diffusion length λ.

one for parallel configuration of the F2, F3 magnetizations (full circles) and another one for antiparallel (open circles). The signal difference between the two curves at $B_z = 0$ corresponds to the spin-valve signal measured with in-plane magnetic field. The parallel configuration measurement is continued for high magnetic fields up to +1.1 T (the measurement is symmetric around the y-axis) and the onset of the signal saturation can be seen.

The curves are fitted with the solutions to the Bloch Equation 4.2 (solid lines in Figure 4.9) and the spin transport parameters are extracted. Aside of a small difference between the curves in the parallel and antiparallel geometry, the fit yields a spin relaxation time $\tau \approx 100$ ps, a diffusion coefficient $D_S \approx 0.02$ m^2/s and a spin diffusion length $\lambda \approx 1.4$ μm. The value, confirmed by measurements on many similar graphene samples [40, 46–48, 69], corresponds to the spin diffusion length range (1–2 μm) extracted from measurements of the spin-valve signal against L we have mentioned earlier. This also indicates that the longitudinal and transverse electron spin relaxation times are the same in graphene as expected ($T_1 = T_2 \equiv \tau$, see Ref. [72]).

An extensive spin precession study was done in the high magnetic field range [73] to determine whether the spin-valve signal saturation value at high field is the same as the zero-field value, that is, is the spin relaxation isotropic in graphene (as it is in metals, see Ref. [38]) or not. For a 2D system where spin–orbit fields in plane (Rashba or Dresselhaus type spin–orbit interactions) dominate the spin relaxation a strong anisotropy is expected up to 50% [72]. The experiments on graphene indicate that $\tau_\perp \approx 0.8\tau_\parallel$; the dominating spin relaxation mechanism seems to be of Elliot–Yafet type. More conclusive studies on this subject will be presented in Section 4.6.

An interesting effect was predicted for spin precession in a clean graphene sheet at the Dirac neutrality point: the equivalent of the Hartman effect (apparent superluminality) in optics. The dynamics of carriers at the neutrality point is "pseudodiffusive," and the authors calculate that the transmission time of spins from injector to detector (distance L) is $\tau = 0.87L/v$, that is, the electrons propagate with an apparent velocity larger than v [74].

4.5.4 DRIFT DIFFUSION UNDER AN ELECTRIC FIELD

Spin drift-diffusion effects in all-electrical n-GaAs-based devices were measured by Lou et al. [75], where the conduction was limited to electrons. Unlike in regular semiconductors, graphene electrostatic gating allows for switching from hole to electron conduction while keeping the carrier mobility, diffusion constant, Fermi velocity, electric conductivity, and other parameters approximately unchanged (see Figure 4.3). The effect of spin drift was studied in metallic (hole- and electron-doped, $n \approx \pm 3.5 \times 10^{12}$ cm^{-2}) graphene as well as in the vicinity of the charge neutrality point under an applied DC electric field in lateral spin valves [76]. The experiments were regular, four-terminal non-local spin-valve measurements using an AC spin injection and detection scheme as in Figure 4.5. The device, however, was extended with a fifth electric contact on the graphene (placed between the inner injector–detector contacts), allowing for a DC electric current and thus a DC

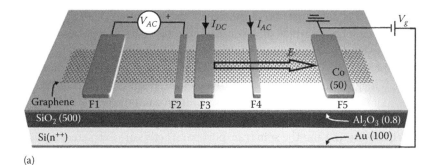

FIGURE 4.10 Sample geometry and contact scheme for the spin drift-diffusion experiments (a), with the additional electrode F3 employed to create a DC electric field E. The measured spin-valve signal is plotted against E in (b) for electrons and holes (density $|n| = 3.5 \times 10^{12}$ cm^{-2}) and for the neutrality region. The solid lines are obtained via the drift-diffusion model.

electric field to be present in the injector–detector region (see Figure 4.10a). Note that there is a major difference between these experiments and the ones presented in the previous section where the AC injection contacts were DC biased to enhance injection efficiency. Here, the spin injection process is undisturbed, only the *transport* of the created spin imbalance is manipulated using the electric field.

Due to the high sheet resistance (≈ 1.5 kΩ in the metallic regime, 4.5 kΩ at the charge neutrality point), sending a 40 µA maximum DC current through the graphene created a 68 kV/m electric field (206 kV/m at the neutrality point). The resulting carrier drift velocity $|v_D| \leq 1.7 \times 10^4$ m/s was sufficient to change the diffusion of the spin imbalance predicted by Equation 4.2 and increase/decrease the spin-valve signals by 50%. These experiments are not possible in metals since achieving a considerable drift velocity in high conductance materials would necessitate unrealistic electrical fields.

Figure 4.10b shows the spin-valve signals R_{SV} as extracted from the multitude of regular non-local measurements, plotted against the applied DC electric field. For high carrier densities, the increase in the spin-valve signal is clearly visible when the electric field is increased to positive (electrons, full circles) or negative values (holes, open circles). At the Dirac neutrality point

($n \approx 0$, stars), the conduction on the large scale happens neither via electrons nor via holes, meaning no spin drift should take place and the spin-valve signal should not depend on the magnitude of E. There is, however, a slight decreasing tendency of the signal for increasingly positive DC fields, similar to the hole conduction regime, indicating that the measurement was done not exactly at the neutrality point.

A quantitative interpretation of the data is possible by adopting the drift-diffusion model introduced by Yu and Flatté [77] for nondegenerate semi-conductors. (The degenerate case is discussed in [78].) The starting equation is, again, Equation 4.2 for steady state, less the precession component:

$$DV^2\vec{n}_s - \frac{\vec{n}_s}{\tau} + \vec{v}_D \nabla \vec{n}_s = 0, \tag{4.7}$$

and we will look at the transport of the spin imbalance n in the x-direction. Similar to the spin diffusion length λ, we define a spin drift length $\lambda_D = D/v_D$. The symmetric diffusion (around the injection point $x = 0$) and asymmetric drift effects add up to form a spin transport characterized by a pair of length scales, the *upstream* and *downstream* spin drift-diffusion lengths:

$$\frac{1}{\lambda_\pm} = \pm \frac{1}{2}\frac{1}{\lambda_D} + \sqrt{\frac{1}{4}\frac{1}{\lambda_D^2} + \frac{1}{\lambda^2}}. \tag{4.8}$$

The general solution to the spin imbalance Equation 4.7 in the direction x is

$$n_s(x) = A \exp\left(+\frac{x}{\lambda_+}\right) + B \exp\left(-\frac{x}{\lambda_-}\right), \tag{4.9}$$

where A and B are determined by the boundary conditions. The dependence of the spin signal versus the applied DC field can, therefore, be written in the form of

$$R_{SV} = R_0 \exp\left(-\frac{L}{\lambda_\pm}\right), \tag{4.10}$$

where:
  R_0 is the spin signal for $x = 0$
  $\lambda_\pm$ carries the electric field dependence through Equation 4.8
  L is the distance between F3 and F4 (the region where spin transport is of drift-diffusion type)

With the parameters calculated for the graphene spin-valve sample in case ($n \approx 3.5 \times 10^{12}\,\text{cm}^{-2}$, $\mu = 2500\,\text{cm}^2/\text{V s}$, $L = 1.5\,\mu\text{m}$, $D = 0.02\,\text{m}^2/\text{s}$, $\lambda = 2\,\mu\text{m}$), the up/downstream lengths are $\lambda_+ = 1.3\,\mu\text{m}$, $\lambda_- = 3\,\mu\text{m}$ for $E \approx \pm 34\,\text{kV/m}$ and $\lambda_+ = 0.9\,\mu\text{m}$, $\lambda_- = 4.4\,\mu\text{m}$ for $E \approx \pm 68\,\text{kV/m}$. At zero electric field $\lambda_\pm \equiv \lambda$ allows us to estimate the value of $R_0 = R_{SV} \exp(L/\lambda)$. The curve $R_{SV}(E_{DC})$ calculated with Equation 4.10 is plotted as solid line both for electrons and holes in Figure 4.10b and shows excellent agreement with the experimental data. There are no free parameters in the calculation and, in fact, these measurements could be used to determine the spin diffusion length λ if it was unknown.

At the Dirac neutrality point, the above model does not apply since the Drude model for calculating the carrier mobility is not valid. In general, the electronic DOS in graphene is

$$\nu(\varepsilon) = \frac{g_S g_V 2\pi |\varepsilon|}{h^2 v_F^2}, \tag{4.11}$$

with g_S and g_V the spin- and valley degeneracies. An expression for the drift velocity can be calculated as $v_D = (\sqrt{\pi} v_F e \tau_d)\big/(\hbar\sqrt{2n})E$ (see Refs. [35, 53]). If the momentum scattering time τ_d is independent of the carrier density n, then at sufficiently small densities the drift velocity drastically increases and reaches the Fermi velocity but at $n=0$ it becomes zero. However, for a $\tau_d \propto \sqrt{n}$ relationship, v_D remains constant at small densities and, again, it becomes zero at $n=0$. It is still an open theoretical question, which of the two situations holds for graphene.

4.6 SPIN RELAXATION PROCESSES

4.6.1 THEORETICAL CONSIDERATIONS

We have seen that experiments done so far on graphene spin valves reveal room temperature spin relaxation times τ of the order of 100–200 ps. Combined with a high carrier mobility characteristic to graphene, this gives an already promising spin diffusion length of 1–2 μm. However, it is still below the theoretically predicted intrinsic limit (~20 μm as estimated by Huertas-Hernando [79]) and the relaxation times measured so far are too low to be attractive for spin qubit type applications in quantum computation (as proposed in Trauzettel et al. [80]; here the authors calculate a spin decoherence time of 20 μs in graphene quantum dots limited by hyperfine interactions with the nuclear spin). Some calculations also consider scattering from the substrate impurities and remote surface phonons and yet predict very long spin relaxation times up to a microsecond [81]. Most models for spin relaxation in graphene are based on a clean, defect-free graphene crystal with ripples [81–87], where the spin–orbit interactions are expected to be weak (carbon being low atomic number) and the hyperfine interactions [80, 88] insignificant since the occurrence of ^{13}C is only 1%. An explanation for the fast spin relaxation must lie therefore in extrinsic mechanisms such as scattering on impurity potential, defects, boundaries, phonons; this was addressed recently by Castro Neto and Guinea [89] considering an sp³ hybridization of carbon due to hydrogen impurities, leading to a spin–orbit coupling enhanced by three orders of magnitude.

4.6.2 EXPERIMENTS: ELLIOT–YAFET VERSUS D'YAKONOV–PEREL' MECHANISM

There are two possible mechanisms discussed in the literature [90, 91] that can be responsible for spin relaxation, scaling differently on the momentum relaxation. In case of the Elliot–Yafet mechanism (spin flip occurs with a finite probability at each momentum scattering center, Figure 4.11a) the spin scattering time τ is proportional to τ_d, the momentum scattering time. The D'yakonov–Perel' mechanism (spins precess under the influence of local spin–orbit fields in between scattering events, Figure 4.11b) is on the other hand, characterized by a $\tau \propto \tau_d^{-1}$ relationship. To identify the scattering mechanism and find the ultimate limit on spin relaxation, one can thus investigate the link between spin transport and the electronic quality of the graphene, in particular the charge diffusion coefficient $D_c = 0.5v_F l$ (or the carrier mobility μ). Ideally, the experiments would be done on a set of graphene spin-valve devices that display a wide spread in their charge transport properties. However, it is experimentally challenging to fabricate consistently good ferromagnetic contacts to the graphene for such a set of samples where parameters like the spin injection efficiency are the same for all contacts.

Another route is to do the experiments on individual devices tuning the charge carrier density from metallic regime down to the lowest values and comparing the behavior of the spin transport to the changes in the charge diffusion coefficient.

At an energy E sufficiently far from the Dirac neutrality point (metallic regime), the charge diffusion coefficient D_c can be calculated from the measured conductivity σ using the Einstein relation $\sigma = e^2 v(E_F) D_c$. Here, $v(E_F)$ is the density of states (DOS) at the Fermi level E_F given by Equation 4.11 from which by integration we can obtain the density $n(E_F) = (g_S g_V \pi E_F^2)/(h^2 v_F^2)$. This yields

$$D_c = \frac{h v_F}{2e^2 \sqrt{g_S g_V \pi}} \frac{\sigma}{\sqrt{n}}. \tag{4.12}$$

For densities $|n| < 0.5 \times 10^{12}$ cm^{-2}, this formula gives unphysical values and results in a singularity at the Dirac neutrality point. This comes from the unrealistic assumption of vanishing carrier density and DOS at the neutrality point. To correct for it, one has to account for a broadened DOS $v^*(E)$ due

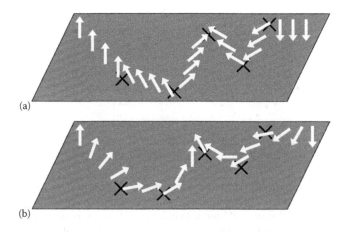

(a)

(b)

FIGURE 4.11   Illustration of the Elliot–Yafet (a) and D'yakonov–Perel' (b) type spin relaxation mechanisms.

to finite temperature, electron–hole puddles, and finite lifetime of electronic states. This can be done by adding a Gaussian broadening energy γ in form of

$$v^*(E) = \frac{1}{\sqrt{2\pi}\gamma} \int_{-\infty}^{\infty} \exp\left(-\frac{(\varepsilon - E)^2}{2\gamma^2}\right) v(\varepsilon)d\varepsilon. \tag{4.13}$$

After integration, we obtain an expression for the broadened DOS that can be used in Equation 4.12 to calculate a modified charge diffusion coefficient D_c^*.

A set of resistivity, spin valve, and Hanle precession measurements was done in a broad range of charge carrier densities to compare spin- and charge transport in SLG [58]. In Figure 4.12a–e, we plot the extracted transport parameters against the applied backgate voltage (top scale) and carrier density (bottom scale). The conductivity (inverse of the measured sheet resistivity) in panel a shows the typical electron and hole conductance regimes separated by the charge neutrality point. Due to trapped charges in the SiO_2 and/or other contaminants on the sample, the neutrality point is usually shifted by a few (tens of) volts from the zero setting of the backgate voltage. This curve is used to calculate and plot (Figure 4.12c) the charge diffusion coefficient using

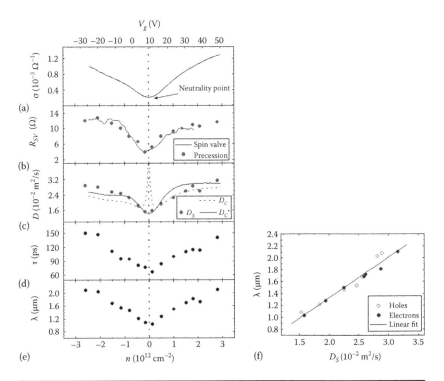

FIGURE 4.12 Charge and spin transport parameters against gate voltage (top scale) and carrier density (bottom scale): conductivity as inverse of sheet resistivity (a), spin-valve signal extracted from spin valve and Hanle precession measurements (b), spin- and charge-diffusion coefficients (c), spin relaxation time (d), spin relaxation length (e). A linear scaling is revealed when the spin relaxation length is plotted against the diffusion coefficient (f).

the unbroadened formula (4.12) (D_c, dashed line) and the broadened version (D_c^*, solid line). A broadening energy of $\gamma \approx 75\,\text{meV}$ was used corresponding to a density variation of $\Delta n \approx \pm\, 0.7 \times 10^{12}\,\text{cm}^{-2}$. The broadening is consistent with the literature values [15] attributed to electron–hole puddles in graphene on a SiO_2 substrate, and takes care of avoiding the singularity at the neutrality point. In the same panel, the spin diffusion coefficient (D_s, circles) is also plotted, as extracted from Hanle precession measurements fitted with the solutions to the Bloch equation (similar to Figure 4.9). The spin- and charge diffusion coefficients are practically the same, although they are obtained from completely different types of experiments. This indicates a minor role of Coulomb-type electron–electron interactions in scattering processes [92, 93] in agreement with the recent results of Li et al. [94] where e–e interactions are detected in high mobility, suspended graphene flakes only.

Figure 4.12b shows the spin-valve signal extracted from spin-valve measurements (solid line, continuously scanning V_g) and from the Hanle precession measurements with $B_z = 0$ (circles), which of course should be the same. We can see a threefold decrease of the signal when the carrier density goes from metallic to neutrality. Since in Equation 4.4 the signal is proportional to the inverse of the conductivity, the strong loss of spin-valve signal must come from a spin diffusion length λ that is considerably lower for low densities.

Figure 4.12d shows the behavior of the spin relaxation time τ against carrier density extracted from the Hanle experiments. Indeed, combining τ with D_s the spin diffusion coefficient from panel c using $\lambda = \sqrt{D_s \tau}$ yields a spin diffusion length that decreases by more than 50% when the carrier density is reduced to the minimum (see panel e). On a device with spin injector–detector spacing $L = 3.1\,\mu\text{m}$, this is also quantitatively consistent with the behavior of the spin-valve signal displayed in panel b.

Finally, let us look at the scaling of the spin diffusion length on the (spin or charge) diffusion coefficient. Plotting λ against D_s (Figure 4.12f) shows a clear linear dependence for both the electron and the hole conduction regime. The spin scattering time is thus directly proportional to the diffusion coefficient, that is, to the momentum scattering time. This is a clear confirmation of a dominating Elliot–Yafet type spin relaxation mechanism, in agreement with the spin relaxation anisotropy studies presented in Section 4.5.3.

Improving the electronic characteristics of the graphene device is, therefore, expected both to enhance the charge transport and prolong the spin scattering time. Assuming the Elliot–Yafet mechanism is still dominating at high carrier mobilities, one can extrapolate the behavior shown in Figure 4.12f to samples displaying charge carrier mobilities in the range of $2 \times 10^5\,\text{cm/V}\,\text{s}$ reported recently, to predict a possible room temperature spin diffusion length up to $100\,\mu\text{m}$.

4.6.3 EXTRINSIC FACTORS FOR SPIN RELAXATION

A number of extrinsic factors were investigated by Popinciuc et al. [48] that could produce the impurity potential for a strong spin (and charge) scattering. One of these is the edges of the graphene flake that, especially in the case of samples etched with plasma, are rough and may be chemically modified, thus could reduce spin diffusion length. A set of spin valve and Hanle experiments on samples with graphene width from 0.1 to 2 μm show a considerable spread in the spin relaxation times (40–200 ps); however, no clear scaling on the width is found.

Another effect that could play a role is the relaxation through the ferromagnetic contacts (i.e., the conductivity mismatch problem); this is contributing to the spin relaxation only for transparent ferromagnet–graphene interfaces, and it can be eliminated by using an oxide barrier of appropriate height as we discussed earlier.

Yet another factor that was excluded as spin relaxation channel is the oxide layer deposited on the graphene. One could imagine a conductor that consists of a single layer of atoms to be strongly affected in its spin (and charge) transport properties by an insulating layer deposited on the top of it, of similar thickness. However, samples were made where the oxide, instead of covering the full graphene flake, was deposited only under the ferromagnetic electrodes (Al_2O_3 [48]; MgO [46]) and no improvement in the spin relaxation times was observed.

One aspect that was so far not studied systematically is how the substrate and contaminants between the substrate and graphene affects spin relaxation. It was shown experimentally that the charge transport is greatly enhanced by (partially) removing the substrate (wet chemical etching of SiO_2) and adsorbed contaminants (annealing with high current densities). Particularly, the carrier mobility is enhanced by a factor of 20 [7–9]. An SEM image of a graphene "bridge" is shown in Figure 4.13 using Cr/Au contacts. Conforming to a dominating Elliot–Yafet spin relaxation mechanism such an increase in charge transport should also come with an improved spin diffusion. Currently, efforts are made to measure spin transport in suspended flakes; however, the fabrication of these devices with ferromagnetic contacts and oxide barriers is technically much more challenging due mainly to the chemical corrosiveness of the materials.

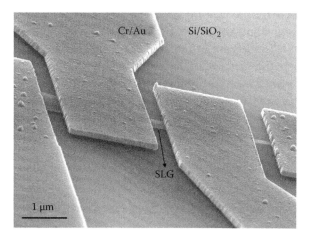

FIGURE 4.13   SEM image of an SLG flake suspended above a Si/SiO₂ (220 nm) substrate, contacted by Cr/Au (5/125 nm) electrodes. (Photo courtesy of J. van den Berg.)

4.7 NEW HORIZONS

Interesting spin transport phenomena can occur at graphene p–n junctions. Such junctions were already made for charge transport experiments using top- and side-gate electrodes that n- or p-dope a limited region of a large graphene flake, yet so far no spintronic measurements were reported. Žutić et al. predicted effects in GaAs-based p–n junctions such as spin amplification (spin density can increase away from the point of spin injection) and spin capacitance (the amount of spin stored in the p–n junctions changes with bias) [95]. Graphene offers the possibility to study these effects with the ease of creating p–n junctions that do not present a depletion region at the interface. One can imagine that an electric field across the graphene flake will cause spin drift toward the p–n junction (electrons and holes having opposite drift velocity for the same electric field), leading to increased spin accumulations at the junction.

Different levels of controlled hydrogenization of graphene, leading to graphane, offer the possibility to study systematically the influence of defects on the spin relaxation. Ultimately, in the strongly disordered regime (variable range hopping), there is an interesting prediction: the magnetoresistance should be exponentially sensitive to an applied magnetic field. This happens because in the localized state a magnetic field reduces the scattering by aligning the electron spin [96, 97].

The idea of a "perfect spin filtering" was proposed by Karpan et al. [98] based on the calculations of the Fermi surfaces for a graphene (or graphite) crystal contacted by (111) fcc or (0001) hcp nickel or cobalt electrodes. Although difficult to realize with the current graphene production technology, such a spin-valve device is predicted to show infinite magnetoresistance due to the fact that the in-plane lattice constants of the materials match almost perfectly while their electronic structures overlap in the reciprocal space for one spin direction only.

Intrinsic ferromagnetism and half-metallic behavior of graphene nanoribbons was predicted to rise from the character of the edges (zigzag versus armchair edge, see Figure 4.1a) that is worthwhile exploring [99]. Furthermore, an extrinsic ferromagnetism is also feasible to be observed, when defects or magnetic adatoms lead to long-range ordered coupling among spins [100, 101]. Ferromagnetic insulators, for example, europium oxide deposited on graphene can induce ferromagnetic correlations as well, through a proximity effect predicted by Haugen et al. [102].

Spin transport inside double- and few-layer-graphene devices on regular substrates, due to the screening effect of the outer layers and/or the onset of the D'yakonov–Perel' scattering mechanism, show improvement when compared to SLG. Recent experiments indicate already a systematic two-fold increase in the relaxation time for few-layer-graphene samples compared to single layers [103] and even stronger in certain bilayer samples [106, 107]. Spin relaxation times up to 6.2 ns (at 20 K temperature) and a peculiar temperature dependence of the spin transport parameters was reported in the bilayer graphene samples described in Ref. [107].

Finally, local studies of spin transmission in *perpendicular* geometry in graphite-based spin valves (NiFe/graphite/Co) were done recently using ballistic electron magnetic microscopy [104]. No measurable loss of spin information was detected when highly (75%) spin-polarized electrons were injected vertically into 17 nm thick graphite nanostructures of 300–500 nm lateral dimensions. The energy/momentum scattering length has been found to be greater than 250 nm at 1.8 eV above the Fermi level, suggesting that the spin relaxation length in graphite would be in the micrometer range at this energy.

4.8 2010 NOBEL PRIZE IN PHYSICS

After finishing this chapter, it was announced that the 2010 Nobel Prize in Physics would be awarded jointly to Andre Geim and Konstantin Novoselov (the University of Manchester, United Kingdom) "for groundbreaking experiments regarding the two-dimensional material graphene." The exceptionally short period (a mere 5 years) from their first electrical measurements on single-layer graphene flakes to winning the most prestigious scientific prize is typical of the rapid progress that this new material itself has made. Whether it will continue to progress as rapidly, whether it will find its way to applications, or whether it will forever remain in research labs has yet to be seen. In any case, it has generated immense experimental and theoretical interest and has made us realize that a simple hexagonal carbon lattice can still surprise us in so many different ways.

ACKNOWLEDGMENTS

We are grateful for the crucial contribution of our collaborators Siemon Bakker, Bernard Wolfs, Johan Holstein, Nikolaos Tombros, Mihaita Popinciuc, Harry Jonkman, Paul Zomer, Thomas Maassen, Alina Veligura, Shinichi Tanabe and for the inspiring discussions with Roland Kawakami and Antonio Castro Neto. We acknowledge the financial support of the Zernike Institute for Advanced Materials, MSC^{plus}, NWO, FOM, and NanoNed.

REFERENCES

1. W. J. M. Naber, S. Faez, and W. G. van der Wiel, Organic spintronics, *J. Phys. D: Appl. Phys.* **40**, R205 (2007).
2. G. Wagoner, Spin resonance of charge carriers in graphite, *Phys. Rev.* **118**, 647 (1960).
3. K. Matsubara, T. Tsuzuku, and K. Sugihara, Electron spin resonance in graphite, *Phys. Rev. B* **44**, 11845 (1991).
4. K. S. Novoselov, D. Jiang, F. Schedin et al., Two-dimensional atomic crystals, *Proc. Natl. Acad. Sci. USA* **102**, 10451 (2005).
5. P. R. Wallace, The band theory of graphite, *Phys. Rev.* **71**, 622 (1947).
6. K. S. Novoselov, A. K. Geim, S. V. Morozov et al., Electric field effect in atomically thin carbon films, *Science* **306**, 666 (2004).
7. X. Du, I. Skachko, A. Barker, and E. Y. Andrei, Approaching ballistic transport in suspended graphene, *Nat. Nanotechnol.* **3**, 491 (2008).
8. K. Bolotin, K. Sikes, Z. Jiang et al., Ultrahigh electron mobility in suspended graphene, *Solid State Commun.* **146**, 351 (2008).

9. S. V. Morozov, K. S. Novoselov, M. I. Katsnelson et al., Giant intrinsic carrier mobilities in graphene and its bilayer, *Phys. Rev. Lett.* **100**, 016602 (2008).
10. J. H. Chen, C. Jang, S. Xiao, M. Ishigami, and M. S. Fuhrer, Intrinsic and extrinsic performance limits of graphene devices on SiO_2, *Nat. Nanotechnol.* **3**, 206 (2008).
11. E. H. Hwang, S. Adam, and S. Das Sarma, Carrier transport in two-dimensional graphene layers, *Phys. Rev. Lett.* **98**, 186806 (2007).
12. Y. Zhang, J. Tan, H. L. Stormer, and P. Kim, Experimental observation of the quantum Hall effect and Berry's phase in graphene, *Nature* **438**, 201 (2005).
13. K. S. Novoselov, A. K. Geim, S. V. Morozov et al., Two-dimensional gas of massless Dirac fermions in graphene, *Nature* **438**, 197 (2005).
14. J. Martin, N. Akerman, G. Ulbricht et al., Observation of electron–hole puddles in graphene using a scanning single-electron transistor, *Nat. Phys.* **4**, 144 (2008).
15. A. Deshpande, W. Bao, F. Miao, C. N. Lau, and B. J. LeRoy, Spatially resolved spectroscopy of monolayer graphene on SiO_2, *Phys. Rev. B* **79**, 205411 (2009).
16. Y. Zhang, V. W. Brar, C. Girit, A. Zettl, and M. F. Crommie, Origin of spatial charge inhomogeneity in graphene, *Nat. Phys.* **5**, 722 (2009).
17. H. Heersche, P. Jarillo-Herrero, J. B. Oostinga, L. M. K. Vandersypen, and A. F. Morpurgo, Bipolar supercurrent in graphene, *Nature* **446**, 56 (2007).
18. A. Cortijo and M. A. H. Vozmediano, Effects of topological defects and local curvature on the electronic properties of planar graphene, *Nucl. Phys. B* **763**, 293 (2007).
19. J. van den Brink, From strength to strength, *Nat. Nanotechnol.* **2**, 199 (2007).
20. A. Rycerz, J. Tworzydlo, and C. W. J. Beenakker, Valley filter and valley valve in graphene, *Nat. Phys.* **3**, 172 (2007).
21. F. Schedin, A. K. Geim, S. V. Morozov et al., Detection of individual gas molecules adsorbed on graphene, *Nat. Mater.* **6**, 652 (2007).
22. T. O. Wehling, K. S. Novoselov, S. V. Morozov et al., Molecular doping of graphene, *Nano Lett.* **8**, 173 (2008).
23. D. C. Elias, R. R. Nair, T. M. G. Mohiuddin et al., Control of graphene's properties by reversible hydrogenation: Evidence for graphene, *Science* **323**, 610 (2009).
24. R. Y. N. Gengler, A. Veligura, A. Enotiadis et al., Large-yield preparation of high-electronic-quality graphene by a Langmuir-Schaefer approach, *Small* **6**, 35 (2010).
25. C. Gmez-Navarro, R. T. Weitz, A. M. Bittner et al., Electronic transport properties of individual chemically reduced graphene oxide sheets, *Nano Lett.* **7**, 3499 (2007).
26. X. L. Li, G. Y. Zhang, X. D. Bai et al., Highly conducting graphene sheets and Langmuir–Blodgett films, *Nat. Nanotechnol.* **3**, 538 (2008).
27. H. A. Becerril, J. Mao, Z. Liu, R. M. Stoltenberg, Z. Bao, and Y. Chen, Evaluation of solution-processed reduced graphene oxide films as transparent conductors, *ACS Nano* **2**, 463 (2008).
28. C. Berger, Z. Song, X. Li et al., Electronic confinement and coherence in patterned epitaxial graphene, *Science* **312**, 1191 (2006).
29. J. Hass, R. Feng, T. Li et al., Highly ordered graphene for two dimensional electronics, *Appl. Phys. Lett.* **89**, 143106 (2006).
30. X. Li, W. Cai, J. An et al., Large-area synthesis of high-quality and uniform graphene films on copper foils, *Science* **324**, 1312 (2009).
31. K. S. Kim, Y. Zhao, H. Jang et al., Large-scale pattern growth of graphene films for stretchable transparent electrodes, *Nature* **457**, 706 (2009).
32. A. Reina, X. Jia, J. Ho et al., Large area, few-layer graphene films on arbitrary substrates by chemical vapor deposition, *Nano Lett.* **9**, 30 (2009).
33. C. Oshima and A. Nagashima, Ultra-thin epitaxial films of graphite and hexagonal boron nitride on solid surfaces, *J. Phys.: Condens. Matter* **9**, 1 (1997).
34. I. Žutić and M. Fuhrer, Spintronics: A path to spin logic, *Nat. Phys.* **1**, 85 (2005).
35. T. Ando, Screening effect and impurity scattering in monolayer graphene, *J. Phys. Soc. Jpn.* **75**, 074716 (2006).

36. M. Johnson and R. H. Silsbee, Interfacial charge-spin coupling: Injection and detection of spin magnetization in metals, *Phys. Rev. Lett.* **55**, 1790 (1985).

37. M. Johnson and R. H. Silsbee, Thermodynamic analysis of interfacial transport and of the thermomagnetoelectric system, *Phys. Rev. B* **35**, 4959 (1987).

38. F. J. Jedema, A. T. Filip, and B. J. van Wees, Electrical spin injection and accumulation at room temperature in an all-metal mesoscopic spin valve, *Nature* **410**, 345 (2001).

39. E. W. Hill, A. K. Geim, K. S. Novoselov, F. Schedin, and P. Blake, Graphene spin valve devices, *IEEE Trans. Magn.* **42**, 2694 (2006).

40. N. Tombros, C. Józsa, M. Popinciuc, H. T. Jonkman, and B. J. van Wees, Electronic spin transport and spin precession in single graphene layers at room temperature, *Nature* **448**, 571 (2007).

41. S. Cho, Y.-F. Chen, and M. S. Fuhrer, Gate-tunable graphene spin valve, *Appl. Phys. Lett.* **91**, 123105 (2007).

42. M. Nishioka and A. M. Goldman, Spin transport through multilayer graphene, *Appl. Phys. Lett.* **90**, 252505 (2007).

43. M. Ohishi, M. Shiraishi, R. Nouchi, T. Nozaki, T. Shinjo, and Y. Suzuki, Spin injection into a graphene thin film at room temperature, *Jpn. J. Appl. Phys. Part 2* **46**, L605 (2007).

44. W. H. Wang, K. Pi, Y. Li et al., Magnetotransport properties of mesoscopic graphite spin valves, *Phys. Rev. B* **77**, 020402(R) (2008).

45. H. Goto, A. Kanda, T. Sato et al., Gate control of spin transport in multilayer graphene, *Appl. Phys. Lett.* **92**, 212110 (2008).

46. W. Han, K. Pi, W. Bao et al., Electrical detection of spin precession in single layer graphene spin valves with transparent contacts, *Appl. Phys. Lett.* **94**, 222109 (2009).

47. M. Shiraishi, M. Ohishi, R. Nouchi et al., Robustness of spin polarization in graphene-based spin valves, *Adv. Funct. Mater.* **19**, 3711 (2009).

48. M. Popinciuc, C. Józsa, P. J. Zomer et al., Electronic spin transport in graphene field effect transistors, *Phys. Rev. B* **80**, 214427 (2009).

49. M. Ishigami, J. H. Chen, W. G. Cullen, M. S. Fuhrer, and E. D. Williams, Atomic structure of graphene on SiO_2, *Nano Lett.* **7**, 1643 (2007).

50. D. Li, W. Windl, and N. P. Padture, Toward site-specific stamping of graphene, *Adv. Mater.* **21**, 1243 (2009).

51. X. Xie, L. Ju, X. Feng et al., Controlled fabrication of high-quality carbon nanoscrolls from monolayer graphene, *Nano Lett.* **9**, 2565 (2009).

52. P. Blake, E. W. Hill, A. H. C. Neto et al., Making graphene visible, *Appl. Phys. Lett.* **91**, 063124 (2007).

53. N. Tombros, Electron spin transport in graphene and carbon nanotubes, PhD dissertation, University of Groningen, the Netherlands, 2008.

54. C. Stampfer, J. Güttinger, F. Molitor, D. Graf, T. Ihn, and K. Ensslin, Tunable Coulomb blockade in nanostructured graphene, *Appl. Phys. Lett.* **92**, 012102 (2008).

55. P. Nemes-Incze, Z. Osváth, K. Kamarás, and L. P. Biró, Anomalies in thickness measurements of graphene and few layer graphite crystals by tapping mode atomic force microscopy, *Carbon* **46**, 1435 (2008).

56. M. Y. Han, B. Ozyilmaz, Y. Zhang, and P. Kim, Energy band-gap engineering of graphene nanoribbons, *Phys. Rev. Lett.* **98**, 206805 (2007).

57. F. Molitor, J. Güttinger, C. Stampfer, D. Graf, T. Ihn, and K. Ensslin, Local gating of a graphene Hall bar by graphene side gates, *Phys. Rev. B* **76**, 245426 (2007).

58. C. Józsa1, T. Maassen, M. Popinciuc et al., Linear scaling between momentum and spin scattering in graphene, *Phys. Rev. B* **80**, 241403 (2009).

59. L. C. Campos, V. R. Manfrinato, J. D. Sanchez-Yamagishi, J. Kong, and P. Jarillo-Herrero, Anisotropic etching and nanoribbon formation in single-layer graphene, *Nano Lett.* **9**, 2600 (2009).

60. J. Moser and A. Bachtold, Fabrication of large addition energy quantum dots in graphene, *Appl. Phys. Lett.* **95**, 173506 (2009).

61. X. Lou, C. Adelmann, S. A. Crooker et al., Electrical detection of spin transport in lateral ferromagnet–semiconductor devices, *Nat. Phys.* **3**, 197 (2007).

62. N. Tombros, S. J. van der Molen, and B. J. van Wees, Separating spin and charge transport in single-wall carbon nanotubes, *Phys. Rev. B* **73**, 233403 (2006).

63. F. J. Jedema, H. B. Heersche, A. T. Filip, J. J. A. Baselmans, and B. J. van Wees, Electrical detection of spin precession in a metallic mesoscopic spin valve, *Nature* **416**, 713 (2002).

64. G. Schmidt, D. Ferrand, L. W. Molenkamp, A. T. Filip, and B. J. van Wees, Fundamental obstacle for electrical spin injection from a ferromagnetic metal into a diffusive semiconductor, *Phys. Rev. B* **62**, 4790(R) (2000).

65. E. I. Rashba, Theory of electrical spin injection: Tunnel contacts as a solution of the conductivity mismatch problem, *Phys. Rev. B* **62**, 16267(R) (2000).

66. A. Fert and H. Jaffrès, Conditions for efficient spin injection from a ferromagnetic metal into a semiconductor, *Phys. Rev. B* **64**, 184420 (2001).

67. P. M. Tedrow and R. Meservey, Spin polarization of electrons tunneling from films of Fe, Co, Ni, and Gd, *Phys. Rev. B* **7**, 318 (1973).

68. T. Kimura, Y. Otani, and J. Hamrle, Enhancement of spin accumulation in a nonmagnetic layer by reducing junction size, *Phys. Rev. B* **73**, 132405 (2006).

69. C. Józsa, M. Popinciuc, N. Tombros, H. T. Jonkman, and B. J. van Wees, Controlling the efficiency of spin injection into graphene by carrier drift, *Phys. Rev. B* **79**, 081402(R) (2009).

70. S. A. Crooker, E. Garlid, A. N. Chantis et al., Bias-controlled sensitivity of ferromagnet/semiconductor electrical spin detectors, *Phys. Rev. B* **80**, 041305(R) (2009).

71. M. Johnson and R. H. Silsbee, Coupling of electronic charge and spin at a ferromagnetic-paramagnetic metal interface, *Phys. Rev. B* **37**, 5312 (1988).

72. J. Fabian, A. Matos-Abiague, C. Ertler, P. Stano, and I. Žutić, Semiconductor spintronics, *Acta Phys. Slov.* **57**, 565 (2007).

73. N. Tombros, S. Tanabe, A. Veligura et al., Anisotropic spin relaxation in graphene, *Phys. Rev. Lett.* **101**, 046601 (2008).

74. R. A. Sepkhanov, M. V. Medvedyeva, and C. W. J. Beenakker, Hartman effect and spin precession in graphene, *Phys. Rev. B* **80**, 245433 (2009).

75. X. Lou, C. Adelmann, M. Furis et al., Electrical detection of spin accumulation at a ferromagnet-semiconductor interface, *Phys. Rev. Lett.* **96**, 176603 (2006).

76. C. Józsa, M. Popinciuc, N. Tombros, H. T. Jonkman, and B. J. van Wees, Electronic spin drift in graphene field-effect transistors, *Phys. Rev. Lett.* **100**, 236603 (2008).

77. Z. G. Yu and M. E. Flatté, Electric-field dependent spin diffusion and spin injection into semiconductors, *Phys. Rev. B* **66**, 201202(R) (2002); ibid 235302.

78. I. D'Amico, Spin injection and electric-field effect in degenerate semiconductors, *Phys. Rev. B* **69**, 165305 (2004).

79. D. Huertas-Hernando, F. Guinea, and A. Brataas, Spin–orbit-mediated spin relaxation in graphene, *Phys. Rev. Lett.* **103**, 146801 (2009).

80. B. Trauzettel, D. V. Bulaev, D. Loss, and G. Burkard, Spin qubits in graphene quantum dots, *Nat. Phys.* **3**, 192 (2007).

81. C. Ertler, S. Konschuh, M. Gmitra, and J. Fabian, Electron spin relaxation in graphene: The role of the substrate, *Phys. Rev. B* **80**, 041405(R) (2009).

82. C. L. Kane and E. J. Mele, Quantum spin Hall effect in graphene, *Phys. Rev. Lett.* **95**, 226801 (2005).

83. H. Min, J. E. Hill, N. A. Sinitsyn et al., Intrinsic and Rashba spin–orbit interactions in graphene sheets, *Phys. Rev. B* **74**, 165310 (2006).

84. D. Huertas-Hernando, F. Guinea, and A. Brataas, Spin–orbit coupling in curved graphene, fullerenes, nanotubes, and nanotube caps, *Phys. Rev. B* **74**, 155426 (2006).

85. Y. Yao, F. Ye, X.-L. Qi, S.-C. Zhang, and Z. Fang, Spin–orbit gap of graphene: First-principles calculations, *Phys. Rev. B* **75**, 041401(R) (2007).

86. K.-H. Ding, G. Zhou, and Z.-G. Zhu, Magnetotransport of Dirac fermions in graphene in the presence of spin–orbit interactions, *J. Phys.: Condens. Matter* **20**, 345228 (2008).

87. M. Gmitra, S. Konschuh, C. Ertler, C. Ambrosch-Draxl, and J. Fabian, Band-structure topologies of graphene: Spin–orbit coupling effects from first principles, *Phys. Rev. B* **80**, 235431 (2009).

88. O. V. Yazyev, Hyperfine interactions in graphene and related carbon nano-structures, *Nano Lett.* **8**, 1011 (2008).

89. A. H. Castro Neto and F. Guinea, Impurity-induced spin–orbit coupling in graphene, *Phys. Rev. Lett.* **103**, 026804 (2009).

90. I. Žutić, J. Fabian, and S. Das Sarma, Spintronics: Fundamentals and applications, *Rev. Mod. Phys.* **76**, 323 (2004).

91. J. Fabian and S. D. Sarma, Band structure effects in the spin relaxation of conduction electrons, *J. Appl. Phys.* **85**, 5075 (1999).

92. Y. Barlas, T. Pereg-Barnea, M. Polini, R. Asgari, and A. H. MacDonald, Chirality and correlations in graphene, *Phys. Rev. Lett.* **98**, 236601 (2007).

93. E. H. Hwang, B. Y.-K. Hu, and S. Das Sarma, Density dependent exchange contribution to $\partial\mu/\partial n$ and compressibility in graphene, *Phys. Rev. Lett.* **99**, 226801 (2007).

94. G. Li, A. Luican, and E. Y. Andrei, Scanning tunneling spectroscopy of graphene on graphite, *Phys. Rev. Lett.* **102**, 176804 (2009).

95. I. Žutić, J. Fabian, and S. Das Sarma, Spin injection through the depletion layer: A theory of spin-polarized p-n junctions and solar cells, *Phys. Rev. B* **64**, 121201(R) (2001).

96. T. G. Rappoport, B. Uchoa, and A. H. Castro Neto, Magnetism and magneto-transport in disordered graphene, *Phys. Rev. B* **80**, 245408 (2009).

97. A. H. Castro Neto, Private communications, 2009.

98. V. M. Karpan, G. Giovannetti, P. A. Khomyakov et al., Graphite and graphene as perfect spin filters, *Phys. Rev. Lett.* **99**, 176602 (2007).

99. Y.-W. Son, M. L. Cohen, and S. G. Louie, Half-metallic graphene nanoribbons, *Nature* **444**, 347 (2006).

100. N. M. R. Peres, F. Guinea, and A. H. Castro Neto, Coulomb interactions and ferromagnetism in pure and doped graphene, *Phys. Rev. B* **72**, 174406 (2005).

101. M. A. H. Vozmediano, M. P. López-Sancho, T. Stauber, and F. Guinea, Local defects and ferromagnetism in graphene layers, *Phys. Rev. B* **72**, 155121 (2005).

102. H. Haugen, D. Huertas-Hernando, and A. Brataas, Spin transport in proximity-induced ferromagnetic graphene, *Phys. Rev. B* **77**, 115406 (2008).

103. T. Maassen, F. K. Dejene, M. H. D. Guimarães et al., Comparison between charge and spin transport in few-layer graphene, *Phys. Rev. B* **83**, 115410 (2011).

104. T. Banerjee, W. G. van der Wiel, and R. Jansen, Spin injection and perpendicular spin transport in graphite nanostructures, *Phys. Rev. B* **81**, 214409 (2010).

105. K. W. Han, K. Pi, K. M. McCreary et al., Tunneling spin injection into single layer graphene, *Phys. Rev. Lett.* **105**, 167202 (2010).

106. T.-Y. Yang, J. Balakrishnan, F. Volmer et al., Observation of long spin relaxation times in bilayer graphene at room temperature, *Phys. Rev. Lett.* **105**, 047206 (2011).

107. K. W. Han and R. K. Kawakami, Spin relaxation in single-layer and bilayer graphene, *Phys. Rev. Lett.* **107**, 047206 (2011).

5
Spintronics in 2D Materials

Wei Han and Roland K. Kawakami

5.1 INTRODUCTION: SPINTRONICS IN 2D MATERIALS

Spintronics aims to utilize the spin degree freedom of electrons for novel information storage and computing devices [1–4]. The potential advantages of spintronic devices include non-volatility, increased data processing speed, increased integration densities, and decreased electric power consumption [1]. Originating from discovery of giant magnetoresistance in metallic ferromagnetic (FM) multilayer systems [5, 6], spintronics is divided into two major research directions: spin-based quantum computing [7, 8] which uses isolated electron or nuclear spins as quantum bits, and spin transport devices which focus on the injection, transport and manipulation of spin-polarized carriers for information storage and spin logic devices [2, 4, 9–12].

The advent of graphene and related van der Waals materials has created new opportunities for electronics. In such 2D materials, the carriers are confined in the 2D plane, which strongly alters their physical properties compared to bulk materials. Graphene is a single layer of carbon atoms that exhibit ultrahigh mobility and room temperature quantum Hall effect with half-integer Hall conductivities due to the Berry phase of the chiral Dirac fermions [13–17]. Single layer transition metal dichalcogenides (TMDCs) have a direct band gap and different properties of K and −K valleys, which gives rise to the valley polarization [18–22]. These TMDCs and black phosphorus could be also used for future field effect transistor applications with low quiescent currents [23–25]. Beyond individual layers of 2D materials, van der Waals heterostrucutres that integrate several layers of different 2D materials could generate novel properties through proximity effects [26, 27].

These 2D materials are also very attractive for spintronics since they exhibit unique spin-dependent and magnetic properties and provide new platforms to investigate the spin degree of freedom. The low spin-orbit coupling in graphene gives rise to long spin lifetimes that could be potentially useful for spin transport devices [28–30] as well as for graphene-based spin qubits [31, 32]. The high spin-orbit coupling in TMDCs produces a strong coupling between the electron spin and valley pseudospin [33, 34], which leads to long spin-valley lifetimes [35]. These 2D materials could also become magnetic through adatom doping and magnetic proximity effects [29, 36–40]. Recently, intrinsic ferromagnetism and antiferromagnetism in 2D materials has been identified and their potential applications have also been explored [41–47].

This chapter on spintronics in 2D materials focuses on the injection, transport, and manipulation of spins in 2D materials as well as on related advances in 2D magnetism. It is organized as follows. The second section presents recent progress in graphene spintronics beyond Chapter 4, Volume 3, by Józsa and van Wees. It includes the current understanding of the spin relaxation mechanisms and the studies towards graphene-based spintronics devices, such as spin logic devices and spin transistors. The third section presents the spin-valley coupling in TMDCs, including the valley Hall effect, long-lived spin-valley polarized states, the manipulation of valleys by spins, the manipulation of spins by valleys, and the spin-valley coupling induced Ising superconductivity. The fourth section discusses the induced extrinsic magnetism in 2D materials, using graphene as an example, by adatom doping with atomic hydrogen, and magnetic proximity effect with a magnetic insulator. The fifth section presents the recent development of intrisinc 2D ferromagnetism/antiferromagnetism, the electric field effect of 2D FM materials, and the magnetic tunnel junctions based on 2D FM materials. The summary and future outlook of spintronics in 2D materials are discussed in the final section. Related topics to this chapter pertain to discussion of nonlocal spin injection in Chapter 6, Volume 1, Chapter 5, Volume 2, and Chapters 7 and 9, Volume 3, spin relaxation in Chapter 1, Volume 2, optical selection rules in Chapter 2, Volume 2, and spin-based logic devices in Chapter 17, Volume 3.

5.2 RECENT PROGRESS IN GRAPHENE SPINTRONICS

For the purpose of manipulation of spins for spin transistors and spin logic devices, a major challenge is to develop a suitable spin transport channel with long spin lifetime and long spin propagation length. Graphene is a very promising material for the spin channel due to the experimental observations of large spin signals, long spin lifetimes, and long spin diffusion lengths at room temperature [29].

In a pioneering study, the van Wees group provided the first demonstration of spin transport in graphene with nonlocal Hanle spin precession [48]. They fabricated graphene spin valve devices, did spin injection and spin transport, and observed nonlocal magnetoresistance (MR) at room temperature. The direct proof of spin transport in graphene was the observation of Hanle spin precession during the spin transport. Subsequently, highly efficient injection of spin-polarized carriers into graphene was achieved (~30% spin injection efficiency) using graphene/TiO_2/MgO/Co tunneling contacts grown by molecular beam epitaxy [49]. High quality graphene has been studied in recent experiments, which identify much longer spin diffusion length over 10 μm and spin lifetimes over 10 ns [50–53].

Beyond the graphene spintronics written by Józsa and van Wees in Chapter 4, Volume 3, of this book and two recent review articles in this field [29, 54], there have been many important breakthroughs in the field of graphene spintronics. Here, we try to summarize them in this section. After a short introduction of the nonlocal spin valve, the spin transport and spin relaxation measurements, we will discuss the recent understanding of the spin relaxation mechanisms and the anistropy of spin relaxation in graphene. There are several overlapping contents of the spin relaxation mechanisms with two recent review articles of graphene spintronics [29, 54]. Then, recent experimental results towards the applications of graphene for spin logic devices and spin transistors will be presented. This includes the demonstration of XOR logic function using modified graphene spin valves [55], and the realization of graphene spin field-effect switches by stacking graphene with TMDC in the spin transport region [56, 57].

5.2.1 MEASUREMENTS OF SPIN TRANSPORT AND RELAXATION IN GRAPHENE

The spin transport and spin relaxation in graphene can be measured by nonlocal MR and nonlocal Hanle spin precession, as shown in Figure 5.1a, b. This nonlocal spin measurement method was invented by Johnson and Silibee in 1985, when they measured the spin injection in an Aluminum bar [58]. Soon after the development of the nanofabrication technique, this measurement has been widely used in the spin injection and detection in metals, semiconductors, and graphene [48, 59–69]. For the nonlocal spin transport measurement (Figure 5.1a), a current source is applied between the electrodes, E1 and E2, where E2 is ferromagnetic (FM) and serves as a spin injector. Then, the spin is detected by measuring the voltage across E3 and E4, where E3 (FM) is the spin detector. The electrodes E1 and E4 are ideally nonmagnetic, but are sometimes FM to simplify the device fabrication process. This measurement is called "nonlocal" because the voltage probes lie outside the charge current loop. The nonlocal voltage probes the spin density at E3 arising from the pure spin current diffused from the graphene underneath E2. The measured voltage (V_{NL}) is positive or negative depending on whether the magnetization configuration of E2 and E3 is parallel or anti-parallel to each other. The difference between these two voltages is the nonlocal spin signal, and it is often converted to units of resistance by dividing out the injection current ($\Delta R_{NL} = (V_{NL}^{P} - V_{NL}^{AP})/I_{inj}$). The nonlocal

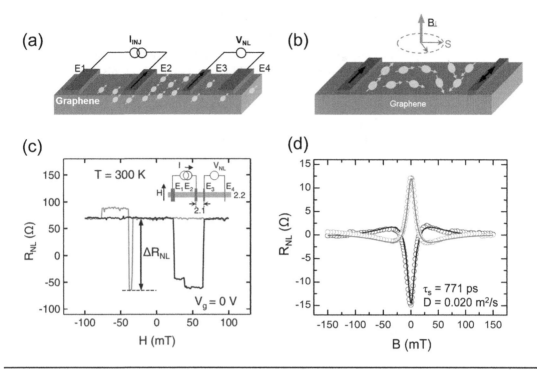

FIGURE 5.1   (a) Nonlocal spin transport in graphene spin valves. E2 and E3 are ferromagnetic electrodes. E1 and E4 could be made of either ferromagnetic or nonmagnetic materials. The spin dependent chemical potential generated by the diffusive spin current from the E2 to E3 is measured by the nonlocal voltage (V_{NL}). (b) Hanle effect: the precession of the spins caused by the torque from a perpendicular magnetic field (T $\propto S \times B_\perp$). (c) A typical nonlocal magnetoresistance measured on a graphene nonlocal spin valve with tunneling contacts. The difference of the two resistance states represents the spin signal and is called nonlocal MR (ΔR_{NL}). (d) Typical Hanle spin precession curves for parallel (red circles) and anti-parallel (black circles) configurations measured in graphene nonlocal spin valve with tunneling contacts. The solid lines represent the best fit results based on Equation 5.1. (After Han, W. et al., *Nat. Nanotechnol.* 9, 794, 2014. With permission.)

resistance ($R_{NL} = V_{NL}/I_{inj}$) is not a resistance in the usual sense; it is a four-terminal resistance which can have positive or negative values depending on the polarity of spin density in graphene. As shown in Figure 5.1c, the nonlocal resistance has multiple values that depend on the magnetization configurations of the FM electrodes. To generate the parallel and anti-parallel magnetization alignments of E2 and E3, an in-plane magnetic field is applied along the long axes of the FM electrodes. As the magnetic field is swept from negative to positive, one electrode, E3, switches the magnetization direction first, resulting in a switching of nonlocal resistance as the magnetization alignment changes from parallel to anti-parallel. Then, the electrode E2 switches the magnetization direction, and the state changes back to the parallel state. The difference in R_{NL} between the parallel and anti-parallel states is called the nonlocal MR (i.e., ΔR_{NL}) and is a result of the spin diffusion from E2 to E3. The minor jumps in the nonlocal MR loop are due to the magnetization switching of the FM electrodes E1 and E4. The value of the nonlocal MR is highly dependent on the spin injection/detection efficiency and the spacing between the spin injector and detector.

For the spin injection/detection efficiency, the FM contacts play the most important role; the tunneling contacts with an insulating barrier between the FM and graphene gives a higher spin injection/detection efficiency and a large MR. Various types of FM contacts have been discussed previously in Ref. [29].

Since the nonlocal MR is a resistance that depends on magnetization of the FM magnetizations, sometimes it shows up along with the same magnetic field-dependent resistance, such as anisotropic magnetoresistance arising from one single FM electrode magnetization switching rather than the spin-dependent chemical potential due to spin diffusion. To unambiguously demonstrate the spin transport in graphene, Hanle spin precession measurement is necessary. To measure this Hanle effect, an in-plane magnetic field is used to set the parallel or anti-parallel configurations first, and set to zero. Then, either the device or the magnetic field is rotated 90°. In this geometry, the magnetic field is perpendicular to the spin polarization direction, since the spin polarization direction is only dependent on the FM magnetization directions, which is fixed in-plane for small out-of-plane magnetic field ($H_\perp$) due to the shape anistropy. As a result of the spin precession induced by $H_\perp$, the spin polarization reaching the detector electrode reduces, resulting a reduced spin signal with increasing magnitude of $H_\perp$. Two typical Hanle curves are shown in Figure 5.1d, of which the red/black open circles represent the parallel/anti-parallel magnetization states of the spin injector and detector. Quantitatively, the Hanle curve can be fitted by [59]:

$$R_{\text{NL}} \propto \pm \int_0^\infty \frac{1}{\sqrt{4\pi Dt}} \exp\left[-\frac{L^2}{4Dt}\right] \cos(\omega_L t) \exp(-t/\tau_s)\, dt, \qquad (5.1)$$

where the + (−) sign is for the parallel (anti-parallel) magnetization state, D is the diffusion constant, and τ_s is the spin lifetime. ω_L is the Larmor frequency of $\omega_L = g\mu_B H_\perp / \hbar$, where g is the g-factor, μ_B is the Bohr magneton, and $\hbar$ is reduced Planck's constant. This numerical process of fitting the experimental results can be used to obtain the spin-dependent properties of the graphene, including the spin lifetimes and the spin diffusion constants. The red/black solid lines in Figure 5.1d are the fitted curves based on Equation 5.1. Note that this fit equation works well for tunneling FM contacts with contact resistance much larger than the graphene spin resistance. However, in the more general case there is spin absorption into the FM contacts [49, 70–73] and more accurate values of the spin lifetimes are obtained using the following equation [74]:

$$R_{\text{NL}}^{\pm} = \pm p_1 p_2 R_N f, \qquad (5.2)$$

where the ± corresponds to the relative alignment of injector/detector magnetizations, p_1 and p_2 are the effective electrode spin polarizations (1 for injector, 2 for detector), $R_N = \dfrac{\lambda}{WL\sigma^N}$ is the spin resistance of graphene, λ

is the spin diffusion length in graphene (related to τ_s by $\lambda = \sqrt{D\tau_s}$), W is the width of the graphene, σ^N is the conductivity of graphene and

$$f = Re\left(2\left[\sqrt{1+i\omega_L\tau_s} + \frac{\lambda}{2}\left(\frac{1}{r_s}+\frac{1}{r_d}\right)\right]e^{\left(\frac{L}{\lambda}\right)\sqrt{1+i\omega_L\tau_s}}\right.$$

$$\left. + \frac{\lambda^2}{r_s r_d}\frac{\sinh\left[\left(\frac{L}{\lambda}\right)\sqrt{1+i\omega_L\tau_s}\right]}{\sqrt{1+i\omega_L\tau_s}}\right)^{-1}. \qquad (5.3)$$

In the function f, $r_i = \frac{R_F + R_C^i}{R_{SQ}}W$ with $i=s,d$ for the injector and detector, respectively, R_C is the contact resistance, R_{SQ} is the graphene sheet resistance, $R_F = \rho_F\lambda_F/A$ is the spin resistance of the ferromagnetic electrode material, ρ_F is the resistivity of the ferromagnet, λ_F is the spin diffusion length in the ferromagnet, and A is the area of the graphene/electrode junction.

5.2.2 Spin Relaxation Mechanisms

Theoretically, several spin relaxation mechanisms have been utilized to explain the spin lifetimes measured in graphene, including the Elliott-Yafet (EY) and Dyakonov-Perel (DP), magnetic resonant scattering, and pseudospin-driven spin relaxation [75–80]. Both EY and DP mechanisms have their roots in metal and semiconductor spintronics, and they rely on spin-orbit coupling and momentum scattering, but their effect is opposite with respect to the latter [2, 81–83], as discussed in Chapter 1, Volume 2, of this book. Magnetic resonant scattering is based on the existence of magnetic moments (from vacancies or adatoms) which provide the spin-flip exchange field [78, 79]. On the other hand, pseudospin-driven spin relaxation mechanism is proposed in graphene without any magnetic impurity. It results from quantum mixing between spin and pseudospin-related Berry's phase and random local Rashba spin-orbit coupling (SOC) due to adatoms [80].

The EY mechanism explains the spin relaxation by spin flips during scattering (Figure 5.2a). In the presence of spin-orbit coupling the Bloch states are an admixture of the Pauli spin up and spin down states. Typically an electron undergoes up to a million scattering events before its spin is flipped. The spin relaxation rate is roughly $1/\tau_s \approx b^2/\tau_p$, where τ_p is the momentum relaxation time and b is the amplitude of the spin admixture. Typically b is determined as the intrinsic spin-orbit coupling divided by the Fermi energy ($b \approx \lambda_I/\varepsilon_F$). Experimental studies have been performed to investigate the relative importance of the EY mechanism in single layer graphene [84–86]. It is observed that the τ_s increases with the diffusion constant D ($\sim \tau_p$), as shown in Figure 5.2c. This linear relationship between D and τ_s is consistent with the EY spin relaxation mechanism. In more detail, the relationship deviates from linearity because b varies with ε_F which also depends on the gate voltage, but the general trend of spin lifetime increasing with momentum scattering time still holds [87].

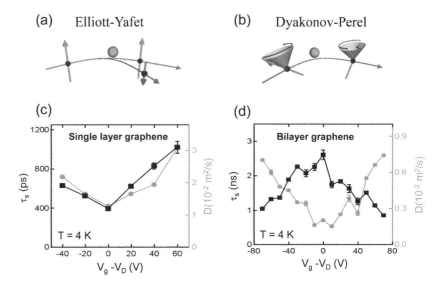

FIGURE 5.2 (a, b) Schematics of Elliott-Yafet (a) and Dyakonov-Perel (b) spin relaxation mechanisms for graphene. (c, d) Spin lifetime (squares) and diffusion coefficient (circles) as a function of gate voltage at 4 K for single layer graphene (c) and bilayer graphene (d). For single layer graphene, the linear relationship between spin lifetime and diffusion coefficient is observed [84]. Similar results are reported in Ref. [85]. For bilayer graphene, the inverse relationship between spin lifetime and diffusion coefficient is observed. Similar results are reported in Ref. [69]. ((a, b) After Han, W. et al., *Nat. Nanotechnol.* 9, 794, 2014. With permission. (c, d) After Han, W. et al., *Phys. Rev. Lett.* 107, 047207, 2011. With permission.)

The DP mechanism is based on the concept of motional narrowing: the more the electron scatters, the less its spin relaxes (Figure 5.2b). Unlike for the EY mechanism, the spins precess between the scattering events. In the absence of space inversion symmetry, the spin-orbit coupling is manifested as a spin-orbit field, say of the Rashba type, and the electron spin precesses about this field. As the electron scatters, the orientation (and/or the value) of the effective magnetic field changes. In this effect, the electron spin precesses in a randomly fluctuating spin-orbit field, where the correlation time of the fluctuations is given by the momentum relaxation time. The spin relaxation rate is then given by $1/\tau_s \approx \lambda_R^2 \tau_p$, where λ_R is the magnitude of the Rashba field, which is the spin precession frequency. Experimental studies have been performed to investigate the relative importance of the DP mechanism in bilayer graphene; it is found that τ_s decreases with increasing D (or τ_p), which suggests a DP spin relaxation mechanism (Figure 5.2d) [69, 84]. The roles of the EY and DP spin relaxation mechanisms remain an open question. For example, it could be hard to distinguish these mechanisms in graphene because both EY-like and DP-like behaviors could be observed due to the randomness of the Rashba field [88, 89].

Magnetic resonant scattering is based on the existence of magnetic moments in graphene. The magnetic scatterers provide an exchange field to induce spin flips. At resonant energies, the scattering electron spends a considerable time at the impurity, allowing the electron spin to precess, say, at least a full circle (Figure 5.3a). As the electron escapes, the spin can be found with equal probability up and down, so the spin flip time equals the spin relaxation time. This happens precisely at the resonant energy. Averaging over resonant energies (due to thermal fluctuations, distributions of resonant energies of different defects, or due to the presence of electron-hole puddles), can yield the spin relaxation times of 100 ps for as little as 1 ppm of magnetic moments (Figure 5.3b) [90].

Recently, another mechanism was proposed to account for the short spin lifetimes, namely pseudospin-driven spin relaxation mechanism. It is proposed in graphene with random Rashba SOC from nonmagnetic impurities, and it results from quantum mixing between spin and pseudospin-related Berry's phase and randomly local Rashba SOC due to adatoms [80]. Figure 5.3c is the schematic of the entangled spin–pseudospin dynamics induced by local Rashba spin-orbit interaction related to adsorbed gold atoms. Figure 5.3d shows the calculated spin lifetimes as a function of the carrier density for several levels of the densities of adatoms.

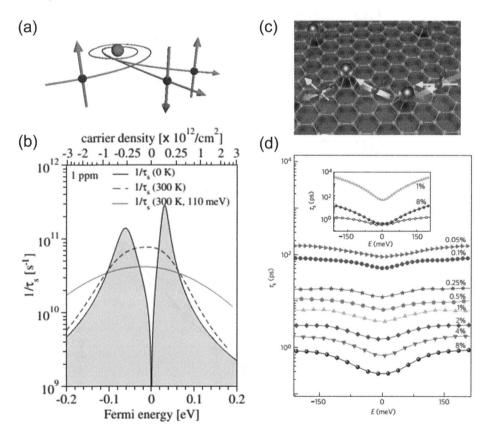

FIGURE 5.3   (a) A schematic of resonant scattering by magnetic impurities in graphene. (b) Spin relaxation rate as a function of energy-carrier density, at 0 K, at 300 K, and at 300 K broadened by puddles with energy fluctuations of 110 meV. Exchange coupling $J = -0.4$ eV and impurity fraction is 1 ppm. (c) Schematic of disordered graphene doped with a random distribution of adatoms. (d) Spin relaxation times (τ_s) in graphene deduced from pseudospin-driven spin relaxation mechanism. Adatom densities are varied between 0.05 and 8% (with $\mu = 0.1\gamma_0$, where μ is the local chemical potential shift and γ_0 is the nearest neighbor hopping). Inset: τ_s values using the continuum model for $\rho = 1\%$ and 8%. A comparison with the microscopic model (with $\mu = 0$) is also given for $\rho = 8\%$ (open circles). ((a) After Han, W. et al., *Nat. Nanotechnol.* 9, 794, 2014. With permission. (b) After Kochan, D., *Phys. Rev. Lett.* 112, 116602, 2014. With permission. (c, d) After Tuan, D.V. et al., *Nat. Phys.* 10, 857, 2014. With permission.)

To study the spin relaxation sources, another approach is to use the anisotropy of spin lifetime, which is the ratio of the lifetime for spins out-of-plane ($\tau_\perp$) to spins in-plane ($\tau_{//}$). For various spin relaxation sources, the anisotropy of the spin lifetimes is expected to exhibit different behaviors: spin-orbit fields pointing in the graphene plane ($\tau_\perp \sim 0.5\,\tau_{//}$), ripple-induced spin relaxation generates spin-orbit fields pointing perpendicular to the graphene plane ($\tau_\perp \gg \tau_{//}$), while random magnetic scattering leads to the relationship of $\tau_\perp \sim \tau_{//}$. This anisotropic behavior and the electrical field tuning have been studied by the van Wees group [52, 91] comparing Hanle spin precession at low and high out-of-plane fields. As shown in Figure 5.4, the spin valve was made based on graphene that was sandwiched between two pieces of hexagonal boron nitride. Without any electric field perpendicular to the graphene, the ratio of $\tau_\perp \sim \tau_{//}$ was obtained to be ~ 0.75. Both top and bottom

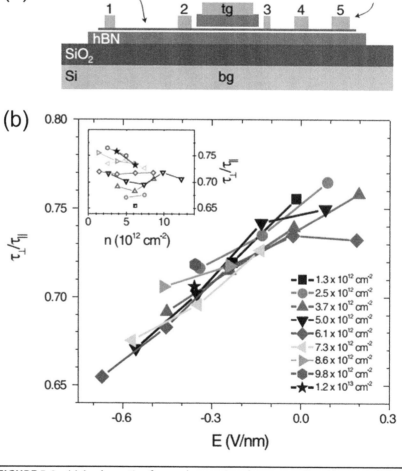

FIGURE 5.4   (a) A schematic of a graphene spin valve encapsulated in two h-BN layers. 1–5 indicate the ferromagnetic contact electrodes (TiO$_x$/Co), and tg (bg) indicate the top (bottom) gate electrodes. (b) The ratio of $\tau_\perp/\tau_\|$ as a function of electric field for different carrier densities. Inset: The ratio of $\tau_\perp/\tau_\|$ as a function of the carrier densities for different electric fields. (After Guimarães, M.H.D. et al., *Phys. Rev. Lett.* 113, 086602, 2014. With permission.)

gates could be applied which can tune the Fermi level of the graphene and can provide a perpendicular electric field. With the perpendicular electric field, the ratio could be tuned down to ~ 0.65. Their results demonstrated that a perpendicular electric field can largely modulate the in-plane Rashba SOC. Recently, another approach to characterize spin lifetime anisotropy by low field oblique Hanle precession has been proposed by Raes et al. and utilized to investigate proximity-induced spin-orbit coupling in graphene/TMDC heterostructures [92], which is discussed later in this chapter.

Finally, another source of spin relaxation is the contact-induced spin relaxation. The main effect is related to spin absorption into the ferromagnetic contacts [49, 70–74], which is taken into account explicitly in Equations 5.2 and 5.3. However, studies indicate that other sources of contact-induced spin relaxation such as effects of fringe fields and spin loss due to interfacial spin flip scattering due to the FM contacts may still be present in graphene spin valves [72].

5.2.3 GRAPHENE SPIN LOGIC DEVICES

The room temperature physical characteristics including long spin diffusion length, large spin signal, and long spin lifetime make graphene one of the most favorable candidates for spin channel material in spin logic applications.

Graphene-based spintronic devices for logic applications have been proposed by Dery and Sham [10, 93]. The building block of graphene spin logic is a magnetologic gate consisting of a graphene sheet contacted by five FM electrodes (Figure 5.5a). Two FM electrodes (A and D) define the input states, two electrodes (B and C) define the operation of the gate, and one electrode (M) is utilized for readout. The logic operation is performed by spin injection/extraction in graphene at the input/operation electrodes (A,B,C,D) followed by mixture and diffusion of spin currents to the output (M). As illustrated in Figure 5.5a, the basic logic operations can be reconfigured based on the FM states of the electrodes B and C, and the magnetization states of A and D represent the logic inputs.

Experimentally, XOR logic application has been achieved on a graphene spin valve at room temperature [55]. The device geometry and measurement circuit are shown in Figure 5.5b. Three FM tunneling contacts and one nonmagnetic Ti/Au contact were fabricated on the graphene. The source current I_S is a combination of I_{ac} (AC current to inject spins) and I_{dc} (DC bias current). The output voltage V_{out} (AC voltage) is measured using standard low-frequency lock-in techniques, and output current I_{out} ($=V_{out}/R_{sen}$) is determined using a current detection scheme by systematically tuning the variable sensing resistor R_{sen}. An offset voltage V_{offs} (AC voltage with phase and frequency locked to I_{ac}) is used to eliminate any background signal unrelated to spin. The experimental demonstration of the XOR logic operation is shown in Figure 5.5c. During the sweeping of the magnetic field, the magnetization of M is kept downward. The magnetic field changes the magnetization of the two input electrodes A and B. Clearly, the measured I_{out} depends on the magnetization states of the input A and B, and the XOR logic operation is achieved. For additional discussion of graphene spin logic devices, please see Chapter 17, Volume 3.

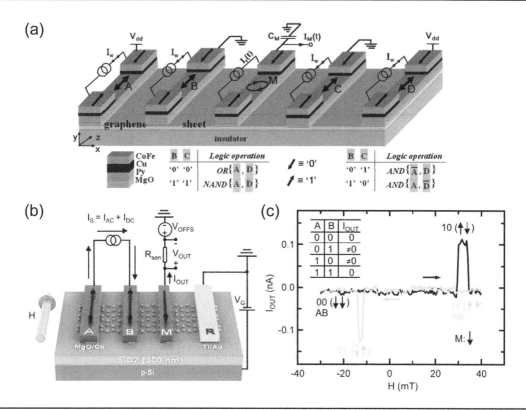

FIGURE 5.5 Graphene spin logic devices. (a) Schematic drawing of graphene-based spin logic gate consisting of a graphene sheet contacted by five ferromagnetic electrodes [29]. Two electrodes (A and D) define the input states, two electrodes (B and C) define the operation of the gate, and one electrode (M) is utilized for read-out; V_{dd} drives the steady-state current, and $I_M(t)$ is the transient current response, which gives the output; C_M is a capacitor; I_w is the writing current to manipulate the magnetization direction of each ferromagnetic electrode; $I_r(t)$ is the current used to perturb the magnetization of the electrode M. (b) Illustration of XOR logic device structure and measurement setup [55]. A, B, and M are ferromagnetic electrodes (Co/MgO). The spin channel is a single-layer graphene. R is a nonmagnetic electrode (Ti/Au) used as ground point. I_{out} and V_{out} are the measured current and voltage, respectively. R_{sen} is a variable resistor, and V_{offs} is an AC voltage source. (c) Experimental demonstration of XOR operation: I_{out} as a function of in-plane magnetic field. Black (red) curve indicates H sweeps upwards (downwards). Vertical arrows indicate the magnetization states of electrodes A and B. Top left inset: truth table of XOR logic operation. ((a) After Han, W. et al., *Nat. Nanotechnol.* 9, 794, 2014. With permission. (b, c) After Wen, H. et al., *Phys. Rev. Appl.* 5, 044003, 2016. With permission.)

5.2.4 GRAPHENE SPIN FIELD-EFFECT SWITCH

In 1990, Datta and Das proposed the idea of a spin field effect transistor, which uses the electric field to generate large Rashba fields [9]. Since the Rashba fields can generate a torque on the spins that rotates the spins, which results a large modulation of the spin signal (ON/OFF). The outstanding spin-dependent properties of graphene at room temperature make it one of the most favorable candidates for spin field effect transistor applications. However, one major challenge is that the perpendicular electric field can only modulate the spin signal a small amount [29, 49, 50, 54].

Recently the graphene spin field-effect switch has been demonstrated based on a van de Waals heterostructure consisting of an atomically thin graphene and a semiconducting MoS_2 [56, 57]. The switch mechanism is different from the electric field generating Rashba fields, as proposed by Datta and Das. This spin field-effect switch can control the spin signal by orders of magnitude, which arises from the gate tuning of the MoS_2 conductivity, which modulates the spin absorption from graphene into the MoS_2. The device structure is illustrated in Figure 5.6a. A MoS_2 layer is put on the top of the graphene in the spin diffusion region between the spin injector and the nonlocal spin detector. The FM contacts are made of Co/TiO_2 by electron-beam lithography to perform the spin injection and detection measurements.

The gate voltage tuning of the spin signal is shown in Figure 5.6b. For large negative V_g, the semiconducting MoS_2 is in the low conductivity state and the nonlocal MR (ΔR_{nl}) is able to be observed, reaching the "ON" state of the spin device. Sweeping the gate voltage towards positive values brings the MoS_2 towards its high conductivity state, reaching the "OFF" state of the spin device. In this case, the resistance of MoS_2 increases by more than six orders of magnitude, as shown in Figure 5.6b. The strong correlation of the spin signal and the sheet conductivity of MoS_2 suggest the switching of spin transport using the graphene/MoS_2 heterostructure relies on the absorption of spins traveling through the graphene by the MoS_2.

5.2.5 PROXIMITY-INDUCED SPIN-ORBIT COUPLING

In addition to inducing spin absorption, the TMDC stacked on top of graphene could generate proximity-induced spin-orbit coupling. Unlike spin absorption which is based on semi-classical spin diffusion, the proximity spin-orbit coupling is a quantum mechanical effect in which the presence

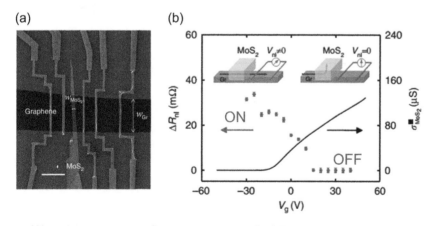

FIGURE 5.6   Graphene spin field-effect switch [56]. Similar results are reported in Ref. [57]. (a) Scanning electron microscope (SEM) image of the graphene spin field-effect switch with MoS_2 intercepting the spin current path. Scale bar, 2 μm. (b) Experimental demonstration of the ON state and the OFF state of the spin signal by gate voltage (blue circles). The black solid line is the sheet conductivity of the MoS_2 as a function of gate voltage. The insets show schematically the spin current path (green arrow) in the OFF state (left inset) and the ON state (right inset) of MoS_2. (After Yan, W. et al., *Nat. Commun.* 7, 13372, 2016. With permission.)

of the TMDC layer directly modifies the band structure of graphene. Theoretical DFT studies of monolayer TMDC on graphene show the presence of spin-split conduction and valence bands and a gap opening at the Dirac point of graphene due to the TMDC overlayer [94]. The proximity spin-orbit coupling is of high interest because it could be used to realize topological states with dissipationless edge channels (e.g. quantum spin Hall, quantum anomalous Hall) [94, 95], and could also be used for efficient spin field-effect switches [96].

Experimentally, the proximity spin-orbit coupling has been demonstrated in monolayer graphene with TMDC by various methods. Weak antilocalization in WS_2/graphene [97–99] and the spin Hall effect in WSe_2/graphene [100] provide evidence for proximity spin-orbit coupling. Most recently, two studies have unambiguously demonstrated the presence of proximity-induced spin-orbit coupling in $MoSe_2$/graphene and WS_2/graphene through spin transport experiments and the measurement of a strong spin lifetime anisotropy [101, 102]. Figure 5.7 highlights the approach of Benitez et al. [102], using oblique Hanle spin precession to determine the spin lifetime anisotropy. As shown in Figure 5.7a, the magnetic field is applied at an oblique angle β measured from the surface ($\beta = 0°$ is the standard spin valve geometry of Figure 5.1c, while $\beta = 90°$ is the standard Hanle geometry of Figure 5.1d). The oblique angle causes the spins to precess about

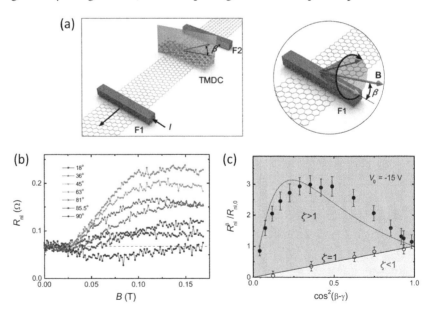

FIGURE 5.7 Proximity spin-orbit coupling in WS_2/graphene [101]. (a) Geometry for the oblique Hanle spin precession experiment, where β is the out-of-plane tilt angle. (b) Nonlocal Hanle curves for different oblique angles β. (c) Plot of normalized nonlocal resistance at $B = 0.15$ T as a function of $\cos^2(\beta - \gamma)$, where γ is the FM tilt angle. The solid green circles are for the WS_2/graphene sample and the green curve is for spin lifetime anisotropy ratio $\zeta = \tau_\perp / \tau_\parallel = 10$. The open black circles are for a reference sample with graphene only, showing no spin lifetime anisotropy. (After Benitez, L.A. et al., 2017. With permission.)

a cone and gain an out-of-plane component. Figure 5.7b shows how the line-shape of the nonlocal spin precession curve (V_{NL} vs. B) depends on the field angle β. When the magnetic field is strong enough, the spin component perpendicular to the field will be fully dephased, and only the parallel component will contribute to the nonlocal signal. The spin lifetime anisotropy ratio $\zeta = \tau_\perp/\tau_\parallel$ can be determined by plotting V_{NL} vs. $\cos^2(\beta-\gamma)$, where γ is the tilt angle of the FM magnetization (small). For $\tau_\perp/\tau_\parallel = 1$ (no anisotropy), V_{NL} increases linearly on $\cos^2(\beta-\gamma)$, while V_{NL} is larger for $\tau_\perp/\tau_\parallel > 1$ and smaller for $\tau_\perp/\tau_\parallel < 1$. The data in Figure 5.7c (solid green circles) shows the V_{NL} vs. $\cos^2(\beta-\gamma)$ for WS_2/graphene, which is substantially above linear dependence. The matching green curve is a simulated curve with spin lifetime anisotropy ratio $\tau_\perp/\tau_\parallel = 10$. This indicates that the proximity spin-orbit coupling has a stronger out-of-plane spin-orbit field (e.g. intrinsic type) than the in-plane spin-orbit field (e.g. Rashba type).

5.3 SPINTRONICS IN TRANSITION METAL DICHALCOGENIDES

With the emergence of graphene, the concept of valleytronics based on manipulating the valley index has attracted great interest [103, 104]. There are two degenerate and inequivalent valleys at the corners of the Brillouin zone in the band structure of graphene. The K and −K valleys are related to each other by time reversal. To manipulate this binary degree of freedom, we need to have measurable physical quantities that distinguish the ±K valleys. In the presence of inversion symmetry, the even parity of pseudo vectors under the inversion operation requires such quantities to take the same value for states related by inversion. Thus, inversion symmetry breaking is a necessary condition for the ±K valleys to exhibit valley contrast properties.

For monolayer TMDC, different from graphene, the inversion symmetry is broken, which automatically generates two inequivalent valleys constituting a binary index for low energy carriers. Furthermore, the strong SOC in TMDCs results in strong valley-spin coupling in the valence band near both K and −K points. The unique spin-valley coupling in TMDCs provides a platform to use valleys/spins for manipulating the spins/valleys, including the manipulation of the valleys by the spin polarization, and the injection of spin polarized carriers by the control of valley polarization.

5.3.1 SPIN-VALLEY COUPLING

Taking MoS_2 as one example, it can be regarded as strongly bonded 2D S-Mo-S layers that are loosely coupled to one another by van der Waals interaction. As shown in Figure 5.8a, bulk MoS_2 has the 2H stacking order with the space group D_{6h}^4, which is inversion symmetric. However, within each layer (Figure 5.8b), the Mo and S atoms form 2D hexagonal lattices, with the Mo atom coordinated by the six neighboring S atoms in a trigonal prismatic geometry. The crystal symmetry reduces to D_{3h}^1, and inversion

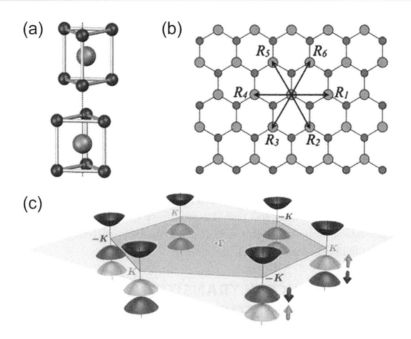

FIGURE 5.8   Spin-valley coupling in monolayers of MoS_2 and other group-VI dichalcogenides [33]. (a) The unit cell of bulk 2H-MoS_2, which has the inversion center located in the middle plane. It contains two unit cells of MoS_2 monolayers, which lacks an inversion center. (b) Top view of the MoS_2 monolayer. R_i ($i=1, 2,...6.$) are the vectors connecting nearest Mo atoms. (c) Schematic drawing of the band structure at the band edges located at the K and −K points. (After Xiao, D. et al., *Phys. Rev. Lett.* 108, 196802, 2012. With permission.)

symmetry is explicitly broken, which leads to valley-dependent optical selection rules for interband transitions at K points at the 2D hexagonal Brillouin zone, as shown in Figure 5.8c. The spin-orbit interaction from the metal d orbitals further leads to strong coupling of spin and valley degrees of freedom, which makes possible selective photon excitation of carriers with various combinations of valley and spin indices. An important consequence of the spin-valley coupling is that the spin index becomes locked with the valley index at the band edges.

The valley contrast is also accompanied by a valley-dependent selection rule for optical excitation with circularly polarized light: the interband transition at K (−K) couples only to σ + (σ −) circularly polarized light. The −K valley is also called K′ in many papers. Experimental breakthroughs in the manipulation of valley pseudospin have been made possible by using monolayer TMDs, which have a direct bandgap at the ±K points in the visible wavelength range [18–22].

5.3.2 Valley Hall Effect

For monolayer TMDC, an applied in-plane electrical voltage bias will generate a usual longitudinal charge current as well as a transverse valley current where electrons in the K and −K valleys flow in opposite transverse directions. This phenomenon is known as the valley Hall effect (VHE) and leads to opposite valley polarization accumulating on the two transverse edges of the sample. Similar to the intrinsic anomalous Hall effect in ferromagnets and the intrinsic spin Hall effect in heavy metal systems, the VHE originates from nonzero Berry curvature in momentum space. Specifically, the Berry curvature is opposite for the K and −K valleys, thus leading to the transverse velocities and Hall conductivities that have opposite signs for the two valleys. In the case of the valence bands where the spin and valleys are strongly coupled, the intrinsic spin Hall effect is equivalent to the intrinsic VHE [33, 104, 105].

Several major experimental breakthoughts have been achieved, demonstrating the valley Hall effect. The materials are single layer MoS_2, single layer graphene and bilayer graphene with inversion symmetry broken property [106–109]. Here, we use the single layer MoS_2 as one example to illustrate the valley Hall effect. As shown in Figure 5.9a, the direct energy gaps are located at the two valleys, which have different spin-dependent properties.

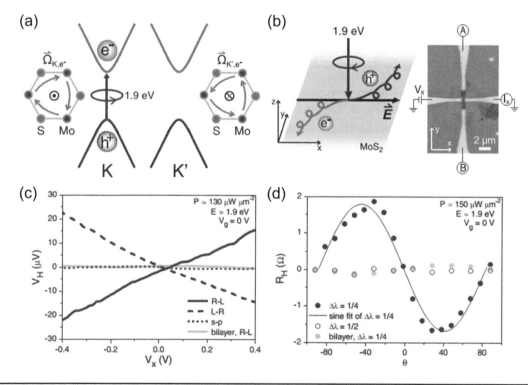

FIGURE 5.9 Valley Hall effect in monolayers of MoS_2 [106]. (a) Schematics of the valley-dependent optical selection rules and the electrons at the K and K′(−K) valleys. (b) Schematic of a photo-induced AHE driven by a net valley polarization (left) and an image of the Hall bar device (right). (c) The sourcedrain bias dependence of the Hall voltage for the monolayer device under R-L (red solid line), L-R (red dashed line), and half-wave s-p modulations (red dotted line). Results for the bilayer device under quarter-wave modulation are also shown (blue circles). (d) The anomalous Hall resistance of the monolayer device as a function of the incidence angle under quarter-wave ($\Delta\lambda = 1/4$, solid red circles) and half-wave ($\Delta\lambda = 1/2$, open red circles) modulations. Results for the bilayer device under quarter-wave modulation are also shown (blue circles). (After Xiao, D. et al., *Phys. Rev. Lett.* 108, 196802, 2012. With permission.)

Due to the spin-valley coupling, the valley-polarized carriers can be optically pumped by circularly polarized photons. The carriers in the two valleys move under the effective magnetic fields that are proportional to the Berry curvature, as shown in Figure 5.9b. In the study by Mak et al., a Hall bar device based on single layer MoS_2 is fabricated, and the time-reversal symmetry is broken explicitly by circularly polarized light. During the measurement, the Hall voltage (V_H, the voltage difference between contacts A and B) is simultaneously recorded as a function of the Fermi level of the MoS_2 channel. A small but finite V_H that scales linearly with V_X (a source-drain voltage between the contacts A and B) is observed under right minus left (R-L) and L-R modulations, as shown in Figure 5.9c. The sign of the signal is opposite for the R-L and L-R modulations. When a linear (s-p) modulation is used, no net Hall voltage could be detected. This is the signature of a photo-induced anomalous Hall effect driven by the net valley polarization.

Furthermore, the anomalous Hall resistance is measured as a function of the incidence angle under quarter-wave and half-wave modulations. The anomalous Hall resistance exhibits a sine dependence under quarter-wave modulation, but no signal is probed under the half-wave modulation, as shown in Figure 5.9d. This observation is consistent with linear relationship between the degrees of valley polarization and the excitation ellipticity [33]. For comparison, no anomalous Hall effect is observed in bilayer MoS_2, which has lattice inversion symmetry. The observation of such photo-induced anomalous Hall effect in monolayer MoS_2 arising from valley polarization demonstrates the valley Hall effect.

5.3.3 LONG-LIVED SPIN-VALLEY POLARIZATION

The presence of strong spin-valley coupling in the valence band on monolayer TMDC was predicted to generate long spin-valley lifetimes. In 2012, photoluminescence (PL) studies using circularly polarized excitation found that the emission retained a high degree of circular polarization [19, 20]. From this, it was inferred that the valley lifetime of the exciton was much longer than its recombination time, and a lower bound of ~ 1 ns for the valley lifetime was proposed. However, it was not until 2015 that long spin-valley lifetimes were reported in monolayer TMDC. In Yang et al. [35], time-resolved Kerr rotation (TRKR) was utilized to measure spin-valley dynamics in n-type monolayer MoS_2 and WS_2. At 5 K, a long-lived Kerr rotation signal of a few nanoseconds was reported for monolayer MoS_2, coming from the spin polarization of resident electrons. Unlike excitons which recombine within a few hundred picoseconds, the resident electrons exist in equilibrium in n-type material so its spin polarization can persist long after the exciton has recombined. Figure 5.10a shows the optical setup, population of spin-polarized resident electrons, and the TRKR time delay scans (for different magnetic fields) measured on monolayer MoS_2 at 5 K. Notably in zero field, the Kerr rotation exhibits long spin lifetimes on the order of a few nanoseconds. With increasing magnetic field, the spin lifetime is reduced.

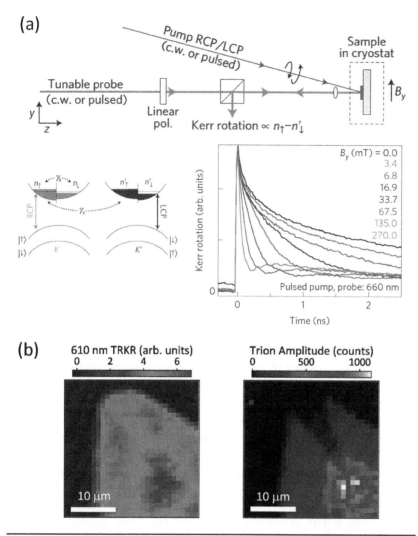

FIGURE 5.10 Long-lived spin-valley polarization in monolayer MoS$_2$ and WS$_2$. (a) Time-resolved Kerr rotation of monolayer MoS$_2$ at 5 K, from Ref. [35]. Circularly polarized optical excitation generates spin polarization of resident electrons in the n-type monolayer MoS$_2$ sheet. Time delay scans show spin lifetimes on the order of a few nanoseconds at 5 K. The magnetic field dependence shows a reduction of spin lifetime with increased magnetic field. (After Yang, L. et al., *Nat. Phys.* 11, 830, 2015. With permission.) (b) Spatial maps of time-resolved Kerr rotation and trion photoluminescence in monolayer WS$_2$ at 6 K, from Ref. [110]. The TRKR image was taken at a pump-probe time delay of 100 ps. The trion emission amplitude was determined by Lorentzian fits of the photoluminescence spectra. (After McCormick, E.J. et al., *2D Mater.* 5, 011010, 2018. With permission.)

Subsequently, longer lifetimes were observed in WS$_2$ and WSe$_2$ monolayers [110–114]. Notable are studies of gate-controlled, encapsulated WSe$_2$ exhibiting spin-valley lifetimes up to 2 μs [114] and studies of MoS$_2$/WSe$_2$ heterostructures reporting lifetimes up to 40 μs [115]. The long spin-valley lifetimes in p-type WSe$_2$ are due to the very large spin orbit splitting (~450 meV), which produces slow valley relaxation due to the requirement of spin-flip during intervalley scattering. While these studies attribute these long lifetimes to the spin-valley polarization of resident carriers, another possible contribution to the long-lived Kerr rotation signal in doped TMDC is the spin-valley polarized dark trion [110, 113]. Currently, the strongest evidence for this is reported in McCormick et al. [110], where TRKR microscopy and PL microscopy measure the spin-valley dynamics at 6 K with high spatial resolution. As shown in Figure 5.10b, spatial maps of the TRKR signal (measured at pump-probe time delay of 100 ps) and the trion PL emission (extracted by Lorentzian fits of the full PL spectrum) exhibit a spatial anticorrelation. Here, the region of high trion PL near the center of the triangular flake has low TRKR amplitude. This is a signature of valley-polarized dark trions as follows: After circularly-polarized optical excitation of a valley-polarized (bright) trion, if it converts to a valley-polarized dark trion, it will simultaneously increase the long-lived Kerr signal and decrease the trion PL emission, thus producing a spatial anticorrelation.

5.3.4 CONTROL OF VALLEY POLARIZATION VIA SPIN INJECTION AND MAGNETIC PROXIMITY EFFECT

Since the valley polarization provides a new degree of freedom for information processing, the control of the valley polarization is crucially important. The spin-valley coupling allows the generation and control of valley polarization via spintronics routes. There are generally two methods to generate spin polarization in one material. The first one is the electrical spin injection, by putting the spin-polarized carriers into the two valleys. The second one is to generate spin polarization by magnetic proximity effect. Recent experimental breakthroughs have been achieved to control the valley polarization via spin-valley coupling by spin injection and magnetic proximity effect [116–118].

For the spin injection route, a ferromagnetic semiconductor is used as the spin source [116], as illustrated in Figure 5.11a. When a current is applied between the (Ga,Mn)As and the single layer TMDC, a spin polarized current is injected, with which the spin polarization direction could be controlled by the magnetization of the ferromagnetic (Ga,Mn)As. Due to the perpendicular anisotropy of the (GaMn)As, a small outward magnetic field of 400 Oe is used to set the magnetization perpendicular to the single layer WS_2 plane. The valley polarization can be probed by detecting the circular polarization of the light emission. As shown in Figure 5.11b, the K valley is populated by spin-up hole injection due to spin–valley locking, resulting in σ− light emission. When the magnetic field is applied along the opposite direction, the K′ valley is populated by spin-down hole injection, resulting in σ+ light emission. Subsequently, a similar study used spin injection from a FM permalloy electrode into WSe_2 and recombination at a WSe_2/MoS_2 heterojunction to emit circularly polarized light [119].

For the magnetic proximity effect, the interfacial magnetic exchange coupling can provide a much larger effect than external magnetic fields. Theoretically, a giant valley splitting has been predicted in monolayer $MoTe_2$ exchange coupled to a ferromagnetic insulator EuO [120]. Experimentally,

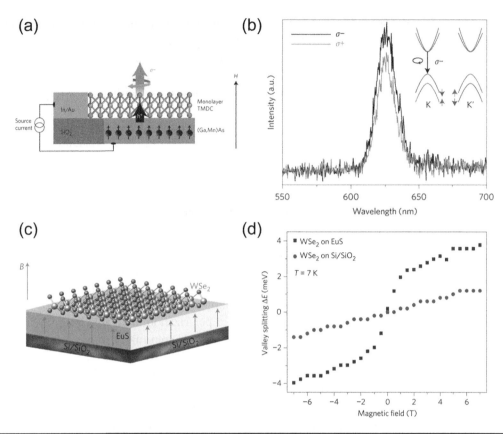

FIGURE 5.11 Control of the valley polarization in monolayer TMDC by electrical spin injection and magnetic proximity effect [116, 117]. (a) Schematic of the monolayer TMDC/(Ga,Mn)As heterojunction for electrical valley polarization devices. (b) Strong valley polarization in the monolayer WS$_2$/(Ga,Mn)As heterojunction as a result of spin-polarized hole injections. The spectra of σ− and σ+ resolved electroluminescence are measureed under an outward magnetic field of 400 Oe perpendicular to the surface with a current of 15 μA. Inset: Schematic representation of electrical excitation and emission processes at the K valley due to spin-valley locking. (c) Proximity effect between TMDC and a ferromangeitc insulator EuS. (d) Large valley splitting due to the magnetic proximity effect. Similar results are observed in Ref. [118]. ((a, b) After Ye, Y. et al., *Nat. Nanotechnol.* 11, 598, 2016. With permission. (c, d) After Zhao, C. et al., *Nat. Nanotechnol.* 12, 757, 2017. With permission.)

two works have independently demonstrated the large valley splitting induced by magnetic proximity effect [117, 118]. As shown in Figure 5.11c, a WSe$_2$ monolayer is put on the top of a EuS layer, which is a ferromagnetic insulator with a Curie temperature of ~ 16 K. As a function of the magnetic field that is applied perpendicular to the WSe$_2$ plane, the valley splitting energy exhibits a much more dramatic effect in WSe$_2$/EuS compared to the WSe$_2$/SiO$_2$, as shown in Figure 5.11d. The valley splitting of WSe2/EuS samples has been enhanced by an order of magnitude to 2.5 meV at 1 T. For WSe$_2$/SiO$_2$, the valley splitting is simply due to the Zeeman splitting of the band edge states by the magnetic fields. The larger valley splitting in WSe$_2$/EuS is an indication of the strong magnetic proximity induced effective magnetic field arising from the large magnetic moment of Eu^{2+} ($S_z = 7$ μ$_B$) and exchange coupling (J ~10 meV). This large valley splitting induced by magnetic proximity effect has also been observed in the van der Waals structure consisting of an ultrathin ferromagnetic semiconductor CrI_3 and a monolayer of WSe_2 [118]. Due to the perpendicular magnetic anisotropy of the CrI_3 layer, a zero-field valley splitting and polarization is observed.

To accurately describe magnetic proximity effect in TMDCs, Coulomb interactions need to be included; neither the energy, nor the spectral weight of excitons is given within the single-particle picture. The corresponding many-body analysis shows that by changing the magnetization direction of a substrate one can transform between dark and bright excitons in TMDCs [121].

5.3.5 OPTO-VALLEYTRONIC SPIN INJECTION

Besides the manipulation of the valley polarization by spin injection and magnetic proximity effect, the reverse process is also possible arising from the spin-valley coupling; the manipulation of spin polarization via the valley degree of freedom. Since the valley polarization can be generated efficiently in TMDC monolayer via circularly polarized light, it can be used as a spin injection source to generate spin polarization in an adjacent material. This valley/spin polarized excitation-induced spin injection has been first theoretically proposed for the graphene/TMDC structures [122].

Recently, the spin injection into graphene by valley polarization has been demonstrated experimentally [123]. The device structure is illustrated in Figure 5.10a. The single layer MoS_2 is used to generate valley polarization via optical excitation by absorbing circularly polarized light. An FM electrode is used in the nonlocal geometry to detect the spins that are injected from the MoS_2 layer to graphene. To detect the spin polarization, an in-plane external magnetic field is used to make the spins precess while diffusing from the MoS_2 to the FM electrode, as shown in Figure 5.12a. The whole spin injection and detection process is described in the following text. First, the circularly polarized photons are absorbed in the monolayer MoS_2 to produce spin/valley-polarized carriers oriented out-of-plane (along +z direction), which subsequently transfer into the adjacent few-layer graphene. The spins (blue arrows) then diffuse within the few-layer graphene toward a ferromagnetic (FM) spin detector with in-plane magnetization. To detect the spin transport, a magnetic field By is applied to induce spin precession. This generates a nonzero component of spin polarization (Sx) along the FM detector's magnetization, which produces a detector voltage (V_{NL}) that is proportional to Sx. The expected and measured results of the detector voltage as a function of the in-plane magnetic field are shown in Figure 5.12a inset and 5.12b. As the magnetization of the FM electrode switches, the opposite spin polarization is detected. Furthermore, tuning the photon energy can be used to adjust the magnitude and even the sign of the injected spin polarization, which is a direct consequence of the spin-valley coupling and the large valley splitting in the valence band of MoS_2. Subsequently, an experiment with WSe_2 and monolayer graphene showed similar results [124].

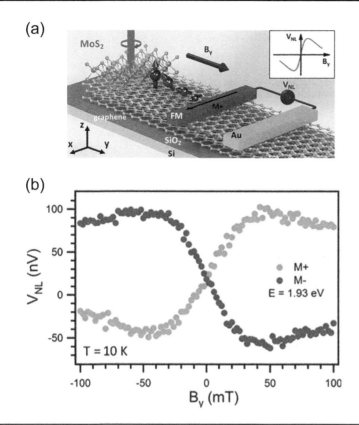

FIGURE 5.12   Opto-valleytronic spin injection into graphene via spin-valley coupling in monolayer MoS_2. (a) Illustration of optical spin injection and electrical spin detection in a monolayer MoS_2/few-layer graphene hybrid spin valve. Inset: expected signal V_{NL} as a function of applied magnetic field B_y, which results in the precession of the spins injected optically. (b) The measured V_{NL} as a function of applied magnetic field B_y at 10 K, which is consistent with theoretical predications. M+ and M− indicate the Co electrodes' magnetization directions along the x axis. The photon energy is tuned to the A excitation resonance at $E = 1.93$ eV. (After Luo, Y.K. et al., *Nano Lett.* 17, 3877, 2017. With permission.)

These experiment results demonstrate the spintronic/valleytronic functionality of a TMDC/graphene device by integrating opto-valleytronic spin injection.

5.3.6 Ising Superconductivity

The spin-valley coupling also gives rise to the Ising pairing in superconducting, evidenced by the recent observation of an extremely large in-plane upper critical field compared to the Pauli paramagnetic limit [125–127].

Figure 5.13a, b illustrate the spin-momentum locking in monolayer $NbSe_2$. Monolayer transition $NbSe_2$ possesses out-of-plane mirror symmetry and broken in-plane inversion symmetry. The electron spin is thus oriented in the out-of-plane direction and in opposite directions for electrons of opposite momenta, basically a spin-momentum locking. The inner and outer

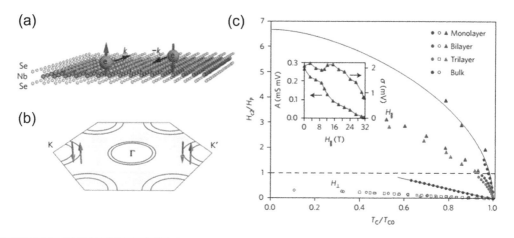

FIGURE 5.13   (a) Spin–momentum locking in monolayer $NbSe_2$, which consists of a layer of Nb atoms sandwiched between two layers of Se atoms. (b) Brillouin zone in the in-plane direction and Fermi surface near the Γ, K and K' points. The Fermi surface is spin split in the monolayer $NbSe_2$. (c) The critical field H_{c2}/H_P as a function of transition temperature T_C/T_{C0} for $NbSe_2$ samples of differing thickness under both out-of-plane $H_\perp$ (open symbols) and in-plane $H_{//}$ (filled symbols) magnetic fields. The dashed line corresponds to the Pauli paramagnetic limit. The inset: the zero-bias peak area and width as a function of $H_{//}$ at 0.36 K obtained from differential conductance measurements in a monolayer $NbSe_2$ device. Similar results are observed in ionic liquid gated MoS_2 [126, 127]. (After Xi, X. et al., *Nat. Phys.* 12, 139, 2016. With permission.)

Fermi surfaces at the K and −K points have out-of-plane spin polarization with up and down directions.

For such systems with spin-valley coupling, the Cooper pairs can only form between the K and −K valleys. Such intervalley pairing is protected by the internal out-of-plane magnetic field. As a result, the Cooper pairs remain robust unless the external in-plane magnetic field exceeds the internal out-of-plane magnetic field. Thus, this qualitative description attributes the anomalous enhancement of $H_{//c2}$ to spin–valley locking, and this mechanism is also referred to as Ising superconductivity. For example, very large in-plane critical magnetic field is observed in monolayer $NbSe_2$, as shown in Figure 5.13c. Furthermore, large H_{c2} is also observed in bi- and trilayers of $NbSe_2$, which can be attributed to spin–layer locking, which means that the spins are locked in each layer and also protects the superconductivity against in-plane external magnetic fields.

The Ising superconductivity has also been observed in thin layers of MoS_2 under ionic liquid gating, which can induce large numbers of carriers in materials to induce superconductivity [126–128].

5.4 INDUCED MAGNETISM IN 2D MATERIALS

The possibility of making 2D materials to be magnetic has attracted much interest from the technological standpoints. Making 2D materials magnetic could potentially give rise to high Curie temperature ferromagnetism and meet the demands of the ever increasing magnetic information storage density. Thus, to generate the magnetic moment in 2D materials, especially graphene, has been one of the most emergent research directions.

5.4.1 Induced Magnetism by Doping and Proximity Effect

Generally speaking, there are two routes to induce ferromagnetism in a non-magnetic material. The first one is the induced ferromagnetic moment by doping another d or f element, such as Mn, Cr, Eu etc. or by creating defects. When the magnetic doping is above a critical concentration, ferromagnetic order can be generated. The most typical example is the diluted magnetic semiconductor, Mn doped GaAs [129, 130]. The exchange interaction between carriers and local moment on Mn generates indirect ferromagnetic exchange coupling, thus introducing magnetism. The Curie temperature has a strong dependence on the Mn doping level [129, 131], as discussed in Chapters 9 and 10, Volume 2, in this book.

The other one is exchange coupling via magnetic proximity effect with adjacent ferromagnetic materials. At the atomic level, when two atoms come into proximity, the highest energy, or valence, orbitals of the atoms change substantially and the electron wavefunctions around the two atoms reorganize. For example, static magnetic proximity effects can occur at the interface of Pt to a ferromagnetic layer due to its close vicinity to the Stoner criterion [132]. The proximity effect has a strong dependence on the overlapping wave functions between the nonmagnetic layer and the adjacent ferromagnetic layer [132]. For example, Cu has a much smaller proximity-induced magnetic moment compared to Pt [133].

In this section, two examples of induced ferromagnetic moment in graphene will be discussed. The first example is ferromagnetic graphene in hydrogen-doped graphene, in which scanning tunneling microscopy is used to identify the ferromagnetic order [40]. The second example is the ferromagnetic graphene in the graphene/YIG heterostructures, where YIG is a ferromagnetic insulator. The ferromagnetic order can be observed up to 250 K, revealed by the anomalous Hall effect [134]. For the case of graphene/YIG, the induced magnetism has been used to fully modulate the spin current in graphene spin valve devices [135, 136].

5.4.2 Magnetism in Hydrogen-Doped Graphene

Hydrogenated graphene is the benchmark case for graphene magnetism, which was theoretically proposed by Yazyev and Helm [137]. Hydrogen chemisorbs reversibly on graphene, forming a strong covalent bond. This effectively removes one p_z orbital (it shifts the bonding state down by several electron volts) from the π band, thus creating a sublattice imbalance. The single hydrogen adatom induces a quasilocalized (resonant) state with magnetic moment of 1 μ_B (Bohr magneton) in accordance with Lieb's theorem [138].

Experimentally, paramagnetic moments were first observed in graphene with hydrogen doping at $\sim$ 15 K [38]. In this experiment, a hydrogen atomic source is used to dope the graphene with hydrogen and the magnetic moments are probed by pure spin current with *in situ* nonlocal spin transport measurements. A dip at zero magnetic field in the nonlocal MR measurement indicates that the spins interact with the local hydrogen-induced magnetic moments via exchanging coupling, thus providing evidence for the formation of magnetic moments. In other words, the graphene spin valve is being utilized as a very sensitive magnetometer of the hydrogen-induced magnetic moments.

Recently, the adsorption of a single hydrogen atom on graphene was shown to induce a magnetic moment by scanning tunneling microscopy [40]. As shown in Figure 5.14a, a scanning tunneling microscopy visualizes a single H atom on graphene as a bright protrusion surrounded by a complex threefold $\sqrt{3} \times \sqrt{3}$ pattern. The atomic H is chemically absorbed onto graphene grown on a SiC (0001) substrate. Figure 5.14b compares the dI/dV spectra measured on pristine graphene and on top of single H atoms. Two narrow peaks, one below and one above E_F, separated in energy by a splitting of $\sim$ 20 meV is observed on top of the H atoms, which could be attributed to the spin-polarized states. This atomically modulated spin texture, which extends several nanometers away from the hydrogen atom, drives the direct coupling between the magnetic moments at unusually

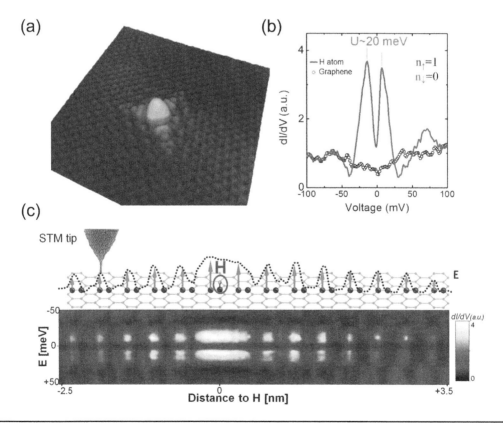

FIGURE 5.14   (a) STM topography of a single H atom chemisorbed on neutral graphene. (b) dI/dV spectra measured on the H atom, showing a fully polarized peak at E_F, and measured on bare graphene far from the H atom. U indicates the energy splitting for the spin up and spin down energy levels. The spectra were acquired at a nominal junction impedance of 2 GΩ (–100 mV, 50 pA). (c) Conductance map [dI/dV(x,E)] along the line across an H atom on graphene. This indiates the spatial extension of the spin-polarized electronic state induced by H atoms in undoped graphene. The spectra were acquired at a nominal junction impedance of 3 gigaohms (100 mV, 33 pA). (After González-Herrero, H. et al., *Science* 352, 437, 2016. With permission.)

long distances. As shown in Figure 5.14c, the magnetic moment is induced on the carbon atoms in the graphene sublattice opposite to the locus of H chemisorption. The large extension of the local magnetic moments associated with H chemisorption suggests that long-range magnetic interactions mediated by direct exchange should take place. Furthermore, the manipulation of the hydrogen atoms with atomic precision is achieved using the STM tip. This paves the way to tailor the magnetism of selected graphene regions.

5.4.3 Proximity-Induced Ferromagnetism in Graphene

The other route to induce ferromagnetism in graphene is by proximity effect in adjacent with ferromagnetic materials with spin polarized d or f orbitals. Since the graphene is a conducting material, anomalous Hall effect can be used to detect the ferromagnetism if it is there. A ferromagnetic insulator is preferred due to its insulating nature, as it will not create parallel conduction paths.

Experimentally, two ferromagnetic insulators have been used to study this magnetic proximity effect at the interface. The first one is yttrium iron garnet (YIG) with a Curie temperature of ~ 560 K. The other one is EuS with a lower Curie temperature of ~ 16 K. For both ferromagnetic insulators, a large proximity effect is observed. Induced ferromagnetic order is detected in graphene on YIG, and a large magnetic exchange field is reported at the interface between graphene and EuS [134, 139].

The proximity-induced ferromagnetism in graphene has been reported by Wang et al. in the graphene/YIG heterostructure [134]. The interface between graphene and YIG is one of the important parameters to achieve the strong magnetic proximity effect and to induce ferromagnetism. In their work, YIG has an atomically flat surface grown by pulsed laser deposition, which provides the high interface quality when graphene is transferred onto the YIG surface. The Hall bar device structure is illustrated in Figure 5.15a. The top gate is used to tune the carrier density of graphene. As shown in Figure 5.15b, anomalous Hall effect is observed up to 250 K in graphene on YIG. One important feature is that the saturating field is identical to that of YIG, which confirms the ferromagnetism is induced from the YIG layer. The observation of anomalous Hall effect is a strong indication of induced ferromagnetism arising from the hybridization between the π orbitals in graphene and the nearby spin-polarized d orbitals in YIG.

One of the important applications of the induced ferromagnetism in graphene is to modulate spin currents in graphene spin valves. This was demonstrated in two experimental studies of graphene on YIG [135, 136]. Figure 5.16 illustrates the full modulation of spin currents in bilayer graphene on YIG [136]. Schematically (Figure 5.16a), when the magnetization of the YIG substrate is collinear with the magnetization of the spin injector electrode, the spin current will traverse the graphene to reach the spin detector and register a nonzero nonlocal spin signal. On the other hand, if the YIG magnetization is rotated to be perpendicular to the spin injector

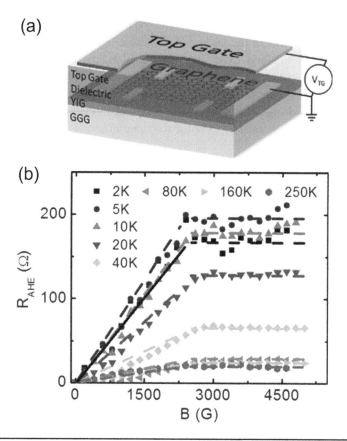

FIGURE 5.15 (a) Schematic drawing (with top gate) of the single layer graphene devices after transferred to a ferromagnetic insulator yttrium iron garnet (YIG). The YIG is grown on gadolinium gallium garnet (GGG) substrates by pulsed laser deposition. (b) Anomalous Hall resistance of ferromagnetic graphene induced by magnetic proximity effect with YIG. (After Wang, Z. et al., *Phys. Rev. Lett.* 114, 016603, 2015. With permission.)

magnetization, then the spin current will exhibit a precessional dephasing in the presence of the exchange field induced by YIG. Thus, the spin current will be dephased before it can reach the spin detector, thereby generating zero nonlocal spin signal. The experiment exhibiting this behavior is shown in Figure 5.16b, where θ is the angle of the applied magnetic field relative to the Co spin injector axis. With a fixed applied field magnitude of 15 mT, this is large enough to set the orientation of the YIG magnetization (YIG saturation field is less than 3 mT) yet small enough to produce minimal changes to the Co magnetization axis as the applied field direction θ is rotated through 360°. The blue (red) curve shows the nonlocal resistance R_{NL} for parallel (anti-parallel) magnetization alignments of the Co spin injector and detector. Thus, the difference of the two curves is R_{NL}, the nonlocal spin signal representing the spin current reaching the detector. As expected, the R_{NL} is maximized when the YIG magnetization is collinear with the Co direction (for $\theta = 0°$, 180°, 360°), while the curves pinch off to yield $R_{NL} = 0$ (no spin

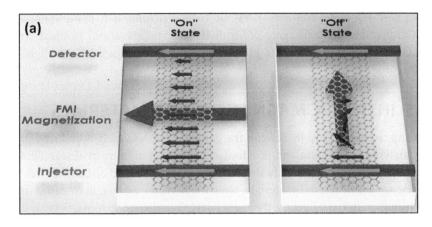

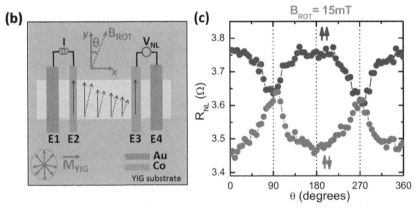

FIGURE 5.16   Full modulation of spin currents in graphene/YIG. (a) Device concept with transmission of spin current with collinear YIG magnetization (left) and dephasing of spin current with perpendicular YIG magnetization (right). (b) Nonlocal spin signal as a function of external magnetic field angle for fixed field magnitude of 15 mT. The blue (red) curve corresponding to parallel (anti-parallel) magnetization alignment of the spin injector and detector magnetizations. The difference between the curves is the nonlocal spin signal. (After Singh, S. et al., *Phys. Rev. Lett.* 118, 187201, 2017. With permission.)

current at detector) when the YIG magnetization is perpendicular to the Co direction (for $\theta = 90°$, 270°).

It is also possible to consider magnetic proximity effect in graphene with ferromagnetic metals [140]. Proximity-induced exchange spin splitting in graphene suggest that spin injection and spin transport should be revisited. Proximity-modified graphene is no longer a nonmagnetic region and by acquiring spin-dependent properties itself should be viewed as a source of spin injection [140, 141]. In Co/h-BN/graphene junction it was predicted that the proximity-induced spin polarization can be reversed by changing the gate voltage [142] and potentially avoid the need for an applied magnetic field in graphene spin logic devices (Figure 5.5). However, in conventional 2D contacts the needed electric field for such spin polarization reversal was very high. Instead, carefully designed 1D edge contacts [143] in Co/encapsulated h-BN/graphene junctions demonstrate gate-controlled spin polarization reversal at much smaller applied field and open novel possibilities with graphene spintronics [144].

5.5 INTRINSIC MAGNETISM IN 2D MATERIALS

During the same time to develop the extrinsic ferromagnetism induced in 2D materials by doping and proximity effect, a great effort by researchers has also been used to identify intrinsically 2D ferromagnetism and antiferromagnetism in layered materials, which exhibit robust magnetic order in their bulk materials. Luckily, there are a large number of ferromagnetic and antiferromagnetic semiconductors/insulators that have a similar layered structure as graphene and predicted to have the magnetic order down to atomic layers, including ternary tri-tellurides $Cr_2X_2Te_6$ (X = Si, Ge), transition-metal chalcogenophosphates MPX_3 (M = V, Cr, Mn, Fe, Co, Ni, Cu, Zn, and X = S, Se, Te), chromium trihalides CrX_3 (X= F, Cl, Br, I), and vanadium-based dichalcogenides [145–153].

Figure 5.17 shows the ground-state magnetic phase diagram of semiconducting transition-metal trichalcogenide monolayers as a function of J_1/J_3 and J_2/J_3. J_1, J_2 and J_3 are the first, second, and third nearest-neighbor exchange interactions. The calculation is performed based on the Heisenberg model using first-principles calculations within density functional theory. Strain could be used to tune the magnetic interactions, thus to cause a phase transition.

Although the long-range magnetic order is prohibited theoretically in the 1D and 2D isotropic Heisenberg model by thermal fluctuations based on the Mermin–Wagner theorem [154], uniaxial magnetic anisotropy can remove

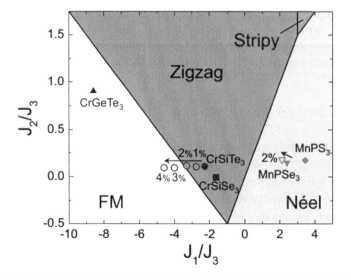

FIGURE 5.17 The ground-state magnetic phase diagram of semiconducting transition-metal trichalcogenide monolayers as a function of J_1/J_3 and J_2/J_3. J_1, J_2 and J_3 are the first, second, and third nearest-neighbor exchange interactions. (After Sivadas, N. et al., *Phys. Rev. B* 91, 235425, 2015. With permission.)

this restriction and induce 2D ferromagnetism or antiferromagnetism. Recently a lot of progress in this field has been achieved both theoretically and experimentally. In this section, the theoretical and experimental identification of the 2D ferromagnetism and 2D antiferromagnetism is first discussed. Then the electric field effect in 2D ferromagnetic material of $Cr_2Ge_2Te_6$ is presented. At the end, the potential applications of the 2D ferromagnetism for spintronics devices, such as magnetic tunnel junctions, will be discussed.

5.5.1 Intrinsic 2D Ferromagnetism

In the search for 2D ferromagnetism, recent experiments have demonstrated that ultrathin exfoliated layers of CrI_3 and $Cr_2Ge_2Te_6$ provide the realizations of 2D Ising and Heisenberg ferromagnets, respectively [42, 43].

The Ising and Heisenberg models are illustrated in Figure 5.18a, b. For the Ising model, the spins on the Cr atoms are constrained to lie perpendicular to the plane of monolayer. For the Heisenberg model, spins are free to point in any spatial direction.

Bulk CrI_3 shows layered Ising ferromagnetism below a Curie temperature of ~ 61 K with an out-of-plane easy axis [151, 155]. The ferromagnetic order, characterized by polar magneto-optical Kerr effect, down to monolayer CrI_3 is shown in Figure 5.18c. The hysteresis curves of monolayer and trilayer CrI_3 indicate ferromagnetic behavior as a function of the external magnetic field perpendicular to the thin layers. For bilayer CrI_3, an antiferromagnetic behavior is observed. At zero magnetic field, the measured magnetic moment is close to zero, indicating that almost perfect compensation of the two CrI_3 layers that are antiferromagnetically coupled.

Bulk $Cr_2Ge_2Te_6$ has an out-of-plane magnetic easy axis with a Curie temperature of ~ 61 K [156–158]. Raman spectra indicate that the $Cr_2Ge_2Te_6$ flakes are potentially ferromagnetic in the 2D limit [159]. Recently, magnetic optical Kerr effect and anomalous Hall effect have been used to demonstrate the ferromagnetic order in $Cr_2Ge_2Te_6$ flakes [43, 45]. Very interestingly, Gong et al. show that the ferromagnetic order could be realized down to 2 layers (~1.1 nm) [43]. The crystal structure of $Cr_2Ge_2Te_6$ is illustrated in Figure 5.18d. The Curie temperature decreases as the layer number decreases (Figure 5.18e), which can be explained theoretically using the Heisenberg model. The inset shows the magnetic moment probed by Kerr rotation as a function of temperature. The Curie temperature is ~ 30 K. The theoretical model also predicts a Curie temperature of 15 K for monolayer $Cr_2Ge_2Te_6$, which needs further experimental demonstration.

5.5.2 Intrinsic 2D Antiferromagnetism

Bulk MPS_3 (M=Fe, Mn, Ni) layered materials exhibit very interesting properties below the Néel temperature for antiferromagnetism [160]. The magnetization axis lies perpendicular to the layers for $MnPS_3$ and $FePS_3$, while it lies in the layer for $NiPS_3$. Besides, $MnPS_3$ is isotropic, $NiPS_3$ shows only a weak anisotropy, and $FePS_3$ exhibits highly anisotropic susceptibility. Furthermore, $MnPS_3$ is best described by the isotropic Heisenberg Hamiltonian, $FePS_3$ is

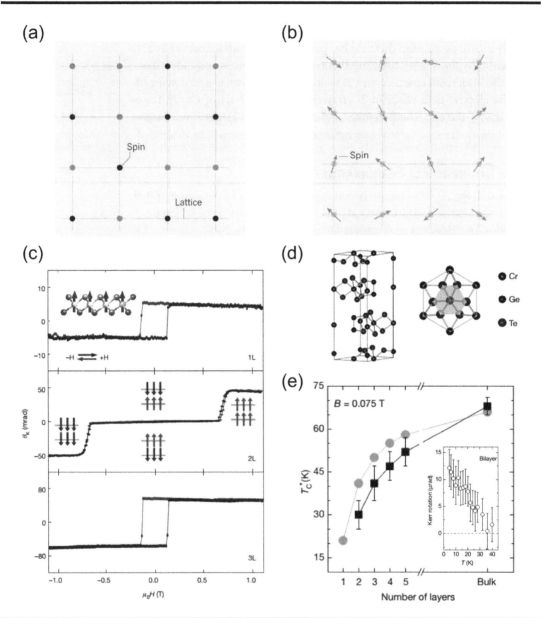

FIGURE 5.18 (a, b) Two-dimensional magnetic systems described by the Ising (a) and Heisenberg (b) models. (c) Polar magneto-optical Kerr effect (MOKE) signal on a monolayer (top), bilayer (middle), and trilayer (bottom) CrI_3 flakes. The hysteresis curves of monolayer and trilayer CrI_3 indicate ferromagnetic behavior. Antiferromagnetic behavior is observed on bilayer CrI_3 flake. Inset: Out-of-plane view of the same CrI_3 structure depicting the Ising spin orientation. (d) Crystal structure (side and top views) of $Cr_2Ge_2Te_6$. (e) The transition temperatures T_C of $Cr_2Ge_2Te_6$ samples of different thickness, obtained from Kerr measurements (blue squares) and theoretical calculations (red circles) under a magnetic field of 0.075 T. Inset: Temperature dependent Kerr rotation intensities of bilayer $Cr_2Ge_2Te_6$ sample under a magnetic field of 0.075 T. ((a, b) After Samarth, N., *Nature* 546, 216, 2017. With permission. (c) After Huang, B. et al., *Nature* 546, 270, 2017. With permission. (d, e) After Gong, C. et al., *Nature* 546, 265, 2017. With permission.)

most effectively treated by the Ising model and $NiPS_3$ by the anisotropic Heisenberg Hamiltonian.

Recently, intrinsic 2D antiferromagnetism has been indicated in $FePS_3$ by Raman studies [44, 161]. The crystal structure is illustrated in Figure 5.19a. Due to magnetic-elastic scattering, the Raman peaks can indicate the antiferromagnetic transition temperature. The transition temperature remains almost independent of the thickness from bulk to the monolayer limit with $T_N \sim 118$ K, indicating the Ising type antiferromagnetism. This also means that the antiferromagnetic ordering is mainly determined by the intralayer exchange interactions, while the interlayer interaction has little effect.

5.5.3 ELECTRIC FIELD EFFECT IN 2D FERROMAGNETIC MATERIALS

2D ferromagnetic materials have several important advantages for the application as information storage, compared to conventional metal and semiconductor materials. The first one is the ultrathin nature, which could be used for high density storage applications. The other one but more important is that the magnetic properties might be manipulated easily by electric field effect, as discussed in Chapter 13, Volume 2, in this book. For example, only ultrathin ferromagnetic metal can be tuned by the electric field effect since the Fermi-Thomas screening length is extremely short arising from the large carrier densities [162, 163]. For ferromagnetic semiconductors, the electric field tuning of the magnetism relies on the carrier densities [164, 165]. While for 2D materials, their physical properties could be largely tuned via the modulation of the carrier density or variation of the band structure itself by a perpendicular electric field [14, 24, 166].

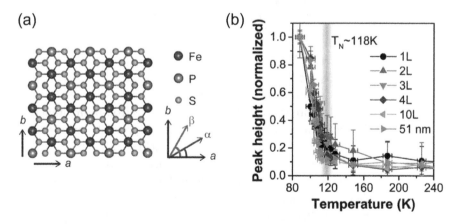

FIGURE 5.19   Ising-type antiferromagnetism in atomically thin $FePS_3$. (a) Crystal structure of $FePS_3$ from the top view. (b) Temperature dependence of peak height for P1a, the low-frequency peak from vibrations including Fe atoms. A Néel temperature (T_N) of ~ 118 K is observed for various $FePS_3$ samples of thickness from monolayer to 51 nm. Similar results but a slightly larger thickness dependence of the Néel temperature are reported in Ref. [161]. (After Lee, J.-U. et al., *Nano Lett.* 16, 7433, 2016. With permission.)

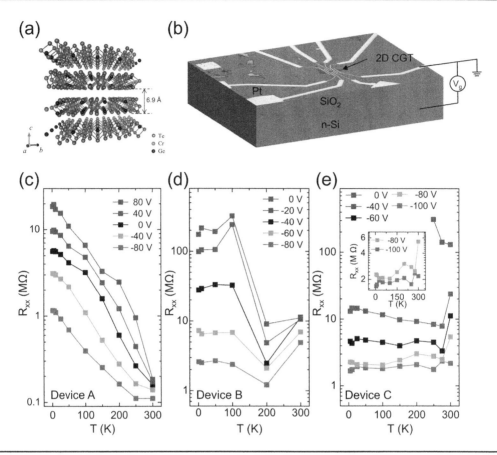

FIGURE 5.20 Electric field effect in 2D ferromagnetic $Cr_2Ge_2Te_6$ (CGT) devices. (a) The crystal structure of $Cr_2Ge_2Te_6$. (b) Schematic of the Hall bar devices fabricated on the 2D CGT flakes on the SiO_2/Si substrate, where the n-doped Si is used as a back gate, the thickness of the SiO_2 layer is 300 nm and Pt electrodes are used for electrical contacts. (c–e) Large modulation of the channel resistances of the 2D CGT devices for thicknesses of ~ 25 nm (c, Device A), ~ 15 nm (d, Device B), and ~ 12 nm (e, Device C) by electric field effect. (After Xing, W. et al., *2D Mater.* 4, 024009, 2017. With permission.)

Recently, Xing et al. demonstrated the large electric field effect on the 2D $Cr_2Ge_2Te_6$ Hall bar devices [45]. Figure 5.20a, b illustrate the crystal structure of $Cr_2Ge_2Te_6$ and the device fabricated based on the 2D flakes. The Hall bar devices are fabricated using standard electron-beam lithography and the electrical contacts are made of ~ 80 nm Pt grown by magnetron sputtering. The substrates are SiO_2/Si, where the thickness of SiO_2 is ~ 300 nm and the highly electron-doped Si is used as a back gate to provide the electric fields. The main results of the electric field effect in the three representative devices (Device A, B, and C) using different thicknesses of $Cr_2Ge_2Te_6$ flakes are shown in Figure 5.20c–e. For Device A (25 nm $Cr_2Ge_2Te_6$), the ungated device exhibits a semiconducting feature. The gate tuning of the channel resistance shows the same trend as that at 300 K in the full temperature range, but the modulation is more dramatic at low temperatures. When the films become thinner, the channel resistances exhibit a large increase. For instance, at 300 K and $V_g = 0$ V, Device B (~15 nm $Cr_2Ge_2Te_6$) has a channel resistance of ~ 10 MΩ, and Device C (~12 nm $Cr_2Ge_2Te_6$) shows channel resistance of ~ 130 MΩ. For even thinner flakes, a giant modulation of the channel resistance is observed. As shown in Figure 5.20e for Device C, the channel resistance changes from very insulating (not measurable below 250 K) at 0 V to quite conducting, which shows a resistance of ~ 1.6 MΩ at −100 V and 1.5 K. From the temperature dependence of the channel resistance, two major conclusions could be drawn. First, there is less gating effect observed in thicker CGT films, which could be attributed to the charge screening effect of the applied electric field. Second and most importantly, for thicker CGT Device A under −80 V, the channel resistances increase as the temperature decreases, still indicating a semiconducting feature. Meanwhile, for thinner CGT Device C, the channel resistance decreases as the temperature decreases below ~ 200 K under $V_g = -100$ V, indicating a metallic behavior, in stark contrast to the semiconducting/insulating properties while ungated.

The electric field effect on the ferromagnetic properties needs further studies.

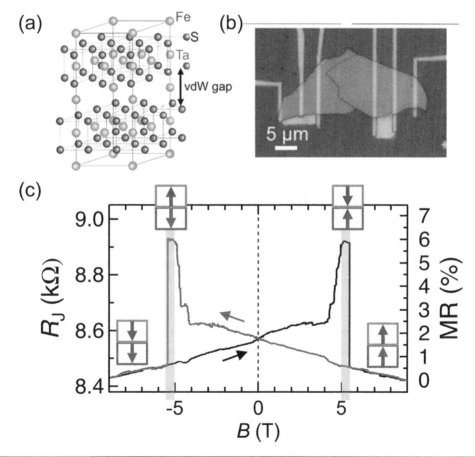

FIGURE 5.21   Van der Waals magnetic tunnel junction using ferromagnetic layered dichalcogenides. (a) Crystal structure of $Fe_{0.25}TaS_2$ from the side view. (b) Optical micrograph of the van der Waals junction based on two $Fe_{0.25}TaS_2$ flakes with a similar thickness of ~ 100 nm. (c) External magnetic field dependence of the four-terminal resistance measured on the magnetic tunnel junction at 5 K. (After Arai, M. et al., *Appl. Phys. Lett.* 107, 103107, 2015. With permission.)

5.5.4 VAN DER WAALS MAGNETIC TUNNEL JUNCTIONS

For applications, 2D ferromagnetic materials have exhibited their potential in magnetic tunnel junctions [46]. Tunneling magnetoresistance is observed in the van der Waals junctions based on ~ 100 nm thick ferromagnetic $Fe_{0.25}TaS_2$, indicating spin-polarized tunneling across a van der Waals gap (Figure 5.21a) between the two ferromagnetic layered materials. $Fe_{0.25}TaS_2$ flakes have a Curie temperature of ~ 160 K. The magnetic tunnel junctions are fabricated using room temperature dry transfer technique, and the cleaved surfaces of different flakes are subsequently connected (Figure 5.21b).

A tunnel magnetoresistance of 6% is observed at 5 K, as shown in Figure 5.21c. Both the top and bottom layers (~100 nm thick bottom $Fe_{0.25}TaS_2$ flakes) have perpendicular magnetic anisotropy. The existence of Ta_2O_5 tunnel barrier at the surface of $Fe_{0.25}TaS_2$ enables spin-polarized tunneling.

5.6 CONCLUSIONS AND OUTLOOK

In summary, there has been a lot of progress in the field of spintronics in 2D materials due to their fantastic spin-dependent properties. Graphene, TMDCs, and extrinsic/intrinsic 2D ferromagnetic materials have been intensively investigated in the last several years.

Looking forward, there is no doubt that 2D materials will continue to be extremely attractive for spintronics. The van der Waals structures that integrate 2D materials with contrasting spin-dependent properties will be particularly interesting. The coupling of spin degree of freedom with valley, pseudospin, layer, and momentum will open up new avenues for 2D spintronics.

ACKNOWLEDGMENTS

We are grateful for the crucial contributions of our collaborators Jaroslav Fabian, Hanan Dery, Jing Shi, Igor Žutić, Ilya Krivorotov, Lu J. Sham, Kathleen McCreary, Berend Jonker, Shuang Jia, Yan Li, Xincheng Xie, Ezekiel Johnston-Halperin, Jay Gupta, Chris Hammel, Michael Flatté, David McComb, Mohit Randeria, Jeanie Lau, Marc Bockrath, Paul Crowell, Steven Koester, Ludwig Bartels, Cengiz Ozkan, Nitin Samarth, Mahesh Neupane, Harry Tom, Umar Mohideen, Fengyuan Yang, Siddharth Rajan, Roberto Myers, Josh Goldberger, Jie Shan, Leonard Brillson, and Wolfgang Windl. Wei Han acknowledges the funding support of National Basic Research Programs of China (973 Grants 2015CB921104 and 2014CB920902), National Natural Science Foundation of China (NSFC Grant 11574006), and the 1000 Talents Program for Young Scientists of China. Roland K. Kawakami acknowledges the funding support from Office of Naval Research (No. N00014-14-1-0350), National Science Foundation (No. DMR-1310661, No. DMR-1420451, No. DMR-1429143), US Department of Energy (No. DE-SC0018172), Army Research Office (No. W911NF-11-1-0182, No. W911NF-14-1-0457), and C-SPIN, one of the six SRC STARnet Centers, sponsored by Microelectronics Advanced Research Corporation (MARCO) and Defense Advanced Research Projects Agency (DARPA).

REFERENCES

1. S. A. Wolf, D. D. Awschalom, R. A. Buhrman et al., Spintronics: A spin-based electronics vision for the future, *Science* **294**, 1488 (2001).
2. I. Žutić, J. Fabian, and S. Das Sarma, Spintronics: Fundamentals and applications, *Rev. Mod. Phys* **76**, 323 (2004).
3. A. Fert, Nobel Lecture: Origin, development, and future of spintronics, *Rev. Mod. Phys.* **80**, 1517 (2008).
4. S. D. Bader and S. S. P. Parkin, Spintronics, *Annu. Rev. Cond. Mat.* **1**, 71 (2010).
5. M. N. Baibich, J. M. Broto, A. Fert et al., Giant magnetoresistance of (001)Fe/ (001)Cr magnetic superlattices, *Phys. Rev. Lett.* **61**, 2472 (1988).
6. G. Binasch, P. Grünberg, F. Saurenbach, and W. Zinn, Enhanced magneto-resistance in layered magnetic structures with antiferromagnetic interlayer exchange, *Phys. Rev. B* **39**, 4828 (1989).
7. R. Hanson, L. P. Kouwenhoven, J. R. Petta, S. Tarucha, and L. M. K. Vandersypen, Spins in few-electron quantum dots, *Rev. Mod. Phys.* **79**, 1217 (2007).
8. C. Kloeffel and D. Loss, Prospects for spin-based quantum computing in quantum dots, *Annu. Rev. Condens. Matter Phys.* **4**, 51 (2013).
9. S. Datta and B. Das, Electronic analog of the electro-optic modulator, *Appl. Phys. Lett.* **56**, 665 (1990).
10. H. Dery, P. Dalal, L. Cywinski, and L. J. Sham, Spin-based logic in semiconductors for reconfigurable large-scale circuits, *Nature* **447**, 573 (2007).
11. B. Behin-Aein, D. Datta, S. Salahuddin, and S. Datta, Proposal for an all-spin logic device with built-in memory, *Nat. Nanotechnol.* **5**, 266 (2010).
12. B. Supriyo and C. Marc, Electron spin for classical information processing: A brief survey of spin-based logic devices, gates and circuits, *Nanotechnol.* **20**, 412001 (2009).
13. K. S. Novoselov, A. K. Geim, S. V. Morozov et al., Two-dimensional gas of massless Dirac fermions in grahene, *Nature* **438**, 197 (2005).
14. K. S. Novoselov, A. K. Geim, S. V. Morozov et al., Electric field effect in atomically thin carbon films, *Science* **306**, 666 (2004).
15. Y. Zhang, Y.-W. Tan, H. L. Stormer, and P. Kim, Experimental observation of the quantum Hall effect and Berry's phase in graphene, *Nature* **438**, 201 (2005).
16. A. K. Geim and K. S. Novoselov, The rise of graphene, *Nat. Mater.* **6**, 183 (2007).
17. S. D. Sarma, S. Adam, E. H. Hwang, and E. Rossi, Electronic transport in two dimensional graphene, *Rev. Mod. Phys.* **83**, 407 (2011).
18. K. F. Mak, C. Lee, J. Hone, J. Shan, and T. F. Heinz, Atomically thin MoS_2: A new direct-gap semiconductor, *Phys. Rev. Lett.* **105**, 136805 (2010).
19. H. Zeng, J. Dai, W. Yao, D. Xiao, and X. Cui, Valley polarization in MoS_2 monolayers by optical pumping, *Nat. Nanotechnol.* **7**, 490 (2012).
20. K. F. Mak, K. He, J. Shan, and T. F. Heinz, Control of valley polarization in monolayer MoS_2 by optical helicity, *Nat. Nanotechnol.* **7**, 494 (2012).
21. T. Cao, G. Wang, W. Han et al., Valley-selective circular dichroism of monolayer molybdenum disulphide, *Nat. Commun.* **3**, 887 (2012).
22. A. M. Jones, H. Yu, N. J. Ghimire et al., Optical generation of excitonic valley coherence in monolayer WSe_2, *Nat. Nanotechnol.* **8**, 634 (2013).
23. B. Radisavljevic, A. Radenovic, J. Brivio, V. Giacometti, and A. Kis, Single-layer MoS_2 transistors, *Nat. Nanotechnol.* **6**, 147 (2011).
24. L. Li, Y. Yu, G. J. Ye et al., Black phosphorus field-effect transistors, *Nat. Nanotechnol.* **9**, 372 (2014).
25. H. Liu, A. T. Neal, Z. Zhu et al., Phosphorene: An unexplored 2D semiconductor with a high hole mobility, *ACS Nano* **8**, 4033 (2014).
26. A. K. Geim and I. V. Grigorieva, Van der Waals heterostructures, *Nature* **499**, 419 (2013).
27. K. S. Novoselov, A. Mishchenko, A. Carvalho, and A. H. Castro Neto, 2D materials and van der Waals heterostructures, *Science* **353** (2016).

28. M. Gmitra, S. Konschuh, C. Ertler, C. Ambrosch-Draxl, and J. Fabian, Band-structure topologies of graphene: Spin-orbit coupling effects from first principles, *Phys. Rev. B* **80**, 235431 (2009).

29. W. Han, R. K. Kawakami, M. Gmitra, and J. Fabian, Graphene spintronics, *Nat. Nanotechnol.* **9**, 794 (2014).

30. D. Pesin and A. H. MacDonald, Spintronics and pseudospintronics in graphene and topological insulators, *Nat. Mater.* **11**, 409 (2012).

31. B. Trauzettel, D. V. Bulaev, D. Loss, and G. Burkard, Spin qubits in graphene quantum dots, *Nat. Phys.* **3**, 192 (2007).

32. P. Recher and B. Trauzettel, Quantum dots and spin qubits in graphene, *Nanotechnology* **21**, 302001 (2010).

33. D. Xiao, G.-B. Liu, W. Feng, X. Xu, and W. Yao, Coupled spin and valley physics in monolayers of MoS_2 and other Group-VI dichalcogenides, *Phys. Rev. Lett.* **108**, 196802 (2012).

34. X. Xu, W. Yao, D. Xiao, and T. F. Heinz, Spin and pseudospins in layered transition metal dichalcogenides, *Nat. Phys.* **10**, 343 (2014).

35. L. Yang, N. A. Sinitsyn, W. Chen, J. Yuan, J. Lou, and S. A. Crooker, Long-lived nanosecond spin relaxation and spin coherence of electrons in monolayer MoS_2 and WS_2, *Nat. Phys.* **11**, 830 (2015).

36. M. M. Ugeda, I. Brihuega, F. Guinea, and J. M. Gómez-Rodríguez, Missing atom as a source of carbon magnetism, *Phys. Rev. Lett.* **104**, 096804 (2010).

37. R. R. Nair, M. Sepioni, I. L. Tsai et al., Spin-half paramagnetism in graphene induced by point defects, *Nat. Phys.* **8**, 199 (2012).

38. K. M. McCreary, A. G. Swartz, W. Han, J. Fabian, and R. K. Kawakami, Magnetic moment formation in graphene detected by scattering of pure spin currents, *Phys. Rev. Lett.* **109**, 186604 (2012).

39. W. Han, Perspectives for spintronics in 2D materials, *APL Mater.* **4**, 032401 (2016).

40. H. González-Herrero, J. M. Gómez-Rodríguez, P. Mallet et al., Atomic-scale control of graphene magnetism by using hydrogen atoms, *Science* **352**, 437 (2016).

41. N. Samarth, Condensed-matter physics: Magnetism in flatland, *Nature* **546**, 216 (2017).

42. B. Huang, G. Clark, E. Navarro-Moratalla et al., Layer-dependent ferromagnetism in a van der Waals crystal down to the monolayer limit, *Nature* **546**, 270 (2017).

43. C. Gong, L. Li, Z. Li, et al., Discovery of intrinsic ferromagnetism in two-dimensional van der Waals crystals, *Nature* **546**, 265 (2017).

44. J.-U. Lee, S. Lee, J. H. Ryoo et al., Ising-type magnetic ordering in atomically thin $FePS_3$, *Nano Lett.* **16**, 7433 (2016).

45. W. Xing, Y. Chen, P. M. Odenthal et al., Electric field effect in multilayer $Cr_2Ge_2Te_6$: A ferromagnetic 2D material, *2D Mater.* **4**, 024009 (2017).

46. M. Arai, R. Moriya, N. Yabuki, S. Masubuchi, K. Ueno, and T. Machida, Construction of van der Waals magnetic tunnel junction using ferromagnetic layered dichalcogenide, *Appl. Phys. Lett.* **107**, 103107 (2015).

47. I. Žutić, A. Matos-Abiague, B. Scharf, H. Dery, and K. Belashchenko, Proximitized materials, *Mater. Today* **22**, 85 (2019).

48. N. Tombros, C. Jozsa, M. Popinciuc, H. T. Jonkman, and B. J. van Wees, Electronic spin transport and spin precession in single graphene layers at room temperature, *Nature* **448**, 571 (2007).

49. W. Han, K. Pi, K. M. McCreary et al., Tunneling spin injection into single layer graphene, *Phys. Rev. Lett.* **105**, 167202 (2010).

50. P. J. Zomer, M. H. D. Guimarães, N. Tombros, and B. J. van Wees, Long-distance spin transport in high-mobility graphene on hexagonal boron nitride, *Phys. Rev. B* **86**, 161416 (2012).

51. B. Dlubak, M.-B. Martin, C. Deranlot et al., Highly efficient spin transport in epitaxial graphene on SiC, *Nat. Phys.* **8**, 557 (2012).

52. M. H. D. Guimarães, P. J. Zomer, J. Ingla-Aynés, J. C. Brant, N. Tombros, and B. J. van Wees, Controlling spin relaxation in hexagonal BN-encapsulated graphene with a transverse electric field, *Phys. Rev. Lett.* **113**, 086602 (2014).

53. M. Drögeler, C. Franzen, F. Volmer et al., Spin lifetimes exceeding 12 ns in graphene nonlocal spin valve devices, *Nano Lett.* **16**, 3533 (2016).

54. S. Roche, J. Åkerman, B. Beschoten et al., Graphene spintronics: The European Flagship perspective, *2D Mater.* **2**, 030202 (2015).

55. H. Wen, H. Dery, W. Amamou, Experimental demonstration of XOR operation in graphene magnetologic gates at room temperature, *Phys. Rev. Applied* **5**, 044003 (2016).

56. W. Yan, O. Txoperena, R. Llopis, H. Dery, L. E. Hueso, and F. Casanova, A two-dimensional spin field-effect switch, *Nat. Commun.* **7**, 13372 (2016).

57. A. Dankert and S. P. Dash, Electrical gate control of spin current in van der Waals heterostructures at room temperature, *Nat. Commun.* **8**, 16093 (2017).

58. M. Johnson and R. H. Silsbee, Interfacial charge-spin coupling: Injection and detection of spin magnetization in metals, *Phys. Rev. Lett.* **55**, 1790 (1985).

59. F. J. Jedema, H. B. Heersche, A. T. Filip, J. J. A. Baselmans, and B. J. van Wees, Electrical detection of spin precession in a metallic mesoscopic spin valve, *Nature* **416**, 713 (2002).

60. F. J. Jedema, A. T. Filip, and B. J. van Wees, Electrical spin injection and accumulation at room temperature in an all-metal mesoscopic spin valve, *Nature* **410**, 345 (2001).

61. T. Kimura and Y. Otani, Large spin accumulation in a permalloy-silver lateral spin valve, *Phys. Rev. Lett.* **99**, 196604 (2007).

62. X. Lou, C. Adelmann, S. A. Crooker et al., Electrical detection of spin transport in lateral ferromagnet-semiconductor devices, *Nat. Phys.* **3**, 197 (2007).

63. T. Sasaki, T. Oikawa, T. Suzuki, M. Shiraishi, Y. Suzuki, and K. Tagami, Electrical spin injection into silicon using MgO tunnel barrier, *Appl. Phys. Expr.* **2**, 53003 (2009).

64. O. M. J. van 't Erve, A. L. Friedman, E. Cobas, C. H. Li, J. T. Robinson, and B. T. Jonker, Low-resistance spin injection into silicon using graphene tunnel barriers, *Nat. Nanotechnol.* **7**, 737 (2012).

65. Y. Zhou, W. Han, L.-T. Chang et al., Electrical spin injection and transport in germanium, *Phys. Rev. B* **84**, 125323 (2011).

66. A. Avsar, J. Y. Tan, M. Kurpas et al., Gate-tunable black phosphorus spin valve with nanosecond spin lifetimes, *Nat. Phys.* **13**, 888 (2017).

67. M. Ohishi, M. Shiraishi, R. Nouchi, T. Nozaki, T. Shinjo, and Y. Suzuki, Spin injection into a graphene thin film at room temperature, *Jpn. J. Appl. Phys* **46**, L605 (2007).

68. W. Han, W. H. Wang, K. Pi et al., Electron-hole asymmetry of spin injection and transport in single-layer graphene, *Phys. Rev. Lett.* **102**, 137205 (2009).

69. T.-Y. Yang, J. Balkrishnna, F. Volmer et al., Observation of long spin relaxation times in bilayer graphene at room temperature, *Phys. Rev. Lett.* **107**, 047206 (2011).

70. T. Maassen, I. J. Vera-Marun, M. H. D. Guimarães, and B. J. van Wees, Contact-induced spin relaxation in Hanle spin precession measurements, *Phys. Rev. B* **86**, 235408 (2012).

71. H. Idzuchi, A. Fert, and Y. Otani, Revisiting the measurement of the spin relaxation time in graphene-based devices, *Phys. Rev. B* **91**, 241407 (2015).

72. W. Amamou, Z. Lin, J. v. Baren, S. Turkyilmaz, J. Shi, and R. K. Kawakami, Contact induced spin relaxation in graphene spin valves with Al_2O_3 and MgO tunnel barriers, *APL Materials* **4**, 032503 (2016).

73. G. Stecklein, P. A. Crowell, J. Li, Y. Anugrah, Q. Su, and S. J. Koester, Contact-induced spin relaxation in graphene nonlocal spin valves, *Phys. Rev. Applied* **6**, 054015 (2016).

74. E. Sosenko, H. Wei, and V. Aji, Effect of contacts on spin lifetime measurements in graphene, *Phys. Rev. B* **89**, 245436 (2014).

75. R. J. Elliott, Theory of the effect of spin-orbit coupling on magnetic resonance in some semiconductors, *Phys. Rev.* **96**, 266 (1954).
76. Y. Yafet, in *Solid State Physics*, F. Seitz and D. Turnbull (Eds.), vol. 14 Academic Press Inc., New York, p. 1, 1963.
77. M. I. Dyakonov and V. I. Perel, Spin relaxation of conduction electrons in noncentrosymmetric semiconductors, *Sov. Phys. Solid State* **13**, 3023 (1972).
78. J. P. Robinson, H. Schomerus, L. Oroszlány, and V. I. Fal'ko, Adsorbate-limited conductivity of graphene, *Phys. Rev. Lett.* **101**, 196803 (2008).
79. T. O. Wehling, S. Yuan, A. I. Lichtenstein, A. K. Geim, and M. I. Katsnelson, Resonant scattering by realistic impurities in graphene, *Phys. Rev. Lett.* **105**, 056802 (2010).
80. D. V. Tuan, F. Ortmann, D. Soriano, S. O. Valenzuela, and S. Roche, Pseudospin-driven spin relaxation mechanism in graphene, *Nat. Phys.* **10**, 857 (2014).
81. M. W. Wu, J. H. Jiang, and M. Q. Weng, Spin dynamics in semiconductors, *Phys. Rep.* **493**, 61 (2010).
82. J. Fabian and S. Das Sarma, Spin relaxation of conduction electrons, *J. Vac. Sci. Technol. B* **17**, 1708 (1999).
83. J. Fabian, A. Matos-Abiague, C. Ertler, P. Stano, and I. Zutic, Semiconductor spintronics, *Acta Phys. Slovaca* **57**, 565 (2007).
84. W. Han and R. K. Kawakami, Spin relaxation in single layer and bilayer graphene, *Phys. Rev. Lett.* **107**, 047207 (2011).
85. C. Józsa, T. Maassen, M. Popinciuc et al., Linear scaling between momentum and spin scattering in graphene, *Phys. Rev. B* **80**, 241403(R) (2009).
86. M. Popinciuc, C. Jozsa, P. J. Zomer et al., Electronic spin transport in graphene field effect transistors, *Phys. Rev. B* **80**, 214427 (2009).
87. H. Ochoa, A. H. Castro Neto, and F. Guinea, Elliot-Yafet mechanism in graphene, *Phys. Rev. Lett.* **108**, 206808 (2012).
88. V. K. Dugaev, E. Y. Sherman, and J. Barnaś, Spin dephasing and pumping in graphene due to random spin-orbit interaction, *Phys. Rev. B* **83**, 085306 (2011).
89. P. Zhang and M. W. Wu, Electron spin relaxation in graphene with random Rashba field: Comparison of the D'yakonov–Perel' and Elliott–Yafet-like mechanisms, *New J. Phys.* **14**, 033015 (2012).
90. D. Kochan, M. Gmitra, and J. Fabian, Spin relaxation mechanism in graphene: Resonant scattering by magnetic impurities, *Phys. Rev. Lett.* **112**, 116602 (2014).
91. N. Tombros, S. Tanabe, A. Veligura et al., Anisotripic spin relaxation in graphene, *Phys. Rev. Lett.* **101**, 046601 (2008).
92. B. Raes, J. E. Scheerder, M. V. Costache et al., Determination of the spin-lifetime anisotropy in graphene using oblique spin precession, *Nat. Commun.* **7**, 11444 (2016).
93. H. Dery, H. Wu, B. Ciftcioglu et al., Nanospintronics based on magnetologic gates, *IEEE Trans. Elec. Dev.* **59**, 259 (2012).
94. M. Gmitra, D. Kochan, P. Hogl, and J. Fabian, Trivial and inverted Dirac bands and the emergence of quantum spin Hall states in graphene on transition-metal dichalcogenides, *Phys. Rev. B* **93**, 155104 (2016).
95. W.-K. Tse, Z. Qiao, Y. Yao, A. H. MacDonald, and Q. Niu, Quantum anomalous Hall effect in single-layer and bilayer graphene, *Phys. Rev. B* **83**, 155447 (2011).
96. M. Gmitra and J. Fabian, Proximity effects in bilayer graphene on monolayer WSe_2: Field-effect spin valley locking, spin-orbit valve, and spin transistor, *Phys. Rev. Lett.* **119**, 146401 (2017).
97. Z. Wang, D.-K. Ki, H. Chen, H. Berger, A. H. MacDonald, and A. F. Morpurgo, Strong interface-induced spin-orbit interaction in graphene on WS_2, *Nat. Commun.* **6**, 8339 (2015).
98. Z. Wang, D.-K. Ki, J. Y. Choo et al., Origin and magnitude of 'designer' spin-orbit interaction in graphene on semiconducting transition metal dichalcogenides, *Phys. Rev. X* **6**, 041020 (2016).
99. B. Yang, M.-F. Tu, J. Kim et al., Tunable spin-orbit coupling and symmetry-protected edge states in graphene/WS_2, *2D Mater.* **3**, 031012 (2016).

100. A. Avsar, J. Y. Tan, T. Taychatanapat et al., Spin-orbit proximity effect in graphene, *Nat. Commun.* **5**, 4875 (2014).

101. T. S. Ghiasi, J. Ingla-Aynés, A. A. Kaverzin, and B. J. van Wees, Large proximity-induced spin lifetime anisotropy in transition metal dichalcogenide/graphene heterostructures, *Nano Lett.* **17**, 7528 (2017).

102. L. A. Benitez, J. F. Sierra, W. Savero Torres et al., Strongly anisotropic spin relaxation in graphene/WS$_2$ van der Waals heterostructures, *arXiv:1710.11568* (2017).

103. A. Rycerz, J. Tworzydlo, and C. W. J. Beenakker, Valley filter and valley valve in graphene, *Nat. Phys.* **3**, 172 (2007).

104. D. Xiao, W. Yao, and Q. Niu, Valley-contrasting physics in graphene: Magnetic moment and topological transport, *Phys. Rev. Lett.* **99**, 236809 (2007).

105. D. Xiao, M.-C. Chang, and Q. Niu, Berry phase effects on electronic properties, *Rev. Mod. Phys.* **82**, 1959 (2010).

106. K. F. Mak, K. L. McGill, J. Park, and P. L. McEuen, The valley Hall effect in MoS$_2$ transistors, *Science* **344**, 1489 (2014).

107. R. V. Gorbachev, J. C. W. Song, G. L. Yu et al., Detecting topological currents in graphene superlattices, *Science* **346**, 448 (2014).

108. M. Sui, G. Chen, L. Ma et al., Gate-tunable topological valley transport in bilayer graphene, *Nat. Phys.* **11**, 1027 (2015).

109. Y. Shimazaki, M. Yamamoto, I. V. Borzenets, K. Watanabe, T. Taniguchi, and S. Tarucha, Generation and detection of pure valley current by electrically induced Berry curvature in bilayer graphene, *Nat. Phys.* **11**, 1032 (2015).

110. E. J. McCormick, M. J. Newburger, Y. K. Luo et al., Imaging spin dynamics in monolayer WS$_2$ by time-resolved Kerr rotation microscopy, *2D Mater.* **5**, 011010 (2018).

111. W.-T. Hsu, Y.-L. Chen, C.-H. Chen et al., Optically initialized robust valley-polarized holes in monolayer WSe$_2$, *Nat. Commun.* **6**, 8963 (2015).

112. X. Song, S. Xie, K. Kang, J. Park, and V. Sih, Long-lived hole spin/valley polarization probed by Kerr rotation in monolayer WSe$_2$, *Nano Lett.* **16**, 5010 (2016).

113. F. Volmer, S. Pissinger, M. Ersfeld, S. Kuhlen, C. Stampfer, and B. Beschoten, Intervalley dark trion states with spin lifetimes of 150 ns in WSe$_2$, *Phys. Rev. B* **95**, 235408 (2017).

114. P. Dey, L. Yang, C. Robert et al., Gate controlled spin-valley locking of resident carriers in WSe$_2$ monolayers, *Phys. Rev. Lett.* **119**, 137401 (2017).

115. J. Kim, C. Jin, B. Chen et al., Observation of ultralong valley lifetime in WSe$_2$/MoS$_2$ heterostructures, *Sci. Adv.* **3**, e1700518 (2017).

116. Y. Ye, J. Xiao, H. Wang et al., Electrical generation and control of the valley carriers in a monolayer transition metal dichalcogenide, *Nat. Nanotechnol.* **11**, 598 (2016).

117. C. Zhao, T. Norden, P. Zhang et al., Enhanced valley splitting in monolayer WSe2 due to magnetic exchange field, *Nat. Nanotechnol.* **12**, 757 (2017).

118. D. Zhong, K. L. Seyler, X. Linpeng et al., Van der Waals engineering of ferromagnetic semiconductor heterostructures for spin and valleytronics, *Sci. Adv.* **3**, e1603113 (2017).

119. O. L. Sanchez, D. Ovchinnikov, S. Misra, A. Allain, and A. Kis, Valley polarization by spin injection in a light-emitting van der Waals heterojunction, *Nano Lett.* **16**, 5792 (2016).

120. J. Qi, X. Li, Q. Niu, and J. Feng, Giant and tunable valley degeneracy splitting in MoTe$_2$, *Phys. Rev. B* **92**, 121403 (2015).

121. B. Scharf, G. Xu, A. Matos-Abiague, and I. Žutić, Magnetic proximity effects in transition-metal dichalcogenides: Converting excitons, *Phys. Rev. Lett.* **119**, 127403 (2017).

122. M. Gmitra and J. Fabian, Graphene on transition-metal dichalcogenides: A platform for proximity spin-orbit physics and optospintronics, *Phys. Rev. B* **92**, 155403 (2015).

123. Y. K. Luo, J. Xu, T. Zhu et al., Opto-valleytronic spin injection in monolayer MoS$_2$/Few-layer graphene hybrid spin valves, *Nano Lett.* **17**, 3877 (2017).

124. A. Avsar, D. Unuchek, J. Liu et al., Optospintronics in graphene via proximity coupling, *ACS Nano*, (2017), doi:10.1021/acsnano.7b06800.

125. X. Xi, Z. Wang, W. Zhao et al., Ising pairing in superconducting $NbSe_2$ atomic layers, *Nat. Phys.* **12**, 139 (2016).

126. J. M. Lu, O. Zheliuk, I. Leermakers et al., Evidence for two-dimensional Ising superconductivity in gated MoS_2, *Science* **350**, 1353 (2015).

127. Y. Saito, Y. Nakamura, M. S. Bahramy et al., Superconductivity protected by spin-valley locking in ion-gated MoS_2, *Nat. Phys.* **12**, 144 (2016).

128. Y. Saito, T. Nojima, and Y. Iwasa, Highly crystalline 2D superconductors, *Nat. Rev. Mater.* **2**, 16094 (2016).

129. T. Dietl, A ten-year perspective on dilute magnetic semiconductors and oxides, *Nat. Mater.* **9**, 965 (2010).

130. T. Dietl and H. Ohno, Dilute ferromagnetic semiconductors: Physics and spintronic structures, *Rev. Mod. Phys.* **86**, 187 (2014).

131. A. H. MacDonald, P. Schiffer, and N. Samarth, Ferromagnetic semiconductors: Moving beyond (Ga,Mn)As, *Nat. Mater.* **4**, 195 (2005).

132. T. Kuschel, C. Klewe, J. M. Schmalhorst et al., Static magnetic proximity effect in $Pt/NiFe_2O_4$ and Pt/Fe bilayers investigated by X-Ray resonant magnetic reflectivity, *Phys. Rev. Lett.* **115**, 097401 (2015).

133. M. G. Samant, J. Stöhr, S. S. P. Parkin et al., Induced spin polarization in Cu spacer layers in Co/Cu multilayers, *Phys. Rev. Lett.* **72**, 1112 (1994).

134. Z. Wang, C. Tang, R. Sachs, Y. Barlas, and J. Shi, Proximity-induced ferromagnetism in graphene revealed by the anomalous hall effect, *Phys. Rev. Lett.* **114**, 016603 (2015).

135. J. C. Leutenantsmeyer, A. A. Kaverzin, M. Wojtaszek, and B. J. van Wees, Proximity induced room temperature ferromagnetism in graphene probed with spin currents, *2D Mater.* **4**, 014001 (2017).

136. S. Singh, J. Katoch, T. Zhu et al., Strong modulation of spin currents in bilayer graphene by static and fluctuating proximity exchange fields, *Phys. Rev. Lett.* **118**, 187201 (2017).

137. O. V. Yazyev and L. Helm, Defect-induced magnetism in graphene, *Phys. Rev. B* **75**, 125408 (2007).

138. E. H. Lieb, Two theorems on the Hubbard model, *Phys. Rev. Lett.* **62**, 1201 (1989).

139. P. Wei, S. Lee, F. Lemaitre et al., Strong interfacial exchange field in the graphene/EuS heterostructure, *Nat. Mater.* **15**, 711 (2016).

140. P. U. Asshoff, J. L. Sambricio, A. P. Rooney et al., Magnetoresistance of vertical Co-graphene-NiFe junctions controlled by charge transfer and proximity-induced spin splitting in graphene, *2D Mater.* **4**, 031004 (2017).

141. P. Lazić, G. M. Sipahi, R. K. Kawakami, and I. Žutić, Graphene spintronics: Spin injection and proximity effects from first principles, *Phys. Rev. B* **90**, 085429 (2014).

142. P. Lazić, K. D. Belashchenko, and I. Žutić, Effective gating and tunable magnetic proximity effects in two-dimensional heterostructures, *Phys. Rev. B* **93**, 241401 (2016).

143. L. Wang, I. Meric, P. Y. Huang et al., One-dimensional electrical contact to a two-dimensional material, *Science* **342**, 614 (2013).

144. J. Xu, S. Singh, J. Katoch et al., Spin inversion in graphene spin valves by gate-tunable magnetic proximity effect at one-dimensional contacts, *Nat. Commun.* **9**, 2869 (2018).

145. T. J. Williams, A. A. Aczel, M. D. Lumsden et al., Magnetic correlations in the quasi-two-dimensional semiconducting ferromagnet $CrSiTe_3$, *Phys. Rev. B* **92**, 144404 (2015).

146. S. Lebègue, T. Björkman, M. Klintenberg, R. M. Nieminen, and O. Eriksson, Two-dimensional materials from data filtering and ab initio calculations, *Phys. Rev. X* **3**, 031002 (2013).

147. N. Sivadas, M. W. Daniels, R. H. Swendsen, S. Okamoto, and D. Xiao, Magnetic ground state of semiconducting transition-metal trichalcogenide monolayers, *Phys. Rev. B* **91**, 235425 (2015).

148. X. Li, T. Cao, Q. Niu, J. Shi, and J. Feng, Coupling the valley degree of freedom to antiferromagnetic order, *Proc. Natl. Acad. Sci. U.S.A.* **110**, 3738 (2013).

149. X. Li, X. Wu, and J. Yang, Half-metallicity in $MnPSe_3$ exfoliated nanosheet with carrier doping, *J. Am. Chem. Soc.* **136**, 11065 (2014).

150. B. L. Chittari, Y. Park, D. Lee et al., Electronic and magnetic properties of single-layer MPX_3 metal phosphorous trichalcogenides, *Phys. Rev. B* **94**, 184428 (2016).

151. M. A. McGuire, H. Dixit, V. R. Cooper, and B. C. Sales, Coupling of crystal structure and magnetism in the layered, ferromagnetic insulator CrI_3, *Chem. Mater.* **27**, 612 (2015).

152. W.-B. Zhang, Q. Qu, P. Zhu, and C.-H. Lam, Robust intrinsic ferromagnetism and half semiconductivity in stable two-dimensional single-layer chromium trihalides, *J. Mater. Chem. C* **3**, 12457 (2015).

153. Y. Ma, Y. Dai, M. Guo, C. Niu, Y. Zhu, and B. Huang, Evidence of the existence of magnetism in pristine VX_2 monolayers (X = S, Se) and their strain-induced tunable magnetic properties, *ACS Nano* **6**, 1695 (2012).

154. N. D. Mermin and H. Wagner, Absence of ferromagnetism or antiferromagnetism in one- or two-dimensional isotropic Heisenberg models, *Phys. Rev. Lett.* **17**, 1133 (1966).

155. J. F. D. Jr. and C. E. Olson, Magnetization, resonance, and optical properties of the ferromagnet CrI_3, *J. Appl. Phys.* **36**, 1259 (1965).

156. V. Carteaux, D. Brunet, G. Ouvrard, and G. Andre, Crystallographic, magnetic and electronic structures of a new layered ferromagnetic compound $Cr_2Ge_2Te_6$, *J. Phys. Condens. Matter* **7**, 69 (1995).

157. H. Ji, R. A. Stokes, L. D. Alegria et al., A ferromagnetic insulating substrate for the epitaxial growth of topological insulators, *J. Appl. Phys.* **114**, 114907 (2013).

158. X. Zhang, Y. Zhao, Q. Song, S. Jia, J. Shi, and W. Han, Magnetic anisotropy of the single-crystalline ferromagnetic insulator $Cr_2Ge_2Te_6$, *Jpn. J. Appl. Phys.* **55**, 033001 (2016).

159. T. Yao, J. G. Mason, J. Huiwen, R. J. Cava, and S. B. Kenneth, Magneto-elastic coupling in a potential ferromagnetic 2D atomic crystal, *2D Mater.* **3**, 025035 (2016).

160. P. A. Joy and S. Vasudevan, Magnetism in the layered transition-metal thiophosphates MPS_3 (M=Mn, Fe, and Ni), *Phys. Rev. B* **46**, 5425 (1992).

161. X. Wang, K. Du, Y. Y. F. Liu et al., Raman spectroscopy of atomically thin two-dimensional magnetic iron phosphorus trisulfide ($FePS_3$) crystals, *2D Mater.* **3**, 031009 (2016).

162. T. Maruyama, Y. Shiota, T. Nozaki, et al., Large voltage-induced magnetic anisotropy change in a few atomic layers of iron, *Nat. Nanotechnol.* **4**, 158 (2009).

163. D. Chiba, S. Fukami, K. Shimamura, N. Ishiwata, K. Kobayashi, and T. Ono, Electrical control of the ferromagnetic phase transition in cobalt at room temperature, *Nat. Mater.* **10**, 853 (2011).

164. H. Ohno, D. Chiba, F. Matsukura et al., Electric-field control of ferromagnetism, *Nature* **408**, 944 (2000).

165. D. Chiba, M. Sawicki, Y. Nishitani, Y. Nakatani, F. Matsukura, and H. Ohno, Magnetization vector manipulation by electric fields, *Nature* **455**, 515 (2008).

166. J. T. Ye, Y. J. Zhang, R. Akashi, M. S. Bahramy, R. Arita, and Y. Iwasa, Superconducting dome in a gate-tuned band insulator, *Science* **338**, 1193 (2012).

Magnetism and Transport in DMS Quantum Dots

Joaquín Fernández-Rossier and R. Aguado

6.1 SUMMARY

The present chapter reviews the electronic structure and transport properties of quantum dots made of diluted magnetic semiconductors (DMS) made of zincblende II-VI and III-V compounds, such as CdTe and GaAs, where a minor fraction of the cation atoms are replaced by magnetic $3d$ transition metal atoms, most often Mn. The resulting DMS compounds are denoted by (Cd,Mn)Te, (Ga,Mn)As, etc. We study how transport electrons and the local spins provided by the transition metal atoms affect each other. As in the case of bulk, electrons and holes exert a very different influence on the local spins because of the very different strength of both spin orbit and carrier-local spin coupling of conduction electrons and valence band holes. A distinctive feature of DMS quantum dots is the fact that the controlled addition of a single carrier can result in a dramatic change in their conductivity and magnetization. This result is derived using both an analytical argument where local spins are treated classically and exact diagonalization of the full quantum Hamiltonian in dots with a few Mn atoms, the magnetic dopant that has been most widely studied in this context. Special attention is devoted to the experimentally relevant case of dots doped with an individual Mn atom, an outstanding example of the so-called *single dopant devices* [1]. We show how the spin of a single Mn can be controlled by changing the number of carriers in the dot. In this limit, we find a rich interplay between Coulomb Blockade, single spin magnetic anisotropy and transport. In particular, we discuss a strong anisotropic magnetoresistance (AMR) effect in quantum dots with 1 Mn atom gated with 1 heavy hole. Some background material for this chapter is discussed in Chapters 9 and 10, Volume 2, on magnetism in III-V semiconductors and dilute oxides. Chapter 13, Volume 2, in this book explains how ideas about electrical control of magnetism in semiconductors can be extended to other material systems.

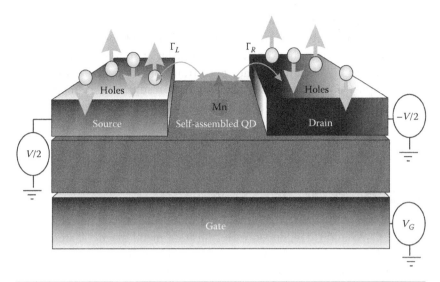

FIGURE 6.1 Schematics of the setup discussed in this chapter for studying transport in a DMS quantum dot. The system is based on a CdTe QD doped with a single Mn^{+2} ion and coupled to metallic reservoirs (tunnelling rates Γ_L and Γ_R and chemical potentials $\mu_{L/R}=\pm V/2$). V_G is the gate voltage.

6.2 INTRODUCTION

6.2.1 MOTIVATION

The electronic spin already plays an important role in transport in quantum dots made of non-magnetic materials, like semiconductor, carbon nanotubes and metallic grains. The standard experimental setup, shown in Figure 6.1, consists of a central region, the quantum dot, weakly coupled to two conducting electrodes, the source and drain. The charge of the dot is controlled by the electric field supplied by the gate voltage. We only consider situations where the charging energy of the dot, inversely proportional to its size, is larger than the temperature. In this Coulomb Blockade regime, the charge in the dot is quantized and whenever the highest energy electron is unpaired, the injection of an additional electron in the dot is only possible when its spin is opposite to that of the residing dot because of the Pauli principle. At low temperatures, these correlations result in the formation of the Kondo singlet, for which non-trivial spin correlations are established between the electron in the dot and those in the electrodes. Non-trivial spin effects in a quantum dot can also happen when the highest energy electrons reside a partially full degenerate shell. Other chapters of this book show how transport in magnetic materials attracts huge attention both for fundamental and applied perspectives. The influence of the magnetic state of the system over the electrical conduction gives rise to a variety of magnetoresistance (MR) effects, like anisotropic MR (AMR) in bulk materials, Giant MR (GMR) in multilayers, Tunneling MR (TMR) and AMR (TAMR) in tunnel junctions with magnetic electrodes and so on. Conversely, the transport electrons can also affect the magnetic state of the system through the so-called spin transfer or spin torque effect, see Chapters 7 and 8, Volume 1. Since both transport in magnetic materials and transport in non-magnetic quantum dots results in very interesting spin-related phenomena, we expect that transport in dots made of magnetic materials leads to new fascinating physics. In this chapter, we focus on transport through quantum dots made of a fascinating class of magnetic materials, the so-called diluted magnetic semiconductors.

6.2.2 DILUTED MAGNETIC SEMICONDUCTORS

As discussed in Chapter 9, Volume 2, conventional semiconductor materials, like GaAs and CdTe, doped with transition metal magnetic atoms like Mn, are known as diluted magnetic semiconductors (DMS). These materials are widely studied [2–6] because they hold the promise of combining in the same sample the useful properties of semiconducting compounds, like electrically tunable conductivity, with those of magnetic materials, like permanent magnetism. Whereas all DMS host magnetic atoms with local spins, only some of them present ferromagnetic order.

The range, strength and sign of exchange interactions between magnetic impurities depend on the density and nature of the states at the Fermi level [5]. Mn-doped semiconductors of the families (II,Mn)VI and (III,Mn)V order ferromagnetically in the presence of carriers that mediate indirect exchange interactions between Mn. In the case of (III,Mn)V compounds like (Ga,Mn) As, Mn acts as an acceptor supplying holes that mediate ferromagnetic indirect exchange interactions responsible of the ferromagnetism below a transition temperature which depends on both Mn and hole densities [6] and can reach 160 K [7]. In contrast, Mn does not supply itinerant carriers in (II,Mn)VI compounds. These materials do not order ferromagnetically [3] unless further doping with acceptors provides holes which produce ferromagnetism below a carrier density–dependent Curie temperature (T_c) of approximately 2 K in modulation-doped structures [8]. In the case of II-VI semiconductors, doping with electrons is also possible, but the Curie temperature are expected to be much lower than p-doped to both to smaller exchange coupling and effective mass.

DMS are also very attractive because low dimensional heterostructures can be made with them using molecular beam epitaxy techniques that have been so successful in non-magnetic semiconductors. Thus, quantum wells [8], delta-doped layers [9], tunnel barriers [10–12], quantum point contacts [11] and quantum dots (see below) have been fabricated, and exciting new physical phenomena, like Tunnel Anisotropic Magnetoresistance [10, 11], have been discovered. A paradigmatic example of the potential of these artificial structures is the reversible electrical control of the Curie temperature (T_c) [13, 14], the coercive field [15, 16] and the magnetic moment orientation, demonstrated experimentally in gated layers of both II-VI and III-V DMS. For instance, the Curie temperature of InAsMn can be reversibly changed from 22 to 23 K and back [14]. The small relative change of T_c is due to the small relative change in carrier density $\delta p \ll p$, because the density of holes before gating is very high in (III,Mn)V materials.

In order to enhance the electric control of the magnetic properties, it seems desirable to reduce the density of carriers or to increase the density of electrically injected carriers. The injection of N carriers in quantum dots of (II,Mn)VI afford both $p = 0$ and a large $\delta p = N/V$ provided that the volume V of the dot is sufficiently small. Additionally, the number of controlled spins can be reduced down to 1 [17], making it possible to control the spin of a single Mn atom by the electrical injection of a single carrier, as demonstrated by Leger and collaborators [18].

6.2.3 DMS QUANTUM DOTS: EXPERIMENTAL STATE OF THE ART

In this section, we summarize the state of the art in the fabrication and probing of DMS quantum dots. According to the fabrication technique, the dots fall into three categories, ordered in increasing size: colloidal nanocrystals [19], self-assembled QD [17, 20–23] and electrically defined islands [24]. In the overwhelming majority of cases, these systems have been probed optically in photoluminescence (PL) experiments under an applied magnetic field. One of the outstanding features in these experiments is the strong shift of the PL spectrum under the application of a magnetic field. This is the so-called Giant Zeeman Splitting [3] and arises from the combination of two physical phenomena: the Mn local spin aligns parallel to the applied field and, because of the exchange coupling to the electron and hole spin, the Mn average magnetization acts as an effective magnetic field on the exciton spin which results in Zeeman splittings as large as 100 meV. The exchange interaction can also give rise to the formation of the so-called magnetic polaron, i.e, the spontaneous alignment at zero magnetic field of the spins of the Mn atoms parallel to the spin of an unpaired carrier or exciton. In bulk, the formation of magnetic polarons results from the competition between exchange energy, which favors the localization of the carrier, and kinetic energy, which opposes it. Quantum dots favor the localization of the carrier and facilitate the formation of magnetic polarons at higher temperatures [23, 25, 26]. This is a particular manifestation of a general rule: quantum dots increase the exchange coupling between itinerant carriers and Mn local spins.

The fabrication of quantum dots doped with an individual transition metal atom has been one of the most striking results in this field. This was achieved first with self-assembled CdTe QD (Besombes 04) and later with InGaAs [22] doped with a single Mn atom. In both cases, the single quantum dot PL presents a characteristic pattern with a well-defined number of narrow peaks (six in CdTe, five in InAs) that evolve under the application of magnetic fields, permitting to associate them to the various spin states of the Mn spin. The addition of a single carrier into the dot results in a dramatic change of the PL in the case of CdTe [18]. More recent reports demonstrate the fabrication CdTe/ZnTe dots doped with a single Cobalt atom [27, 28], a single Fe atom [29] and a single Cr atom [30]. In all cases, optical spectroscopy reveals multiple photoluminescence peaks that reveal the spin interactions between the photoexcited electron-hole pair and the local spin of the individual transition metal. Single Mn doping in other II-VI hosts, such as CdSe/ZnSe, has also been demonstrated [31].

The model Hamiltonians presented below are able to provide a good account of these non-trivial PL spectra both in CdTe [32] and InGaAs quantum dots [33]. Thus, a single Mn-doped semiconductor quantum dot provides an ideal test bench for our theoretical understanding of the various spin couplings in DMS. Additionally, this system has made it possible to measure the spin relaxation time, T_1, of a single Mn atom, both in the presence and absence of carriers in the dot [34] as well as to polarize its spin using optical pumping at zero applied field [35, 36]. An important step in the scaling up of this approach has been the demonstration of a CdTe/ZnTe quantum dot doped with *two* Mn atoms [37]. The addition of individual Mn atoms in spontaneously coupled quantum dot pairs has also been demonstrated [38].

Thus, quantum dots have made it possible to measure and manipulate the spin of individual 3d atoms through their interaction with optically injected carriers, most notably, individual Mn atoms. This naturally leads to one of the central themes of this chapter: how would transport electrons and Mn spin affect each other in a hypothetical experiment in which a DMS quantum dot plays the role of the central island in a single electron transistor? Theory already provides some answers to that question, as we discuss below [39–41]. We first review the experimental progress in that direction.

To the best of our knowledge, two different transport experiments in DMS quantum dots have been reported. The group of Mollenkamp has been able to fabricate planar tunnel junctions with self-assembled CdSe quantum dots and Mn atoms in the barriers [21]. This device features resonant tunneling through the dots. The evolution of the resonant energies as a function of the applied magnetic field shows zero-field spin splitting that might arise from the formation of a magnetic polaron in the dot. However, in this experimental configuration, it is not possible to control the number of carriers in the dot with a gate voltage. The second system was an electrically defined GaMnAs island [24]. In this case, the island is in the Coulomb Blockade regime, but there is no evidence of quantum confinement. This peculiar single hole transistor behaves very differently as the direction of an applied magnetic field is changed. This Coulomb Blockade AMR arises from the strong magnetic anisotropy of GaMnAs island, related to the spin orbit of holes [4, 42].

The fabrication of a single electron transistor based on a self-assembled quantum dot with magnetic impurities still remains a challenge. Whereas gating self-assembled dots is relatively simple and has been achieved both in single Mn-doped CdTe [18] and InGaAs [22], the main technological obstacle is the coupling of the quantum dot to the conducting electrodes, but very rapid progress is being achieved in this regard. The group of professor

Arakawa has been able to fabricate single electron transistors based on self-assembled InAs quantum dots both with ferromagnetic [43] and superconducting electrodes.

Experimental setups that make it possible to study transport across individual molecules that host a single magnetic atom, and permit to probe its spin properties, have been reported [44–46]. These systems hold an obvious similarity with quantum dots doped with an individual magnetic atom: a gapped non-magnetic host with electronic states that connect both to the transport electrodes, via tunneling, and to the atomic spin, via exchange. Tsukahara et al. probed Fe phthalocyanine molecules deposited on a surface using a scanning tunneling microscope (STM), more specifically, an inelastic electron tunneling spectroscopy (IETS) that permits to probe spin excitations with atomic scale resolution [47]. The same technique was later used to probe the collective spin excitations of engineered Mn spin linear spin chains [48] as well as Mn_{12} molecular magnets [49]. STM transport experiments are in a two terminal configuration, with limited capability to gate the charge state of the molecules.

Gated break junctions, fabricated by electromigration, permit transport measurements through a single bis(phthalocyaninato)terbium(III) single molecule magnet [46] in a three-terminal geometry that permits to precisely control the charge state of the molecule, which acts as a single electron transistor. In the Kondo regime, the linear conductance of the molecule turns out to be sensitive to the electronic spin state of the molecule. Remarkably, in this specific system, the coercive field of the single molecule magnet is controlled by the nuclear spin state of the Terbium atom, which makes it possible to perform its readout [46]. In addition, taking advantage of the hyperfine Stark effect, coherent manipulation of the nuclear spin was also demonstrated in this system [44]. These experiments permit to infer a very long spin relaxation lifetime of the nuclear spin and illustrate very well the very interesting physical phenomena that could be studied using the setup of Figure 6.1.

Single hole transistors have been fabricated using CdSe nanocrystals [50]. This provides another route to have DMS single electron transistors in the quantum confinement regime. Whether or not it is possible to dope colloidal nanocrystals in general, and with Mn atoms in particular, has been the subject of a longstanding controversy which has been recently settled with clear experimental signatures that Mn has successfully been incorporated into these systems [19, 51], like the observation of magnetic circular dichroism and exciton magnetic polarons in Mn-doped CdSe colloidal nanocrystals. Finally, nanowire-based quantum dots offer another promising route towards the fabrication of DMS single electron transistor.

The rest of this chapter is organized as follows. In Section 6.3, we describe the electronic structure of DMS quantum dots. In Section 6.4, we discuss sequential transport in quantum dots doped with a single magnetic atom, a system which, in spite of its simplicity, shows the rich interplay between magnetism and transport in this system.

6.3 SPIN INTERACTIONS IN DMS

In this section, we briefly review the spin interactions relevant to Mn-doped quantum dots. We first consider a II-VI material and then discuss briefly the differences in the case of a III-V material. As a substitutional impurity of the group II atom, Mn is an isoelectronic impurity. The electrons in the half-filled d shell can be considered localized in the Mn atom and, in agreement with the Hunds rule, they form a spin $S = 5/2$. In the model, electrons and holes in the confined levels of the quantum dot are described in the k_p effective mass approximation. We consider quantum dots small enough as to have a sizable quantization of their energy spectrum.

As in the case of bulk, the electrons and holes are exchanged and coupled to the Mn spins through a local interaction [32, 40, 52–56]:

$$H_{e\_Mn} + H_{h\_Mn} = \sum_I \vec{M}_I \cdot \left(\alpha \vec{S}_e(\vec{r}_I) + \beta \vec{S}_e(\vec{r}_I) \right), \tag{6.1}$$

where the spin 5/2 operator is associated with the Mn atom located in $\vec{r}_I$; $\vec{S}_e(\vec{r}_I)$ and $\vec{S}_h(\vec{r}_I)$ are the spin densities of conduction band electrons and valence band holes, respectively, and α and β are the corresponding exchange coupling constants, which have dimensions of energy times volume [3]. These constants differ in sign and magnitude because the conduction band is mainly made of s orbitals of the group II, whereas the valence band is mainly composed of p orbitals of the group VI atom. In the case of the conduction band, the exchange is direct and ferromagnetic, whereas, in the case of the valence band, the exchange is dominated by the kinetic exchange that arises from pd hybridization and results in antiferromagnetic Mn-hole coupling.

Even in the absence of carriers, the Mn spins interact with each other via an antiferromagnetic superexchange which decays exponentially with distance. This term is important when two Mn spins are located in the first neighbor unit cells, i.e., when they are separated by a single atom. The Hamiltonian for this coupling reads [3]:

$$H_{AF} = \sum_{I,I'} J(\vec{r}_I - \vec{r}_{I'}) \vec{M}_I \cdot \vec{M}_{I'}. \tag{6.2}$$

The function $J(\vec{r}_I - \vec{r}_{I'})$ decays very quickly as the interatomic distance increases. For first neighbor coupling, the exchange interaction is of the order of 1 meV. The spin 5/2 of the Mn feels the presence of the crystal environment of a Zinc Blende lattice through the combination of the different electrostatic energy felt by the different d orbitals and the spin–orbit interaction. This results in a single-ion Hamiltonian whose symmetry depends dramatically on the symmetry of the crystal environment. For instance, in the case of bulk, the single-ion Hamiltonian for Mn in zinc blende is respectful with the cubic symmetry of the lattice and reads [57]:

$$H_{\text{cubic}} = a\left(M_x^4 + M_y^4 + M_z^4 \right). \tag{6.3}$$

The otherwise six degenerate spin levels of the Mn split in a doublet and a quartet. The splitting is proportional to a and is so small (below 10 μeV) that it can only be measured with electron paramagnetic resonance (EPR). The small value is due to the fact that Mn^{2+} has a half-full shell, the orbital moment is zero and spin–orbit interaction is small.

In the case of strained quantum wells, closer to the situation expected for the CdTe quantum dots, the single-ion Hamiltonian has the growth direction as a preferred easy axis, keeping the in-plane symmetry in the layer [58]. In the case of quantum dots, the in-plane symmetry might be broken so that the single-ion Hamiltonian reads:

$$H_{\text{strain}} = DM_z^2 + E\left(M_x^2 - M_y^2\right). \tag{6.4}$$

In the case of the layers, $E = 0$ and the reported value of D are distributed around a mean value of 10 microeV [58], much smaller than the temperature at which experiments are done. Thus, even if $M_z = \pm 1/2$ is the ground state, all the six levels are almost equally occupied at 4 K. In the case of the dots, E is probably non-zero and the sign of D could change, but their values are not known and they probably condition the single Mn dynamics observed experimentally [34, 35]. In addition, dynamical phonons result in a dynamical fluctuation of the strainfield, which results in spin-phonon interactions that have an influence on the spin dynamics of CdTe quantum dots doped with 1 Mn atom [59].

Finally, in EPR experiments, the hyperfine coupling between the nuclear spin (which is also 5/2) and the electronic spin is clearly observed [58] and might play some role in the Mn spin dynamics. For the remainder of this chapter, the effect of single-ion terms on anisotropy and hyperfine coupling will be neglected. Whereas this approximation is very good at describing the main features of the system, the single-ion terms are more relevant when it comes to describing single-ion dynamics [35].

6.4 ELECTRON-DOPED (II,Mn)V QUANTUM DOT

We now consider a Mn-doped QD where conduction band electrons are electrically injected [40, 53, 55, 56, 60–63]. We denote the single particle confined wave functions and their corresponding energies by $\psi_n(\vec{r})$ and $\varepsilon_{n,\sigma} = \varepsilon_n + g\mu_B \dfrac{\sigma}{2}B$ respectively. This notation implies that the spin quantization axis is parallel to the applied magnetic field B and also that the very small spin-orbit coupling is neglected. The second quantization operator that creates an electron with spin σ in the state n is denoted by $f_{n\sigma}$:

$$H_{oe} + H_{e-Mn} = \sum_{n,\sigma} \varepsilon_{n,\sigma} f_{n,\sigma} f_{n,\sigma} + \alpha \sum_{I,n,n'} \vec{M}_I \cdot \frac{\vec{\tau}_{\sigma,\sigma'}}{2} \psi_n^*(\vec{r}_I)\psi_{n'}(\vec{r}_I), \tag{6.5}$$

where $\bar{\tau}$ are the spin 1/2 Pauli matrices. Attending to the level index n, the exchange interaction has both intra- and inter-sublevel terms. Attending to the spin index, the exchange interaction has both spin-conserving and spin-flip terms.

In CdTe, the exchange coupling constant α takes values of -30 eVÅ^3. The negative sign implies ferromagnetic coupling. The strength of the carrier-Mn exchange depends both on α, a property of the material, and on the amplitude of spin density on the Mn location, a sample property. In order to build some intuition about the latter, we make use of a very simple model for the confinement potential: a hard wall cuboid with dimensions L_x, L_y and L_z. The wave functions for an electron in the non-degenerate effective mass model reads:

$$\psi_{nx,ny,nz}(x,y,z) = \sqrt{\frac{8}{L_x L_y L_z}} \sin\left(\frac{\pi n_x x}{L_x}\right) \sin\left(\frac{\pi n_y y}{L_y}\right) \sin\left(\frac{\pi n_z z}{L_z}\right), \quad (6.6)$$

where *nx*, *ny* and *nz* are the positive integer numbers. For a dot of volume *V*, the strength of the maximal value of interaction between Mn-electron coupling can be estimated as α/V. In a self-assembled dot of dimensions 5*5*2 nm, this yields 0.6 meV. The maximal Zeeman splitting of the electron in a dot with those dimensions and *N* atoms of Mn would be 0.6*N meV.

When more than one carrier is present in the dot, the effect of Coulomb interactions needs to be considered (see for instance [55, 56, 60, 62, 64]). In most instances, the Coulomb interaction results only in a shift of the many body levels that do not change their total spin. The only exception to this rule occurs in quantum dots with a partially filled degenerate shell [55, 56]. In that situation, Coulomb repulsion favors electronic ground states with large spin even in the absence of magnetic impurities. Degenerate shells occur in dots with cylindrical or spherical symmetry as well as in the cuboid model above, provided that at least two of the three lateral dimensions are identical. We now analyze the properties of the Mn-e Hamiltonian for situations of increasing complexity. The fact that both the number of electrons and the number of Mn atoms can be varied in the dot justifies the detailed account of the properties that follow.

6.4.1 ONE ELECTRON

We consider first a quantum dot with only one Mn atom, like the ones studied by Besombes, Mariette and collaborators, and with 2 Mn atoms. In Figure 6.2, we show the spectra of a dot with dimensions ($L_x = 14$, $L_y = 12$, $L_z = 3$) nm with 1 and 2 Mn atoms, respectively, and with one and two electrons. The spectrum features three groups of levels corresponding to the three lowest energy single particle confined levels. The fine structure within those groups is associated with the spin interactions and is of the order of $\Delta \simeq \alpha/V$. The distinctive feature of these zero dimensional systems is $\delta \gg \Delta$.

In the dot with 1 Mn (Figure 6.2), the low energy group is split into a ground state manifold with seven states and an excited state manifold with five states, as seen in the inset. Since the Hamiltonian commutes with the

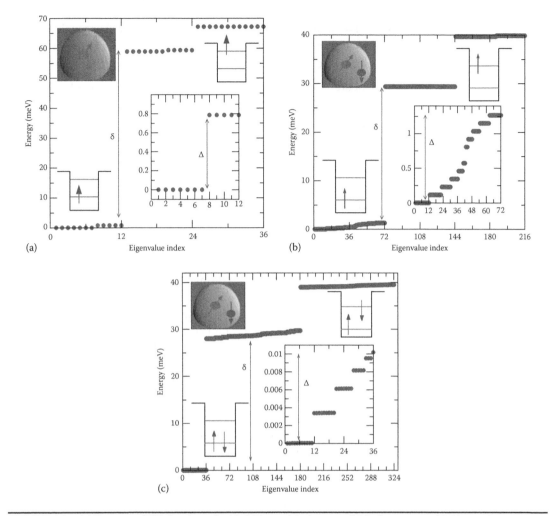

FIGURE 6.2   (a) Spectra of QD dot doped with 1 Mn and 1 electron, (b) 2 electrons and 2 Mn, and (c) 1 electron and 2 Mn (center). Insets: low energy sector of the spectra.

total spin operator, we can label the states according to the total spin S, with multiplicity $2S + 1$. Thus, we can assign to the ground state manifold of the dots with 1 a total spin $S = 6/2 = 3$, and the excited state in the low energy group a spin of $S = 2$. Notice that $S = 3$ is the maximal total spin that the Mn spin 5/2 and the electron spin $s = 1/2$ can form.

The low energy groups seen in the left and central panels of Figure 6.2 correspond basically to the electron in the lowest energy-confined state exchange coupled to the Mn spin(s) through an effective Heisenberg model because the inter-level terms of the Hamiltonian have a very small effect when $\delta \gg \Delta$ (and levels are not doubly occupied). When the higher energy single particle levels are neglected, the effective Hamiltonian can be simplified to the effective Heisenberg model

$$H_{1e} = \sum_I j_e(I)\vec{M}_I \cdot \vec{S}_e, \tag{6.7}$$

where $j_e(I) = \alpha |\psi_0(\vec{r}_I)|^2$, $\vec{S}_e$ is the spin 1/2 operator of the electron. In the case of a single Mn atom, this is the Heisenberg Hamiltonian for two spins, which can be solved exactly by taking advantage of the fact that the total spin and the Heisenberg coupling are related: $S_T^2 = (\vec{S}_1 + \vec{S}_2)^2 = \vec{S}_1^2 + \vec{S}_2^2 + 2\vec{S}_1 \cdot \vec{S}_2$. The square of the spin operators $\vec{S}_1^2$ are $S_i(S_i + 1)$ times the unit matrix, whereas the total spin can take the values between $S_T = S_1 + S_2$ and $S_T = |S_1 - S_2|$. In this case, the allowed values are $S_T = 3$ and $S_T = 2$. Thus, in the case of the Mn spin 5/2 and the electron spin 1/2, we can write $J\vec{S}_1 \cdot \vec{S}_2 = \frac{J}{2}\left(S_T(S_T+1) - \frac{3}{4} - \frac{35}{4}\right)$, with eigenvalues ($10J/8$) and for S = 3 and $-14J/8$ for S = 2. The splitting is 3J. Importantly, this intra-level splitting Δ scales linearly with the exchange coupling constant α.

In the dot with 2 Mn, taken sufficiently apart so that their direct coupling J_{AF} is zero, the low energy group has 72 states, whose fine structure is shown in the inset. The ground state manifold within these 72 states is a group with 12 states which can be identified with a total spin S = 11/2, the maximal possible spin that 2 Mn atoms and one electron can have. Thus the electron favors a large spin ground state, inducing a ferromagnetic spin correlation between the 2 Mn spins. This is a clear example of indirect or carrier-mediated exchange coupling. The strength of the coupling can be estimated from the energy splitting of the low energy group of the spectrum, Δ, which is of the order of 1 meV. Importantly, it also scales linearly with α. The splitting Δ sets the temperature scale below to where the Mn spins are not independent. This analysis implies also that one electron will induce ferromagnetic correlations between more than 2 Mn. This suggests that the addition of a single carrier to the quantum dot significantly changes its spin properties. Without carriers, the Mn spins do not interact significantly, unless they are first neighbors.

6.4.2 Two and More Electrons

We now discuss how the spin of the ground state of the dot evolves as additional carriers are injected into the dot by means of a gate voltage (as opposed to the injection of non-equilibrium transport carriers). When several Mn spins interact with a single electron, they tend to orient their spins parallel to that of the electron, which can take two orientations. The addition of a second electron in the same single particle level results in the formation of an electron spin singlet which should have very little influence on the Mn spins. That is exactly what is obtained from the numerical diagonalization of the Hamiltonian. In the case of the dot with 1 Mn and two electrons, the spin of the ground state is S = 5/2, the same of that in a dot without extra carriers. The case of 2 Mn and two electrons is more interesting because, in the absence of carriers and interactions, the spin of this system can take values within S = 0 and S = 5, all of them with the same energy. The presence of the two electrons breaks this degeneracy through virtual inter-level excitations (see inset of right panel in Figure 6.2. The lowest energy configuration has the two electrons with total spin 0 in the lowest energy level. This configuration is coupled to higher energy configurations with two unpaired electrons, one of them in the excited quantum dot level. This coupling enables spin density fluctuations of the two electron droplet, involving as intermediate or virtual state an excitation of one of the electrons to an excited quantum dot level. As we show below, these processes are responsible for indirect exchange interaction between different Mn atoms whose strength scales as Δ^2/δ, which is much smaller than the interaction mediated by an unpaired electron, which is order Δ. This is the quantum dot version of RKKY interaction and, as in the bulk case, it can be either positive or negative, depending on the position of the two Mn atoms in the quantum dot.

The addition of a third electron leaves the electron with two paired electrons in the low energy level and one unpaired electron in the second energy level [53]. Not surprisingly, the behavior of such dot is very similar to that with a single electron. As a general rule, there are two types of behavior. When there are unpaired electrons, there are ferromagnetic correlations like those of magnetic polarons, enhanced compared to bulk because confinement favors the localization of the itinerant electron. In the absence of unpaired electrons, the carrier-mediated interactions are much weaker than in the polaron case and can be either ferromagnetic or antiferromagnetic. In the next section, we provide an analytical background for these statements, which so far are backed up by numerics.

6.4.3 MANY IMPURITIES: CLASSICAL APPROXIMATION

Here we provide an analytical argument [53] that permits us to go beyond the numerical approach of the previous section, which is limited to a few magnetic atoms (see also [65]). The generalization is made possible by an approximation: we consider the Mn spins as classical variables. By so doing, the size of the Hilbert space is reduced by a factor 6N, where N is the number of Mn spins. The partition function of the dot is written, in the canonical ensemble, as:

$$Z_N = \int d\vec{M}_1 \cdots d\vec{M}_N e^{-\beta H_{AF}(\vec{M}_I)} \sum_{\Sigma} e^{-\beta E_{\sigma,N}(\vec{M}_I)}, \qquad (6.8)$$

where Σ labels a Slater determinant eigenstate of the fermionic Hamiltonian with eigen-energy, which depends parametrically on the spin arrangement. The integrals sum over all the possible classical spin configurations. We make further progress by doing some approximations. First, in the case of $\delta \gg k_B T$, we can keep the lowest energy configuration in the sum. Second, in the case of an unpaired electron configuration, the dominant contribution to the multielectronic energy comes from the exchange coupling of the unpaired to the classical spins. This term is formally identical to the energy of an electron interacting with an effective magnetic field $g\mu_B \vec{B}_{\text{eff}} = \sum_I j_e(I)\vec{M}_I$, where $j_e(I) \equiv \alpha |\psi_n(\vec{r}_I)|^2$, where n is the quantum dot level where occupied by the unpaired electron. By aligning its spin parallel to this effective magnetic field, the unpaired electron lowers its energy by an amount

$$g\mu_B \left| \vec{B}_{\text{eff}} \right| = \sqrt{\sum_{I,I'} j_e(I) j_e(I') \vec{M}_I \cdot \vec{M}_{I'}}. \tag{6.9}$$

It is apparent that this term favors collinear Mn spin arrangements and it scales linearly with the Mn-electron interaction constant α. Interestingly, the resulting effective model is the square root of a long-range Heisenberg model.

In the case of paired ground state configurations, the contribution above has opposite signs for the two electrons and vanishes. Thus, the leading order contribution comes from the inter-level exchange. Since this term is small, we can obtain it by perturbation theory using an unperturbed Hamiltonian without exchange coupling. The sum of the second order perturbation theory corrections of the occupied one-particle levels read:

$$\varepsilon_e = \sum_{I,I'} \left[\frac{J_c^2}{2} \sum_{n,n'} \gamma_{n,n'}(I,I') \frac{f_n - f_{n'}}{\varepsilon_n^0 - \varepsilon_{n'}^0} \right] \vec{M}_I \cdot \vec{M}_{I'}, \tag{6.10}$$

where $\gamma_{n,n'}(I,I') \equiv \phi_n^*(\vec{x}_I) \phi_{n'}(\vec{x}_I) \phi_n^*(\vec{x}_{I'}) \phi_{n'}(\vec{x}_{I'})$ and $f_n = 0,1$ are the occupation of the unperturbed dot in the ground state electronic configuration. Importantly, $\gamma_{n,n'}(I,I')$ can be both positive or negative, so that the effective coupling can be either ferromagnetic or antiferromagnetic depending on the location of the Mn atoms and the number of electrons in the dot. This full shell coupling is quite similar to bulk RKKY interaction and is much smaller than the unpaired electron or polaronic contribution.

In summary, the addition/removal of a single carrier can quite drastically change the effective interactions between the Mn spins. Dots with an unpaired electron tend to form a magnetic polaron. Dots with closed shell have residual carrier-induced interactions that can be either ferromagnetic or antiferromagnetic. Thus, they should behave like nanometric-sized spin glasses at very low temperatures, and like standard paramagnets above a few Kelvin. In both situations, the very small size of the spin–orbit interaction for both half-full d shell (Mn^{2+}) and conduction band electrons results in a very small magnetic anisotropy for the types of states. As we see in the next section, the situation is radically different in the case of Mn interacting with quantum dot holes.

6.5 HOLE-DOPED (II,Mn)VI QUANTUM DOT

In the case of holes, both the degeneracy of the valence band and the very strong spin–orbit interaction play an important role [3, 42]. In bulk, the top of the valence band is formed by two degenerate bands, light and heavy holes (LH, HH), well above in energy from the split-off band. In the case of CdTe, this splitting is larger than 0.8 eV, reflecting the strength of the spin–orbit interaction of Te. Within the effective mass approximation, the states of these bands can be described as linear combinations of the states at the Γ

point, which have the same symmetry as the atomic states and are strongly affected by spin–orbit interaction. The light and heavy hole wave functions transform like the eigenstates of the total angular momentum J formed from the sum of a $l = 1$ and a spin $s = 1/2$ object. The eigenstates of the $J = 3/2$ sector are written as linear combinations of the product basis in which s_z and l_z are good quantum numbers [66]:

$$\left| J = \frac{3}{2}, J_z = \frac{+3}{2} \right\rangle = \left| l_z = 1, \uparrow \right\rangle$$

$$\left| J = \frac{3}{2}, J_z = \frac{-1}{2} \right\rangle = \frac{1}{\sqrt{3}} \left| l_z = -1, \uparrow \right\rangle + \sqrt{\frac{2}{3}} \left| l_z = 0, \downarrow \right\rangle$$

$$\left| J = \frac{3}{2}, J_z = \frac{+1}{2} \right\rangle = \frac{1}{\sqrt{3}} \left| l_z = 1, \downarrow \right\rangle + \sqrt{\frac{2}{3}} \left| l_z = 0, \uparrow \right\rangle$$

$$\left| J = \frac{3}{2}, J_z = \frac{-3}{2} \right\rangle = \left| l_z = -1, \downarrow \right\rangle.$$

$$(6.11)$$

It is apparent that spin is not a good quantum number in these basis states and it will not be a good quantum number for holes in general. In the quantum dot, confinement and strain split the light and heavy holes. In most instances, the lowest energy hole states (i.e., the highest energy states in the valence band) are made almost entirely of heavy hole states.

In order to describe the single particle states for the quantum dot holes, we use the same hard wall cuboid approximation as in the case of electrons. But in this case, the resulting effective mass problem leads to six coupled differential equations, due both to spin–orbit interaction and the degeneracy of the top of the valence band. This model has been proposed by Kyrychenko and Kossut [67] and used extensively in the study of single Mn-doped quantum dots [32, 68]. For given dimensions of the dot, L_x, L_y and L_z, and given principal quantum numbers for the envelope function, n_x, n_y and n_z, the six differential equations are mapped into a range six matrix problem, given by the Kohn-Luttinger kp Hamiltonian. The $k = 0$ eigenstates of this Hamiltonian transform like the eigenstates of $J = 3/2$ ($J_z = \pm 3/2, \pm 1/2$) and $J = 1/2$ ($J_z = \pm 1/2$), the two values of the angular momentum that can be formed with $l = 1$, the orbital momentum of p orbitals and $s = 1/2$, the spin of the electrons. The finite $\vec{k}$ states are a linear combination of these states, and, therefore, J and J_z are no longer good quantum numbers. In the approach of Kyrychencko and Kossut [67], the confined quantum dot states are obtained by the diagonalization of the bulk Kohn-Luttinger Hamultonian where k_a^2 is replaced by $\left(\frac{\pi n_a}{L_a} \right)^2$ where $a = x, y, z$, and the terms linear in $\vec{k}$ vanish. In this model, the lowest energy bands are made mostly of light and heavy holes (states within the $J = 3/2$ manifold). In bulk, the wave vector determines the degree of mixing between light ($J = 3/2$, $J_z = \pm 1/2$) and heavy ($J = 3/2$, $J_z = \pm 3/2$) holes. In the quantum dot, effective wave vector of the state is determined by the dimensions of the dot. Within this model, the ground state of a quantum dot with $L_z \ll L_x, L_y$ is mostly made of heavy holes, and the degree of mixing to light holes depends on the aspect ratio $r = L_x/L_y$. For $r = 1$, the ground state is purely made of heavy holes.

6.5.1 ONE HOLE

We are now in a position to study the problem of quantum dot holes interacting with Mn spin in a quantum dot. We denote the confined quantum dot hole states by the label ν, which combines orbital and spin degrees of freedom. The Hamiltonian for holes exchanged coupled to Mn atoms in a quantum dot reads:

$$\mathcal{H}_{0h} + \mathcal{H}_{h-Mn} = \sum_\nu \varepsilon_\nu f_\nu^\dagger f_\nu + \beta \sum_{I,\nu,\nu'} \vec{M}_I \cdot \left\langle \nu \left| \frac{\vec{\tau}_{\sigma,\sigma'}}{2} \delta(\vec{r} - \vec{r}_I) \right| \nu' \right\rangle. \quad (6.12)$$

The first case we consider is a quantum dot doped with one hole, which can occupy to states that we denote 2006 $\Uparrow$ and $\Downarrow$. This has been studied experimentally [18, 69]. The competition between spin orbit and exchange becomes apparent in the evaluation of the spin matrix elements in the sub-space formed by the lowest energy doublet. If there is absolutely no LH-HH mixing, the spin-flip process by which the heavy hole with spin +1/2 goes to −1/2, increasing in one unit the spin of a Mn spin, is forbidden because the orbital part of initial and final heavy hole states are orthogonal:

$$\left\langle \frac{3}{2}, \frac{-3}{2} \left| \tau^- \right| \frac{3}{2}, \frac{+3}{2} \right\rangle = \langle l = 1, m = -1 | l = 1, m = +1 \rangle \times \left\langle \downarrow \left| \tau^- \right| \uparrow \right\rangle = 0. \quad (6.13)$$

In contrast, the spin-conserving channel is totally open. Thus, Mn-heavy hole spin exchange results in an Ising coupling:

$$\mathcal{H}_{I\,sing} = \frac{\beta}{2} \left(f_\Uparrow^\dagger f_\Uparrow - f_\Downarrow^\dagger f_\Downarrow \right) \sum_I |\psi_0(\vec{r}_I)|^2 M_I^z. \quad (6.14)$$

This Hamiltonian describes the Mn spins of the dot Ising coupled to the spin of the hole. The quantization axis is given by the growth direction which determines the shape of the dot. We define the effective coupling $j_h(I) \equiv \beta |\psi_0(\vec{r}_I)|^2$ which is positive because the hole–Mn coupling is antiferromagnetic. In the case of a CdTe dot doped with a single Mn, the spectrum of this Hamiltonian is given by six doublets and can take six values between $E(M_z, \pm) = \pm \frac{J_h}{2} M^z$, where M_z is the Mn spin projection along the growth axis 2 and +5/2 (see Figure 6.3). This spectrum can be rationalized as two Zeeman split families associated to the two possible orientations of the hole spin. The ground state doublet is given by the states ($\Uparrow$,−5/2) and ($\Downarrow$,+5/2).

This apparently simplistic model is validated by the experimentally observed PL of CdTe quantum dots doped with 1 Mn [17] that features six peaks. Basically, the dominant exciton–Mn coupling is given by the

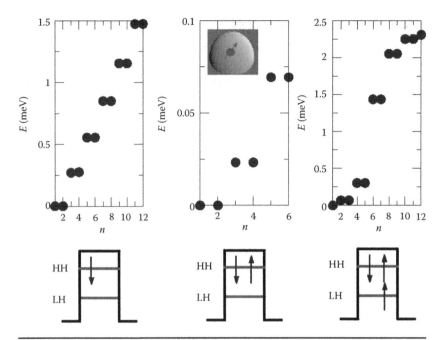

FIGURE 6.3 Spectra of single Mn QD dot doped with 1, 2 and 3 holes.

hole–Mn coupling since we have $\beta \simeq 4\alpha$ [3]. Thus, the Mn is Ising coupled to the exciton and both the Mn and the exciton spin are conserved. Thus, the energy of the exciton depends on which of the six Mn spin projections along the growth axis. Thus, in the presence of the heavy hole, the spin of the Mn acquires a strong anisotropy with the growth direction as the easy axis. This anisotropy arises from the strong spin orbit of the hole, rather than from the very weak single-ion anisotropy of the Mn atom. Thus, the quantum dot affords the unique opportunity of controlling the anisotropy of the Mn atom(s) by the injection of a single hole [18, 52].

In general, the quantum dot confined states have some mixing between heavy and light holes, determined by the shape of the dot and the strain. This mixing opens the spin-flip channel a little, resulting in strongly anisotropic Heisenberg coupling:

$$H_{XXY} = \sum_I jh(1)\left[\left(f_\Uparrow^\dagger f_\Uparrow - f_\Downarrow^\dagger f_\Downarrow\right)M_I^z + \frac{\varepsilon}{2}\left(f_\Uparrow^\dagger f_\Downarrow M_I^{(-)} + f_\Downarrow^\dagger f_\Uparrow M_I^{(+)}\right)\right], \quad (6.15)$$

where ε is a dimensionless factor linearly proportional to the overlap of both $\Uparrow$ and parameter $\Downarrow$ states with the light hole states ($J = 3/2\ J_z = 1/2$). As long as the anisotropy parameter ε is different from 1 (which would retrieve the isotropic Heisenberg anisotropy coupling), the spin of the holes and the spin of the Mn will not be good quantum numbers. Interestingly, the PL spectra of some single Mn doped can only be modeled if we take a finite value of ε, typically smaller than 0.2 [32]. The fact that different dots have different ε is understood within the theory above because of the variation of shapes, sizes and strains from dot to dot.

6.5.2 TWO AND MORE HOLES

As we did in the case of conduction band electrons, we consider now the effect of increasing the number of holes in the dot one by one. On top of the even-odd effect found in the case of electrons, we can also expect strong variations in the anisotropy of the hole–Mn coupling depending on the light or heavy nature of the holes. We restrict the study to dots doped with 1 Mn, which are already interesting. We start with the case of two holes. The single particle energy spacing is smaller for holes than for electrons, both because the larger effective mass and, in the case of CdTe/ZnTe, because the energy barriers are smaller for the valence band. On the other side, exchange interaction with the Mn is larger for holes than for electrons. However, it is still true that the inter-level spacing is much larger than the exchange energy. Thus, the ground state with two holes is mostly made of configurations where the two holes occupy the lowest energy doublet and behave almost like a spin zero object, like in the case of the dot with two electrons. Thus, the ground state is similar to that of the free Mn spin. But in contrast with the electrons, the total spin S is not a good quantum number and inter-level coupling has some observable effect on the spectrum of the system. In particular, the otherwise degenerate six states of the quasi-free Mn spin acquire a fine structure, as seen in the figure, which can be fitted according to the equation $D_{2h}M_z^2$, where $D_{2h} \propto \dfrac{j_2^h}{\delta}$. This fine-structure splitting has been observed in the PL spectra of biexciton to exciton transitions [70], which is slightly different from the exciton to ground PL. The size of the splitting is 0.12 meV, the same order of magnitude than the one in Figure 6.3 and significantly larger than the single-ion anisotropy.

The case of a QD with three holes is also different from the case with three electrons. The ground state wave function is mostly composed configurations with two holes in the lowest energy doublet and one unpaired hole in the LH level which dominates the behavior of the system. The low energy sector spectrum of the system, shown in the right panel of Figure 6.3, can be rationalized as follows: there are two main groups of states: five low energy states and seven high energy states which, in absence of spin–orbit interaction, would merge into a $S = 2$ low energy quintuplet and a $S = 3$ high energy manifold that reveals the AF Mn-hole coupling. The splitting inside the groups arises from the anisotropic exchange interaction which results from SO coupling.

6.6 ELECTRICALLY TUNABLE NANOMAGNET

The results of the previous sections could be summarized as follows: it is possible to reversibly tune the magnetic properties of a given quantum dot–doped Mn by changing the charge state of the dot Q. Focusing on the case of single Mn-doped dots, electron doping permits to have two types of states, free Mn with $S = 5/2$ for $Q = 0, -2, -4$, in closed-shell configurations and, in open-shell configurations with one unpaired electron ($Q = -1, -3, -5$), "polaronic" state where the Mn and the unpaired electron spin have strong ferromagnetic spin correlations behaving like a spin $S = 3$ object. On top of this open- versus closed-shell dichotomy, holes add two additional ingredients. First, the $2S + 1 = 6$ degeneracy of closed-shell configurations is split due to inter-level exchange that produces a fine-structure splitting identical to that of single-ion anisotropy. Second, in configurations with one unpaired hole, the symmetry properties of the magnetic state are very different depending on the LH/HH character of the highest occupied confined level.

The fact that the low energy spectra of the single Mn-doped quantum dot can be changed reversibly for different charge states, as shown in Figure 6.4, has lead us to propose this system as an *electrically tunable nanomagnet* [40], and is related to the idea of voltage control of the a DMS quantum dot [53, 62, 71]. Hole-doped single-Mn quantum dots definitely have one of the two main properties of a nanomagnet, namely, a finite magnetic moment. Whether or not the system displays remanence and coercivity, like in the case of single molecule magnets, remains to be determined experimentally. Theoretically, remanence and coercivity require two ingredients: magnetic anisotropy and blocking of the spin relaxation. Single

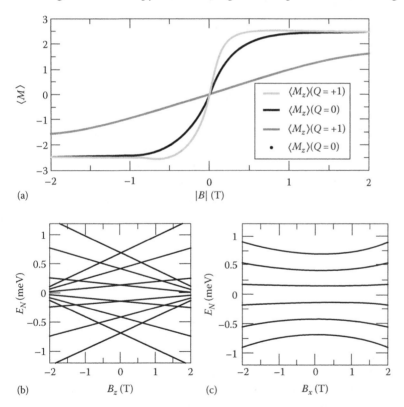

FIGURE 6.4 (a) Equilibrium magnetization of the Mn spin for a dot with $Q = 0$ and $Q = +1$ as a function of the intensity of the applied field for two orientations. (b) and (c) Energy spectra of a single Mn-doped quantum dot with 1 hole as a function of the magnetic field for two directions: parallel to growth axis (b) and perpendicular to the growth axis (c). (After Fernández-Rossier, J. et al., *Phys. Status Solidi C* 3, 3734, 2006. With permission.)

Mn-doped CdTe quantum dot with one resident hole presents a very large magnetic anisotropy: the response of the system to an applied magnetic field along the growth (z) axis (Farady configuration) is very different to the one with the field applied in the plane (Voigt configuration). Experimentally, this is inferred from the very different Voigt and Faraday magneto-PL spectra [32, 72]. In Figure 6.4, we plot the average magnetization of the Mn spin for a dot with $Q = 0$ and $Q = +1$ as a function of the intensity of the applied field for two orientations. The magnetization is calculated using the equilibrium ensemble. In the $Q = 0$ case, there is a very small difference caused by the very weak crystalline anisotropy of the Mn ion. In contrast, the magnetic anisotropy caused by the hole in the $Q = +1$ case is much larger. It can be understood by the inspection of the evolution of the 12 low energy states of the QD with 1 Mn and 1 hole as a function of the applied field. In the Faraday case, the field splits the levels with well-defined M_z and S_z. In the Voigt case, there is a competition between Zeeman coupling and Ising coupling which decreases the susceptibility.

6.7 Mn-Doped III-V Quantum Dots

As opposed to the case of Mn-doped II-VI quantum dots, for which there is a wide range of experimental work, Mn-doped III-V quantum dots are much less studied. In this section, we briefly comment on the microscopic model adequate to understand the PL spectra of a single Mn-doped (In,Ga) As quantum dot reported by Kudelsky et al. [22].

In stark contrast with the II-VI, substitutional Mn in bulk GaAs and InAs behaves like a shallow acceptor. EPR [73] and photoemission [74] experiments indicate that Mn retains the five d electrons when doping concentrations are small. Hence, Mn keeps an oxidation state of +2, resulting in an effective charge of −1 that repels the electrons nearby. The Mn impurity remains charge neutral, at the scale of a few unit cells, by binding a hole. The binding energy of the hole is of 110 meV in Ga(Mn)As and 28 meV in In(Mn) As. The acceptor hole state has a radius of 1 nm and has been probed by STM experiments both in GaAs and InAs [75]. If the size of the dot is much larger than the acceptor hole state and the Mn is far from the dot interface, we can expect that the dot has a weak influence on the acceptor state. This view is consistent with the PL experiments [22, 33].

In bulk, the acceptor hole has a four-fold degeneracy inherited from the top of the valence band, which is lifted by quantum confinement and/ or strain [76]. Because of the strong spin–orbit interaction, it is convenient to treat the acceptor hole as a spin $J = 3/2$ object, exchanged coupled to the Mn spin $M = 5/2$. It is also convenient to think of the acceptor states as if there was an unpaired electron with quantum numbers $J = 3/2$, $j_z = \pm 3/2$, $\pm 1/2$ interacting antiferromagnetically with the Mn spin. The operators acting upon this object are the four by four $J = 3/2$ angular momentum matrices, $\vec{j}$. In the spherical approximation the Mn-acceptor hole, spin coupling reads [76]: $HG_0 = \varepsilon \vec{M} \cdot \vec{j}$ where $\varepsilon = +5$ meV is the antiferromagnetic coupling between the hole and the Mn, $\vec{M}$ are the $S = 5/2$ spin matrices of the Mn and $\vec{j}$ are the $J = 3/2$ matrices corresponding to the total angular momentum of the valence band states. The Hamiltonian HG_0 is readily diagonalized in the basis of the total spin $F = M + J$ the spin of the Mn plus hole complex. F can take integer values between one and four. The eigenvalues are $E(F) = \varepsilon F(F + 1)/2 + E_0$. Since the coupling is antiferromagnetic, the ground state has $F = 1$, separated from the $F = 2$ states by a sizable energy barrier of 2ε. Hence, as long as the Mn-hole complex is not distorted by perturbations that couple different F manifolds, it is a good approximation to think of it as a composite object with total spin $F = 1$, whose wave functions read $|F = 1, F_z = \pm 1, 0\rangle = \sum_{M_z, j_z} C_{M, j_z}(F, F_z)|M_z, j_z\rangle$. We see that, because of the very strong exchange interaction, the spins of the Mn and the acceptor hole are strongly correlated and they are not good quantum numbers separately.

The cubic symmetry of the ideal crystal and the presence of quantum confinement and strain result in additional terms in the Hamiltonian that need to be summed to HG_0. Both quantum dot confinement and strain can result in a splitting of the light and heavy hole bands that can be modeled as a $-Dj_z^2$ term. This term would be present in thin film layers with strain and still preserves rotational invariance in the xy plane. The presence of the quantum dot potential will break this in-plane symmetry. To the lowest order, this can be modeled by an additional term in the Hamiltonian, $E(j_x^2 - j_y^2)$ [77]. It is assumed that these perturbations act on the acceptor state only and not on the Mn d electrons. This is justified since the hole is spread over tens of unit cells, whereas the Mn d states are confined in a single unit cell. Hence, the following model for the ground state Hamiltonian results: $H_G = \varepsilon \vec{M} \cdot \vec{j} - Dj_z^2 + E(j_x^2 - j_y^2)$ Since $\varepsilon \gg D \gg E$ this Hamiltonian can be projected on the $F = 1$ lowest energy manifold, resulting in the Hamiltonian [33, 77]:

$$H_G = -D_F F_z^2 + E_F(F_x^2 - F_y^2) + g_F \mu_B F_z B, \tag{6.16}$$

where:

$$D_F = (3/10)D$$
$$E_F = (3/10)E$$
$$g_F = ((7/4)g_M - (3/4)g_j)$$

The D_F term splits the $F = 1$ triplet into a singlet $|1, 0\rangle$ and a doublet, $|1, \pm 1\rangle|$. The E_F term mixes the two states in the ± 1 doublet, resulting in a bonding and antibonding states along the Y and X axis respectively.

A III-V quantum dot doped with a Mn atom could undergo two types of low energy charge fluctuation. First, removal of the acceptor hole, which would liberate the spin $S = 5/2$ of the Mn from its strong coupling with the acceptor hole. Second, the addition of a hole. From the interpretation [33] of the PL experiments [22], we know that the second hole goes into a quantum dot state, rather than to doubly occupy the acceptor hole. The experimental results are accounted for if the interaction between the QD hole and the $F = 1$ composite spin of the Mn-acceptor complex is modeled as an Ising coupling.

6.8 SEQUENTIAL TRANSPORT IN Mn-DOPED QUANTUM DOTS

6.8.1 INTRODUCTION

The experimental demonstration of electrical control of DMS quantum dots down to the limit of a single Mn atom [18] paves the way to electric transport studies in these systems. Although such challenging experiments have not hitherto been performed, transport through individual self-assembled non-magnetic QDs has been achieved. The main technical difficulty in performing such measurements through individual self-assembled QDs, i.e., the attachment of electrodes, has been overcome by developing specific techniques such as planar single electron transistors [78], gated submicrometer mesa devices [79, 80] and nanolithographically defined metallic nanogap electrodes [81]. Recent successes in making electrical contact to a single self-assembled quantum dot beyond the latter case (normal metallic contacts) include the coupling to ferromagnetic leads [43] and to superconducting leads [82]. Quite remarkably, this technique even allows one to reach the strong tunneling limit where the Kondo effect for both normal [83] and superconducting regimes [84] has been measured. All this experimental progress suggests that our theoretical proposal of electrical transport measurements with Mn-doped self-assembled quantum dots should be within reach in the next few years. Before explaining in detail transport through Mn-doped quantum dots, we summarize, for completeness, the main concepts and ideas of sequential transport in non-magnetic quantum dots [85].

Arguably, the most relevant transport phenomenon in quantum dots is Coulomb Blockade. When tunneling occurs, the charge on the quantum dot changes by the quantized amount e. This event changes the electrostatic energy of the quantum dot by the charging energy $E_C = e^2/C$. Charging effects become important when E_C exceeds the thermal energy $k_B T$. Another important requirement is that the tunneling resistance of the barrier $R_t = 1/(e^2\Gamma)$, with Γ being the typical level width due to tunneling, should be much larger than the resistance quantum $h/e^2 = 25.813$ kΩ such that quantum fluctuations of charge due to tunneling are strongly suppressed. At the lowest order in tunneling, transport occurs if the electrochemical potential of the quantum dot occupied by N electrons $\mu_{dot}(N)$ lies between the electrochemical potentials of the reservoirs $\mu_L = E_F + V_L$ and $\mu_R = E_F + V_R$, where $V = \mu_L - \mu_R = V_L - V_R$ is the applied bias voltage. The electrochemical potential of the dot is, by definition, the minimum energy needed to add the Nth electron to the dot: $\mu_{dot}(N) = U(N) - U(N-1)$, where $U(N)$ is the total ground state energy for N electrons on the dot at zero temperature. When at fixed gate voltage, the number of electrons changes by one, the change in electrochemical potential is given by the addition energy

$\mu_{dot}(N+1) - \mu_{dot}(N) = \Delta\varepsilon + E_C$, which is large for small capacitances, and/or large energy splittings $\Delta\varepsilon$ between confined states. Typical energy scales for both $\Delta\varepsilon$ and E_C in self-assembled quantum dots are in the range of a few tens of meV [81] such that Coulomb Blockade is observed for temperatures below $T \approx 5\,K$. Note that if two electrons are added to the same spin, the degenerate level $\Delta\varepsilon = 0$ and their energy difference is given by the charging energy only. A lot of theoretical papers consider this situation in which transport is supposed to occur through only one spin-degenerate state. When $E_C \gg k_B T$, the charging energy dominates transport. If $\mu_{dot}(N+1) > \mu_L, \mu_R > \mu_{dot}(N)$, transport is blocked and the quantum dot is in the Coulomb Blockade regime. The Coulomb Blockade can be removed by changing the gate voltage to align $\mu_{dot}(N+1)$ between the chemical potentials of the reservoirs such that an electron can tunnel from the left reservoir to the dot and from the dot to the right reservoir, which causes the electrochemical potential to drop back to $\mu_{dot}(N)$. A new electron can now enter the dot such that the cycle $N \Rightarrow N+1 \Rightarrow N$ is repeated. This process is called single-electron tunneling and quantum dots exhibiting these physics are often called single electron transistors (SETs). By changing the gate voltage, the linear conductance oscillates between zero (Coulomb Blockade) and non-zero. In the regions of zero conductance, the number of electrons inside the quantum dot is fixed. At finite bias voltages, the large regions of nearly zero conductance form characteristic Coulomb diamonds in differential conductance plots. Excited states (e.g. electronic, vibrational, or spin degrees of freedom) appear as visible lines in the diamonds, making this kind of transport measurement a useful form of level spectroscopy (Figure 6.5).

As the tunneling to the reservoirs becomes larger, namely as the resistance of the barriers approaches the quantum of resistance, higher-order tunneling events become relevant. In this situation, quantum fluctuations dominate transport and electrons are allowed to tunnel via intermediate virtual states where first-order tunneling would be suppressed. Thus, the intrinsic width of the energy levels of the quantum dot does not only include contributions from direct elastic tunneling but also tunneling via virtual states. These higher-order tunneling events are referred to as co-tunneling processes. These higher-order tunneling events lead to spectacular effects when the spin of the electrons is also involved. Importantly, a quantum dot with a net spin coupled to electron reservoirs resembles a magnetic impurity coupled to itinerant electrons in a metal and, thus, can exhibit the Kondo effect. In this regime, quantum fluctuations induce an effective exchange coupling $J \approx \dfrac{\Gamma E_c}{|\Delta\varepsilon||\Delta\varepsilon + E_c|}$ between the quantum dot spin and the reservoir ones. The spin and the accompanying exchange result in a complete screening of the quantum dot singly, the ground state becomes a singlet between confined and itinerant carriers. This mechanism gives rise to a new many-body resonance in the density of states of the quantum dot around the Fermi energy of the reservoirs. The most remarkable manifestation of this new "Kondo" resonance is the transition from near-zero conductance due to Coulomb Blockade to perfect transmission as the temperature is

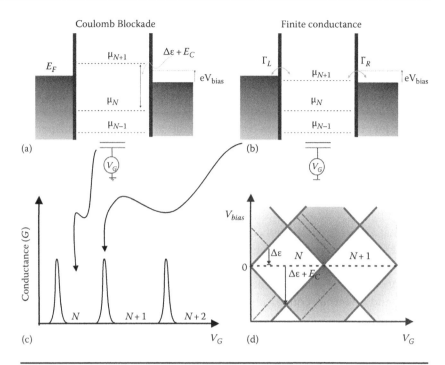

FIGURE 6.5 Schematic representation of the energy configuration of Quantum Dots in the Coulomb Blockade regime (a) and transport (b) configurations; (c) Schematics of the linear conductance oscillations through such device; (d) At finite bias voltages, the large regions of nearly zero conductance form characteristic Coulomb diamonds in differential conductance plots. Excited states appear as dashed lines.

lowered well below the so-called Kondo temperature $T_K \sim e^{\frac{-\pi|\Delta\varepsilon||\Delta\varepsilon+E_C|}{2\Gamma E_C}}$. Clear experimental signatures of Kondo physics have been reported in various types of quantum dots as well as other systems which can be understood in terms of a SET setup. The latter includes molecules, break junctions, etc. For a short review on the Kondo effect in quantum dots [86].

6.8.2 MODEL

All the above physics is captured by the Hamiltonian

$$H = H_{QD} + H_{\text{res}} + H_{\text{tunn}}, \tag{6.17}$$

which describes a quantum dot coupled to reservoirs. $H_{QD} = \sum_n \varepsilon_n \mathbf{f}_n^\dagger \mathbf{f}_n + H_C + H_{\text{int}}$ is the Hamiltonian of the isolated QD. The first term defines the energy of confined carriers within the QD. Coulomb effects are described within the constant interaction model $H_C = \left(\hat{Q} - Q_{\text{gate}}\right)^2 / 2C \cdot \hat{Q} = e\hat{N}$ is the extra charge in the QD (where $\hat{N} = \sum_n \mathbf{f}_n^\dagger \mathbf{f}_n - N_0$ is the number of extra carriers on top of the background N_0), $C = C_L + C_R + C_{\text{gate}}$ is the total capacitance and $Q_{\text{gate}} = V_L C_L + V_R C_R + V_{\text{gate}} C_{\text{gate}}$ is the total induced gate charge. Finally, H_{int} contains extra interactions not included in the above charging model.

In our case, this would correspond to the Hamiltonian of confined carriers interacting with the Mn that we have described in previous sections. The Hamiltonian of the metallic electrodes is $H_{res} = \sum_{k\alpha \in L,R} \varepsilon_{k\alpha} \mathbf{c}_{k\alpha}^\dagger \mathbf{c}_{k\alpha}$, with $\mathbf{c}_{k\alpha}^\dagger$ the creation operator of an electron/hole in the left/right reservoir (α is a channel index that takes into account spin, orbital degeneracies, etc.). The coupling between the QD and the reservoirs reads

$$H_T = \sum_{\substack{k\alpha \in L,R \\ n}} V_{k\alpha,n} \mathbf{c}_{k\alpha}^\dagger \mathbf{f}_n + V_{k\alpha,n}^* \mathbf{f}_n^\dagger \mathbf{c}_{k\alpha}, \qquad (6.18)$$

which is the standard tunneling Hamiltonian describing electron/hole tunneling in/out of the QD.

The transport properties of a QD whose dynamics is governed by the Hamiltonian in Equation 6.17 cannot be solved exactly and one has to adopt some approximation scheme. Most of the calculations employed for the description of transport through an interacting region fall into two different families: nonequilibrium (Keldysh) Green's functions techniques and Quantum Master equations. To clarify the advantages/disadvantages of each approach, and their connection, we briefly present them in the next two sections. Those readers who only want to acquire new information about the transport properties of magnetic QDs are encouraged to skip these two technical sections and go directly to Section 6.8.5.

6.8.3 EXACT EXPRESSION FOR THE CURRENT USING THE KELDYSH METHOD

In the first approach, more "condensed-matter" oriented, one writes physical averages with the help of single-particle non-equilibrium Keldysh Green's functions. Formally, this approach is very powerful as it allows one to derive an exact expression for the current through the interacting region [87]. Using

$$J_L = -e\langle \dot{N} \rangle = -\frac{ie}{\hbar} \langle [H, N_L] \rangle = \frac{ie}{\hbar} \sum_{\substack{k,\alpha \in L \\ n}} \left[V_{k\alpha,n} \langle \mathbf{c}_{k\alpha}^\dagger \mathbf{f}_n \rangle - V_{k\alpha,n}^* \langle \mathbf{f}_n^\dagger \mathbf{c}_{k\alpha} \rangle \right], \qquad (6.19)$$

and assuming that the leads are non-interacting, the current can be written as

$$J_L = \frac{ie}{\hbar} \sum_{\substack{\alpha \in L \\ n,m}} \int \frac{d\varepsilon}{2\pi} v_\alpha(\varepsilon) V_{\alpha,n}(\varepsilon) V_{\alpha,m}^*(\varepsilon)$$

$$\qquad (6.20)$$

$$\left\{ f_L(\varepsilon) G_{nm}^>(\varepsilon) - \left[1 - f_L(\varepsilon)\right] G_{nm}^<(\varepsilon) \right\},$$

where $v_\alpha(\varepsilon) = \sum_k \delta(\varepsilon - \varepsilon_{k\alpha})$ is the density of states in channel α and $f(\varepsilon)$ is the Fermi function. $G_{nm}^{\gtrless}(\varepsilon)$ is the Fourier transform of the so-called lesser (greater) Green's functions defined as $G^<(t,t') = \frac{i}{\hbar} \langle \mathbf{f}_m^\dagger(t') \mathbf{f}_n(t) \rangle$ and $G^>(t,t') = -\frac{i}{\hbar} \langle \mathbf{f}_n(t) \mathbf{f}_m^\dagger(t') \rangle$, respectively. Then it is useful to define the level-broadening functions as $\left[\Gamma^L(\varepsilon) \right]_{mn} = 2\pi \sum_{\alpha \in L} v_\alpha(\varepsilon) V_{\alpha,n}(\varepsilon) V_{\alpha,m}^*(\varepsilon)$. When these expressions are substituted to Equation 6.20, the current from the left (right) contact to the central region becomes

$$J_{L(R)} = \frac{ie}{\hbar} \int \frac{d\varepsilon}{2\pi} \text{Tr}\left\{ \mathbf{\Gamma}^{L(R)}(\varepsilon) \left(f_{L(R)}(\varepsilon) \mathbf{G}^{>}(\varepsilon) + \left[1 - f_{L(R)}(\varepsilon)\right] \mathbf{G}^{<}(\varepsilon) \right) \right\}, \quad (6.21)$$

where the bold notations denote matrices. $G^{\gtrless}(\varepsilon)$ is the many-body densities of states for the addition (removal) of an extra particle of energy ε which suggests the intuitive interpretation of the current in Equation 6.21 as "current-in" minus "current-out." In fact, using the definition of the full many-body spectral function as $\mathbf{A} = i\left[\mathbf{G}^{>}(\varepsilon) - \mathbf{G}^{<}(\varepsilon)\right]$, one can write the total current $J = (J_L - J_R)/2$ as a generalized Landauer equation

$$J = \frac{e}{\hbar} \int \frac{d\varepsilon}{2\pi} \left[f_L(\varepsilon) - f_R(\varepsilon) \right] \text{Tr}\left\{ \frac{\mathbf{\Gamma}^{L}(\varepsilon)\mathbf{\Gamma}^{R}(\varepsilon)}{\mathbf{\Gamma}^{L}(\varepsilon) + \mathbf{\Gamma}^{R}(\varepsilon)} \left(\mathbf{A}(\varepsilon) \right) \right\}. \quad (6.22)$$

The term in brackets can be interpreted as a many-body transmission coefficient. Equation 6.22 contains broadening to arbitrary order and is, of course, equivalent to the Scattering Matrix method in the absence of interactions. However, its usefulness depends on whether one is able to make some suitable perturbative expansion for the interacting Green's functions. This Keldysh method, combined with ab-initio quantum chemistry calculations, is extensively used in molecular electronics transport calculations (see Chapter 11, Volume 3). Another useful route, fully equivalent to real time Green's functions, is the so-called imaginary-time formulation for steady-state non-equilibrium [88] which has its roots in Hershfield's non-equilibrium steady state density matrix [89]. This formalism has the advantage of having the mathematical structure of equilibrium theory and, in principle, can be easily used in combination with established equilibrium techniques.

6.8.4 The Quantum Master Approach

The second approach, more "quantum-optics" oriented, focuses on the reduced dynamics of the QD density operator after the reservoirs are integrated out. In this approach, interaction effects, such as magnetic exchange, are fully taken into account from the beginning (the density matrix is written in the exact eigenbasis which diagonalizes the isolated problem). The drawback of this approach is that it is very difficult to include the broadening of many-body states beyond low orders in tunneling. This results in a Quantum Master Equation (QME) which describes the approximate dissipative dynamics of exact many-body states.

QMEs can be derived under many different approximations and assumptions. These include various Markovian schemes, such as the standard Bloch-Redfield approach, and more sophisticated superoperator methods such as the Nakajima-Zwanzig (NZ) projection operator technique or diagrammatic expansions in Liouville space, just to mention a few. A comprehensive review of all these techniques can be found in [90].

The dynamics of the density operator in the interaction picture is governed by the Liouville-von Neumann equation $\dfrac{d\hat{\rho}^I}{dt} = -i\left[H_T, \hat{\rho}^I\right]$. Integrating this equation with the initial condition $\hat{\rho}(t_0) = \hat{\rho}^{QD}(t_0) \otimes \hat{\rho}^{res}_{eq}$, which assumes a product state at initial times between the QD density matrix and the reservoirs at equilibrium, tracing over reservoir degrees of freedom and iterating once yields $\dfrac{d\hat{\rho}^I(t)}{dt} = -\displaystyle\int_{t_0}^{t} dt' \mathrm{Tr}_{res}\left\{\left[H_T(t), \left[H_T(t'), \hat{\rho}^I(t')\right]\right]\right\}$. If we now assume a product state at all times $\hat{\rho}(t) = \hat{\rho}^{QD}(t) \otimes \hat{\rho}^{res}_{eq}$ (Born approximation), we end up with the equation

$$\frac{d\hat{\rho}^I(t)}{dt} = -\int_{t_0}^{t} dt' \mathrm{Tr}_{res}\left\{\left[H_T(t), \left[H_T(t'), \hat{\rho}^{QD,I}(t') \otimes \hat{\rho}^{res}_{eq}\right]\right]\right\}. \tag{6.23}$$

Equation 6.23, which is still non-Markovian, is the starting point of most weak coupling QMEs describing transport through an interacting region. In order to convert Equation 6.23 into a QME local in time, one makes one further approximation $\hat{\rho}^{QD,I}(t') \approx \rho^{\wedge QD,I}(t)$ such that Equation 6.23 becomes

$$\frac{d\hat{\rho}^I(t)}{dt} = -\int_{t_0}^{t} dt' \mathrm{Tr}_{res}\left\{\left[H_T(t), \left[H_T(t'), \hat{\rho}^{QD,I}(t) \otimes \hat{\rho}^{res}_{eq}\right]\right]\right\}. \tag{6.24}$$

This equation is called the Redfield equation. The final form of the Markovian QME is obtained by making $\tau = t - t'$ and $t_0 \to -\infty$, such that one gets

$$\frac{d\hat{\rho}(t)}{dt} = -\int_0^{\infty} d\tau \mathrm{Tr}_{res}\left\{\left[H_T(t), \left[H_T(t-\tau), \hat{\rho}^{QD,I}(t) \otimes \hat{\rho}^{res}_{eq}\right]\right]\right\}. \tag{6.25}$$

Equation 6.25 is usually termed the Born-Markov Quantum Master equations and can be also obtained using the Nakajima-Zwanzig superoperator technique at the lowest order.

By projecting Equation 6.25 onto the eigenstates $|N\rangle$ that diagonalize the many-body Hamiltonian H_{QD}, one obtains

$$\frac{d\rho^I_{NM}(t)}{dt} = \sum_{KL} e^{i(\omega_N - \omega_M - \omega_K + \omega_L)t} R_{NMKL} \rho^I_{KL}(t), \tag{6.26}$$

with $R_{NMKL} = -\delta_{L,M} \sum_R \Gamma^+_{NRRK} - \delta_{N,K} \sum_R \Gamma^-_{LRRM} + \Gamma^+_{LMNK} + \Gamma^-_{LMNK}$ is the so-called Redfield tensor. All the information about the reservoirs is encapsulated in the rates

$$\Gamma^+_{LMNK} = \int_0^{\infty} d\tau e^{-i(\omega_N - \omega_K)\tau} \mathrm{Tr}_{res}\left\{\langle L|H_T(\tau)|M\rangle\langle N|H_T|K\rangle \hat{\rho}^{res}_{eq}\right\}$$

$$\Gamma^-_{LMNK} = \int_0^{\infty} d\tau e^{-i(\omega_N - \omega_K)\tau} \mathrm{Tr}_{res}\left\{\langle L|H_T|M\rangle\langle N|H_T(\tau)|K\rangle \hat{\rho}^{res}_{eq}\right\}, \tag{6.27}$$

which contain integrals of the form $\int_0^\infty d\tau e^{iz\tau} = \pi\delta(z) + iP\dfrac{1}{z}$. The reals parts which are proportional to delta functions give rise to dissipative effects, whereas the imaginary principal parts induce shifts in the spectrum known as Stark and Lamb shifts in Quantum Optics.

One further rotating wave approximation, which involves averaging over rapidly oscillating terms approximation where only terms such that $e^{i(\omega_N - \omega_M - \omega_K + \omega_L)t} \approx 1$ leads to the so-called secular approximation where only terms such that $(\omega_N - \omega_M - \omega_K + \omega_L) = 0$ are considered in the summations of Equation 6.26. This approximation guarantees that the QME has a Lindblad form, namely that it preserves the positivity of the density matrix. After all the above approximations, the QME can be written as

$$\frac{d\rho(t)}{dt} = \mathbf{L}\rho(t),\tag{6.28}$$

where $\rho(t)$ is a vector containing both populations and coherences. The latter are important because of the intrinsic many-body degeneracies of the QD spectra. The matrix L contains tunneling rates of the form

$$\Gamma_{NM}^+ = \sum_{\alpha\in LR} \Gamma_\alpha f_\alpha \left(E_N - E_M\right) \sum_n \left|\left\langle N|\mathbf{f}_n^\dagger|M\right\rangle\right|^2$$
$$\Gamma_{NM}^- = \sum_{\alpha\in LR} \Gamma_\alpha \left[1 - f_\alpha\left(E_N - E_M\right)\right] \sum_n \left|\left\langle N|\mathbf{f}_n|M\right\rangle\right|^2.\tag{6.29}$$

The notation $\Gamma_{NM}^\pm$ implies that states M with charge Q are connected with states N with charge $Q+1$ (or $Q-1$) via the matrix elements $\left\langle N|\mathbf{f}_n^\dagger|M\right\rangle$ (or $\left\langle N|\mathbf{f}_n|M\right\rangle$). In our case, these matrix elements depend on the spin property system and lead to remarkable transport properties as we will discuss in the following sections. The tunneling rates also depend on the Fermi functions $f_{L/R}$ of the left/right reservoir. As before, the tunneling coupling is parameterized as $\Gamma^\alpha(\varepsilon) = 2\pi \sum_{k,\alpha\in L,R} |V_{k,\alpha}|^2 \delta(\varepsilon - \varepsilon_{k\alpha}) = 2\pi \sum_{\alpha\in L,R} \mathbf{v}_\alpha(\varepsilon)|V_\alpha(\varepsilon)|^2$ which is usually assumed to be constant. The steady state density matrix is obtained from the solution of $\mathbf{L}\rho^{\text{stat}} = 0$ and is used to compute average quantities such as charge, magnetization and current. The most general expression for the latter is given by:

$$J_{L/R} = \frac{e}{\hbar}\Gamma_{L/R} \sum_{KLM}\sum_n \left\{ f_{L/R}\left(E_K - E_L\right)\left\langle L|\mathbf{f}_n|M\right\rangle\left\langle M|\mathbf{f}_n^\dagger|K\right\rangle\right.$$
$$\left. -\left[1 - f_{L/R}\left(E_K - E_M\right)\right]\left\langle L|\mathbf{f}_n^\dagger|M\right\rangle\left\langle M|\mathbf{f}_n|K\right\rangle\right\}\rho_{KL}^{\text{stat}}.\tag{6.30}$$

The key observation to relate the above expression to the current obtained within the Quantum Master Equation approach and the one in Equation 6.21 which we obtained using the Keldysh formalism is to note that Equation 6.21 already contains an explicit $\Gamma_{L/R}$ in front of the Green's functions. Therefore, in order to obtain the current to lowest (sequential) order we have to

use Green's functions to *zeroth order in the tunneling coupling*. Using
$G^<(t) = \frac{i}{\hbar}\langle \mathbf{f}_m^\dagger \mathbf{f}_n(t)\rangle = \frac{i}{\hbar}Tr\{\mathbf{f}_m^\dagger \mathbf{f}_n(t)\hat{\rho}^{QD}\} = \frac{i}{\hbar}\sum_{MKL}\langle K|\hat{\rho}^{QD}|L\rangle\langle L|\mathbf{f}_m^\dagger|M\rangle\langle M|\mathbf{f}_n(t)|K\rangle$, the zeroth order
is obtained by using Heisenberg operators with respect to the *isolated* many-
body Hamiltonian, namely $\mathbf{f}_n(t) = e^{iH_{QD}t}\mathbf{f}_n e^{-iH_{QD}t}$, such that the lesser
Green's function is $G^<(\varepsilon) = \frac{2\pi i}{\hbar}\sum_{MKL}\delta(\varepsilon - E_K + E_M)\langle K|\hat{\rho}^{QD}|L\rangle\langle L|\mathbf{f}_m^\dagger|M\rangle\langle M|\mathbf{f}_n(t)|K\rangle$.
Substituting this expression in Equation 6.21 one immediately finds
Equation 6.30.

6.8.5 TRANSPORT IN DMS QUANTUM DOTS

In the following, we show results for two dots of CdTe with $L_z = 60$ Å,
$L_x = 80$ Å and different $L_y = 80$ Å (dot A) and $L_y = 75$ Å (dot B), both doped with
1 Mn atom. As we explained in Section 6.8.1, the steady state of a standard
SET is uniquely characterized by external voltages. In the linear response
regime, a new charge is accommodated in the dot at precise values of the
gate voltage, when the electrochemical potential of the dot falls within the
bias window, for example, $\mu_L > \mu_{dot}(N+1) > \mu_R>$. Importantly, for DMS quantum dots, the charge and the conductance of the SET *depend also on the
quantum state of the Mn spin*. This is demonstrated in Figure 6.6, where we
show the charge and linear conductance of a single dot A as the gate injects
either one electron (left) or one hole (right). In both cases, the initial gate is
chosen so that only the $Q = 0$ states are occupied. This initial condition is
described by a thermal equilibrium DM with six equally populated Mn spin
states, $M_z = \pm\frac{5}{2}, \pm\frac{3}{2}, \pm\frac{1}{2}$. We increase the gate and solve the master equation

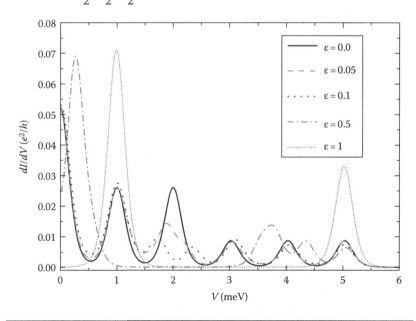

FIGURE 6.6 $G_0(V_G)$ (upper panels), charge (middle panels) and diagonal terms of
the ρ (lower panels) for QD A as a function of V_G around the $Q = -1 \leftrightarrow 0$ transition
(left) and $Q = 0 \leftrightarrow +1$ (right). Lower panels: solid (dashed) lines correspond to $Q = 0$
($|Q| = 1$) states. Results obtained with $\Gamma_L = \Gamma_R = 0.01$ meV and $k_B T = 0.05$ meV.

to obtain the steady state density matrix, which is used as an initial condition for the next run with higher V_G. Importantly, in some cases, the steady state is not equal to the thermal density matrix, which has dramatic consequences for linear transport. This is the case of the positively charged dot A for which the transverse spin interaction is zero, which results in anomalous linear conductance curves that exhibit a three-peak structure in regions of gate voltage, corresponding to the charge degeneracy region. This is in stark contrast with a normal quantum dot which would instead exhibit one single Coulomb Blockade peak. The key to an understanding of this unusual Coulomb Blockade is the fact that M_z does not relax during transport: in the absence of holes, the spin of the Mn ion is free and therefore all the six projections are degenerate. Sweeping the gate voltage towards the charge degeneracy region, the only allowed transitions are those which conserve M_z. This results in three possible charge degeneracy points corresponding to the transition $Q = 0 \Leftrightarrow Q = 1$ at different values of M_z. For example, the first CB peak corresponds to charge degeneracy between $|-5/2,0\rangle$ and $|-5/2,\Uparrow_h\rangle$ $\left(\text{or} |-5/2,0\rangle \text{and} |-5/2,\Downarrow_h\rangle \right)$. Importantly, at this gate voltage, the density matrix has populations in both charge sectors, namely states with $Q = +1, |M_z| = 5/2$ and states with $Q = 0, |M_z| \neq 5/2$ *coexist*. This is true, provided that a quasi steady-state limit for populations (and current) can be reached at time scales smaller than the spin relaxation time, see below. By further increasing V_G, states with $|M_z| = 3/2$ become degenerate and a new peak in the conductance occurs (reflecting the population of the $Q = +1, |M_z| = 3/2$ states). The third peak corresponds to the transition with $|M_z| = 1/2$. Therefore, as a result of this mechanism, the Coulomb Blockade is spin-dependent.

The discharge simulation is done analogously. If the initial V_G is such that there is one hole in the QD, the Ising interaction removes the degeneracy among states with different $|M_z|$. The ground state of the $Q = +1$ sector is now the lowest energy doublet with $|M_z| = 5/2$, which is the only state occupied in thermal equilibrium. Clearly, this initial state is different from the steady state obtained during the charging procedure which produces a hysteretic linear conductance vs. gate voltage curve: as the gate is ramped to discharge the dot, a single peak in the conductance is obtained, corresponding to the resonance condition with the $Q = 0 |M_z| = 5/2$ states.

In contrast, the charge and discharge curves for the conduction band case are equal to each other and display a single peak in the linear conductance curve, as in a standard SET. Of course, the difference between electrons and holes arises from the presence of spin-flip terms in the Hamiltonian, which sets a new time scale in the system: the Mn-spin relaxation time, T_1. In the case of dot A (Ising coupling), T_1 is infinite for holes, which makes the steady state different from the thermal state. In the case of electrons (left panels), T_1 is comparable to the charge relaxation time ($\Gamma_{L,R}^{-1}$) so that steady and thermal DM are identical. In real positively charged dots, T_1 may be long but not infinite. Two independent mechanisms, missing in the simulations shown in the right panels of Figure 6.6, yield a finite T_1. For neutral dots, the Mn T_1, due to superexchange with other Mn spins, scales exponentially with the Mn density [91]. For bulk $Cd_{0.995}Mn_{0.005}$ Te, we have $T_1 = 10^{-3}s$, which is a lower limit estimate for T_1 of the QD with a single Mn. The second mechanism is the transverse spin interaction, which is proportional to the light-hole–heavy-hole mixing and is small according to recent experiments [18]. We have simulated a QD with finite T_1. If we integrate the master equation for $\Gamma^{-1} \ll t \ll T_1$, the $G_0(V_G)$ curve displays two peaks. In contrast, if we integrate the master equation for $\Gamma^{-1} \ll T_1 \ll t$, the $G_0(V_G)$ has a single peak, in agreement with the physical picture of the previous paragraph.

The finite bias conductance of the device also depends strongly on the charge state (=magnetic state) of the dot. In general, current flows whenever the addition of a fermion is permitted by energy conservation and spin selection rules. This provides a link between the dI/dV curve and the spectral function of the fermion in the dot. Thus, the dI/dV curve yields information about the elementary excitation of the Mn spin coupled to the carriers. In the case of a dot doped with 1 Mn whose charge fluctuates between 0 and 1 carrier, the relevant spectral function can be derived from Hamiltonian (6.15), changing ε and J_h. For instance, for conduction band electrons, we would have $\varepsilon = 1$ and for pure heavy holes $\varepsilon = 0$. To illustrate how the tunneling spectra might be used to infer the degree of anisotropy, we calculate the $\dfrac{dI}{dV_B}$ curves corresponding to the effective model of Equation 6.15 and study systematically the above changes as a function of the parameter ε (Figure 6.7). As we have explained, if valence band mixing is non-zero, $|\pm 3/2\rangle$ heavy holes couple with $|\pm 1/2\rangle$ light holes. This mixing

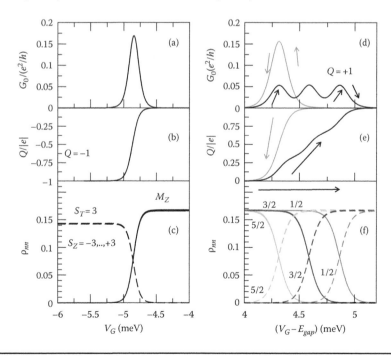

FIGURE 6.7 Differential conductance as a function of bias for holes A (a,b,c) and different anisotropy parameters. V_G was chosen to yield finite $V=0$ conductance. (After Fernández-Rossier, J. et al., *Phys. Rev. Lett.* 98, 106805, 2007. With permission.)

allows simultaneous spin flips between the hole and the Mn spins which results in split states which are bonding and antibonding combinations of $\left|M_z = -1/2, \Uparrow\right\rangle$ and $\left|M_z = +1/2, \Downarrow\right\rangle$. This splitting can be extracted directly from the dI/dV curves as we illustrate in Figure 6.7. For $\varepsilon = 0$, and V_G such that the conductance at zero bias is finite, we obtain six equally spaced peaks corresponding to the Ising spectrum of a QD with a single Mn and doped with a hole (solid curve). Increasing ε, a clear splitting shows up centered at $V_B = 2$(dashed and dotted lines). This situation, $\varepsilon \leq 0.1$, corresponds to realistic parameters (see the discussion of Section 6.5.1). For larger ε, the splitting and the energy shifts are larger, which results in overlapping peaks that are not evenly spaced (dashed-dotted curve). For $\varepsilon = 1$, we recover the fully isotropic case which, as we have discussed in Section 6.4.1, presents two peaks with a zero field splitting, corresponding to the energy difference between a spin $S = 2$ ground state and the $S = 3$ excited state septuplet. The differential conductance, in this case, is, therefore, a direct measure of the spectrum of Figure 6.2.

Interestingly, the strongly anisotropic character of the exchange can be exploited at finite magnetic fields to produce a strong AMR which can be tuned both via gate and bias voltages. This concept is illustrated in Figure 6.8. The upper plots in Figure 6.8 show the linear conductance as a function of gate voltage for 3 situations: $B = 0$, $\vec{B} = (0,0,B)$ (Faraday)

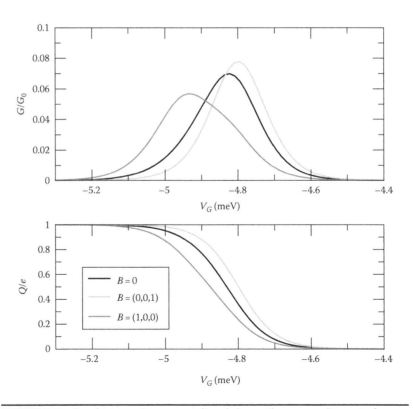

FIGURE 6.8 Conductance (upper panel) and charge (lower panel) vs gate for $\vec{B} = 0, \vec{B} = B_0 \hat{x}$ and $\vec{B} = B_0 \hat{z}$.

and $\vec{B} = (B,0,0)$ (Voigt), whereas the lower plot shows the average charge of the dot. We see how both the charging and G curves depend on $|\vec{B}|$, since they are different from zero and finite field and also on the orientation of $\vec{B}$. At a fixed gate voltage, the change in conductance with the orientation of the applied field can be very large. The origin of the anisotropy lies in the different spectra for the dot doped with 1 hole (Figure 6.4). Since the $Q = 0$ spectrum is quite isotropic, the addition energy becomes anisotropic. Interestingly, compared to the $B = 0$ case, the charging energy is smaller in the Faraday configuration and larger in the Voigt configuration. In the Faraday case, exchange and Zeeman cooperate, whereas, in the Voigt configuration, they compete. The resulting AMR effect is maximized in regions where the charge in one of the configurations begins to saturate. This large AMR, predicted for a device with a single magnetic dopant, could have practical applications.

6.9 OUTLOOK

Semiconductor quantum dots with a few Mn atoms combine two very different length scales, the radius of the confined carrier and the atomic scale where the d electrons lie. In this unique system, the magnetic properties of the Mn atoms are thus dominated by a single or a few carriers whose wave functions extend over thousands of atoms nearby, in stark contrast with magnetic atoms in insulating crystals where only the first neighbor atoms affect the single-ion anisotropy. Conversely, the spectrum of a semiconductor quantum dot with thousands of atoms is dramatically modified by the presence of a single magnetic dopant, as observed experimentally [17, 18, 22]. Thus, probing the dot with excitons or in transport permits one to access the spin state of a single or a few magnetic atoms. Here we have reviewed the theoretical understanding of the electronic structure and transport of this promising system.

REFERENCES

1. J. Fernández-Rossier, Single-atom devices: Quantum engineering, *Nat. Mater.* **12**, 480 (2013).
2. T. Dietl, Ferromagnetic semiconductors, *Semicond. Sci. Technol.* **17**, 377392 (2002).
3. J. K. Furdyna, Diluted magnetic semiconductors, *J. Appl. Phys.* **64**, R29 (1988).
4. T. Jungwirth, J. Sinova, J. Mašek, J. Kučera, and A. H. MacDonald, Theory of ferromagnetic (III,Mn)V semiconductors, *Rev. Mod. Phys.* **78**, 809 (2006).
5. P. Kacman, Spin interactions in diluted magnetic semiconductors and magnetic semiconductor structures, *Semicond. Sci. Technol.* **16**, R25 (2001).
6. H. Ohno, Making nonmagnetic semiconductors ferromagnetic, *Science* **281**, 951 (1998).
7. K. W. Edmonds, P. Bogusławski, K. Y. Wang et al., Mn interstitial diffusion in (Ga,MnAs), *Phys. Rev. Lett.* **92**, 037201 (2004).
8. A. Haury, A. Wasiela, A. Arnoult et al., Observation of a ferromagnetic transition induced by two-dimensional hole gas in modulation-doped CdMnTe quantum wells, *Phys. Rev. Lett.* **79**, 511 (1997).

9. R. K. Kawakami, E. Johnston-Halperin, L. F. Chen et al., (Ga,Mn)As as a digital ferromagnetic heterostructure, *Appl. Phys. Lett.* **77**, 2379 (2000).

10. C. Gould, C. Räuster, T. Jungwirth et al., Tunneling anisotropic magnetoresistance: A spin valve-like tunnel magnetoresistance using a single magnetic layer, *Phys. Rev. Lett.* **93**, 117203 (2004).

11. C. Rüster, T. Borzenko, C. Gould et al., Very large magnetoresistance in lateral ferromagnetic (Ga,Mn)As wires with nanoconstrictions, *Phys. Rev. Lett.* **91**, 216602 (2003).

12. M. Tanaka and Y. Higo, Large tunneling magnetoresistance in GaMnAs/AlAs/GaMnAs ferromagnetic semiconductor tunnel junctions, *Phys. Rev. Lett.* **87**, 026602 (2001).

13. H. Boukari, P. Kossacki, M. Bertolini et al., Light and electric field control of ferromagnetism in magnetic quantum structures, *Phys. Rev. Lett.* **88**, 207204 (2002).

14. H. Ohno, D. Chiba, F. Matsukura et al., Electric-field control of ferromagnetism, *Nature* **408**, 944 (2000).

15. D. Chiba, K. Takamura, F. Matsukura, and H. Ohno, Effect of low-temperature annealing on (Ga,Mn)As trilayer structures, *Appl. Phys. Lett.* **82**, 3020 (2003a).

16. D. Chiba, M. Yamanouchi, F. Matsukura, and H. Ohno, Electrical manipulation of magnetization reversal in ferromagnetic semiconductor, *Science* **301**, 943 (2003b).

17. L. Besombes, Y. Léger, L. Maingault, D. Ferrand, and H. Mariette, Probing the spin state of a single magnetic ion in an individual quantum dot, *Phys. Rev. Lett.* **93**, 207403 (2004).

18. Y. Léger, L. Besombes, J. Fernández-Rossier, L. Maingault, and H. Mariette, Electrical control of a single Mn atom in a quantum dot, *Phys. Rev. Lett.* **97**, 107401 (2006).

19. R. Beaulac, P. Archer, S. T. Ochsenbein, and D. Gamelin, Mn^{2+}-doped CdSe quantum dots: New inorganic materials for spin-electronics and spin-photonics, *Adv. Func. Mater.* **18**, 3873 (2008).

20. G. Bacher, A. A. Maksimov, H. Schomig et al., Monitoring statistical magnetic fluctuations on the nanometer scale, *Phys. Rev. Lett.* **89**, 127201 (2002).

21. C. Gould, A. Slobodskyy, D. Supp et al., Remanent zero field spin splitting of self-assembled quantum dots in a paramagnetic host, *Phys. Rev. Lett.* **97**, 017202 (2006).

22. A. Kudelski, A. Lemaître, A. Miard et al., Optically probing the fine structure of a single Mn atom in an InAs quantum dot, *Phys. Rev. Lett.* **99**, 247209 (2007).

23. A. A. Maksimov, G. Bacher, A. McDonald et al., Magnetic polarons in a single diluted magnetic semiconductor quantum dot, *Phys. Rev. B* **62**, R7767 (2000).

24. J. Wunderlich, T. Jungwirth, B. Kaestner et al., Coulomb blockade anisotropic magnetoresistance effect in a (Ga,Mn)As single-electron transistor, *Phys. Rev. Lett.* **97**, 077201 (2006).

25. T. Gurung, S. Mackowski, H. E. Jackson, and L. M. Smith, Optical studies of zero-field magnetization of CdMnTe quantum dots: Influence of average size and composition of quantum dots, *J. Appl. Phys.* **96**, 7407 (2004).

26. I. R. Sellers, R. Oswaldowski, V. R. Whiteside et al., Robust magnetic polarons in type-II (Zn,Mn)Te/ZnSe magnetic quantum dots, *Phys. Rev. B* **82**, 195320 (2010).

27. J. Kobak, T. Smoleński, M. Goryca et al., Designing quantum dots for solotronics, *Nat. Commun.* **5**, 3191 (2014).

28. J. Kobak, A. Bogucki, T. Smoleński et al., Direct determination of zero-field splitting for single Co^{2+} ion embedded in a CdTe/ZnTe quantum dot, *Phys. Rev.* **97**, 045305 (2016).

29. T. Smoleński, T. Kazimierczuk, J. Kobak et al., Magnetic ground state of an individual Fe^{2+} ion in strained semiconductor nanostructure, *Nat. Commun.* **7**, 10484 (2016).

30. A. Lafuente-Sampietro, H. Utsumi, H. Boukari, S. Kuroda, and L. Besombes, Individual Cr atom in a semiconductor quantum dot: Optical addressability and spin-strain coupling, *Phys. Rev. B* **93**, 161301(R) (2016).
31. T. Smoleński, W. Pacuski, M. Goryca, M. Nawrocki, A. Golnik, and P. Kossacki, Optical spin orientation of an individual Mn^{2+} ion in a CdSe/ZnSe quantum dot, *Phys. Rev. B* **91**, 045306 (2015).
32. J. Fernández-Rossier, Single exciton spectroscopy in semimagnetic quantum dots, *Phys. Rev. B* **73**, 045301 (2006).
33. J. Van Brie, P. M. Koenraad, and J. Fernández-Rossier, Single-exciton spectroscopy of single Mn doped InAs quantum dots, *Phys. Rev. B* **78**, 165414 (2008).
34. L. Besombes, Y. Leger, J. Bernos et al., Optical probing of spin fluctuations of a single paramagnetic Mn atom in a semiconductor quantum dot, *Phys. Rev. B* **78**, 125324 (2008).
35. C. Le Gall, L. Besombes, H. Boukari, R. Kolodka, J. Cibert, and H. Mariette, Optical spin orientation of a single manganese atom in a semiconductor quantum dot using quasi resonant photoexcitation, *Phys. Rev. Lett.* **102**, 127402 (2009).
36. C. Le Gall, R. S. Kolodka, C. L. Cao et al., Optical initialization, readout, and dynamics of a Mn spin in a quantum dot, *Phys. Rev. B* **81**, 245315 (2010).
37. L. Besombes, C. L. Cao, S. Jamet, H. Boukari, and J. Fernández-Rossier, Optical control of the spin state of two Mn atoms in a quantum dot, *Phys. Rev. B* **86**, 165306 (2012).
38. M. Koperski, M. Goryca, T. Kazimierczuk et al., Introducing single Mn^{2+} ions into spontaneously coupled quantum dot pairs, *Phys. Rev. B* **89**, 075311 (2014).
39. Al. L. Efros, E. I. Rashba, and M. Rosen, Paramagnetic ion-doped nanocrystal as a voltage-controlled spin filter, *Phys. Rev. Lett.* **87**, 206601 (2001).
40. J. Fernández-Rossier and R. Aguado, Single electron transport in electrically tunable nanomagnets, *Phys. Rev. Lett.* **98**, 106805 (2007).
41. F. Qu and P. Vasilopoulos, Spin transport across a quantum dot doped with a magnetic ion, *Appl. Phys. Lett.* **89**, 122512 (2006).
42. T. Dietl, H. Ohno, F. Matsukura, J. Cibert, and D. Ferrand, Zener model description of ferromagnetism in Zinc-Blende magnetic semiconductors, *Science* **287**, 1019 (2000).
43. K. Hamaya, Spin-related current suppression in a semiconductor quantum dot spin-diode structure, *Phys. Rev. Lett.* **102**, 236806 (2009).
44. S. Thiele, F. Balestro, R. Ballou, S. Klyatskaya, M. Ruben, and W. Wernsdorfer, Electrically driven nuclear spin resonance in single-molecule magnets, *Science* **344**, 1135 (2014).
45. N. Tsukahara, K. Noto, M. Ohara et al., Adsorption-induced switching of magnetic anisotropy in a single iron (II) phthalocyanine molecule on an oxidized Cu (110) surface, *Phys. Rev. Lett.* **102**, 167203 (2009).
46. R. Vincent, S. Klyatskaya, M. Ruben, W. Wernsdorfer, and F. Balestro, Electronic read-out of a single nuclear spin using a molecular spin transistor, *Nature* **488**, 357 (2012).
47. A. J. Heinrich, J. A. Gupta, C. P. Lutz, and D. M. Eigler, Single-atom spin-flip spectroscopy, *Science* **306**, 466 (2004).
48. C. Hirjibehedin, C. P. Lutz, and A. J. Heinrich, Spin coupling in engineered atomic structures, *Science*, **312**, 1021 (2006).
49. S. Kahle, Z. Deng, N. Malinowski et al., The quantum magnetism of individual manganese-12-acetate molecular magnets anchored at surfaces, *Nano Lett.* **12**, 518 (2011).
50. D. L. Klein, R. Roth, A. K. L. Lim, A. P. Alivisatos, and P. L. McEuen, A single-electron transistor made from a cadmium selenide, *Nature* **389**, 699 (1997).
51. S. T. Ochsenbein, Y. Feng, K. M. Whitaker et al., Charge-controlled magnetism in colloidal doped semiconductor nanocrystals, *Nat. Nanotech.* **4**, 681 (2009).

52. J. Fernández-Rossier and R. Aguado, Mn doped II-VI quantum dots: Artificial single molecule magnets, *Phys. Status Solidi C* **3**, 3734 (2006).

53. J. Fernández-Rossier and L. Brey, Ferromagnetism mediated by few electrons in semimagnetic quantum dots, *Phys. Rev. Lett.* **93**, 1172001 (2004).

54. A. O. Govorov, Optical and electron properties of quantum dots with magnetic impurities, *C. R. Phys.* **9**, 857 (2008).

55. F. Qu and P. Hawrylak, Magnetic exchange interactions in quantum dots containing electrons and magnetic ions, *Phys. Rev. Lett.* **95**, 217206 (2005).

56. F. Qu and P. Hawrylak, Theory of electron mediated Mn-Mn interactions in quantum dots, *Phys. Rev. Lett.* **96**, 157201 (2006).

57. M. Blume and R. Orbach, Spin-lattice relaxation of S-State ions: Mn 2+ in a cubic environment, *Phys. Rev.* **127**, 1587 (1962).

58. M. Qazzaz, Electron paramagnetic resonance of Mn^{2+} in strained-layer semiconductor superlattices, *Solid State Commun.* **96**, 405 (1995).

59. C. Cao, L. Besombes, and J. Fernández-Rossier, Spin-phonon coupling in single Mn doped CdTe quantum dots, *Phys. Rev. B.* **84**, 205305 (2011).

60. R. Abolfath, P. Hawrylak, and I. Žutić, Tailoring magnetism in quantum dots, *Phys. Rev. Lett.* **98**, 207203 (2007).

61. R. Abolfath, A. G. Petukhov, and I. Žutić, Piezomagnetic quantum dots, *Phys. Rev. Lett.* **101**, 207202 (2008).

62. J. I. Climente, M. Korkusiński, P. Hawrylak, and J. Planelles, Voltage control of the magnetic properties of charged semiconductor quantum dots containing magnetic ions, *Phys. Rev. B* **71**, 125321 (2005).

63. A. O. Govorov and A. V. Kalameitsev, Optical properties of a semiconductor quantum dot with a single magnetic impurity: Photoinduced spin orientation, *Phys. Rev. B* **71**, 035338 (2005).

64. N. T. Nguyen and F. M. Peeters, Correlated many-electron states in a quantum dot containing a single magnetic impurity, *Phys. Rev. B* **76**, 045315 (2007).

65. J. M. Pientka, R. Oszwałdowski, A. G. Petukhov, J. E. Han, and I. Žutić, Magnetic ordering in quantum dots: Open versus closed shells, *Phys. Rev. B* **92**, 155402 (2015).

66. M. Cardona, *Fundamentals of Semiconductors*, Springer, Dordrecht, 1999.

67. F. V. Kyrychenko and J. Kossut, Diluted magnetic semiconductor quantum dots: An extreme sensitivity of the hole Zeeman splitting on the aspect ratio of the confining potential, *Phys. Rev. B* **70**, 205317 (2004).

68. D. E. Reiter, T. M. Kuhn, and V. M. Axt, All-optical spin manipulation of a single manganese atom in a quantum dot, *Phys. Rev. Lett.* **102**, 177403 (2009).

69. B. Varghese, H. Boukari, and L. Besombes, Dynamics of a Mn spin coupled to a single hole confined in a quantum dot, *Phys. Rev. B* **90**, 115307 (2014).

70. L. Besombes, Y. Leger, L. Maingault, D. Ferrand, and H. Mariette, Carrier-induced spin splitting of an individual magnetic atom embedded in a quantum dot, *Phys. Rev. B* **71**, 161307 (2005).

71. A. O. Govorov, Voltage-tunable ferromagnetism in semimagnetic quantum dots with few particles: Magnetic polarons and electrical capacitance, *Phys. Rev. B* **72**, 075359 (2005).

72. Y. Léger, L. Besombes, L. Maingault, D. Ferrand, and H. Mariette, Hole spin anisotropy in single Mn-doped quantum dots, *Phys. Rev. B* **72**, 241309 (2005).

73. J. Szczytko, A. Twardowski, M. Palczewska, R. Jabłoński, J. Furdyna, and H. Munekata, Electron paramagnetic resonance of Mn in $In_{1-x}Mn_xAs$ epilayers, *Phys. Rev. B* **63**, 085315 (2001).

74. J. Okabayashi, T. Mizokawa, D. D. Sarma et al., Electronic structure of In1-xMnxAs studied by photoemission spectroscopy: Comparison with $Ga_{1-x}Mn_xAs$, *Phys. Rev. B* **65**, 161203(R) (2002).

75. A. M. Yakunin, A. Yu. Silov, P. M. Koenraad et al., Spatial structure of an individual Mn acceptor in GaAs, *Phys. Rev. Lett.* **92**, 216806 (2004).

76. A. K. Bhattarcharjee and C. Benoit a la Guillaume, Model for the Mn acceptor in GaAs, *Solid State Commun.* **113**, 17 (2000).

77. A. O. Govorov, Optical probing of the spin state of a single magnetic impurity in a self-assembled quantum dot, *Phys. Rev. B* **70**, 035321 (2004).

78. K. H. Schmidt, M. Versen, U. Kunze, D. Reuter, and A. D. Wieck, Electron transport through a single InAs quantum dot, *Phys. Rev. B* **62**, 15879 (2000).

79. T. Ota, K. Ono, M. Stopa et al., Single-dot spectroscopy via elastic single-electron tunneling through a pair of coupled quantum dots, *Phys. Rev. Lett.* **93**, 066801 (2004).

80. T. Ota, M. Rontani, S. Tarucha et al., Few-electron molecular states and their transitions in a single InAs quantum dot molecule, *Phys. Rev. Lett.* **95**, 236801 (2005).

81. M. Jung, K. Hirakawa, Y. Kawaguchi, S. Komiyama, S. Ishida, and Y. Arakawa, Lateral electron transport through single self-assembled InAs quantum dots, *Appl. Phys. Lett.* **86**, 033106 (2005).

82. K. Shibata, C. Buizert, A. Oiwa, K. Hirakawa, and S. Tarucha, Lateral electron tunneling through single self-assembled InAs quantum dots coupled to superconducting nanogap electrodes, *Appl. Phys. Lett.* **91**, 112102 (2007).

83. Y. Igarashi, M. Jung, M. Yamamoto et al., Spin-half Kondo effect in a single self-assembled InAs quantum dot with and without an applied magnetic field, *Phys. Rev. B* **76**, 081303R (2007).

84. C. Buizert, A. Oiwa, K. Shibata, K. Hirakawa, and S. Tarucha, Kondo universal scaling for a quantum dot coupled to superconducting leads, *Phys. Rev. Lett.* **99**, 136806 (2007).

85. L. P. Kouwenhoven, C. M. Marcus, P. L. McEuen, S. Tarucha, R. M. Westervelt, and N. S. Wingreen, Electron transport in quantum dots, in *Mesoscopic Electron Transport*, L. L. Sohn, L. P. Kouwenhoven, and G. Schon (Eds.), Kluwer, Dordrecht, pp. 105, 1997.

86. L. P. Kouwenhoven and L. Glazman, Revival of the Kondo effect, *Phys. World* **14**, 33 (2001).

88. J. E. Han and R. J. Heary, Imaginary-time formulation of steady state nonequilibrium: Application to strongly correlated transport, *Phys. Rev. Lett.* **99**, 236808 (2007).

87. Y. Meir and N. S. Wingreen, Landauer formula for the current through an interacting electron region, *Phys. Rev. Lett.* **68**, 2512 (1992).

89. S. Hershfield, Reformulation of steady state nonequilibrium quantum statistical mechanics, *Phys. Rev. Lett.* **70**, 2134 (2007).

90. C. Timm, Tunneling through molecules and quantum dots: Master equation approaches, *Phys. Rev. B* **77**, 195416 (2008).

91. J. Lambe and Ch. Kikuchi, Paramagnetic resonance of CdTe: Mn and CdS: Mn, *Phys. Rev.* **119**, 1256 (1960).

7

Spin Transport
in Hybrid
Nanostructures

Saburo Takahashi and Sadamichi Maekawa

The spin-dependent transport in hybrid nanostructures is currently of great interest, particularly in the emergence of new phenomena as well as the potential applications to spintronic devices [1–6]. Recent experimental and theoretical studies have demonstrated that the spin-polarized carriers injected from a ferromagnet (F) into a nonmagnetic material (N), such as a normal conducting metal, semiconductor, and superconductor, create nonequilibrium spin accumulation and spin current over the spin diffusion length in the range from nanometers to micrometers. Efficient spin injection and detection and creation of large spin current, spin accumulation, and spin transfer are key factors in utilizing the spin degrees of freedom of carriers as a new functionality in spintronic devices [7, 8].

In this chapter, we discuss the basic aspects for the spin transport in hybrid nanostructures containing normal conducting metals and transition-metal ferromagnets by focusing on the spin current and spin accumulation in a nonlocal spin device of F1/N/F2 structure, where F1 is a spin injector and F2 a spin detector connected to N. We derive basic spin-dependent transport equations for the electrochemical potentials of up-spin and down-spin electrons, and apply them to the F1/N/F2 structure with arbitrary interface resistance ranging from an ohmic contact to a tunneling regime. The injection of spin-polarized electrons and the detection of spin current and accumulation depend strongly on the nature of the junction interface (ohmic contact or tunnel barrier) in the structure. By analyzing the spin-dependent transport in the structure, we show that the relative magnitude of the electrode resistance to the interface resistance plays a crucial role for the spin transport in the F1/N/F2 structure. When a tunnel barrier is inserted into the junction interfaces, spin injection and detection are most effective. When the N/F2 interface is an ohmic contact, the injected spin current from F1 is strongly absorbed by F2 (spin sink) owing to the small spin resistance of ferromagnets. The spin absorption effect plays a vital role in nonlocal spin manipulation. We present the spin Hall effect in nonmagnetic metals caused by the spin-orbit scattering of conducting electrons, which enables the interconversion between the spin (charge) current and charge (spin) current using a nonlocal spin device. Spin injection into a ferromagnetic insulator is briefly mentioned. This chapter provides useful background for Chapters 8 and 9, Volume 3.

7.1 SPIN INJECTION AND DETECTION

Johnson and Silsbee [9, 10] first reported that spins are injected from a ferromagnet into an Al film and diffuse over the spin diffusion length of the order of 1 μm (or even several hundred μm for pure Al), see also Chapter 5,

Volume 1. This rather long spin diffusion length led to the proposal of a spin injection and detection technique in three terminal devices of F1/N/F2 structure (F1 is an injector and F2 a detector) [11], in which the output voltage at F2 depends on the relative orientation of the magnetizations of F1 and F2. In 2001, Jedema et al. performed the spin injection and detection experiments with a nonlocal measurement in a lateral structure of permalloy/copper/permalloy (Py/Cu/Py), and observed a clear spin accumulation signal at room temperature [12]. Subsequently, they measured a large spin accumulation signal in a cobalt/aluminum/cobalt (Co/I/Al/I/Co) structure with tunnel barriers (I = Al_2O_3) [13]. Nonlocal spin injection and detection experiments have been conducted by many groups using normal metals [14–26], graphenes [27–33], superconductors [34–39], semiconductors [40–43], and carbon nanotubes [44].

A basic structure of spin injection and detection device which consists of a nonmagnetic metal N connected to the ferromagnets of the injector F1 and detector F2 is shown in Figure 7.1a, b. F1 and F2 are ferromagnetic electrodes with width w_F and thickness d_F, which are separated by distance L. N is a normal-metal electrode with width w_N and thickness d_N. The magnetizations of F1 and F2 are oriented either parallel or antiparallel. In this

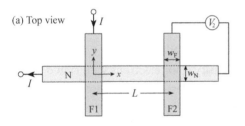

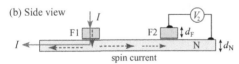

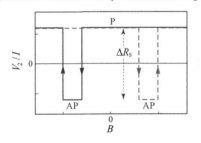

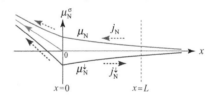

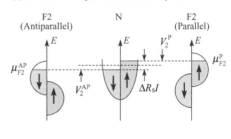

FIGURE 7.1 Nonlocal spin injection and detection device of F1/N/F2. (a) Top view and (b) Side view. Current I is sent from F1 to the left end of N. The spin accumulation at $x = L$ is detected by F2 by measuring voltage V_2. (c) Nonlocal resistance V_2 / I as a function of in-plane magnetic field B, where P and AP represent the parallel and antiparallel orientations of magnetizations in F1 and F2. (d) Spatial variation of the electrochemical potential (ECP) for up-spin and down-spin electrons in N. (e) Densities of states diagram for detection of spin accumulation in N (center) by F2 in the parallel (left) and antiparallel (right) orientations. (After Takahashi, S. et al., *Phys. Soc. Jpn.* 77, 031009, 2008. With permission.)

device, spin-polarized electrons are injected into N from F1 by sending the bias current I from F1 to the left side of N, and the spin accumulation is detected by F2 at distance L from F1, by measuring the voltage V_2 between F2 and N. Since any voltage source is absent in the right side of the device, there is no charge current in the electrodes that lie in the right side of F1. By contrast, owing to the diffusive nature of polarized spins, the injected spins are equally diffused in the left and right directions in N, creating spin current and accumulation on both sides. In this way, the spin and charge transports are separated to realize the region with purely spin transport, which is of great advantage to the nonlocal measurement.

7.2 SPIN TRANSPORT IN NONLOCAL SPIN DEVICES

We describe the nonlocal spin transport in a nanostructured F1/N/F2 device and reveal how the spin transport depends on interface resistance, electrode resistance, spin polarization, and spin diffusion length, and obtain the conditions for efficient spin injection, spin accumulation, and spin current in the device.

7.2.1 BASIC FORMULATION

In the presence of electric field $\mathbf{E}$ and gradient of carrier density n_σ in a conductor, the electrical current density $\mathbf{j}_\sigma$ for electrons with spin σ ($\sigma = \uparrow, \downarrow$) is given by $\mathbf{j}_\uparrow = \sigma_\uparrow \mathbf{E} - eD_\uparrow \nabla n_\uparrow$ and $\mathbf{j}_\downarrow = \sigma_\downarrow \mathbf{E} - eD_\downarrow \nabla n_\downarrow$, where σ_σ is the electrical conductivity, D_σ is the diffusion constant, and e is the electron charge ($e < 0$). Using $\nabla n_\sigma = N_\sigma \nabla \mu_c^\sigma$, where N_σ is the density of states of spin-σ electrons and μ_c^σ is the chemical potential, and the Einstein relation $\sigma_\sigma = e^2 N_\sigma D_\sigma$, we have the spin-dependent currents

$$\mathbf{j}_\uparrow = -\left(\sigma_\uparrow / e\right) \nabla \mu_\uparrow, \quad \mathbf{j}_\downarrow = -\left(\sigma_\downarrow / e\right) \nabla \mu_\downarrow, \tag{7.1}$$

where:

$\mu_\sigma = \mu_c^\sigma + e\phi$ is the electrochemical potential (ECP)
ϕ is the electric potential ($\mathbf{E} = -\nabla \phi$)

The continuity equations for charge and spin in the steady state are

$$\nabla \cdot \left(\mathbf{j}_\uparrow + \mathbf{j}_\downarrow\right) = 0, \quad \nabla \cdot \left(\mathbf{j}_\uparrow - \mathbf{j}_\downarrow\right) = -2e\frac{\delta n_\uparrow}{\tau_{\uparrow\downarrow}} + 2e\frac{\delta n_\downarrow}{\tau_{\downarrow\uparrow}}, \tag{7.2}$$

where:

$\delta n_\sigma = n_\sigma - \bar{n}_\sigma$ is the deviation of carrier density n_σ from equilibrium one $\bar{n}_\sigma$
$\tau_{\sigma\sigma'}$ is the scattering time of an electron from spin state σ to σ'

Making use of the continuity equations and detailed balance $N_\uparrow / \tau_{\uparrow\downarrow} = N_\downarrow / \tau_{\downarrow\uparrow}$, which ensures the absence of spin imbalance in equilibrium, we obtain the basic equations for ECP that describe the charge and spin transport in each electrode [45–50]

$$\nabla^2 \left(\sigma_\uparrow \mu_\uparrow + \sigma_\downarrow \mu_\downarrow \right) = 0, \quad \nabla^2 \left(\mu_\uparrow - \mu_\downarrow \right) = \frac{1}{\lambda^2} \left(\mu_\uparrow - \mu_\downarrow \right), \tag{7.3}$$

where:

$\lambda = \sqrt{D\tau_S}$ is the spin diffusion length

$D = (N_\uparrow + N_\downarrow)/(N_\uparrow D_\downarrow^{-1} + N_\downarrow D_\uparrow^{-1})$ is the effective diffusion constant

$\tau_S = (\tau_{\uparrow\downarrow}^{-1} + \tau_{\downarrow\uparrow}^{-1})^{-1}$ is the spin relaxation time [48]

In the N electrode, the physical quantities are spin-independent, e.g., the electrical conductivity is $\sigma_N^\uparrow = \sigma_N^\downarrow = \frac{1}{2}\sigma_N$, while in the F electrode, those are spin-dependent, e.g., $\sigma_F^\uparrow \neq \sigma_F^\downarrow$ ($\sigma_F = \sigma_F^\uparrow + \sigma_F^\downarrow$). The spin diffusion lengths (λ_F) of transition-metal ferromagnets and alloys, such as NiFe (Py) and CoFe, are on the nanometer scale [51], whereas the spin diffusion lengths (λ_N) of non-magnetic metals, such as Cu, Ag, and Al, are on the micrometer scale. The large difference between λ_F and λ_N plays a crucial role in the spin-dependent transport in nonlocal devices.

The interfacial current across the junctions is described by using the CPP-GMR (current-perpendicular-plane giant magnetoresistance) theory developed by Valet and Fert [46]. In the presence of spin-dependent interface resistance R_i^σ at junction i ($i = 1,2$), the ECP changes discontinuously at the interface when the current flows across the junction. The spin-dependent interfacial current I_1^σ (I_2^σ) from F1 (F2) to N is given by the ECP difference at the interface [46–48]:

$$I_1^\sigma = \frac{1}{eR_1^\sigma} \left(\mu_{F1}^\sigma - \mu_N^\sigma \right), \quad I_2^\sigma = \frac{1}{eR_2^\sigma} \left(\mu_{F2}^\sigma - \mu_N^\sigma \right), \tag{7.4}$$

where the distribution of the current is assumed to be uniform over the interface. The total charge and spin currents across the ith interface are $I_i = I_i^\uparrow + I_i^\downarrow$ and $I_i^s = I_i^\uparrow - I_i^\downarrow$. The above interfacial currents can be applicable from tunnel junctions to transparent ohmic contacts. In a transparent ohmic contact ($R_i^\sigma \rightarrow 0$), the ECP of each spin channel is continuous at the interface, which puts a constraint for spin accumulation on the N side, because the spin accumulation on F is very small due to the short spin diffusion length. In a tunnel junction, spin accumulation on the N side is free from such constraint owing to a large discontinuous change in the ECP at the junction.

In a real device, the distribution of the current across the interface depends on the relative magnitude of the interface resistance to the electrode resistance [52]. When the interface resistance is much larger than the electrode resistance as in tunnel junctions, the current distribution is uniform over the contact area [53], which validates the assumption of uniform interface current assumed in Equation 7.4. However, when the interface resistance is comparable to or smaller than the electrode resistance as in ohmic-contact junctions, the interface current may have inhomogeneous distribution with higher current density around a corner of the contact [54]. In this case, the effective contact area through which the current passes is smaller than the actual contact area $A_J = w_N w_F$ of the junction.

When current I is sent from F1 to the left side of N ($I_1 = I$), the solution of Equation 7.3 takes the form

$$\mu_N^\sigma = \bar{\mu}_N + \sigma\Big(a_1 e^{-|x|/\lambda_N} - a_2 e^{-|x-L|/\lambda_N}\Big). \tag{7.5}$$

Here, the first term describes the charge transport and is $\bar{\mu}_N = -[eI/(\sigma_N A_N)]x$ ($A_N = d_N w_N$) for $x < 0$ and $\bar{\mu}_N = 0$ (ground level of ECP) for $x > 0$, and the second term is the shift in ECP of up-spin ($\sigma = +$) and down-spin ($\sigma = -$) electrons, where the a_1 term represents the spin accumulation due to spin injection from F1, whereas the a_2 term is the spin depletion due to spin leakage into F2. Note that, in the region of $x > 0$, the charge current ($j_N = j_N^\uparrow + j_N^\downarrow$) is absent and only the spin current ($j_N^s = j_N^\uparrow - j_N^\downarrow$) flows (Figure 7.1d), implying the pure spin current is created in this region.

In the F1 and F2 electrodes, the thicknesses are much larger than the spin diffusion length ($d_F \gg \lambda_F$), as in the case of Py or CoFe, so that the solutions close to the interfaces may take the forms of vertical transport along the z direction:

$$\mu_{F1}^\sigma = \bar{\mu}_{F1} + \sigma b_1\Big(\sigma_F/\sigma_F^\sigma\Big)e^{-z/\lambda_F}, \quad \mu_{F2}^\sigma = \bar{\mu}_{F2} - \sigma b_2\Big(\sigma_F/\sigma_F^\sigma\Big)e^{-z/\lambda_F}, \tag{7.6}$$

where:

$\bar{\mu}_{F1} = -[eI/(\sigma_F A_J)]z + eV_1$ describes the charge current flow in F1
$\bar{\mu}_{F2} = eV_2$ has a constant potential with no charge current in F2
V_1 and V_2 are the voltage drops across junctions 1 and 2

The coefficients a_i, b_i, and V_i in Equations 7.5 and 7.6 are determined by the matching conditions that the spin and charge currents are continuous at the interfaces of junctions 1 and 2, and are listed in Section 7.6. The voltages, V_2^P and V_2^{AP}, detected by F2 in the parallel (P) and antiparallel (AP) alignments of magnetizations (Figure 7.1e) are used to calculate the nonlocal resistances V_2^P/I and V_2^{AP}/I, the difference of which is the spin accumulation signal ΔR_s (Figure 7.1c).

7.2.2 SPIN ACCUMULATION SIGNAL

The spin accumulation signal $\Delta R_s = (V_2^P - V_2^{AP})/I$ in the nonlocal spin injection detection device is given by [49]

$$\Delta R_s = R_N \frac{\left(P_1 \dfrac{2R_1^*}{R_N} + p_F \dfrac{2R_F^*}{R_N}\right)\left(P_2 \dfrac{2R_2^*}{R_N} + p_F \dfrac{2R_F^*}{R_N}\right)e^{-L/\lambda_N}}{\left(1 + \dfrac{2R_1^*}{R_N} + \dfrac{2R_F^*}{R_N}\right)\left(1 + \dfrac{2R_2^*}{R_N} + \dfrac{2R_F^*}{R_N}\right) - e^{-2L/\lambda_N}}, \tag{7.7}$$

where $R_N = (\rho_N \lambda_N)/A_N$ is the resistance of N electrode with resistivity ρ_N, length λ_N, and cross-sectional area $A_N = w_N d_N$, and

$$R_i^* = R_i/(1 - P_i^2), \quad R_F^* = R_F/(1 - p_F^2), \tag{7.8}$$

are, respectively, the renormalized version of interface resistance R_i $(1/R_i = 1/R_i^\uparrow + 1/R_i^\downarrow)$ at junction i and that of resistance $R_F = (\rho_F \lambda_F)/A_J$ of F electrode with resistivity ρ_F, length λ_F, and contact area $A_J = w_N w_F$ of the junctions. Hereafter, we call R_N and R_F^* the *spin resistance* of N and F, respectively. In Equation 7.8, $P_i = |R_i^\uparrow - R_i^\downarrow|/(R_i^\uparrow + R_i^\downarrow)$ is the interfacial current spin polarization of junction i and $p_F = |\rho_F^\uparrow - \rho_F^\downarrow|/(\rho_F^\uparrow + \rho_F^\downarrow)$ is the spin polarization of F1 and F2, where $\rho_F^\sigma = 1/\sigma_F^\sigma$ is the spin-dependent resistivity of F. In ohmic-contact junctions, P_i and p_F are in the range of around 40%–70%, as determined from GMR experiments [55] and point-contact Andreev-reflection experiments [56]. In tunnel junctions, P_i is in the range of around 30%–55% for alumina (Al_2O_3) tunnel barriers [57–59], and very large (up to ~85%) for crystalline MgO barriers [60, 61], as determined from superconducting tunneling spectroscopy experiments.

The spin accumulation signal ΔR_s strongly depends on the relative magnitude between the junction resistances (R_1, R_2) and the electrode resistances (R_F, and R_N). Since R_F is much smaller than R_N ($R_N \gg R_F$), as in a device with Cu and Py, Equation 7.7 is simplified as follows. When junctions 1 and 2 are both tunnel junctions ($R_1^*, R_2^* \gg R_N \gg R_F^*$) [11, 13],

$$\Delta R_s/R_N = P_T^2 e^{-L/\lambda_N}, \tag{7.9}$$

where $P_T = P_1 = P_2$ is the tunnel spin polarization. When junction 1 is a tunnel junction and junction 2 is a transparent ohmic contact ($R_1^* \gg R_N \gg R_F^* \gg R_2$) [49],

$$\Delta R_s/R_N = 2p_F P_T \left(\frac{R_F^*}{R_N}\right) e^{-L/\lambda_N}. \tag{7.10}$$

When junctions 1 and 2 are both transparent metallic contacts ($R_N \gg R_F \gg R_1^*, R_2^*$) [12, 47, 48, 62],

$$\Delta R_s/R_N = 4p_F^2 \left(\frac{R_F^*}{R_N}\right)^2 \frac{e^{-L/\lambda_N}}{\left[1 + 2\left(R_F^*/R_N\right)\right]^2 - e^{-2L/\lambda_N}} \approx \frac{2p_F^2}{\sinh(L/\lambda_N)}\left(\frac{R_F^*}{R_N}\right)^2. \tag{7.11}$$

Note that ΔR_s in the above limiting cases is independent of interface resistance. In the intermediate regime ($R_F \ll R_i^* \ll R_N$), however, ΔR_s depends on the interface resistance as

$$\Delta R_s/R_N = \frac{2P_1 P_2}{\sinh(L/\lambda_N)}\left(\frac{R_1^* R_2^*}{R_N^2}\right). \tag{7.12}$$

Figure 7.2a shows the spin accumulation signal ΔR_s for $R_F/R_N = 0.03$, $p_F = 0.7$, and $P_1 = P_2 = 0.4$. We see that ΔR_s increases by one order of magnitude by replacing an ohmic contact with a tunnel barrier, since the resistance mismatch, which is represented by $(R_F/R_N) \ll 1$, is removed by replacing an ohmic contact with a tunnel junction. Note that the mismatch originates from a large difference in the spin diffusion lengths between N and F $(\lambda_F \ll \lambda_N)$. When a nonmagnetic semiconductor is used for N, the resistance mismatch arises from the resistivity mismatch $(\rho_N \gg \rho_F)$ [62–64].

Figure 7.2b shows the experimental data of ΔR_s as a function of distance L in devices of CoFe/I/Al/I/CoFe [66] and Py/Mg/Py [25]. In the tunnel device of CoFe/I/Al/I/CoFe $(I = Al_2O_3)$, fitting Equation 7.9 to the data of Valenzuela and Tinkham [66] leads to $\lambda_N = 705$ nm at 4.2 K, $P_T = 0.3$, and $R_N = 4\Omega$.* The relation $\lambda_N^2 = D\tau_S$ with $\lambda_N = 705$ nm and $D = 1/[2e^2 N(0)\rho_N] \sim 45$ cm^2/s leads to $\tau_S = 82$ ps at 4.2 K, which is consistent with the value of the spin-orbit parameter $b = \hbar/(3\tau_S \Delta_{Al}) \sim 0.01$ obtained by superconducting tunneling spectroscopy [59, 60]. In the ohmic-contact device of Py/Mg/Py, fitting Equation 7.11 to the data of Idzuchi et al. leads to $\lambda_N = 720$ nm, $R_N = 1.7\Omega$,† $p_F = 0.41$, $R_F = 0.1\Omega$,‡ and $\tau_S = 58$ ps at 10 K.

Efficient spin injection has been achieved in nonlocal spin devices with NiFe/MgO/Ag junctions, which leads to a giant enhancement of detected voltage increasing 100-fold with high-applied current, and the crossover of ΔR_s between an ohmic contact to a tunneling regime was demonstrated by changing the junction resistance [65].

In the tunneling regime, the spin splitting of ECP at position x in N is given by

$$2\delta\mu_N(x) = P_T e R_N I e^{-|x|/\lambda_N}. \tag{7.13}$$

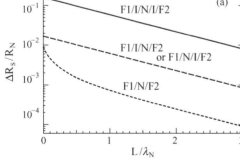
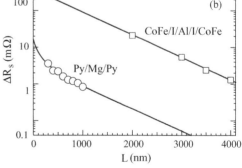

FIGURE 7.2 (a) Spin accumulation signal ΔR_s as a function of distance L between F1 and F2. (b) Spin accumulation signal ΔR_s as a function of distance L in tunnel and metallic-contact devices. The symbols (□) are the experimental data of CoFe/I/Al/I/CoFe tunnel devices [66] at 4.2 K, and (○) are those of Py/Mg/Py metallic-contact devices [25] at 10 K. (After Takahashi, S. et al., *Phys. Rev. B* 67, 052409, 2003. With permission.)

---

* $\rho_N = 5.88$ µΩcm, $w_N = 400$ nm, and $d_N = 25$ nm for Al [66].
† $\rho_N = 4$ µΩcm, $w_N = 170$ nm, and $d_N = 100$ nm for Mg [25].
‡ $\rho_F = 35$ µΩcm, $w_F = 130$ nm, and $d_F = 20$ nm for Py [25].

For a device with $P_T = 0.3$, $R_N = 4\Omega$, and $I = 100\,\mu A$, the values of $\delta\mu_N(x)$ is about 60 μeV at $x = 0$ and 20 μeV at $x = \lambda_N$. This indicates that $\delta\mu_N(x)$ is much smaller than the superconducting gap $\Delta \sim 200\,\mu eV$ of Al films. In a device with a superconductor (Al), it has been predicted that the spin accumulation signal ΔR_s is dramatically enhanced due to the increase of the spin splitting of ECP in the superconducting state [49], and has been observed experimentally [34, 35, 38].

If a half-metallic ferromagnet ($p_F = 1$, $R_F^* \to \infty$) is used for F1 and F2, a largest signal $\Delta R_s \approx R_N e^{-L/\lambda_N}$ is expected without tunnel barriers, which is the advantage of using a half-metallic ferromagnet with 100% spin polarization.

7.2.3 SPIN PRECESSION AND DEPHASING (HANLE EFFECT)

When a magnetic field $\mathbf{B} = (0,0,B)$ is applied perpendicular to the device plane, the injected spins in the N electrode precess around the z axis parallel to B as shown in Figure 7.3a. This precession changes the direction of accumulated spins by angle $\varphi = \omega_L t$, where $\omega_L = \gamma_e B$ is the Larmor frequency and $\gamma_e = 2\mu_B / \hbar$ is the gyromagnetic ratio of conduction electrons [13, 66]. The motion of the magnetization $\mathbf{m}_N$ due to the spin accumulation under the magnetic field is governed by the Bloch-Torrey equation [67]

$$\frac{\partial \mathbf{m}_N(x,t)}{\partial t} = -\gamma_e \mathbf{m}_N(x,t) \times \mathbf{B} - \frac{\mathbf{m}_N(x,t)}{\tau_S} + D_N \nabla^2 \mathbf{m}_N(x,t). \tag{7.14}$$

In weak magnetic fields, the out-of-plane component m_z of magnetization is small and is disregarded for simplicity. If a spin accumulation polarized in the y direction is created at position $x = 0$ and time $t = 0$ in a delta-function form $m_y(x,0) \propto \delta(x)$, Equation 7.14 gives a time evolution of magnetization as $\tilde{m}(x,t) \propto \wp(x,t)e^{-t/\tau_S}e^{i\omega_L t}$, where $\tilde{m} = m_y + im_x$ is a complex representation of magnetization and $\wp(x,t) = (4\pi D_N t)^{-1/2} e^{-x^2/4D_N t}$ is the diffusion kernel in one-dimension.

In a nonlocal F1/N/F2 device with tunnel barriers, spins are steadily injected from F1 to N to create the stationary magnetization at position x:

$$\tilde{m}(x) = \frac{\mu_B}{e} \frac{P_T I}{A_N} \int_0^\infty dt \wp(x,t) e^{-t/\tau_S} \exp(i\omega_L t)$$

$$= \frac{\mu_B}{e} \frac{P_T I}{2 D_N A_N} \lambda_\omega \exp(-|x|/\lambda_\omega), \tag{7.15}$$

where $\lambda_\omega = \lambda_N/\sqrt{1 + i\omega_L \tau_S}$. Since the detected voltage by F2 at $x = L$ is given by $P_T m_y(L)/(\mu_B N(0))$, the nonlocal resistances R_s^P and R_s^{AP} in the parallel and antiparallel orientations become

$$R_s^P = -R_s^{AP} = \frac{1}{2} P_T^2 R_N \mathrm{Re}\left[\left(\lambda_\omega/\lambda_N\right) e^{-L/\lambda_\omega} \right]. \tag{7.16}$$

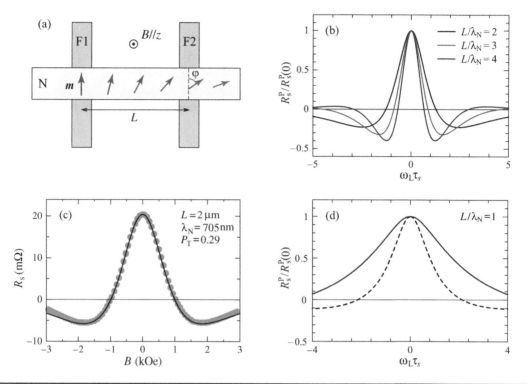

FIGURE 7.3 (a) Spin precession in which electron spins rotate by an angle φ during electron travel from the injector to the detector. (b) Nonlocal resistance as a function of Larmor frequency $\omega_L (= \gamma_e B)$ for distances $L/\lambda_N = 2$, 3, and 4 between F1 and F2 in the parallel alignment of magnetizations in a tunnel device. (c) Nonlocal resistance as a function of perpendicular magnetic field B for $L/\lambda_N = 2.84$ in the parallel alignment of magnetizations. The symbol ($\bullet$) is the experimental data of CoFe/I/Al/I/CoFe [68]. (d) Nonlocal resistance for distance $L/\lambda_N = 1$ in the parallel alignment of magnetizations in a device with a tunnel injector and an ohmic detector. The dashed curve corresponds to that in a tunnel device.

Figure 7.3b shows the normalized nonlocal resistance for distances $L/\lambda_N = 2$, 3, and 4 between F1 and F2 in the parallel alignment of the magnetizations in a nonlocal spin device with tunnel barriers. As seen from the figure, R_s^p shows a damped oscillation due to the spin precession and dephasing (Hanle effect). Figure 7.3c shows the experimental data of R_s^p as a function of perpendicular magnetic field B for $L = 2\,\mu m$ in the parallel alignment of magnetizations in CoFe/I/Al/I/CoFe at 4.2 K [68]. Fitting Equation 7.16 to the data together with $\lambda_N = 705\,nm$ and $R_N = 4\,\Omega$ gives the values of $\tau_S = 52\,ps$ and $P_T = 0.29$.

When F1/N is tunnel contact and F2/N is ohmic contact, the nonlocal resistances are obtained as

$$R_s^p = -R_s^{AP} = P_T p_F R_F^* \frac{\left(\mathrm{Re}\,\lambda_\omega\right)\mathrm{Re}\left[\lambda_\omega e^{-L/\lambda_\omega}\right] + \left(\mathrm{Im}\,\lambda_\omega\right)\mathrm{Im}\left[\lambda_\omega e^{-L/\lambda_\omega}\right]}{\left(\mathrm{Re}\,\lambda_\omega\right)^2 + \left(\mathrm{Im}\,\lambda_\omega\right)^2}, \quad (7.17)$$

which reduces to Equation 7.10 in the absence of the perpendicular magnetic field. When both are ohmic contact, the expression is somewhat complicated [69]. Figure 7.3d shows the normalized R_s^p for distance $L/\lambda_N = 1$

between F1 and F2 in the case of ohmic contacts. With increasing field, R_s^p in the ohmic-contact junctions decays much more slowly than in the tunnel junctions, which may reflect the strong absorption of the spin current by F2 in the ohmic-contact junctions as discussed in Section 7.2.4. The results indicate that the Hanle effect depends strongly on whether the junctions are tunnel or ohmic-contact junctions, in qualitative agreement with recent experiments [69].

7.2.4 NONLOCAL SPIN-CURRENT INJECTION AND MANIPULATION

We next discuss how the spin current is nonlocally injected from F1 to F2 (Figure 7.4a) through N, because of an interest in magnetization switching [70–72] caused by the absorption of pure spin current by F2 [73–75].

The magnitude and distribution of the spin accumulation and spin current in a nonlocal device strongly depend on the relative magnitudes between the interface resistances (R_i) and the electrode spin resistances (R_F, R_N). Figure 7.4b shows the spatial variation of spin accumulation $\delta\mu_N$ in the N electrode in the F1/I/N/F2 structure, in which the first junction is a tunnel junction and the second junction is an ohmic-contact junction. In the absence of F2, the spin accumulation has a symmetric distribution around F1, as shown by the dashed curves of $L/\lambda_N = \infty$. By contrast, when F2 is placed at distance $L/\lambda_N = 0.5$, the spin accumulation is strongly suppressed by the ohmic contact of F2, leaving a small amount of spin accumulation on the right side of F2, as shown by the solid curve. This behavior is caused by

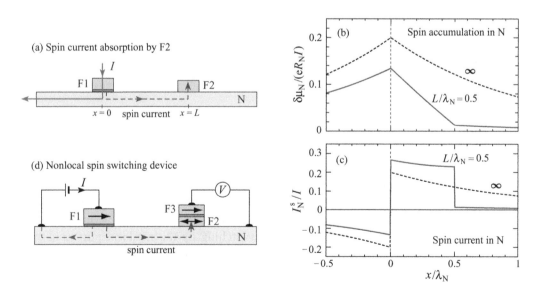

FIGURE 7.4 (a) Nonlocal spin-current injection device of F1/I/N/F2, where junction 1 is a tunnel junction and second 2 is an ohmic contact. Spatial variations of (b) spin accumulation $\delta\mu_N$ and (c) spin current I_N^s in N for $L/\lambda_N = 0.5$ and ∞. The parameter values are the same as those in Figure 7.2. The discontinuous change of spin current at $L/\lambda_N = 0.5$ indicates that most of the spin current flows out of N through the N/F2 interface. (d) Spin switching device utilizing nonlocal spin-current injection from F1 into F2 [73]. The magnetization direction in F2 is detected by F3. (After Takahashi, S. et al., *Sci. Technol. Adv. Mater.* 9, 014105, 2008. With permission.)

the absorption of spins by F2 with small spin resistance R_F^*, and has been observed in a nonlocal device with three Py electrodes [16, 17]. Note that the curve between F1 and F2 ($0 < x < L$) has a steeper slope than that of the dashed curve, indicating an enhanced spin current I_N^s between F1 and F2 compared with that in the absence of F2, as seen in Figure 7.4c. The large discontinuous drop of I_N^s at $x = L$ indicates that most of the spin current flows out of N into F2 through the N/F2 interface, leaving very small spin current on the right side of F2 ($x > L$). This result clearly demonstrates strong absorption of the spin current by ferromagnets.

The spin current I_2^s across the N/F2 interface is calculated as [49, 74]

$$I_2^s = \frac{2\left(P_1\dfrac{R_1^*}{R_N} + p_F\dfrac{R_F^*}{R_N}\right)e^{-L/\lambda_N}}{\left(1+2\dfrac{R_1^*}{R_N}+2\dfrac{R_F^*}{R_N}\right)\left(1+2\dfrac{R_2^*}{R_N}+2\dfrac{R_F^*}{R_N}\right)-e^{-2L/\lambda_N}}I, \tag{7.18}$$

which leads to the spin-current injection from N into F2 being the largest efficiency in the case where the first junction is a tunnel junction and the second junction is an ohmic contact with F2 working as a strong spin absorber, such as Py or CoFe. In this case ($R_2^* \ll R_F^* \ll R_N \ll R_1^*$), the spin current injected into F2 becomes

$$I_2^s \approx P_T I e^{-L/\lambda_N}. \tag{7.19}$$

When both junctions are ohmic-contact junctions ($R_i^* \ll R_F^* \ll R_N$), the spin current injected into F2 becomes

$$I_2^s \approx \frac{p_F I}{\sinh\left(L/\lambda_N\right)}\frac{R_F^*}{R_N}. \tag{7.20}$$

When a small F2 island or disk is placed on N with the contact area of $(50\text{nm})^2$ at distance L within the spin diffusion length $\sim \lambda_N$, the injected pure spin-current density into F2 becomes of the order of $I_2^s \sim 10^6 - 10^7$ A/cm^2 for $I = 1$ mA and $P_T = 0.3$, suggesting that a large pure spin current can be injected, and hence, the spin angular momentum is efficiently transferred from F1 to a small F2. This result provides a method for manipulating the orientation of magnetization due to spin transfer torque in nonlocal spin devices. The magnetization switching by the nonlocal pure spin-current injection has been observed in lateral spin-valve Py/Cu/Py devices [75, 76].

7.3 SPIN HALL EFFECT

Conduction electrons in metals or semiconductors are scattered by local potentials created by impurities or defects in a crystal. The relativistic interaction between the spin and orbital motion of electrons (spin-orbit interaction) at local potentials causes a spin-asymmetric scattering of conduction electrons [77]. In ferromagnetic materials, the electrical current is carried by up-spin (majority) and down-spin (minority) electrons, in which the flow of up-spin electrons are slightly deflected in a transverse direction while the down-spin (minority) electrons in the opposite direction, resulting in the electron flow in the direction perpendicular to both the applied electric field and the magnetization directions. Since up-spin (majority) and down-spin (minority) electrons are strongly imbalanced in ferromagnets, both spin and charge currents are generated in the transverse direction, the latter of which is probed as the Hall voltage and is called the anomalous Hall effect (AHE) [78].

Nonlocal spin injection in nanostructured devices provides a new opportunity for observing the Hall effect originating from spin-orbit interaction in nonmagnetic conductors, the so-called spin Hall effect (SHE) [79–85]. When a pure spin current without accompanying charge current is created in N via nonlocal spin injection, the up- and down-spin currents flowing in opposite directions are deflected in the same direction to induce a charge current in the transverse direction, resulting in a charge accumulation on the edges of N. Inversely, when an unpolarized charge current flows in N as a result of an applied electric field, the up- and down-spin currents flowing in the same direction are deflected in the opposite direction to induce a spin current in the transverse direction, resulting in a spin accumulation near the edges of N. As a consequence, the spin (charge) degrees of freedom are converted to charge (spin) degrees of freedom owing to spin-orbit scattering in nonmagnetic conductors. SHE has been observed using nonlocal spin injection in metal-based nanostructured devices [66, 86–91], which paves the way for future spin electronic applications. In addition to these extrinsic SHEs, intrinsic SHEs have been intensively studied in semiconductors which do not require impurities or defects [92–97].

In the following, we consider the effect of spin-orbit scattering on the spin and charge transports in nonmagnetic metals (N) such as Al, Au, and Pt, and discuss SHE by taking into account the *side jump* (SJ) and *skew scattering* (SS) mechanisms [77, 98–103], and derive formulas for the SHE induced by spin-orbit scattering in nonmagnetic metals and superconductors [83, 104].

7.3.1 Spin and Charge Currents Induced by SHE

The spin-orbit interaction in the presence of nonmagnetic impurities in a metal is derived as follows [105]. The impurity potential $V(\mathbf{r})$ gives rise to an additional electric field $\mathbf{E} = -(1/e)\nabla V(\mathbf{r})$. When an electron passes through the field with velocity $\hat{\mathbf{p}}/m = (\hbar/i)\nabla/m$, the electron feels an effective magnetic field $\mathbf{B}_{\text{eff}} = -(1/mc)\hat{\mathbf{p}} \times \mathbf{E}$, which leads to the spin-orbit coupling $V_{\text{so}} = \mu_B \boldsymbol{\sigma} \cdot \mathbf{B}_{\text{eff}} = \eta_{so}\boldsymbol{\sigma} \cdot [\nabla V(\mathbf{r}) \times \nabla/i]$, where $\boldsymbol{\sigma}$ is the Pauli spin operator and η_{so} is the spin-orbit coupling parameter. Though the value of $\eta_{so} = (\hbar/2mc)^2$ in the free-electron model is too small to account for SHE as well as AHE observed in experiments, the value of η_{so} in real metals may be enhanced by several orders of magnitude for Bloch electrons [77]. In the following, η_{so} is treated as a phenomenological parameter. The total impurity potential $U(\mathbf{r})$ is the sum of the ordinary impurity potential and the spin-orbit potential: $U(\mathbf{r}) = V(\mathbf{r}) + V_{so}(\mathbf{r})$.

The one-electron Hamiltonian in the presence of the impurity potential $U(\mathbf{r})$ is given by

$$\mathcal{H} = \sum_{k,\sigma} \xi_k a_{k\sigma}^{\dagger} a_{k\sigma} + \sum_{k,k'} \sum_{\sigma,\sigma'} \langle \mathbf{k'}\sigma' | U | \mathbf{k}\sigma \rangle a_{k'\sigma'}^{\dagger} a_{k\sigma}. \quad (7.21)$$

Here, the first term is the kinetic energy of conduction electrons with energies $\xi_k = (\hbar k)^2/2m - \mu_c$ measured from the chemical potential, and the second term describes the scattering of conduction electrons between states $|\mathbf{k}\sigma\rangle$ with momentum $\mathbf{k}$ and spin σ with the scattering amplitude

$$\langle \mathbf{k'}\sigma' | U | \mathbf{k}\sigma \rangle = \left(V_{imp}/V \right) \left[\delta_{\sigma'\sigma} + i\eta_{so}\boldsymbol{\sigma}_{\sigma'\sigma} \cdot (\mathbf{k'} \times \mathbf{k}) \right] \sum_i e^{i(\mathbf{k}-\mathbf{k'})\cdot \mathbf{r}_i}, \quad (7.22)$$

where the first and second terms represent the matrix elements of $V(\mathbf{r})$ and $V_{so}(\mathbf{r})$, respectively, V_{imp} is the strength of impurity potential, V is the volume, $\boldsymbol{\sigma}$ is the Pauli matrix, and $\mathbf{r}_i$ is impurity position.

The velocity $\mathbf{v}_k^{\sigma}$ of an electron in the presence of spin-orbit potential is calculated as follows. By taking the matrix element $\mathbf{v}_k^{\sigma} = \langle \mathbf{k}^+ \sigma | \hat{\mathbf{v}} | \mathbf{k}^+ \sigma \rangle$ of the velocity operator $\hat{\mathbf{v}} = d\mathbf{r}/dt = [\mathbf{r}, \mathcal{H}]/(i\hbar)$ [106] between the scattering state $|\mathbf{k}^+ \sigma\rangle$ of an electron with momentum $\mathbf{k}$ and spin σ in the Born approximation, we obtain $\mathbf{v}_k^{\sigma} = v_k + \omega_k^{\sigma}$ with the usual velocity $v_k = \hbar \mathbf{k}/m$ and *anomalous velocity* $\omega_k^{\sigma} = \theta_{SH}^{SJ} (\boldsymbol{\sigma}_{\sigma\sigma} \times \hbar \mathbf{k}/m)$, where θ_{SH}^{SJ} is so-called the spin Hall angle representing the strength of side jump

$$\theta_{SH}^{SJ} = \frac{\hbar \bar{\eta}_{so}}{2\mu_c \tau_{tr}^0} = \frac{\bar{\eta}_{so}}{k_F l}, \quad (7.23)$$

where $\tau_{tr}^0 = 1/[(2\pi/\hbar)n_{imp}N(0)V_{imp}^2]$ is the scattering time due to impurities, n_{imp} is the impurity concentration, $\bar{\eta}_{so} = k_F^2 \eta_{so}$ is the dimensionless spin-orbit coupling parameter, k_F is the Fermi momentum, and l is the mean-free path.

Introducing the current operator $\hat{\mathbf{J}}_{\sigma} = e \sum_k \left(v_k + \omega_k^{\sigma} \right) a_{k\sigma}^{\dagger} a_{k\sigma}$ for conduction electrons with spin σ, the total charge current $\mathbf{J}_q = \mathbf{J}_{\uparrow} + \mathbf{J}_{\downarrow}$ and the total spin current $\mathbf{J}_s = \mathbf{J}_{\uparrow} - \mathbf{J}_{\downarrow}$ are expressed as

$$\mathbf{J}_q = \mathbf{J}_q' + \theta_{SH}^{SJ} \left[\hat{\mathbf{z}} \times \mathbf{J}_s' \right], \quad \mathbf{J}_s = \mathbf{J}_s' + \theta_{SH}^{SJ} \left[\hat{\mathbf{z}} \times \mathbf{J}_q' \right], \quad (7.24)$$

where $\mathbf{J}_q' = e\sum_k v_k \left(f_{k\uparrow} + f_{k\downarrow} \right)$, $\mathbf{J}_s' = e\sum_k v_k \left(f_{k\uparrow} - f_{k\downarrow} \right)$, $f_{k\sigma} = \langle a_{k\sigma}^{\dagger} a_{k\sigma} \rangle$ is the distribution function of an electron with energy ξ_k and spin σ, and $\hat{\mathbf{z}}$ is the polarization axis ($\boldsymbol{\sigma}_{\sigma\sigma} = \sigma\hat{\mathbf{z}}$). The second terms in Equation 7.24 are the charge and spin Hall currents due to side jump. In addition to the side-jump contribution, there is the skew-scattering contribution which originates from the modification of the distribution function due to anisotropic scattering by the spin-orbit interaction.

The distribution function $f_{\mathbf{k}\sigma}$ is calculated based on the Boltzmann transport equation in the steady state,

$$\mathbf{v}_{\mathbf{k}} \cdot \nabla f_{\mathbf{k}\sigma} + \frac{e\mathbf{E}}{\hbar} \cdot \nabla_{\mathbf{k}} f_{\mathbf{k}\sigma} = \left(\frac{\partial f_{\mathbf{k}\sigma}}{\partial t} \right)_{\text{scatt}}, \tag{7.25}$$

where $\mathbf{v}_k = \hbar \mathbf{k} / m$, E is the external electric field, and the collision term due to impurity scattering is written as

$$\left(\frac{\partial f_{\mathbf{k}\sigma}}{\partial t} \right)_{\text{scatt}} = \sum_{\mathbf{k}'\sigma'} \left[P^{\sigma\sigma'}_{\mathbf{k}\mathbf{k}'} f_{\mathbf{k}'\sigma'} - P^{\sigma'\sigma}_{\mathbf{k}'\mathbf{k}} f_{\mathbf{k}\sigma} \right], \tag{7.26}$$

where the first term in the brackets is the scattering-in term ($\mathbf{k}'\sigma' \to \mathbf{k}\sigma$) and the second term is the scattering-out term ($\mathbf{k}\sigma \to \mathbf{k}'\sigma'$), $P^{\sigma'\sigma}_{\mathbf{k}'\mathbf{k}} = (2\pi / \hbar)$ $n_{\text{imp}} |\langle \mathbf{k}'\sigma' | \hat{T} | \mathbf{k}\sigma \rangle|^2 \delta(\xi_{\mathbf{k}} - \xi_{\mathbf{k}'})$ is the scattering probability from state $|\mathbf{k}\sigma\rangle$ to state $|\mathbf{k}'\sigma'\rangle$, $\hat{T}$ is the scattering matrix whose matrix elements are calculated within the second-order Born approximation. Thus, we find that the scattering probability has the symmetric and asymmetric contributions:

$$P^{\sigma'\sigma(1)}_{\mathbf{k}'\mathbf{k}} = \frac{2\pi}{\hbar} \frac{n_{\text{imp}}}{V} V^2_{\text{imp}} \left(\delta_{\sigma\sigma'} + \eta^2_{\text{so}} |(\mathbf{k}' \times \mathbf{k}) \cdot \boldsymbol{\sigma}_{\sigma\sigma'}|^2 \right) \delta(\xi_{\mathbf{k}'} - \xi_{\mathbf{k}}), \tag{7.27}$$

$$P^{\sigma'\sigma(2)}_{\mathbf{k}'\mathbf{k}} = -\frac{(2\pi)^2}{\hbar} \eta_{\text{so}} \frac{n_{\text{imp}}}{V} V^3_{\text{imp}} N(0) \delta_{\sigma\sigma'} \left[(\mathbf{k}' \times \mathbf{k}) \cdot \boldsymbol{\sigma}_{\sigma\sigma} \right] \delta(\xi_{\mathbf{k}'} - \xi_{\mathbf{k}}). \tag{7.28}$$

In solving the Boltzmann equation, it is convenient to separate $f_{\mathbf{k}\sigma}$ into three parts [107] as

$$f_{\mathbf{k}\sigma} = f^0_{\mathbf{k}\sigma} + g^{(1)}_{\mathbf{k}\sigma} + g^{(2)}_{\mathbf{k}\sigma}, \tag{7.29}$$

where $f^0_{\mathbf{k}\sigma}$ is a non-directional distribution function defined by the average of $f_{\mathbf{k}\sigma}$ with respect to the solid angle $\Omega_{\mathbf{k}}$ of $\mathbf{k}$: $f^0_{\mathbf{k}\sigma} = \int f_{\mathbf{k}\sigma} d\Omega_{\mathbf{k}} / (4\pi)$, and $g^{(1)}_{\mathbf{k}\sigma}$ and $g^{(2)}_{\mathbf{k}\sigma}$ are both directional distribution functions, i.e., $\int g^{(i)}_{\mathbf{k}\sigma} d\Omega_{\mathbf{k}} = 0$, and are related with the symmetric and asymmetric contributions, respectively.

We first determine $g^{(1)}_{\mathbf{k}\sigma}$ from the Boltzmann equation with the collision term

$$\sum_{\mathbf{k}'\sigma'} \left[P^{\sigma\sigma'(1)}_{\mathbf{k}\mathbf{k}'} f_{\mathbf{k}'\sigma'} - P^{\sigma'\sigma(1)}_{\mathbf{k}'\mathbf{k}} f_{\mathbf{k}\sigma} \right] = -\frac{g^{(1)}_{\mathbf{k}\sigma}}{\tau_{\text{tr}}} - \frac{f^0_{\mathbf{k}\sigma} - f^0_{\mathbf{k}-\sigma}}{\tau_{\text{sf}}(\theta)}, \tag{7.30}$$

where τ_{tr} is the transport relaxation time and $\tau_{\text{sf}}(\theta)$ is the spin-flip relaxation time, which are given by $1 / \tau_{\text{tr}} = (1 / \tau^0_{\text{tr}})\left(1 + 2\bar{\eta}^2_{\text{so}} / 3\right)$ and $1 / \tau_{\text{sf}}(\theta) = (\bar{\eta}^2_{\text{so}} / 3\tau^0_{\text{tr}})$ $(1 + \cos^2 \theta)$ with angle θ between k and the z axis. Then, the Boltzmann Equation 7.25 with the collision term 7.30 is [82, 108]

$$\mathbf{v}_{\mathbf{k}} \cdot \nabla f_{\mathbf{k}\sigma} + \frac{e\mathbf{E}}{\hbar} \cdot \nabla_{\mathbf{k}} f_{\mathbf{k}\sigma} = -\frac{g^{(1)}_{\mathbf{k}\sigma}}{\tau_{\text{tr}}} - \frac{f^0_{\mathbf{k}\sigma} - f^0_{\mathbf{k}-\sigma}}{\tau_{\text{sf}}(\theta)}, \tag{7.31}$$

where the first term in the r.h.s. describes the momentum relaxation due to impurity scattering and the second term the spin relaxation due to spin-flip scattering. Since $\tau_{tr} \ll \tau_{sf}$, the momentum relaxation occurs first, followed by slow spin relaxation.

The distribution function $f_{k\sigma}^0$ describes the spin accumulation by the shift $\sigma\delta\mu_c(\mathbf{r})$ in the chemical potential from the equilibrium one μ_c, and may be expanded as

$$f_{k\sigma}^0 \approx f_0(\xi_k) + \sigma\left(-\frac{\partial f_0}{\partial \xi_k}\right)\delta\mu_c(\mathbf{r}), \tag{7.32}$$

where $f_0(\xi_k)$ is the Fermi distribution function. Using $f_{k\sigma}^0$ for $f_{k\sigma}$ in Equation 7.31 and $(-\partial f_0 / \partial \xi_k) \approx \delta(\xi_k)$, we obtain

$$g_{k\sigma}^{(1)} \approx -\tau_{tr}\left(-\frac{\partial f_0}{\partial \xi_k}\right)\nu_k \cdot \nabla\mu_N^\sigma(\mathbf{r}), \tag{7.33}$$

where:

$\mu_N^\sigma(\mathbf{r}) = \mu_c + \sigma\delta\mu_c + e\phi$ is the electrochemical potential (ECP)
ϕ is the electric potential ($\mathbf{E} = -\nabla\phi$)

By substituting Equations 7.32 and 7.33 into the Boltzmann Equation 7.31 and summing over $\mathbf{k}$, one obtains the spin diffusion equation

$$\nabla^2(\mu_N^\uparrow - \mu_N^\downarrow) = \lambda_N^{-2}(\mu_N^\uparrow - \mu_N^\downarrow), \tag{7.34}$$

with $\lambda_N = \sqrt{D\tau_{sf}/2}$, $D = (1/3)\tau_{tr}\nu_F^2$, and $\tau_{sf}^{-1} = \langle\tau_{sf}^{-1}(\theta)\rangle_{av} = (4/9)\bar{\eta}_{so}^2/(\tau_{tr}^0)$. Note that $\tau_{sf} = 2\tau_S$. The spin-flip relaxation time τ_{tr} is related to the transport relaxation τ_{sf} time through the spin-orbit coupling parameter η_{so} as $\tau_{tr}/\tau_{sf} = (4/9)\bar{\eta}_{so}^2/[1 + (2/3)\bar{\eta}_{so}^2]$.

The distribution function $g_{k\sigma}^{(2)}$ contributed from *skew scattering* is determined by the asymmetric terms in the Boltzmann equation $\sum_{k'\sigma'}[-P_{k'k}^{\sigma'\sigma(1)}g_{k\sigma}^{(2)} + P_{k'k}^{\sigma'\sigma(2)}g_{k'\sigma'}^{(1)}] = 0$ with Equations 7.27, 7.28, and 7.33, yielding

$$g_{k\sigma}^{(2)} = \theta_{SH}^{SS}\tau_{tr}\left(-\frac{\partial f_0}{\partial \xi_k}\right)(\boldsymbol{\sigma}_{\sigma\sigma} \times \nu_k) \cdot \nabla\mu_N^\sigma(\mathbf{r}), \tag{7.35}$$

where θ_{SH}^{SS} is the so-called spin Hall angle representing the strength of skew scattering

$$\theta_{SH}^{SS} = -(2\pi/3)\bar{\eta}_{so}N(0)V_{imp}. \tag{7.36}$$

Using the distribution function $f_{k\sigma}$ obtained above, we can calculate the first terms in Equation 7.24 as $\mathbf{J}_s' = \mathbf{j}_s + \theta_{SH}^{SS}(\hat{\mathbf{z}} \times \mathbf{j}_q)$ and $\mathbf{J}_q' = \mathbf{j}_q + \theta_{SH}^{SS}(\hat{\mathbf{z}} \times \mathbf{j}_s)$, where the first terms are the longitudinal spin and charge currents:

$$\mathbf{j}_s = -(\sigma_N/2e)\nabla\delta\mu_N, \quad \mathbf{j}_q = \sigma_N\mathbf{E}, \tag{7.37}$$

where $\sigma_N = 2e^2 N(0)D$ is the electrical conductivity and $\delta\mu_N = \mu_N^\uparrow - \mu_N^\downarrow$ is the chemical potential shift, and the second terms are the Hall spin and charge currents induced by the charge and spin currents, respectively. Therefore, the total spin and charge currents in Equation 7.24 are written as

$$\mathbf{J}_q = \mathbf{j}_q + \theta_{SH}\left(\hat{\mathbf{z}} \times \mathbf{j}_s\right), \quad \mathbf{J}_s = \mathbf{j}_s + \theta_{SH}\left(\hat{\mathbf{z}} \times \mathbf{j}_q\right), \qquad (7.38)$$

where $\theta_{SH} = \theta_{SH}^{SJ} + \theta_{SH}^{SS}$ with $\theta_{SH}^{SJ} = \bar{\eta}_{so}/(k_F l)$ and $\theta_{SH}^{SS} = -(2\pi/3)\bar{\eta}_{so}N(0)V_{imp}$. Equation 7.38 indicates that the spin current $\mathbf{j}_s$ induces the transverse charge current $\mathbf{j}_q^{SH} = \theta_{SH}(\hat{\mathbf{z}} \times \mathbf{j}_s)$, while the charge current $\mathbf{j}_q$ induces the transverse spin current $\mathbf{j}_s^{SH} = \theta_{SH}(\hat{\mathbf{z}} \times \mathbf{j}_q)$, as shown in Figure 7.5.

The spin Hall conductivity σ_{SH} is given by the sum of the side-jump contribution $\sigma_{SH}^{SJ} = \theta_{SH}^{SJ}\sigma_N$ and the skew-scattering contribution $\sigma_{SH}^{SS} = \theta_{SH}^{SS}\sigma_N$. We note that the SJ conductivity has the form $\sigma_{SH}^{SJ} = (e^2/\hbar)\eta_{so}n_e$, n_e being the carrier density, and is independent of the impurity concentration, while the SS conductivity depends on the strength and distribution of impurities. When the impurities have a narrow distribution of potentials with definite sign (either positive or negative), as in doped impurities, the SS contribution is dominant for SHE, whereas when the impurity potentials are distributed with positive and negative sign such that the average of V_{imp} over the impurity distribution vanishes ($\langle V_{imp}\rangle \approx 0$), then SJ contribution is dominant. The spin Hall resistivity $\rho_{SH} \approx \sigma_{SH}/\sigma_N^2$, has linear and quadratic terms in ρ_N representing the contributions from side jump and skew scatterings, respectively: $\rho_{SH} \approx a_{SS}\rho_N + b_{SJ}\rho_N^2$, where $a_{SS} = -(2\pi/3)\bar{\eta}_{so}N(0)V_{imp}$ and $b_{SJ} = (2/3\pi)\bar{\eta}_{so}(e^2/h)k_F$.

7.3.2 Spin-Orbit Coupling Parameter

The electrical resistivity and the spin diffusion length are fundamental quantities for the charge and spin transports. By multiplying the resistivity and the spin diffusion length, we obtain $\rho_N \lambda_N = (\sqrt{3}\pi R_K/2k_F^2)(\tau_{sf}/\tau_{tr})^{1/2}$, where $R_K = h/e^2 \approx 25.8$ kΩ is the quantum resistance. Since the spin-orbit

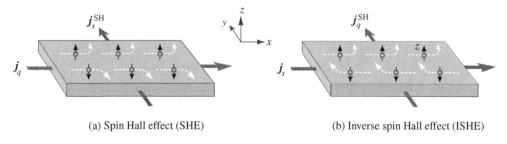

(a) Spin Hall effect (SHE)   (b) Inverse spin Hall effect (ISHE)

FIGURE 7.5   (a) Spin Hall effect (SHE) in which the charge current j_q along the x direction induces the spin current j_s^{SH} in the y direction with the polarization parallel to the z axis. (b) Inverse spin Hall effect (SHE) in which the spin current j_s flowing along the x direction with the polarization parallel to the z axis induces the charge current j_q^{SH} in the y direction. (After Takahashi, S. et al., *Sci. Technol. Adv. Mater.* 9, 014105, 2008. With permission.)

coupling parameter is given by $\bar{\eta}_{so} = (3/2)(\tau_{tr}/\tau_{sf})^{1/2}$ for $\eta_{so}^2 \ll 1$, we have the relation [74, 84, 109]:

$$\bar{\eta}_{so} \approx \frac{3R_K}{k_F^2 \rho_N \lambda_N}. \tag{7.39}$$

Equation 7.39 implies that the spin-orbit coupling parameter $\bar{\eta}_{so}$ is expressed in terms of the product $\rho_N \lambda_N$, providing a useful method of evaluating the spin-orbit coupling in nonmagnetic metals. Using the experimental data of ρ_N and λ_N for various metals in Equation 7.39, we obtain the values of the spin-orbit coupling parameter $\bar{\eta}_{so}$ in those metals, as listed in Table 7.1. We note that $\bar{\eta}_{so}$ is small for Al (light metal), large for Pt (heavy metal), and intermediate for Cu, Ag, and Au. In the case of Al, the spin-orbit coupling parameters for different samples are very close to each other, despite the scattered values of λ_N and ρ_N between them. The values of $\bar{\eta}_{so}$ estimated from the spin injection method are several orders of magnitude larger than the value of $\bar{\eta}_{so} = (\hbar k_F/2mc)^2$ in the free-electron model.

The spin Hall angle due to side-jump contribution (7.23) has the simple form $\theta_{SH}^{SJ} = (3/8)^{1/2}/(k_F \lambda_N)$ with the aid of the relation $l/\lambda_N = (6\tau_{tr}/\tau_{sf})^{1/2} = (8/3)^{1/2}\bar{\eta}_{so}$. This formula is useful to estimate θ_{SH}^{SJ} and σ_{SH}^{SJ} from the measured values of λ_N and σ_N. Using the data in Table 7.1, one can obtain the magnitude of θ_{SH}^{SJ} and σ_{SH}^{SJ}. For example, $\theta_{SH}^{SJ} = 5 - 8 \times 10^{-5}$ and $\sigma_{SH}^{SJ} = 8 - 9 \, (\Omega cm)^{-1}$ for Al [66]. For Pt, $\theta_{SH}^{SJ} = 4.4 \times 10^{-3}$ and $\sigma_{SH}^{SJ} = 340 \, (\Omega cm)^{-1}$ [87], which are much larger than those of Al, since Pt is a heavy metal element with large $\bar{\eta}_{so}$ and short λ_N.

7.3.3 NONLOCAL SPIN HALL EFFECT

In nonlocal spin injection devices, a pure spin current is created in a nonmagnetic metal electrode. It is fundamentally important to verify the existence of the spin current flowing in a nonmagnetic metal. A most simple and direct verification of the spin current is made by using a nonlocal spin Hall device shown in Figure 7.6a [74, 84, 109, 114]. In this device, the magnetization of the ferromagnet (F) is in the z direction perpendicular to the plane. Spin injection is made by applying the current I from F to the left end of N, while the Hall voltage (V_{SH}) is measured by the Hall bar at distance L, where the pure spin current $\mathbf{j}_s = (j_s, 0, 0)$ flows in the x direction. Thus it follows from Equation 7.38 that

$$\mathbf{J}_q = \sigma_N \mathbf{E} + \theta_{SH}(\hat{\mathbf{z}} \times \mathbf{j}_s), \tag{7.40}$$

where the second term is the Hall current induced by the spin current. In an open circuit condition in the transverse direction, the ohmic current builds up in the transverse direction as opposed to the Hall current such that the y component of $\mathbf{J}_q$ in Equation 7.40 vanishes, resulting in the relation between the Hall electric field E_y and the spin current j_s, $E_y = -\theta_{SH}\rho_N j_s$, which is integrated over the width w_N of N to yield the Hall voltage

$$V_{SH} = \theta_{SH} w_N \rho_N j_s, \tag{7.41}$$

TABLE 7.1
Spin-Orbit Coupling Parameter $\bar{\eta}_{so}$ for Al, Mg, Cu, Ag, Au, and Pt. here, the Fermi Momenta, $k_F = 1.75 \times 10^8$ cm^{-1} (Al), 1.36×10^8 cm^{-1} (Mg, Cu), 1.20×10^8 cm^{-1} (Ag), and 1.21×10^8 cm^{-1} (Au) from [110] are Taken, and 1×10^8 cm^{-1} for Pt is Assumed

	λ_N (nm)	ρ_N ($\mu\Omega$cm)	τ_{sf}/τ_{tr}	$\bar{\eta}_{so}$	References
Al (4.2 K)	650	5.90	5.6×10^4	0.0063	[13]
Al (4.2 K)	455	9.53	7.2×10^4	0.0056	[66]
Al (4.2 K)	705	5.88	6.5×10^4	0.0059	[66]
Mg (10 K)	720	4.00	1.2×10^4	0.014	[25]
Cu (4.2 K)	1000	1.43	2.8×10^3	0.028	[12]
Cu (4.2 K)	1500	1.00	3.1×10^3	0.027	[111]
Cu (4.2 K)	546	3.44	4.9×10^3	0.021	[20]
Cu (<4 K)	520	3.5	4.6×10^3	0.022	[112]
Ag (4.2 K)	162	4.00	3.7×10^2	0.080	[21]
Ag (4.2 K)	195	3.50	4.1×10^2	0.076	[21]
Ag (<4 K)	750	3.0	5.3×10^3	0.021	[112]
Ag (77 K)	3000	1.1	9.2×10^3	0.015	[24]
Au (4.2 K)	168	4.00	3.8×10^2	0.078	[22]
Au (<4 K)	58.5	3.3	31.4	0.27	[112]
Au (10 K)	63	1.36	6.2	0.60	[18]
Au (<10 K)	40	4	13.4	0.32	[91]
Pt (4.2 K)	14	4.2	1.4	1.27	[113]
Pt (5 K)	14	12.8	13	0.42	[87]
Pt (<10 K)	10	10	4	0.75	[91]

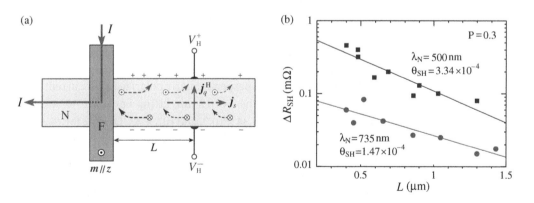

FIGURE 7.6 (a) Nonlocal spin Hall device. The magnetization of F is pointed perpendicular to the plane. The injected pure spin current generates the spin Hall voltage $V_{SH} = V_{SH}^+ - V_{SH}^-$ in the transverse direction, which is measured by the Hall bar at distance L. (b) Nonlocal spin Hall resistance ΔR_{SH} as a function of distance L between the spin injector and the Hall bar for two different devices in high magnetic fields perpendicular to the device plane. The symbols (●, ■,) are the experimental data of CoFe/I/Al at 4.2 K [66]. (After Valenzuela, S.O. et al., *Nature* 442, 176, 2006. With permission.)

indicating that the induced Hall voltage is proportional to the spin current. The spin current at $x = L$ is given by

$$j_s \approx \frac{1}{2} P_{\text{eff}} \left(I/A_N \right) e^{-L/\lambda_N}, \tag{7.42}$$

where P_{eff} is the effective spin polarization which takes the tunnel spin polarization $P_{\text{eff}} = P_T$ for a tunnel junction and $P_{\text{eff}} = p_F(R_F^*/R_N)$ for an ohmic-contact junction. Therefore, the nonlocal Hall resistance $R_{\text{SH}} = V_{\text{SH}}/I$ becomes [74, 84, 109, 114]

$$R_{\text{SH}} = \frac{1}{2} P_{\text{eff}} \theta_{\text{SH}} \frac{\rho_N}{d_N} e^{-L/\lambda_N}. \tag{7.43}$$

For typical values of device parameters ($P_{\text{eff}} \sim 0.3$, $d_N \sim 10$ nm, and $\rho_N \sim 5\,\mu\Omega$ cm), and $\theta_{\text{SH}} \sim 0.01 - 0.0001$ for $\bar{\eta}_{\text{so}} = 0.5 - 0.005$ (Table 7.1), $k_F l \sim 100$, and $V_{\text{imp}} N(0) \sim 0.1 - 0.01$, the expected value of R_{SH} at $L = \lambda_N/2$ is of the order of 0.05–5 mΩ, indicating that SHE is measurable using nonlocal Hall devices. Using the finite element method in three dimensions, a more realistic and quantitative calculation is possible to investigate the spin Hall effect and reveal the spatial distribution of spin and charge currents in a nonlocal device by taking into account the device structure and geometry [115].

The spin Hall effect was observed by using nonlocal spin injection devices: CoFe/I/Al (I = Al$_2$O$_3$) under high magnetic fields perpendicular to the device plane by Valenzuela and Tinkham [66, 116], Py/Cu/Pt using strong spin absorption by Pt by Kimura et al. [86, 87], and FePt/Au with a perpendicularly magnetized FePt by Seki et al. [88], and by using ferromagnetic resonance (FMR) in Py/Pt bilayer films by Saitoh et al. [89, 117].

Figure 7.6b shows the experimental results of CoFe/Al [66] for the nonlocal spin Hall resistance $\Delta R_{\text{SH}} = R_{\text{SH}}(B_s) - R_{\text{SH}}(-B_s)$ as a function of distance L between the spin injector CoFe and the Hall bar in high perpendicular magnetic fields ($\pm B_s$) to saturate the magnetization of CoFe in the perpendicular direction, producing perpendicularly polarized spin current in Al. In the device with the Al thickness of 12 nm, fitting Equation 7.43 to the data leads to $\lambda_N = 500$ nm, $\theta_{\text{SH}} = 3.34 \times 10^{-4}$, and $\sigma_{\text{SH}} = 43$ (Ωcm)$^{-1}$ for $P_T = 0.3$ and $\rho_N/d_N = 8\Omega^*$. In the device with the Al thickness of 25 nm, fitting Equation 7.43 to the data leads to $\lambda_N = 735$ nm, $\theta_{\text{SH}} = 1.5 \times 10^{-4}$, and $\sigma_{\text{SH}} = 25$ (Ωcm)$^{-1}$ for $P_T = 0.3$, and $\rho_N/d_N = 2.4\Omega$. These estimated values for the spin Hall angle and the spin Hall conductivity are in rather satisfactory agreement with those presented in Section 7.3.2.

It has been pointed out that a large spin Hall effect is caused by the skew scattering mechanism in nonmagnetic metal with magnetic impurities [118]. Based on a first principle band structure calculation [119] and a quantum Monte Carlo simulation [120] for Fe impurities in an Au host metal, a novel type of Kondo effect due to strong electron correlation at iron impurities

---

* $\rho_N = 9.53\,\mu\Omega$cm and $d_N = 12$ nm for Al [66].

tremendously enhances the spin-orbit interaction of the order of the hybrid-ization energy (~eV), leading to a large spin Hall angle comparable to that observed in experiments of giant spin Hall effect [88].

It has been demonstrated that, using a spin absorption technique in a nonlocal Py/Cu/NbN device, the inverse spin Hall signal mediated by quasiparticles in a superconducting NbN shows a huge enhancement and becomes more than a thousand times larger than that in the normal state as the spin injection current decreases [38].

7.4 SPIN INJECTION INTO A FERROMAGNETIC INSULATOR

The recent observation that a spin current in a Pt wire is injected into an insulator magnet $Y_3Fe_5O_{12}$ and an electric signal is detected by another Pt wire [121] has stimulated new research interests in spintronics. Magnetic insulators are unique in that they are electrically inactive with frozen charge degrees of freedom, but magnetically active due to the spins of localized elec-trons. The low-lying excitation of spin wave (magnon) carries spin angular momentum. The spin current carried by spin waves are called a spin-wave (magnon) spin current. Utilizing purely magnetic excitation without charge excitation is crucial for developing energy-saving spintronic devices.

In this section, we demonstrate that the spin exchange interaction between local moments and conduction electrons at the interface of normal metal and ferromagnetic insulator plays a vital role for the spin current across the interface. Making use of spin correlation functions and fluctuation-dissipation theorem, we derive a formula for the spin current through the interface, and discuss the conversion efficiency of the spin currents.

7.4.1 SPIN CURRENT THROUGH A FI/N INTERFACE

We consider a junction which consists of a normal metal (N) and a ferro-magnetic insulator (FI) as shown in Figure 7.1. The magnetization of FI is oriented in the z direction. For simplicity, we assume a spin accumulation in N, in which the up-spin and down-spin chemical potentials are split by $\delta\mu_N = \mu_N^\uparrow - \mu_N^\downarrow$, and consider a collinear situation, in which the accumulation direction is either parallel or antiparallel to the magnetization of FI depend-ing on the sign of $\delta\mu_N$. This situation is realized by spin injection from other ferromagnets connected to N or by the spin Hall effect by applying a current in the N layer.

When conduction electrons in N are incident on FI, the electrons are reflected back at the FI/N interface since electrons are prohibited to enter FI due to the large energy gap at the Fermi energy. At the scattering, there is a spin-flip process in which an electron reverses its spin, thereby emitting or absorbing a magnon in FI. The spin-flip scattering with magnon excita-tion gives rise to the transfer of spin angular momentum between FI and N. When the up-spin chemical potential is higher than the down-spin chemical potential (see Figure 7.7), the spin-flip process from up-spin to down-spin

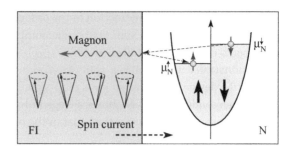

FIGURE 7.7 Schematic diagram of a junction with a ferromagnetic insulator (FI) and a normal metal (N) with spin accumulation $\delta\mu_N = \mu_N^\uparrow - \mu_N^\downarrow$, which plays a role of spin voltage. Magnon emission associated with scattering of an up-spin electron to a down-spin electron by the exchange interaction at the interface.

state dominates over the reversed process, so that the spin current flows from N to FI across the FI/N interface.

We describe the electron-magnon interaction at the interface by using the *s-d* exchange interaction between local moments and conduction electrons at the interface:

$$H_{sd} = J_{sd}v_e \sum_{n=1}^{N_I} S_n \cdot \psi_{\sigma'}^\dagger(\mathbf{r}_n)\hat{\sigma}_{\sigma'\sigma}\psi_\sigma(\mathbf{r}_n), \tag{7.44}$$

where:

J_{sd} is the exchange interaction energy between a local moment and a conduction electron

$\mathbf{S}_n$ are local moments at position $\mathbf{r}_n$ at the interface

$\psi_\sigma^\dagger(\mathbf{r}_n)$ and $\psi_\sigma(\mathbf{r}_n)$ are the creation and annihilation operator of an incident electron with spin σ at position $\mathbf{r}_n$, respectively

σ is the Pauli spin operator

v_e is the volume per electron

N_I is the number of local moments that interact with conduction electrons at the interface

Using the Fourier transformation $\psi_\sigma(\mathbf{r}) = \sum_k c_{k\sigma} e^{ik\cdot r}$ and $\mathbf{S}_n = N_S^{-1/2} \sum_q \mathbf{S}_q e^{iq\cdot r_n}$, where $c_{k\sigma}$ is the annihilation operator of an incident electron with momentum $\mathbf{k}$ and spin σ and N_S is the number of localized moments in FI, Equation 7.44 is written as

$$H_{sd} = J_{eff} \sum_{\mathbf{k},\mathbf{k}',q} \left[S_q^- c_{k'\uparrow}^\dagger c_{k\downarrow} + S_q^+ c_{k'\downarrow}^\dagger c_{k\uparrow} + S_q^z (c_{k'\uparrow}^\dagger c_{k\uparrow} - c_{k'\downarrow}^\dagger c_{k\downarrow}) \right] \rho_{q-k'+k}, \tag{7.45}$$

where $J_{eff} = J_{sd}/(N_N N_S^{1/2})$, $S_q^\pm = S_q^x \pm i S_q^y$, and $\rho_{q-k'+k} = \sum_{n=1}^{N_I} e^{i(q-k'+k)\cdot r_n}$.

The spin current through the interface is calculated from $I_{FI/N}^s = e(d/dt) \langle N_e^s \rangle$, where $N_e^s = (N_e^\uparrow - N_e^\downarrow)$ and $N_e^\sigma = \sum_k c_{k\sigma}^\dagger c_{k\sigma}$ is the number operator of electrons with spin σ. The average value of $(d/dt)\langle N_e^s \rangle$ is obtained by following the procedure to derive the current operator [122], yielding the spin-current density across the interface

$$j_{FI/N}^{s} = \frac{2e}{\hbar^2} n_I J_{eff}^2 \sum_{p,q} \int_{-\infty}^{\infty} dt e^{-i\delta\mu_N t/\hbar} \left[C_{FI}^{+-}(\mathbf{q},t) C_N^{-+}(\mathbf{p},t) - C_{FI}^{-+}(\mathbf{q},-t) C_N^{+-}(\mathbf{p},-t) \right], \quad (7.46)$$

with the interface density n_I of localized moments and the spin correlation functions in FI and N:

$$C_{FI}^{-+}(\mathbf{q},t) = \left\langle S_{-\mathbf{q}}^{-}(t) S_{\mathbf{q}}^{+}(0) \right\rangle, \quad C_{FI}^{+-}(\mathbf{q},t) = \left\langle S_{\mathbf{q}}^{+}(t) S_{-\mathbf{q}}^{-}(0) \right\rangle, \quad (7.47)$$

$$C_N^{-+}(\mathbf{p},t) = \left\langle \sigma_{-\mathbf{p}}^{-}(t) \sigma_{\mathbf{p}}^{+}(0) \right\rangle, \quad C_N^{+-}(\mathbf{p},t) = \left\langle \sigma_{\mathbf{p}}^{+}(t) \sigma_{-\mathbf{p}}^{-}(0) \right\rangle, \quad (7.48)$$

where $\sigma_p^+ = \sum_k c_{k\uparrow}^\dagger c_{k+p\downarrow}$ and $\sigma_p^- = \sum_k c_{k\downarrow}^\dagger c_{k+p\uparrow}$. Using the fluctuation-dissipation theorem which relates the spin correlation function to the imaginary part of the spin susceptibility, we obtain the spin-current density across the interface [121, 123–125]

$$j_{FI/N}^{s} \approx \frac{8\pi e}{\hbar} n_I \langle S_z \rangle \left[J_{sd} v_e N(0) \right]^2 \frac{1}{N_S} \sum_q (\hbar\omega_q + \delta\mu_N)$$

$$\left[n(\hbar\omega_q + \delta\mu_N) - n(\omega_q) \right], \quad (7.49)$$

where:

$N(0)$ is the density of states of conduction electrons at the Fermi level
$f(\xi_k)$ is the Fermi distribution function
$n(\omega_q)$ is the Bose distribution function

Note that $j_{FI/N}^{s}$ vanishes for $\delta\mu_N \to 0$ as expected.

To examine how the spin accumulation is converted to the magnon spin current and how the spin current depends on the magnon energy and temperature, we consider a simple model in which all the magnons have the same energy $\hbar\omega_0$, for which Equation 7.49 becomes

$$j_{FI/N}^{s} = \frac{4\pi e}{\hbar} n_I \langle S_z \rangle \left[J_{sd} v_e N(0) \right]^2 (\hbar\omega_0 + \delta\mu_N)$$

$$\left[\coth\left(\frac{\hbar\omega_0 + \delta\mu_N}{2k_B T} \right) - \coth\left(\frac{\hbar\omega_0}{2k_B T} \right) \right]. \quad (7.50)$$

Figure 7.8 shows the normalized spin current $j_{FI/N}^{s}/j_{FI/N}^{s0}$ versus spin accumulation $\delta\mu_N$ for several values of $\hbar\omega_0/k_B T$, where $j_{FI/N}^{s0} = (8\pi/\hbar) n_I \langle S_z \rangle$ $[J_{sd} v_e N(0)]^2 \hbar\omega_0$. In the high temperature regime ($\hbar\omega_0/k_B T \ll 1$), the spin current depends linearly on $\delta\mu_N$ as $j_{FI/N}^{s}/j_{FI/N}^{s0} \approx -(k_B T/\hbar\omega_0)(\delta\mu_N/\hbar\omega_0)$. In the low temperature regime ($\hbar\omega_0/k_B T \gg 1$), $j_{FI/N}^{s}$ exhibits a strong asymmetry in the $j_{FI/N}^{s} - \delta\mu_N$ curve; $j_{FI/N}^{s} \approx 0$ for $\delta\mu_N > -\hbar\omega_0$ and increases linearly as

$j_{FI/N}^{s} \approx -j_{FI/N}^{s0}[(\delta\mu_N / \hbar\omega_0)+1]$ for $\delta\mu_N < -\hbar\omega_0$ due to magnon excitation. This result shows that the FI/N junction with a ferromagnetic insulator has a rectification effect for the spin current (Figure 7.8).

7.4.2 Spin Injection into FI via Spin Hall Effect

When an external current is applied in the N layer along the y direction in a FI/N bilayer, the spin current is induced by the spin Hall effect (SHE) to flow along the x direction perpendicular to the N layer [7, 74, 117],

$$\mathbf{j}_s\left(x\right) = -\frac{\sigma_N}{2e}\nabla\delta\mu_N\left(x\right) + \theta_{SH}\left(\boldsymbol{\sigma}_s \times \mathbf{j}\right), \quad \left(0 < x < d_N\right), \quad (7.51)$$

where:

σ_N is the electrical conductivity

$\delta\mu_N(x) = \mu_N^{\uparrow}(x) - \mu_N^{\downarrow}(x)$ represents the spin accumulation polarized in the z direction

θ_{SH} is the spin Hall angle

$\boldsymbol{\sigma}_s$ is the polarization vector of spin current induced by SHE ($\boldsymbol{\sigma}_s \parallel \mathbf{z}$ in the present setup)

$\mathbf{j} = (0, j, 0)$ is the applied current density

We look for a solution for the spin accumulation of the form $\delta\mu_N(x) = ae^{-x/\lambda_N} + be^{x/\lambda_N}$ ($0 < x < d_N$), where λ_N is the spin diffusion length, and a and b are determined by the boundary conditions $j_s(0) = j_{FI/N}^s$ at the interface and $j_s(d_N) = 0$ at the outer surface. The resulting spin accumulation and spin current are

$$\delta\mu_N\left(x\right) = 2e\rho_N\lambda_N\left[\frac{\cosh\left[\left(x-d_N\right)/\lambda_N\right]}{\sinh\left(d_N/\lambda_N\right)}j_{FI/N}^s - \theta_{SH}\frac{\sinh\left[\left(x-d_N/2\right)/\lambda_N\right]}{\cosh\left(d_N/2\lambda_N\right)}j\right], \quad (7.52)$$

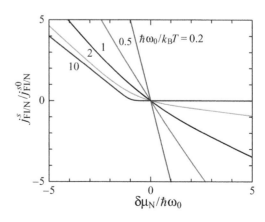

FIGURE 7.8   Spin current $j_{FI/N}^s$ as a function of spin accumulation $\delta\mu_N$ for the ratio $\hbar\omega_0/k_BT$ of magnon energy to temperature.

$$j_s(x) = -\frac{\sinh[(x-d_N)/\lambda_N]}{\sinh(d_N/\lambda_N)} j^s_{FI/N} - \theta_{SH}\left[1-\frac{\cosh[(x-d_N/2)/\lambda_N]}{\cosh(d_N/2\lambda_N)}\right]j. \quad (7.53)$$

In the high temperature regime, the interface spin current depends linearly on $\delta\mu_N(0)$ as

$$j^s_{FI/N} = -\frac{G_{\uparrow\downarrow}}{2eA_J}\delta\mu_N(0), \quad (7.54)$$

where:

$G_{\uparrow\downarrow}$ is the interface spin conductance
A_J is the junction area

In the model of all magnons with the same energy $\hbar\omega_0$, the area spin conductance is given by

$$G_{\uparrow\downarrow}/A_J = \frac{16\pi^2}{R_K}\frac{n_I\langle S_z\rangle[J_{sd}v_e N(0)]^2}{(\hbar\omega_0/k_BT)}, \quad (7.55)$$

where $R_K = h/e^2$ ($R_K \approx 25.8\,\text{k}\Omega$) is the quantum resistance. Using Equation 7.52 in Equation 7.54, we have the interface spin-current density

$$j^s_{FI/N} = -\theta_{SH}\frac{\left(G_{\uparrow\downarrow}/G^s_N\right)\tanh\left(d_N/2\lambda_N\right)}{1+\left(G_{\uparrow\downarrow}/G^s_N\right)\coth\left(d_N/\lambda_N\right)}j, \quad (7.56)$$

and the interface spin accumulation

$$\delta\mu_N(0) = \theta_{SH}\frac{2e(\rho_N\lambda_N)\tanh\left(d_N/2\lambda_N\right)}{1+\left(G_{\uparrow\downarrow}/G^s_N\right)\coth\left(d_N/\lambda_N\right)}j, \quad (7.57)$$

where $G^s_N = A_J/(\rho_N\lambda_N)$ is the spin conductance of N. As seen in Equation 7.56, the efficiency of spin injection is governed by the spin Hall angle θ_{SH}, the conductance ratio ($G_{\uparrow\downarrow}/G^s_N$), and thickness d_N. Using the parameter values $\rho_N = 20\,\mu\Omega\text{cm}$, $\lambda_N = 10\,\text{nm}$, and $d_N = 10\text{nm}$ for a Pt film, and $J_{sd}v_e N(0) \approx J_{sd}/\varepsilon_F$, $\langle S_z\rangle = 5/2$, $a_S = 0.6\text{nm}$, $T = 300\text{K}$, and $(\hbar\omega_0/k_BT) = 0.5$ for a ferromagnetic insulator, Equation 7.55 yields $g_{\uparrow\downarrow} = (G_{\uparrow\downarrow}/A_J)/(2e^2/h) \sim 2.2\times10^{17}(J_{sd}/\varepsilon_F)^2\,\text{cm}^{-2}$. For $J_{sd}/\varepsilon_F = 0.1$, $g_{\uparrow\downarrow} \sim 10^{15}\,\text{cm}^{-2}$, which is comparable to experimental results of YIG/Pt bilayers [126, 127] and even those of metallic bilayers [128], indicating that the spin Hall effect is useful for spin injection into FI [129–132].

7.5 SUMMARY

In this chapter, we have discussed the spin-dependent transport phenomena caused by spin injection from ferromagnets into normal metals in hybrid nanostructure devices, and clarified the conditions for efficient spin injection, accumulation, and transport in these devices. In particular, the nonlocal spin injection creates a flow of "pure" spin current in nonmagnetic conductors (N), so that we have the opportunity to observe novel spin-current-induced phenomena, such as the spin Hall effect in N and nonlocal spin manipulation by pure spin-current injection. The observation of SHE provides a direct verification of the existence of spin current flowing in N. In a reversible way, the electrical current creates the spin current and spin accumulation by the SHE, which provides a source for spin generation without the use of magnetic materials. In addition, the nonlocal spin injection and absorption makes it possible to realize the nonlocal spin manipulation, by which the magnetization direction of a small ferromagnet connected to N can be switched by the spin transfer torque due to nonlocal spin-current absorption. Nonlocal lateral spin devices have advantages in flexibility of layout device geometry and the relative ease of fabricating multiterminal devices with different functionalities. Magnetic insulators with purely magnetic excitations will be useful for developing energy-saving spintronic devices. The development of nonlocal spin devices as well as the exploration of new phenomena open a new avenue in the research of spintronics.

7.6 APPENDIX: ELECTROCHEMICAL POTENTIALS IN F1/N/F2

The coefficients (a_i, b_i) and voltages V_i $(i = 1,2)$ in the electrochemical potentials (ECP) of the N and F electrodes are determined by the matching conditions that the charge and spin currents are continuous at the interfaces of F1/N/F2. The resulting coefficients a_1 and a_2 in Equation 7.5 in the N electrode are

$$a_1 = \frac{\left(P_1 r_1 + p_F r_F\right)\left(1 + 2r_2 + 2r_F\right)}{\left(1 + 2r_1 + 2r_F\right)\left(1 + 2r_2 + 2r_F\right) - e^{-2L/\lambda_N}} R_N eI, \tag{7.58}$$

$$a_2 = \frac{\left(P_1 r_1 + p_F r_F\right)e^{-L/\lambda_N}}{\left(1 + 2r_1 + 2r_F\right)\left(1 + 2r_2 + 2r_F\right) - e^{-2L/\lambda_N}} R_N eI, \tag{7.59}$$

where r_i $(i = 1,2)$ and r_F are the normalized resistances: $r_i = R_i^*/R_N$ and $r_F = R_F^*/R_N$ with $R_i^* = R_i/(1 - P_i^2)$ and $R_F^* = R_F/(1 - p_F^2)$. The spin accumulation voltage V_2 detected by F2 is

$$V_2 = \pm \frac{2\left(P_1 r_1 + p_F r_F\right)\left(P_2 r_2 + p_F r_F\right)e^{-L/\lambda_N}}{\left(1 + 2r_1 + 2r_F\right)\left(1 + 2r_2 + 2r_F\right) - e^{-2L/\lambda_N}} R_N I, \tag{7.60}$$

where "+" and "−" correspond to the parallel and antiparallel alignments of magnetizations, respectively. The coefficients b_1 and b_2 in ECP of Equation 7.6 in F1 and F2 are given in terms of a_1 and a_2 as

$$b_1 = \left(1/2\right)p_F R_F eI - \left(R_F/R_N\right)a_1, \quad b_2 = \left(R_F/R_N\right)a_2. \tag{7.61}$$

Thus the interfacial spin currents across junctions 1 and 2 are

$$I_1^s = \left(2/eR_N\right)a_1, \quad I_2^s = -\left(2/eR_N\right)a_2. \tag{7.62}$$

In the tunneling regime, $I_1^s = P_1 I$ and $I_2^s \approx 0$, as expected.

ACKNOWLEDGMENT

The authors would like to thank M. Ichimura, R. Sugano, J. Ieda, J. Ohe, H. Adachi, Bo Gu, H. Imamura, H. Idzuchi, Y. Fukuma, Y. Niimi, T. Kimura, Y. Otani, T. Seki, S. Mitani, K. Takanashi, K. Ando, K. Uchida, Y. Kajiwara, K. Harii, and E. Saitoh for their helpful discussions and their collaboration. This work was supported by MEXT, and the Next Generation Supercomputer Project of MEXT, Japan.

REFERENCES

1. S. Maekawa and T. Shinjo (Eds.), *Spin Dependent Transport in Magnetic Nanostructures*, Taylor & Francis, London, 2002.
2. S. Maekawa (Ed.), *Concepts in Spin Electronics*, Oxford University Press, Oxford, 2006.
3. H. Kronmuller and S. Parkin (Eds.), *Handbook of Magnetism and Advanced Magnetic Materials*, vol. 1–5, John Wiley & Son, Hoboken, NJ, 2007.
4. S. Maekawa, S. O. Valenzuela, E. Saitoh, and T. Kimura (Eds.), *Spin Current*, Oxford University Press, Oxford, 2012.
5. Y. Xu, D. Awschalom, and J. Nitta (Eds.), *Handbook in Spintronics*, Springer, Netherlands, 2016.
6. I. Žutić, J. Fabian, and S. Das Sarma, Spintronics: Fundamentals and applications, *Rev. Mod. Phys.* **76**, 323 (2004).
7. S. Takahashi and S. Maekawa, Spin current in metals and superconductors, *J. Phys. Soc. Jpn.* **77**, 031009 (2008).
8. S. Takahashi and S. Maekawa, Spin current, spin accumulation and spin Hall effect, *Sci. Technol. Adv. Mater.* **9**, 014105 (2008).
9. M. Johnson and R. H. Silsbee, Interfacial charge-spin coupling: Injection and detection of spin magnetization in metals, *Phys. Rev. Lett.* **55**, 1790 (1985).
10. M. Johnson and R. H. Silsbee, Spin-injection experiment, *Phys. Rev. B* **37**, 5326 (1988).
11. M. Johnson, Spin accumulation in gold films, *Phys. Rev. Lett.* **70**, 2142 (1993).
12. F. J. Jedema, A. T. Filip, and B. J. van Wees, Electrical spin injection and accumulation at room temperature in an all-metal mesoscopic spin valve, *Nature* **410**, 345 (2001).
13. F. J. Jedema, H. B. Heersche, A. T. Filip, J. J. A. Baselmans, and B. J. van Wees, Electrical detection of spin precession in a metallic mesoscopic spin valve, *Nature* **416**, 713 (2002).
14. M. Zaffalon and B. J. van Wees, Zero-dimensional spin accumulation and spin dynamics in a mesoscopic metal island, *Phys. Rev. Lett.* **91**, 186601 (2003).
15. M. Urech, J. Johansson, V. Korenivski, and D. B. Haviland, Spin injection in ferromagnet-superconductor/normal-ferromagnet structures, *J. Magn. Magn. Mater.* **272–276**, e1469 (2004).
16. T. Kimura, J. Hamrle, Y. Otani, K. Tsukagoshi, and Y. Aoyagi, Spin-dependent boundary resistance in the lateral spin-valve structure, *Appl. Phys. Lett.* **85**, 3501 (2004).

17. T. Kimura, J. Hamrle, and Y. Otani, Estimation of spin-diffusion length from the magnitude of spin-current absorption: Multiterminal ferromagnetic/non-ferromagnetic hybrid structures, *Phys. Rev. B* **72**, 14461 (2005).

18. Y. Ji, A. Hoffmann, J. S. Jiang, and S. D. Bader, Spin injection, diffusion, and detection in lateral spin-valves, *Appl. Phys. Lett.* **85**, 6218 (2004).

19. K. Miura, T. Ono, S. Nasu, T. Okuno, K. Mibu, and T. Shinjo, Electrical spin injection in $Ni_{81}Fe_{19}/Al/Ni_{81}Fe_{19}$ with double tunnel junctions, *J. Magn. Magn. Mater.* **286**, 142 (2005).

20. S. Garzon, I. Žutić, and R. A. Webb, Temperature-dependent asymmetry of the nonlocal spin-injection resistance: Evidence for spin nonconserving interface scattering, *Phys. Rev. Lett.* **94**, 176601 (2005).

21. R. Godfrey and M. Johnson, Spin injection in mesoscopic silver wires: Experimental test of resistance mismatch, *Phys. Rev. Lett.* **96**, 136601 (2006).

22. J. H. Ku, J. Chang, H. Kim, and J. Eom, Effective spin injection in Au film from Permalloy, *Appl. Phys. Lett.* **88**, 172510 (2006).

23. T. Kimura, Y. Otani, and P. M. Levy, Electrical control of the direction of spin accumulation, *Phys. Rev. Lett.* **99**, 166601 (2007).

24. T. Kimura and Y. Otani, Large spin accumulation in a permalloy-silver lateral spin valve, *Phys. Rev. Lett.* **99**, 196604 (2007).

25. H. Idzuchi, Y. Fukuma, L. Wang, and Y. Otani, Spin diffusion characteristics in magnesium nanowires, *Appl. Phys. Exp.* **3**, 063002 (2010).

26. Y. Niimi, D. Wei, H. Idzuchi, T. Wakamura, T. Kato, and Y. Otani, Experimental verification of comparability between spin-orbit and spin-diffusion lengths, *Phys. Rev. Lett.* **106**, 126601 (2011).

27. N. Tombros, C. Jozsa, M. Popinciuc, H. T. Jonkman, and B. J. van Wees, Electronic spin transport and spin precession in single graphene layers at room temperature, *Nature* **448**, 571 (2007).

28. M. Ohishi, M. Shiraishi, R. Nouchi, T. Nozaki, T. Shinjo, and Y. Suzuki, Spin injection into a graphene thin film at room temperature, *Jpn. J. Appl. Phys.* **46**, L605 (2007).

29. W. Han, W. H. Wang, K. Pi et al., Electron-hole asymmetry of spin injection and transport in single-layer graphene, *Phys. Rev. Lett.* **102**, 137205 (2009).

30. K. Pi, W. Han, K. M. McCreary, A. G. Swartz, Y. Li, and R. K. Kawakami, Manipulation of spin transport in graphene by surface chemical doping, *Phys. Rev. Lett.* **104**, 187201 (2009).

31. M. Popinciuc, C. Jozsa, P. J. Zomer et al., Electronic spin transport in graphene field-effect transistors, *Phys. Rev. B* **80**, 214427 (2009).

32. H. Idzuchi, A. Fert, and Y. Otani, Revisiting the measurement of the spin relaxation time in graphene-based devices, *Phys. Rev. B* **91**, 241407(R) (2015).

33. K. Vaklinova, A. Hoyer, M. Burghard, and K. Kern, Current-induced spin polarization in topological insulator-graphene heterostructures, *Nano Lett.* **16**, 2595 (2016).

34. M. Urech, J. Johansson, N. Poli, V. Korenivski, and D. B. Haviland, Enhanced spin accumulation in a superconductor, *J. Appl. Phys.* **99**, 08M513 (2006).

35. K. Miura, S. Kasai, K. Kobayashi, and T. Ono, Non local spin detection in ferromagnet/superconductor/ferromagnet spin-valve device with double-tunnel junctions, *Jpn. J. Appl. Phys.* **45**, 2888 (2006).

36. K. Ohnishi, T. Kimura, and Y. Otani, Nonlocal injection of spin current into a superconducting Nb wire, *Appl. Phys. Lett.* **96**, 192509 (2010).

37. C. H. L. Quay, D. Chevallier, C. Bena, and M. Aprili, Spin imbalance and spin-charge separation in a mesoscopic superconductor, *Nat. Phys.* **9**, 84 (2013).

38. T. Wakamura, H. Akaike, Y. Omori et al., Quasiparticle-mediated spin Hall effect in a superconductor, *Nat. Mater.* **14**, 675 (2015).

39. D. Beckmann, Spin manipulation in nanoscal superconductors, *J. Phys. Condens. Matter.* **28**, 163001 (2016).

40. X. Lou, C. Adelmann, S. A. Crooker et al., Electrical detection of spin transport in lateral ferromagnet-semiconductor devices, *Nat. Phys.* **3**, 197 (2007).

41. Y. Ando, K. Hamaya, K. Kasahara et al., Electrical injection and detection of spin-polarized electrons in silicon through an Fe₃Si/Si Schottky tunnel barrier, *Appl. Phys. Lett.* **94**, 182105 (2009).

42. K. Ando, S. Takahashi, J. Ieda et al., Electrically tunable spin injector free from the impedance mismatch problem, *Nat. Mater.* **10**, 655 (2011).

43. T. Sasaki, T. Oikawa, M. Shiraishi, Y. Suzuki, and K. Noguchi, Comparison of spin signals in silicon between nonlocal four-terminal and three-terminal methods, *Appl. Phys. Lett.* **98**, 012508 (2011).

44. C. Feuillet-Palma, T. Delattre, P. Morfin et al., Conserved spin and orbital phase along carbon nanotubes connected with multiple ferromagnetic contacts, *Phys. Rev. B* **81**, 115414 (2010).

45. P. C. van Son, H. van Kempen, and P. Wyder, Boundary resistance of the ferromagnetic-nonferromagnetic metal interface, *Phys. Rev. Lett.* **58**, 2271 (1987).

46. T. Valet and A. Fert, Theory of the perpendicular magnetoresistance in magnetic multilayers, *Phys. Rev. B* **48**, 7099 (1993).

47. A. Fert and S. F. Lee, Theory of the bipolar spin switch, *Phys. Rev. B* **53**, 6554 (1996).

48. S. Hershfield and H. L. Zhao, Charge and spin transport through a metallic ferromagnetic-paramagnetic-ferromagnetic junction, *Phys. Rev. B* **56**, 3296 (1997).

49. S. Takahashi and S. Maekawa, Spin injection and detection in magnetic nanostructures, *Phys. Rev. B* **67**, 052409 (2003).

50. M. Johnson and J. Byers, Charge and spin diffusion in mesoscopic metal wires and at ferromagnet/nonmagnet interfaces, *Phys. Rev. B* **67**, 125112 (2003).

51. J. Bass and W. P. Pratt Jr., Spin-diffusion lengths in metals and alloys, and spin-flipping at metal/metal interfaces: An experimentalist's critical review, *J. Phys. Condens. Matter* **19**, 183201 (2007).

52. R. J. M. van de Veerdonk, J. Nowak, R. Meservey, J. S. Moodera, and W. J. M. de Jonge, Current distribution effects in magnetoresistive tunnel junctions, *Appl. Phys. Lett.* **71**, 2839 (1997).

53. M. Ichimura, S. Takahashi, K. Ito, and S. Maekawa, Geometrical effect on spin current in magnetic nanostructures, *J. Appl. Phys.* **95**, 7255 (2004).

54. J. Hamrle, T. Kimura, T. Yang, and Y. Otani, Current distribution inside Py/Cu lateral spin-valve devices, *Phys. Rev. B* **71**, 094402 (2005).

55. J. Bass and W. P. Pratt Jr., Version 7/21/01 current-perpendicular-to-plane (CPP) magnetoresistance, *Physica B* **321**, 1 (2002).

56. R. J. Soulen Jr., J. M. Byers, M. S. Osofsky et al., Measuring the spin polarization of a metal with a superconducting point contact, *Science* **282**, 85 (1998).

57. R. Meservey and P. M. Tedrow, Spin-polarized electron tunneling, *Phys. Rep.* **238**, 173 (1994).

58. J. S. Moodera and G. Mathon, Spin polarized tunneling in ferromagnetic junctions, *J. Magn. Magn. Mater.* **200**, 248 (1999).

59. D. J. Monsma and S. S. P. Parkin, Spin polarization of tunneling current from ferromagnet/Al₂O₃ interfaces using copper-doped aluminum superconducting films, *Appl. Phys. Lett.* **77**, 720 (2000).

60. S. S. P. Parkin, C. Kaiser, A. Panchula et al., Giant tunnelling magnetoresistance at room temperature with MgO (100) tunnel barriers, *Nat. Mater.* **3**, 862 (2004).

61. S. Yuasa, T. Nagahama, A. Fukushima, Y. Suzuki, and K. Ando, Giant room-temperature magnetoresistance in single-crystal Fe/MgO/Fe magnetic tunnel junctions, *Nat. Mater.* **3**, 868 (2004).

62. E. I. Rashba, Theory of electrical spin injection: Tunnel contacts as a solution of the conductivity mismatch problem, *Phys. Rev. B* **62**, 16267(R) (2000).

63. A. Fert H. Jaffrès, Conditions for efficient spin injection from a ferromagnetic metal into a semiconductor, *Phys. Rev. B* **64**, 184420 (2001).

64. G. Schmidt, G. Richter, P. Grabs, C. Gould, D. Ferrand, and L. W. Molenkamp, Fundamental obstacle for electrical spin injection from a ferromagnetic metal into a diffusive semiconductor, *Phys. Rev. B* **62**, 4790(R) (2000).

65. Y. Fukuma, L. Wang, H. Idzuchi, S. Takahashi, S. Maekawa, and Y. Otani, Giant enhancement of spin accumulation and long-distance spin manipulation in metallic lateral spin valves, *Nat. Mater.* **10**, 527 (2011).

66. S. O. Valenzuela and M. Tinkham, Direct electronic measurement of the spin Hall effect, *Nature* **442**, 176 (2006).

67. M. Johnson and R. H. Silsbee, Coupling of electronic charge and spin at a ferromagnetic-paramagnetic metal interface, *Phys. Rev. B* **37**, 5312 (1988).

68. S. O. Valenzuela, Nonlocal electronic spin detection, spin accumulation and the spin Hall effect, *Int. J. Mod. Phys. B* **23**, 2413 (2009).

69. H. Idzuchi, Y. Fukuma, S. Takahashi, S. Maekawa, and Y. Otani, Effect of anisotropic spin absorption on the Hanle effect in lateral spin valves, *Phys. Rev. B* **89**, 081308(R) (2014).

70. J. C. Slonczewski, Current-driven excitation of magnetic multilayers, *J. Magn. Magn. Mater.* **159**, L1 (1996).

71. L. Berger, Emission of spin waves by a magnetic multilayer traversed by a current, *Phys. Rev. B* **54**, 9353 (1996).

72. J. A. Katine, F. J. Albert, R. A. Buhrman, E. B. Myers, and D. C. Ralph, Current-driven magnetization reversal and spin-wave excitations in Co/Cu/Co pillars, *Phys. Rev. Lett.* **84**, 3149 (2000).

73. S. Maekawa, K. Inomata, and S. Takahashi, Japan Patent No. 3818276 (23 June, 2006).

74. S. Takahashi and S. Maekawa, Spin injection and transport in magnetic nanostructures, *Physica C* **437–438**, 309 (2006).

75. T. Kimura, Y. Otani, and J. Hamrle, Switching magnetization of a nanoscale ferromagnetic particle using nonlocal spin injection, *Phys. Rev. Lett.* **96**, 037201 (2006).

76. T. Yang, T. Kimura, and Y. Otani, Giant spin-accumulation signal and pure spin-current-induced reversible magnetization switching, *Nat. Phys.* **4**, 851 (2008).

77. C. L. Chien and C. R. Westgate (Eds.), *The Hall Effect and Its Applications*, Plenum, New York, 1980.

78. N. Nagaosa, J. Sinova, S. Onoda, A. H. MacDonald, and N. P. Ong, Anomalous Hall effect, *Rev. Mod. Phys.* **82**, 1539 (2010).

79. J. Sinova, S. O. Valenzuela, J. Wunderlich, C. H. Back, and T. Jungwirth, Spin Hall effect, *Rev. Mod. Phys.* **87**, 1213 (2015).

80. M. I. Dyakonov and V. I. Perel, Current induced spin orientation of electrons in semiconductors, *Phys. Lett. A* **35**, 459 (1971).

81. J. E. Hirsch, Spin Hall effect, *Phys. Rev. Lett.* **83**, 1834 (1999).

82. S. Zhang, Spin Hall effect in the presence of spin diffusion, *Phys. Rev. Lett.* **85**, 393 (2001).

83. S. Takahashi and S. Maekawa, Hall effect induced by a spin-polarized current in superconductors, *Phys. Rev. Lett.* **88**, 116601 (2002).

84. S. Takahashi, H. Imamura, and S. Maekawa, Spin injection and spin transport in hybrid nanostructures, in *Concept in Spin Electronics*, S. Maekawa (Ed.), Oxford University Press, Oxford, 2006.

85. R. V. Shchelushkin and A. Brataas, Spin Hall effects in diffusive normal metals, *Phys. Rev. B* **71**, 045123 (2005).

86. T. Kimura, Y. Otani, T. Sato, S. Takahashi, and S. Maekawa, Room-temperature reversible spin Hall effect, *Phys. Rev. Lett.* **98**, 156601 (2007).

87. L. Vila, T. Kimura, and Y. Otani, Room-temperature reversible spin Hall effect, *Phys. Rev. Lett.* **99**, 226604 (2007).

88. T. Seki, Y. Hasegawa, S. Mitani et al., Giant spin Hall effect in perpendicularly spin-polarized FePt/Au devices, *Nat. Mater.* **7**, 125 (2008).

89. E. Saitoh, M. Ueda, H. Miyajima, and G. Tatara, Conversion of spin current into charge current at room temperature: Inverse spin-Hall effect, *Appl. Phys. Lett.* **88**, 182509 (2006).

90. A. Hoffmann, Spin Hall effect in metals, *IEEE Trans. Magn.* **49**, 5172 (2013).

91. Y. Niimi, H. Suzuki, Y. Kawanishi et al., Extrinsic spin Hall effects measured with lateral spin valve structures, *Phys. Rev. B* **89**, 054401 (2014).

92. R. Karplus and J. M. Luttinger, Hall effect in ferromagnetics, *Phys. Rev.* **95**, 1154 (1954).

93. S. Murakami, N. Nagaosa, and S.-C. Zhang, Dissipationless quantum spin current at room temperature, *Science* **301**, 1348 (2003).

94. J. Sinova, D. Culcer, Q. Niu, N. A. Sinitsyn, T. Jungwirth, and A. H. MacDonald, Universal intrinsic spin Hall effect, *Phys. Rev. Lett.* **92**, 126603 (2002).

95. J. Inoue, G. E. W. Bauer, and L. W. Molenkamp, Suppression of the persistent spin Hall current by defect scattering, *Phys. Rev. B* **70**, 041303 (2004).

96. Y. K. Kato, R. C. Myers, A. C. Gossard, and D. D. Awschalom, Observation of the spin Hall effect in semiconductors, *Science* **306**, 1910 (2004).

97. J. Wunderlich, B. Kaestner, J. Sinova, and T. Jungwirth, Experimental observation of the spin-Hall effect in a two-dimensional spin-orbit coupled semiconductor system, *Phys. Rev. Lett.* **94**, 047204 (2005).

98. J. Smit, The spontaneous Hall effect in ferromagnetics II, *Physica* **24**, 39 (1958).

99. L. Berger, Side-jump mechanism for the Hall effect of ferromagnets, *Phys. Rev. B* **2**, 4559 (1970).

100. A. Crépieux and P. Bruno, Theory of the anomalous Hall effect from the Kubo formula and the Dirac equation, *Phys. Rev. B* **64**, 14416 (2001).

101. W.-K. Tse and S. Das Sarma, Spin Hall effect in doped semiconductor structures, *Phys. Rev. Lett.* **96**, 56601 (2006).

102. H. A. Engel, E. I. Rashba, and B. I. Halperin, Theory of spin Hall effects in semiconductors, in *Handbook of Magnetism and Advanced Magnetic Materials*, vol. 5, Wiley, New York, 2006.

103. N. A. Sinitsyn, Semiclassical theories of the anomalous Hall effect, *J. Phys. Condens. Matter* **20**, 023201 (2008).

104. S. Takahashi and S. Maekawa, Spin Hall effect in superconductors, *Jpn. J. Appl. Phys.* **51**, 010110 (2012).

105. J. J. Sakurai and S. F. Tuan, *Modern Quantum Mechanics*, Addison-Wesley, Redwood City, CA, 1985.

106. S. K. Lyo and T. Holstein, Side-jump mechanism for ferromagnetic Hall effect, *Phys. Rev. Lett.* **29**, 423 (1972).

107. J. Kondo, Anomalous Hall effect and magnetoresistance of ferromagnetic metals, *Prog. Theor. Phys.* **27**, 772 (1964).

108. J.-P. Ansermet, Perpendicular transport of spin-polarized electrons through magnetic nanostructures, *J. Phys. Condens. Matter* **10**, 6027 (1998).

109. S. Takahashi and S. Maekawa, Nonlocal spin Hall effect and spin-orbit interaction in nonmagnetic metals, *J. Magn. Magn. Mater.* **310**, 2067 (2007).

110. N. W. Ashcroft and D. Mermin, *Solid State Physics*, Saunders College, Philadelphia, PA, 1976.

111. T. Kimura, J. Hamrle, and Y. Otani, Spin accumulation and Hall effect measured in non-local configuration, *J. Magn. Soc. Jpn.* **29**, 192 (2005).

112. A. B. Gougam, F. Pierre, H. Pothier, D. Esteve, and N. O. Birge, Comparison of energy and phase relaxation in metallic wires, *J. Low Temp. Phys.* **118**, 447 (2000).

113. H. Kurt, R. Loloee, K. Eid, W. P. Pratt, Jr., and J. Bass, Spin-memory loss at 4.2 K in sputtered Pd and Pt and at Pd/Cu and Pt/Cu interfaces, *J. Appl. Phys.* **81**, 4787 (2002).

114. S. Takahashi and S. Maekawa, Effect of spin injection and spin accumulation, in *Spinelectronics–Basic and Applications*, K. Inomata (Ed.), CMC Publishing Co., Ltd., Tokyo, p. 28, 2004.

115. R. Sugano, M. Ichimura, S. Takahashi, and S. Maekawa, Three dimensional simulations of spin Hall effect in magnetic nanostructures, *J. Appl. Phys.* **103**, 07A715 (2008).

116. S. O. Valenzuela and M. Tinkham, Electrical detection of spin currents: The spin-current induced Hall effect, *J. Appl. Phys.* **101**, 09B103 (2007).

117. K. Ando, S. Takahashi, K. Harii et al., Electric manipulation of spin relaxation using the spin Hall effect, *Phys. Rev. Lett.* **101**, 036601 (2008).

118. A. Fert and O. Jaoul, Left-right asymmetry in the scattering of electrons by magnetic impurities, and a Hall effect, *Phys. Rev. Lett.* **28**, 303 (1972).

119. G.-Y. Guo, S. Maekawa, and N. Nagaosa, Enhanced spin Hall effect by resonant skew scattering in the orbital-dependent Kondo effect, *Phys. Rev. Lett.* **102**, 036401 (2009).

120. B. Gu, J.-Y. Gan, N. Bulut et al., Quantum renormalization of the spin Hall effect, *Phys. Rev. Lett.* **105**, 086401 (2010).

121. Y. Kajiwara, K. Harii, S. Takahashi et al., Transmission of electrical signals by spin-wave interconversion in a magnetic insulator, *Nature* **464**, 262 (2010).

122. G. D. Mahan, *Many Particle Physics*, Kluwer Academic/Plenum Publisher, New York, 2000.

123. S. Takahashi, E. Saitoh, and S. Maekawa, Spin current through a normal-metal/insulating-ferromagnet junction, *J. Phys. Conf. Ser.* **200**, 062030 (2010).

124. S. S.-L. Zhang and S. Zhang, Spin current through a normal-metal/insulating-ferromagnet junction, *Phys. Rev. B* **86**, 214424 (2012).

125. S. A. Bender, L. A. Duine, and Y. Tserkovnyak, Electron pumping of quasi-equilibrium Bose-Einstein-condensed magnons, *Phys. Rev. Lett.* **108**, 246601 (2012).

126. F. D. Czeschka, L. Dreher, M. S. Brandt et al., Scaling behavior of the spin pumping effect in ferromagnet-platinum bilayers, *Phys. Rev. Lett.* **107**, 046601 (2011).

127. B. Heinrich, C. Burrowes, E. Montoya et al., *Phys. Rev. Lett.* **107**, 066604 (2011).

128. Y. Tserkovnyak, A. Brataas, G. E. Bauer, and B. I. Halperin, Nonlocal magnetization dynamics in ferromagnetic heterostructures, *Rev. Mod. Phys.* **77**, 1375 (2005).

129. Y.-T. Chen, S. Takahashi, H. Nakayama et al., Theory of spin Hall magnetoresistance, *Phys. Rev. B.* **87**, 144411 (2013).

130. T. Chiba, G. E. W. Bauer, and S. Takahashi, Current-induced spin torque resonance for magnetic insulators, *Phys. Rev. Appl.* **2**, 034003 (2014).

131. S. Chatterjee and S. Sachdev, Probing excitations in insulators via injection of spin currents, *Phys. Rev. B.* **92**, 165113 (2015).

132. W. Chen, M. Sigrist, and D. Manske, Spin Hall effect induced spin transfer through an insulator, *Phys. Rev. B.* **94**, 104412 (2016).

8

Spin Caloritronics

Rafael Ramos and Eiji Saitoh

8.1 INTRODUCTION

Spin caloritronics [1–3] focuses on the study of the interaction between the charge and spin degrees of freedom with heat currents, this area bridges two very active fields of research: thermoelectricity and spintronics, which have the potential to harvest and reduce the energy consumption of modern logic devices. Thermoelectric phenomena emerge from the interaction between heat and charge, manifesting itself as a coupled transport of heat and electricity in electrically conductive materials [4]. Meanwhile, spintronics deals with the fundamental role of the spin of the electron in solid state physics and its potential applications [5, 6]. The pioneering work of Johnson and Silsbee in 1987 [7] started the field of spin caloritronics; they performed a theoretical study to include spin transport in the description of the thermoelectric effect at the interface of a heterostructure comprising a junction of a ferromagnetic and a normal metal layer (a detailed description of their thermodynamic theory is given in Chapter 5, Volume 1). Despite this initial effort, the activity in the spin caloritronics field remained low for many years with only a few experimental studies in metallic magnetic multilayers [8, 9], mainly related to the study of giant magnetoresistive effects [10, 11]. It was not until recently that the field gained renewed interest, mainly after the discovery of the spin Seebeck effect (SSE) by Uchida and co-workers in 2008 [12, 13], demonstrating that spin currents can be thermally generated from a ferromagnetic film into a non-magnetic metal attached and electrically detected by the inverse spin Hall effect, this effect is also regarded as a thermal spin pumping effect [14, 15].

In this chapter we will mainly focus on the SSE, from its initial measurement to recent experimental developments. In Section 8.3 we introduce the spin Seebeck effect and discuss its basic physical mechanism and experimental measurement configuration. In Section 8.4 we introduce the enhancement of the SSE in multilayer systems; it is shown that in structures formed by multiple repetitions of F/N bilayers, where F is a ferromagnet and N is a normal metal, the heat-to-electricity conversion efficiency of the spin Seebeck effect is greatly increased as a result of an unexpected spin Seebeck voltage enhancement in these structures. In Section 8.5 we present the magnetic field–induced suppression of the SSE, which is explained by the magnetic field dependence of the magnon dispersion in the ferromagnetic material: as a result of an applied magnetic field, the Zeeman splitting induces a gap in the magnon spectrum which can be observed due to the magnonic nature of the SSE in ferromagnets. In Section 8.6 we briefly review recent experimental developments of the SSE in materials with different types of magnetic order: compensated ferrimagnets, antiferromagnets and paramagnetic materials. Finally, in Section 8.7 other spin caloritronics phenomena are discussed, particularly the observation of spin-dependent Seebeck and Peltier effects in metallic structures; these are the spin-dependent versions of conventional thermoelectric effects due to conduction electrons.

8.2 SPIN CURRENTS

Spin currents [16, 17] are ubiquitous to all spintronics phenomena [5, 6, 18]; they describe a flow of angular momentum carried by the spin degree of freedom in magnetic materials. This angular momentum can be carried by the spin of conduction electrons in electrically conductive ferromagnets or by the elementary excitations of the spin system in insulating ferromagnets (i.e., spin waves or magnons); as a result we can identify two types of spin currents: conduction-electron spin currents (J_S) and magnon spin currents (J_M). Conduction-electron spin currents can be described as the rate of change of the spin accumulation, defined as the difference between electrochemical potentials for spin-up and spin-down electrons in metals; a more detailed analysis about this type of spin currents is shown in Chapters 7 and 9, Volume 3.

In this section, we will briefly introduce the concept of spin-wave spin currents. We start by considering the Landau-Lifshitz-Gilbert (LLG) equation:

$$\frac{\partial \mathbf{M}(\mathbf{r},t)}{\partial t} = -\gamma \mathbf{M} \times \mathbf{H}_{\text{eff}} + \frac{\alpha}{M} \mathbf{M} \times \frac{\partial}{\partial t} \mathbf{M}, \tag{8.1}$$

where γ is the gyromagnetic ratio and $\mathbf{H}_{\text{eff}}$ is an effective magnetic field that describes general interactions acting on the spin, such as external magnetic fields, magnetic anisotropy and exchange interaction. The last term is the Gilbert damping term which describes the relaxation of the spin due to the interaction with conduction electrons or the crystal lattice and it will be neglected for simplicity. The effective magnetic field can be expressed in terms of the spin-dependent free energy $F(\mathbf{M})$ of the system as: $-\delta F / \delta \mathbf{M}$. Then, after considering the exchange interaction contribution to the effective field [19], the LLG equation can be rewritten as (for a detailed derivation see [16]):

$$\frac{\partial \mathbf{M}}{\partial t} = -\nabla J_M, \tag{8.2}$$

which has the form of a continuity equation, representing spin angular momentum conservation, when the Gilbert damping term is neglected ($\alpha = 0$). J_M is an exchange spin current with components given by:

$$J^{\beta}_{M_\alpha} = \frac{D}{M_S}[\mathbf{M} \times \nabla_i M]_\alpha. \tag{8.3}$$

The exchange spin current can be driven by elementary excitations of magnetically ordered states, known as spin waves or magnons. These can be generated by application of microwave, acoustic or heat excitations [12, 15, 20–23]. Now, we consider an exchange spin current carried when a spin wave is excited. We introduce a spin-wave wave function $\Psi(\mathbf{r},t) = M_+(\mathbf{r},t) = M_x(\mathbf{r},t) + iM_y(\mathbf{r},t)$ and its conjugate complex $\Psi^*(\mathbf{r},t)$. The z-component of the exchange spin current can be written as

$$J^{\beta}_{M_z} = \frac{1}{2i}\frac{D}{M_S}[\Psi^*(\mathbf{r},t)\nabla_\beta \Psi(\mathbf{r},t) - \Psi(\mathbf{r},t)\nabla_\beta \Psi^*(\mathbf{r},t)]. \tag{8.4}$$

By introducing creation and annihilation operators $(b_q^\dagger, b_q)$ of spin-wave excitations (magnons) with the frequency ω_q and the wave number q by $\Psi = M_+ = \sqrt{2/M_s} \sum_q b_q e^{iq\cdot r}$ and $\Psi^* = M_- = \sqrt{2/M_s} \sum_q b_q e^{-iq\cdot r}$, the exchange spin current is expressed as

$$J_{M_z}^\beta = \sum_{p,q} v_q n_q, \tag{8.5}$$

where $v_q = \partial\omega_q / \partial q = 2Dq$ is the spin-wave group velocity and $n_q = \langle b_q^\dagger b_q^\dagger \rangle$ is the number of spin waves. The above equation implies that when the number of spin waves in k space is different between q and $-q$, a non-zero net exchange spin current is carried by the spin waves: a spin-wave spin current.

The spin-wave spin currents have been observed in References. [24–27]; one of the main advantages of these type of spin currents are their spin coherence length, which is orders of magnitude larger than that in conduction-electron spin currents [28]. This fact, together with its lesser dissipation in magnetic insulators, makes spin-wave spin currents very promising for applications in the implementation of spin-based logic devices [29–33].

8.3 SPIN SEEBECK EFFECT

The SSE was first reported by Uchida and co-workers in a ferromagnetic metal $Ni_{81}Fe_{19}$ film [12]. The basic mechanism of the SSE is schematically shown in Figure 8.1; it consists on the generation of a spin current from a ferromagnetic material (F) into a normal metal (N) driven by a temperature gradient via the thermal magnetization dynamics. The thermally injected spin current is detected in the N layer as an electric voltage (SSE voltage) by means of the inverse spin Hall effect (ISHE) as a result of the spin-orbit interaction [34–38] (see Chapter 7, Volume 3, for a more detailed discussion about

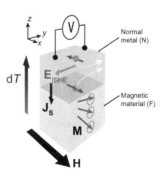

FIGURE 8.1 Schematic representation depicting the basic mechanism of the spin Seebeck effect (SSE) with the induced inverse spin-Hall effect (ISHE) in a normal metal attached to a ferromagnetic or ferrimagnetic material (the longitudinal SSE configuration is shown). E_{ISHE} denote the electric field generated by the ISHE. **H**, **M**, ∇T, and J_S denote the magnetic field, magnetization, temperature gradient and spatial direction of the spin current, respectively.

the spin Hall effect). The ISHE electric field ($\mathbf{E}_{ISHE}$) in N can be expressed according to the following relation:

$$\mathbf{E}_{ISHE} = \theta_{SH}\rho(\mathbf{J}_S \times \sigma), \tag{8.6}$$

where θ_{SH} and ρ denote the spin Hall angle and electric resistivity of the N layer, respectively. $\mathbf{J}_S$ is the spatial direction of the injected spin current into N, perpendicular to the F/N interface (parallel to ∇T) and σ is the spin-polarization vector (parallel to the magnetization of F).

The SSE was initially detected in metals using the transverse SSE configuration (more details about the measurement will be given in Section 8.3.2). It was first measured in $Ni_{81}Fe_{19}$ film [12], then it was subsequently reported in other metallic systems, particularly: Fe, Ni films [39, 40] and in a Co_2MnSi half-metal Heusler alloy film [41]. Then Jaworski and co-workers also reported the observation of the SSE at low temperatures in the ferromagnetic semiconductor GaMnAs [42, 43]. The observation of the SSE in ferrimagnetic insulating oxides: $LaY_2Fe_5O_{12}$ (La:YIG) [23], demonstrated that the effect was still present even in the absence of charge carriers (YIG has a band gap of 2.85 eV [24]), challenging the interpretation based on the conduction electrons [44]. This result pointed to the SSE originating from the thermally driven magnetization dynamics of the ferromagnet and opened the possibility of using electrically insulating materials for thermoelectric conversion [45–47]. Experimentally, it also allowed them to perform the SSE measurements using the longitudinal SSE configuration [48, 49] (see Section 8.3.2), which makes interpretation of the results easier. The effect has also been subsequently reported in ferrimagnetic spinel ferrites [50–55], among other oxides [56–61], thus establishing the spin Seebeck effect as a general non-equilibrium transport phenomenon in magnetic materials. Furthermore, the Onsager reciprocal of the SSE, the spin Peltier effect, has also been experimentally confirmed in YIG [62]: a charge current in a Pt layer drives a magnon excitation in a YIG ferrimagnet and, as a consequence, the travelling magnons transport heat, inducing a measurable temperature difference in the material.

One main advantage of the spin Seebeck is that it provides a simple and robust method for the generation of spin currents; it has been actually shown that the spin current densities that can be achieved are about two orders of magnitude larger than those obtained by resonant excitation methods [63]. Furthermore, the effect bridges the area of spintronics and thermoelectric energy conversion, with potential advantages over conventional thermoelectric devices, such as the possibility of lower dissipation energy conversion using insulating ferromagnets. However, the current values of the detected electric voltages are still far from potential applications, and efforts are currently devoted to further understanding the physics of the SSE in order to gain insight and develop possible routes to increase the magnitude of the SSE voltage. The basic mechanism, experimental measurement geometry, and recent developments in spin Seebeck effect are subsequently described.

8.3.1 BASIC MECHANISM FOR THE SPIN SEEBECK EFFECT

The spin Seebeck effect was initially formulated in terms of conduction electrons [12, 44]. However, the observation of the SSE in magnetic insulators upset this interpretation and pointed to the thermal excitation of localized spins in the ferromagnet (magnons) as the possible origin of the observed effect. The theoretical model for the magnon-driven SSE was first proposed by Xiao et al. [64] using the scattering formalism, and subsequently developed by Adachi et al. [65, 66] using linear response theory.

In this model the SSE is explained as a result of the thermal non-equilibrium between the magnon and the electron systems in F and N, respectively. The thermal excitation of the magnon and electron systems in F and N can be described in terms of effective temperatures of magnons in the ferromagnet (T_F) and the electrons in the normal metal (T_N). The fluctuation dissipation theorem connects these effective temperatures to thermal fluctuations of the magnetization in F and thermal noise in N, which are described in terms of the random fields h and l, respectively. The random field in the ferromagnet satisfies the white noise position-time correlator of the form:

$$\left\langle h_i^\mu(t) h_j^\nu(t') \right\rangle = \frac{2 k_B T_F \alpha}{\gamma M_S} \delta_{ij} \delta_{\mu\nu} \delta(t - t'), \tag{8.7}$$

where:

k_B is the Boltzmann constant
γ is the gyromagnetic ratio
M_s is the saturation magnetization
α the Gilbert damping constant

The thermal noise field in the ferromagnet (h) injects a spin current from F to N proportional to T_F, this is known as the spin pumping component (J_s^{pump}) [15]. A similar expression of the above described white noise correlator also relates the effective electron temperature to the noise field l in N, this field generates a spin current flowing back from N to F proportional to T_N, known as the backflow component (J_s^{back}) [67]. The total spin current (J_s) at the F/N interface is given by the difference between the spin pumping and backflow components and is proportional to the temperature difference between the magnon and electron systems, therefore:

$$J_s = J_s^{\text{pump}} - J_s^{\text{back}} \propto T_F - T_N. \tag{8.8}$$

In the case of thermal equilibrium: $T_F = T_N$, and no spin current is generated. Then if an external temperature gradient is applied, an effective magnon-electron temperature difference is induced [68] and a spin current is generated across the F/N interface.

The SSE has also been formulated using other approaches. A SSE theory using Landau-Lifshitz-Gilbert formalism was developed by Hoffman et al. [69] to study the thickness dependence and length scale of the SSE.

More recently, Rezende et al. [70, 71] and Zhang et al. [72, 73] have independently formulated the SSE in terms of a bulk magnon spin current thermally induced in the ferromagnet, not at the F/N interface. Here, we will briefly introduce the details of their model. When a thermal gradient is applied to a ferromagnetic material, a number of magnons are excited out-of-thermal equilibrium: $\delta n_k(\mathbf{r}) = n_k(\mathbf{r}) - n_k^0$, where n_k^0 is the number of magnons in thermal equilibrium, described by a Bose-Einstein distribution: $n_k^0 = 1/[\exp(\varepsilon_k/k_BT) - 1]$, where $\varepsilon_k = \hbar\omega_k$ is the k-magnon energy. The density of magnons in excess of equilibrium defines the magnon accumulation: $\delta n_m(\mathbf{r}) = 1/(2\pi)^3 \int d^3k[n_k(\mathbf{r}) - n_k^0]$ [72, 73]. Then, the bulk spin current of magnons propagating with velocity $\mathbf{v}_k$ can be defined as [70–74]:

$$\mathbf{J}_M = \frac{\hbar}{(2\pi)^3} \int d^3k \mathbf{v}_k \left[n_k(\mathbf{r}) - n_k^0 \right]. \tag{8.9}$$

In order to estimate the magnon spin current in the equation above, one needs to know the magnon distribution under an applied temperature gradient. This can be calculated using the Boltzmann transport equation; in the relaxation approximation and in the absence of external forces, we obtain the following solution in the steady state:

$$n_k(\mathbf{r}) - n_k^0 = -\tau_k \mathbf{v}_k \cdot \nabla n_k(\mathbf{r}), \tag{8.10}$$

where τ_k is the k-magnon relaxation time. Considering the expression of the magnons excited out-of-thermal equilibrium ($\delta n_k(\mathbf{r}) = n_k(\mathbf{r}) - n_k^0$), the above solution to the Boltzmann equation can be written as:

$$n_k(\mathbf{r}) - n_k^0 = -\tau_k \mathbf{v}_k \cdot \left[\nabla \delta n_k(\mathbf{r}) + \frac{\partial n_k^0}{\partial T} \nabla T \right], \tag{8.11}$$

by substitution of Equation 8.11 into Equation 8.9, they obtained the magnon spin current as the sum of two terms, $\mathbf{J}_M = \mathbf{J}_{M\nabla T} + \mathbf{J}_{M\delta n}$. The first one is the magnon spin current contribution due to the flow of magnons driven by an applied temperature gradient:

$$\mathbf{J}_{M\nabla T} = -\frac{\hbar}{(2\pi)^3} \int d^3k\tau_k \frac{\partial n_k^0}{\partial T} \mathbf{v}_k \left[\mathbf{v}_k \cdot \nabla T \right], \tag{8.12}$$

which is proportional to the applied temperature gradient $\mathbf{J}_{M\nabla T} = -C\nabla T$. The second term is due to the spatial variation of the magnon accumulation:

$$\mathbf{J}_{M\delta n} = -\frac{\hbar}{(2\pi)^3} \int d^3k\tau_k \mathbf{v}_k \left[\mathbf{v}_k \cdot \nabla \delta n_k(\mathbf{r}) \right]. \tag{8.13}$$

The equation above can be expressed as a magnon diffusion current (see [71] for a detailed derivation). For magnons propagating in the z direction:

$$J_M(z) = -\hbar D_m \frac{\partial}{\partial y} \delta n_m(z), \tag{8.14}$$

where D_m is the magnon diffusion coefficient. The relaxation of the magnon accumulation into de lattice can be described by a magnon-phonon relaxation time τ_{mp} and by conservation of angular momentum we have: $\frac{\partial J_M}{\partial z} = -\hbar \frac{\delta n_m(z)}{\tau_{mp}}$. Using this relation in the above equation, we obtain a diffusion equation for the magnon accumulation,

$$\frac{\partial^2 \delta n_m(z)}{\partial z^2} = \frac{\delta n_m(z)}{\Lambda^2}, \tag{8.15}$$

where $\Lambda = \sqrt{D_m \tau_{mp}}$ is the magnon diffusion length. Then the spatial variation of the magnon accumulation has solutions of the form $\delta n_m(z) = A e^{z/\Lambda} + B e^{-z/\Lambda}$. Inserting this expression in Equation 8.14, we can see that the magnon spin current in the F/N structure is given by:

$$J_M(z) = -C \nabla_z T - \hbar \frac{D_m}{\Lambda} A e^{z/\Lambda} + \hbar \frac{D_m}{\Lambda} B e^{-z/\Lambda}. \tag{8.16}$$

The coefficients A and B can be obtained by considering the boundary conditions in the F/N structure, which are given by the conservation of angular momentum flow, implying continuity of the spin current at the F/N interface [15, 70–74] and vanishing spin current at the top and bottom surfaces of the F/N structure. With these boundary conditions, and considering that the physics of the magnon thermal properties are mostly described by the dispersion relation in the acousitc branch [75], then the magnon spin current density at the F/N interface can be described as (more details of this derivation can be found in References [70, 71]):

$$J_M(0) = -L \frac{B_1 B_S}{\sqrt{B_0 B_2}} \rho g_{\text{eff}}^{\uparrow\downarrow} \nabla T, \tag{8.17}$$

where L is a prefactor dependent on material parameters, $\rho = \frac{\cosh(t_F/\Lambda) - 1}{\sinh(t_F/\Lambda)}$ describes the effect of the ferromagnetic layer thickness (t_F), $g_{\text{eff}}^{\uparrow\downarrow}$ is the real part of effective spin mixing conductance that accounts for spin pumping and backflow components and $B_i(q, \eta_q)$ ($i = 0,1,2,S$) are integrals dependent on the normalized wave number $q = k/k_m$, and on the magnon lifetime through the relaxation rate $\eta_q = \tau_0/\tau_k$, where τ_0 is the lifetime of magnons near $k \approx 0$. The expression for η_q can be obtained from fits to the calculated relaxation rates due to 3- and 4-magnon scattering processes as described in References [70, 71, 76].

Equation 8.17 describes the essential features of the spin Seebeck effect in ferromagnets, and can be used to explain a variety of recent experimental observations in the SSE, such as: the dependence of the SSE on the F thickness [77], the temperature dependence [70, 78], the SSE suppression at high magnetic fields [79–81], the SSE in magnetic multilayers [82] and the observation of the SSE in antiferromagnetic materials [83–86].

Other models have also been developed to explain different aspects of the SSE [87–99]. A mechanism for the phonon-mediated SSE has also been described [100, 101].

8.3.2 MEASUREMENT OF THE SPIN SEEBECK EFFECT

There are two main experimental geometries employed for the measurement of the spin Seebeck effect, these are: the transverse and the longitudinal SSE configurations. The transverse SSE was originally employed in the observation of the SSE of $Ni_{81}Fe_{19}$ film [12], in this arrangement the directions of injected spin current and applied thermal gradient are perpendicular to each other. Figure 8.2a shows a schematic illustration of the transverse SSE geometry, consisting of a ferromagnetic sample with an N wire attached, having its long axis (y) perpendicular to that of the ferromagnet (x). The thermal gradient and the magnetic field are applied parallel to each other along the x direction. The thermal excitation of the magnetization of the ferromagnet results in a spin current injection across the F/N interface (parallel to the z direction). The SSE is detected by means of the ISHE by measuring the voltage along the length of the wire (y) direction (see Equation 8.6). The transverse measurement geometry is suitable for the detection of the SSE in all type of magnetic systems: metals, semiconductors and insulators. However, this configuration requires the careful thermal design of the sample and measurement system. For instance, inappropriate choice of substrate can result in out-of-plane thermal gradients (∇T_z) arising from thermal conductivity mismatch between the substrate and F/N layers. As a result, spurious voltage signals due to conventional spin-dependent thermoelectric effects can be detected, such as the anomalous [102–106] and planar Nernst effects [107], which are the thermal equivalents of the charge induced anomalous and planar Hall effects [108].

Furthermore, there are still some unanswered questions about the transverse SSE, such as the length scale (in the mm range) over which the effect can be observed, much longer than the spin diffusion length of ferromagnetic metals (in the nm range) [28]. There is experimental evidence that suggest this unconventional length scale might be related to phonon transport through the substrate, as shown in experiments where the SSE can still be observed after the ferromagnetic thin film is cut (suppressing any possible spin transport) and only phonon transport through the single crystal substrate is possible [42] or the disappearance of the transverse SSE in amorphous substrates [20]. These experiments suggest that the microscopic mechanisms to explain the length dependence of the transverse SSE can possibly be related to heat transfer by phonons [101] and magnon-phonon drag [100]. However, this is still a matter of debate [20, 101, 106].

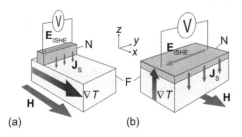

(a)   (b)

FIGURE 8.2 Schematic representation of the different configurations used for the detection of the spin Seebeck effect (SSE). (a) Transverse SSE. (b) Longitudinal SSE.

Due to the above mentioned complications with the transverse SSE, the longitudinal SSE is the configuration mainly employed for the investigation of the SSE, due to its simplicity. All the results shown in this chapter were obtained employing the longitudinal SSE. Figure 8.2b shows the experimental geometry for the longitudinal SSE measurement: the applied thermal gradient is parallel to the injected spin current (perpendicular to the F/N interface, along z), the magnetic field is directed parallel to the x direction and the voltage is detected along the y direction. This experimental configuration is the same as the one employed for the measurement of the anomalous Nernst effect (ANE) in electrically conductive magnetic materials [109–113], where the ANE-induced electric field is given by:

$$\mathbf{E}_{\mathrm{ANE}} = Q_S \mu_0 \nabla T \times \mathbf{M}, \tag{8.18}$$

where Q_S, μ_0 and $\mathbf{M}$ are the ANE coefficient, vacuum permeability and the magnetization of the sample, respectively. The longitudinal configuration is ideally suited for the detection of the SSE in magnetic insulators [48]. In the case of electrically conductive ferromagnets, the SSE and ANE-induced voltage signals are entangled. However, it has been shown that in ferromagnetic systems where the electrical conductivity of F is comparatively lower than that of the N layer, the SSE dominates the measured voltage. This is due to the suppression of the ANE-induced voltage of the ferromagnet by the larger electrical conductivity of the N layer; the degree of suppression can be estimated considering the expression for the charge current density (J_m^i) in the linear response region, given by $J_m^i = \sigma_m^{ij} E^j - \alpha_m^{ik} \nabla_k T$, where E^j is the electric field, $\nabla_k T$ is the thermal gradient, and σ_m^{ij} (α_m^{ik}) are the elements of the electrical conductivity (thermopower) tensor. Under the conditions of the longitudinal SSE experiment, this expression can be expanded as:

$$J_I^z = \sigma_I^{zz} E^z + \sigma_I^{zy} E^y - \alpha_I^{zz} \left(\nabla_z T \right)_I,$$

$$J_I^y = \sigma_I^{yz} E^z + \sigma_I^{yy} E^y - \alpha_I^{zy} \left(\nabla_z T \right)_I,$$

$$J_{II}^z = \sigma_{II}^{zz} E^z - \alpha_{II}^{zz} \left(\nabla_z T \right)_{II}, \tag{8.19}$$

$$J_{II}^y = \sigma_{II}^{yy} E^y,$$

where $m = I$ and $m = II$ stand for the ferromagnet and the normal metal, respectively. The SSE measurement is performed in the open circuit condition, which can be described as: $I^z = A J_I^z = A J_{II}^z = 0$ and $I^y = S_I J_I^y + S_{II} J_{II}^y = 0$, here $A = L_x L_y$ is the area of the junction normal to the film surface and $S_m = L_x t_m$ is the area along the y direction, where t_m ($m = I, II$) is the thickness of each of the layers comprising the F/N junction. After solving the above equation, the electric field in the non-magnetic conductor due to the ANE in the ferromagnet is expressed as follows:

$$E^y = \left(\frac{r}{1+r} \right) E_{\mathrm{ANE}}, \tag{8.20}$$

where $r = \dfrac{\rho_{II}\, t_I}{\rho_I\, t_{II}}$ is the shunting factor which accounts for the degree of suppression of the ANE-induced electric field in the ferromagnet $\left(E_{\mathrm{ANE}} = \left[\dfrac{\alpha^{yz}}{\sigma^{zz}} - \left(\dfrac{\sigma^{yz}}{\sigma^{zz}}\dfrac{\alpha^{zz}}{\sigma^{zz}}\right)\right](\nabla_z T)_I\right)$ and $\rho_m = 1/\sigma_m$ being the electrical resistivity of the ferromagnet ($m = I$) and normal metal ($m = II$).

To illustrate this model, we consider the case of a magnetite/platinum (Fe_3O_4/Pt) system. Magnetite is a well-known ferrimagnetic oxide with a moderate electrical conductivity, due to electron hopping between the Fe cations at the octahedral sites of the spinel structure [114]. In Figure 8.3b we show the SSE measurement of a Fe_3O_4/Pt bilayer with layer thicknesses of $t_{Fe_3O_4} = 34$ nm ($t_{Pt} = 17$ nm) and resistivities $\rho_{Fe_3O_4} = 6.94 \times 10^{-5}\,\Omega\mathrm{m}$ ($\rho_{Pt} = 6.94 \times 10^{-7}\,\Omega\mathrm{m}$). Considering the phenomenological model described above, we can estimate the contribution from the ANE-induced voltage of the Fe_3O_4 film, which only accounts for about 0.4% of the SSE voltage

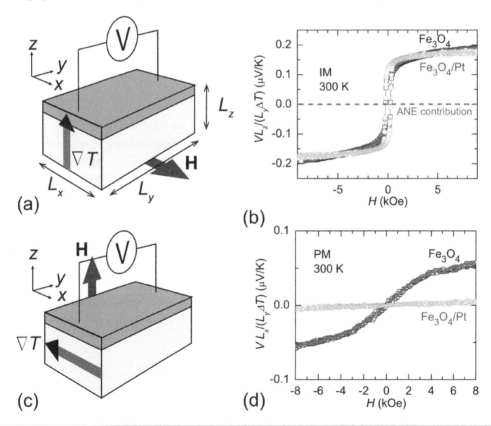

FIGURE 8.3   (a) Schematic illustration of the in-plane magnetized configuration for the measurement of the SSE. (b) H dependence for the ANE and SSE measured on a MgO(001)//Fe_3O_4(34) and MgO(001)//Fe_3O_4(34)/Pt(17) sample respectively, in the IM configuration (thicknesses in nm). The dashed line shows the estimated contribution of the ANE of Fe_3O_4 to the SSE voltage. (c) Schematic illustration of the perpendicular magnetized configuration where only the ANE can be observed and the SSE vanishes due to ISHE symmetry (spin current injected across the Fe_3O_4/Pt interface is parallel to the applied magnetic field). (d) H dependence of the ANE voltage measured in the PM configuration. The ANE signal in Fe_3O_4 film is still present; while the SSE vanishes due to measurement geometry.

measured in the Fe_3O_4/Pt bilayer, clearly showing that the transversal thermoelectric voltage in this system is dominated by the SSE.

Even in electrically insulating ferromagnetic systems, another possible source of a spurious transverse thermoelectric voltage is due to magnetic proximity effects (MPE) in Pt, which can result in an ANE-induced voltage in the Pt layer [115, 116]. This is due to the fact that Pt is close to the Stoner instability criterion for ferromagnetic order [117] and when placed in contact with a ferromagnetic material it can become weakly magnetized in a region close to the interface with the ferromagnet. To evaluate the MPE-induced ANE, a phenomenological model similar to the one explained above is used. In this model the Pt layer is considered to be divided in a magnetic and a non-magnetic region, the MPE-induced ANE voltage of the magnetized region is suppressed due to short-circuit effect in the Pt non-magnetic region, therefore we can estimate the degree of suppression of the ANE-induced electric field in the Pt layer. Since the material properties are essentially the same between the magnetic and non-magnetic regions, the degree of suppression is basically described by the ratio of the thickness of the magnetized region (t_I) to the total Pt thickness (t_{Pt}) and expressed as follows:

$$E^y = \left(\frac{t_I}{t_{Pt}}\right) E_{ANE}^{MPE}, \tag{8.21}$$

where the prefactor (t_I / t_{Pt}) accounts for the reduction of the output voltage due to short-circuit effect and E_{ANE}^{MPE} is the proximity effect–induced electric field in the magnetized region of the Pt film, which can be expressed as follows: $E_{ANE}^{MPE} = S_{ANE}^{MPE} \nabla_z T_{Pt}$ where S_{ANE}^{MPE} is the proximity ANE coefficient of the magnetized Pt layer. In order to evaluate the MPE-induced ANE, the temperature gradient across the Pt film can be estimated by considering heat current conservation across the thickness direction of the bilayer structure, $-\kappa_{Subs}\nabla_z T_{Subs} = -\kappa_{Fe_3O_4}\nabla_z T_{Fe_3O_4} = -\kappa_{Pt}\nabla_z T_{Pt}$, here we have neglected interfacial thermal resistance effects [118]. Since, $t_{Subs} \gg t_{Fe_3O_4}, t_{Pt}$, the gradient in the Pt layer can be expressed as $\nabla_z T_{Pt} = (\kappa_{Subs} / \kappa_{Pt})\nabla_z T_{Subs}$, then the MPE-induced ANE can be written in terms of the measured temperature difference across the sample $\nabla_z T_{Subs} \sim \Delta T / t_{Subs}$ and the MPE-induced voltage is given by the relation:

$$\frac{V_{ANE}^{MPE} t_{Subs}}{L_y \Delta T} = \left(\frac{\kappa_{Subs}}{\kappa_{Pt}}\right)\left(\frac{t_{Pt(I)}}{t_{Pt}}\right) S_{ANE}^{MPE}, \tag{8.22}$$

taking the reported values for $\kappa_{Pt} = 72$ W/mK and $\kappa_{MgO} = 53$ W/mK [119, 120], considering an upper value for the magnetized Pt region of $t_I = 1$ nm and the value of the MPE ANE coefficient calculated for a Pt film $S_{ANE}^{MPE} = 0.058$ μV/K [116], we obtain a maximum value of the proximity-induced ANE voltage $\frac{V_{ANE} L_{Subs}}{L_y \Delta T} \sim 3$ nV/K for 17 nm of Pt; this is two orders of magnitude lower than the measured SSE voltages, therefore showing that the ANE induced by MPE is negligibly small.

We have previously shown that the ANE contribution can come either from the ferromagnet, in electrically conductive materials, or by a static MPE in Pt. A method for clear experimental separation of the SSE from the ANE effect consists of performing transverse thermoelectric measurements in the F/N system with the applied magnetic field oriented within the plane of the films; in-plane magnetized configuration (IM) or with the magnetic field applied parallel to their surface normal; perpendicularly magnetized configuration (PM). These can be simply attained by interchanging the directions of the magnetic field (**H**) and applied thermal gradients (∇T) while keeping the directions of the thermal gradient, magnetic field and direction of measured voltage orthogonal to each other. The IM configuration (Figure 8.3a) is the conventional longitudinal SSE setup, while in the PM configuration (Figure 8.3c) the thermal gradient and the magnetic field are applied in the in-plane ($\| x$) and out-of-plane directions of the film ($\| z$), respectively. In the PM configuration a voltage due to the ANE can be detected, but the observation of the spin Seebeck effect is forbidden due to the symmetry of ISHE (see Equation 8.6), note that the direction of injected spin current ($J_S \| z$) is parallel to that of the spin-polarization vector ($\sigma \| z$). Figure 8.3d shows the results obtained for Fe_3O_4 (34) film and Fe_3O_4 (34)/Pt(17) bilayer systems; it can be clearly seen that in the PM configuration the ANE can be measured in the Fe_3O_4 layer, but is strongly suppressed for the Fe_3O_4/Pt bilayer, therefore showing that the proximity-induced ANE in Fe_3O_4/Pt system is negligibly small.

8.4 ENHANCEMENT OF THE SSE IN MULTILAYERS

The SSE offers a potential alternative to conventional thermoelectric conversion by means of heat-induced spin transport. The fact that it has been observed in electrically insulating magnetic materials is attractive due to the possibility of thermoelectric generation free from Joule heating in the thermal part of the device and it also opens the possibility to study other oxide materials not conventionally used for thermoelectric investigations. Moreover, since SSE devices comprise a F/N junction, the advantage is that the thermally excited (ferromagnet) and electrical voltage generating parts (normal metal) can be optimized independently. This is in contrast to conventional thermoelectrics [121–123] where the thermal conductivity and electrical conductivity are strongly interrelated due to Widemann-Franz law. Another advantage lies in the SSE measurement geometry in which the heat (parallel to spin current) and electric current paths are perpendicular to each other, increasing the versatility of SSE devices, which can be easily implemented by just a thin film coating over conventional surfaces [124] as in the recent demonstration of flexible SSE devices [55].

However, the main roadblock for the application of the SSE is the low magnitude of the generated electric voltages, with values two orders of magnitude lower than observed in conventional thermoelectrics [123, 125], strongly limiting its applicability for thermoelectric devices. Therefore, current efforts are devoted to studying the SSE mechanism in order to improve its conversion efficiency. One approach is aimed at understanding the spin-to-electric current detection mechanism in order to further exploit the spin Hall angle characteristics of the N layer [126–131]. In this sense, large voltage magnitude increases have been demonstrated by the use of spin Hall thermopiles in which metals with spin Hall angles of opposite sign are deposited forming an array of alternate wires electrically connected in series forming a zig-zag structure [132]. However, the reduction of the effective device area, together with the increase of the internal resistance, strongly limits the output power of the spin Hall thermopiles [13], therefore other approaches are necessary.

Recently, we have shown that in magnetic multilayers formed by repeated growth on n number of Fe_3O_4/Pt bilayers (see Figure 8.4a, b for a schematic and a scanning transmission electron microscope image of a multilayer with $n=6$), it is possible to obtain both an enhancement of the SSE voltage and a reduction in sample resistance [82], which translates into an enhancement of the electric output power of the devices. The reduction in sample resistance

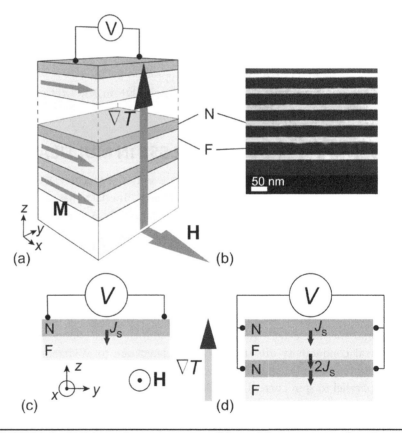

FIGURE 8.4   (a) Schematic representation of the multilayer structures for spin Seebeck effect measurements formed by repeated growth of F/N bilayers, the experimental measurement geometry is also shown. (b) Scanning transmission electron microscopy image of a $[Fe_3O_4/Pt]_6$ system with thicknesses for Fe_3O_4 (Pt) layers of 34 nm (17 nm). Schematic illustration of the spin currents (J_s) injected in the N layer for a (c) single bilayer (F/N) and a (d) double bilayer considering the simple picture of thermal spin current injection at the F/N interface.

is expected due to parallel contact of all the layers forming the multilayer structure. However, the voltage enhancement is unexpected and cannot be explained by the simple picture of thermal spin injection at the F/N interface, shown in Figure 8.4c. According to this picture, if we treat the multilayer system as a set of independent Fe_3O_4/Pt bilayers electrically connected in parallel with a constant heat flow and therefore the same magnitude of thermally injected spin current at the interfaces, a voltage enhancement is only expected for the inner Pt layers due to the additive spin current injection from top and bottom Pt/Fe_3O_4 interfaces, as it is depicted in Figure 8.4d, which shows a representative example for the case of a double bilayer $n = 2$. In the case of a very large number of bilayers ($n \gg 2$), a maximum voltage of $V \approx 2V_{BL}$ can be expected under this picture (here V_{BL} is the SSE voltage of a single F/N bilayer).

Figure 8.5a shows the H dependence of the measured SSE voltage in the multilayer system (V_{ML}), normalized by ΔT, for a different number of Fe_3O_4/Pt bilayers, n. The SSE voltage significantly increases with the number of bilayers n; this voltage enhancement was unforeseen, with a magnitude significantly larger than previously expected (see Figure 8.5b). Even for the multilayer case with $n = 2$, the measured voltage clearly exceeds the maximum expected value according to the simple picture described above ($V_{ML} \gg 2V_{BL}$). Similar voltage enhancement was also recently reported in other multilayer systems based on all-metallic [133, 134] and all-oxide layers [61], although in the case of all-metallic systems the SSE and ANE of the F layers are strongly entangled. In the case of Fe_3O_4/Pt-based multilayers, as shown in the previous section, the SSE is expected to be the dominant contribution, due to the electrical resistivity of Fe_3O_4 being much greater than that of Pt and the ANE contribution of Fe_3O_4 being strongly suppressed due to short-circuit effects [51].

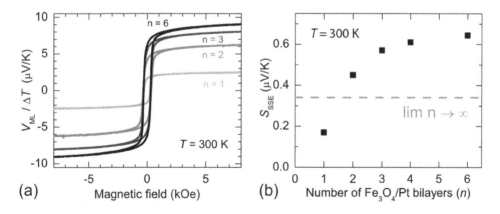

FIGURE 8.5   Magnetic field dependence of $V_{ML}/\Delta T$ in [Fe_3O_4(34)/Pt(17)]$_n$ multilayers as a function of the number of Fe_3O_4/Pt bilayers n, measured at 300 K. The sample dimensions are $L_x = 2$ mm, $L_y = 7$ mm and $L_z = 0.5$ mm (b) n dependence of the SSE coefficient [$S_{SSE} = (V_{ML}^{sat}/\Delta T)(L_z/L_y)$], where V_{ML}^{sat} is the measured SSE voltage at 7 kOe. The grey dashed line represents the upper limit of the SSE voltage considering the simple picture of spin injection at the Fe_3O_4/Pt interfaces and the parallel electrical connection between all the layers of the multilayer system.

A possible modulation of the thermal conductivity of the multilayer system with the increased number of interfaces as a possible origin of an increased thermoelectric voltage can be ruled out by thermal transport measurements of the $[Fe_3O_4/Pt]_n$ multilayers, showing that the thermal conductivity is independent of the number of Fe_3O_4/Pt interfaces, suggesting that another mechanism must be at the origin of the observed SSE voltage enhancement. Another possibility lies at the spin transport across the multilayer thickness, through the propagation of spin currents. In this sense, we studied the effect of the suppression of the spin current in the multilayers by measuring a multilayer system in which the inner Pt interlayers of the sample were replaced by thick MgO layers, leaving only the topmost Pt layer. In this structure, heat transport across the multilayer thickness is maintained (with MgO being a good thermally conductive material), while the electron and spin transport across MgO is suppressed. Figure 8.6 shows the comparative results obtained for Fe_3O_4/Pt, $[Fe_3O_4/Pt]_3$ and $[Fe_3O_4/MgO]_2/Fe_3O_4/Pt$ heterostructures. The SSE voltage of the multilayer system with MgO interlayers shows a strong reduction as a consequence of the absence of spin current propagation through the MgO interlayers, with a measured voltage comparable to the one obtained for a single Fe_3O_4/Pt bilayer. This result suggests that spin transport across the multilayer is important for the observed SSE voltage enhancement in which the existence of multiple Fe_3O_4/Pt interfaces is a relevant factor.

Now we will introduce the details of the model used to qualitatively explain the physics behind the observed SSE voltage enhancement in multilayers. In our model, we consider two types of spin currents: a magnon spin current in the F layer (J_M) [24, 70–73] and a conduction-electron spin current in the N layer (J_S). Then, we consider the following boundary conditions for the spin currents propagating normal to the F/N interfaces: (i) the continuity of the magnon and conduction-electron spin current at the various

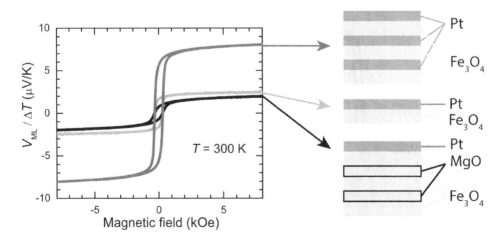

FIGURE 8.6 Suppression of the spin Seebeck effect voltage enhancement in multilayers by insertion of an electrically insulating 8 nm thick MgO layer between the Fe_3O_4 layers, only the topmost Pt layer is kept for electrical detection of the SSE. The graph shows the comparison between the $V_{ML}/\Delta T$ for Fe_3O_4/Pt, $[Fe_3O_4/Pt]_3$ and $[Fe_3O_4/MgO]_2/Fe_3O_4/Pt$ heterostructures. The thickness of Fe_3O_4 and Pt is 34 and 17 nm, respectively.

F/N (N/F) interfaces of the multilayer [70–73, 135] and (ii) the cancellation of the conduction-electron and magnon spin currents at the top and bottom surfaces of the multilayer structure. As a consequence of these boundary conditions, the magnitude and spatial profile of the spin current is affected; the continuous and non-vanishing spin current at the inner F/N interfaces of the multilayer structure results in an enhancement of the spin current in the multilayers, with maximum values at the inner layers of the multilayer. A schematic of the model's prediction of the spin current profile across the multilayer thickness is shown in Figure 8.7a. The effect can be explained by the appearance of a new length scale that characterizes the overall variation of the spin current in the multilayer and the presence of N within the multilayer that electrically detect the largest spin current at the middle of the multilayer sample.

The measured spin Seebeck voltage in the multilayer structure (V_{ML}) can be regarded as an average value over all the N layers, due to the parallel contact between all the layers, and therefore proportional to the average spin current over all N layers ($< J_S >$). Figure 8.7b shows the predicted values of the average spin current according to our model; we can clearly see that the model can qualitatively reproduce the observed SSE dependence vs the number of $(F/N)_n$ multilayers previously shown in Figure 8.5b.

Now we would briefly like to mention on the improvement of the thermoelectric performance of the spin Seebeck effect in multilayers [136]. Figure 8.8a shows the SSE voltage measured for $[Fe_3O_4/Pt]_n$ multilayers with Pt layers of thickness 7 nm, we can see that the SSE voltage is further enhanced reaching almost one order of magnitude increase with respect to the single bilayer. The main advantage of the multilayers is that the voltage enhancement is accompanied by a reduction of the sample resistance and an increase in the device electric power output is expected. The thermoelectric power generated by these SSE multilayers was determined by performing measurements of the SSE voltage on a variable resistive load attached to the

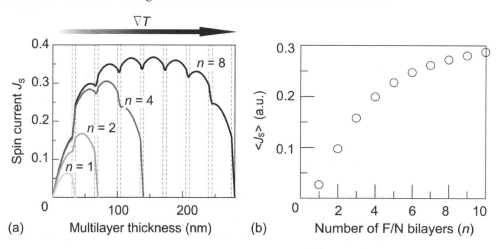

(a)   (b)

FIGURE 8.7   (a) Calculated spin current J_S profile for a $[F/N]_n$ multilayer system for several values of the number of F/N bilayers n. (b) n dependence of the spin current magnitude averaged over all the N layers $< J_S >$. The details of the model used for the calculation can be found in [82].

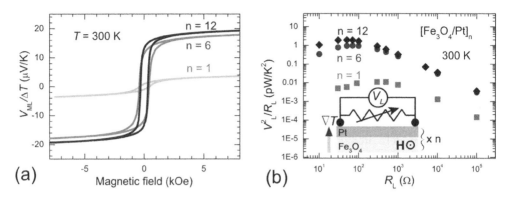

FIGURE 8.8 (a) Magnetic field dependence of $V_{ML} / \Delta T$ of $[Fe_3O_4/Pt]_n$ multilayers with $n = 1$, 6 and 12. The Fe_3O_4 and Pt thicknesses are 23 and 7 nm, respectively. (b) Electrical power (V_L^2 / R_L) extracted from the multilayers as a function of the load resistance R_L. The inset shows the schematic illustration of the measurement configuration to determine the electrical power output characteristics of the SSE.

multilayers (inset of Figure 8.8b). Figure 8.8b shows that the power output characteristics of the SSE multilayers are largely improved with an enhancement of more than two orders of magnitude between a single bilayer ($n = 1$) and a multilayer with $n = 12$. Therefore demonstrating the increase of the SSE efficiency in the multilayer system.

According to the above described model, the enhancement of the SSE in multilayers is expected to be strongly dependent on spin transport characteristics of the F and N layers and suggests that the voltage might be further enhanced by careful choice of material parameters and layer thickness.

8.5 SUPPRESSION OF THE SSE BY MAGNETIC FIELDS

Recently, the importance of magnon spin currents within the ferromagnet in the SSE has been theoretically highlighted [70–73], this is also supported by experimental reports demonstrating: the dependence of the SSE voltage on the thickness of the ferrimagnetic material [77], the SSE enhancement in magnetic multilayers described in the previous section [82] and the suppression of the SSE by strong magnetic fields [78]. In this section we will focus on the magnetic field dependence of the SSE, which has been independently reported by several groups [79–81]. The study of the magnetic field dependence offers an important clue in order to understand the role of magnons in the SSE, here we will focus on the recently reported results by Kikkawa et al. [78, 79] in YIG-bar/N junction systems (N: Pt or Au). Figure 8.9a, b show the magnetic field dependence of the SSE coefficient, defined as $S_{SSE} = (V / \Delta T)(L_z / L_y)$ in a YIG-bar/Pt sample measured at 300 and 5 K, respectively. The SSE effect monotonically decreases upon application of a large magnetic field after having reached its maximum value at low fields, while the magnetization value remains constant. It has been previously shown that the SSE suppression is irrelevant to the normal Nernst effect of the Pt layer [79] and is still present after changing the thickness of

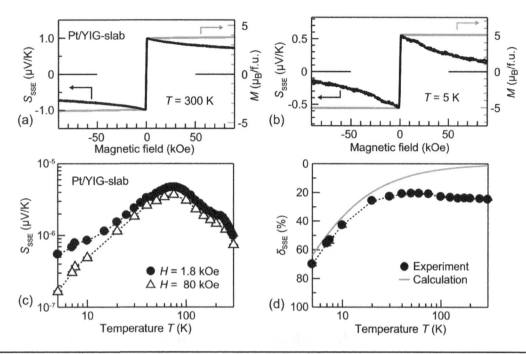

FIGURE 8.9   Magnetic field H dependence of the SSE in the YIG-slab/Pt sample and the magnetization M of the YIG slab measured at (a) 300 K and (b) 5 K, showing the SSE suppression at high magnetic fields. (c) Temperature dependence of the spin Seebeck coefficient S_{SSE} measured at $H=1.8$ kOe (filled circles) and 80 kOe (empty triangles). (d) Temperature dependence of the SSE suppression δ_{SSE}, the grey line shows calculated temperature dependence of δ_{SSE} considering the conventional SSE model as detailed in [78].

Pt or even using a different material as spin detector (Au), indicating that the phenomenon is solely dependent on the magnetic properties of YIG. The SSE suppression by magnetic field is explained by the gap opening in the magnon dispersion due to the Zeeman effect ($g\mu_B H$, where g and μ_B are the g-factor and Bohr magneton) demonstrating that the SSE magnitude is affected by the modulation of magnon excitation by an applied magnetic field, and pointing to the dependence of the SSE on the bulk magnon properties of the ferrimagnet.

Figure 8.9c represents S_{SSE} measured as a function of temperature at magnetic field values of 1.8 and 80 kOe with and without SSE voltage suppression, respectively. The overall temperature dependence of the SSE is similar to that previously observed [14]. Remarkably, the suppression of the SSE δ_{SSE} shows temperature-dependent behavior as it is also shown in Figure 8.9d, where $\delta_{SSE} = (S_{max} - S_{80kOe})/S_{max}$ with S_{max} and S_{80kOe} being the SSE coefficients at the maximum position and at $H=80$ kOe, respectively. The SSE suppression remains almost constant down to $T=30$ K, then it monotonically increases as the temperature is reduced reaching $\delta_{SSE} \sim 70\%$ at 5 K. Figure 8.9d also shows estimation of the SSE suppression based on the conventional SSE model (see details in [78]); at $T<30$ K we can see that it is in good agreement with the experimental results. However, at $T>30$ K it starts to deviate from the measured values; this is due to the fact that as the temperature increases the thermal fluctuations start to overcome the energy gap due to Zeeman energy.

Note that at $H = 80$ kOe, the magnon energy gap is $g\mu_B H / k_B = 10.7$ K and in order to influence the magnon excitation by magnetic fields, the magnon energy has to be of similar or lower magnitude than the Zeeman energy. The fact that the SSE suppression can be observed at room temperature suggests that the low-frequency magnons of energy comparable to the Zeeman energy are relevant to SSE even at room temperature.

This result shows that the SSE has a magnon spectral dependence, which can be understood in terms of the frequency dependence of the magnon thermalization lengths, with the lower energy magnons generally presenting longer lengths [25, 26, 64, 68, 71–73, 137, 138]. This interpretation is also in agreement with recent experimental observations showing that the magnitude of the SSE voltage monotonically decreases with a reduction of the thickness of YIG [77]. The above results provide an important piece of information about the physics of the SSE and also show that in order to maximize the SSE voltage in thin film–based structures, the thickness of the magnetic insulator has to be longer than the low-frequency magnons which provide the dominant contribution to the SSE.

8.6 SPIN SEEBECK EFFECT IN MATERIALS WITH DIFFERENT TYPE OF MAGNETIC ORDER

The SSE comprises a F/N junction system, where F is usually a ferromagnetic or ferrimagnetic material with a net magnetic moment and N is a normal metal, generally a paramagnetic (i.e: Pt) or diamagnetic (i.e: Au) material. In this subsection, we will briefly review recent observations of the SSE in F/N junctions using materials with different magnetic properties.

The SSE in insulating materials usually comprises a ferrimagnetic oxide, consisting of two non-compensated magnetic sublattices resulting in a non-vanishing total magnetic moment and therefore the material is effectively treated as a conventional ferromagnet. The dependence of the SSE on the magnetization of the sublattices was recently investigated by studying the SSE in a bilayer junction system formed by $Gd_3Fe_5O_{12}$ (GdIG) and Pt [139] layers. GdIG is a ferrimagnet with a magnetic compensation point at a temperature of around $T_C \sim 288$ K [140]; it was observed that the SSE shows a sign reversal upon cooling with a vanishing effect at T_C, in agreement with theoretical predictions [89].

The SSE in electrically insulating materials with antiferromagnetic order was also recently reported, both theoretically [85] and experimentally [83, 84]. It was shown that upon application of a magnetic field along the antiferromagnet easy axes, a spin-flop transition is induced in the antiferromagnet, therefore generating a net magnetic moment that can be electrically detected by means of the SSE voltage in an attached Pt layer; this observation has also been theoretically explained [85].

The SSE is theoretically explained in terms of thermal excitation of the magnon states of the magnetic materials. However, recently the SSE has also been reported in insulating paramagnetic single crystals, specifically in gadolinium gallium garnet ($Ga_3Ga_5O_{12}$) and dysprosium scandium oxide

(DyScO$_3$) [141]. These results suggest that short range magnetic interactions can still generate a measurable spin current at the interface with the normal metal, challenging the current theoretical understanding of the SSE. It is also interesting to highlight that the SSE has also been observed in a non-magnetic semiconductor [142] at low temperatures and high magnetic fields, although the origin of the effect is fundamentally different from the SSE phenomena discussed in this chapter.

As for the materials used for electrical detection, the thermally induced spin currents in the SSE are usually detected by means of the inverse spin Hall effect in a normal metal with large spin-orbit interaction, where Pt is the most common material used as the spin current detector due to its relatively large spin Hall angle and fabrication simplicity [130, 131]. Recently, the ISHE has also been investigated in antiferromagnetic [143–146] and ferromagnetic materials [79, 147–152]; these studies open the possibility of exploring a wider range of materials which are fundamentally interesting and can also possibly lead to increased thermoelectric conversion efficiency of spin Seebeck devices, due to hybrid thermoelectric generation by means of additive contributions from the SSE and ANE of the F and N layers, respectively.

8.7 OTHER SPIN CALORITRONIC PHENOMENA

In this chapter, we have mainly focused on the physics of the SSE, a phenomenon driven by heat-induced collective excitations of the spin in magnetic materials. There are more demonstrations of collective phenomena in spin caloritronics such as: thermally induced spin transfer torque [153–158], domain wall motion [159–160] and heat flow control by spin waves [161] among others [162]. In this section, we will focus on spin caloritronic phenomena related to heat excitation of incoherent spin currents, these are spin currents carried by conduction electrons, typically in magnetic tunnel junctions [163] or metallic spin valve systems [119, 164]; in the following subsection, we briefly discuss the latter.

8.7.1 SPIN-DEPENDENT THERMOELECTRIC TRANSPORT IN METALLIC STRUCTURES

In metallic systems with ferromagnetic order, the electrical transport properties are usually described by Mott's two current model in which spin-up ($\uparrow$) and spin-down ($\downarrow$) electron currents are considered as two independent parallel channels [165]. The electrical transport properties can be described in terms of spin-dependent electrical conductivities ($\sigma_\uparrow$, $\sigma_\downarrow$) and Seebeck coefficients ($S_\uparrow, S_\downarrow$) [166], by the following expression:

$$J_{\uparrow,\downarrow} = -\sigma_{\uparrow,\downarrow}\left(\frac{1}{e}\nabla\mu_{\uparrow,\downarrow} + S_{\uparrow,\downarrow}\nabla T\right), \qquad (8.23)$$

where $\mu_{\uparrow,\downarrow}$ is the spin-dependent chemical potential and e is the charge of the electron. In the absence of an electric current, an applied thermal gradient generates a spin current $J_S = J_\uparrow - J_\downarrow = -\sigma_F(1-P^2)S_S\nabla T/2$, which is proportional to the spin-dependent Seebeck coefficient: $S_S = S_\uparrow - S_\downarrow$. Here, σ_F is the electrical conductivity of the ferromagnet and $P = (\sigma_\uparrow - \sigma_\downarrow)/(\sigma_\uparrow + \sigma_\downarrow)$ its spin polarization.

The possibility of thermoelectric generation of pure spin currents has been previously discussed by Kovalev et al. [154, 167]. Then, A. Slachter et al. [164] experimentally demonstrated the spin dependence of the Seebeck coefficient using a lateral spin valve arrangement, such as the one schematically illustrated in Figure 8.10a. By passing an electric current through a ferromagnetic metal (F1), the Joule heating induces a thermal gradient over the F/N interface, which results in a thermal injection of a spin current from the ferromagnet (F1) into the normal metal (N) and a spin accumulation appears in N. Then, by placing an analyzing ferromagnetic contact (F2) within the range of the spin diffusion length of N, the thermally induced spin accumulation can be detected. The thermally injected spin current is detected as a voltage which is dependent on the relative orientation between F1 and F2 (as it is schematically shown in Figure 8.10c).

The Onsager reciprocal of the spin-dependent Seebeck effect, the spin-dependent Peltier effect, has also been observed by the same group [119]. The effect is based on the ability of majority and minority charge carriers to transport heat independently [8]. When a pure spin current is injected into N, the spin-up and spin-down electrons travel in opposite directions and the

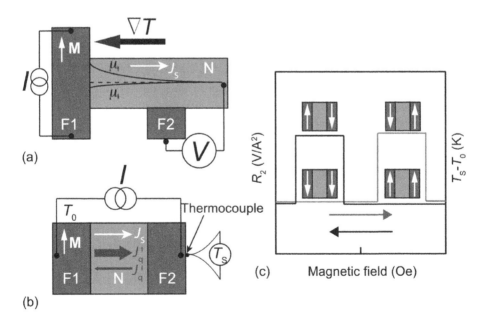

FIGURE 8.10 Schematic illustration of the experimental geometry used for the measurement of (a) the spin-dependent Seebeck effect [164] and (b) the spin-dependent Peltier effect [119]. (c) Schematic representation of the magnetic field dependence of the spin-dependent Seebeck and Peltier effects. The left axis shows the second armonic signal R_2 (proportional to thermal spin injection by spin-dependent Seebeck effect), where $V = R_1 I + R_2 I^2 + \dots$ with I and V being the charge current applied to F1 and voltage difference between F2 and N. The left axis represents the temperature difference, induced by the spin-dependent Peltier effect, obtained from the first harmonic component of the thermocouple voltage (second harmonic is used to remove conventional Peltier effect contribution).

net charge flow is zero. Then a net heat current is only possible if the heat carried by spin-up and spin-down channels is different, as it is described by the spin-dependent Peltier coefficient: $\Pi_S = \Pi_\uparrow - \Pi_\downarrow$, where $\Pi_\uparrow$ ($\Pi_\downarrow$) is the Peltier coefficients for the spin-up (spin-down) conduction channels. If the thickness of N is lower than its spin relaxation length, the spin-dependent Peltier effect will generate a temperature difference between the N/F interface and the ferromagnet, which is given by the following expression:

$$\Delta T \propto \sigma_F \left(1 - P^2\right) \Pi_S \mu_S^0, \tag{8.24}$$

where we can see that the spin-induced temperature difference is directly proportional to the spin accumulation at the F/N interface, $\mu_S^0 = \mu_\uparrow^0 - \mu_\downarrow^0$.

To measure the spin-dependent Peltier effect, a metallic spin valve structure was used: one ferromagnetic electrode (F1) is anchored at T_0 and a thermocouple is attached to the other ferromagnetic electrode (F2) (see Figure 8.10b). The spin current was driven by applying a charge current between F1 and F2, then the charge and spin parts of the Peltier effect are separated [119, 168]. When the magnetization of the ferromagnets are anti-parallel, the spin accumulation at each of the F/N interfaces generates an additional spin temperature difference ΔT, which is effectively measured as a dependence of the thermocouple voltage on the relative orientation between the magnetizations of F1 and F2 (as schematically shown in Figure 8.10c). The authors have also proved the Kelvin-Onsager relation between the spin-dependent Seebeck and Peltier coefficients ($\Pi_S = S_S T$), therefore demonstrating the expected reciprocity between these spin-dependent thermoelectric effects.

8.8 SUMMARY

In this chapter, we have reviewed the field of spin caloritronics with a special emphasis on the spin Seebeck effect, from its basic mechanism and measurement geometry to recent experimental findings. Particularly, we have shown that in magnetic multilayers formed by repeated growth of F/N bilayers, the SSE voltage is substantially enhanced. This result represents a significant leap in the thermoelectric efficiency of the SSE, reaching two orders of magnitude larger electric power output than conventional F/N bilayer structures. The observed voltage enhancement is expected to depend on the spin transport properties of the multilayer materials, so we can anticipate that further improvements in the SSE can be achieved by careful choice of material properties and geometry.

We have also introduced the results on the suppression of the SSE at high magnetic fields, which directly point to the dependence of the SSE on the bulk magnon properties of the magnetic material and the fact that the effect can be observed even at room temperatures indicates that low energy magnons play a relevant role in SSE.

Finally, the spin-dependent versions of conventional thermoelectric effects, measured in metallic spin valve structures, were briefly reviewed.

ACKNOWLEDGMENTS

The authors would like to thank K. Uchida, T. Kikkawa, H. Adachi, S. Maekawa, A. Anadon, I. Lucas, M. H. Aguirre, P. A. Algarabel, L. Morellon and M. R. Ibarra for helpful discussions and their collaboration.

BIBLIOGRAPHY

1. G. E. W. Bauer, A. H. MacDonald, and S. Maekawa, Spin caloritronics, *Solid State Commun.* **150**, 459 (2010).
2. G. E. W. Bauer, E. Saitoh, and B. J. van Wees, Spin caloritronics, *Nat. Mater.* **11**, 391 (2012).
3. S. R. Boona, R. C. Myers, and J. P. Heremans, Spin caloritronics, *Energy Environ. Sci.* **7**, 885 (2014).
4. H. J. Goldsmith, *Introduction to Thermoelectricity*, Springer-Verlag, Berlin, Germany, 2009.
5. S. A. Wolf, D. D. Awschalom, R. A. Buhrman et al., Spintronics: A spin-based electronics vision for the future, *Science* **294**, 1488 (2001).
6. I. Žutić, J. Fabian, and S. Das Sarma, Spintronics: Fundamentals and applications, *Rev. Mod. Phys.* **76**, 323 (2004).
7. M. Johnson and R. H. Silsbee, Thermodynamic analysis of interfacial transport and of the thermomagnetoelectric system, *Phys. Rev. B* **35**, 4959 (1987).
8. L. Gravier, S. Serrano-Guisan, F. Reuse, and J.-P. Ansermet, Spin-dependent Peltier effect of perpendicular currents in multilayered nanowires, *Phys. Rev. B* **73**, 052410 (2006).
9. L. Gravier, S. Serrano-Guisan, F. Reuse, and J.-P. Ansermet, Thermodynamic description of heat and spin transport in magnetic nanostructures, *Phys. Rev. B* **73**, 024419 (2006).
10. M. N. Baibich, J. M. Broto, A. Fert et al., Giant magnetoresistance of (001)Fe/(001)Cr magnetic superlattices, *Phys. Rev. Lett.* **61**, 2472 (1988).
11. G. Binasch, P. Griinberg, F. Saurenbach, and W. Zinn, Enhanced mag-netoresistance in layered magnetic structures with antiferromagnetic interlayer exchange, *Phys. Rev. B* **39**, 4828 (1989).
12. K. Uchida, S. Takahashi, K. Harii et al., Observation of the spin Seebeck effect, *Nature* **455**, 778 (2008).
13. K. Uchida, M. Ishida, T. Kikkawa, A. Kirihara, T. Murakami, and E. Saitoh, Longitudinal spin Seebeck effect: From fundamentals to applications, *J. Phys. Condens. Matter* **26**, 343202 (2014).
14. K. Uchida, T. Ota, H. Adachi et al., Thermal spin pumping and magnon-phonon-mediated spin-Seebeck effect, *J. Appl. Phys.* **111**, 103903 (2012).
15. Y. Tserkovnyak, A. Brataas, G. E. W. Bauer, and B. I. Halperin, Nonlocal magnetization dynamics in ferromagnetic heterostructures, *Rev. Mod. Phys.* **77**, 1375 (2005).
16. S. Maekawa, E. Saitoh, S. O. Valenzuela, and T. Kimura, *Spin Current*, Oxford University Press, Oxford, 2012.
17. S. Maekawa, H. Adachi, K. Uchida, J. Ieda, and E. Saitoh, Spin current: Experimental and theoretical aspects, *J. Phys. Soc. Jpn.* **82**, 102002 (2013).
18. I. Zutic and H. Dery, Spintronics: Taming spin currents, *Nat. Mater.* **10**, 647 (2011).
19. D. D. Stancil and A. Prabhakar, *Spin Waves*, Springer US, New York, NY, 2009.
20. K. Uchida, H. Adachi, H. An et al., Long-range spin Seebeck effect and acoustic spin pumping, *Nat. Mater.* **10**, 737 (2011).
21. K. Uchida, Z. Qiu, T. Kikkawa, and E. Saitoh, Pure detection of the acoustic spin pumping in Pt/YIG/PZT structures, *Solid State Commun.* **198**, 26 (2014).

22. K. Uchida, H. Adachi, T. An et al., Acoustic spin pumping: Direct generation of spin currents from sound waves in $Pt/Y_3Fe_5O_{12}$ hybrid structures, *J. Appl. Phys.* **111**, 053903 (2012).

23. K. Uchida, J. Xiao, H. Adachi et al., Spin Seebeck insulator, *Nat. Mater.* **9**, 894 (2010).

24. Y. Kajiwara, K. Harii, S. Takahashi et al., Transmission of electrical signals by spin-wave interconversion in a magnetic insulator, *Nature* **464**, 262 (2010).

25. L. J. Cornelissen, J. Liu, R. A. Duine, J. Ben Youssef, and B. J. van Wees, Long-distance transport of magnon spin information in a magnetic insulator at room temperature, *Nat. Phys.* **11**, 1022 (2015).

26. B. L. Giles, Z. Yang, J. S. Jamison, and R. C. Myers, Long-range pure magnon spin diffusion observed in a nonlocal spin-Seebeck geometry, *Phys. Rev. B* **92**, 224415 (2015).

27. J. Li, Y. Xu, M. Aldosary et al., Observation of magnon-mediated current drag in Pt/yttrium iron garnet/Pt(Ta) trilayers, *Nat. Commun.* **7**, 10858 (2016).

28. J. Bass and W. P. Pratt Jr., Spin-diffusion lengths in metals and alloys, and spin-flipping at metal/metal interfaces: An experimentalist's critical review, *J. Phys. Condens. Matter* **19**, 183201 (2007).

29. V. V. Kruglyak, S. O. Demokritov, and D. Grundler, Magnonics, *J. Phys. D Appl. Phys.* **43**, 264001 (2010).

30. A. A. Serga, A. V. Chumak, and B. Hillebrands, YIG magnonics, *J. Phys. D Appl. Phys.* **43**, 264002 (2010).

31. B. Lenk, H. Ulrichs, F. Garbs, and M. Munzenberg, The building blocks of magnonics, *Phys. Rep.* **507**, 107 (2011).

32. A. V. Chumak, A. A. Serga, and B. Hillebrands, Magnon transistor for all-magnon data processing, *Nat. Commun.* **5**, 4700 (2014).

33. A. V. Chumak, V. I. Vasyuchka, A. A. Serga, and B. Hillebrands, Magnon spin-tronics, *Nat. Phys.* **11**, 453 (2015).

34. A. Azevedo, L. H. Vilela Leao, A. B. Oliveira, and S. M. Rezende, Dc effect in ferromagnetic resonance: Evidence of the spin-pumping effect? *J. Appl. Phys.* **97**, 10C715 (2005).

35. E. Saitoh, M. Ueda, H. Miyajima, and G. Tatara, Conversion of spin current into charge current at room temperature: Inverse spin-Hall effect, *Appl. Phys. Lett.* **88**, 182509 (2006).

36. M. V. Costache, M. Sladkov, S. M. Watts, C. H. van der Wal, and B. J. van Wees, Electrical detection of spin pumping due to the precessing magnetization of a single ferromagnet, *Phys. Rev. Lett.* **97**, 216603 (2006).

37. S. O. Valenzuela and M. Tinkham, Direct electronic measurement of the spin Hall effect, *Nature* **442**, 176 (2006).

38. T. Kimura, Y. Otani, T. Sato, S. Takahashi, and S. Maekawa, Room-temperature reversible spin hall effect, *Phys. Rev. Lett.* **98**, 156601 (2007).

39. T. Ota, K. Uchida, Y. Kitamura, T. Yoshino, H. Nakayama, and E. Saitoh, Electric detection of the spin-Seebeck effect in Ni and Fe thin films at room temperature, *J. Phys. Conf. Ser.* **200**, 062020 (2010).

40. K. Uchida, T. Ota, K. Harii, K. Ando, H. Nakayama, and E. Saitoh, Electric detection of the spin-Seebeck effect in ferromagnetic metals (invited), *J. Appl. Phys.* **107**, 09A951 (2010).

41. S. Bosu, Y. Sakuraba, K. Uchida et al., Spin Seebeck effect in thin films of the Heusler compound Co_2MnSi, *Phys. Rev. B* **83**, 224401 (2011).

42. C. M. Jaworski, J. Yang, S. Mack, D. D. Awschalom, J. P. Heremans, and R. C. Myers, Observation of the spin-Seebeck effect in a ferromagnetic semiconductor, *Nat. Mater.* **9**, 898 (2010).

43. C. M. Jaworski, J. Yang, S. Mack, D. D. Awschalom, R. C. Myers, and J. P. Heremans, Spin-Seebeck effect: A phonon driven spin distribution, *Phys. Rev. Lett.* **106**, 186601 (2011).

44. K. Uchida, S. Takahashi, J. Ieda et al., Phenomenological analysis for spin-Seebeck effect in metallic magnets, *J. Appl. Phys.* **105**, 07C908 (2009).

45. K. Uchida, H. Adachi, T. Kikkawa et al., Thermoelectric generation based on spin Seebeck effects, *Proc. IEEE* **104**, 1946 (2016).
46. A. B. Cahaya, O. A. Tretiakov, and G. E. W. Bauer, Spin Seebeck power conversion, *IEEE Trans. Magn.* **51**, 0800414 (2015).
47. A. B. Cahaya, O. A. Tretiakov, and G. E. W. Bauer, Spin Seebeck power generators, *Appl. Phys. Lett.* **104**, 042402 (2014).
48. K. Uchida, H. Adachi, T. Ota, H. Nakayama, S. Maekawa, and E. Saitoh, Observation of longitudinal spin-Seebeck effect in magnetic insulators, *Appl. Phys. Lett.* **97**, 172505 (2010).
49. K. Uchida, T. Nonaka, T. Kikkawa, Y. Kajiwara, and E. Saitoh, Longitudinal spin Seebeck effect in various garnet ferrites, *Phys. Rev. B* **87**, 104412 (2013).
50. K. Uchida, T. Nonaka, T. Ota, and E. Saitoh, Longitudinal spin-Seebeck effect in sintered polycrystalline $(Mn,Zn)Fe_2O_4$, *Appl. Phys. Lett.* **97**, 262504 (2010).
51. R. Ramos, T. Kikkawa, K. Uchida et al., Observation of the spin Seebeck effect in epitaxial Fe_3O_4 thin films, *Appl. Phys. Lett.* **102**, 072413 (2013).
52. D. Meier, T. Kuschel, L. Shen et al., Thermally driven spin and charge currents in thin $NiFe_2O_4/Pt$ films, *Phys. Rev. B* **87**, 054421 (2013).
53. T. Niizeki, T. Kikkawa, K. Uchida et al., Observation of longitudinal spin-Seebeck effect in cobalt-ferrite epitaxial thin films, *AIP Adv.* **5**, 053603 (2015).
54. E.-J. Guo, A. Herklotz, A. Kehlberger, J. Cramer, G. Jakob, and M. Klaui, Thermal generation of spin current in epitaxial $CoFe_2O_4$ thin films, *Appl. Phys. Lett.* **108**, 022403 (2016).
55. A. Kirihara, K. Kondo, M. Ishida et al., Flexible heat-flow sensing sheets based on the longitudinal spin Seebeck effect using one-dimensional spin-current conducting films, *Sci. Rep.* **6**, 23114 (2016).
56. H. Asada, A. Kuwahara, N. Sakata et al., Longitudinal spin seebeck effect in $Nd_2BiFe_{5-x}Ga_xO_{12}$ prepared on gadolinium gallium garnet (001) by metal organic decomposition method, *J. Appl. Phys.* **117**, 17C724 (2015).
57. A. Aqeel, N. Vlietstra, J. A. Heuver et al., Spin-Hall magnetoresistance and spin Seebeck effect in spin-spiral and paramagnetic phases of multiferroic $CoCr_2O_4$ films, *Phys. Rev. B* **92**, 224410 (2015).
58. P. Li, D. Ellsworth, H. Chang et al., Generation of pure spin currents via spin Seebeck effect in self-biased hexagonal ferrite thin films, *Appl. Phys. Lett.* **105**, 242412 (2014).
59. R. Takagi, S. Seki, Y. Tokunaga, T. Ideue, Y. Taguchi, and Y. Tokura, Thermal generation of spin current in a helimagnetic multiferroic hexaferrite, *APL Mater.* **4**, 032502 (2016).
60. Y. Shiomi and E. Saitoh, Paramagnetic spin pumping, *Phys. Rev. Lett.* **113**, 266602 (2014).
61. Y. Shiomi, Y. Handa, T. Kikkawa, and E. Saitoh, Transverse thermoelectric effect in $La_{0.67}Sr_{0.33}MnO_3|SrRuO_3$ superlattices, *Appl. Phys. Lett.* **106**, 232403 (2015).
62. J. Flipse, F. K. Dejene, D. Wagenaar, G. E. W. Bauer, J. Ben Youssef, and B. J. van Wees, Observation of the spin Peltier effect for magnetic insulators, *Phys. Rev. Lett.* **113**, 027601 (2014).
63. M. Weiler, M. Althammer, M. Schreier et al., Experimental test of the spin mixing interface conductivity concept, *Phys. Rev. Lett.* **111**, 176601 (2013).
64. J. Xiao, G. E. W. Bauer, K. Uchida, E. Saitoh, and S. Maekawa, Theory of magnon-driven spin Seebeck effect, *Phys. Rev. B* **81**, 214418 (2010).
65. H. Adachi, J. Ohe, S. Takahashi, and S. Maekawa, Linear-response theory of spin Seebeck effect in ferromagnetic insulators, *Phys. Rev. B* **83**, 094410 (2011).
66. H. Adachi, K. Uchida, E. Saitoh, and S. Maekawa, Theory of the spin Seebeck effect, *Rep. Prog. Phys.* **76**, 36501 (2013).
67. J. Foros, A. Brataas, Y. Tserkovnyak, and G. E. W. Bauer, Magnetization noise in magnetoelectronic nanostructures, *Phys. Rev. Lett.* **95**, 016601 (2005).
68. D. J. Sanders and D. Walton, Effect of magnon-phonon thermal relaxation on heat transport by magnons, *Phys. Rev. B* **15**, 1489 (1977).

69. S. Hoffman, K. Sato, and Y. Tserkovnyak, Landau-Lifshitz theory of the longitudinal spin Seebeck effect, *Phys. Rev. B* **88**, 064408 (2013).

70. F. S. M. Rezende, R. L. Rodriguez-Suarez, R. O. Cunha et al., Magnon spin-current theory for the longitudinal spin-Seebeck effect, *Phys. Rev. B* **89**, 014416 (2014).

71. S. M. Rezende, R. L. Rodriguez-Suarez, R. O. Cunha, J. C. Lopez Ortiz, and A. Azevedo, Bulk magnon spin current theory for the longitudinal spin Seebeck effect, *J. Magn. Magn. Mater.* **400**, 171 (2016).

72. S. S.-L. Zhang and S. Zhang, Magnon mediated electric current drag across a ferromagnetic insulator layer, *Phys. Rev. Lett.* **109**, 096603 (2012).

73. S. S.-L. Zhang and S. Zhang, Spin convertance at magnetic interfaces, *Phys. Rev. B* **86**, 214424 (2012).

74. J. Ren, T. Liu, J. Ren, and J. Zhang, Spin and heat transport in ferromagnetic insulators (FI)/normal metal (NM) mesoscopic devices – A Boltzmann approach, *Spin* **5**, 1540010 (2015).

75. S. M. Rezende, R. L. Rodriguez-Suarez, J. C. Lopez Ortiz, and A. Azevedo, Thermal properties of magnons and the spin Seebeck effect in yttrium iron garnet/normal metal hybrid structures, *Phys. Rev. B* **89**, 134406 (2014).

76. S. M. Rezende, R. L. Rodriguez-Suarez, and A. Azevedo, Magnetic relaxation due to spin pumping in thick ferromagnetic films in contact with normal metals, *Phys. Rev. B* **88**, 014404 (2013).

77. A. Kehlberger, U. Ritzmann, D. Hinzke et al., Length scale of the spin Seebeck effect, *Phys. Rev. Lett.* **115**, 096602 (2015).

78. T. Kikkawa, K. Uchida, S. Daimon, Z. Qiu, Y. Shiomi, and E. Saitoh, Critical suppression of spin Seebeck effect by magnetic fields, *Phys. Rev. B* **92**, 064413 (2015).

79. T. Kikkawa, K. Uchida, S. Daimon et al., Separation of longitudinal spin Seebeck effect from anomalous Nernst effect: Determination of origin of transverse thermoelectric voltage in metal/insulator junctions, *Phys. Rev. B* **88**, 214403 (2013).

80. H. Jin, S. R. Boona, Z. Yang, R. C. Myers, and J. P. Heremans, The effect of the magnon dispersion on the longitudinal spin Seebeck effect in yttrium iron garnets (YIG), *Phys. Rev. B* **92**, 054436 (2015).

81. U. Ritzmann, D. Hinzke, A. Kehlberger, E.-J. Guo, M. Klaui, and U. Nowak, Magnetic field control of the spin Seebeck effect, *Phys. Rev. B* **92**, 174411 (2015).

82. R. Ramos, T. Kikkawa, M. H. Aguirre et al., Unconventional scaling and significant enhancement of the spin Seebeck effect in multilayers, *Phys. Rev. B* **92**, 220407(R) (2015).

83. S. Seki, T. Ideue, M. Kubota et al., Thermal generation of spin current in an antiferromagnet, *Phys. Rev. Lett.* **115**, 266601 (2015).

84. S. M. Wu, W. Zhang, K. C. Amit et al., Antiferromagnetic spin Seebeck effect, *Phys. Rev. Lett.* **116**, 097204 (2015).

85. S. M. Rezende, R. L. Rodriguez-Suarez, and A. Azevedo, Theory of the spin Seebeck effect in antiferromagnets, *Phys. Rev. B* **93**, 014425 (2016).

86. S. M. Rezende and R. L. Rodriguez-Suarez, Diffusive magnonic spin transport in antiferromagnetic insulators, *Phys. Rev. B* **93**, 054412 (2016).

87. J. Ohe, H. Adachi, S. Takahashi, and S. Maekawa, Numerical study on the spin Seebeck effect, *Phys. Rev. B* **83**, 115118 (2011).

88. S. A. Bender, R. A. Duine, and Y. Tserkovnyak, Electronic pumping of quasi-equilibrium Bose-Einstein-condensed magnons, *Phys. Rev. Lett.* **108**, 246601 (2012).

89. Y. Ohnuma, H. Adachi, E. Saitoh, and S. Maekawa, Spin Seebeck effect in anti-ferromagnets and compensated ferrimagnets, *Phys. Rev. B* **87**, 014423 (2013).

90. L. Chotorlishvili, Z. Toklikishvili, V. K. Dugaev, J. Barnas, S. Trimper, and J. Berakdar, Fokker-Planck approach to the theory of the magnon-driven spin Seebeck effect, *Phys. Rev. B* **88**, 144429 (2013).

91. J. Ren, Predicted rectification and negative differential spin Seebeck effect at magnetic interfaces, *Phys. Rev. B* **88**, 220406 (2013).

92. S. A. Bender and Y. Tserkovnyak, Interfacial spin and heat transfer between metals and magnetic insulators, *Phys. Rev. B* **91**, 140402 (2015).

93. I. I. Lyapilin, M. S. Okorokov, and V. V. Ustinov, Spin effects induced by thermal perturbation in a normal metal/magnetic insulator system, *Phys. Rev. B* **91**, 195309 (2015).

94. S. R. Etesami, L. Chotorlishvili, and J. Berakdar, Spectral characteristics of time resolved magnonic spin Seebeck effect, *Appl. Phys. Lett.* **107**, 132402 (2015).

95. A. Brataas, H. Skarsvag, E. G. Tveten, and E. L. Fjrbu, Heat transport between antiferromagnetic insulators and normal metals, *Phys. Rev. B* **92**, 180414(R) (2015).

96. L. Chotorlishvili, Z. Toklikishvili, S. R. Etesami, V. K. Dugaev, J. Barnas, and J. Berakdar, Magnon-driven longitudinal spin Seebeck effect in F|N and N|F|N structures: Role of asymmetric in-plane magnetic anisotropy, *J. Magn. Magn. Mater.* **396**, 254 (2015).

97. Y. Tserkovnyak, S. A. Bender, R. A. Duine, and B. Flebus, Bose-Einstein condensation of magnons pumped by the bulk spin Seebeck effect, *Phys. Rev. B* **93**, 100402(R) (2016).

98. L. J. Cornelissen, K. J. H. Peters, G. E. W. Bauer, R. A. Duine, and B. J. van Wees, Magnon spin transport driven by the magnon chemical potential in a magnetic insulator, *Phys. Rev. B* **94**, 014412 (2016).

99. B. Scharf, A. Matos-Abiague, I. Žutić, and J. Fabian, Theory of thermal spin-charge coupling in electronic systems, *Phys. Rev. B* **85**, 085208 (2012).

100. H. Adachi, K. Uchida, E. Saitoh, J. Ohe, S. Takahashi, and S. Maekawa, Gigantic enhancement of spin Seebeck effect by phonon drag, *Appl. Phys. Lett.* **97**, 252506 (2010).

101. K. S. Tikhonov, J. Sinova, and A. M. Finkel'stein, Spectral non-uniform temperature and non-local heat transfer in the spin Seebeck effect, *Nat. Commun.* **4**, 1945 (2013).

102. S. Bosu, Y. Sakuraba, K. Uchida et al., Thermal artifact on the spin Seebeck effect in metallic thin films deposited on MgO substrates, *J. Appl. Phys.* **111**, 07B106 (2012).

103. S. H. Wang, L. K. Zou, J. W. Cai, B. G. Shen, and J. R. Sun, Transverse thermoelectric effects in platinum strips on permalloy films, *Phys. Rev. B* **88**, 214304 (2013).

104. M. Schmid, S. Srichandan, D. Meier et al., Transverse spin Seebeck effect versus anomalous and planar Nernst effects in permalloy thin films, *Phys. Rev. Lett.* **111**, 187201 (2013).

105. M. Weiler, M. Althammer, F. D. Czeschka et al., Local charge and spin currents in magnetothermal landscapes, *Phys. Rev. Lett.* **108**, 106602 (2012).

106. D. Meier, D. Reinhardt, M. van Straaten et al., Longitudinal spin Seebeck effect contribution in transverse spin Seebeck effect experiments in Pt/YIG and Pt/NFO, *Nat. Commun.* **6**, 8211 (2015).

107. A. D. Avery, M. R. Pufall, and B. L. Zink, Determining the planar Nernst effect from magnetic-field-dependent thermopower and resistance in nickel and permalloy thin films, *Phys. Rev. B* **86**, 184408 (2012).

108. N. Nagaosa, J. Sinova, S. Onoda, A. H. MacDonald, and N. P. Ong, Anomalous Hall effect, *Rev. Mod. Phys.* **82**, 1539 (2010).

109. W.-L. Lee, S. Watauchi, V. L. Miller, R. J. Cava, and N. P. Ong, Anomalous Hall heat current and Nernst effect in the $CuCr_2Se_{4-x}Br_x$ Ferromagnet, *Phys. Rev. Lett.* **93**, 226601 (2004).

110. T. Miyasato, N. Abe, T. Fujii et al., Crossover behavior of the anomalous Hall effect and anomalous Nernst effect in itinerant ferromagnets, *Phys. Rev. Lett.* **99**, 86602 (2007).

111. Y. Pu, D. Chiba, F. Matsukura, H. Ohno, and J. Shi, Mott relation for anomalous Hall and Nernst effects in $Ga_{1-x}Mn_xAs$ ferromagnetic semiconductors, *Phys. Rev. Lett.* **101**, 117208 (2008).

112. M. Mizuguchi, S. Ohata, K. Uchida, E. Saitoh, and K. Takanashi, Anomalous Nernst effect in an L10-ordered epitaxial FePt thin film, *Appl. Phys. Express* **5**, 093002 (2012).

113. R. Ramos, M. H. Aguirre, A. Anadon et al., Anomalous Nernst effect of Fe_3O_4 single crystal, *Phys. Rev. B* **90**, 054422 (2014).

114. F. Walz, The Verwey transition – A topical review, *J. Phys. Condens. Matter* **14**, R285 (2002).

115. S. Y. Huang, X. Fan, D. Qu et al., Transport magnetic proximity effects in platinum, *Phys. Rev. Lett.* **109**, 107204 (2012).

116. G. Y. Guo, Q. Niu, and N. Nagaosa, Anomalous Nernst and Hall effects in magnetized platinum and palladium, *Phys. Rev. B* **89**, 214406 (2014).

117. H. Ibach and H. Luth, *Solid-State Physics: An Introduction to the Principles of Materials Science*, Springer-Verlag, Berlin, Germany, 2009.

118. M. Schreier, A. Kamra, M. Weiler et al., Magnon, phonon, and electron temperature profiles and the spin Seebeck effect in magnetic insulator/normal metal hybrid structures, *Phys. Rev. B* **88**, 094410 (2013).

119. J. Flipse, F. L. Bakker, A. Slachter, F. K. Dejene, and B. J. van Wees, Direct observation of the spin-dependent Peltier effect, *Nat. Nanotechnol.* **7**, 166 (2012).

120. D. A. Dalton, W.-P. Hsieh, G. T. Hohensee, D. G. Cahill, and A. F. Goncharov, Effect of mass disorder on the lattice thermal conductivity of MgO periclase under pressure, *Sci. Rep.* **3**, 2400 (2013).

121. M. S. Dresselhaus, G. Chen, M. Y. Tang et al., New directions for low-dimensional thermoelectric materials, *Adv. Mater.* **19**, 1043 (2007).

122. F. J. DiSalvo, Thermoelectric cooling and power generation, *Science* **285**, 703 (1999).

123. G. J. Snyder and E. S. Toberer, Complex thermoelectric materials, *Nat. Mater.* **7**, 105 (2008).

124. A. Kirihara, K. Uchida, Y. Kajiwara et al., Spin-current-driven thermoelectric coating, *Nat. Mater.* **11**, 686 (2012).

125. L.-D. Zhao, S.-H. Lo, Y. Zhang et al., Ultralow thermal conductivity and high thermoelectric figure of merit in SnSe crystals, *Nature* **508**, 373 (2014).

126. L. Liu, C.-F. Pai, Y. Li, H. W. Tseng, D. C. Ralph, and R. A. Buhrman, Spin-torque switching with the giant spin Hall effect of tantalum, *Science* **336**, 555 (2012).

127. C.-F. Pai, L. Liu, Y. Li, H. W. Tseng, D. C. Ralph, and R. A. Buhrman, Spin transfer torque devices utilizing the giant spin Hall effect of tungsten, *Appl. Phys. Lett.* **101**, 122404 (2012).

128. Y. Niimi, Y. Kawanishi, D. H. Wei et al., Giant spin hall effect induced by skew scattering from bismuth impurities inside thin film CuBi alloys, *Phys. Rev. Lett.* **109**, 156602 (2012).

129. P. Laczkowski, J.-C. Rojas-Sanchez, W. Savero-Torres et al., Experimental evidences of a large extrinsic spin Hall effect in AuW alloy, *Appl. Phys. Lett.* **104**, 142403 (2014).

130. A. Hoffmann, Spin Hall effects in metals, *IEEE Trans. Magn.* **49**, 5172 (2013).

131. J. Sinova, S. O. Valenzuela, J. Wunderlich, C. H. Back, and T. Jungwirth, Spin Hall effects, *Rev. Mod. Phys.* **87**, 1213 (2015).

132. K. Uchida, T. Nonaka, T. Yoshino, T. Kikkawa, D. Kikuchi, and E. Saitoh, Enhancement of spin-Seebeck voltage by spin-hall thermopile, *Appl. Phys. Express* **5**, 093001 (2012).

133. K. Uchida, T. Kikkawa, T. Seki et al., Enhancement of anomalous Nernst effects in metallic multilayers free from proximity-induced magnetism, *Phys. Rev. B* **92**, 094414 (2015).

134. K.-D. Lee, D.-J. Kim, H. Y. Lee et al., Thermoelectric signal enhancement by reconciling the spin Seebeck and anomalous Nernst effects in ferromagnet/non-magnet multilayers, *Sci. Rep.* **5**, 10249 (2015).

135. S. Takahashi, E. Saitoh, and S. Maekawa, Spin current through a normal-metal/insulating-ferromagnet junction, *J. Phys. Conf. Ser.* **200**, 062030 (2010).

136. R. Ramos, A. Anadon, I. Lucas et al., Thermoelectric performance of spin Seebeck effect in Fe_3O_4/Pt-based thin film heterostructures, *APL Mater.* **4**, 104802 (2016).

137. S. T. B. Goennenwein, R. Schlitz, M. Pernpeintner et al., Non-local magnetoresistance in YIG/Pt nanostructures, *Appl. Phys. Lett.* **107**, 172405 (2015).

138. L. J. Cornelissen and B. J. van Wees, Magnetic field dependence of the magnon spin diffusion length in the magnetic insulator yttrium iron garnet, *Phys. Rev. B* **93**, 020403 (2016).

139. S. Geprags, A. Kehlberger, F. D. Coletta et al., Origin of the spin Seebeck effect in compensated ferrimagnets, *Nat. Commun.* **7**, 10452 (2016).

140. R. Pauthenet, Spontaneous magnetization of some garnet ferrites and the aluminum substituted garnet ferrites, *J. Appl. Phys.* **29**, 253 (1958).

141. S. M. Wu, J. E. Pearson, and A. Bhattacharya, Paramagnetic spin Seebeck effect, *Phys. Rev. Lett.* **114**, 186602 (2015).

142. C. M. Jaworski, R. C. Myers, E. Johnston-Halperin, and J. P. Heremans, Giant spin Seebeck effect in a non-magnetic material, *Nature* **487**, 210 (2012).

143. L. Frangou, S. Oyarzun, S. Auffret, L. Vila, S. Gambarelli, and V. Baltz, Enhanced Spin pumping efficiency in antiferromagnetic IrMn thin films around the magnetic phase transition, *Phys. Rev. Lett.* **116**, 077203 (2016).

144. W. Zhang, M. B. Jungfleisch, W. Jiang, J. E. Pearson, and A. Hoffmann, Spin Hall effects in metallic antiferromagnets, *Phys. Rev. Lett.* **113**, 196602 (2014).

145. D. Qu, S. Y. Huang, and C. L. Chien, Inverse spin Hall effect in Cr: Independence of antiferromagnetic ordering, *Phys. Rev. B* **92**, 020418 (2015).

146. P. Wadley, B. Howells, J. Zelezny et al., Electrical switching of an antiferromagnet, *Science* **351**, 587 (2016).

147. B. F. Miao, S. Y. Huang, D. Qu, and C. L. Chien, Inverse spin Hall effect in a ferromagnetic metal, *Phys. Rev. Lett.* **111**, 066602 (2013).

148. A. Azevedo, O. Alves Santos, G. A. Fonseca Guerra, R. O. Cunha, R. Rodriguez-Suarez, and S. M. Rezende, Competing spin pumping effects in magnetic hybrid structures, *Appl. Phys. Lett.* **104**, 052402 (2014).

149. S. M. Wu, J. Hoffman, J. E. Pearson, and A. Bhattacharya, Unambiguous separation of the inverse spin Hall and anomalous Nernst effects within a ferromagnetic metal using the spin Seebeck effect, *Appl. Phys. Lett.* **105**, 092409 (2014).

150. C. Du, H. Wang, F. Yang, and P. C. Hammel, Systematic variation of spin-orbit coupling with d-orbital filling: Large inverse spin Hall effect in 3d transition metals, *Phys. Rev. B* **90**, 140407 (2014).

151. D. Tian, Y. Li, D. Qu, X. Jin, and C. L. Chien, Separation of spin Seebeck effect and anomalous Nernst effect in Co/Cu/YIG, *Appl. Phys. Lett.* **106**, 212407 (2015).

152. T. Seki, K. Uchida, T. Kikkawa, Z. Qiu, and E. Saitoh, Observation of inverse spin Hall effect in ferromagnetic FePt alloys using spin Seebeck, *Appl. Phys. Lett.* **107**, 092401 (2015).

153. M. Hatami, G. E. W. Bauer, Q. Zhang, and P. J. Kelly, Thermal spin-transfer torque in magnetoelectronic devices, *Phys. Rev. Lett.* **99**, 066603 (2007).

154. A. A. Kovalev and Y. Tserkovnyak, Thermoelectric spin transfer in textured magnets, *Phys. Rev. B* **80**, 100408 (2009).

155. K. M. D. Hals, A. Brataas, and G. E. W. Bauer, Thermopower and thermally induced domain wall motion in (Ga, Mn)As, *Solid State Commun.* **150**, 461 (2010).

156. H. Yu, S. Granville, D. P. Yu, and J.-Ph. Ansermet, Evidence for thermal spin-transfer torque, *Phys. Rev. Lett.* **104**, 146601 (2010).

157. J. C. Slonczewski, Initiation of spin-transfer torque by thermal transport from magnons, *Phys. Rev. B* **82**, 054403 (2010).

158. X. Jia, K. Xia, and G. E. W. Bauer, Thermal spin transfer in Fe-MgO-Fe tunnel junctions, *Phys. Rev. Lett.* **107**, 176603 (2011).

159. D. Hinzke and U. Nowak, Domain wall motion by the magnonic spin Seebeck effect, *Phys. Rev. Lett.* **107**, 027205 (2011).

160. A. A. Kovalev and Y. Tserkovnyak, Thermomagnonic spin transfer and Peltier effects in insulating magnets, *Europhys. Lett.* **97**, 67002 (2012).

161. T. An, V. I. Vasyuchka, K. Uchida et al., Unidirectional spin-wave heat conveyer, *Nat. Mater.* **12**, 549 (2013).

162. M. V. Costache, G. Bridoux, I. Neumann, and S. O. Valenzuela, Magnon-drag thermopile, *Nat. Mater.* **11**, 199 (2012).

163. M. Walter, J. Walowski, V. Zbarsky et al., Seebeck effect in magnetic tunnel junctions, *Nat. Mater.* **10**, 742 (2011).

164. A. Slachter, F. L. Bakker, J.-P. Adam, and B. J. van Wees, Thermally driven spin injection from a ferromagnet into a non-magnetic metal, *Nat. Phys.* **6**, 879 (2010).

165. N. F. Mott, Electrons in transition metals, *Adv. Phys.* **13**, 325 (1964).

166. A. A. Tulapurkar and Y. Suzuki, Contribution of electron-magnon scattering to the spin-dependent Seebeck effect in a ferromagnet, *Solid State Commun.* **150**, 466 (2010).

167. A. A. Kovalev and Y. Tserkovnyak, Magnetocaloritronic nanomachines, *Solid State Commun.* **150**, 500 (2010).

168. F. K. Dejene, J. Flipse, G. E. W. Bauer, and B. J. van Wees, Spin heat accumulation and spin-dependent temperatures in nanopillar spin valves, *Nat. Phys.* **9**, 636 (2013).

Nonlocal Spin Valves in Metallic Nanostructures

**Yoshichika Otani, Takashi Kimura,
Yasuhiro Niimi, and Hiroshi Idzuchi**

9.1 INTRODUCTION

The main focus of this chapter is recent experimental developments of nonlocal spin injection in metallic systems and the related phenomena after the year 2000. Recent related theoretical work is discussed in Chapter 7, Volume 3, while the early developments of spin injection in metals are discussed in Chapter 5, Volume 1. The pioneering work on spin injection in metals was established in 1985 by Johnson and Silsbee [1, 2]. Their measurements were performed in pure "bulk wire" of aluminum with permalloy leads used for injection and detection up to ~50 K and yielded ~pV signals. By revisiting these ideas in micro-fabricated lateral ferromagnet/nonmagnet/ferromagnet (F/N/F) structures Jedema et al. [3, 4] demonstrated room temperature spin injection and detection with much larger signals. All of these experiments have spurred very intensive research efforts to investigate diffusive spin transport in lateral devices with important implications for the emerging field of spintronics [5].

In Section 9.2, we describe the basic aspect of spin injection, spin accumulation, and spin current in lateral devices comprised of non-magnetic and ferromagnetic metals. We introduce the concept of the spin resistance, that is, a measure of impedance against spin relaxation; thereby lateral spin valve structures can be expressed by equivalent spin-resistance circuits in analogy with cascaded transmission lines. This certainly facilitates designing the lateral spin valves exhibiting efficient spin injection and detection. In Section 9.3, we put our focus on the effect of the device structure, including the junction size, on the magnitude of the spin accumulation stored in the nonmagnet and also the magnitude of the spin current flowing in it. This implies that the spin injection efficiency can be tuned by connecting materials with appropriate structures and spin resistances. Thereby the optimized spin injection enables us to demonstrate pure spin current–induced phenomena such as magnetization switching, and *direct* and *inverse* spin Hall effects (SHEs). This chapter provides some background material to understand the field of spin caloritronics in Chapter 8, Volume 3.

9.2 SPIN INJECTION, ACCUMULATION, AND SPIN CURRENT

Spin injection is a key technology to give rise to spin-dependent transport phenomena, which leads to spin accumulation over the spin-diffusion length in the vicinity of the interface of the junction consisting of ferromagnetic (F) and non-magnetic (N) wires as in Figure 9.1, showing a spatial distribution of the spin-dependent electrochemical potential in the transparent single F/N junction device together with a schematic illustration.

When the spin-polarized electrons are injected through the interface from F to N as indicated by the dashed arrow in the illustration in Figure 9.1, the electron spins accumulate in both F and N metals across the interface over the spin-diffusion length λ_{sf}^{F} for F and λ_{sf}^{N} for N. This spin accumulation results in the spin current, which can be described by considering up ($\uparrow$) and down ($\downarrow$) diffusive spin channels based on the two-current model [6–8] as follows. The spin-dependent current density $j_{\uparrow,\downarrow}$ in both F and N is given by

$$j_{\uparrow,\downarrow} = \sigma_{\uparrow,\downarrow} E + e D_{\uparrow,\downarrow} \nabla \delta n_{\uparrow,\downarrow}. \tag{9.1}$$

Here the first term corresponds to the ordinary Ohm's low, so-called drift current for each spin channel with the spin-dependent conductivities $\sigma_{\uparrow,\downarrow}$,

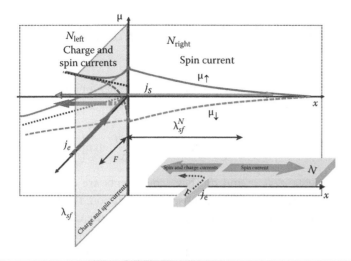

FIGURE 9.1   Spatial distribution of the spin dependent electrochemical potential in the transparent single F/N junction device together with its schematic illustration.

and the second term describes the diffusive contribution of the accumulated spin density $\delta n_{\uparrow,\downarrow}$ with the spin-dependent diffusion constant $D_{\uparrow,\downarrow}$, satisfying Einstein's relation $D_{\uparrow,\downarrow} = \sigma_{\uparrow,\downarrow}/e^2 N_{\uparrow,\downarrow}$ with the density of states $N_{\uparrow,\downarrow}$ at the Fermi level for each spin. The relation between the spin-dependent current density and the electrochemical potential can thus be expressed as

$$j_{\uparrow,\downarrow} = \frac{\sigma_{\uparrow,\downarrow}}{e} \nabla \mu_{\uparrow,\downarrow}, \tag{9.2}$$

where $\mu_{\uparrow,\downarrow} = e\phi + \delta n_{\uparrow,\downarrow}/N_{\uparrow,\downarrow}(E_F)$ with ϕ the electric potential. One should note here that the total electric current density is given by $j_e \equiv j_\uparrow + j_\downarrow = \nabla(\mu_\uparrow \sigma_\uparrow + \mu_\downarrow \sigma_\downarrow)/e$ while the spin current density is $j_s \equiv j_\uparrow - j_\downarrow = \nabla(\mu_\uparrow \sigma_\uparrow - \mu_\downarrow \sigma_\downarrow)/e$, of which spin polarization P is defined as $P = j_s/j_e = (j_\uparrow - j_\downarrow)/(j_\uparrow + j_\downarrow)$. These relations imply that both the drift and diffusive contributions take place in the left-hand side of N with respect to the junction whereas only the diffusive contribution does in the right-hand side; that is, only the spin current flows in the right-hand side in N but no charge current (Figure 9.1). It is important to note here that this spin current is called "pure spin current" to distinguish from the spin (polarized) current flowing in the left-hand side and is only present in the length scale of the spin-diffusion length λ_{sf} along the x-axis. This implies that the spin current decays through spin relaxation process via momentum scattering by phonons or impurities in the presence of the spin-orbit coupling that mixes spin up $|\uparrow\rangle$ and down $|\downarrow\rangle$ states [9]. The diffusive spin current is proportional to the slope of the spin accumulation $\Delta\mu(=\mu_\uparrow - \mu_\downarrow)$, which is well described by the diffusion equation

$$\nabla^2\left(\mu_\uparrow - \mu_\downarrow\right) = \frac{\left(\mu_\uparrow - \mu_\downarrow\right)}{\lambda_{sf}^2}, \tag{9.3}$$

where:

$\lambda_{sf} = \sqrt{D\tau_{sf}}$ with $D = D_\uparrow D_\downarrow (N_\uparrow + N_\downarrow)/(N_\uparrow D_\uparrow + N_\downarrow D_\downarrow)$
τ_{sf} is the spin relaxation time

By applying the low conservation and continuity for charge current, one obtains general solutions for both up and down spin channels

$$\mu_{\uparrow,\downarrow} = a + bx \pm \frac{c}{\sigma_{\uparrow,\downarrow}} \exp\left(-\frac{x}{\lambda_{sf}}\right) \pm \frac{d}{\sigma_{\uparrow,\downarrow}} \exp\left(\frac{x}{\lambda_{sf}}\right), \quad (9.4)$$

with coefficients *a*, *b*, *c*, and *d*. These coefficients are determined by the boundary conditions at the junctions. In the solution, the first linear term $a + bx$ corresponds to a linear voltage drop due to the charge current with Ohmic resistance; thus the spin accumulation $\Delta\mu(=\mu_\uparrow - \mu_\downarrow)$ is given as a sum of the third and the fourth terms in the above equation.

The experimentally accessible quantity, the induced spin accumulation voltage, is thus given by

$$\Delta V_S = \frac{\Delta\mu}{e} = V_+ \exp\left(-\frac{|x|}{\lambda_{sf}}\right) \pm V_- \exp\left(\frac{|x|}{\lambda_{sf}}\right). \quad (9.5)$$

It should be remarked here that the coefficients V_+ and V_- in Equation 9.5 are attributable to the weight of the tunnel-like character and that of the ohmic-like character of the junction.

From Equations 9.2 and 9.5, the spin current $J_S(\equiv j_S S_N)$ in the *N* wire can be obtained as

$$J_S = \frac{1}{R_S^N}\left(V_+ \exp\left(-\frac{x}{\lambda_{sf}^N}\right) \pm V_- \exp\left(\frac{x}{\lambda_{sf}^N}\right)\right), \quad (9.6)$$

where $R_S^N = \lambda_{sf}^N/\sigma_N S_N$ with the conductivity σ_N and the cross-sectional area S_N of the *N* wire.

Similar considerations for the *F* wire also yield the same expression for the spin current J_S in Equation 9.6 with $R_S^F = \lambda_{sf}^F/(1-P^2)\sigma_F S_F$ instead of $R_S^N = \lambda_{sf}^N/\sigma_N S_N$, where σ_F and S_F are, respectively, the conductivity and the junction size of the *F/N* interface. One should notice that Equations 9.5 and 9.6 mimic the electrical transmission line of the characteristic impedance R_S^N with attenuation constant of $1/\lambda_{sf}^N$ [10, 11]. We call R_S^N and R_S^F the spin resistance, indicating a measure of impedance against spin relaxation. Therefore, the *F/N* junction in Figure 9.1 can be regarded as an equivalent spin-resistance circuit (Figure 9.2a), where Equations 9.5 and 9.6 hold in each subsection of the circuit. The spin accumulation decays exponentially along two *N* arms and one *F* arm. By taking into account these three decaying paths, the voltage ΔV_S^i induced by the spin current J_S is

$$\Delta V_S^I = J_S \frac{R_S^N R_S^F}{R_S^N + 2R_S^F}. \quad (9.7)$$

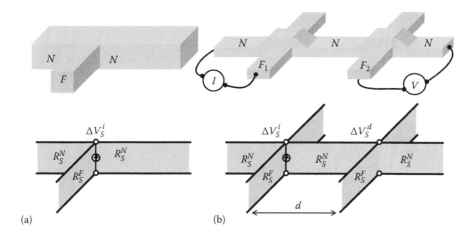

FIGURE 9.2 Schematic illustrations of (a) single F/N junction and (b) lateral spin valve together with an equivalent spin resistance circuits.

In the following section, we discuss the lateral spin valve based on the earlier discussion.

9.3 NONLOCAL SPIN INJECTION EXPERIMENTS

9.3.1 NONLOCAL LATERAL SPIN VALVE

The nonlocal spin valve (NLSV) measurement is usually performed to determine the spin splitting $\Delta\mu$ $(=\Delta eV_S)$ in the chemical potential by using the lateral spin valve structure typically consisting of two ferromagnetic wires $F_{1,2}$ bridged by a non-magnetic wire N, as schematically shown in Figure 9.2b. There are two families of lateral spin valves, one is fully based on the ohmic junctions and the other is based on the tunnel junctions. Valet et al. [8] first introduced the spin-dependent interface resistance in the F/N junction to describe its tunnel- or ohmic-like character, that is, discontinuity appearing in the spin-dependent electrochemical potential $\mu_{\uparrow,\downarrow}$ when the current passes through it. Following their definition, the tunnel junction here means the discontinuity in $\mu_{\uparrow,\downarrow}$ much larger than the spin splitting $\Delta\mu$ $(=\Delta eV_S)$, while the ohmic junction means no discontinuity in $\mu_{\uparrow,\downarrow}$. Here we limit our discussion mostly to the ohmic junctions. A series of reports on the ohmic lateral spin valves performed in the 2000s can be found in Refs. [10, 12–14]. Some representative experimental demonstrations using tunnel junctions are shown in Refs. [3, 15–18]. The F_1 wire is the spin injector through which the spin-polarized current is injected into the N wire. The spin accumulation builds up in the vicinity of the F_1/N junction as discussed in the previous section. The spatially decaying spin accumulation then gives rise to a spin current flowing from the F_1/N interface to the F_2/N interface, and relaxing into the F_2 detector. Since $\sigma_\uparrow$ and $\sigma_\downarrow$ are different in ferromagnet, measurable nonequilibrium voltage ΔV_P or ΔV_{AP} is produced according to the relative magnetization orientation of F_1 and F_2 either parallel (P) or antiparallel (AP), as in Figure 9.3. This situation can also be well described by the equivalent

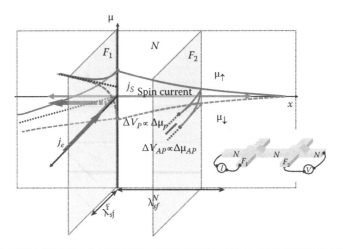

FIGURE 9.3 Spatial distribution of the spin dependent electrochemical potential in the lateral spin valve device and a schematic illustration of the device structure.

spin-resistance circuit in Figure 9.2b. The total spin resistance of the device R_S^{tot} can be calculated similarly to that of the single F/N junction as [10].

$$
\begin{aligned}
R_S^{\text{tot}} &= \left(\frac{1}{R_S^N} + \frac{1}{R_S^F} + \frac{1}{R_S^i} \right)^{-1} \\
&= \frac{QR_S^N\left[Q\exp\left(d/\lambda_{sf}\right) + 4\sinh\left(d/\lambda_{sf}\right) \right]}{2\left[Q(4+Q)\exp\left(d/\lambda_{sf}\right) + 4\sinh\left(d/\lambda_{sf}\right) \right]},
\end{aligned}
\tag{9.8}
$$

where:

R_S^i is the injector spin resistance consisting of N and F wires

Q is the spin resistance ratio R_S^F/R_S^N

d is the center-to-center spacing between the injector and detector wires

Hence the spin accumulation voltage ΔV_S^d at the detector junction is given by $\Delta V_S^d = TJ_S R_S^{\text{tot}}$, where $T = Q/\left(Q\exp\left(d/\lambda_{sf}\right) + 4\sinh\left(d/\lambda_{sf}\right)\right)$. Finally we obtain the overall change ΔR_{NL}^d in NLSV signal in ohm between P and AP states as

$$
\Delta R_{NL}^D \equiv \frac{P\Delta V_S^d}{J_e} \approx \frac{2P^2 Q^2 R_S^N}{2Q\exp\left(d/\lambda_{sf}\right) + \sinh\left(d/\lambda_{sf}\right)}.
\tag{9.9}
$$

Figure 9.4a and b shows a scanning electron micrograph (SEM) of a typical lateral spin valve consisting of $F_{1,2}$ permalloy (Py: $Ni_{19}Fe_{79}$) wires bridged by a N (Cu) wire and its NLSV signal ΔR_{NL} measured at room temperature as a function of applied magnetic field. The devices are usually fabricated by means of lift-off or shadow evaporation methods combined with e-beam lithography [19]. Due to the shape anisotropy, the magnetization directions

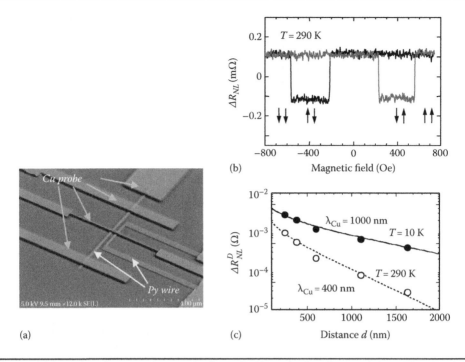

FIGURE 9.4 (a) Scanning electron micrograph of a typical lateral spin valve consisting of Py wires bridged by a Cu wire; (b) the nonlocal spin valve signal at room temperature, and (c) the overall change ΔR_{NL}^D as a function of separating distance at various temperatures.

of $F_{1,2}$ wires are parallel to the long axes. The F_2 wire having domain wall reservoirs at both edges exhibits a smaller switching field than the straight F_1 wire. As the magnetic field is swept from positive to negative (black curve) after having set both magnetizations in P configuration at a large positive magnetic field of 1.0 kOe, a sharp drop in the signal ΔR_{NL} is observed at about -0.2 kOe when the $F_{1,2}$ magnetizations turn into AP configuration. Further decrease down to -0.55 kOe recovers the high ΔR_{NL} value in P configuration. A similar process takes place when the field sweep is reversed (gray curve).

The spin-diffusion lengths of Py and Cu wires can be evaluated by performing above NLSV measurements as a function of the separation d. Fitting the measured overall change ΔR_{NL}^D versus d to Equation 9.9 (Figure 9.4c) yields the spin-diffusion length of Cu as $\lambda_{sf}^{Cu} = 1000$ nm at 10 K and $\lambda_{sf}^{Cu} = 400$ nm at 290 K. Careful attention has to be paid when the cross-sectional areas S_N and S_F for Cu and Py wires are estimated. As mentioned earlier λ_{sf}^{Cu} is of the order of several hundred nanometers, whereas λ_{sf}^{Py} a few nanometers [18]. This means that the spin current flows homogeneously along the Cu wire but not in the Py wire where the spin current only exists in the vicinity of the junction. Therefore S_N for Cu should be the cross-sectional area of the wire, while S_F for Py should be the junction area. This tendency is clearly assured by the numerical calculation of spin current distribution [20].

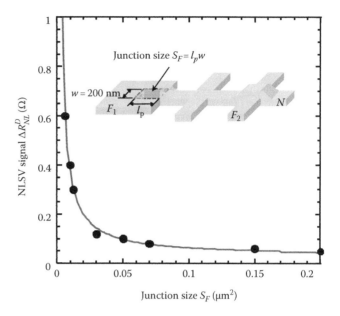

FIGURE 9.5 Overall nonlocal spin valve signal ΔR_{NL}^{D} as a function of the junction size S. The gray curve is the best fit to the experimental data using the spin resistance circuit model.

9.3.2 Junction Size Dependence

As discussed earlier, the cross-sectional area for the spin current is represented by the effective junction size for the case of a ferromagnet such as Py with a very small λ_{sf}^{Py} of about a few nanometers. This implies that the spin accumulation in the vicinity of the Py/Cu junction is readily suppressed by the progressive spin relaxation in Py. This fact has to be well considered in order to realize efficient spin injection and detection. One of the ways to suppress the spin relaxation in the vicinity of the interface is to increase its spin resistance, equivalently meaning that the junction size has to be diminished. This was experimentally demonstrated as in Figure 9.5, showing drastic improvement in the NLSV signal ΔR_{NL}^{D} as the junction size S_F is minimized [21]. If one seeks the best detection efficiency the point contact should be the one as well as the tunnel junction. However the current density for spin injection cannot be maximized for both cases. It should be noted here again that the NLSV signal greatly varies not only with the injector-detector spacing d but also with the junction size S. Alternative solution for injection and detection efficiency could rely on the interface engineering. By introducing the spin-dependent interface layer, the efficiency can be significantly optimized [22].

9.3.3 Spin Absorption (Sink) Effect

We now apply our spin-resistance circuit model to a triple junction shown schematically in Figure 9.6. In this case, the spin accumulation voltage ΔV_S^d induced between the F_2 and N wires is expected to be influenced by the M

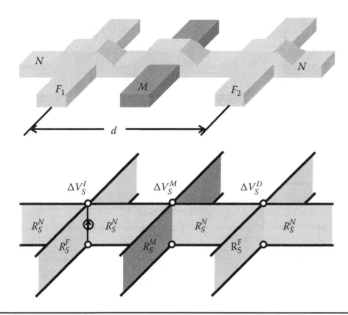

FIGURE 9.6 (top) Schematic illustration of a triple junction device consisting of two F/N and a middle M/N junctions and (bottom) an equivalent spin resistance circuit.

wire inserted right in the middle at $d/2$ with the spin resistance of R_S^M and the spin-diffusion length of λ_{sf}^M, providing an additional spin relaxation volume. The magnitude of ΔR_{NL}^D with M, ΔR_{NL}^{Dwith} can be calculated by using the same formalism given in Section 9.3.1

$$\Delta R_{NL}^{Dwith} \approx \frac{4P^2Q^2Q_MR_S^N}{\cosh\left(d/\lambda_N\right)-1+2Q_M\sinh\left(d/\lambda_N\right)+2Qe^{d/\lambda_N}-2Q}, \quad (9.10)$$

where:

Q_M is defined as R_S^M/R_S^N

d is the spacing between F_1 and F_2

Figure 9.7 shows typical lateral spin valve structures consisting of Py $F_{1,2}$ wires bridged by a Cu N wire with and without an inserted Pt M wire and corresponding NLSV signals. The overall change in NLSV signal $\Delta R_{NL}^{Dwithout}$ amounts to 0.6 mΩ. However, the signal ΔR_{NL}^{Dwith} is considerably lowered to about 0.05 mΩ by the inserted Pt wire. This indeed indicates that the Pt wire absorbs most of the spins accumulated in between the Py injector and detector wires; in other words, the Pt wire acts as a spin sink.

Putting into Equation 9.10 the values of the spin resistances, $R_S^{Cu} = 2.60\,\Omega$ and $R_S^{Py} = 0.34\,\Omega$ separately determined by the double junction experiments, and the experimental value of ΔR_{NL}^{Dwith} for Pt, we obtain the value of $R_S^{Pt} = 0.15\,\Omega$ and then the spin-diffusion length $\lambda_{sf}^{Pt} = 10\,nm$. In the similar manner, the spin resistance and the spin-diffusion length for Au are determined, respectively, as $R_S^{Au} = 0.62\,\Omega$ and $\lambda_{sf}^{Au} = 60\,nm$. In this

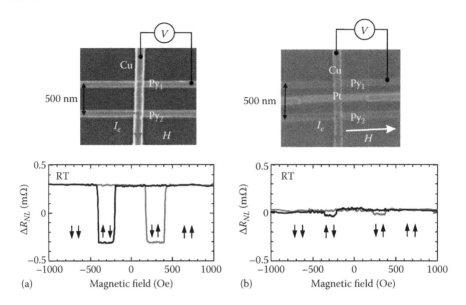

FIGURE 9.7 Typical lateral spin valves consisting of Py wires bridged by a Cu wire with (left) and without (right) an inserted Pt wire and corresponding nonlocal spin valve signal as a function of external field measured at room temperature.

way, the value of the spin resistance R_S^M of the inserted wire can be directly determined from the measured value of $\Delta R_{NL}^{D\text{with}}$, thereby the spin-diffusion length λ_{sf}^M can be calculated. Figure 9.8 shows the R_S^M versus $\Delta R_{NL}^{D\text{with}}$ plot for spin valves with $d = 500$ and 650 nm calculated by using Equation 9.10. This plot is cross-checking that the values of the spin resistances obtained by using the triple junction are equivalent to those determined by using the double junction. The values obtained with the inserted Cu and Py wires are, respectively, 2.67 and 0.33 Ω, which are indeed in good agreement with those determined by using the double junction. Important to note is that the spin-diffusion length can be simply determined from the ratio of $\Delta R_{NL}^{D\text{with}}$ to $\Delta R_{NL}^{D\text{without}}$ without varying the injector-detector spacing d, and that all the experimental data points lie on the same curve only when the spacing d and the junction size S are the same, as can be seen in Figure 9.8.

One should also note here that the spin current prefers to flow into the additionally connected wire when the spin resistance R_S^M is much smaller than R_S^N, meaning that this can be used as a technique to inject a pure spin current into a target material with a small R_S^M. We will show some of the examples such as the pure spin current induced switching, and SHEs in Sections 9.4 and 9.5.

9.3.4 Collective Spin Precession Hanle Effect

The Hanle effect is the manifestation of dephasing process in precessional motion of diffusively traveling collective electronic spins. In the measurement, the external field is usually applied perpendicular to the axis of the

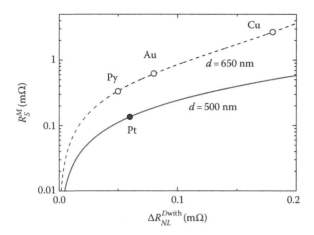

FIGURE 9.8 Calculated spin resistance as a function of the overall change ΔR_{NL}^{D} for the triple junction devices with $d=500$ nm and $d=650$ nm. All the measured values of ΔR_{NL}^{D} for the devices with middle junctions of Py, Au and Cu lie on the same calculated curve (dashed curve) when the value of d is the same. The calculated solid curve corresponds to the device with $d=500$ nm with ΔR_{NL}^{D} for the device with a Pt middle junction.

accumulated spins to induce the precessional motion about the field [23]. The averaged orientation of spins detected at the position of the detector thus results in a modulation in the NLSV signal as a function of the external field [4]. The orientation is modulated by the Larmor precession frequency, therefore this is a powerful means to characterize transport properties of the injected spins, particularly spin relaxation time, and is a counter part of nonlocal spin valve measurement, which characterizes the spin-diffusion length (see Figure 9.7). In spintronics dealing with mesoscopic systems, Hanle effect is widely used to characterize spin transport properties (Al: [4], Cu: [16], Ag: [23], Graphene: [24], GaAs: [25], more references can be found in [26]). However, since the spin accumulation is strongly affected by the Ohmic contact as we have discussed in the previous section, the transit-time of diffusing electronic spins from the injector to the detector should also be affected. This results in that the apparent spin relaxation time characterized without taking account of the spin absorption is different from the intrinsic spin relaxation time of the material.

Before describing Hanle effect, we show how the junction property affects the spin absorption properties. Figure 9.9a shows spin signal ΔR_S for a lateral spin valve (LSV) with Py/Ag Ohmic contact and LSVs with Py/MgO/Ag junctions with different MgO thicknesses. With no MgO layer, Py/Ag metallic contact shows transparent property (low resistance) and the junction resistance of Py/MgO/Ag junction increases with thickness of the inserted MgO [23]. As an increase of the contact resistance, the spin signal increases and levels off. This behavior can be explained by considering the spin absorption effect i.e., with an increase of contact resistance, the strength of spin absorption by the contacts decreases and becomes negligible. Interestingly the crossover resistance between too fast spin absorption

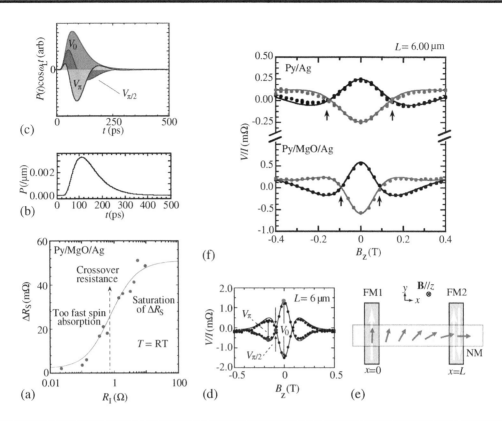

FIGURE 9.9 (a) Nonlocal spin signal ΔR_S as a function of contact resistance R_I for NiFe(Py)/Ag lateral spin valves with MgO layer between Py and Ag, for the separation $L = 0.30$ μm [23]. The crossover resistance between the conductivity mismatch regime (too fast spin absorption by the contacts) and saturation (intrinsic spin relaxation in Ag), $R_{crossover} = R_N = \rho_{Ag}\lambda_{Ag}/t_{Ag}w_{Ag}$ is the scale governing the variation with R_I [27] and is of the order of 1 Ω for a metal as Ag (resistivity ρ_{Ag} in the μΩcm range, spin diffusion length λ_{Ag} around 1 μm and the thickness t_{Ag} (width w_{Ag}) in the 10^1 nm (10^2 nm) range. (After Idzuchi, H. et al., *Phys. Rev. B* 91, 241407, 2015. With permission.) (b) Transit-time distribution of spin current for lateral spin valve with Py/MgO/Ag junctions. (c) Transit-time distribution of spin current for lateral spin valve with Py/MgO/Ag junctions. Red, blue and green lines show $P(t)\cos(\omega_L t)$ with magnetic field in which $\int_0^\infty dt P(t)\cos(\omega_L t)$ shows maximum ($B = 0$), zero ($B = B_{\pi/2}$) and minimum ($B = B_\pi$) and correspond to 0, $\pi/2$ and π rotation, respectively. (After Idzuchi, H. et al., *Physica E* 68, 239, 2015. With permission.) (d) Hanle signal for lateral spin valve with dual injector Py/MgO/Ag junctions with $L = 6$ μm. (After Idzuchi, H. et al., *Sci. Rep.* 2, 628, 2012. With permission.) Red, blue and green dots are proportional to the colored area in (c). (e) Precession of accumulated spins in lateral spin valve in the presence of perpendicular magnetic field **B** where the spin accumulation rotates during the travel of distance L between the injector FM1 and the detector FM2. (f) Hanle signal in LSVs with Py/Ag junctions and Py/MgO/Ag junctions with various separations $L = 6$ μm. Black and red circles show spin valve signal V/I of parallel and antiparallel magnetic configurations of the injector and detector electrodes, respectively at $T = 10$ K. Curves are obtained by the formula of Hanle effect (Equations 9.14 and 9.16) with adjusting parameters shown in Table 9.1. Arrows show first crosspoints of the Hanle signal for the parallel and antiparallel configurations corresponding to the collective $\pm\pi/2$ rotation of diffusive spins. (After Idzuchi, H. et al., *Phys. Rev. B* 89, 081308, 2014. With permission.)

and saturation corresponds to the spin resistance of the nonmagnet, in this case Ag. In this way, we can control spin absorption property by changing the junction resistance, which enables us to study the influence of the spin absorption on the Hanle effect.

Now, we depict intuitive picture of Hanle effect. If the spin absorption by the contacts can be neglected, the probability $P(t)$ of the spin reached at the detector position $x = L$ after the spin is injected at the time $t = 0$ is expressed as

$$P(t) = \frac{1}{\sqrt{4\pi D\, t}} \exp\!\left(-\frac{L^2}{4D\, t} - \frac{t}{\tau_{sf}}\right), \tag{9.11}$$

where τ_{sf} is the spin relaxation time. As shown in Figure 9.9b, it gives a characteristic feature, in which the peak indicates the crossover from the spin diffusion in short time scale, to the spin-flip characters in long time scale. The nonlocal voltage is proportional to the overall area of $P(t)\cos(\omega_L t)$ and the spin precession causes a decrease in (or even sign reversal of) nonlocal voltage as depicted in Figure 9.9c and d.

Now in order to formulate Hanle effect for both cases of with and without spin absorption, we start from "equation of motion" for nonequilibrium magnetization $\mathbf{m}_N$ (Bloch-Torrey equation). In case of without spin absorption

$$\frac{\partial \mathbf{m}_N}{\partial t} = -\gamma_e \left[\mathbf{m}_N(\mathbf{r}) \times \mathbf{B}\right] - \frac{\mathbf{m}_N}{\tau_{sf}} + D\, \nabla^2 \mathbf{m}_N(\mathbf{r}), \tag{9.12}$$

where γ_e is the gyromagnetic ratio, and each term of right-hand side gives the effect of spin precession, spin relaxation and spin diffusion. Now we extend the chemical potential to the complex value so that the real and imaginary parts describe the spin polarization in x- and y-direction, respectively (see Figure 9.9e for the definition of axis).

$$\delta\tilde{\mu}_N = \frac{I_m}{eN(\varepsilon_F)S_N} \frac{\tau_{sf}}{2\lambda_{sf}} \left[\frac{\lambda_\omega}{\lambda_{sf}} \exp(-|x|/\lambda_\omega)\right], \tag{9.13}$$

where $\lambda_\omega = \lambda_{sf}\big/\sqrt{1 + i\omega_L \tau_{sf}}$.

$V = \mathrm{Re}[\delta\tilde{\mu}_N(x)]/e$, i.e.,

$$\frac{V}{I} = \frac{1}{2} P_{I1} P_{I2} R_N \, \mathrm{Re}\!\left[(\lambda_\omega/\lambda_{sf})\exp(-L/\lambda_\omega)\right], \tag{9.14}$$

where:

V is nonlocal voltage

P_{I1}, P_{I2} and are the spin polarization of junctions (injector and detector)

Under the influence of the spin absorption, the Bloch-Torrey equation is

$$\begin{aligned}
\frac{\partial \mathbf{m}_N}{\partial t} =\; & -\gamma_e \left[\mathbf{m}_N(\mathbf{r}) \times \mathbf{B}\right] - \frac{\mathbf{m}_N}{\tau_{sf}} + D_N \nabla^2 \mathbf{m}_N(\mathbf{r}) \\
& + \frac{\hbar\gamma_e}{2e} \frac{I_{s1}^x}{S_N} \mathbf{e}_x \delta(x) + \frac{\hbar\gamma_e}{2e} \frac{I_{s1}^y}{S_N} \mathbf{e}_y \delta(x) \\
& + \frac{\hbar\gamma_e}{2e} \frac{I_{s2}^x}{S_N} \mathbf{e}_x \delta(x - L) + \frac{\hbar\gamma_e}{2e} \frac{I_{s2}^y}{S_N} \mathbf{e}_y \delta(x - L),
\end{aligned} \tag{9.15}$$

where:

$\hbar$ is Reduced Planck constant

I_{si}^{j} is the spin current through i-th contact with j-polarization, and the additional term gives the flow of spin current in each junction

Then we obtain

$$\frac{V}{I} = -2R_N \left(\frac{P_{F1}}{1-P_{F1}^2} \frac{R_{F1}}{R_N} + \frac{P_{I1}}{1-P_{I1}^2} \frac{R_{I1}}{R_N} \right) \left(\frac{P_{F2}}{1-P_{F2}^2} \frac{R_{F2}}{R_N} + \frac{P_{I2}}{1-P_{I2}^2} \frac{R_{I2}}{R_N} \right) \frac{C_{12}}{\det(\hat{X})}, \quad (9.16)$$

where:

$$\hat{X} = \begin{pmatrix} r_{1\|} + \mathrm{Re}[\overline{\lambda}_\omega] & \mathrm{Re}[\overline{\lambda}_\omega e^{-L/\hat{\lambda}_\omega}] & -\mathrm{Im}[\overline{\lambda}_\omega] & -\mathrm{Im}[\overline{\lambda}_\omega e^{-L/\hat{\lambda}_\omega}] \\ \mathrm{Re}[\overline{\lambda}_\omega e^{-L/\hat{\lambda}_\omega}] & r_{2\|} + \mathrm{Re}[\overline{\lambda}_\omega] & -\mathrm{Im}[\overline{\lambda}_\omega e^{-L/\hat{\lambda}_\omega}] & -\mathrm{Im}[\overline{\lambda}_\omega] \\ \mathrm{Im}[\overline{\lambda}_\omega] & \mathrm{Im}[\overline{\lambda}_\omega e^{-L/\hat{\lambda}_\omega}] & r_{1\perp} + \mathrm{Re}[\overline{\lambda}_\omega] & \mathrm{Re}[\overline{\lambda}_\omega e^{-L/\hat{\lambda}_\omega}] \\ \mathrm{Im}[\overline{\lambda}_\omega e^{-L/\hat{\lambda}_\omega}] & \mathrm{Im}[\overline{\lambda}_\omega] & \mathrm{Re}[\overline{\lambda}_\omega e^{-L/\hat{\lambda}_\omega}] & r_{2\perp} + \mathrm{Re}[\overline{\lambda}_\omega] \end{pmatrix},$$

$$r_{k\|} = \left(\frac{2}{1-P_{Ik}^2} \frac{R_{Ik}}{R_N} + \frac{2}{1-P_{Fk}^2} \frac{R_{Fk}}{R_N} \right), \quad r_{k\perp} = \frac{1}{R_N S_{Ik} G_k^{\uparrow\downarrow}}, \quad (k=1,2).$$

$$C_{12} = -\det \begin{pmatrix} \mathrm{Re}[\overline{\lambda}_\omega e^{-L/\hat{\lambda}_\omega}] & -\mathrm{Im}[\overline{\lambda}_\omega e^{-L/\hat{\lambda}_\omega}] & -\mathrm{Im}[\overline{\lambda}_\omega] \\ \mathrm{Im}[\overline{\lambda}_\omega] & r_{1\perp} + \mathrm{Re}[\overline{\lambda}_\omega] & \mathrm{Re}[\overline{\lambda}_\omega e^{-L/\hat{\lambda}_\omega}] \\ \mathrm{Im}[\overline{\lambda}_\omega e^{-L/\hat{\lambda}_\omega}] & \mathrm{Re}[\overline{\lambda}_\omega e^{-L/\hat{\lambda}_\omega}] & r_{2\perp} + \mathrm{Re}[\overline{\lambda}_\omega] \end{pmatrix}.$$

$\overline{\lambda}_\omega = \lambda_\omega / \lambda_{sf}$, $R_F = \lambda_{sf}^F / \sigma_F S_F$, $R_N = \lambda_{sf}^N / \sigma_N S_N$, and S_{Ik} is the area of k-th junction. The junction is generally anisotropic and can be characterized by the real part of spin mixing conductance of k-th junction $G_k^{\uparrow\downarrow}$ [28]. We note that magnetic field could affect the direction of ferromagnetic injector and detector, which gives the background similar to parabolic curve [29]. The detail formulation of Hanle effect can be found in [30, 31].

Now we show the influence of spin absorption on the Hanle effect on Ag based LSVs with Py/Ag contact and Py/MgO/Ag junction. Figure 9.9f shows Hanle effect for NLSV with Py/Ag contact and Py/MgO/Ag junction. Hanle signal for Py/Ag contact is wider than the one for Py/MgO/Ag junctions, which means apparent velocity is faster for Py/Ag junctions. By using equation above we can obtain intrinsic spin and consistent transport parameters as shown in Table 9.1. Therefore this effect is caused by spatial distribution of spin-dependent chemical potential by spin absorption effect.

This effect can be observed in different class of material. As a matter of fact, spin absorption effect is more severe for the materials with larger R_S^N. Therefore semiconductive (semi-metallic) system is expected to be largely affected by contact effect. Interesting case is graphene where theoretical spin relaxation time and experimental one is largely different [32]. Spin absorption effect in graphene is numerically analyzed but reported as the spin relaxation is not limited by spin absorption and mainly limited by other

TABLE 9.1
Adjusting parameters for Hanle signals which are shown in Figure 9.9f

Junction	P_F	$P_{I(Py/MgO/Ag)}$	$P_{I(Py/Ag)}$	τ_{sf} (ps)	$G_{\uparrow\downarrow}$ (m$^{-2}\Omega^{-1}$)
Py/Ag	0.55 ± 0.12	N/A	0.76 ± 0.06	42.9 ± 7.9	$(3.6 \pm 8.4) \times 10^{14}$
Py/MgO/Ag	N/A	0.26 ± 0.07	N/A	45.0 ± 10.2	N/A

Source: From [30].

effect [33]. Recently some of us and a collaborator showed that the experimental report of different spin relaxation time of graphene, is caused by contact. Co/Graphene and Co/MgO/TiO$_2$/Graphene, can be quantitatively explained by the formula above introduced [31]. That demonstrates that considering spin absorption at contacts are important to correctly characterize spin transport properties in Hanle effect.

9.4 PURE SPIN CURRENT INDUCED MAGNETIZATION REVERSAL

Spin transfer torques arise whenever the flow of spin angular momentum is throwing a portion of its angular momentum while passing through a ferromagnet. This happens typically in F$_1$/N/F$_2$ tri-layered nano-pillars where a spin-polarized current generated by spin filtering of the first ferromagnetic F$_1$ layer enters the second ferromagnetic F$_2$ layer through the N layer, and its polarization rotates to follow the magnetization direction of the F$_2$. This is equivalent to the situation where a pure spin current is absorbed by a magnetic wire whose magnetic moment is not collinear with the polarization direction of the pure spin current.

Kimura et al. [34, 35] demonstrated for the first time the pure spin current induced magnetization switching by using the structure similar to the lateral spin valve in Figure 9.9b but F$_2$ wire is replaced with a small Py nanoscale particle to which the pure spin current was injected *via* spin absorption process. The above mentioned behavior can be understood that a flow of up-spin electrons penetrate diffusively into a magnetic element and transfers their transverse component of spin angular momentum, while an equal number of down-spin electrons with an averaged spin component collinear with the magnet come out with generating no spin torque.

More recently [36] succeeded in improving the spin injection and detection efficiency by a factor of 10 by employing in-situ shadow evaporation technique for fabricating a lateral spin valve consisting of 2 separated Py (20 nm)/Au nano-pillars structured on a Cu wire as shown in Figures 9.10b, d and 9.11a. Remarkable is that well defined lateral giant magneto resistance effect as well as the NLSV effect are observed (Figure 9.10a, c) and the overall change in ΔR_L of 32 mΩ is about twice as much as that in the NLSV signal ΔR_{NL} of 18.5 mΩ. In addition these values are almost an order of magnitude larger than the reported values [3, 20, 10] and also in good agreement with the theoretical values calculated by using Equation 9.9. These facts assure

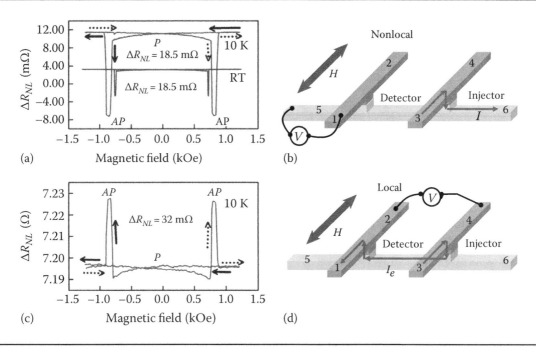

FIGURE 9.10   (a) Nonlocal spin valve signal ΔR_{NL} as a function of magnetic field together with (b) a schematic device configuration. (c) Local spin valve signal ΔR_L as a function of magnetic field together with (d) a schematic device configuration.

that clean interfaces of both spin injector and detector well contributes to produce the giant NLSV signal.

For the purpose of realizing efficient pure spin current induced switching, the detector Py/Au nano-pillar thickness was reduced to 4 nm to match the spin-diffusion length so that the spin current necessary for the switching can be minimized. The value of the NLSV signal ΔR_{NL} for the device is about 4 mΩ, smaller than that in Figure 9.10a but still larger than previously reported values. This device enables to switch reversibly the magnetization of a Py particle $75 \times 170 \times 4$ nm^3 in size at the local spin current density in the range 2–6×10^{10} A/m^2 which is reasonable compared with the conventional spin injection experiments using nano-pillars [37, 38].

9.5 INVERSE AND DIRECT SPIN HALL EFFECTS

As discussed in Section 9.4, spin accumulation, absorption or spin pumping enables us to inject a pure spin current into a nonmagnetic material. If the N material possesses the scattering centers such as impurities causing spin-orbit interaction, the injected spins are scattered asymmetrically according to their orientation [19, 39–41]. Thereby the spin current is converted to the charge current flowing perpendicular to the spin current. The direction of the induced charge current J_e is given by the vector product $\mathbf{J_S} \times \mathbf{s}$, where s is the spin direction, yielding the flow of J_e normal to both J_S and $\mathbf{s}$. This behavior is called *inverse spin Hall effect* (SHE). In reverse, a charge current flowing in the nonmagnetic material can be converted to a spin current

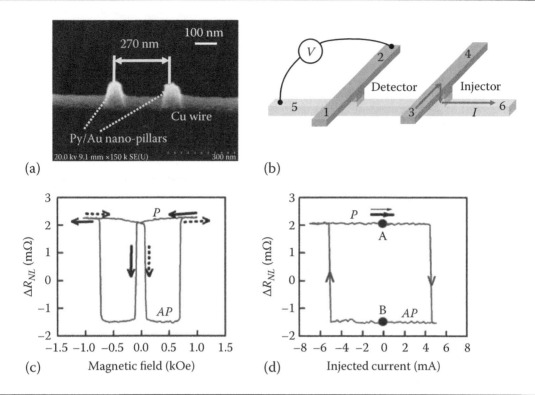

(a)   (b)

(c)   (d)

FIGURE 9.11   Nonlocal spin injection results for a sample with 4 nm thick Py detector nano-magnet. (a) Scanning electron micrograph of two Py/Au nano-pillars bridged by a Cu wire. (b) a schematic device configuration. (c) Nonlocal spin valve signal as a function of magnetic field. (d) The nonlocal spin valve signal as a function of injected d.c. current. The loop starts at the initial parallel (P) state A. The switching from P to anti-parallel (AP) states occurs when the injected d.c. current reaches 5 mA. The reverse switching from AP to P states takes place when the current polarity is reversed.

flowing perpendicular to the charge current and up and down electronic spins accumulate along the opposing edges of the N material. This is called (*direct*) *spin Hall effect* (SHE), which has been observed in semiconductors [42, 43] and also metals [19].

For the *inverse* SHE measurements, the spin absorption technique explained in Section 9.3.3 was used for injecting a pure spin current into a Pt wire. The device structure of the SHE experiments is as shown in the inset of Figure 9.12a, typically consisting of a large Py injector pad 30 nm thick connected to a 80 nm wide Pt wire with thickness of about 5 nm *via* a 100 nm wide 80 nm thick Cu cross. The charge current is then injected from the Py pad into the Cu cross and drained into one of the 2 arms at the center of the cross. Induced spin current is diffusively transferred and absorbed into the Pt wire perpendicular to the device plane. When the spin direction s is oriented perpendicular to the Pt wire, the induced charge current J_e is directed along the wire, and its magnitude is detected as the Hall voltage V_C appearing between the edges of the Pt wire. The Hall signal in ohm $\Delta V_C / J_e$ varies as a function of the external field, reflecting the magnetization process of the Py pad as shown in Figure 9.12a since the averaged spin direction is defined by the magnitude of the magnetization vector.

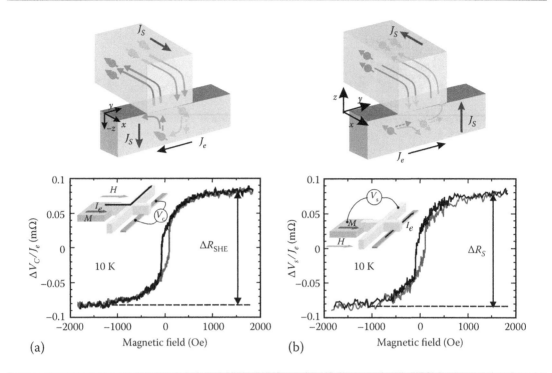

(a)

(b)

FIGURE 9.12 (a) Schematic explanation of the inverse spin Hall effect, conversion process in the Cu (light orange)/Pt (orange) junction from the spin current J_S to the charge current J_e together with the measured Hall voltage as a function of magnetic field. The inset schematically shows a device configuration. (b) Schematic explanation of the direct spin Hall effect, reversed conversion process from J_e to J_S, together with the spin accumulation signal as a function of magnetic field.

The same device can also be used for the *direct* SHE measurements. The large Py pad is now used as a detector instead of injector. It should be remarked that in the above experiment, the Pt wire is used as a spin current absorber, but the Pt wire now acts as a spin current source, which may provide a new means to produce the spin current for spintronics applications. The induced spin accumulation at the surface of the Pt wire is transferred through the Cu wire and detected by the Py pad as the spin accumulation signal in ohm $\Delta V_S/J_e$ as in Figure 9.12b which varies almost the same as $\Delta V_C/J_e$ in Figure 9.12a corresponding to Onsager's reciprocal relation between spin and charge currents. This is equivalent to say that the spin current induced SH conductivity σ_{SHE} is equal to the charge current-induced SH conductivity σ'_{SHE}. The spin Hall conductivity can be expressed by using the experimentally determined overall change ΔR_{SHE} in $\Delta V_C/J_e$ [44];

$$\sigma_{SHE} = \frac{w\sigma^2}{x}\left(\frac{J_C}{J_S}\right)\Delta R_{SHE}, \tag{9.17}$$

where x is a correction factor taking into account the fact that the horizontal current in the Pt wire is partially shunted by the Cu wire above the Pt/Cu interface [45]. J_C/J_S is the ratio of the injected charge current to the spin current contributing to the SHE, of which magnitude can be estimated by Equation 9.18 with the spin resistance of Pt R_S^{Pt} determined by the spin absorption experiment discussed in Section 9.3.3,

$$\frac{J_C}{J_S} \approx \frac{t_{Pt}}{\lambda_{sf}^{Pt}} \frac{\left\{1-\exp\left(-2t_{Pt}/\lambda_{sf}^{Pt}\right)\right\}}{\left\{1-\exp\left(-t_{Pt}/\lambda_{sf}^{Pt}\right)\right\}^2} \frac{R_S^{Cu}\{\cosh\{d/\lambda_{sf}^{Cu}\}-1\}+2R_S^{Py}\{\exp(d/\lambda_{sf}^{Cu})-1\}+2R_S^{Pt}\sinh(d/\lambda_{sf}^{Cu})}{2P_{Py}R_S^{Py}\sinh\left(d/2\lambda_{sf}^{Cu}\right)},$$

(9.18)

where t_{Pt} is the thickness of Pt and λ_{sf}^{Pt} is the spin-diffusion length of Pt. Equations 9.17 and 9.18 yield the value of σ_{SHE} of 1.7×10^5 $(\Omega m)^{-1}$ and the spin Hall angle, that is the ratio of the SH conductivity to the electrical conductivity of Pt $\alpha_{SHE} = \sigma_{SHE}/\sigma$ of 2.1×10^{-2} at 10 K [44].

REFERENCES

1. M. Johnson and R. H. Silsbee, Interfacial charge-spin coupling: Injection and detection of spin magnetization in metals, *Phys. Rev. Lett.* **55**, 1790 (1985).
2. M. Johnson and R. H. Silsbee, Coupling of electronic charge and spin at a ferromagnetic-paramagnetic metal interface, *Phys. Rev. B* **37**, 5312 (1988).
3. F. J. Jedema, A. T. Filip, and B. J. van Wees, Electrical spin injection and accumulation at room temperature in an all-metal mesoscopic spin valve, *Nature (London)* **410**, 345 (2001).
4. F. J. Jedema, H. B. Heershe, A. T. Filip, J. J. A. Baselmans, and B. J. van Wees, Electrical detection of spin precession in a metallic mesoscopic spin valve, *Nature (London)* **416**, 713 (2002).
5. A. Brataas, Y. Tserkovnyak, G. Bauer, and B. Halperin, Spin battery operated by ferromagnetic resonance, *Phys. Rev. B* **66**, 060404(R) (2002).
6. N. F. Mott, The electrical conductivity of transition metals, *Proc. Roy. Soc.* **153**, 699 (1936).
7. P. C. van Son, H. van Kempen, and P. Wyder, Boundary resistance of the ferromagnetic-nonferromagnetic metal interface, *Phys. Rev. Lett.* **58**, 2271 (1987).
8. T. Valet and A. Fert, Theory of the perpendicular magnetoresistance in magnetic multilayers, *Phys. Rev. B* **48**, 7099 (1993).
9. R. J. Elliott, Theory of the effect of spin-orbit coupling on magnetic resonance in some semiconductors, *Phys. Rev.* **96**, 266 (1954).
10. T. Kimura, J. Hamrle, and Y. Otani, Estimation of spin-diffusion length from the magnitude of spin-current absorption: Multiterminal ferromagnetic/nonferromagnetic hybrid structures, *Phys. Rev. B* **72**, 014461 (2005).
11. S. Ramo, J. R. Winnery, and T. V. Duzer, *Fields and Waves in Communication Electronics*, 3rd ed., Wiley, New York, p. 247, 1994.
12. Y. Ji, A. Hoffmann, J. S. Jiang, and S. D. Bader, Spin injection, diffusion, and detection in lateral spin-valves, *Appl. Phys. Lett.* **85**, 6218 (2004).
13. T. Kimura, J. Hamrle, Y. Otani, K. Tsukagoshi, and Y. Aoyagi, Spin-dependent boundary resistance in the lateral spin-valve structure, *Appl. Phys. Lett.* **85**, 3501 (2004).
14. F. Casanova, A. Sharoni, M. Erekhinsky, and I. K. Schuller, Control of spin injection by direct current in lateral spin valves, *Phys. Rev. B* **79**, 184415 (2009).
15. S. O. Valenzuela and M. Tinkham, Spin-polarized tunneling in room-temperature mesoscopic spin valves, *Appl. Phys. Lett.* **85**, 5914 (2004).
16. S. Garzon, I. Žutić, and R. A. Webb, Temperature-dependent asymmetry of the nonlocal spin-injection resistance: Evidence for spin nonconserving interface scattering, *Phys. Rev. Lett.* **94**, 176601 (2005).

17. N. Poli, M. Urech, V. Korenivski, and D. B. Haviland, Spin-flip scattering at Al surfaces, *J. Appl. Phys.* **99**, 08H701 (2006).

18. J. Bass and W. P. Pratt Jr., Spin-diffusion lengths in metals and alloys, and spin-flipping at metal/metal interfaces: An experimentalist's critical review, *J. Phys. Condens. Matter* **19**, 183201 (2007).

19. T. Kimura, Y. Otani, T. Sato, S. Takahashi, and S. Maekawa, Room temperature reversible spin Hall effect, *Phys. Rev. Lett.* **98**, 156601 (2007); Erratum: *Phys. Rev. Lett.* **98**, 249901 (2007).

20. J. Hamrle, T. Kimura, Y. Otani, K. Tsukagoshi, and Y. Aoyagi, Three-dimensional distribution of spin-polarized current inside (Cu/Co) pillar structures, *Phys. Rev. B* **71**, 094402 (2005).

21. T. Kimura and Y. Otani, Spin transport in lateral ferromagnetic/nonmagnetic hybrid structures, *J. Phys. Condens. Matter* **19**, 16521 (2007).

22. Y. Fukuma, H. Wang, H. Idzuchi, and Y. Otani, Enhanced spin accumulation obtained by inserting low-resistance MgO interface in metallic lateral spin valves, *Appl. Phys. Lett.* **97**, 012507 (2010).

23. Y. Fukuma, L. Wang, H. Idzuchi, S. Takahashi, S. Maekawa, and Y. Otani, Giant enhancement of spin accumulation and long-distance spin precession in metallic lateral spin valves, *Nature Materials* **10**, 527 (2011).

24. N. Tombros, C. Jozsa, M. Popinciuc, H. T. Jonkman, and B. J. van Wees, Electronic spin transport and spin precession in single graphene layers at room temperature, *Nature (London)*, **448**, 571 (2007).

25. X. Lou, C. Adelmann, S. A. Crooker et al., Electrical detection of spin transport in lateral ferromagnet–semiconductor devices, *Nat. Phys.* **3**, 197 (2007).

26. H. Idzuchi, A. Fert, and Y. Otani, Revisiting the measurement of the spin relaxation time in graphene-based devices, *Phys. Rev. B* **91**, 241407 (2015).

27. S. Takahashi and S. Maekawa, Spin injection and detection in magnetic nanostructures, *Phys. Rev. B* **67**, 052409 (2003).

28. A. Brataas, Y. V. Nazarov, and G. E. Bauer, Spin-transport in multi-terminal normal metal-ferromagnet systems with non-collinear magnetizations, *Eur. Phys. J. B* **22**, 99 (2001).

29. H. Idzuchi, Y. Fukuma, and Y. Otani, Towards coherent spin precession in pure-spin current, *Sci. Rep.* **2**, 628 (2012).

30. H. Idzuchi, Y. Fukuma, S. Takahashi, S. Maekawa, and Y. Otani, Effect of anisotropic spin absorption on the Hanle effect in lateral spin valves, *Phys. Rev. B* **89**, 081308 (2014).

31. H. Idzuchi, Y. Fukuma, and Y. Otani, Spin transport in non-magnetic nanostructures induced by non-local spin injection, *Physica E* **68**, 239 (2015).

32. W. Han, R. K. Kawakami, M. Gmitra, and J. Fabian, Graphene spintronics, *Nat. Nanotechnol.* **9**, 794 (2014).

33. T. Maassen, I. J. Vera-Marun, M. H. D. Guimarães, and B. J. van Wees, Contact-induced spin relaxation in Hanle spin precession measurements, *Phys. Rev. B* **86**, 235408 (2012).

34. T. Kimura, Y. Otani, and J. Hamrle, Enhancement of spin accumulation in a nonmagnetic layer by reducing junction size, *Phys. Rev. B* **73**, 132405 (2006).

35. T. Kimura, Y. Otani, and J. Hamrle, Switching magnetization of a nanoscale ferromagnetic particle using nonlocal spin injection, *Phys. Rev. Lett.* **96**, 037201 (2006).

36. T. Yang, T. Kimura, and Y. Otani, Giant spin-accumulation signal and pure spin-current-induced reversible magnetization switching, *Nat. Phys.* **4**, 851 (2008).

37. I. N. Krivorotov, N. C. Emley, J. C. Sankey et al., Time-domain measurements of nanomagnet dynamics driven by spin-transfer torques, *Science* **307**, 228 (2005).

38. H. Kurt, R. Loloee, W. P. Pratt, Jr., and J. Bass, Current-induced magnetization switching in permalloy-based nanopillars with Cu, Ag, and Au, *J. Appl. Phys.* **97**, 10C706 (2005).

39. E. Saitoh, M. Ueda, H. Miyajima, and G. Tatara, Conversion of spin current into charge current at room temperature: Inverse spin-Hall effect, *Appl. Phys. Lett.* **88**, 182509 (2006).

40. S. O. Valenzuela and M. Tinkham, Direct electronic measurement of the spin Hall effect, *Nature* **442**, 176 (2006).

41. T. Seki, Y. Hasegawa, S. Mitani et al., Giant spin Hall effect in perpendicularly spin-polarized FePt/Au devices, *Nat. Mater.* **7**, 125 (2008).

42. Y. Kato, R. C. Myers, A. C. Gossard, and D. D. Awschalom, Observation of the spin Hall effect in semiconductors, *Science* **306**, 1910 (2004).

43. J. Wunderlich, B. Kaestner, J. Sinova, and T. Jungwirth, Experimental observation of the spin-Hall effect in a two-dimensional spin-orbit coupled semiconductor system, *Phys. Rev. Lett.* **94**, 047204 (2005).

44. M. Morota, Y. Niimi, K. Ohnishi et al., Indication of intrinsic spin Hall effect in 4d and 5d transition metals, *Phys. Rev. B* **83**, 174405 (2011).

45. Y. Niimi and Y. Otani, Reciprocal spin Hall effects in conductors with strong spin-orbit coupling: A review, *Rep. Prog. Phys.* **78**, 124501 (2015).

10

Magnetic Skyrmions
on Discrete Lattices

Elena Y. Vedmedenko and Roland Wiesendanger

Magnetic skyrmions have recently been in the focus of intense worldwide research activities, because they exhibit intriguing properties such as small lateral size and expected high stability due to their topological protection and the Dzyaloshinskii–Moriya (DM) interaction energy involved. A broad range of theoretical investigations on magnetic skyrmionic systems have been conducted in the framework of field theory, micromagnetics, or other theoretical approaches based on continuum media. There is also a considerable body of publications on magnetic skyrmions using discrete atomistic simulations or analytical calculations for atomistic models. While both theoretical approaches agree on the basic properties of the skyrmionic systems, there are certain subtleties that make one or another model more preferable for a certain problem. Particularly, the discrete treatment is very important for the interfacial skyrmionic systems in ultrathin magnetic films and for the confined geometries at the nanoscale.

In this chapter, we discuss these two approaches with respect to known skyrmionic systems. The emphasis is put on ultrathin films with DM interactions induced by an interface between magnetic and non-magnetic metals. Particular attention is paid to the differences in the description of topological properties between continuum and atomistic systems. We make a classification of the interfacial skyrmionic systems based on the involved magnetic interactions as well as the structure of the underlying atomistic lattice. Using this classification, we review the multi-dimensional phase diagrams of the magnetic chiral systems, discuss the conditions for the appearance of skyrmions in each system class, and review their physical properties. Particular attention is paid to the review of investigations devoted to the stability of skyrmions and to skyrmion manipulation, because these two aspects are decisive for skyrmion-based memory and logic device applications.

10.1 FIELD-THEORETICAL APPROACH

The Skyrme model was formulated in the framework of non-linear field theory and describes the SU(2) valued fields embedded into three-dimensional spatial dimensions [1]. Later, a two-dimensional version of the Skyrme model was considered [2]. In this field theoretical model, the two-dimensional flat space was mapped onto a three-dimensional field manifold. The two-dimensional Skyrme model is known as the "baby Skyrme model." The framework of the baby Skyrme model leads to the following energy density of the field:

$$\varepsilon = \frac{1}{2}\left(\partial_1\vec{\phi}(\vec{r})^2 + \partial_2\vec{\phi}(\vec{r})^2\right) + \frac{1}{2}\left(\partial_1\vec{\phi}(\vec{r}) \cdot \partial_2\vec{\phi}(\vec{r})\right)^2 + \mu^2(1 - n \cdot \vec{\phi}(\vec{r})), \quad (10.1)$$

where ∂_i, $i = 1,2$ denotes the partial derivatives with respect to two Cartesian coordinates, the third term is typically denoted as potential with $n = (0,0,1)$, and μ is a constant with the dimension of inverse length and can be interpreted as the inverse Compton length. This potential term is crucial, because it reduces the symmetry of the model. The energy functional can be obtained by integration of the energy density given in Equation 10.1:

$$E[\phi] = \int d^2r \, \varepsilon. \tag{10.2}$$

One is then interested in field configurations of finite energy; that is, in stationary points. One of the most important features of the Skyrme models is that the expansion of the energy density given in Equation 10.2 possesses terms linear in the first spatial derivatives of the order parameter $\vec{\phi}(\vec{r})$, such as $\left(\dfrac{\partial}{\partial \vec{r}}\right) \times \vec{\phi}(\vec{r})$. This is important, because these linear invariants are responsible for the existence of stable solutions belonging to the class of solitons. Particularly, it is known that a configuration $\vec{\phi}(\vec{r})$ corresponds to a stationary point of $E[\phi]$, if the first variation of $E[\phi]$ under transformation $\vec{\phi}(\vec{r}) \mapsto \vec{\phi}(\vec{r}) + \xi(\vec{r}) \times \vec{\phi}(\vec{r})$ vanishes for any function $\xi(\vec{r})$ [2, 3]. If such a stable configuration $\vec{\phi}(\vec{r})$ exists for $n = 1$ (see Equation 10.1), it is called a π-*soliton* or *baby skyrmion*; for $n > 1$, the solutions are called *multi-dimensional solitons* or *multi-skyrmions*.

The smooth mapping between the field and spatial variables described earlier corresponds to the mapping between classical vector fields and two-dimensional Euclidean space. Already in the early publications on baby Skyrme models, an alternative physical interpretation of the field $\vec{\phi}(\vec{r})$ was suggested [2, 3]. Particularly, the baby Skyrme model has been proposed to describe complex magnetization textures in two dimensions. The potential term then describes the coupling of the magnetization to an external magnetic field. However, the energies considered by the Skyrme models possess higher-order derivative terms, which might not be appropriate for magnetic systems, and therefore, they were not immediately considered in solid state physics.

10.2 MICROMAGNETICS

Micromagnetism describes systems with continuous magnetic degrees of freedom [4], which can be regarded as a vector field. In contrast to nonlinear field theory addressing general fields taking values on a manifold, the fields in micromagnetism are based on micromagnetic energy terms. The most important of these terms is related to the exchange interaction between electrons. The most general expression for a two-site electronic system reads

$$E_{ex} = \vec{S}_i \, \vec{\vec{A}}_{ij} \vec{S}_j, \tag{10.3}$$

where $\ddot{A}_{ij} = \{A_{ij}^{\alpha\beta}\}$ with $\alpha, \beta = x, y, z$ is a generalized exchange tensor, which can be decomposed into three contributions:

$$\ddot{A}_{ij} = J_{ij}\ddot{I} + \ddot{A}_{ij}^s + \ddot{A}_{ij}^a, \tag{10.4}$$

corresponding to an isotropic part of the exchange tensor

$$J_{ij} = \frac{1}{3}Tr(\ddot{A}_{ij}), \tag{10.5}$$

the traceless symmetric anisotropic exchange

$$\ddot{A}_{ij}^s = \frac{1}{2}\left(\ddot{A}_{ij} + \ddot{A}_{ij}^+\right), \tag{10.6}$$

and the antisymmetric exchange tensor

$$\ddot{A}_{ij}^a = \frac{1}{2}\left(\ddot{A}_{ij} - \ddot{A}_{ij}^+\right).$$

The isotropic part of the exchange tensor, also known as the *exchange integral*, can reach values of several hundreds of millielectron volts. In the continuous representation of micromagnetics, the neighboring spins are only allowed to vary by a small angle, and the corresponding exchange energy is given by

$$E_i^{exch} = \int_V \frac{A}{M_s^2}\left(\frac{\partial M_i}{\partial x_j}\right)^2 dV, \tag{10.7}$$

where A is the exchange stiffness constant and M_s is the saturation magnetization, as described in many excellent reviews [5]. The symmetric contribution for slowly varying spin density, which is responsible for the anisotropy in the exchange interactions, is a correction to the isotropic term of Equation 10.7:

$$E_s^{exch} = \int_V \frac{A}{M_s^2}\left(|\nabla M_x|^2 + |\nabla M_y|^2 + |\nabla M_z|^2\right) dV, \tag{10.8}$$

while the antisymmetric part takes the form

$$E_{as}^{exch} = D\int_V \vec{M} \cdot \left(\vec{\nabla} \times \vec{M}\right) dV. \tag{10.9}$$

The antisymmetric exchange interaction becomes important in magnetic systems without inversion symmetry, such as the antiferromagnetic insulating compounds CrF_3, $\alpha\text{-}Fe_2O_3$, or the chiral bulk magnet MnSi [6–9]. The energy term of Equation 10.9 possesses a single gradient in the magnetization field similarly to the energy density of the field in the Skyrme model described in Section 10.1. This term, which is linear in the first spatial derivatives in the expansion of the non-equilibrium thermodynamic potential, was first introduced by Dzyaloshinskii for weak ferromagnets [7], and nowadays it is known as the *Dzyaloshinskii–Moriya energy density term*, with D being the DM parameter. It has already been realized quite some time ago that the antisymmetric DM interaction is particularly important at interfaces, where inversion symmetry is always broken [10, 11]. In this case, gradients along the surface normal can be neglected, and the DM energy reads

$$E_{DM} = t \int_S \frac{D}{M_s} \left[\left(M_x \frac{\partial M_x}{\partial x} - M_z \frac{\partial M_x}{\partial x} \right) \right.$$

$$\left. + \left(M_y \frac{\partial M_z}{\partial y} - M_z \frac{\partial M_y}{\partial y} \right) \right] d^2\vec{r},$$

(10.10)

where:

    D   has the units of joules per meter^2

    t   is the film thickness [12]

The summands of Equation 10.10, which are linear with respect to the first spatial derivatives of the magnetization, are also known as *Lifshitz invariants* [13] in magnetism. Depending on the crystallographic symmetry of the system, different combinations of Lifshitz invariants are present in the DM energy functional.

The introduction of a single gradient term, such as the one given in Equation 10.10, into the energy density potential favors the formation of stable magnetic domain walls with a unique rotational sense, which can be described as one-dimensional soliton solutions [14]. Specific combinations of Lifshitz invariants lead to specific symmetries of the soliton solutions. Particularly, based on Equation 10.10, two-dimensional metastable magnetic structures identical to baby skyrmions were predicted even before the first papers on baby skyrmions [14–16] (see Figure 10.1). Similarly to the

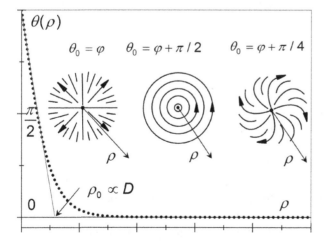

FIGURE 10.1 Typical magnetization profile of a baby skyrmion as a function of the spatial coordinate ρ. This profile corresponds to a one-dimensional soliton with the boundary conditions θ (0) = π and θ (∞) = 0. The schematics of three possible classes of skyrmionic solutions with different phases are given in the insets.

baby Skyrme model, these non-trivial magnetic structures were denoted as π-*vortices*. It has been also recognized that these quasiparticle-like, countable states may appear *spontaneously* in condensed matter systems with chiral interactions [17], and the magnetic π-vortices predicted theoretically in the late 1980s got the name *skyrmions*. Detailed field-theoretical analysis has demonstrated that magnetic skyrmions can appear in the form of isolated entities as well as skyrmion lattices (SkX). The main reason for the formation of these particle-like topological solitons is the interplay between the exchange and the DM interaction. This is the main difference between these particle-like topological solitons or chiral skyrmions and various multiple-Q states. The chiral magnetic skyrmions have been found experimentally in a variety of systems [18–24].

There are several secondary energy contributions, which may be important for the formation of skyrmions. One of these is the stray field energy; that is, the magnetic field energy produced by the body itself. The stray field $\vec{H}_{st}$ is generated by the divergence of the magnetization:

$$\nabla \times \vec{H}_{st} = -\nabla \times \vec{M}, \tag{10.11}$$

and has the energy

$$E_{\text{dip}} = -\frac{\mu_0}{2} \int \vec{H}_{st} \cdot \vec{M} \ dV. \tag{10.12}$$

The Zeeman energy describes the interaction of the magnetization with external magnetic fields:

$$E_H = -\int \vec{H} \cdot \vec{M} \ dV. \tag{10.13}$$

The magnetic anisotropy energy may have different origins and symmetries in various systems. However, for the skyrmionic systems, the most important term can be approximated by the so-called *uniaxial anisotropy*:

$$E_a = -\int \left(K_1 \cdot M_z^2 + K_2 \cdot M_z^4 \right) \ dV. \tag{10.14}$$

The continuum theory of micromagnetism is very successful for the description of low-temperature chiral systems, where temperature fluctuations do not play a significant role. For skyrmionic systems on discrete lattices, however, fluctuations are particularly important, because their strength depends on the interplay between the lattice symmetry and the symmetry of interactions. The effect of the lattice becomes pronounced at temperatures that are close to the Curie temperature. To take the temperature effects into account, atomistic simulations such as Langevin dynamics or Monte-Carlo simulations are required. Most of the temperature-dependent phase diagrams from the discrete systems stem from Monte-Carlo simulations. The energy terms for the atomistic simulations exhibit certain differences with respect to Equations 10.8 through 10.14 and will be discussed in Section 10.4.

10.3 TOPOLOGICAL PROPERTIES OF CONTINUOUS SYSTEMS

The phenomenon of topological stability is simple and fundamental. Any object with topological order is robust against local perturbations. The notion of topological stability can be applied to a variety of physical systems that can be described by continuous vector fields. Magnetic materials have always played a significant role among these systems. Particularly, magnetic solitons and skyrmions, introduced earlier, belong to the class of topologically non-trivial objects.

The non-uniform magnetization distribution of magnetic solitons and skyrmions is typically defined by the spherical coordinates (θ, ψ) on the unit sphere, while for the spatial variables, cylindrical coordinates (ρ, φ, z) are introduced:

$$\langle \vec{M}(\rho,\varphi) \rangle = M\big(\sin\theta(\rho,\varphi)\cos\psi(\rho,\varphi), \sin\theta(\rho,\varphi)\sin\psi(\rho,\varphi), \cos\theta(\rho,\varphi)\big).$$

The z coordinate for the two-dimensional case is just a constant. The variational problem for the total energy-containing contributions described in the previous chapter admits, then, the solution $\{\psi(\rho,\varphi) = \varphi, \ \theta(\rho,\varphi) = \theta\}$. This solution corresponds to the extremum of the energy functional. The function $\theta(0,\varphi) = \pi$ and $\theta(\infty,\varphi) = 0$ defines an isolated soliton (skyrmion). Other classes of solutions, for example $\theta(0,\varphi) = \pi$, $\theta(\infty,\varphi) = 0$, and $\psi(\rho,\varphi) = \varphi \pm \pi/2$, shown in Figure 10.1, can also be realized, but it is more difficult to treat them. Recently, it has been analytically shown that skyrmionic solutions with arbitrary helicity can be realized in a two-dimensional magnet by assuming an appropriate form of DM interaction [25].

Mathematically, any continuous flow on a sphere must always have at least one point of discontinuity (zero magnitude of the vector field). On the torus, this is not necessary. Since a skyrmion can be described by a mapping to a sphere, the angle variable $\theta(\rho,\varphi)$ in all cases is continuous everywhere except in the subspace of dimensionality d_s less than d. For $d=2$, $\theta(\rho,\varphi) = \pi + \theta_0$ and $\psi(\rho,\varphi) = \varphi \pm \varphi_0$, the magnetization $\langle \vec{M}(\rho,\varphi) \rangle$ is continuous, and $\nabla\theta = 1/\rho$ is finite everywhere except at the origin (core of the skyrmion in Figure 10.1). This mathematical singularity can be removed by cutting a hole out of the material.

The magnetization defines a topological space. Once a topological space has been defined, one can define a path through it as a mapping $f(\rho,\varphi)$ of the real-line segment $[(\rho_1,\varphi_1),(\rho_2,\varphi_2)]$ $((\rho_1,\varphi_1) < (\rho,\varphi) < (\rho_2,\varphi_2))$ to the topological space. If $(\rho_1,\varphi_1) = (\rho_2,\varphi_2)$, the path becomes closed. If the space has holes in it, these loops can be divided into classes, each one characterized by the number of times the loop winds round the hole. This quantity is known as *winding number*. The winding number in all cases of Figure 10.1 is unity, because $\theta(\rho,\varphi)$ and $\psi(\rho,\varphi)$ change only once by 2π in one circuit along any contour enclosing the core [26]. Such configurations are important, because they cannot disappear by any continuous deformation of the order parameter (in our case, the magnetization) because of the singularity at the origin.

In contrast, if such a loop does not enclose a core, it can disappear and is topologically trivial with a winding number equal to zero.

Since the winding number is an integer value, it also has been given the name of a *topological charge, Q*. For two-dimensional magnetization fields, it can easily be calculated using the following expression [27]:

$$Q = \frac{1}{4\pi M_s} \int_A \vec{M} \cdot \left(\frac{\partial \vec{M}}{\partial x} \times \frac{\partial \vec{M}}{\partial y} \right) \mathrm{d}x \mathrm{d}y. \tag{10.15}$$

For an isolated skyrmion, the local magnetization directions cover the surface of the unit sphere contributing 4π to the integral. If there is a large region of parallel magnetization, that is, there is no change in the magnetization, this part will not contribute to the integral. Consequently, the integral of Equation 10.15 is equal to the surface area of a unit sphere. The Q number divided by 4π is an integer, and each skyrmion contributes ± 1 to the total Q in the case of continuous fields. Hence, the topological number is quantized.

Within field theory, states with different topological charges cannot continuously be deformed into one another. In other words, once created, the skyrmions are indestructible because of the quantized singularity at the origin. Physically, if an external magnetic field is applied parallel to the magnetization in the center of a skyrmion, its radius will become divergently small, but the skyrmion will not disappear. This property is typically denoted as *topological protection*. The field dependence of the radius of continuous magnetic skyrmions can be found in Bogdanov and Hubert [12, 16].

10.4 TOPOLOGICAL PROPERTIES OF DISCRETE SYSTEMS

The spin texture in magnetic materials is restricted to atomistic crystalline lattices. For an atomistic description of chiral solitons or skyrmions, the continuous function defined in Equation 10.15 has to be partitioned into values on a per-atom basis. This partition is not evident *a priori*, because it might depend on the lattice symmetry. It is also not clear *a priori* what happens with the phenomenon of topological stability, because in atomistic magnetism, the discrete spin variables at the lattice sites are coupled by pair/many particle potentials instead of continuum energy functionals. In contrast to the functionals, the atomistic potentials allow abrupt changes in the magnetization. Hence, some local excitation, such as local fields or a short-time increase in temperature due to, for example, a laser pulse, might overcome the interparticle potentials and cause a magnetization reversal in a small spatial region. This short excitation might then destroy a particle-like skyrmionic state, if this region coincides with the skyrmion center. A detailed understanding of chiral magnetism at the atomistic level is very important for fundamental research as well as for realistic applications and, therefore, lies at the center of broad scientific activity. In the following, several important aspects of the atomistic description will be addressed and compared with the continuum approach.

The topological charge of a lattice spin texture differs from that of a continuum. Quite generally, it is the sum of solid angles of elementary atomistic triangles $\Delta\left(\vec{S}_1, \vec{S}_2, \vec{S}_3\right)$:

$$Q = \frac{1}{4\pi} \sum_{\Delta} \Omega_{\Delta}, \tag{10.16}$$

where $-2\pi < \Omega_{\Delta} < 2\pi$. Evidently, the triangularization is lattice dependent. Therefore, the definition of Ω_{Δ} differs for different symmetries. First, it has been derived [28] for a square lattice as

$$e^{i\Omega_{\Delta}/2} = \frac{1 + \vec{S}_1 \cdot \vec{S}_2 + \vec{S}_2 \cdot \vec{S}_3 + \vec{S}_1 \cdot \vec{S}_3 + i \cdot \vec{S}_1 \cdot \left(\vec{S}_2 \times \vec{S}_3\right)}{\left[2\left(1 + \vec{S}_1 \cdot \vec{S}_2\right)\left(1 + \vec{S}_2 \cdot \vec{S}_3\right)\left(1 + \vec{S}_1 \cdot \vec{S}_3\right)\right]^{1/2}}. \tag{10.17}$$

The corresponding expression for Ω_{Δ} on a hexagonal lattice can be found in Van Oosterom and Strackee [29]:

$$\tan\left(\Omega_{\Delta}/2\right) = \frac{\vec{S}_1 \cdot \left(\vec{S}_2 \times \vec{S}_3\right)}{S_1 \cdot S_2 \cdot S_3 + \left(\vec{S}_1 \cdot \vec{S}_2\right)S_3 + \left(\vec{S}_2 \cdot \vec{S}_3\right)S_1 + \left(\vec{S}_1 \cdot \vec{S}_3\right)S_2}. \tag{10.18}$$

In contrast to the topological number, which can be regarded as an order parameter for magnetic configurations in chiral systems, the magnetic pair interactions governing the formation of these configurations do not depend on the lattice symmetry. Discrete analogs of micromagnetic potentials of Equations 10.11 through 10.14 are introduced in the following.

The exchange interaction in atomistic systems is generally described by Equation 10.3. For the isotropic and symmetric parts of the tensor A in Equation 10.3, corresponding to the continuum Equations 10.7 and 10.8, however, one often uses the atomistic form

$$E_{ex} = -\sum_{i<j} J_{ij}(\vec{r}) \, \vec{S}_i \vec{S}_j, \tag{10.19}$$

where $\vec{S}_i$ could be vectorial magnetic moments $\vec{S}_i = \vec{\mu}_i/\mu_s$ ($\mu_s = |\vec{\mu}_i|$) or spin operators at sites i and j. This term is known as the *Heisenberg exchange Hamiltonian* and has many possible applications and variations. If the scalar factor $J_{ij}(\vec{r})$ is direction independent and acts between nearest neighbors only, then it describes the direct isotropic exchange interactions. If several or even all neighbors are taken into account, then Equation 10.19 may describe the indirect or super-exchange mediated by other sorts of atoms/orbitals. If the parameter $J_{ij}(\vec{r})$ is of long-range nature, and its amplitude and sign are distance dependent, that is, $J(\vec{r}) = f(k_F)/(|\vec{r}|^n)$, then Equation 10.19 is applicable to systems with Ruderman–Kittel–Kasuya–Yoshida (RKKY) interactions mediated by itinerant electrons. Eventually, if the amplitude and sign of $J_{ij}(\vec{r})$ depend on the direction, the very same expression can be used for the description of the anisotropic exchange. The strength and form of $J_{ij}(\vec{r})$ can be quantified by experiments [30, 31], different kinds of density-functional methods [32–34], or tight-binding calculations [35] or derived from symmetry considerations. Often, phase diagrams using numerical methods can be constructed, which can then be used for applications to different classes of particular material systems.

For chiral systems, antisymmetric contributions to the exchange interaction given in Equations 10.9 and 10.10 are described by the atomistic DM term

$$E_{DM} = -\sum_{i<j} \vec{D}_{ij}(\vec{r}) \cdot \left(\vec{S}_i \times \vec{S}_j \right). \tag{10.20}$$

The strength and direction of the DM vector ($\vec{D}_{ij}(\vec{r})$) for each atomic pair can rigorously be derived in the framework of fully relativistic Korringa–Kohn–Rostocker (KKR) methods [36]. However, there are also many other effective approximations on the basis of other DFT frameworks [37–39], the three-site model by Lévy and Fert [40, 41], symmetry considerations [42], or experimental data [43]. Typically, $\vec{D}_{ij}(\vec{r})$ is perpendicular to the connecting bond and has fixed rotational sense around the atomic site i, as shown in Figure 10.2 (see also Ref. [44]).

There are several other terms, which are important to describe skyrmionic systems with significant contributions of itinerant electrons or with long-range frustrated exchange interactions [45]. In materials with itinerant character of the charge carriers many-body terms gain importance in addition to the standard pairwise interactions. To account for the itinerant nature of *3d* magnets, an extension of the Hamiltonian given in Equations 10.3 and 10.19 by terms going beyond the nearest-neighbor Heisenberg coupling is required. These higher-order terms can be derived by means of the fourth-order perturbation expansion of the Hubbard model [46]. Apart from the effective, longer-range Heisenberg-like terms readily described by Equation 10.19, the most important higher-order contributions are the biquadratic and the four-spin exchange interaction resulting from the hopping of electrons over four neighboring sites [47, 48].

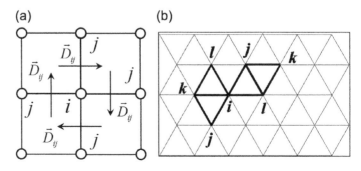

FIGURE 10.2 (a) Typical orientations of atomistic DM vectors $\vec{D}_{ij}$ on a square lattice. (b) Schematics of minimal parallelograms of the four-spin interaction on a hexagonal lattice discussed in the text.

The biquadratic term is given by the expression

$$E_{bi} = \sum_{i<j} J_{ij}^{bi}(\vec{r}) \cdot \left(\vec{S}_i \cdot \vec{S}_j\right)^2, \tag{10.21}$$

while the four-particle interaction reads

$$E_{4\text{-spin}} = J_{ij}^{4\text{-spin}} \sum_{\langle i,j,k,l \rangle} \left[\left(\vec{S}_i \cdot \vec{S}_j\right)\left(\vec{S}_k \cdot \vec{S}_l\right) + \left(\vec{S}_i \cdot \vec{S}_k\right)\left(\vec{S}_j \cdot \vec{S}_l\right) \right.$$
$$\left. - \left(\vec{S}_i \cdot \vec{S}_l\right)\left(\vec{S}_j \cdot \vec{S}_k\right) \right], \tag{10.22}$$

with $J_{ij}^{bi}(\vec{r})$ and $J_{ij}^{4\text{-spin}}$ being the biquadratic and four-spin exchange parameters, respectively. The four involved sites i, j, k, and l in Equation 10.22 form a minimal parallelogram, where each side is a line connecting two nearest neighbors. In the case of a square lattice, this parallelogram is just a square. For a triangular lattice, it is a rhombus, as shown in Figure 10.2a. For the latter case, 12 minimal parallelograms exist for each central magnetic moment i. Two examples of minimal parallelograms are shown in Figure 10.2b.

The secondary terms described in Equations 10.12 through 10.14 also have their discrete analogs. They are: the dipolar interaction leading to the shape anisotropy and many other interesting lattice effects [49]

$$E_{dip} = \frac{\mu_0 \mu_s^2}{4\pi a^3} \sum_{i<j} \left(\frac{\mathbf{S}_i \cdot \mathbf{S}_j}{r_{ij}^3} - 3 \frac{\left(\mathbf{S}_i \cdot \mathbf{r}_{ij}\right)\left(\mathbf{S}_j \cdot \mathbf{r}_{ij}\right)}{r_{ij}^5} \right), \tag{10.23}$$

where the interaction strength is defined via saturated magnetic moments of individual atoms μ_s and a is the lattice constant; the uniaxial second-order anisotropy [50]

$$E_{an} = -\sum_i \left[D_{z1}\left(S_i^z\right)^2 + D_{z2}\left(S_i^z\right)^4 \right], \tag{10.24}$$

and the Zeeman energy

$$E_B = -\mu_s \sum_i \left[\vec{B} \cdot \vec{S}_i \right]. \tag{10.25}$$

10.5 CLASSIFICATION OF MAGNETIC SKYRMIONS

Phase diagrams showing conditions under which multiple phases co/exist in equilibrium are indispensable for a systematic description of any physical system. For skyrmionic materials, they are particularly important, because this exotic phase of matter appears in a rather narrow range of thermodynamic parameters. This can be seen in a universal phase diagram for bulk chiral material, which can be found, for example, in Figure 1 of Bauer and

Pfleiderer [51]. The SkX in this diagram occupies a tiny pocket between much larger spin spiral (SS) and ferromagnetic (FM) phases; that is, for very specific temperature (T) and magnetic field (B) values.

While all phase diagrams of skyrmionic matter distinguish between SS, SkX, and FM states, there is a large diversity in the coordinates of those phases. There are three main reasons for this strong diversity. (i) There are several different mechanisms for the formation of skyrmionic phases involving different sets of interactions. (ii) Single phase regions are separated by lines of non-analytical behavior where phase transitions occur. These so-called *phase boundaries* in one and the same skyrmionic system are multi-dimensional. They might be driven by temperature, strength of DM interaction, strength of applied magnetic field, pressure, and so on. (iii) There are also differences in the thermodynamic behavior for different lattice symmetries, thereby making the diversity of phases even larger.

10.5.1 CLASSIFICATION USING ENERGY FUNCTIONAL

We propose the following classification of the skyrmionic systems on the basis of the corresponding energy functionals.

10.5.1.1 Skyrmion Class I (SC-I)

$$H = -\sum_{i<j} J_{ij}(\vec{r})\, \vec{S}_i \vec{S}_j - \sum_{i<j} \vec{D}_{ij}(\vec{r}) \cdot \left(\vec{S}_i \times \vec{S}_j \right) - \mu_s \sum_i \left[\vec{B} \cdot \vec{S}_i \right] + O. \quad (10.26)$$

Skyrmions appear in systems described by the Hamiltonian given in Equation 10.26, where is the distance $J_{ij}(\vec{r})$ dependent exchange integral, $D_{ij}(\vec{r})$ distance dependent DM coupling and $\vec{B}$ an external magnetic field; $\vec{S}_i$ and $\vec{S}_j$ denote magnetic moments of magnitude μ_s at sites i and j. This Hamiltonian can lead to the formation of skyrmions in bulk crystals as well as in interfacial systems. The only difference between these two cases is the origin of the DM interaction. The classical ingredients for the formation of skyrmions formulated in the late 1980s are the ferromagnetic, nearest-neighbor exchange interaction (Equation 10.19), the DM interaction (Equation 10.20), and the Zeeman interaction (Equation 10.25). The dipolar interaction and anisotropies of different kinds (secondary O-terms in Equation 10.26) will make the phase diagram richer but are not indispensable. The energy functional given in Equation 10.26 governs the formation of skyrmions in B20 and B2 materials such as MnSi [19], FeGe [20, 52], and FeCoSi [53] or GaV$_4$S$_8$ [54], as well as in interfacial magnets such as Pd/Fe bilayers on Ir(111) [24] and Co/Pt/Ir multilayers [55]. The SC-I skyrmions can appear spontaneously [17]. Depending on the ratio $\vec{D}_{ij}/\vec{B}$, individual skyrmions (Sk) as well as SkX can be observed, as will be discussed in Section 10.5.3. The isolated skyrmions of this class are metastable, because they appear at the borderline separating the ferromagnetic phase and SkX. Both phases here have similar energies but are separated from each other by an energy barrier. Importantly, the skyrmions of Class I arise at finite values of magnetic field only. The ground state at zero field corresponds to the spin spiral even in the case of strong spin–orbit coupling [56, 57]. Hence, to create stable skyrmions of this sort, an externally applied magnetic field is needed. The SC-I skyrmions may be rather large. Their size differs for different systems from several hundred (e.g., of the order of 150 nm for $Co_8Mn_{10}Zn_2$ [58]) to several nanometers (e.g., of the order of 3 nm for Pd/Fe/Ir(111) [59]).

10.5.1.2 Skyrmion Class II (SC-II)

$$H = -\sum_{i<j} J_{ij}(\vec{r})\,\vec{S}_i\vec{S}_j - \sum_{i<j} \vec{D}_{ij}(\vec{r})\cdot\left(\vec{S}_i\times\vec{S}_j\right) + \sum_{i<j} J_{ij}^{bi}(\vec{r})\cdot\left(\vec{S}_i\cdot\vec{S}_j\right)^2$$

$$+ J_{ij}^{4\text{-}spin}\sum_{\langle i,j,k,l\rangle}\left[\left(\vec{S}_i\cdot\vec{S}_j\right)\left(\vec{S}_k\cdot\vec{S}_l\right) + \left(\vec{S}_i\cdot\vec{S}_k\right)\left(\vec{S}_j\cdot\vec{S}_l\right)\right. \tag{10.27}$$

$$\left. - \left(\vec{S}_i\cdot\vec{S}_l\right)\left(\vec{S}_j\cdot\vec{S}_k\right)\right] - \sum_i D_{z1}\left(S_i^z\right)^2 + O.$$

Recently, another class (SC-II) of skyrmions in interfacial magnetic systems has been found [21]. The size of these skyrmions is very small (of the order of 1 nm). Therefore, they are known as *nanoskyrmions*. It has been realized that the Hamiltonian responsible for the formation of SC-II skyrmions is different from that of Equation 10.26. The main ingredients of the SC-II Hamiltonian (Equation 10.27) are medium-range exchange interactions of alternating sign (up to eighth neighbor) (Equation 10.19), the four-spin interaction (Equation 10.22), the DM interaction (Equation 10.20), the biquadratic exchange interaction (Equation 10.21), and a perpendicular anisotropy (Equation 10.24) with corresponding coupling parameters. Magnetic skyrmions of the second class appear spontaneously and, in contrast to the SC-I, do not need any magnetic field for their formation. The SkX phase of square symmetry corresponds to the ground state. Until now, nanoskyrmions have been found in Fe monolayers on bulk Ir(111) substrates [21, 60] or on Ir deposited on multilayers with (111) symmetry; e.g., Ir/YSZ/Si(111) [61].

10.5.1.3 Skyrmion Class III (SC-III)

$$H = -\sum_{i<j} J_{ij}(\vec{r})\,\vec{S}_i\vec{S}_j - \sum_i D_{z1}\left(S_i^z\right)^2 - \mu_s\sum_i\left[\vec{B}\cdot\vec{S}_i\right] + O. \tag{10.28}$$

It is well known that multiple periodic states appear in a wide variety of physical systems due to competing exchange interactions [48]. The physical reason for such behavior is quite well known [46, 48]: the fundamental solutions of the Heisenberg exchange model on a periodic lattice are spiral spin structures. The reciprocal vector $\vec{Q}$ of a spin spiral depends on the strength of the exchange parameters and on the crystal structure. Recently, it has been shown that the addition of magnetic anisotropy and magnetic field to competing exchange interactions may lead to the formation of skyrmions [62, 63]. These systems with frustrated magnetic interactions constitute a third class of magnetic skyrmions (SC-III). In contrast to chiral skyrmions, which require the presence of the DM interaction, their counterparts originating from magnetic frustration arise in systems with ferro/antiferromagnetic exchange interactions, Zeeman energy, and perpendicular magnetic anisotropy. For a triangular lattice, the specific ratio of exchange constants between first and second nearest neighbors required for the appearance of modulated configurations is $|J_1|/|J_2| = 3$. As the DM energy fixing the sense of magnetic rotation is not present in this case, the chirality of frustrated skyrmions can be changed by an external magnetic field.

10.5.1.4 Skyrmion Class IV (SC-IV)

$$H = -\sum_{i<j} \vec{D}_{ij}(\vec{r}) \cdot (\vec{S}_i \times \vec{S}_j) - \sum_i D_{z1}(S_i^z)^2$$

$$+ \frac{\mu_0 \mu_s^2}{4\pi a^3} \sum_{i<j} \left(\frac{\vec{S}_i \cdot \vec{S}_j}{r_{ij}^3} - 3 \frac{(\vec{S}_i \cdot \vec{r}_{ij})(\vec{S}_j \cdot \vec{r}_{ij})}{r_{ij}^5} \right) - \mu_s \sum_i [\vec{B} \cdot \vec{S}_i] + O. \tag{10.29}$$

There is a fourth class of materials (SC-IV) in which skyrmion-like states are ubiquitous. These are nanostructured materials with pronounced dipolar interactions, which are comparable to or even stronger than the DM interaction. Skyrmions are close relatives of magnetic bubble domains, and the transformation of the spiral state into the skyrmion state is similar to the field-induced transition between the stripe domain state and the bubble array [64]. These magnetic configurations, initially called *skyrmions* [65], are now named *bubble skyrmions* or *skyrmionic bubble domains* (see, e.g., Bar'yakhtar et al. [66] and Dai et al. [67]). The reasons for this distinction have been discussed, for example, in Kiselev et al. [68] Ideal bubble domains, without any defects in the cylindrical domain walls surrounding any bubble, are topologically equivalent to skyrmions ($Q = \pm 1$). However, the dipolar interaction is perfectly symmetric. Therefore, bubbles of both chiralities can be found in magnetic samples with equal probability. The introduction of the DM interaction, however, immediately breaks this symmetry. However, the magnetization profile of a skyrmion and that of a skyrmionic bubble are different.

10.5.2 *B-T* DIAGRAMS

All bulk SC-I skyrmions share a universal magnetic phase diagram, discussed earlier, as thoroughly reviewed in Bauer and Pfleiderer [51]. The SkX state occupies a single phase pocket, and the entire magnetic phase diagram is well accounted for by a Ginzburg–Landau approach including the effects of thermal fluctuations. While the first *B-T* diagram for this kind of Hamiltonian was found experimentally in the 1990s [69], it has been confirmed theoretically for bulk systems much later [70–72] by means of the Landau–Ginzberg theory and Monte-Carlo simulations (MC). The corresponding phase diagram for the interfacial skyrmionic systems in the framework of micromagnetics demonstrates the existence of additional meron states at high temperatures [73]. In this numerical approach, one also has distinguished between the SkX and the skyrmion gas (SkG).

On a discrete lattice, three ordered states—conical SS, hexagonal SkX, and the FM state—have been generally addressed [68, 74]. The meron phase has not been addressed so far, while the SkG phase has recently also been identified [75]. In Dupé [74], the fields needed for skyrmion creation were found to be much larger than in experimental investigations [59], whereas better agreement has been found in Rózsa [71]. For the system of Pd/Fe/Ir(111) ($0.05 < D_{ij}/J_{ij} < 0.19$), the field ranges are cycloidal SS for $B < 1.4$ T, hexagonal SkX for 1.4 T $< B < 3$ T, and FM state for $B > 3$ T. The SkX pocket has been found to be somewhat larger than for bulk B20 systems (see Figure 10.3a), which is interesting for possible applications. In contrast to the fluctuation-induced first-order phase transition in bulk systems [77], the new phase, that is, the fluctuation disordered state, has been found between the ordered phases described earlier and the paramagnetic disordered region [71] (see Figure 10.3b). In this state, the skyrmion lifetime is finite, and the topological charge starts to fluctuate, while the average number of skyrmions remains the same. Hence, this fluctuation disordered state is a topologically non-trivial analog to the superparamagnetic switching of topologically trivial single-domain magnetic nanoislands [78].

While the temperature-dependent investigations on SC-II systems revealed that the critical temperature of the transition from the SkX to the paramagnetic state is on the order of 27 K for Fe/Ir(111) [79] at zero field, the complete, systematic B-T phase diagram is still missing in the literature. The same is true for SC-III skyrmionic systems. Interesting B-T data for specific systems, however, are available in the literature. An example is given by the B-T phase diagram for $D_{z1} = 0.5J_1$ and $D_{z2} = 0$ in Hayami et al. [62] These data, however, cannot directly be compared with those of Leonov and Mostavoy [63] because of different parameters and a different number of nearest-neighbor bonds taken into account. Quite generally, the SC-II and SC-III systems show much richer phase diagrams encompassing multiple-Q states and different kinds of SS. They are strongly dependent on the lattice symmetry.

The SC-VI systems, which are dominated by dipolar interactions, are always very challenging for calculations of thermodynamic properties, because of the long-range character of those interactions. Nevertheless, valuable information about the B-T phase diagram studied within the micromagnetic approach has recently appeared in the literature [80]. Similarly to SC-I systems, the limits of the three phases—stripes corresponding to SS, bubbles corresponding to SkX, and FM phases—have been found. The phase diagram of dipolar skyrmionic materials, in contrast to the SC-I case, has been found to be qualitatively different for different $\mu_0\mu_s^2/(J_1 4\pi a^3)$ ratios. The transitions between different phases in thin films seem to show reentrant behavior rather than abrupt first-order transitions [81].

10.5.3 *B-D* DIAGRAMS

This type of phase diagram is very important for studies of skyrmionic systems, because it permits skyrmion formation to be characterized in different materials. Each material class has a characteristic value of the D/J

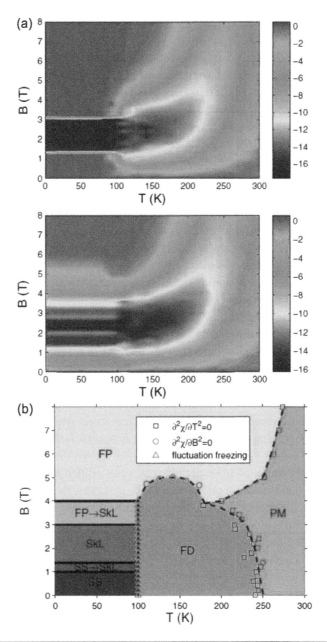

FIGURE 10.3   B-T phase diagrams for an atomistic SC-I skyrmionic system for (a) increasing and (b) decreasing temperature. (c) B-T phase diagram for the same system constructed from the inflection points of magnetic susceptibility. Solid lines denote first-order transitions, while dashed lines correspond to the second-order transitions. SkL corresponds to the skyrmion lattice (SkX in the rest of the chapter), while FP denotes the fully polarized ferromagnetic (FM) state. The region FD corresponds to the fluctuation disordered states discussed in the text. (After Rózsa, L. et al., *Phys. Rev. B* 93, 024417, 2016. With permission.)

ratio. Hence, plotting the value of the order parameter in *B-D* coordinates yields an overview of fields at which the skyrmionic phase is observable. Typically, the winding number *Q* (see Equations 10.15 and 10.16) is used as the order parameter [57, 82, 83]. The winding number provides information about the number of skyrmions in a sample. By that means, one can distinguish between the ferromagnetic phase and the regime of skyrmion lattices. However, two skyrmion lattices of different density and with different skyrmion sizes but with identical numbers of skyrmions have identical winding numbers *Q*. That is, they are indistinguishable. Additionally, it is difficult to distinguish between the regimes of isolated skyrmions and that of the skyrmion lattice because of the absence of a strict criterion for how many skyrmions are needed to form a lattice.

It is known, however, that the skyrmion size and the properties of skyrmion lattices are parameter dependent [12]. Therefore, any specific information about the skyrmion size or about the density of a skyrmion lattice might be important for the design and characterization of experimental systems. In a recent investigation [75], a *B-D* phase diagram of the interfacial chiral systems belonging to the class SC-I, including information about the skyrmion size and density, has been derived. This phase diagram provides access to important information about the size dependence of isolated skyrmions and skyrmion lattices as a function of material parameters and the strength of an applied external magnetic field.

For that purpose, additionally to *Q*, the density order parameter ρ has been defined in Siemens et al. [75] The density parameter gives the ratio between an effective area occupied by skyrmions and the area occupied by a closed packed lattice of skyrmions having the same radius *R*. The skyrmion radius *R* has been determined numerically by a two-dimensional surface approximation of the azimuthal angle Θ of the local magnetization texture with a formula that was originally used to describe 360° Néel domain walls in Kubetzka et al. [84].

To obtain the phase diagrams, very large samples (on the order of 10^6 sites) of rectangular shape, with a *D/J* gradient along the *x*-axis, have been equilibrated at various magnetic fields *B*. This theoretical procedure is similar to experimental setups with wedge geometry of a sample. The magnetocrystalline anisotropy was set to zero for simplicity. The obtained results have been cross-checked via standard calculations without any gradient as explained in the previous sections. Figure 10.4 shows data corresponding to $d(D/J)/dr < 0.005/a$ corresponding to $dD/dr < 0.035$ meV/*a*.

The advantage of the method of gradients is the possibility of a direct visualization of the magnetic microstructure as a function of the interaction strength. Figure 10.4a shows the equilibrium micromagnetic structure of a sample subject to a magnetic field of strength $\mu B/J = 0.7$ at a temperature $k_B T/J \approx 8.5 \cdot 10^{-3} J$. The sample geometry permits the observation of the transitions from the FM to the SkX and eventually, to the SS phase, driven by an increasing DM interaction. In addition to this information, one immediately recognizes a variation of the skyrmion size and the skyrmion density with

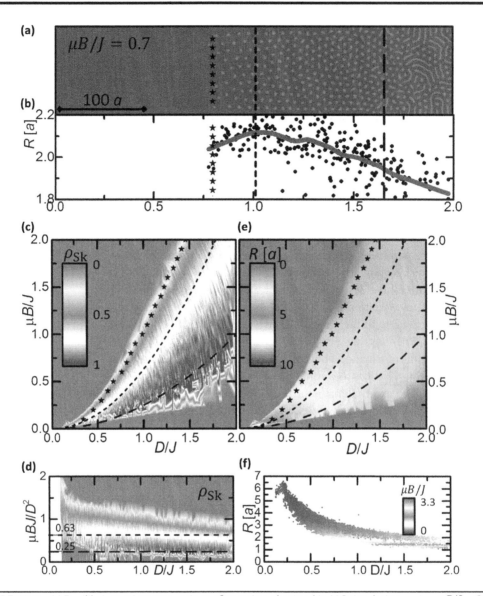

FIGURE 10.4 (a) Equilibrium micromagnetic configuration obtained in MC simulations using $\mu B/J = 0.7$ at a temperature $k_B T/J \approx 8.5 \cdot 10^{-3} J$ with a gradient of the DM interaction strength along the x-axis enforcing a transition from the FM state to the spiral state. (b) Visualization of the corresponding magnetic textures as a function of the D/J. (c, d) Phase diagrams using the skyrmion density ρ as an order parameter. (e, f) Phase diagrams using the skyrmion radius as an order parameter. (After Siemens, A. et al., *New J. Phys.* 18, 045021, 2016. With permission.)

increasing D/J. Figure 10.4b exhibits the corresponding D/J dependence of the skyrmion radius. Here, the numerical values are represented by dots, while the mean skyrmion radius R is given by the solid red line.

Interestingly, one can observe a drastic change in the behavior of the skyrmion radius at $D/J \approx 1.05$. According to our calculations, the skyrmion size increases with the strength of the DM interaction for $D/J < 1.05$ in the single skyrmion regime, while it decreases with the DM interaction for $D/J > 1.05$ in the skyrmion lattice regime. There are several descriptions in the literature of the connection between the skyrmion size and the strength of the DM interaction. While these predictions concern continuous systems, they are in good agreement with our data on discrete skyrmions shown in Figure 10.4a and b. For single interfacial skyrmions, the analytic dependence is given in Rohart and Thieville [85] and Kiselev et al. [86]. The quantitative comparison, however, is difficult because of the lacking field dependence. According to the literature mentioned earlier, the skyrmion radius should be hyperbolic in B and linear in D ($R \propto D/B$) for isolated skyrmions but hyperbolic in D and independent of B ($R \propto J/D$) in skyrmion lattices. The calculations using the discrete model [75] confirm this qualitative behavior (increase or decrease of the skyrmion radius with the DM interaction) for interfacial skyrmions in lattices. To identify the position of the transition from the FM phase with occasional isolated metastable skyrmions to the dense skyrmion lattice, the maximum R value is taken.

The resulting phase diagrams are shown in Figure 10.4c and e. In both cases, the phase diagrams are colored using the density order parameter ρ but with two different scaling schemes ($\rho = f(J/D, \mu BJ/D^2)$ and $\rho = f(J/D, \mu BJ/D^2)$). The short-dashed lines in all panels of Figure 10.4 indicate the transitions between the phase of isolated skyrmions and that of the skyrmion lattice, while the long-dashed lines represent transitions between the SkX and the SS phase. Both dashed lines correspond to the functions of the type $\mu B_c J/D^2$ first defined in Banerjee et al. [57] and Dupé et al. [83], where B_c represents the critical field of the transition between two phases. The described data show that at finite temperatures, the isolated skyrmions survive up to the boundary indicated by the stars. The transition from the skyrmion lattice to the spin-spiral state appears to be diffuse: a rather broad region exists in which skyrmions and elongated spiral-like textures coexist, as can also be observed in Figure 10.4a.

Hence, the density order parameter introduced earlier permits us to identify the phase of isolated skyrmions additionally to the commonly addressed SkX and SS phases. The phase diagrams presented in Figure 10.4 are in good agreement with investigations [57, 83], which show that the functions $\mu B_c J/D^2$ should be constant at the FM–SkX and SkX–SS phase boundaries. The strong D/J dependence of the skyrmion radius can be appreciated in Figure 10.4d and f. Furthermore, the skyrmion radius decreases with increasing magnetic field for a given set of D/J and vanishes at a finite minimal size defined in Figure 10.4f.

The last observation is particularly interesting, because it is in contrast to calculations using a continuous description of skyrmions. The skyrmions in a continuum can have any non-vanishing size due to the topological protection, which does not allow transitions between the skyrmionic and a field-polarized state. The calculations [75] show that skyrmions with a radius less than a certain critical value of the order of one lattice constant (diameter of the order of $2a$) do not exist in discrete systems.

10.6 SIZE AND SHAPE OF SKYRMIONS

10.6.1 SIZE OF SKYRMIONS

A more detailed analysis of the radius of individual skyrmions of type SC-I at the phase boundary between the skyrmion lattice and the diluted phase has been studied theoretically in Siemens et al. [75] and Kravchuk et al. [89] Figure 10.5a shows the skyrmion radius at $\mu B_c J/D^2 \approx 0.63$. According to Siemens et al. [75], the skyrmion radius decreases with an increasing D/J ratio and may be well approximated by a hyperbolic function with an offset of $(0.525 \pm 0.008)a$, which gives the smallest possible radius for a metastable skyrmion at the phase boundary between the FM and the skyrmionic phase at large *D/J*.

Close to absolute zero, a skyrmion will possess an infinite lifetime as long as it rests in a local energy minimum protected by a finite energy barrier against a transition to the energetically lower FM state. Therefore, the interesting question is: at which skyrmion size does the system overcome the separating energy barrier and inevitably relax into the FM state? In other words, the question is whether a single reversed magnetic moment can be stabilized by an external magnetic field. This issue was studied by means of Landau–Lifshitz–Gilbert (LLG) simulations at zero temperature in Siemens et al. [75].

The results of this study are presented in Figure 10.5b. In contrast to the MC simulations of Figure 10.5a concerning the size of equilibrium skyrmions, the dynamical simulations of Figure 10.5b show the radius of unstable skyrmions at the very moment when a skyrmion is dynamically annihilated. In both cases, an SC-I system with energy parameters typical for the Pd/Fe/Ir(111) system ($J \approx 7$ meV and $J \approx 2.2$ meV) is addressed. Both investigations demonstrate that the statical and dynamical *D/J* dependencies have the same functionality. The minimal skyrmion radii are very small and lie in the region of $(0.5-1)a$ for the investigated parameter range of $D/J \in [0.3, 2.0]$. These data are in very good agreement with a recent experimental study [59].

The size of the SC-II type skyrmions depends much less on the external magnetic field [87], because in this case, the nanoskyrmion lattice

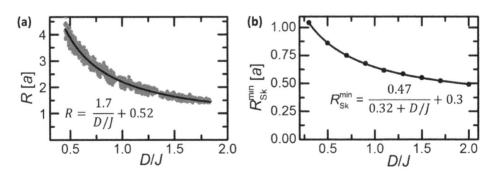

FIGURE 10.5 (a) The equilibrium skyrmion radius at the phase boundary separating the SkX and FM phases. (b) The dynamical critical skyrmion radius derived from LLG simulations. For smaller radii, the system inevitably relaxes to the FM state. (After Siemens, A. et al., *New J. Phys.* 18, 045021, 2016. With permission.)

corresponds to the ground state of a system. Any external perturbation of the system does not alter this robust two-dimensional magnetic ground state but only leads to a change of the rotational domain structure [87]. The diameter of nanoskyrmions, however, is comparable to the minimal size of the SC-I skyrmions and is of the order of 1 nm only.

The sizes of SC-III and SC-IV type skyrmions vary in a much broader range and have not systematically been investigated yet. However, it is known that not only the applied field but also changes in material parameters, lattice strain, and alloy composition play an important role in the skyrmion size [88, 89].

10.6.2 Shape of Skyrmions

Until now, we have addressed spherically symmetric skyrmions, as only in this case does the skyrmion radius unambiguously define its geometrical properties. However, various experimentally feasible material systems naturally exhibit spatially anisotropic behavior. This phenomenon is particularly strong at interfaces [90] and in systems with significant spin–orbit coupling. As shown in a recent investigation [25], the mixture of Rashba and Dresselhaus spin–orbit coupling in combination with a uniaxial anisotropy can stabilize skyrmions of elliptical shape. The main reason for the unusual skyrmion shape in this case is the incompatibility of the sixfold symmetry of the SkX phase with structural deformations in crystals with C_{2v} symmetry.

Another example of interfacial anisotropic skyrmionic systems is given by double [91] and triple [92] atomic layers of Fe on Ir(111). A systematic theoretical investigation of skyrmion formation in systems with an anisotropic environment is presented in Hagemeister et al. [93]. This investigation shows that spatial modulations of the exchange interaction and the anisotropy energy in combination with an isotropic DM interaction lead to the formation of deformed skyrmionic objects.

The shape and size of deformed skyrmions strongly depend on the particular energy landscape. An example of a non-trivial deformed skyrmion obtained with the help of MC simulations is analyzed in Figure 10.6. In this case, a spatial modulation of magnetocrystalline anisotropy between the in-plane and the out-of-plane orientation has been considered to account for the skyrmion deformation (see Figure 10.6a). Such a modulation might occur, for instance, due to stress-induced surface reconstructions. Additionally, it is known that the exchange interaction parameters J_{ij} might be modulated as a function of the orientation of the respective bond and also of the bond position in the lattice [93, 94]. Figure 10.6b–c show the skyrmionic structures for two different magnetic fields with such a spatial modulation of the exchange parameters but without any anisotropy modulation. One observes ordered, bent non-collinear spin states with a non-vanishing topological charge. Figure 10.6d and e show the corresponding equilibrium structure if a spatial modulation of anisotropy has been taken into account. The distorted skyrmionic objects remain but become ordered along linear tracks. The detailed spin structure of the deformed skyrmions is shown in Figure 10.6 f.

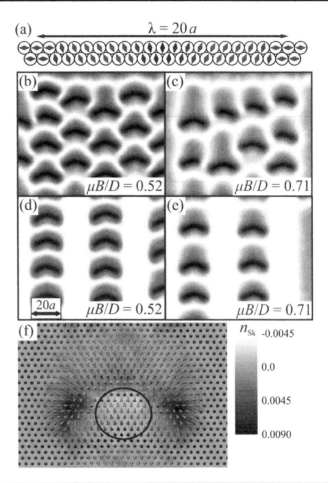

FIGURE 10.6 (a) Sketch of the atomic lattice indicating the spatial modulation of the direction of the easy anisotropy axis. (b, c) Color maps of the perpendicular component of the magnetization of equilibrium skyrmionic states for different strengths of magnetic field and spatial modulation of the exchange interaction corresponding to 3 monolayers Fe on Ir(111) [92]. (d, e) The same as in (b, c) with an additional modulation of the anisotropy according to the scheme outlined in panel (a). (f) Spin structure and local density of the topological charge (grey scale) for a deformed magnetic skyrmion. (After Hagemeister, J. et al., *Phys. Rev. B* 94, 104434, 2016. With permission.)

10.7 STABILITY OF SKYRMIONS

In many cases, isolated magnetic skyrmions correspond to metastable states, which can be deleted or created by fields or currents. This metastability permits the use of topologically distinct skyrmionic and ferromagnetic states as information bits. The critical parameter for any bit of information is its stability. The stability of any state can be quantified by measuring the state's lifetime. The lifetimes of metastable states in turn depend on temperature, external magnetic field, and other intrinsic or extrinsic parameters. At zero temperature, for example, a skyrmion will possess an infinite lifetime as long as it rests in a local energy minimum protected by a finite energy barrier against a transition to the energetically lower field-polarized (FM) state. At higher temperatures, the thermal energy has to be compared with the height of the energy barrier between the two states. Therefore, the interesting question is how the lifetimes of the skyrmionic and ferromagnetic states depend on the field and temperature. This important issue has been recently addressed in several theoretical investigations [71, 95, 96] for the case of SC-I type skyrmions.

In these investigations, the lifetimes of the skyrmionic states have been studied by means of MC simulations [96] as well as atomistic spin dynamics [71]. In both cases, it has been found that the skyrmion lifetime follows the Arrhenius law:

$$\tau_{sk} = \tau_0 e^{\Delta E/k_B T}. \tag{10.30}$$

Calculation of the lifetimes as a function of field led to the interesting conclusion that—despite the similar energy of the skyrmionic and ferromagnetic states—a skyrmion is much more stable than the FM state. This can be seen in Figure 10.7a, where the mean lifetimes of an isolated skyrmion and the polarized FM state are plotted as a function of applied magnetic field. While the system is found with equal probabilities in the Sk and FM states at the critical field B_c, the Sk state becomes preferred for fields $B_0 < B_c$, and the FM state is preferred for $B_0 > B_c$. The lifetimes of a skyrmion, however, are up to three orders of magnitude larger than those of the FM state for identical $|B_0 - B_c|$. This trend is true for any temperature.

Surprisingly, this asymmetric behavior of the mean lifetimes cannot be explained by the height of the energy barrier between the skyrmionic and the FM states. The energy of both states and the corresponding activation barriers are shown in Figure 10.7b. In the entire plotted field range, the activation barrier for a skyrmion was larger than that for an FM. The skyrmionic state was, nevertheless, more stable. To shed light on this finding, the prefactors in the Arrhenius law—known also as the *attempt frequencies*—have been investigated. The ratio of the attempt frequencies for the FM and the Sk states obtained from the Arrhenius fit is on the order of 8×10^3 inverse time units (Monte-Carlo steps) throughout the whole range of explored magnetic fields. This large difference has been identified as the reason for the high stability of skyrmions and results from the higher entropy of the skyrmionic state [96]. To understand why the increased entropy S leads to higher attempt frequencies, one has to consider the Eyring equation Eyring [97] and Laidler and King [98] as a more general form of the Arrhenius law:

$$\tau = \tau_0 \cdot e^{\Delta S/k_B} \cdot e^{\Delta E/k_B T} = \tau_{\text{eff}} \cdot e^{\Delta E/k_B T}. \tag{10.31}$$

Hence, if the entropy of one of the states is larger, then the effective prefactor of the Arrhenius expression will also be larger. On the other hand, the difference in entropy can also be interpreted in the framework of the Kramer's theory describing the shape of the energy minimum. Both kinds of analysis reveal that the stability of skyrmions can be traced back to the dynamics and the geometry of the energy landscape. Particularly, the attempt frequency of

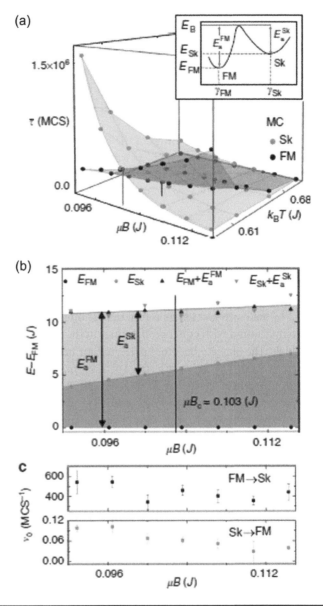

FIGURE 10.7 (a) Mean lifetimes of the Sk and FM states as a function of the magnetic field and the temperature. The red and black points are the results of MC calculations. The points of intersection are marked by green spheres. The inset of (a) shows a sketch of the energy landscape. The energy minima E_{Sk} and E_{FM} of the Sk and FM states are separated by activation energies given by E^{Sk}_a and E^{Sk}_{FM}, respectively. (b) The energies E_{Sk} and E_{FM} of the Sk and FM states, respectively, as a function of the magnetic field for the temperature $k_BT = 0.61$ J alongside the energy barrier E_B, which is given by the sum of the energy levels and activation energies. (c) The attempt frequencies. (After Hagemeister, J. et al., *Nat. Commun.* 6, 8455, 2015. With permission.)

the skyrmion annihilation is orders of magnitude smaller than that of the skyrmion creation due to a more shallow shape of the potential. To make the skyrmions stable even at room temperature, the activation energies have to be increased while leaving the D/J ratio constant. Recent studies showed that this can be achieved by, for example, multilayers [55], thereby increasing the number of nearest neighbors and at the same time the number of interfaces, both being necessary for a strong DM coupling.

10.8 SKYRMION ANNIHILATION

As has been discussed in the previous sections, the lifetimes of metastable skyrmions depend on temperature. At low temperatures, a skyrmion will possess long lifetimes as long as it rests in a local energy minimum protected by a finite energy barrier against a transition to the energetically lower FM state. Therefore, the interesting question is: at which critical skyrmion size does the system inevitably relax into the ferromagnetic state? In other words, can a single reversed magnetic moment be stabilized by an external magnetic field? Another exciting question is: what is the dynamical path of the skyrmion annihilation? These issues were studied analytically and by means of LLG simulations at zero temperature in Siemens et al. [75] and Rohart et al. [99].

While the MC simulations of Figure 10.5 concern the equilibrium skyrmion sizes, the dynamical simulations of Siemens et al. [75] are concentrated on unstable isolated skyrmions at the very moment when a skyrmion is dynamically annihilated. Here, a typical SC-I type skyrmion with energy parameters of the Pd/Fe/Ir(111) system ($J \approx 7$ meV and $J \approx 2.2$ meV) is considered. The skyrmions in that system are rotationally symmetric. This suggests that their centers might relax to a lattice point that possesses the same symmetry. The possible positions of the skyrmion center surrounded by two, three, or six spins are depicted in the upper part of Figure 10.8 for a hexagonal lattice. The atomistic LLG simulations show very good agreement with this symmetry consideration. Starting with a random configuration and using the magnetic fields from the corresponding part of the phase diagram of Figure 10.5, the two- and six-spin equilibrium skyrmion center configurations were obtained. The three-neighbor-spin configuration was not observed. This result can be understood on the basis of the Peierls condition [100], according to which the center of a magnetic non-collinear structure preferably occurs between the atomic sites.

The annihilation of a single skyrmion was achieved using a magnetic field B greater than or equal to the critical field Bc (see Figure 10.5), at which the lifetimes of the skyrmion and ferromagnetic phases are identical. The dynamical evolution of the skyrmion radius as a function of time for different field strengths is shown in Figure 10.8. According to atomistic simulations at zero temperature, the dynamics of the skyrmion annihilation follows the same scenario for any field strength. It consists of two dynamical stages: continuous contraction and discontinuous annihilation of isolated skyrmions. The discontinuous phase starts as soon as the skyrmion radius reaches its

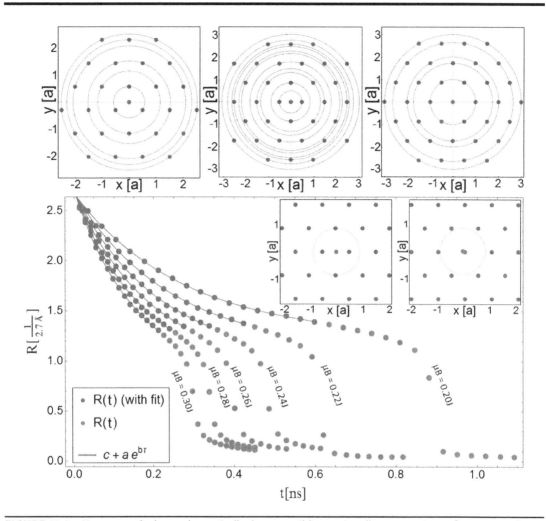

FIGURE 10.8 Upper panels show schematically three possible rotationally symmetric configurations of a skyrmion on a hexagonal lattice corresponding to Pd/Fe/Ir(111). Bottom panel shows the numerically calculated time evolution (points) and fits (solid lines) of the skyrmion radius at different applied magnetic fields B at zero temperature. The insets in the bottom panel show the smallest skyrmions obtained in atomistic spin dynamics. (After Siemens, A. et al., *New J. Phys.* 18, 045021, 2016. With permission.)

minimal value R_{min} as visualized in Figure 10.8. While the duration of the contraction phase decreases with increasing field, the duration of the annihilation phase is practically independent of the field strength. This two-phase transition has also been observed for skyrmions annihilated by the use of spin-polarized currents [101]. The inset in the bottom panel of Figure 10.8 shows the minimal skyrmion radius. In accordance with MC simulations discussed earlier, the ultimately small skyrmion consists of four or seven atomic spins only. This value corresponds to $R_{min} \approx 2.35\text{Å}$, which is close to recent experimental results [59]. Fit functions for the field dependence of the skyrmion dynamics during the annihilation process as well as an estimation of the corresponding energy barriers can be found in Siemens et al. [75].

The dynamical behavior of the energy barriers and the annihilation of isolated skyrmions have also been studied using a path method and atomistic simulations [99]. The set of material parameters was typical for a Co/Pt(111) monolayer. The interesting observation of this investigation is that the most obvious path of the skyrmion collapse presented in Siemens et al. [75] and Fert and Lévy [101], featuring a homogeneous shrinking, gives the largest energy, while the lowest energy barrier, corresponding to a destabilizing field of 0.25 T at around 80 K, corresponds to another path. Along this path, the skyrmion destabilization occurs before any topology change. It is suggested in this investigation that topology plays a minor role in the skyrmion stability for this system, whereas the most important contribution to the skyrmion stability is coming from the DM interaction.

10.9 SKYRMION CONFINEMENT

As can be seen in phase diagrams discussed in previous sections, skyrmions can either form lattices within certain field ranges or be created and manipulated individually. The individual manipulation of skyrmions offers great potential for data storage, transfer, and processing [102, 103]. Until now, the properties of infinitely large lattices or isolated skyrmions in an infinite ferromagnetic background have been discussed. However, in view of the application aspects of skyrmionic systems, theoretical investigations exploring the effect of boundaries and confinement become more and more important, as can be seen by the contemporary literature. Particularly, it has recently been shown [85] that the confinement of a skyrmion in a circular nanoisland leads to its isolation because of the specific boundary conditions induced by the DM interaction. More specifically, the magnetization is bent at the boundaries with respect to the ferromagnetic axis, confining by that means the diameter of a skyrmion. For small D_{ij}, the skyrmion diameter is independent of the dot diameter and coincides with the infinite film solution. At some critical strength of the DM interaction, however, the skyrmions are confined in the island, which limits the diameter increase. Later, it was demonstrated that this confinement can even favor the skyrmion over the spiral phase when compared with infinitely large systems [83]. These properties can be used for the manipulation of single confined skyrmions of type SC-I and for their applications in logic devices [104].

All the abovementioned publications concentrated on individual skyrmions. A first investigation of the interplay between a skyrmion lattice of type SC-II [21], as found in the Fe/Ir(111) system, and the confined geometry of nanoscale islands has recently been reported [105]. A strong coupling of one diagonal of the square magnetic unit cell to the close-packed edges of Fe nanostructures has been observed experimentally by means of SP-STM and theoretically by means of Monte-Carlo simulations. The details of the theoretically calculated micromagnetic structure are presented in Figure 10.9. A clear trend of close-packed edges favoring one of the three rotational domains of the skyrmionic lattice can be seen in Figure 10.9c. However, in an island of triangular shape, it is impossible to orient the diagonal of a square nanoskyrmion lattice along all three edges of the island simultaneously. The mismatch of the symmetries of the skyrmionic lattice

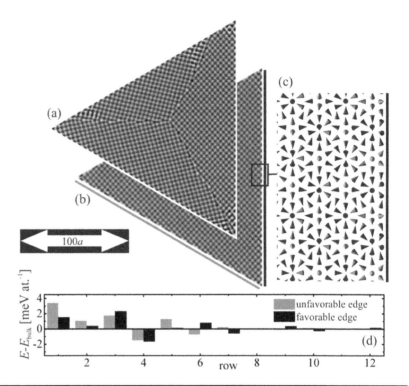

FIGURE 10.9 Triangular Fe/Ir(111) islands with open boundary conditions exhibiting (a) multi- and (b) single-domain states at low temperatures. The displayed out-of-plane sensitive SP-STM images have been calculated based on the Monte-Carlo spin configurations. (c) Spin structure of the nanoskyrmion lattice at the energetically favorable edge, taken from (b). (d) Average energy cost per atom in the nth atomic row parallel to the edge for the two sides marked in (b) with respect to the corresponding value in the interior of a very large sample. The energetically (un)favorable edge is marked in (red) black in (d). (After Hagemeister, J. et al., *Phys. Rev. Lett.* 117, 207202, 2016. With permission.)

and the shape of the island leads to frustration and triple-domain states, as visualized in Figure 10.9a. On the other hand, the formation of domain walls (Figure 10.9a) is accompanied by an energy increase with respect to a monodomain state (Figure 10.9b).

To determine the energy at edges and within domain walls, and to identify the contributions of the different energy terms to the total energy of the system, spatially resolved energy maps of the triangular islands have been analyzed [105]. The result of this analysis can be appreciated in Figure 10.9, where the average energy cost per atom with respect to the corresponding value in the interior of a very large sample in the nth atomic row parallel to the edge for favorable and unfavorable edges is plotted. Surprisingly, the monodomain state has lower internal energy than the triple-domain state. However, despite the lower energy of the single-domain state, multi-domain configurations show up in experiments and numerical simulations due to the combined effect of entropy and an intrinsic domain wall pinning, which results from the skyrmionic character of the spin texture [105].

10.10 MANIPULATION OF SKYRMIONS

Recent exciting investigations toward skyrmionic storage and logic devices are mainly based on the current- or temperature-driven motion of magnetic skyrmions along tracks formed by nanostructuring or by structural lattice distortions [92, 102, 106–113]. To read or write a bit of information, a skyrmion has to be moved toward the reading or writing element. There, each skyrmion has to be addressed individually using local currents or field gradients. It has recently been proposed to manipulate magnetic skyrmions individually using the spin torque induced by spin-polarized electrons originating from the sharp magnetic tip of an SP-STM [24]. The influence of an SP-STM tip on a magnetic skyrmion has been studied by means of atomistic MC simulations using the *s-d* model as first applied to describe current-induced magnetic domain wall motion [108, 114].

The focus has been put on the mobility of skyrmions in the system of three monolayers of Fe on Ir(111) as presented in Figure 10.6. The skyrmions were aligned along tracks due to energy barriers resulting from a spatial modulation of the anisotropy landscape. The skyrmion confinement was restricted to the $[10\bar{1}]$ direction, while the skyrmions were able to move by a driving force along the $[\bar{1}2\bar{1}]$ direction.

The driving mechanism, that is, the spin-polarized current from a magnetic SP-STM tip, has been described by an additional term H_T in the Hamiltonian

$$H_T = -g\sum_i T_i \cdot S_i,\qquad(10.32)$$

for the Monte-Carlo calculations. Therein, g is a coupling constant, and T_i takes the spin-polarized current into account:

$$T_i = -I_0\exp(-2\kappa\sqrt{(x_i - x_{tip})^2 + (y_i - y_{tip})^2 + h^2})\cdot P\cdot m_{tip},\qquad(10.33)$$

where:

P is the tip polarization

m_{tip} is a unity vector parallel to the magnetization direction of the tip

κ is the inverse decay length in vacuum

$r_i = (x_i,y_i,0)$ and $r_{tip} = (x_{tip},y_{tip},h)$ are the positions of the lattice sites and the tip

I_0 is the spin-polarized current

Typical values for the decay length and the current $\kappa = 3\text{Å}^{-1}$ and $I_0 = 10^5 \dfrac{\mu_s}{\gamma D}$ were used. The tip velocity was set to 1.5×10^{-5} lattice constants per Monte-Carlo step, while the position of the tip was updated in intervals of 5000 Monte-Carlo steps. The magnetization direction of the tip is chosen parallel to the skyrmion center, and $g = 1$ and $h = 1a$, with a being the lattice constant. It has been found that the tip is able to move the skyrmions shown in Figure 10.6 (see also [92, 93]) along the given tracks at temperatures on the

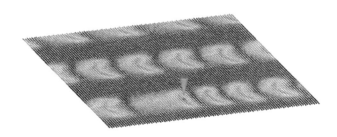

FIGURE 10.10   Schematic representation of skyrmion manipulation by the magnetic tip of a spin-polarized scanning tunneling microscope.

order of $k_B T/D = 0.086$. The tip was positioned above a skyrmion, and when the tip was moved along the track, it moved the skyrmion along. This caused the other skyrmions to move as well. Periodic boundary conditions in the direction of the tracks have been applied in this investigation. Therefore, the skyrmions that leave the sample to one side reappear at the opposite side of the sample. A sketch of this computer experiment is shown in Figure 10.10.

10.11 CONCLUSIONS

Several important aspects of the contemporary research on magnetic skyrmions have been presented. Particularly, a classification of the existing skyrmionic systems according to energy functionals has been provided, and corresponding phase diagrams have been described. The physical properties of magnetic skyrmions have been put into the context of the first field theoretical investigations of these exciting objects. The role of the discreteness of an atomic lattice for the properties of these topological objects has been discussed with an emphasis on interfacial skyrmions. Particular attention has been paid to the stability and manipulation of skyrmionic objects, as they are regarded as hot candidates for future magnetic data storage and logic devices.

REFERENCES

1. T. H. R. Skyrme, A non-linear field theory, *Proc. Roy. Soc. A* **260**, 127 (1961).
2. B. M. A. G. Piette, H. J. W. Miller-Kirsten, D. H. Tchrakian, and W. J. Zakrzewski, Dynamics of baby Skyrmions, *Phys. Lett. B* **320**, 294 (1994).
3. B. M. A. G. Piette, B. J. Schroers, and W. J. Zakrzewski, Multisolitons in a two-dimensional Skyrme model, *Phys. C* **65**, 165 (1995).
4. W. F. Brown, Jr., *Micromagnetism*, John Wiley & Sons, New York, 1963.
5. J. Fidler and T. Schrefl, Micromagnetic modelling - The current state of the art, *Phys. D Appl. Phys.* **33**, R135 (2000).
6. V. G. Bar'yakhtar, A. E. Borovik, and V. A. Popov, Theory of intermediate states in antiferromagnets, *Sov. Phys. Solid State* **11**, 1566 (1970).
7. I. E. Dzyaloshinskii, Theory of helicoidal structures in antiferromagnets, *Sov. Phys. JETP* **19**, 960 (1964).
8. K. P. Bak and M. H. Jensen, Theory of helical magnetic structures and phase transitions in MnSi and FeGe, *J. Phys. C* **13**, L881 (1980).
9. M. I. Ishida, Y. Endoch, S. Mitsuda, Y. Ishikawa, and M. Tanka, Crystal chirality and helicity of the helical spin density wave in MnSi. II. Polarized neutron diffraction, *J. Phys. Soc. Jpn* **54**, 2975 (1985).

10. A. Fert and P. M. Lévy, Role of anisotropic exchange interactions in determining the properties of spin-glasses, *Phys. Rev. Lett.* **44**, 1538 (1980).

11. A. Fert, Magnetic and transport properties of metallic multilayers, *Mater. Sci. Forum* **59–60**, 439 (1990).

12. A. Bogdanov and A. Hubert, Thermodynamically stable magnetic vortex states in magnetic crystals, *J. Magn. Magn. Mater.* **138**, 255 (1994).

13. L. D. Landau and E. M. Lifshitz, *Course of Theoretical Physics, vol. 5, Statistical Physics*, Pergamon, Oxford, 1997.

14. A. N. Bogdanov and D. A. Yablonskii, Thermodynamically stable "vortices" in magnetically ordered crystals. The mixed state of magnets, *Sov. Phys. JETP* **95**, 178 (1989).

15. A. N. Bogdanov, M. V. Kudinov, and D. A. Yablonskii, To the theory of magnetic vortices in easy axis ferromagnets, *Fizika Tverdogo Tela* **31**, 99 (1989).

16. A. N. Bogdanov and A. Hubert, The stability of vortex-like structures in uniaxial ferromagnets, *J. Magn. Magn. Mater.* **195**, 182 (1999).

17. U. K. Rößler, A. N. Bogdanov, C. Pfleiderer, Spontaneous skyrmion ground states in magnetic metals, *Nature (London)* **442**, 797 (2006).

18. G. E. Brown and M. Rho (Eds.), *The Multifaceted Skyrmion*, World Scientific, Singapore, 2010.

19. S. Mühlbauer, B. Binz, F. Jonietz, C. Pfleiderer, A. Rosch, A. Neubauer, R. Georgii, and P. Böni, Skyrmion lattice in a chiral magnet, *Science* **323**, 915 (2009).

20. X. Z. Yu, N. Kanazawa, Y. Onose et al., Near room-temperature formation of a skyrmion crystal in thin-films of the helimagnet FeGe, *Nat. Mater.* **10**, 106 (2011).

21. S. Heinze, K. von Bergmann, M. Menzel et al., Spontaneous atomic-scale magnetic skyrmion lattice in two dimensions, *Nat. Phys.* **7**, 713 (2011).

22. A. Neubauer, C. Pfleiderer, B. Binz et al., Topological Hall effect in the A phase of MnSi, *Phys. Rev. Lett.* **102**, 186602 (2009).

23. A. Tonomura, X. Yu, K. Yanagisawa et al., Real-space observation of skyrmion lattice in helimagnet MnSi thin samples, *Nano Lett.* **12**, 1673 (2012).

24. N. Romming, C. Hanneken, M. Menzel et al., Writing and deleting single magnetic skyrmions, *Science* **341**, 636 (2013).

25. U. Güngördü, R. Nepal, O. A. Tretiakov, K. Belashchenko, and A. A. Kovalev, Stability of skyrmion lattices and symmetries of quasi-two-dimensional chiral magnets, *Phys. Rev. B* **93**, 064428 (2016).

26. P. M. Chaikin and T. C. Lubensky, *Principles of Condensed Matter Physics*, Cambridge University Press, Cambridge, 1995.

27. R. Rajaraman, *Solitons and Instantons*, North-Holland, Amsterdam, 1987.

28. B. Berg and M. Lüscher, Definition and statistical distributions of a topological number in the lattice O(3) σ-model, *Nucl. Phys. B* **190**, 412 (1981).

29. A. van Oosterom and J. Strackee, The solid angle of a plane triangle, *IEEE Trans. Biomed. Eng.* **30**, 125 (1983).

30. L. Zhou, J. Wiebe, S. Lunis, E. Y. Vedmedenko, F. Meier, S. Blügel, P. H. Dederichs, and R. Wiesendanger, Strength and directionality of surface Ruderman-Kittel-Kasuya-Yosida interaction mapped on the atomic scale, *Nat. Phys.* **6**, 187 (2010).

31. A. Stupakiewicz, E. Y. Vedmedenko, A. Fleurence, T. Maroutian, P. Beauvillain, A. Maziewski, and R. Wiesendanger, Atomic-level control of the domain wall velocity in ultrathin magnets by tuning of exchange interactions, *Phys. Rev. Lett.* **103**, 137202 (2009).

32. A. Antal, B. Lazarovits, and L. Balogh, Multiscale studies of complex magnetism of nanostructures based on first principles, *Phil. Mag.* **88**, 2715 (2008).

33. H. J. Gotsis, N. Kioussis, and D. A. Papaconstantopoulos, Evolution of magnetism of Cr nanoclusters on Au(111): First-principles electronic structure calculations, *Phys. Rev. B* **73**, 014436 (2006).

34. C. Etz, J. Zabloudil, P. Weinberger, and E.Y. Vedmedenko, Magnetic properties of single atoms of Fe and Co on Ir(111) and Pt(111), *Phys. Rev. B* **77**, 184425 (2008).

35. R. B. Muniz and D. L. Mills, Theory of spin excitations in Fe (110) monolayers, *Phys. Rev. B* **66**, 174417 (2002).

36. A. Antal, B. Lazarovits, L. Udvardi et al. First-principles calculations of spin interactions and the magnetic ground states of Cr trimers on Au(111), *Phys. Rev. B* **77**, 174429 (2008).

37. R. Takeda, S. Yamanaka, M. Shoji et al. Ab initio calculation of the Dzyaloshinskii-Moriya parameters: Spin-orbit GSO-HF, HFT, and CI approaches, *Int. J. Quant. Chem.* **107**, 1328 (2007).

38. B. Dupé, G. Bihlmayer, M. Boettcher et al. Engineering skyrmions in transition-metal multilayers for spintronic, *Nat. Commun.* **7**, 11779 (2016).

39. K. V. Shanavas and S. Satpathy, Electronic structure and the origin of the Dzyaloshinskii-Moriya interaction in MnSi, *Phys. Rev. B* **93**, 195101 (2016).

40. A. Fert and P. M. Lévy, Anisotropy induced by nonmagnetic impurities in Cu Mn spin-glass alloys, *Phys. Rev. B* **23**, 4667 (1981).

41. A. Crepieux and C. Lacroix, Dzyaloshinsky-Moriya interactions induced by symmetry breaking at a surface, *J. Magn. Magn. Mater.* **182**, 341 (1998).

42. E. Y. Vedmedenko, L. Udvardi, P. Weinberger, and R. Wiesendanger, Chiral magnetic ordering in two-dimensional ferromagnets with competing Dzyaloshinsky-Moriya interactions, *Phys. Rev. B* **75**, 104431 (2007).

43. D.-S. Han, N.-H. Kim, J.-S. Kim et al. Asymmetric hysteresis for probing Dzyaloshinskii -Moriya interaction, *Nano Lett.* **16**, 4438 (2016).

44. L. Udvardi, A. Antal, L. Szunyogh et al. Magnetic pattern formation on the nanoscale due to relativistic exchange interactions, *Physica B* **403**, 402 (2008).

45. R. Wieser, E. Y. Vedmedenko, and R. Wiesendanger, Entropy driven phase transition in itinerant antiferromagnetic monolayers, *Phys. Rev. B* **77**, 064410 (2008).

46. A. H. MacDonald, S. M. Girvin, and D. Yoshioka, t/U expansion for the Hubbard model, *Phys. Rev. B* **37**, 9753 (1988).

47. Ph. Kurz, G. Bihlmayer, S. Blügel, K. Hirai, and T. Asada, Comment on Ultrathin Mn films on Cu(111) substrates: Frustrated antiferromagnetic order, *Phys. Rev. B* **63**, 096401 (2001).

48. E. Y. Vedmedenko, *Competing Interactions and Pattern Formation in Nanoworld*, Wiley-VCH, Weinheim, 2007.

49. E. Y. Vedmedenko, H. P. Oepen, and J. Kirschner, Size-dependent magnetic properties in nanoplatelets, *J. Magn. Magn. Mater.* **256**, 237 (2003).

50. E. Y. Vedmedenko, H. P. Oepen, and J. Kirschner, Microstructure of the spin reorientation transition in second-order approximation of magnetic anisotropy, *Phys. Rev. B* **66**, 214401 (2002).

51. A. Bauer and C. Pfleiderer, Generic aspects of skyrmion lattices in chiral magnets, in *Topological Structures in Ferroic Materials: Domain Walls, Vortices and Skyrmions*, J. Seidel (Ed.), *Book Series: Springer Series in Material Sciences*, **228**, 1, 2016.

52. H. Wilhelm, M. Baenitz, M. Schmidt, U. K. Rössler, A. A. Leonov, and A. N. Bogdanov, Precursor phenomena at the magnetic ordering of the cubic helimagnet FeGe, *Phys. Rev. Lett.* **107**, 127203 (2011).

53. W. Münzer, A. Neubauer, T. Adams et al., Skyrmion lattice in the doped semiconductor $Fe_{1-x}Co_xSi$, *Phys. Rev. B* **81**, 041203(R) (2010).

54. I. Kézsmárki, S. Bordács, P. Milde et al., Néel-type skyrmion lattice with confined orientation in the polar magnetic semiconductor GaV_4S_8, *Nat. Mater.* **14**, 1116 (2015).

55. C. Moreau-Luchaire, C. Moutafis et al., Skyrmions at room temperature: From magnetic thin films to magnetic multilayers, *Nat. Nanotechnol.* **11**, 444 (2016).

56. S.-Z. Lin, A. Saxena, and C. D. Batista, Skyrmion fractionalization and merons in chiral magnets with easy-plane anisotropy, *Phys. Rev. B* **91**, 224407 (2015).

57. S. Banerjee, J. Rowland, O. Erten, and M. Randeria, Enhanced stability of skyrmions in two-dimensional chiral magnets with rashba spin-orbit coupling, *Phys. Rev. X* **4**, 031045 (2014).

58. Y. Tokunaga, X. Z. Yu, J. S. White, H. M. Rønnow, D. Morikawa, Y. Taguchi, and Y. Tokura, A new class of chiral materials hosting magnetic skyrmions beyond room temperature, *Nat. Commun.* **6**, 7638 (2015).

59. N. Romming, A. Kubetzka, C. Hanneken, K. von Bergmann, and R. Wiesendanger, Field-dependent size and shape of single magnetic skyrmions, *Phys. Rev. Lett.* **114**, 177203 (2015).

60. J. Brede, N. Atodiresei, V. Caciuc et al., Long-range magnetic coupling between nanoscale organic-metal hybrids mediated by a nanoskyrmion lattice, *Nat. Nanotechnol.* **9**, 1018 (2014).

61. A. Schlenhoff, P. Lindner, J. Friedlein, S. Krause, R. Wiesendanger, M. Weinl, M. Schreck, and M. Albrecht, Magnetic nano-skyrmion lattice observed in a si-wafer-based multilayer system, *ACS Nano* **9**, 5908 (2015).

62. S. Hayami, S.-Z. Lin, and C. D. Batista, Bubble and skyrmion crystals in frustrated magnets with easy-axis anisotropy, *Phys. Rev. B* **93**, 184413 (2016).

63. A. O. Leonov and M. Mostovoy, Multiply periodic states and isolated skyrmions in an anisotropic frustrated magnet, *Nat. Commun.* **6**, 8275 (2015).

64. E. Y. Vedmedenko, A. Ghazali, and J-C. S. Levy, Magnetic structures of Ising and vector spins monolayers by Monte-Carlo simulations, *Surf. Sci.* **402**, 391 (1998).

65. M. Ezawa, Giant skyrmions stabilized by dipole-dipole interactions in thin ferromagnetic films, *Phys. Rev. Lett.* **105**, 197202 (2010).

66. F. Büttner, C. Moutafis, M. Schneider, B. Krüger, C. M. Günther, J. Geilhufe, C. V. Schmising, C. V. Korff, J. Mohanty, B. Pfau, S. Schaffert, A. Bisig, M. Foerster, T. Schulz, C. A. F. Vaz, J. H. Franken, H. J. M. Swagten, M. Kläui, and S. Eisebitt, Dynamics and inertia of skyrmionic spin structures, *Nat. Phys.* **11**, 225 (2015).

67. Y. Y. Dai, H. Wang, P. Tao, T. Yang, W. J. Ren, and Z. D. Zhang, Skyrmion ground state and gyration of skyrmions in magnetic nanodisks without the Dzyaloshinsky-Moriya interaction, *Phys. Rev. B* **88**, 054403 (2013).

68. N. S. Kiselev, A. N. Bogdanov, R. Schäfer, and U. K. Rössler, Comment on Giant skyrmions stabilized by dipole-dipole interactions in thin ferromagnetic films, *Phys. Rev. Lett.* **107**, 179701 (2011).

69. B. Lebech, P. Harris, J. Skov Pedersen, K. Mortensen, C. I. Gregory, N. R. Bernhoeft, M. Jermy, and S. A. Brown, Magnetic phase diagram of MnSi, *J. Magn. Magn. Mater.* **140–144**, 119 (1995).

70. J. H. Han, J. Zang, Z. Yang, J.-H. Park, and N. Nagaosa, Skyrmion lattice in a two-dimensional chiral magnet, *Phys. Rev. B* **82**, 094429 (2010).

71. L. Rózsa, E. Simon, K. Palotás, L. Udvardi, and L. Szunyogh, Complex magnetic phase diagram and skyrmion lifetime in an ultrathin film from atomistic simulations, *Phys. Rev. B* **93**, 024417 (2016).

72. S. Buhrandt and L. Fritz, Skyrmion lattice phase in three-dimensional chiral magnets from Monte Carlo simulations, *Phys. Rev. B* **88**, 195137 (2013).

73. M. Ezawa, Compact merons and skyrmions in thin chiral magnetic films, *Phys. Rev. B* **83**, 100408 R (2011).

74. B. Dupé, M. Hoffmann, C. Paillard, and S. Heinze, Tailoring magnetic skyrmions in ultra-thin transition metal films, *Nat. Commun.* **5**, 4030 (2014).

75. A. Siemens, Y. Zhang, J. Hagemeister, E. Y. Vedmedenko, and R. Wiesendanger, Minimal radius of magnetic skyrmions: Statics and dynamics, *New J. Phys.* **18**, 045021 (2016).

76. E. Simon, K. Palotás, L. Rózsa, L. Udvardi, and L. Szunyogh, Formation of magnetic skyrmions with tunable properties in PdFe bilayer deposited on Ir(111), *Phys. Rev. B* **90**, 094410 (2014).

77. M. Janoschek, M. Garst, A. Bauer et al., Fluctuation-induced first-order phase transition in Dzyaloshinskii-Moriya helimagnets, *Phys. Rev. B* **87**, 134407 (2013).

78. E.Y. Vedmedenko, N. Mikuszeit, T. Stapelfeldt et al., Spin-spin correlations in ferromagnetic nanosystems, *Eur. Phys. J. B* **80**, 331 (2011).

79. A. Sonntag, J. Hermenau, S. Krause, and R. Wiesendanger, Thermal stability of an interface-stabilized skyrmion lattice, *Phys. Rev. Lett.* **113**, 077202 (2014).

80. A. Mendoza-Coto, O. V. Billoni, S. A. Cannas, and D. A. Stariolo, Modulated systems in external fields: Conditions for the presence of reentrant phase diagrams, *Phys. Rev. B* **94**, 054404 (2016).

81. A. Abanov, V. Kalatsky, V. L. Pokrovsky, and W. M. Saslow, Phase diagram of ultrathin ferromagnetic films with perpendicular anisotropy, *Phys. Rev. B* **51**, 1023 (1995).

82. S. D. Yi, S. Onoda, N. Nagaosa, and J. H. Han, Antiferromagnetically driven electronic correlations in iron pnictides and cuprates, *Phys. Rev. B* **80**, 2009 (2009).

83. R. Keesman, A. O. Leonov, P. van Dieten, S. Buhrandt, G. T. Barkema, L. Fritz, and R. Duine, Degeneracies and fluctuations of Néel skyrmions in confined geometries, *Phys. Rev. B* **92**, 134405 (2015).

84. A. Kubetzka, O. Pietzsch, M. Bode, and R. Wiesendanger, Spin-polarized scanning tunneling microscopy study of 360° walls in an external magnetic field, *Phys. Rev. B* **67** 020401 (2003).

85. S. Rohart and A. Thiaville, Skyrmion confinement in ultrathin film nanostructures in the presence of Dzyaloshinskii-Moriya interaction, *Phys. Rev. B* **88**, 184422 (2013).

86. A. B. Butenko, A. A. Leonov, A. N. Bogdanov, and U. K. Rößler, Theory of vortex states in magnetic nanodisks with induced Dzyaloshinskii-Moriya interactions, *Phys. Rev. B* **80**, 134410 (2009).

87. K. von Bergmann, A. Kubetzka, O. Pietzsch, and R. Wiesendanger, Interface-induced chiral domain walls, spin spirals and skyrmions revealed by spin-polarized scanning tunneling microscopy, *J. Phys. Condens. Matter* **26**, 394002 (2014).

88. K. Shibata, X. Z. Yu, T. Hara, D. Morikawa, N. Kanazawa, K. Kimoto, S. Ishiwata, Y. Matsui, and Y. Tokura, Towards control of the size and helicity of skyrmions in helimagnetic alloys by spin-orbit coupling, *Nat. Nanotechnol.* **8**, 723 (2013).

89. V. P. Kravchuk, U. K. Rößler, O. M. Volkov, D. S. Scheka, J. van den Brink, D. Makarov, H. Fuchs, H. Fangohr, and Y. Gaididei, Topologically stable magnetization states on a spherical shell: Curvature-stabilized skyrmions, *Phys. Rev. B* **94**, 144402 (2016).

90. S. Woo, K. Litzius, B. Krüger et al., Observation of room-temperature magnetic skyrmions and their current-driven dynamics in ultrathin metallic ferromagnets, *Nat. Mater.* **15**, 501 (2016).

91. P. J. Hsu, A. Finco, L. Schmidt, A. Kubetzka, K. von Bergmann, and R. Wiesendanger, Guiding spin spirals by local uniaxial strain relief, *Phys. Rev. Lett.* **116**, 017201 (2016).

92. P. J. Hsu, A. Kubetzka, A. Finco, N. Romming, K. von Bergmann, and R. Wiesendanger, Electric-field-driven switching of individual magnetic skyrmions, *Nat. Nanotechnol.* **12**, 123 (2017).

93. J. Hagemeister, E. Y. Vedmedenko, and R. Wiesendanger, Pattern formation in skyrmionic materials with anisotropic environments, *Phys. Rev. B* **94**, 104434 (2016).

94. E. Y. Vedmedenko, K. von Bergmann, H. Oepen, and R. Wiesendanger, Lattice-dependent anisotropy in the orientation of magnetic domain walls, *J. Magn. Magn. Mater.* **290–291**, 746 (2005).

95. C. Schütte, J. Iwasaki, A. Rosch, and N. Nagaosa, Inertia, diffusion, and dynamics of a driven skyrmion, *Phys. Rev. B* **90**, 174434 (2014).

96. J. Hagemeister, N. Romming, K. von Bergmann, E. Y. Vedmedenko, and R. Wiesendanger, Stability of single skyrmionic bits, *Nat. Commun.* **6**, 8455 (2015).

97. H. Eyring, The activated complex in chemical reactions, *J. Chem. Phys.* **3**, 107 (1935).

98. K. J. Laidler and M. C. King, Development of transition-state theory, *J. Phys. Chem.* **87**, 2657 (1983).

99. S. Rohart, J. Miltat, and A. Thiaville, Path to collapse for an isolated Néel skyrmion, *Phys. Rev. B* **93**, 214412 (2016).

100. K. S. Novoselov, A. K. Geim, S. V. Dubonos, E. W. Hill, and I. V. Grigorieva, Subatomic movements of a domain wall in the Peierls potential, *Nature* **426,** 812 (2003).

101. A. D. Verga, Skyrmion to ferromagnetic state transition: A description of the topological change as a finite-time singularity in the skyrmion dynamics, *Phys. Rev. B* **90**, 174428 (2014).

102. J. Sampaio, V. Cross, S. Rohart, A. Thiaville, A. Fert, Nucleation, stability and current-induced motion of isolated magnetic skyrmions in nanostructures, *Nat. Nanotechnol.* **8**, 839 (2013).

103. R. Wiesendanger, Nanoscale magnetic skyrmions in metallic films and multilayers: A new twist for spintronics, *Nat. Rev. Mater.* **1**, 16044 (2016).

104. X. Zhang, M. Ezawa, and Y. Zhou, Magnetic skyrmion logic gates: Conversion, duplication and merging of skyrmions, *Sci. Rep.* **5**, 9400 (2015).

105. J. Hagemeister, D. Iaia, E. Y. Vedmedenko, K. von Bergmann, A. Kubetzka, and R. Wiesendanger, Skyrmions at the edge: Confinement effects in Fe/Ir(111), *Phys. Rev. Lett.* **117**, 207202 (2016).

106. S. Woo, K. Litzius, B. Krüger et al., Observation of room-temperature magnetic skyrmions and their current-driven dynamics in ultrathin metallic ferromagnets, *Nat. Mater.* **15**, 501 (2016).

107. O. Heinonen, J. Wanjun, and H. Somaily, Generation of magnetic skyrmion bubbles by inhomogeneous spin Hall currents, *Phys. Rev. B* **93**, 094407 (2016).

108. W. Jiang, P. Upadhyaya, W. Zhang et al., Blowing magnetic skyrmion bubbles, *Science* **349**, 283 (2015).

109. K. Evershor, M. Garst, B. Binz et al., Rotating skyrmion lattices by spin torques and field or temperature gradients, *Phys. Rev. B* **86**, 054432 (2012).

110. L. Kong and J. Zang, Dynamics of an insulating skyrmion under a temperature gradient, *Phys. Rev. Lett.* **111**, 067203 (2013).

111. S.-Z. Lin, C. D. Batista, C. Reichhardt, and A. Saxena, ac current generation in chiral magnetic insulators and skyrmion motion induced by the spin Seebeck effect, *Phys. Rev. Lett.* **112**, 187203 (2014).

112. A. A. Kovalev, Erratum: Skyrmionic spin Seebeck effect via dissipative thermomagnonic torques [*Phys. Rev. B* **89**, 241101(R) (2014)], *Phys. Rev. B* **91**, 239903 (2015).

113. C. Schütte and M. Garst, Magnon-skyrmion scattering in chiral magnets, *Phys. Rev. B* **90**, 094423 (2014).

114. T. Stapelfeldt, R. Wieser, E. Y. Vedmedenko, and R. Wiesendanger, Domain wall manipulation with a magnetic tip, *Phys. Rev. Lett.* **107**, 027203 (2011).

Molecular Spintronics

Stefano Sanvito

11.1 INTRODUCTION

While spintronics is consolidating as the leading technology in the magnetic data storage industry and is emerging into both the sensors and the random access memories arena, the interest of the scientific community has grown into investigating new materials different from conventional metals and semiconductors. Ferromagnetic insulators presenting a spontaneous electric polarization and magnetic semiconductors have been both proposed as electronic barriers for magnetic tunnel junctions (MTJ) [1, 2] raising the expectation for multifunctional devices. Likewise, the search for new materials and material combinations to be used for generating highly spin-polarized currents, but at the same time having a small magnetic moment, has witnessed a substantial boost because of their potential for delivering applications based on spin-transfer torque [3, 4].

Among these new materials, organic molecules seem to occupy a rather unique position. These, in fact, not only can be prepared in a practically infinite range of types and combinations at a usually modest cost, but also span an equally large horizon in terms of electronic properties and functionalities. For instance, electron transport in carbon-based materials goes from virtually scattering-free band conduction in graphene and C nanotubes to hopping-assisted conduction in molecular semiconductors, with the conductivity of these latter that can be tuned over 15 orders of magnitude [5]. As they stand, organic molecules appear as an incredibly versatile playground for exploring new spintronics concepts and/or for implementing existing ones.

As a first application, molecules have been used as nonmagnetic spacers in a conventional spin-valve architecture (two magnetic electrodes separated by a nonmagnetic material) with both planar [6] and vertical geometries [7]. The first experiments have been followed by many others over the years (for a review see Ref. [8]) and, despite the present lack of a satisfactory rigorous metrology [9], a few indisputable facts have now emerged about the spin-scattering properties of organic molecules. These are summarized in Figure 11.1, where we report the spin relaxation time, τ_S (the average time that an electron travels before its spin direction is changed), as a function of the spin diffusion length, l_S (the average distance travelled in a time τ_S), for different materials.

Although the figure is only illustrative, since it compares data taken at different temperatures and reports values for τ_S and l_S measured/inferred in different ways, the clear message is that organic materials occupy the top left corner of the τ–l diagram, that is, they are characterized by long relaxation times but short relaxation lengths. Long τ_S's in organics are well known to the electron paramagnetic resonance community [19], and they are rooted in the lack of an efficient mechanism for relaxing spins in organics. In fact, spin–orbit interaction is generally weak (it scales as Z^4 with Z the atomic number) and hyperfine coupling is hampered by the lack of nuclei with nuclear spins in the upper part of the periodic table and by the fact that the molecular orbitals relevant for electron transport are generally delocalized [20]. Likewise also, short l_S's can be easily accounted for, since the typical mobility, μ, of an organic semiconductor is orders of magnitude smaller than that of standard semiconductors. For instance, μ in rubrene (the most conductive among the organic semiconductors) is just $10\,\mathrm{cm^2\,V^{-1}\,s^{-1}}$ compared to $450\,\mathrm{cm^2\,V^{-1}\,s^{-1}}$ for p-type Si.

One may then ask, what can organic molecules do for spintronics? Of course a natural answer is to think about applications where it is necessary to keep the spin coherence for long times, but not to drag the electron spins very far from the point of injection. Quantum information

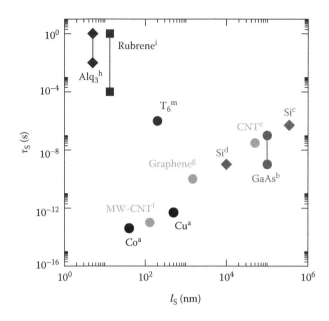

FIGURE 11.1   Spin relaxation time, τ_S, against spin diffusion length, l_S, for different classes of materials. Note that nonmagnetic organic materials occupy the upper left corner, that is, they are characterized by long τ_S and short l_S. The picture is adapted from Ref. [9]. Data in the figure correspond to the following references: a [10], b [11], c [12], d [13], e [14], f [15], g [16], h [7, 17], i [18], and m [6]. (After Szulczewski, G. et al., *Nat. Mater.* 8, 693, 2009. With permission.)

technology appears as one prospect [21] and other new concepts are now emerging [22]. A second possibility is to conceive devices where the organic layer is shorter than its rather short spin-diffusion length so that spin information is not lost in the transfer between the injector and the collector. Crucial to this second strategy is the design of structures where spins can be manipulated while diffusing in the organic layer. A third strategy, which departs quite drastically from conventional spintronics, is that of focusing on single-molecule transport, that is, that of looking at situations where the active part of the device is an individual molecule [23, 24]. This is the main focus of the present chapter. For those readers interested in spin transport in extended organic semiconductors, we suggest a review [8] and Chapter 5, Volume 2, of this book. Furthermore, here we will primarily look at small molecules and will not consider carbon-only inorganic macromolecules, such as carbon nanotubes, graphene, and graphene nanoribbons. Also in this case, we refer the reader to the recent literature [25].

There are three main conceptual device structures that one can envision with single molecules. First, one can consider metallic magnetic electrodes sandwiching nonmagnetic molecules [23]. This is the molecular version of conventional giant magnetoresistance (GMR) or tunneling magnetoresistance (TMR) spin valves, and one can achieve one or the other depending on the conducting nature of the molecule. The important observation here is that the bonding structure of the molecule to the magnetic surface can drastically alter the spin selectivity of the interface, so that one can convert bond engineering in spin engineering. The second class of structures is that where the molecule itself carries a magnetic degree of freedom, that is, it is a single molecular magnet (SMM) [26]. These can be used both in combination with magnetic electrodes or with nonmagnetic ones, and the idea is to use the SMM to manipulate or store information. Such structures are often characterized by complex many body effects such as Coulomb and spin blockade, as well as by a rich inelastic spectrum of magnetic nature. Finally, the last class of devices is populated by molecules that possess non-trivial internal degrees of freedom (magnetic, vibrational, etc.) and that can be switched between different metastable configurations characterized by different transport properties. This is the case of spin-crossover [27] or optically switchable molecules [28].

Of course this is not a rigid classification and several proposals for combining different classes have been already made (for a complete review see Ref. [29]). For instance, one can fabricate spin valves with carbon nanotube spacers and then graft magnetic molecules on the tubes. Even more ambitious is the prospect of having devices where all the elements (electrodes and spacers) are organics, although such an idea at the moment seems more remote because of the lack, with a few exceptions [30], of room temperature organic magnets.

11.2 ESTABLISHING A SIMPLE CONCEPTUAL FRAMEWORK

Electron transport in single molecules is rather different from standard diffusion in metals and semiconductors and here we briefly review the main concepts. Consider the schematic diagram of Figure 11.2 [31]. Here, a generic molecular orbital of wave function ψ_α and single-particle level ε_α is coupled

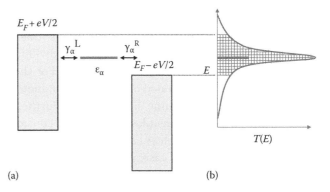

(a)   (b)

FIGURE 11.2 Schematic electronic transport through the single-particle level ε_α associated with a molecular orbital ψ_α. (a) The molecular orbital is electronically coupled to two current/voltage electrodes via the hopping parameters, so that electrons can move from an occupied state in the left electrode to an empty state in the right electrode. The electrodes are kept at a potential difference of V by a battery, which effectively shifts the position of their Fermi level, E_F. (b) The energy-integral of the resulting transmission coefficient, $T(E)$, is the current.

to two featureless (constant density of states—DOS) electrodes via the hopping parameters $\gamma_\alpha^m (m = \text{L,R})$. The transmission rates from/to the two electrodes are $\gamma_\alpha^m / \hbar$ so that the state lifetime is $\frac{1}{\hbar}(\gamma_\alpha^L + \gamma_\alpha^L)$.

When ψ_α is weakly coupled to the electrodes, the associated DOS, $D_\alpha(E)$, is approximated by

$$D_\alpha(E) = \frac{1}{2\pi} \frac{\gamma_\alpha}{(E - \varepsilon_\alpha)^2 + (\gamma_\alpha/2)^2}, \tag{11.1}$$

where $\gamma_\alpha = \gamma_\alpha^L + \gamma_\alpha^R$. Note that the inelastic electronic interaction with the molecule internal degrees of freedom (phonon, spin waves, etc.) can be incorporated in the description by adding an additional broadening, γ_α^{in}, whose details depend on the microscopic nature of the interaction and on how the inelastic excitations decay into the electrodes. Generally, this requires the solution of additional equations of motion, but more simply γ_α^{in} can be interpreted as a "virtual" electrode with no net electron flux [32].

The actual position of the single-particle energy level ε_α is a function of the molecule charging state. At a simple mean field level, we can assume that ε_α scales linearly with the orbital occupation [33]. This is given by

$$n_\alpha = \int_{-\infty}^{+\infty} dE \, D_\alpha(E) \frac{\gamma_\alpha^L f^L(E) + \gamma_\alpha^R f^R(E)}{\gamma_\alpha}, \tag{11.2}$$

where f^L (f^R) is the Fermi function evaluated at $E - \mu_L (E - \mu_R)$ with μ_L (μ_R) the chemical potential of the left- (right-) hand side electrode ($\mu_L - \mu_R = eV$). The transmission coefficient associated to the molecular level ψ_α is

$$T_\alpha(E) = \frac{\gamma_\alpha^L \gamma_\alpha^R}{(E - \varepsilon_\alpha)^2 + (\gamma_\alpha/2)^2}, \tag{11.3}$$

that is simply a resonance centered around ε_α. Finally, the current carried by ψ_α can be calculated by simply integrating $T_\alpha(E)$ within the bias window, that is, it is zero if ε_α is outside the energy interval $\mu_L - \mu_R$ (ψ_α is either completely empty or fully occupied), while for ε_α in the bias window we have

$$I_\alpha = 2\pi \frac{e}{h} \frac{\gamma_\alpha(1 - r_\alpha^2)}{4}, \tag{11.4}$$

where:

    h is the Planck constant
    e is the electron charge

and we have introduced the asymmetry parameter $r_\alpha = (\gamma_\alpha^L - \gamma_\alpha^R)/(\gamma_\alpha^L + \gamma_\alpha^R)$.

The model presented is based on the assumption that the steady state current is essentially a balance between the electron flux into and out from the molecule. If ψ_α is more strongly coupled to one of the electrodes, as in a scanning tunnel microscopy (STM) geometry, then $|r_\alpha| \sim 1$ and ε_α will be pinned to the chemical potential of that electrode. As a consequence, $T(E)$ will have a peak at $E = \varepsilon_\alpha$ moving as a function of bias together with the chemical potential that pins ψ_α and in general the current will be small. In contrast, if $\gamma_\alpha^L \sim \gamma_\alpha^R (r_\alpha \sim 0)$, then the dynamics of the energy level is given only by its charging properties and the current will be maximized.

Finally, it is important to mention that when no molecular levels are inside the bias window, current can still flow due to tunneling. In this case, an envelope function/effective mass picture of the tunneling process is inadequate and one simply writes the transmission coefficient as $T(E) = T_0(E)$ $e^{-\beta(E)L}$ with T_0 being a prefactor depending on the bonding structure between the molecule and the electrodes and the energy-dependent decay coefficient $-\beta(E)$ being a molecular specific quantity (L is the molecule length). In this tunneling limit, one is usually interested in linear response transport only so that the conductance is $G = \left(2e^2/h\right)T_0(E_F)e^{-\beta(E_F)L}$.

Spin can modify this description in two ways. First, the electrodes can be spin-polarized so that both their DOS and the electronic coupling to ψ_α become spin dependent. This implies $\gamma_\alpha^m \rightarrow \gamma_\alpha^{m\sigma}$, where we have introduced the spin index σ ($\sigma = \uparrow,\downarrow$). Second, the molecular level itself can be spin-polarized (for instance, if the molecule is magnetic and/or if it is spin-polarized by proximity to the electrodes), that is, $\varepsilon_\alpha \rightarrow \varepsilon_\alpha^\sigma$. Of course, both the effects can be present, so that one in general may have $\gamma_\alpha^m \rightarrow \gamma_\alpha^{m\sigma\sigma'}$, with the first spin index associated to the electrodes and the second to the molecule. In general, it is a good approximation to neglect spin-flip events so that $\sigma = \sigma'$, and all the transport quantity depends on a single spin index. A more complete description of electron and spin transport in molecules and of how this can be calculated from quantitative ab initio methods can be found in the more specialized literature [34, 35].

11.3 MAGNETIC ELECTRODES AND NONMAGNETIC MOLECULES

The simpler and first device architecture investigated both experimentally and theoretically consists in a nonmagnetic molecule attached to magnetic electrodes. This is the single-molecule version of the spin valve, but it is considerably more experimentally challenging than its inorganic counterpart. In addition to the usual difficulties associated with molecular electronics (mechanical stability of the contacts, control over the bonding geometry, etc.), the use of transition metal electrodes adds the complication of the high reactivity of the surface. This effectively excludes the use of a large class of experimental protocols based on breaking junctions in wet conditions, which was proved rather successful for nonmagnetic noble metal electrodes [36]. This means that chemically stable interfaces must be made, indeed a difficult challenge.

In an early pioneering work, nanoconstrictions were constructed incorporating self-assembled monolayers of octanethiol (an insulating molecule) [37] and Ni electrodes. Remarkably, a low temperature magnetoresistance was found, although both the amplitude and the sign were extremely dependent on the specific sample, with most of the devices being either too resistive or not displaying any magnetoresistive signal at all. This rang an alarm bell toward the difficulties with maintaining clean interfaces and avoiding unwanted surface chemical reaction (oxidation of the transition metal for instance). To date, in fact, no other single-molecule MTJ device has been made, and the only other experiment available for spin-polarized transport in single molecules in a spin-valve architecture is a pump-probe optical experiment on quantum dots connected by organic linkers [38].

A better strategy appears to be that of operating with molecules that "fly," that is, molecules that can be evaporated over a magnetic surface in ultrahigh vacuum conditions and can be probed by STM (either polarized or not). Along these lines, a somehow less direct evidence that spin-polarized electrons can be injected across molecules is provided by two-photon photoemission experiments for copper phthalocyanine molecules deposited over Co surfaces [39], although no detailed microscopic information on how the various molecular features affecting the transport can be provided. Despite the fact that the experimental activity in this area is still rather limited (a much larger experimental effort is currently focused on organic tunnel junctions, where the organic layer is several nanometers in size), single-molecule spin transport has a huge potential, if not for making real devices, at least for developing an understanding of the details of spin injection across an organic/inorganic interface. Let us briefly discuss how large spin polarization (and TMR) can be achieved in molecular junctions.

First, consider the case of tunneling. Here the situation, at a first glance, may look similar to that of standard MTJs, where the symmetry of the complex band structure of the insulating barrier and its matching to the Fermi surface of the magnetic electrodes determines the spin polarization of the current [40]. However, when dealing with single molecules there are three main differences. First, the problem of electron transport through single molecules does not have translational invariance, so that attributing a symmetry to the evanescent wave function in the barrier is meaningless. Second, because of the lack of translational invariance, one expects the transmission to be maximized for electrons with very large longitudinal momentum, that is, at around the Γ point in the 2D Brillouin zone of the electrodes in the plane perpendicular to the molecule. Finally, one expects that the bonding structure at the interface and the molecular orbital character of the molecule may also affect drastically the transmission (T_0 may depend strongly on bonding structure). Overall, however, one expects the spin polarization of the current to be related to the spin polarization of a reduced DOS [41], constructed from those energy levels with a large longitudinal momentum. Since this is never too large in transition metals, we do not expect spectacularly large MR in the tunneling limit. First principle calculations for alkanethiol on Ni confirm this conjecture [23, 42].

A different situation is attainable when the transport is resonant across a specific molecular level ψ_α. The main mechanism is illustrated in Figure 11.3. In this case, electrons tunnel from the electrodes to a delocalized molecular orbital, which in turns provides high transmission, that is, it supports a

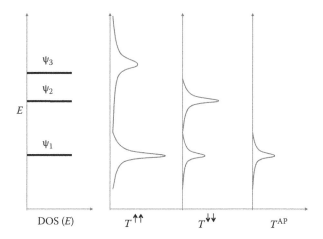

FIGURE 11.3 Illustration of the concept of orbital selectivity applied to a molecular spin valve. Assume a molecule has three molecular levels relevant for the electron transport within the bias window of interest, ψ_1, ψ_2, and ψ_3. One of them, ψ_1, is electronically coupled to both majority and minority spin electrons in the electrodes, while ψ_2 and ψ_3 are only coupled to minority and majority spins, respectively. In this case, $T^{\uparrow\uparrow}$ ($T^{\downarrow\downarrow}$) will present resonances associated to ψ_1 and ψ_3 (ψ_1 and ψ_2). In contrast, T^{AP} will be nonvanishing only where both $T^{\uparrow\uparrow}$ and $T^{\downarrow\downarrow}$ do not vanish, i.e., for ψ_1.

conductance reaching up to the quantum conductance. The total transmission of the junction thus is

$$T_\alpha^{\sigma\sigma'}(E) = \frac{\gamma_\alpha^{L\sigma}\gamma_\alpha^{R\sigma'}}{(E-\varepsilon_\alpha)^2 + (\gamma_\alpha/2)^2}, \tag{11.5}$$

with $\gamma_\alpha = \gamma_\alpha^{L\sigma} + \gamma_\alpha^{R\sigma'}$. Since for transport across small organic molecules spin-flip events can be neglected, we obtain two simple expressions for the total transmission across the molecular orbital ψ_α, when the magnetization vectors of the electrodes are either parallel (P, $\sigma = \sigma'$) or antiparallel (AP, $\sigma = -\sigma'$,) to each other. These are, respectively,

$$T_\alpha^P = T_\alpha^{\uparrow\uparrow} + T_\alpha^{\downarrow\downarrow} \quad \text{and} \quad T_\alpha^{AP} = 2\sqrt{T_\alpha^{\uparrow\uparrow}T_\alpha^{\downarrow\downarrow}}. \tag{11.6}$$

Equation 11.6 establishes the interesting result that the transmission in the AP configuration is the convolution of the transmission for the two spin channels in the P one [23], as illustrated in Figure 11.3. As such, large GMR can be obtained in situations where the transmission in one of the spin channels is strongly suppressed.

We now look at what determines the γ's in the case of resonant tunneling across a molecular orbital. This is rooted in the concept of *orbital selectivity*. Let us assume that the molecular orbital relevant for the transport, ψ_α, has a given well-defined symmetry. It is then expected that only those states in the electrodes that share the same symmetry with ψ_α can efficiently couple to that molecular orbital [43]. As an example, let us consider the situation where the molecule is attached in a straight position to an adatom (along the z direction), so that only the bonding structure between the molecule and the adatom determines the transport. Let us also assume that the relevant molecular orbital for the transport has a π symmetry parallel to the plane of the electrodes, pointing in the x direction. Then, there are only a handful of atomic orbitals that can electronically couple to ψ_α, for instance p_x and d_{xz}, while for the other, for instance s or p_y, the electron hopping is suppressed by symmetry (in standard tight-binding theory, the matrix elements between orbitals not sharing the same angular momentum about the bond axis vanish). Therefore, if for energies $E \sim \varepsilon_\alpha$ no symmetry-allowed orbitals of the electrode are present, then $\gamma_\alpha(E \sim \varepsilon_\alpha) \sim 0$ and the transmission will be highly suppressed. In contrast, if symmetry-allowed orbitals are present, then the transmission will be large. Note that for this argument, we have in practice introduced an energy dependence of the hopping parameter $\gamma = \gamma(E)$, which originates from the nontrivial DOS of the electrodes.

Several demonstrations that orbital selectivity can drastically affect the device transmission and hence the I–V can be found in the literature for nonmagnetic electrodes [44, 45]. In the case of magnetic electrodes, such a property can be used to enhance the spin polarization of the current. In general, in fact, magnetic materials have different Fermi surfaces for spins of different sign [46]. In particular, the orbital content at E_F (which atomic orbitals contribute to the DOS at E_F) can be quite different. This means that one can translate the orbital selectivity of the atomic bond into a spin selectivity via an appropriate choice of the magnetic material for the electrodes, the molecule, and the bonding/anchoring structure.

An example of orbital selectivity is provided in Figure 11.4 (see also Ref. [47]), where we plot the spin polarization of the current through a benzene-thiol-thiolate molecule attached to an Ni surface and probed with an Ni STM tip. This has been calculated by using the nonequilibrium Green's function method implemented with density functional theory electronic structure [34]. The spin polarization of the current at a bias V is defined for the parallel alignment of the magnetization of the substrate and the tip as

$$P(V) = \frac{I^\uparrow - I^\downarrow}{I^\uparrow + I^\downarrow}, \tag{11.7}$$

where $I^\uparrow$ ($I^\downarrow$) is the majority (minority) electrons contribution to the current. From the figure, one can immediately observe two important features. First, $P(V)$ is rather a complicate function of the applied bias, reflecting the fact that different molecular orbitals contribute to the current differently. Second, and most important for the concept of orbital selectivity, the spin polarization appears to be sensitive to the exact position of the STM tip with respect to the molecule. Thus, for instance at the positive voltage of 250 meV, P is positive if the current is collected from the π orbitals of the benzene, but it turns negative when collected from the S atom of the SH group.

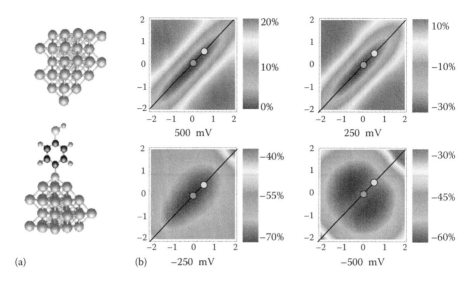

FIGURE 11.4 Spin-polarized transport through a benzene-thiol-thiolate molecule attached to an Ni surface and probed with an Ni tip. In (a), we present a sketch of the calculated system and in (b) a two-dimensional map of the spin polarization P of the current as a function of the position of the STM tip and for different voltages. The diagonal black line in the panels in (b) indicates the plane of the benzene ring, the green circle the position of S, and the blue one that of H in the SH group.

11.4 MAGNETIC MOLECULE DEVICES

The second class of devices is made by contacting molecules comprising magnetic centers (usually transition metal ions) with either nonmagnetic or magnetic electrodes. This is a rather large class, which includes both STM-studied single spins on surfaces and more complex two- and three-terminal devices incorporating single-molecule magnets. Furthermore, depending on the interaction between the molecule and the substrate and on the nature of the substrate itself, a variety of phenomena can be observed, ranging from Coulomb and spin blockade, to Kondo resonant transport, to magnetic inelastic tunnel spectroscopy, to exchange coupling. In what follows, we will briefly review the main related concepts.

11.4.1 STM EXPERIMENTS

As for the case of nonmagnetic molecules, STM-type measurements appear simpler in terms of device fabrication than two- and three-terminal devices. In fact, there is a good choice of standard planar molecules incorporating magnetic centers (porphyrins, phthalocyanines, salens, etc.) that can be evaporated in ultrahigh vacuum conditions, that is, they are fully compatible with STM. In addition to the ease of fabrication, STM measurements for single spins can provide detailed information on the magnetic excitation spectrum and on how this is affected by magnetic coupling, anisotropy, and magnetic fields. In particular, the enabling experimental technique is spin-flip inelastic tunnel spectroscopy (SF-IETS) [48].

SF-IETS is based on a principle common to all inelastic spectroscopy techniques: a change in the electric conductance occurs every time the bias voltage reaches the critical value characteristic of a given excitation. The main concept is illustrated in Figure 11.5 and it is valid for molecules possessing internal degrees of freedom of some kind. An electron in the left electrode at a given energy E has a probability $T(E)$ of being transmitted. Let us assume that there is a molecular orbital of energy ε_α pinned at the highest of the two chemical potentials. In this situation, the transmission coefficient will show a resonance at $E = \varepsilon_\alpha$. As the bias is increased, a second transport channel becomes available. This is produced by the emission of a molecular excitation of energy $\hbar\omega$. If the energy of the final electronic state (after emission) $E = \varepsilon_\alpha - \hbar\omega$ is lower than the chemical potential, μ_2, of the right (collector) electrode, then the electrons cannot be absorbed since the final state in the electrode is occupied. However, if $\varepsilon_\alpha - \hbar\omega > \mu_2$, then the final state in the electrode is empty and the electron can be absorbed. This second inelastic channel contributes to the current and generates a satellite peak in $T(E)$ at $\varepsilon_\alpha - \hbar\omega$.

Whether or not the opening of an inelastic channel increases the current depends strongly on the details of the junction. In the case the molecular excitations are phonons; propensity rules have been established correlating the symmetry of the various vibrational modes and of their coupling to the electrodes to the transmission [49, 50]. As a rule of thumb, in junctions where the transmission is approaching the quantum limit one expects inelastic scattering to reduce the conductance, since the contributions to the current given by the additional inelastic channels are surpassed by an enhancement of the elastic backscattering [51]. In contrast, if the elastic transmission is low, inelastic channels are likely to increase the current. In the first case, one will observe a drop in the differential conductance as a function of bias and a deep in the $\mathrm{d}^2I/\mathrm{d}V^2(V)$, while the second case corresponds to an

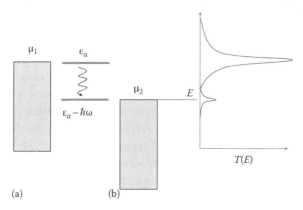

(a) (b)

FIGURE 11.5 Graphical illustration of an inelastic scattering process. (a) An electron at energy ε_α emits an excitation of energy $\hbar\omega$ and lowers down its electronic energy to $\varepsilon_\alpha - \hbar\omega$. If the chemical potential μ_2 of the collector electrode is lower than $\varepsilon_\alpha - \hbar\omega$, such an inelastically scattered electron can be absorbed by the leads and will contribute to the current. (b) Such a scattering event produces a shoulder in the transmission coefficient at $E = \varepsilon_\alpha - \hbar\omega$.

enhancement of $dI/dV(V)$ and a peak in $d^2I/dV^2(V)$. The details of the IETS spectrum are, however, quite complicated to predict, since phonon dissipation after emission can play an important role [52].

In the case where molecular excitations have spin origin, that is, they involve the flipping of the spin direction of the moving electrons, the technique takes the name of spin-flip inelastic tunnel spectroscopy [48]. This presents some additional features to the phonons case, since for spin transition selections rules must be satisfied while it is not the case for phonons. This indeed makes the theory more informative. Other main differences with respect to the phonon IETS is that the energy scale of the magnetic excitations is typically in the sub-meV range, that is, approximately one order of magnitude lower than that of the molecular vibrations. As such, SF-IETS requires severely low temperatures.

SF-IETS was pioneered at IBM with a number of experiments all conducted on magnetic ions deposited over metallic surfaces covered by a thin insulating film [48]. Landmark results include the measurement of the magnetic coupling [53] and of the magnetic anisotropy [54] of single atom chains. The same technique has been employed for planar magnetic molecules deposited on surfaces. The investigations of the details of the super-exchange mechanism in Co phthalocyanine atomic layers deposited on Pb [55] and of the charging state of the same molecules [56] are some examples of this research. It is important to note in this context that SF-IETS works best in the limit of weak electronic coupling between the magnetic center and the electrodes. In the opposite situation, the transport, at least at low bias, is dominated by the Kondo effect. This should deserve a review in itself [57]; here we wish to mention that the Kondo effect in magnetic molecules is well established and can be tuned by STM manipulation of the molecule itself [58]. Also, it is important to remark that a quantitative parameter-free theory of the Kondo effect has begun to emerge [59].

Notably, SF-IETS does not require magnetic electrodes. However, the possibility to perform STM measurements by using spin-polarized currents opens a completely new prospect for the investigation of magnetism at the nano- and atomic scale. This is the growing area of spin-polarized STM [60], where both the probing STM tip and the sample to measure are magnetic (and in some cases also the substrate). A typical measurement then proceeds essentially as a spatially resolved TMR experiment, where the tunneling current is determined by the mutual orientation of the magnetization vectors of the tip and of the sample (in first approximation by their spin-polarized density of states). These, however, are rather delicate experiments, in particular if one aims at atomic scale spatial resolution. For instance, antiferromagnetic tips must be used to avoid stray fields that can affect the magnetic order of the sample to measure, that is, to avoid that the STM data depend on the tip-to-sample interaction. Furthermore, since the decay of the wave function in vacuum is different for different orbital components, one expects that the spatial mapping of the spin polarization may depend on the tip-to-sample separation.

In any case, the amount and quality of information that can be extracted from spin-polarized STM (SP-STM) have probably no equal at present in the field of atomic-scale spintronics. For instance, the fact that different energy (molecular) levels have a different spin polarization allows us to map accurately the spin-polarized spectrum of a nanoscaled object. Although no experiments are available at present for molecules, such a capability makes SP-STM a potentially major tool for the understanding of spin injection into molecular materials. Most of the applications of atomic-scale SP-STM to date have been focused on atoms on surfaces. For instance, the detection of hysteresis signals from Co adatoms of Pt [61] has allowed the investigation of Ruderman-Kittel-Kasuya-Yosida (RKKY)-interaction at the atomic level. Similarly, SP-STM experiments for O adsorbates on Fe have proved the presence of spin-polarized standing waves in thin magnetic films [62]. In the near future, we can envision new experiments where SF-IETS is combined with SP-STM to investigate in detail the transitions between different spin states and spin excitations.

11.4.2 SINGLE MOLECULAR MAGNETS DEVICES

An alternative to studying planar magnetic molecules by STM is that of constructing two- and three-terminal junctions incorporating magnetic molecules. This of course will be the ultimate geometry for fully functional devices, although many hurdles connected to nanofabrications and characterization must be overcome. One over many is the fact that despite many classes of magnetic molecules can be synthesized [26], most of them are extremely fragile away from solution [63, 64]. This means that most likely the molecule entering a device is not the same as the one that was designed for the device. Rapid progress, however, has been made and in 2009 it was demonstrated that molecules belonging to the Fe_4 family can survive on surfaces and preserve both their spin state and most of the anisotropy [65].

Assuming a device can be made, the next important question is how this should operate, that is, what is the physical principle under which the device should function? The first option that comes to mind is that of using spin-valve-like architecture where one magnetic electrode serves as polarizer and the molecule itself as analyzer. This concept, however, lacks a fundamental ingredient; the mutual orientation of the magnetic moment of the molecule with that of the electrode should be stable. Unfortunately, the magneto-crystalline anisotropy in magnetic molecules is usually small (typical energy barriers are in the few Kelvin range), mostly because only a handful of atoms carry the magnetic moment. As a consequence, the molecule is likely to behave as a paramagnetic object at any reasonable temperature and the resulting current not to be spin-polarized. For the same reason, the idea of using magnetic molecules as spin polarizers [66] at the moment appears difficult to realize. An interesting proposal along this direction is that of using magnetic molecules, where the anisotropy results in little or no magnetic moment but a non-zero toroidal moment [67]. Such molecules are predicted to be able to switch the spin polarization of an injected current [68] and still be protected from dipolar interaction (one of the main sources of spin relaxation, at least in molecular crystals).

A second and in our opinion more promising strategy for making devices out of single-molecule magnets is that of relying on the exchange interaction instead of the anisotropy. This essentially means to use the spin state of the molecule as physical property to read, write, and manipulate. The question then becomes, can one address (read, write, and manipulate) the spin state of the molecule without the need of maintaining the spin-quantization axis fixed in time, that is, without the need of strong magneto-crystalline anisotropy? In this section, we will consider the problem of reading the magnetic state, while we will discuss the strategy for writing and manipulating in the last section.

The idea is to be able to convert spin information into molecular orbital information to be detected by an electrical current. In practice, one wants to demonstrate that the different spin states of a molecule present different frontier molecular orbitals or different coupling to the electrodes, both features that might affect the electric current. Again, because of the unlikely possibility of fixing the molecule anisotropy axis, the ideal situation would be to be able to read the molecule electrically without the need of using a spin-polarized current, that is, by using nonmagnetic electrodes.

Such an idea was explored theoretically [69] and it was demonstrated that in a two-terminal device incorporating an Mn_{12} molecule, one can distinguish between the $S = 10$ ground state (GS) and a spin-flip state (SFS) obtained by flipping the magnetic moment of both a Mn^{3+} and a Mn^{4+} ion (the total spin projection of this configuration is 9). Most importantly, this is done by a single non-spin-polarized electrical readout, made possible because of magnetic state–specific molecular orbital re-hybridization under bias. In order to understand this concept, let us first observe that the two spin configurations investigated have a rather similar DOS. The main difference between the GS and the SFS at around E_F is that a molecular orbital singlet belonging to the Mn_{12} highest occupied molecular orbital (HOMO) multiplet has different spin orientation, namely, it is a majority state for the GS and a minority one for the SFS. One then expects such a difference not to be relevant if nonmagnetic electrodes are used, since different spin orientations cannot be distinguished by a non-spin-polarized current. This, however, is no longer true at finite bias as demonstrated by the calculated $I–V$ curve shown in Figure 11.6.

The main feature of the $I–V$ is a number of pronounced negative differential resistances (NDRs) specific for the given spin state. These arise because of the dynamic electron coupling of the various molecular orbital to the electrodes under bias, that is, $\gamma_\alpha^m = \gamma_\alpha^m(E,V)$. Such a dependence of γ_α^m over bias is a consequence of the polarization in an electric field of the wave function, as also illustrated in Figure 11.6. In fact, as the wave function polarizes under the effect of an electric field, its overlap with the electrodes' extended states changes and consequently γ_α^m also changes. If one now considers Equation 11.4, it is clear that a change in the symmetry of the electronic

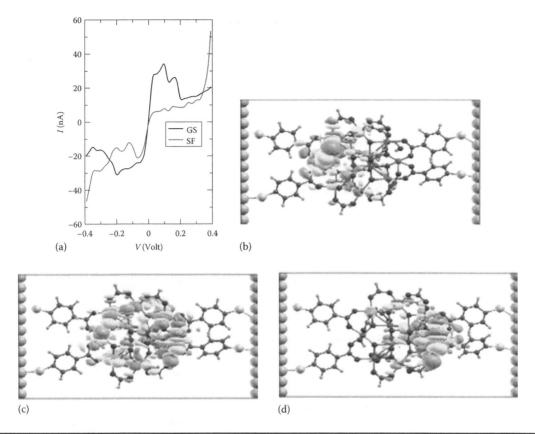

FIGURE 11.6   Transport properties of an Mn_{12} two-probe device. The I–V curves for both the GS and the SFS are presented in (a). The three small plots show isosurfaces of the HOMO wave function for the GS at different bias: (b) 300 mV, (c) = 0, and (d) –300 mV. Note the drastic polarization of the wave function in the electric field.

coupling (change in r_{α}) may reduce the current carried by a given molecular orbital, leading to an NDR. The situation, however, is slightly more complicated because the molecular orbital polarization and then $\gamma_{\alpha}(E,V)$ is associated to the re-hybridization of molecular levels closely spaced in energy. This enables us to read the magnetic state of the molecule. In fact, the molecular orbital that gets spin-flipped when going from the GS to the SFS participates to re-hybridization only in the GS, since in absence of spin flip no hybridization can occur for states of different spins. Therefore, the re-hybridization of the HOMO multiplet in an electric field is different for the GS and the SFS, since the multiplicity of the HOMO for a given spin is different. This means that the response of a spin state to an electric field is driven by the dielectric response of its molecular orbitals, that is, spin information are translated in orbital information.

A mechanism similar to that described here can explain, at least in a qualitative way, some of the low-energy features (NDRs satellite to Coulomb Blockade not involving different charging states) of the I–V curves measured in recent experiments on an Mn_{12} incorporated in a three-terminal device [70, 71]. Alternative explanations ascribe such features to selection rule–forbidden transitions between different spin states [72, 73]. Certainly more work is needed in this area, in particular in the characterization and in general in gathering more data for the various devices. In summary, we wish to remark that the demonstration that different spin states of a molecule can be distinguished electrically has been provided both experimentally and theoretically. Whether or not one will be able to assign with certainty a given spin state to a specific fingerprint in the transport and whether this will be controllable still remains an open question.

11.5 MANIPULATION AND CONTROL

Having demonstrated that the spin state of a molecule can be detected from a transport measurement, the next question is how to read and eventually manipulate such a state. This essentially means how to manipulate the exchange interaction in a controllable way. Note that this rather fundamental aspect underpins not only the potential for new logic devices and sensors, but also the use of magnetic molecules as elements for quantum computation [21, 74]. Demonstrations that the spin state of a molecule can be changed by an external stimulus are abundant. In fact, there exists an entire class of molecules named spin crossover, whose magnetic GS can be altered by light, temperature, or pressure [27]. Typically, such molecules display a high-spin to low-spin transition, which is driven by a change in the geometric configuration between the magnetic center and the surrounding ligands, with the change initiated by the external stimulus. Importantly, spin-crossover activity has been demonstrated only in molecular crystals and to date there is little evidence that the same can happen at the single-molecule level on a surface.

A second strategy for altering the magnetic properties of a molecule is that of acting chemically. For instance, it was demonstrated that the magnetic exchange between two Cr_7Ni rings can be modified by changing the molecular linker coupling the rings [75]. Thus, a linker containing a single $S = 1/2$ Cu^{2+} ion couples the rings antiferromagnetically, while no coupling is detected when the linker is made from a $S = 0$ Cu_2 complex. An intriguing aspect of this protocol is that the electronic structure of the constituent magnetic elements (the Cr_7Ni rings) is little affected by the linker, so that the freedom in the design of the desired compound is rather large. For example, one can envision the use of active linkers, that is, of linkers whose electronic structure can be changed by a local probe such as a gate. Examples of these are the one-electron reducible $[Ru^{2+}Ru^{3+}(OCCMe_3)_4]^+$ linker still for Cr_7Ni rings [76] and $[PMo_{12}O_{40}(VO)_2]q-$ molecules [76]. Finally, it is also worth mentioning that the chemical strategy has also been successfully employed for manipulating the magnetic anisotropy of planar magnetic molecules on surfaces [77].

The important advantage of using active linkers in order to manipulate the exchange interaction between magnetic ions is that the manipulation may be driven by an electric stimulus. A second possibility for manipulating electrically the exchange interaction is based on the electrostatic spin-crossover effect (ESCE), first proposed for high electron density molecular wires [78] and then for polar molecules [79]. In general, when an electric field, $\vec{E}$, is applied to a quantum mechanical system, its energy will change (Stark effect). Such a change is proportional to the first order to $\vec{p} \cdot \vec{E}$, with $\vec{p}$ the permanent electric dipole, and to the second order to $\frac{1}{2} E_i \alpha_{ij} E_j$, where α_{ij} is the polarizability tensor. The idea behind the ESCE is that the high-spin and the low-spin state of a molecule in general have different polarizabilities and, in the case of polar molecules, also different permanent electric dipoles. For this reason, the electric field-driven energy shift depends on the magnetic state of the molecule, so that one can speculate that there exist particular conditions where a high-spin to low-spin crossover is possible. The fundamental question is how large is the crossover field.

If one considers molecules with inversion symmetry, then the first order contribution to the Stark effect vanishes and the difference in the polarizability alone induces the crossover [78]. In order to obtain realistic crossover electric fields, E_{cross}, one should consider molecules with a large spin contrast in the polarizability (i.e., the polarizability of different spin states must be largely different). This translates into a large charge density and a small HOMO-lowest unoccupied molecular orbital (LUMO) gap, and brings two serious drawbacks. First, one needs an extremely challenging chemistry to produce the desired molecules (the ones proposed in Ref. [78] were 10π benzene) and it is not clear whether such a chemical route will ever be available. Second and most importantly, the small HOMO-LUMO gap prevents the onset of a large electric field in a real device, so that even if 10π benzene are made, one will probably never be able to produce an electric field intense enough to switch the molecule.

The use of polar molecules circumvents these problems. The crucial idea is that the permanent electric dipole can effectively "bias" the crossover field to smaller fields. In fact, the energy change as a function of field, Δ_{GS}, is a parabolic function, centered at $E = 0$ for nonpolar molecules (see Figure 11.7). Then, E_{cross} is determined by the interception between the parabolas associated to the different spin states (a spin singlet and a spin triplet in Figure 11.7). The addition of a linear term shifts the center of the parabolas bringing their interception closer to $E = 0$ at least for one of the two field polarities. Hence, the molecule electric dipole effectively introduces a bias field, E_{bias}, which reduces the external electric field needed for the crossover. This appears to be a much more promising strategy since one can first engineer the magnetic molecule and then the specific electric dipole, as demonstrated for acetylene-bridged di-Cobaltocene($CoCp_2$) molecules functionalized with different substituents [79].

Such a field-induced crossover has important consequences for the electric transport. If one of these molecules is sandwiched in a two-terminal device in such a way that there is a potential drop between the magnetic centers, then the strength of the magnetic coupling will change with bias. Interestingly, for biases corresponding to the crossover field the different

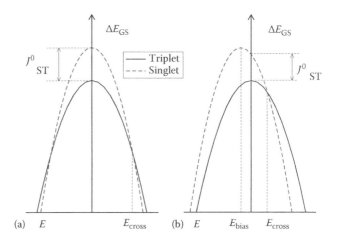

FIGURE 11.7   Stark energy gain, ΔE_{GS}, for the singlet and triplet state of a magnetic molecule as a function of the applied electric field, E. In panel (a), we represent a molecule with inversion symmetry, while in (b) one where the symmetry is broken by an electric dipole. Note that the shift of the $\Delta E_{GS}(E)$ parabola by E_{bias} generates a shift of the crossover field to lower fields. J^0_{ST} is the exchange interaction energy at zero field.

spin states becomes degenerate, that is, on average the molecule will spend an identical amount of time in any one of them. This essentially means that there is no magnetic energy scale at E_{cross} and the current is predicted to become temperature independent [80]. Intriguingly, the electric control of a high-spin to low-spin transition has been recently demonstrated experimentally in a single Mn^{2+} ion coordinated by two tetrapyridine ligands in a three-terminal device geometry [81].

We wish to close this section by mentioning that some theoretical work has been devolved to understand how the magnetization direction of a magnetic molecule can be manipulated with an external current, by essentially translating the concept of spin-transfer torque to the molecular magnets world [82]. Importantly it was shown that a spin-polarized current can switch the magnetic moment of the molecule despite the molecule-intrinsic spin relaxation.

11.6   CONCLUSION AND OUTLOOK

Although the first demonstration of spin transport in an organic material is not even two decades old [6], the field has already evolved very quickly and branched out in several subfields. Here, we have examined only one of them, namely, that of single-molecule spintronics. This appears to be particularly fascinating since it explores the interaction between current electrons and spins at the most fundamental level. It is also of great technological appeal since new multifunctional devices can be envisioned to replace existing inorganic ones or to implement completely different new concepts. A particularly welcome evolution of molecular spintronics is that of addressing molecules with internal degrees of freedom, in particular magnetic molecules.

This seems to be a natural evolution that can bring important consequences on other frontier areas of science such as quantum logic.

However, as the fields mature so do the challenges. At present, there is still no clear way to produce stable and reproducible single-molecule devices. This is a common problem to standard molecular electronics but the situation is even more problematic when the molecules are magnetic, because of their fragility on surfaces. Likewise, magnetic electrodes are more reactive than noble metals and by large incompatible with wet chemical deposition. Then, it is important to develop new useful molecules that are compatible with vacuum deposition, or that can be deposited on magnetic surfaces without oxidation. These need to be characterized thoroughly when in the device. For instance, one should be able to establish and control the bonding structure with the electrodes. Furthermore, it is desirable to engineer new molecules that can be switched between different magnetic states by external stimuli compatible with standard device architecture such as an electrostatic gate.

Finally, we believe it is crucial to find new applications for molecular spin devices. Simply translating concepts from inorganic spintronics does not seem to be a winning strategy. Molecules are not just small, but they are also quantum objects. Therefore, exploiting quantum properties appears like a natural choice. At the same time, it is important to realize that the electronic properties of molecules are always the result of a delicate interplay between different interactions. These can be used for constructing multi-functional devices where more than one property is addressed at the same time. We believe that this is a field that has a lot to offer in the future. In particular, the construction of functional, reproducible, and useful devices may act as a catalytic center for focusing otherwise disconnected disciplines. It is from this synergy of different fields that we expect new fascinating science and revolutionary applications to emerge.

ACKNOWLEDGMENT

This work is sponsored by Science Foundation of Ireland, CRANN and the EU under the FP7 program.

REFERENCES

1. J. S. Moodera, T. S. Santos, and T. Nagahama, The phenomena of spin-filter tunneling, *J. Phys. Condens. Matter* **19**, 165202 (2007).
2. M. Gajek, M. Bibes, S. Fusil et al., Tunnel junctions with multiferroic barriers, *Nat. Mater.* **6**, 296 (2006).
3. J. C. Slonczewski, Current-driven excitation of magnetic multilayers, *J. Magn. Magn. Mater.* **159**, L1 (1996).
4. L. Berger, Emission of spin-waves by a magnetic multilayer traversed by a current, *Phys. Rev. B* **54**, 9353 (1996).
5. C. K. Chiang, C. R. Fincher Jr., Y. W. Park et al., Electrical conductivity in doped polyacetylene, *Phys. Rev. Lett.* **39**, 1098 (1977).
6. V. Dediu, M. Murgia, F. C. Matacotta, C. Taliani, and S. Barbanera, Room temperature spin polarized injection in organic semiconductor, *Solid State Commun.* **122**, 181 (2002).

7. Z. H. Xiong, D. Wu, Z. Valy Vardeny, and J. Shi, Giant magnetoresistance in organic spin-valves, *Nature* (London) **427**, 821 (2004).

8. V. A. Dediu, L. E. Hueso, I. Bergenti, and C. Taliani, Spin routes in organic semiconductors, *Nat. Mater.* **8**, 707 (2009).

9. G. Szulczewski, S. Sanvito, and J. M. D. Coeys, A spin of their own, *Nat. Mater.* **8**, 693 (2009).

10. J. Bass and W. P. Pratt Jr., Spin-diffusion lengths in metals and alloys, and spin-flipping at metal/metal interfaces: An experimentalist's critical review, *J. Phys. Condens. Matter* **19**, 183201 (2007).

11. J. Kikkawa and D. D. Awschalom, Lateral drag of spin coherence in gallium arsenide, *Nature (London)* **397**, 139 (1999).

12. B. Huang, D. Monsma, and I. Appelbaum, Coherent spin transport through a 350 micron thick silicon wafer, *Phys. Rev. Lett.* **99**, 177209 (2007).

13. I. Appelbaum, B. Huang, and D. Monsma, Electronic measurement and control of spin transport in silicon, *Nature (London)* **447**, 295 (2007).

14. L. E. Hueso, J. M. Pruneda, V. Ferrari et al., Transformation of spin information into large electrical signals using carbon nanotubes, *Nature (London)* **445**, 410413 (2007).

15. K. Tsukagoshi, B. W. Alphenaar, and H. Ago, Coherent transport of electron spin in a ferromagnetically contacted carbon nanotube, *Nature (London)* **401**, 572574 (1999).

16. N. Tombros, C. Jozsa, M. Popinciuc, H. T. Jonkman, and B. J. van Wees, Electronic spin transport and spin precession in single graphene layers at room temperature, *Nature (London)* **448**, 571 (2007).

17. S. Pramanik, C.-G. Stefanita, S. Patibandla et al., Observation of extremely long spin relaxation times in an organic nanowire spin valve, *Nat. Nanotechnol.* **2**, 216219 (2007).

18. J. H. Shim, K. V. Raman, Y. J. Park et al., Large spin diffusion length in an amorphous organic semiconductor, *Phys. Rev. Lett.* **100**, 226603 (2008).

19. V. I. Krinichnyi, 2-mm Waveband electron paramagnetic resonance spectroscopy of conducting polymers, *Synth. Met.* **108**, 173 (2000).

20. S. Sanvito and A. R. Rocha, Molecular-spintronics: The art of driving spin through molecules, *J. Comput. Theor. Nanosci.* **3**, 624 (2006).

21. M. Affronte, Molecular nanomagnets for information technologies, *J. Mater. Chem.* **19**, 1731 (2009).

22. H. Agarwal, S. Pramanik, and S. Bandyopadhyay, Single spin universal Boolean logic gate, *New J. Phys.* **10**, 015001 (2008).

23. A. R. Rocha, V. M. Garcia Suarez, S. W. Bailey, C. J. Lambert, J. Ferrer, and S. Sanvito, Towards molecular spintronics, *Nat. Mater.* **4**, 335 (2005).

24. S. Sanvito, Injecting and controlling spins in organic materials, *J. Mater. Chem.* **17**, 4455 (2007).

25. W. J. M. Naber, S. Faez, and W. G. Van der Wiel, Organic spintronics, *J. Phys. D: Appl. Phys.* **40**, R205 (2007).

26. D. Gatteschi, R. Sessoli, and J. Villain, *Molecular Nanomagnets*, Oxford University Press, Oxford, UK, 2006.

27. P. Gütlich and H. A. Goodwin (Eds.), *Spin Crossover in Transition Metal Compounds*, Springer, New York, 2004.

28. M. Alemani, M. V. Peters, S. Hecht, K.-H. Rieder, F. Moresco, and L. Grill, Electric field-induced isomerization of azobenzene by STM, *J. Am. Chem. Soc.* **128**, 14446 (2006).

29. L. Bogani and W. Wernsdorfer, Molecular spintronics using single-molecule magnets, *Nat. Mater.* **7**, 179 (2008).

30. K. I. Pokhodnya, A. J. Epstein, and J. S. Miller, Thin-film V[TCNE]$_x$ magnets, *Adv. Mater.* **12**, 410 (2000).

31. S. Datta, Electrical resistance: An atomistic view, *Nanotechnology* **15**, S433 (2004).

32. M. Büttiker, Role of quantum coherence in series resistors, *Phys. Rev. B* **33**, 3020 (1986).

33. C. Toher, A. Filippetti, S. Sanvito, and K. Burke, Self-interaction errors in density-functional calculations of electronic transport, *Phys. Rev. Lett.* **95**, 146402 (2005).

34. A. R. Rocha, V. M. Garcá-Suárez, S. Bailey, C. J. Lambert, J. Ferrer, and S. Sanvito, Spin and molecular electronics in atomically generated orbital landscapes, *Phys. Rev. B* **73**, 085414 (2006).

35. I. Rungger and S. Sanvito, Algorithm for the construction of self-energies for electronic transport calculations based on singularity elimination and singular value decomposition, *Phys. Rev. B* **78**, 035407 (2008).

36. N. J. Tao, Electron transport in molecular junctions, *Nat. Nanotechnol.* **1**, 173 (2006).

37. J. R. Petta, S. K. Slater, and D. C. Ralph, Spin-dependent transport in molecular tunnel junctions, *Phys. Rev. Lett.* **93**, 136601 (2004).

38. M. Ouyang and D. D. Awschalom, Coherent spin transfer between molecularly bridged quantum dots, *Science* **301**, 1074 (2003).

39. M. Cinchetti, K. Heimer, J.-P. Wüstenberg et al., Determination of spin injection and transport in a ferromagnet/organic semiconductor heterojunction by two-photon photoemission, *Nat. Mater.* **8**, 115 (2009).

40. W. H. Butler, X.-G. Zhang, T. C. Schulthess, and J. M. MacLaren, Spin-dependent tunneling conductance of Fe–MgO–Fe sandwiches, *Phys. Rev. B* **63**, 054416 (2001).

41. M. Julliere, Tunneling between ferromagnetic films, *Phys. Lett. A* **54**, 225 (1975).

42. Z. Ning, Y. Zhu, J. Wang, and H. Guo, Quantitative analysis of nonequilibrium spin injection into molecular tunnel junctions, *Phys. Rev. Lett.* **100**, 056803 (2008).

43. A. P. Sutton, *Electronic Structure of Materials*, Oxford University Press, Oxford, UK, 1993.

44. J. Ning, Z. K. Qian, R. Li, S. M. Hou, A. R. Rocha, and S. Sanvito, Effect of the continuity of the π-conjugation on the conductance of ruthenium-octene-ruthenium molecular junctions, *J. Chem. Phys.* **126**, 174706 (2007).

45. S. M. Hou, Y. Q. Chen, X. Shen et al., High transmission in ruthenium-benzene-ruthenium molecular junctions, *Chem. Phys.* **354**, 106 (2008).

46. T.-S. Choy, J. Naset, J. Chen, S. Hershfield, and C. Stanton, A database of Fermi surface in virtual reality modeling language (vrml), *Bull. Am. Phys. Soc.* **45**, L36 (2000).

47. A. R. Rocha and S. Sanvito, Resonant magnetoresistance in organic spin valves, *J. Appl. Phys.* **101**, 09B102 (2007).

48. A. J. Heinrich, J. A. Gupta, C. P. Lutz, and D. M. Eigler, Single-atom spin-flip spectroscopy, *Science* **306**, 466 (2004).

49. A. Troisi and M. A. Ratner, Molecular transport junctions: Propensity rules for inelastic electron tunneling spectra, *Nano Lett.* **6**, 1784 (2006).

50. M. Paulsson, T. Frederiksen, H. Ueba, N. Lorente, and M. Brandbyge, Unified description of inelastic propensity rules for electron transport through nanoscale junctions, *Phys. Rev. Lett.* **100**, 226604 (2008).

51. N. Jean and S. Sanvito, Inelastic transport in molecular spin valves, *Phys. Rev. B* **73**, 094433 (2006).

52. H. Nakamura, K. Yamashita, A. R. Rocha, and S. Sanvito, Efficient ab initio method for inelastic transport in nanoscale devices: Analysis of inelastic electron tunneling spectroscopy, *Phys. Rev. B* **78**, 235420 (2008).

53. C. F. Hirjibehedin, C. P. Lutz, and A. J. Heinrich, Spin coupling in engineered atomic structures, *Science* **312**, 1021 (2006).

54. C. F. Hirjibehedin, C.-Y. Lin, A. F. Otte et al., Large magnetic anisotropy of a single atomic spin embedded in a surface molecular network, *Science* **317**, 1199 (2007).

55. X. Chen, Y.-S. Fu, S.-H. Ji et al., Probing superexchange interaction in molecular magnets by spin-flip spectroscopy and microscopy, *Phys. Rev. Lett.* **101**, 197208 (2008).

56. Y.-S. Fu, T. Zhang, S.-H. Ji et al., Identifying charge states of molecules with spin-flip spectroscopy, *Phys. Rev. Lett.* **103**, 257202 (2009).

57. G. D. Scott and D. Natelson, Kondo resonances in molecular devices, *ACS Nano* **4**, 3560 (2010).

58. A. Zhao, Q. Li, L. Chen et al., Controlling the kondo effect of an adsorbed magnetic ion through its chemical bonding, *Science* **309**, 1542 (2005).

59. P. Lucignano, R. Mazzarello, A. Smogunov, M. Fabrizio, and E. Tosatti, Kondo conductance in atomic nanocontact from first principles, *Nat. Mater.* **8**, 563 (2009).

60. R. Wiesendanger, Spin mapping at the nanoscale and atomic scale, *Rev. Mod. Phys.* **81**, 1495 (2009).

61. F. Meier, L. Zhou, J. Wiebe, and R. Wiesendanger, The magnetization of individual adatoms, *Science* **320**, 8286 (2008).

62. K. von Bergmann, M. Bode, A. Kubetzka, M. Heide, S. Blügel, and R. Wiesendanger, Spin-polarized electron scattering at single oxygen adsorbates on a magnetic surface, *Phys. Rev. Lett.* **92**, 046801 (2004).

63. S. Voss, M. Fonin, U. Rüdiger, M. Burgert, U. Groth, and Yu. S. Dedkov, Electronic structure of Mn_{12} derivatives on the clean and functionalized Au surface, *Phys. Rev. B* **75**, 045102 (2007).

64. M. Mannini, P. Sainctavit, R. Sessoli et al., XAS and XMCD investigation of Mn_{12} monolayers on gold, *Chem. Eur. J.* **14**, 7530 (2008).

65. M. Mannini, F. Pineider, P. Sainctavit et al., Magnetic memory of a single-molecule quantum magnet wired to a gold surface, *Nat. Mater.* **8**, 194 (2009).

66. S. Barraza-Lopez, K. Park, V. García-Suárez, and J. Ferrer, First-principles study of electron transport through the single-molecule magnet Mn_{12}, *Phys. Rev. Lett.* **102**, 246801 (2009).

67. A. Soncini and L. F. Chibotaru, Toroidal magnetic states in molecular wheels: Interplay between isotropic exchange interactions and local magnetic anisotropy, *Phys. Rev. B* **77**, 220406(R) (2008).

68. A. Soncini and L. F. Chibotaru, Molecular spintronics using noncollinear magnetic molecules, *Phys. Rev. B* **81**, 132403 (2010).

69. C. D. Pemmaraju, I. Rungger, and S. Sanvito, Ab initio calculation of the bias-dependent transport properties of Mn_{12} molecules, *Phys. Rev. B* **80**, 104422 (2009).

70. H. B. Heersche, Z. de Groot, J. A. Folk et al., Electron transport through single Mn_{12} molecular magnets, *Phys. Rev. Lett.* **96**, 206801 (2006).

71. M.-H. Jo, J. E. Grose, K. Baheti et al., Signatures of molecular magnetism in single-molecule transport spectroscopy, *Nano Lett.* **6**, 2014 (2006).

72. C. Romeike, M. R. Wegewijs, and H. Schoeller, Spin quantum tunneling in single molecular magnets: Fingerprints in transport spectroscopy of current and noise, *Phys. Rev. Lett.* **96**, 196805 (2006).

73. C. Romeike, M. R. Wegewijs, M. Ruben, W. Wenzel, and H. Schoeller, Charge-switchable molecular magnet and spin blockade of tunneling, *Phys. Rev. B* **75**, 064404 (2007).

74. M. N. Leuenberger and D. Loss, Quantum computing in molecular magnets, *Nature* **410**, 789 (2001).

75. G. A. Timco, S. Carretta, F. Troiani et al., Engineering the coupling between molecular spin qubits by coordination chemistry, *Nat. Nanotechnol.* **4**, 173 (2009).

76. J. Lehmann, A. Gaita-Ariño, E. Coronado, and D. Loss, Spin qubits with electrically gated polyoxometalate molecules, *Nat. Nanotechnol.* **2**, 312 (2007).

77. P. Gambardella, S. Stepanow, A. Dmitriev et al., Supramolecular control of the magnetic anisotropy in two-dimensional high-spin Fe arrays at a metal interface, *Nat. Mater.* **8**, 189 (2009).

78. M. Diefenbach and K. S. Kim, Towards molecular magnetic switching with an electric bias, *Angew. Chem. Int. Ed.* **46**, 7784 (2007).

79. N. Baadji, M. Piacenza, T. Tugsuz, F. Della Sala, G. Maruccio, and S. Sanvito, Electrostatic spin crossover effect in polar magnetic molecules, *Nat. Mater.* **8**, 813 (2009).

80. S. K. Shukla and S. Sanvito, Electron transport across electrically switchable magnetic molecules, *Phys. Rev. B* **80**, 184429 (2009).

81. E. A. Osorio, K. Moth-Poulsen, H. S. J. van der Zant et al., Electrical manipulation of spin states in a single electrostatically gated transition-metal complex, *Nano Lett.* **10**, 105 (2009).

82. M. Misiorny and J. Barnaś, Effects of intrinsic spin-relaxation in molecular magnets on current-induced magnetic switching, *Phys. Rev. B* **77**, 172414 (2008).

Section VII
Applications

12

Magnetoresistive Sensors Based on Magnetic Tunneling Junctions

Gang Xiao

12.1 INTRODUCTION

The age of electronically based semiconductor devices has been with us for six decades. With more and more electronic devices being packed into smaller and smaller spaces, the limit of physical space may prevent further expansion in the direction the microelectronics industry is currently going. However, a new breed of electronics, dubbed *spintronics* [1, 2],* may alleviate the challenge. Instead of relying on the electron's charge to manipulate electron motion or to store information, spintronics devices would further rely on the electron's spin or its magnetic moment. In spintronics, electrons in a device can be easily manipulated by internal and external magnetic fields or spin-polarized electrical currents. The advantage of spin-based electronics or spintronics is that they are nonvolatile [1] compared with charge-based electronics, and quantum mechanical computing based on spintronics could achieve speeds unheard of with conventional computing. Spintronics will also lead to the emergence of ultrasensitive magnetic sensors, creating new and exciting applications.

Nearly every physical object generates magnetic fields, some strong and some extremely weak. A human heart generates picotesla-scale (10^{-12} T) magnetic pulses, revealing critical cardiac information. A spinning disk inside a hard drive emits magnetic signals with frequency approaching 1 GHz, making the information age possible. The Earth's magnetic field can be a useful navigation tool, particularly where global positioning systems (GPS) are not accessible (e.g., underground and deep sea). Magnetic sensors have been used pervasively in industrial and consumer products [3]. Ultrasensitive magnetic sensors find increasing utility in a number of emerging applications [3]. Magnetocardiography (MCG) [4] uses magnetic sensors to measure the weak electrical signals from the beating heart, allowing the diagnostics of cardiac functions. Magnetoencephalography (MEG) [5], on the other hand, is the magnetic measurement of the electrical activities in the brain. The information obtained from MEG can be used to pinpoint problem regions in the brain of a patient to minimize the invasiveness of

---

* See Chapter 1, Volume 1.

brain surgery. Ultrasensitive magnetic sensors used in MCG and MEG are expensive superconducting quantum interference devices (SQUIDs), which require low-temperature operation.

Thin-film solid-state magnetic sensors enjoy the advantages of small physical size and low cost, though they lack the sensitivity of SQUIDs. Significant progress has been made in improving the sensitivity of thin-film sensors. In data storage applications, magnetic sensors have allowed computers to store massive amounts of information cheaply and reliably [6]. Magnetic sensors are quintessential in the emerging technology of magnetic random access memory (MRAM) [1].* For semiconductor manufacturers, magnetic sensors have made the diagnostics of integrated circuits (ICs) possible by sensing the high-frequency magnetic fields emitted from transistors and interconnects [7–10]. Magnetic sensors are able to detect cracks below the surface of an aircraft or engine turbine by sensing electromagnetic waves originating from the defective area. Ultrasensitive magnetic sensors also have applications in a number of defense-related applications, such as in the detection of submerged submarines or as part of a distributed sensor network [11] ("smart dust") with the eventual goal of monitoring a battlefield remotely. Magnetic sensors are an enabling technology in many areas of science and engineering.

In this chapter, the development of magnetic sensors based on magnetic tunneling junctions (MTJs) is reviewed. Older sensing technologies suffer from varying deficiencies, such as large size, poor sensitivity, and slow speed. MTJ is a disruptive sensing device. It is a tiny "dot" on silicon, invisible to the naked eye, and can be integrated with other thin-film devices such as an IC. The physics of magnetic quantum tunneling [12] dictates that MTJ can function up to multiple gigahertz. MTJ is truly nature's gift, containing a treasure trove of rich physics and promises of applications; so much so that shortly after its discovery, MTJ has taken a large market share in the read/write heads of the data storage industry. The large magnitude of the magnetoresistance (MR) effect in MTJ has created an excellent opportunity for the creation of a new class of high-performance sensor devices.

12.1.1 Magnetic Sensor Types

It is helpful to put MTJ sensors in the context of other available magnetic sensors. There are two broad categories of magnetic sensors: one based on thin-film technology and the other either not based on thin-film technology or requiring special operating conditions [13–16]. Under each category, there are numerous types of magnetic sensors. The two most common in the latter category are SQUIDs and fluxgates. SQUIDs require a cryogenic system to be functional, while fluxgates are expensive to manufacture. Both suffer the disadvantages of high prices, large power consumption, and relatively large physical size. SQUIDs are extremely sensitive, capable of sensing field

---

* See Chapter 13, Volume 3.

strength on the order of femtotesla. Low-temperature SQUIDs work at 4.2 K, whereas high-temperature SQUIDs typically function at 77 K. Each SQUID can cost thousands of dollars or more. Fluxgates consist of two coils of wires wrapped around a soft magnetic cylindrical core. By driving these coils in a particular way and using the resulting magnetic field to saturate the core, the strength of the external magnetic field can be accurately measured. Fluxgates are sensitive, capable of detecting field on the order of $10–100$ pT/Hz$^{1/2}$ over a limited dynamic range. The typical cost of a fluxgate is about $1000.

Sensors that can be fabricated based on standard thin-film methods can be made in large quantities and are generally priced at less than $10 each. Sensors in this category are less sensitive than SQUIDs and fluxgates. However, they enjoy the advantages of small size and low cost. They consume less power than SQUIDs and fluxgates. Thin film–based sensors include Hall-effect (HE), anisotropic magnetoresistance (AMR), giant magnetoresistance (GMR), giant magnetoinductance (GMI), and MTJ sensors.

HE sensors [17] are based on the physics of Lorentz force on an electron inside a magnetic field. They have a simple structure and are easy to manufacture. A typical Hall sensor is very linear within a wide dynamic range, has low sensitivity and moderate to high power consumption, and is among the least expensive thin film–based sensors. Hall sensors account for 98.5% of all magnetic sensors sold worldwide [3].

AMR sensors [13] are based on the physics of electron spin–orbit coupling inside a soft ferromagnetic metal. They use a small resistivity change ($\sim$2%–3%) experienced by certain magnetic materials in the presence of an applied field. AMR sensors are moderately priced, have good sensitivity, and have a low dynamic range.

GMR [6, 18, 19]* sensors are based on the spin-dependent scattering of electrons by the interfaces inside a ferromagnetic and metallic multilayer, first discovered in 1988 [18, 19]. The maximum useful MR in commercial GMR sensors is $\sim$10%–20%. GMR sensors are more sensitive than Hall sensors and cost about the same as AMR sensors.

GMI sensors [20] are based on a large change in impedance of an amorphous metal film inside a magnetic field. This is a relatively new sensor technology. GMI sensors require complicated electronics to detect the magnetic field. GMI sensor development is still in an early stage, and the sensors do not seem to provide a compelling advantage over MTJ sensor technology.

12.1.2 MAGNETIC TUNNELING JUNCTION SENSORS

MTJ sensors represent the latest magnetic sensor technology with several intrinsic advantages. Because these devices rely on tunneling processes, they have a magnetoresistive response, which is fundamentally larger than those of AMR and GMR sensors. The use of a properly fabricated MgO tunnel barrier increases the device MR to over 200%. This leads to a 10-fold increase in device voltage response. In addition, because they are devices

---

* See Chapter 4, Volume 1.

where the current flow is perpendicular to the film plane, their resistance can be varied by five to six orders of magnitude by simply varying the thickness of the insulating barrier layer. These allow MTJ sensors to have a larger response, higher sensitivity, and lower power consumption than HE, AMR, and GMR sensors.

When subjected to an external magnetic field, certain materials undergo a change in their electrical resistivity. The relative resistance change is called *magnetoresistance* (MR), defined as

$$\frac{\Delta R}{R} = \frac{R(H) - R(0)}{R_S},\qquad (12.1)$$

where $R(H)$, $R(0)$, and R_S are the resistance values at field H, at zero field, and at saturation field, respectively. The MR property has been used in magnetoresistive sensors. The field of spintronics was given a huge boost in the late 1980s, when a GMR effect was discovered in layered structures [18, 19]. The commercially developed GMR spin valve, a trilayer structure (NiFe/Cu/Co), first commercialized by IBM around 1997, has been widely used as the sensing element in the read/write heads of computer hard disks [6]. The two scientists who discovered the GMR effect received the 2007 Nobel Prize in physics [21].

The discovery of GMR sparked the search for new devices that could exploit the spin of the electron. The year 1995 saw the introduction of a new type of MR called *tunneling magnetoresistance* (TMR) [22, 23].* TMR occurs in MTJ trilayer structures composed of two ferromagnetic layers separated by a thin insulating barrier (FM/I/FM) [22, 23]. The resistance of an MTJ is a function of the tunneling probability, which depends on the relative orientations of the magnetization (M) vectors of the magnetic layers [12]. When the M vectors are parallel, there is a maximal match between the numbers of occupied electron states in one electrode and available states in the other. Thus, the electron tunneling probability is maximized, and the tunneling resistance (R) is at a minimum. In the antiparallel configuration, electron tunneling occurs between majority states in one electrode and minority states in the other. This mismatch reduces tunneling, causing R to increase. The maximum MR of an MTJ is given by the Julliere model [24]

$$MR = \frac{R_{\uparrow\downarrow} - R_{\uparrow\uparrow}}{R_{\uparrow\uparrow}} = \frac{2P_1 P_2}{1 - P_1 P_2},\qquad (12.2)$$

where:

$R_{\uparrow\downarrow}\ (R_{\uparrow\uparrow})$ is the resistance in the antiparallel (parallel) configuration
P_1 and P_2 are the spin polarizations of the ferromagnets (two electrodes in an MTJ)

In 2001, theoretical work predicted the possibility of extremely large MR values in certain new trilayer structures, which used MgO as the barrier

---

* See Chapter 9, Volume 3.

material [25, 26].* Due to a coherent tunneling effect, it was calculated that a properly prepared MTJ structure using an MgO barrier might exhibit up to 5000% MR. In 2004, two groups independently reported record MR ratios of over 100% in MgO-based MTJ devices [27, 28].[†]

In the case of Al_2O_3-based junctions, the only condition required for the barrier was that it be sufficiently nonconductive. In contrast, MgO-based junctions rely on the quantum mechanical properties of the MgO layer for the realization of ultrahigh MR. Because of this, creating successful tunnel junctions using MgO as the barrier layer is significantly more complicated. One major reason for this is that the coherent tunneling processes required for ultrahigh MR can only occur if the MgO layer has a certain crystal orientation. Achieving this crystal orientation requires tight control of deposition and annealing parameters, and only a handful of groups worldwide have reported success in fabricating MgO-based tunnel junction devices.

Figure 12.1 is a schematic of a typical MTJ layer structure. The two FM electrodes are fabricated from CoFeB, and the insulating barrier is composed of MgO. The remaining layers in the structure are chosen to enhance the material and magnetic characteristics of the device. Typically, to achieve a linear, bipolar operation, one of the two magnetic electrodes (the "pinned layer") in each sensor has its magnetization fixed by the exchange-biasing phenomenon. In Figure 12.1, the bottom FM layer is the pinned layer, whose magnetization is fixed by the adjacent IrMn antiferromagnetic layer. The other magnetic layer, the top FM layer in Figure 12.1, is designed to "freely" respond to an external magnetic field. This magnetic top layer is often called

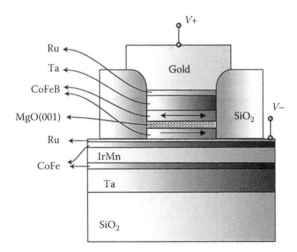

FIGURE 12.1   Schematic of a typical MTJ multilayer structure. The CoFeB layer on top of the MgO barrier is the "free layer," and the CoFeB layer below the MgO is the pinned layer, whose magnetization is fixed by the antiferromagnetic IrMn layer. In a sensor, the magnetization of the free layer is set to be orthogonal to that of the pinned layer at zero external field. In a memory element, the two magnetization vectors are designed to be parallel, as shown.

---

* See Chapter 12, Volume 3.
† See Chapter 11, Volume 1.

the *free layer*. The resulting structure has an electrical resistance that varies linearly as a function of the magnetic field strength over a substantial field range. These devices can be fabricated using conventional semiconductor methods, making them potentially very cheap in large quantities. In addition, because of their two-terminal nature, they are customizable and easy to use.

MTJ magnetic sensors have been commercialized as read heads in hard disk drives and are the most important application of magnetic sensors [29, 30]. The first MTJ heads based on a TiO tunnel barrier were introduced by Seagate in 2004, soon followed by AlO-based MTJ heads by TDK and other manufacturers. Since 2007, these first-generation MTJ heads have been replaced by MgO-based MTJ heads.

12.2 DEVELOPMENT OF MAGNETIC TUNNELING JUNCTION SENSORS

12.2.1 Basic Performance Parameters of MTJ Sensors

The performance parameters used to characterize any magnetic sensor are also applicable to MTJ sensors, although the latter have some unique parameters. What distinguishes MTJ sensors from other types of thin film–based sensors is the parameter of *sensitivity*, which is the largest for MTJ sensors. For an MTJ resistor sensor, the sensitivity is defined as the resistance change (%) per unit field Oersted (%/Oe) at zero field. For an MTJ bridge sensor, sensitivity takes the form of mV/V/Oe (voltage output per 1 volt of applied voltage per Oe). These numbers can be converted from one to the other.

The sensitivity of an MTJ sensor depends on the externally applied voltage (bias) to the sensor, as a high bias voltage will lead to a lower MR due to tunneling physics. Often, an MTJ sensor consists of many MTJ elements connected in series, so that the total applied voltage is divided into only a small voltage to a single element. This allows the MTJ sensor to operate under a large applied voltage. This method also enhances the protection of MTJ sensors against electrostatic discharge (ESD).

The *detectability* of a magnetic sensor is the intrinsic magnetic noise, which is dependent on frequency. For low-frequency applications, the detectability has the units of nanotesla/hertz$^{1/2}$ at $f = 1$ Hz. Typical MTJ sensors have detectability of 1–10 nT, and the goal in the research community is to lower this to a few picotesla.

A linear and bipolar magnetic sensor is true only within a field *dynamic range* (±Oe). In general, the push for high sensitivity leads to the narrowing of dynamic range. A magnetic sensor should have as low a *hysteresis* as possible. Hysteresis is in general measured as a percentage of the dynamic range.

The *thermal coefficients* result from the temperature dependence of the tunneling resistance (thermal coefficient of resistance [TCR]) and that of the magnetoresistance (thermal coefficient of magnetoresistance [TCMR]). The effect of TCR can be compensated by using the design of a bridge, but the effect of TCMR cannot. One needs to rely on electronics and calibration data to take TCMR into consideration. At very high temperatures, MTJs can break down due to interlayer diffusion or decaying magnetization of the free layer. It is useful to know the operating temperature range of an MTJ sensor.

One of the advantages of MTJ sensors is that their *resistance* can be easily varied over many orders of magnitude by adjusting the tunneling barrier thickness. High resistance is important for applications requiring low power consumption (mobile devices), whereas low resistance benefits high-frequency (MRAM, read/write head) or low-noise applications.

Another advantage of MTJ sensors is that they can be made at a very small *physical size*. The size is defined by the area or volume of the free layer, which can have dimensions as small as tens of nanometers. As the resistance is concentrated near the barrier, the lead resistances can often be ignored. Nanoscaled MTJ sensors benefit the applications of scanning magnetic microscopy [7–9, 31]. Here, small size means high spatial resolution in field measurement.

MTJ sensors enjoy a broad *frequency response*. An MTJ sensor is always a good sensor for a direct current (dc) field. Intrinsically, the energy scale of magnetic tunneling is set by the Fermi energy, and a large MR effect remains intact over a broad frequency range. A well-designed MTJ sensor can be ultrafast in response. There are two frequency limits. One is the ferromagnetic resonance frequency of the free layer, which can be as high as tens of gigahertz. The other is the resistance-capacitance (RC) time constant of the MTJ sensor. Properly designed, the upper limit of an MTJ sensor is many gigahertz.

12.2.2 PHYSICAL MODEL OF MTJ SENSOR

An MTJ stack is a multilayer deposited on thermally oxidized silicon substrates using dc and radiofrequency (rf) sputtering. A representative layer structure has the sequence of (thickness in nanometers) 5 Ta/30 Ru/5 Ta/2 CoFe/12 IrMn/2 CoFe/1 Ru/3 CoFeB/2 MgO/3 CoFeB/2 Ta/2 Ru (see Figure 12.1). The 12 IrMn/2 CoFe/1 Ru/3 CoFeB layer assembly just below the MgO barrier is a synthetic antiferromagnetic (SAF) pinned layer. The net magnetization vector of the SAF layer is fixed in the plane of the layer and only weakly perturbed by an external field. The CoFeB layer above the MgO barrier is patterned into an oval shape having a certain aspect ratio and dimensions from tens of nanometers to tens of micrometers.

To make a sensor, the pinning direction of the pinned layer is set to be perpendicular to the easy axis of the free layer. The external field to be sensed is applied along the hard axis of the free layer, causing the magnetization vector of the free layer to rotate through a small angle from the quiescent state. As a consequence, the MTJ sensor experiences a change in resistance, which is proportional to the external field in a bipolar fashion. In an ideal case, the free layer should behave like a single domain "particle" with a uniaxial magnetic anisotropy having a very small anisotropy constant and a very small magnetic hysteresis. The free layer should be magnetically "decoupled" from the pinned layer. Also, the pinning force should be very large, so that the pinning direction will not be affected by the external field. The objective in the development of MTJ sensors is to achieve conditions as close to the ideal sensor as possible by performing research on magnetic materials, layer structures and lateral geometries, and fabrication processing.

12.2.3 MAGNETORESISTANCE RATIO AND SATURATION FIELD: KEYS TO LARGE SENSITIVITY

For an ideal sensor, the field sensitivity is given by

$$S = \frac{R_{\uparrow\downarrow} - R_{\uparrow\uparrow}}{R_{\uparrow\downarrow} + R_{\uparrow\uparrow}} \cdot \frac{1}{B_{sat}}, \tag{12.3}$$

where B_{sat} is the magnetic saturation field of the free layer. Note that the MR ratio, $2(R_{\uparrow\downarrow} - R_{\uparrow\uparrow})/(R_{\uparrow\downarrow} + R_{\uparrow\uparrow})$, is different from the commonly used MR ratio in the literature, $(R_{\uparrow\downarrow} - R_{\uparrow\uparrow})/R_{\uparrow\uparrow}$. According to Equation 12.3, the key to increasing the field sensitivity is to obtain the largest possible MR and to reduce B_{sat}, which can be achieved by using available methods in micromagnetics. Obtaining large MR requires the discovery of new materials and new physics.

Since the discovery of MTJs at room temperature in 1995, there have been continuing efforts in developing MTJ magnetic sensors. The gradual increase in MR has benefited sensor performance. Earlier on, most research was focused on MTJs with an Al_2O_3 insulating tunneling barrier. Typically, an Al sputtering target was used and deposited in an oxidizing environment of Ar and oxygen plasma. The amorphous Al_2O_3 barrier thus formed is dense and of high quality. Between 1995 and 2004, the MR ratio steadily increased from about 10% to 70% at room temperature [22, 23, 32]. Figure 12.2 shows some sensor parameters of a series of FeNi/Al_2O_3/FeNi MTJs after a systematic thermal annealing treatment [33]. After an optimal annealing at 170 °C, the maximum MR reaches about 35% with a maximal field sensitivity of 5%/Oe. The junction area (free layer) is $100 \times 150\,\mu m^2$. In a CoFe/Al_2O_3/NiFeCo junction with a total MR of 20%, a sensitivity of 3%/Oe was obtained by applying a biasing field along the hard axis of the free layer [34].

The Al_2O_3-based MTJs typically have a polycrystalline structure, and the Al_2O_3 barrier is amorphous. Electrons with different wavefunction symmetries are allowed to tunnel, and the overall spin polarization is limited. To increase MR, it was concluded theoretically that single-crystal MTJ structures can enhance MR substantially [25, 26]. This is because tunneling efficiency depends on wavefunction symmetry. Some particular bands having high spin polarization can lead to very high MR if these electrons become the dominant contributors to tunneling in a certain crystalline orientation. Soon, experiments on epitaxial [001] Fe/MgO/Fe and FeCoB/MgO/FeCoB MTJs confirmed the theoretical conclusions [27, 28, 35]. Today, MR has reached over 600% to as high as 1056% in this type of [001] MgO-based MTJs [36, 37], which have become the primary MTJs for sensor development. With high MR ratios, the sensitivity of MTJ sensors has also steadily increased.

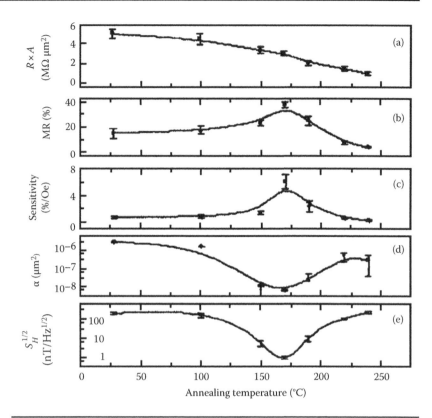

FIGURE 12.2   Annealing temperature dependence of (a) junction resistance, (b) TMR ratio, (c) sensitivity, (d) normalized $1/f$ noise, and (e) magnetic field sensitivity. (After Liu, X. et al., *J. Appl. Phys.* 94, 6218, 2003. With permission.)

Equally importantly, reducing the saturation field B_{sat} will also increase the sensitivity, according to Equation 12.3. Fe and CoFeB are not very soft ferromagnets, but they are required to form the MTJ free layer because of their special band structures. However, one can deposit a thicker ferromagnetic soft film on top of the CoFeB free layer to make a compound free layer. B_{sat} can be reduced, because it is now dominated by the thicker "soft" film. Recently, Egelhoff et al. [38] have used 100 nm of $Ni_{77}Fe_{14}Cu_5Mo_4$ as the thicker "soft" film on top of the 5 nm CoFeB free layer of a working MgO-based MTJ. The low-field hysteresis loop was found to be linear and nonhysteretical and extrapolates to saturation at 0.8 Oe. The resulting sensitivity is about 70%/Oe, which is the highest sensitivity ever reported for any MTJ structure on the wafer level. Processing the MTJ structure into smaller sensor sizes will increase the B_{sat} somewhat due to shape anisotropy. Still, this extremely high sensitivity sets a limit for the MTJ sensors.

There is another approach to increasing the sensitivity by incorporating a flux concentrator near the free layer. A flux concentrator is a soft ferromagnetic film with high magnetic susceptibility and with a certain geometry. By focusing the external magnetic flux toward the free layer, the magnetic field can be increased by a "gain" factor, and so can the sensitivity. Chaves et al.

demonstrated this approach by adding 500 nm thick $Co_{93}Zr_3Nb_4$ flux concentrators with different shapes to CoFeB/MgO/CoFeB MTJ sensors [39, 40]. They were able to increase the sensitivity substantially, with some hysteresis even under a hard-axis biasing field. One disadvantage of this method is that depositing and patterning the flux concentrators increases the manufacturing cost and the physical size of the final sensor products. For this reason, sensors with flux concentrators are not preferred in applications where high spatial resolution is required, such as in scanning magnetic microscopy.

12.2.4 NOISE CHARACTERISTICS AND FIELD DETECTABILITY

The field detectability of a magnetic sensor is determined by the intrinsic noise of the MTJ. Noise characterization can reveal physical behaviors that may be difficult to detect in other experiments. For both sensing and MRAM applications, it is important to understand noise in relation to signal levels. Earlier on, Ingvarsson et al. [41, 42] measured the bias and magnetic field dependence of voltage noise in micron-scale Al_2O_3-based MTJs. At low frequency (f), they found that the $1/f$ noise has two sources: one due to the charge traps in the tunnel barrier and the other due to quasi-equilibrium thermal magnetization fluctuation in the ferromagnetic electrode. The $1/f$ noise was shown to be proportional to the slope dR/dH, tracking the dc susceptibility for the top free layer [42]. This finding provides a good quantitative test of the validity of the fluctuation–dissipation theorem. It was concluded that to curtail the low-frequency noise in MTJs, the thermal stability of the magnetization must be enhanced by subduing the activated motion of domain walls. Nowak et al. [43] studied the low-frequency noise in similar junctions with large areas and attributed the $1/f$ noise only to charge trapping processes.

The field detectability is dependent on two quantities: the signal, expressed typically in change of output voltage per unit of magnetic field (V/Oe), and the noise, expressed as the noise voltage per unit frequency (V/Hz$^{1/2}$). The field detectability of any sensor is simply given by the ratio of these two outputs.

To obtain good detectability, one needs to minimize the noise of a sensor. Many types of noise inhibit the field-sensing capability of magnetic sensors. $1/f$ noise, intrinsic to sensors and amplifiers, is typically dominant at low frequencies. Beyond a knee frequency, white frequency noises dominate. Johnson–Nyquist noise is intrinsic and dependent on sensor resistance and temperature; shot noise is caused by current fluctuations and scales with current. In addition, sensors can have Barkhausen noise, caused by domain wall motion. Mathematically, a typical noise spectrum in MTJs can be described as

$$S_V(f) = \alpha \frac{V^2}{Af} + 4kRT + 2eVR, \tag{12.4}$$

where:

V is the voltage across the sensor

R is the sensor resistance

A is the junction area

α is a constant (Hooge parameter) based on material parameters

k is the Boltzmann constant

e is the electron charge

The first term represents a sensor's intrinsic $1/f$ noise, and the second and third terms are the Johnson and shot noise, respectively. Once the voltage noise is experimentally measured, the magnetic field noise (detectability) can be calculated from [44]

$$S_H = \left(\frac{\Delta H}{\Delta V}\right)^2 \quad S_V = \frac{1}{\left(\dfrac{1}{R}\cdot\dfrac{\Delta R}{\Delta H}\right)^2}\cdot\frac{S_V}{V^2}. \tag{12.5}$$

The term $(1/R)\cdot(\Delta R/\Delta H)$ is simply the sensitivity of a sensor (measured in relative change of resistance per Oe).

Ren et al. [44] studied in detail the low-frequency noise in FeNi/Al$_2$O$_3$/FeNi MTJ sensors. They confirmed that the $1/f$ noise component scales with sensor bias voltage according to Equation 12.4, as shown in Figure 12.3. They further confirmed that the $1/f$ noise is dominated by magnetic noise. A strong correlation between the $1/f$ noise and the sensitivity $(1/R)\cdot(\Delta R/\Delta H)$ was observed. Figure 12.4 shows the transfer curve (R vs. H) and the associated sensitivity variable as a function of H. It also shows that $1/f$ noise level at $f = 1$ Hz is a function H. At fields where the sensor is most sensitive, so is the $1/f$. Figure 12.5 shows the normalized noise value versus the sensitivity. A linear correlation was observed. This was explained by the thermally

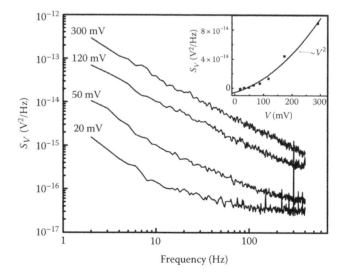

FIGURE 12.3   Voltage power spectral density S_V in frequency domain under different bias voltage V on an MTJ sensor. Inset is S_V measured at frequency $f = 10$ Hz as a function of bias voltage. (After Ren, C. et al., *Phys. Rev. B* 69, 104405, 2004. With permission.)

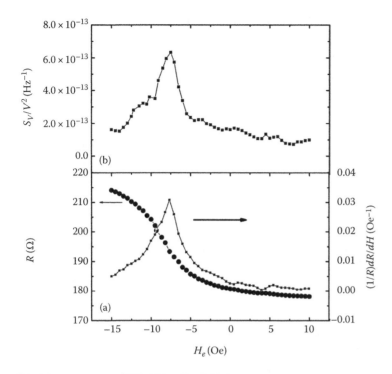

FIGURE 12.4 The voltage-normalized noise in field spectra (a) and the corresponding $R(H)$ curve and its derivative (b) for a 36×24 μm^2 MTJ with biasing field of 8 Oe. The noise value is taken from extrapolating low-frequency noise to 1 Hz. (After Ren, C. et al., *Phys. Rev. B* 69, 104405, 2004. With permission.)

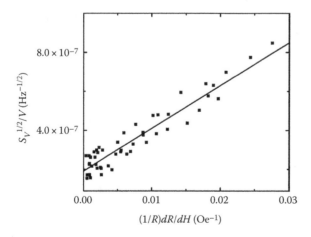

FIGURE 12.5 The voltage noise $S_V^{1/2}/V$ versus the sensitivity (derivative of MR) in an MTJ sensor. (After Ren, C. et al., *Phys. Rev. B* 69, 104405, 2004. With permission.)

activated magnetization fluctuations in the free layer. It can be concluded that a large MR ratio or sensitivity would increase $1/f$ magnetic noise and therefore, would not enhance field detectability as described by Equation 12.5. The only way to lower the $1/f$ magnetic noise is to reduce the magnetization fluctuations. The slope in the linear correlation in Figure 12.5 is a measure of the strength of magnetization fluctuations in the free layer. The smaller the slope, the lower the $1/f$ magnetic noise. The noise component at $(1/R)\cdot(\Delta R/\Delta H) = 0$ (the intercept on the vertical axis in Figure 12.5) results from the nonmagnetic contribution to $1/f$ noise. This component is attributed to defects and localized charge traps in the barrier.

Liu and Xiao [33] found that proper thermal annealing on FeNi/Al$_2$O$_3$/FeNi MTJs reduces $1/f$ magnetic noise, as shown in Figure 12.2. By varying the annealing temperature up to 250 °C, they found that at 170 °C, MR doubles to 35% from 17% without annealing. Most importantly, the field detectability is improved by two orders of magnitude, from over 100 nT/Hz$^{1/2}$ to 1 nT/Hz$^{1/2}$ at 1 Hz. Liou et al. [45] have also found that high-temperature annealing in hydrogen gas can reduce the $1/f$ noise by about 50% in Al$_2$O$_3$-based MTJ sensors.

The $1/f$ characteristics in MgO-based MTJ sensors are similar to those observed in the Al$_2$O$_3$-based MTJ sensors. The voltage noise was again found to be linear in the sensitivity of a sensor [46]. Even if the sensitivity is increased, the increased noise will cancel out any gains in sensor performance.

The figure of 1 nT/Hz$^{1/2}$ at $f = 1$ Hz for MTJ sensors is similar to the detectability of commercial AMR and GMR sensors. One major difference is that commercial AMR and GMR sensors are equipped with flux concentrators, but the MTJ sensors studied are not. Stutzke et al. [47] surveyed the low-frequency noise in commercial sensors. Most of the GMR and AMR sensors with flux concentrators have a noise figure of 3–10 nT/Hz$^{1/2}$ at 1 Hz. The best AMR sensor has a noise level of 0.3 nT/Hz$^{1/2}$ at 1 Hz. This further demonstrates that better sensitivity in MTJ sensors than in GMR and AMR sensors does not mean better low-frequency noise level. The only way to reduce the low-frequency noise in magnetoresistive sensors is to enhance the thermal magnetic stability of the free layer. It is beneficial to use a ferromagnetic material with a large magnetic exchange constant and to configure a stable and uniform single domain structure free of minor domains and excessive domain wall motions. It is also useful to make a high-quality tunnel barrier free of detects and charge trappings to reduce the nonmagnetic $1/f$ noise level.

Sensitivity and $1/f$ noise are strongly dependent on biasing magnetic fields applied along the easy and hard axes of the free layer. Mazumdar et al. [48] have developed a characterization procedure to map the sensitivity, $1/f$ voltage noise, and field noise in an orthogonal magnetic field arrangement (±100 Oe in both axes), as shown in Figure 12.6, on CoFeB/MgO/CoFeB MTJ sensors. The normalized voltage noise map (in units of volts/hertz$^{1/2}$ at 100 Hz) of Figure 12.6b resembles the sensitivity map (in units of %/Oersted) of Figure 12.6a, implying the scaling of noise and sensitivity

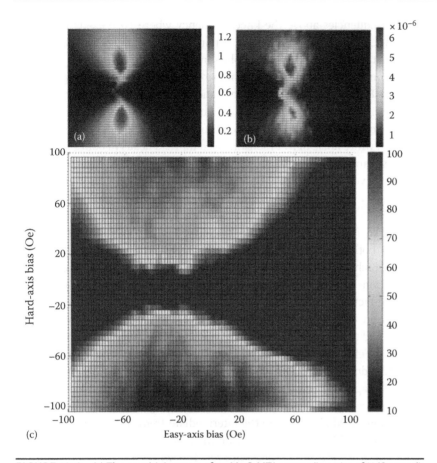

FIGURE 12.6 (a) The sensitivity map of an MgO MTJ sensor (in units of %/Oersted); (b) the corresponding normalized voltage noise map in units of hertz$^{-1/2}$, both measured in a 2D orthogonal field arrangement; (c) magnetic noise map. The color bars are in units of nanotesla/hertz$^{1/2}$ and set to 10–100 nT/Hz$^{1/2}$. (After Mazumdar, D. et al., *J. Appl. Phys.* 103, 113911, 2008. With permission.)

and the dominance of magnetization fluctuation. Using Equation 12.5, the magnetic field noise map is generated as in Figure 12.6c. It is seen that magnetic field noise can be reduced by applying a hard-axis biasing field. One can achieve this by depositing a thin-film magnet structure around the free layer of an MTJ sensor.

By incorporating a hard-axis field bias and a flux concentrator, Chaves et al. [39, 40] demonstrated a noise level of about 0.25 nT/Hz$^{1/2}$ at 1 Hz and 5 pT/Hz$^{1/2}$ at 500 kHz in CoFeB/MgO/CoFeB MTJ sensors. Without the flux concentrator, the noise level increased to 16 nT/Hz$^{1/2}$ at 1 Hz for a very large sensor [49] with a total area of 10 mm^2. As shown in Equation 12.4, a sensor's $1/f$ noise scales inversely with junction area. Using large areas, flux concentrators, and on-chip magnets presents challenges to cost-effective manufacturing. More research is needed to develop better materials and sensor configurations as means of reducing the intrinsic noise level of MTJ sensors.

At frequencies above the knee frequency where $1/f$ noise is no longer dominant, Johnson and shot (white) noise set the limit of field detection (see Equation 12.4). To reduce the Johnson noise, low resistance, for example, tens of ohms, is preferred. Johnson noise does not depend on voltage bias to a sensor. The shot noise, on the other hand, does depend on the voltage bias. Therefore, increasing the voltage bias, which generates a larger signal, does not lead to increased signal-to-noise ratio. Klaassen et al. [50] measured the electrical white noise of MTJ sensors used for read heads in data storage and with resistances ranging from a few ohms to over 100 Ω. They found that at low bias voltages $\ll 2$ kT/e, the noise is dominated by Johnson noise, and at $\gg 2$ kT/e, noise is dominated by shot noise for high-quality ("healthy") junctions.

Tserkovnyak and Brataas [51] developed a semiclassical theory of shot noise in both GMR and MTJ physical systems and calculated the Fano factor, which is the ratio of real shot noise to the full shot noise defined in Equation 12.4. In the cases of ballistic, diffusive, and tunnel junctions, the Fano factor is 0, 1/3, and 1, respectively. Furthermore, the Fano factor depends on the relative orientation between the magnetization vectors of the free and the pinned layer. Guerrero et al. [52] measured the shot noise of $Co/Al_2O_3/FeNi$ MTJs and found that the Fano factor is about 0.7. They attribute the reduced Fano factor to sequential tunneling through nonmagnetic and paramagnetic impurity levels inside the tunnel barrier. Understanding how the Fano factor depends on materials and junction quality will benefit applications of MTJ sensors in the intermediate-frequency range.

For small MR sensors (GMR or MTJ), there exists another kind of white noise in addition to Equation 12.4. This white resistance noise, often called the *mag-noise* [53, 54], results from the thermal magnetization fluctuations (spin waves) in the free and pinned layers. Unlike electronic thermal noise (Johnson), the mag-noise scales with field sensitivity and inversely with the volumes of the ferromagnetic electrodes. In MR read heads used in hard disk drives, mag-noise can become the dominant white noise because of the sensors' small sizes. For MTJ sensors with large junction area (hundreds of micrometers), mag-noise is less of an issue. Finite element micromagnetic simulations [55] show that the mag-noise, which increases linearly with temperature, can be suppressed significantly by an easy-axis bias field.

The current goal in the development of MR sensors is to increase field detectability from roughly 1 nT to 1 pT in sensors that are compact, ultra-sensitive, not power hungry, and inexpensive to manufacture [56]. Judging from the progress in the last 10 years, such a goal is within reach in the future, but challenges abound. Research is ongoing by focusing both *intrinsically* and *extrinsically* on the MTJs. The *intrinsic* way is to improve barrier and junction quality and to search for "good" ferromagnetic materials and the optimal micromagnetic design and device processing. One of the *extrinsic* ways is to incorporate MTJ sensors with microelectromechanical system (MEMS) actuators [57–59]. By using MEMS to create small-amplitude, high-frequency oscillation of the sensor–sample separation, $1/f$ noise can be suppressed. Such an approach tends to increase the manufacturing and integration cost.

12.2.5 SENSOR LINEARIZATION AND MAGNETIC HYSTERESIS

In an ideal magnetic sensor, the transfer curve (output vs. field) should be linear within its intended field dynamic range, and magnetic hysteresis in the transfer curve should be vanishingly small. Figure 12.7 shows a transfer curve of a commercial MgO-based MTJ sensor over a dynamic range of ±1 G. The sensitivity is 0.71%/G, and the sensor is very linear (the small amount of hysteresis in this measurement is actually an artifact of the electromagnets, which supply the field).

To achieve linearization, MTJs should be designed such that the free and pinned layer magnetizations are at a 90° angle in the absence of an applied field. The pinned layer magnetization direction is set by annealing MTJ wafers in a uniform strong magnetic field (more than a few kilogauss). After cooling back to room temperature, the AFM/FM exchange coupling keeps the pinned layer magnetization direction firmly fixed. To make the free layer magnetization direction orthogonal to that of the pinned layer, sensor designers would typically take advantage of the magnetic shape anisotropy and pattern the free layer into an oval shape. The longer (easy) axis is perpendicular to the pinned layer magnetization direction. The larger the eccentricity, the larger the field dynamic range, and the lower the field sensitivity. This method, called the *orthogonal sensor*, uses the intrinsic uniaxial anisotropy of the free layer to align its orientation in the desired direction without the need for external magnetic fields. The intrinsic uniaxial anisotropy is not limited to shape but also can result from magnetocrystalline, magnetostrictive, and other types of anisotropy. Figure 12.8 shows the schematic transfer curve for an orthogonal sensor. Lu et al. [60] have used shape anisotropy to control the transfer curves in MTJs.

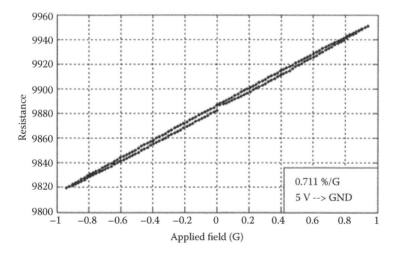

FIGURE 12.7   Transfer curve of an MgO MTJ sensor (resistance sensitivity is 0.711%/G).

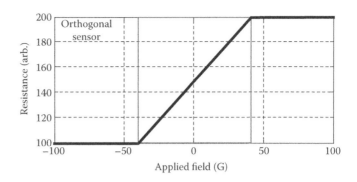

FIGURE 12.8   Schematic transfer curve for an orthogonal sensor.

In reality, making an orthogonal sensor is challenging. If the junction area is large, the free layer often breaks into a multi-domain state, making magnetization rotation in unison impossible. Such a sensor would become nonlinear and hysteretic. On the other hand, if the junction area is small, the edge roughness and other structural defects may hinder the free rotation of local magnetizations. This would cause Barkhausen jumps in the transfer curve. To eliminate the hysteresis and Barkhausen jumps, sensor designers can add a constant magnetic field that forces the free layer magnetization into the desired state. This can be done by an external field or a local field via a thin-film magnet or current-carrying stripline. Such a sensor is called the *offset sensor*, whose schematic transfer curve is shown in Figure 12.9. GMR and MTJ read heads in hard drive disks are examples of the offset sensor.

Liu et al. [61] have studied the method of the offset sensor on FeNi/ Al_2O_3/FeNi MTJs. The junctions have the shape of rectangles with dimensions of 100×150 μm². Figure 12.10 shows the transfer curves under different biasing fields, which are orthogonal to the sensing field direction. At 3 Oe, the hysteresis nearly disappears, and the sensor is linearized. The collapse of coercivity is further demonstrated in Figure 12.11. Pong et al. [62] have also demonstrated hysteresis collapse for linear response by sensor rotation and applying an external field on $Co_{50}Fe_{50}/Al_2O_3/Co_{50}Fe_{50}/Ni_{77}Fe_{14}Cu_5Mo_4$ MTJ wafers.

Recently, a new method to linearize an MTJ sensor has been found [63, 64]. The hysteresis property in CoFeB/MgO/CoFeB is sensitively dependent

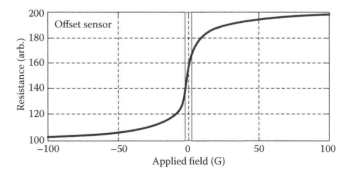

FIGURE 12.9   Schematic transfer curve for an offset sensor.

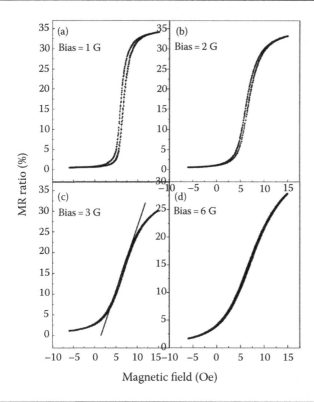

FIGURE 12.10 Transfer curves under different biasing fields (a through d) that are orthogonal to the sensed field direction. (After Liu, X. et al., *J. Appl. Phys.* 92, 4722, 2002. With permission.)

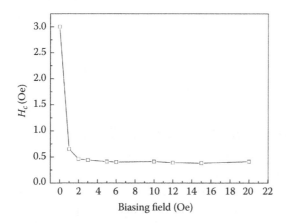

FIGURE 12.11 Coercivity of an MTJ sensor versus biasing field. Coercivity collapses under a biasing field of 2 Oe. (After Liu, X. et al., *J. Appl. Phys.* 92, 4722, 2002. With permission.)

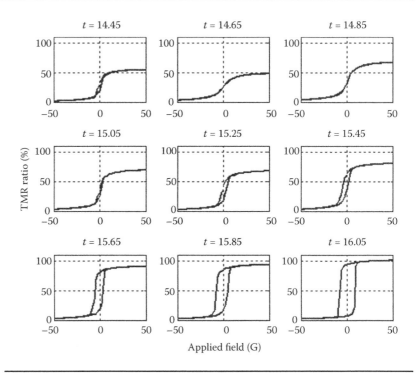

FIGURE 12.12 Representative transfer curves (resistance versus field) for a number of MgO MTJs with increasing free layer (CoFeB) thickness, ranging from 14.45 to 16.05 Å. Below the critical thickness of 15 Å, the free layer becomes superparamagnetic at room temperature, and the MTJ device behaves like a hysteresis-free linear sensor. (After Shen, W. et al., *Phys. Rev. B* 79, 014418, 2009. With permission.)

on the thickness of the free layer. Linear and hysteresis-free transfer curves are obtained simply by reducing the free layer thickness to below a critical thickness of 1.5 nm (see Figure 12.12). Detailed analysis on the transfer curves shows that they agree well with the Langevin equation describing superparamagnetism [64]. In this magnetic state, the free layer enjoys zero hysteresis and still large magnetic susceptibility. It is not clear whether sensor performance will be strongly dependent on temperature, as is the relaxation rate of a superparamagnet.

12.2.6 INTERLAYER COUPLING AND SENSOR OFFSET

Because of the close proximity, there exists a magnetic interlayer coupling between the free layer and the pinned layer. Such a coupling generates an internal field along the sensing axis of the free layer. As a result, the center of the transfer curve is shifted by this internal coupling field (offset field), as shown in Figure 12.13. Since the highest sensitivity typically occurs near the center of the transfer curve, this sensor offset will lower the sensor's sensitivity and reduce its dynamic range. Two effects produce extraneous magnetic fields in the plane of the free layer [65–68]: Néel "orange-peel" coupling

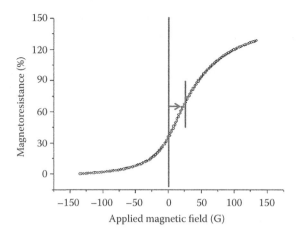

FIGURE 12.13   The center of a transfer curve of an MgO MTJ sensor is shifted by an offset field due to an internal coupling field between the free layer and the pinned layer.

(H_N), due to interface roughness, and magnetostatic coupling (H_M) due to

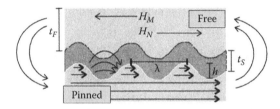

FIGURE 12.14   Internal magnetic coupling between the free layer and the pinned layer. H_N is Néel "orange-peel" coupling due to interface roughness, and H_M is magnetostatic coupling due to uncompensated magnetic poles on the edges of the pinned layer. (After Schrag, B.D. et al., *Appl. Phys. Lett.* 77, 2373, 2000. With permission.)

uncompensated poles on the edges of the pinned layer. Figure 12.14 shows schematically the origins of these two fields.

Assuming a sinusoidal roughness profile, the Néel field [65, 66] is given by

$$H_N = \frac{\pi}{\sqrt{2}}\left(\frac{h^2}{\lambda t_F}\right)M_S \exp\left(-\frac{2\pi\sqrt{2}t_s}{\lambda}\right),\qquad(12.6)$$

where:

h is the amplitude of the roughness profile
λ is the wavelength of the roughness profile
t_F is the thickness of the free layer
t_s is the thickness of the barrier
M_S is the magnetization of the free layer

Equation 12.6 has been confirmed in various studies on MTJs based on the measurement of H_N as a function of t_F and t_s and transmission electron microscopy imaging of the roughness profile [65–68]. According to Equation 12.6, the most effective way to reduce the Néel field is to reduce the interface roughness (reducing h and increasing λ). Shen et al. [69] have developed a deposition process to reduce the interface roughness. They found that low (Ar) pressure sputtering can lead to interface smoothness in MgO-based MTJs. Also, when possible, using a thinner free layer and/or a thicker barrier will reduce the Néel field.

The magnetostatic coupling field, H_M, can be expressed by $H_M = B/L$, with B as an adjustable constant dependent on junction shape and L as the dimension of the pinned layer along the magnetization direction [65–67]. Studies on multiple MTJ samples [55–57] have shown that for L of ~20 μm, H_M is on the order of 1–2 Oe (barrier thickness 1–1.5 nm) for permalloy pinned layers. For smaller junctions, H_M can exceed 10 Oe. To eliminate H_M, an antiferromagnetic synthetic pinned layer is often adopted. In this case, the free layer consists of two magnetic layers antiferromagnetically coupled through a thin Ru layer [46]. The total magnetization of the pinned layer is, therefore, compensated and incapable of generating H_M.

12.2.7 THERMAL STABILITY AND TEMPERATURE DEPENDENCE

There are applications requiring magnetic sensors to function in a broad temperature range or at high temperatures. Thermal stability is an important issue. For example, magnetic sensors used in oil and gas well drilling often encounter up to 200 °C and harsh conditions. Sensors used for monitoring engines face similar situations. In MTJs, high temperatures can cause interlayer diffusion and degradation of the barrier, affecting sensor performance [46, 70]. Furthermore, temperature can lead to varying sensor resistance and field sensitivity. It is, therefore, necessary to characterize the temperature coefficients of both the resistance and the sensitivity.

Mazumdar and coworkers [46] have investigated the thermal stability and temperature dependence in CoFeB/MgO/CoFeB magnetic sensors. They used IrMn, which is widely used due to its high thermal stability, as the antiferromagnet for the pinned layer. They also used an antiferromagnetically coupled synthetic trilayer structure (CoFe/Ru/CoFeB) on top of the IrMn film as the pinned layer. The large separation between the InMn and the MgO barrier reduces potential diffusion of Mn ions, which can degrade the barrier quality at high temperatures. Figure 12.15a shows a sensor's transfer curves at various temperatures. The MTJ sensor remains functioning at 375 °C. The temperature dependence of sensor resistance at zero field is approximately linear with a TCR (TCR = $(1/R)(\Delta R/\Delta T)$) of -7.1×10^{-4} °C^{-1}. Figure 12.15b shows the temperature dependence of the field sensitivity at zero field ($S = (1/R)(\Delta R/\Delta H)$) and detectability. Between room temperature and 275 °C, S varies between 1.0 and 1.14%/Oe, remaining relatively constant over such a wide temperature range.

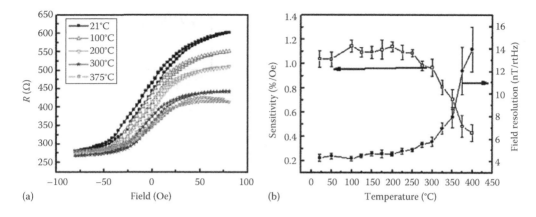

FIGURE 12.15   (a) Temperature dependence of the transfer curves of an MgO MTJ sensor. (b) Sensitivity (open squares) and estimated magnetic field resolution (solid circles) of the same MTJ sensor as a function of temperature. (After Mazumdar, D. et al., *J. Appl. Phys.* 101, 09B502, 2007. With permission.)

These results demonstrate that properly designed and processed MgO-based MTJ sensors enjoy a high degree of thermal stability with low TCR and stable field sensitivity. There are a few factors contributing to the good thermal properties. First of all, MgO is a robust tunnel barrier. Second, the physics of magnetic tunneling persists to high temperature. Finally, the ferromagnetic and antiferromagnetic materials used in the MTJs have high magnetic ordering temperatures.

In the bridge configuration, if all four arms are made of MTJ resistors, the effect of TCR is canceled. Since all arms have the same TCR, the bridge will remain balanced in a varying temperature environment. The thermal coefficient of sensitivity, on the other hand, cannot be canceled in a bridge. Fortunately, the sensitivity for the MgO-based MTJ sensors can be made relatively insensitive to temperature.

12.2.8 FREQUENCY RESPONSE

The evolution of data storage systems requires more sensitive and faster magnetic sensors. Noninvasive diagnostics of semiconductor ICs based on magnetic current density imaging has been made possible by the availability of high-performance MTJ sensors. In these and other applications, it is important that these magnetic devices can respond to signals with frequencies exceeding 1 GHz. The speed of modern electronics has been driven into the deep subnanosecond regime, while the dynamics of today's magnetic materials remains slower.

The physics of magnetic tunneling does allow ultrafast MTJ sensors to work at extremely high frequencies. This is because the energy scale of tunneling electrons is on the order of electronvolts. However, there are many factors that influence the frequency response of MTJ sensors: (1) the effective RC time constant, (2) the ferromagnetic resonance frequency of the free layer, and (3) the magnetic damping characteristics of the free layer.

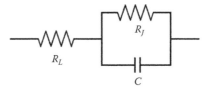

FIGURE 12.16   Effective MTJ circuit model.

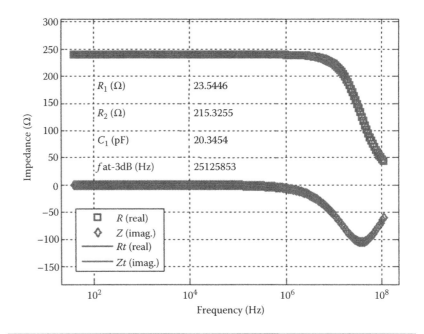

FIGURE 12.17   Fitting results (based on the *RC* circuit in Figure 12.16) for an MTJ sensor, with the extracted parameters of R_1 (lead resistance), R_2 (junction resistance), and C_1 (junction capacitance). Note that the top curve is the real part of the impedance, and the bottom curve is the imaginary part of the impedance.

MTJ sensors can be described using a simplified circuit model for the sensor [71–73], including a lead resistance term (R_L) in series with the traditional parallel *RC* circuit (Figure 12.16). In this case, the maximum operating frequency f_{RC} of the sensor (the –3 dB point) is given by $f_{RC} = 1/2\pi RC$. Figure 12.17 shows the impedance data (real and imaginary) of an MgO-based MTJ sensor. The lines are fittings based on the simplified circuit model, with the fitting results of resistance and capacitance overlaid in the figure. In this sensor, the junction resistance is 215 Ω, and the capacitance is 20.3 pF. This particular sensor works from dc to 25 MHz. With the proper choice of resistance-area (RA) product and junction capacitance, it is possible to produce sensors with $R \sim 20\ \Omega$ and $C \sim 4$ pF for an ultimate f_{RC} of ~ 2 GHz.

If the effective R and C of the packaged device are reduced to small enough values, the limiting factor on the speed of the MTJ sensor becomes the magnetic dynamics of the free layer. To respond to a change in the external magnetic field, the magnetic orientation of the free layer must change, and this process happens at a speed dictated by the Landau–Lifshitz–Gilbert (LLG) equation:

$$\frac{\partial \vec{M}}{\partial t} = -|\gamma| \vec{M} \times \vec{H}_{\text{eff}} - \frac{\alpha}{M_S} \vec{M} \times \frac{\partial \vec{M}}{\partial t}, \tag{12.7}$$

where:

M is the magnetization vector
H_{eff} is the effective magnetic field
M_S is the saturation magnetization
γ is the gyromagnetic ratio
α (damping constant) is a measure of the amount of damping experienced by M

The LLG equation indicates that if a magnetic layer without damping ($\alpha \to 0$) experiences a slight perturbation, such that M and H_{eff} are not quite parallel, then M will simply precess about its lowest energy state indefinitely. α plays a key role in the high-frequency response: if α is too small, it will take a long time for M to settle into its new orientation; if it is too large, the transition between states may also become undesirably slow. α is dependent on the composition of the free layer. By changing the dc field, it is possible to vary the magnitude of H_{eff}, which will in turn vary the precession frequency. Sensors normally work up to the precession frequency, which can be tuned above a few gigahertz. More studies are needed to understand the magnetic dynamics in MTJ sensors.

12.2.9 SENSOR CHARACTERIZATIONS

Modern MTJ devices, as magnetic sensors or used for nonvolatile magnetic memory cells, have complicated structures with performance dependent on more than a dozen material and structural parameters. Accurate characterization of the governing properties of MTJs is critical to understanding the physics and the performance in these devices. Since the discovery of large MR in MTJs, some novel measuring techniques have been introduced, improving the efficiency of measurement.

In MTJ research, MR and RA product are two of the most basic parameters. In the past, obtaining the two parameters was a tedious process. One must pattern the MTJ wafers into vertical structures and attach electrical contacts. Worledge and Trouilloud [74] invented the technique of current-in-plane tunneling (CIPT), which allowed the measurement of MR and RA of unpatterned MTJ wafers. In this technique, a series of micro-point probes are gently placed on the top surface of a wafer. By measuring the four-point resistance at various probe spacings, one can calculate the RA using a formula that is an exact analytical solution of the CIPT conceptual model. By repeated measurement of RA as a function of magnetic field, the MR can be obtained. The CIPT method is purely a wafer-level characterization tool and highly efficient and fast. It is a very useful tool in designing and testing the materials and structures used in MTJs.

Researchers at Brown University and Micro Magnetics, Inc. have developed the techniques of *sensitivity map* [48], *magnetic noise map* [48], and *circle transfer curves* [75] in characterizing MTJ devices. In the *sensitivity map* method, an MTJ device's field sensitivity, $S = (1/R)(\Delta R/\Delta H)$, is measured automatically in a two-dimensional (2D) magnetic field (bipolar) over a large range of $\pm Bx$ and $\pm By$. A map of the S parameter is then generated in the 2D field space. An example of an S map [48] is shown in Figure 12.16a. With the map, one can locate the most sensitive region for a particular sensor design. In most cases, the most sensitive region is not centered around the ambient state of $Bx = By = 0$. By the use of hard- and easy-axis field biasings (on-chip magnets or striplines), the best sensitivity can be achieved.

In the *magnetic noise map* method [48], an MTJ device's voltage noise at a given frequency (e.g., 1 Hz) is measured over a 2D magnetic field space as in the sensitivity map. A high-resolution map consists of hundreds of data points. Automation in data taking and analysis is necessary to generate the map. Once the voltage noise map is obtained, the magnetic noise map can be plotted by combining the voltage noise map and the sensitivity map according to Equation 12.5. In sensor development, the magnetic noise map and the sensitivity map will help the designer to select the optimal operating regions for a magnetic sensor. Figure 12.6b and c show examples of a voltage noise map and a magnetic noise map.

The *circle transfer curves* method [75] is an efficient technique to measure a plethora of important parameters, such as magnetic anisotropy strength, magnetostatic coupling, antiferromagnetic exchange coupling, and the absolute magnetic orientation of both the free and the pinned layers across a whole wafer. Figure 12.18 shows some of the parameters of an MTJ multilayer that can be determined using this method. Traditional magnetometry methods can only measure bulk films or devices with relatively large volumes. The circle transfer curves method can handle devices of any size. In this technique, the MTJ resistance (or conductance) is measured as a function of the applied field angle with a fixed field strength (R vs. angle). Two pairs of orthogonal Helmholtz coils can be used to generate a rotating magnetic field. Only three circle transfer curves at different fixed fields are needed to extract all the desired information about the junction (see detailed analysis description in Safron et al. [75]). The circle transfer curves method is fast and can be used to measure the variation or uniformity parameters across a whole wafer with patterned MTJ devices.

Figure 12.19 shows three experimental circle curves (conductance vs. angle) of a representative ±MTJ device [75]. The deviation from a pure cosine function is due to the free layer anisotropy, which is more pronounced at the lower applied field strength (e.g., 40 Oe). As the field strength increases, the circle curves will be more cosine-like. The solid lines are best-fit curves based on the circle transfer curves model, from which many parameters (see Figure 12.18) can be derived. This technique has been used to map six parameters of MTJs across a whole wafer.

Figure 12.20 shows the distribution of the junction anisotropy angle and pinned layer direction across a wafer [75]. This particular wafer

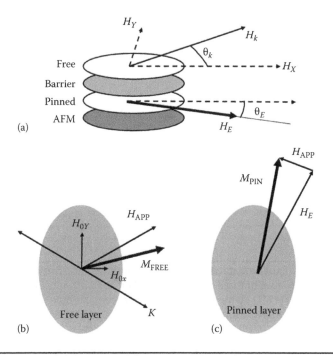

FIGURE 12.18   (a) Schematic of the MTJ multilayer showing the relevant physical quantities, which dictate the junction behavior. (b) Simplified model of the free layer, where the magnetization is determined by the two offset field components (H_{0X} and H_{0Y}), the sample's uniaxial anisotropy, and the external applied field (H_{APP}). (c) Model of the pinned layer magnetization, which is assumed to be the vector sum of two forces: the applied field and the exchange biasing field (H_E). (After Safron, N.S. et al., *J. Appl. Phys.* 103, 033507, 2008. With permission)

consists of hundreds of different MTJ sensors. Figure 12.21 shows the distribution of the junction anisotropy angle and magnitude and the pinned layer direction for a wafer consisting of hundreds of MTJ memory cells. These results demonstrate that circle curves can be used as a new tool, complementary to other methods, for researchers to design better MTJ sensors and devices.

12.2.10 APPLICATIONS: EXAMPLES

12.2.10.1 Magnetic Current Density Microscope

The advent of MTJ sensor development has facilitated applications in different fields. Here, two examples are presented. The first example is a magnetic current density microscope. Magnetic imaging techniques have found numerous practical applications in many areas of science and technology. In addition, recent advances in spintronics have created new and powerful magnetic devices. GMR sensors have become universally adopted in the computer hard disk drive market. MTJ sensors have several significant advantages over GMR technology. Scientists have built on these sensing technologies to develop a new spintronics metrology imaging tool

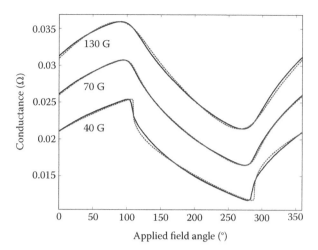

FIGURE 12.19 Experimental data (solid lines) and theoretical fitting results (dashed lines) for a set of three circle curves (taken at 40, 70, and 130 G) taken on a representative MTJ element. All fits are made using a single set of junction parameters. (After Safron, N.S. et al., *J. Appl. Phys.* 103, 033507, 2008. With permission.)

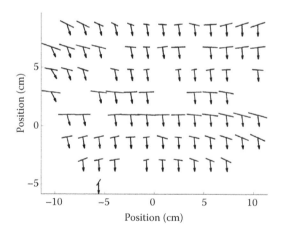

FIGURE 12.20 Plot showing the distribution of junction anisotropy angle and pinned layer direction as a function of MTJ wafer position. The arrows indicate the pinned layer orientation, while the length and orientation of the solid lines indicate the strength and direction of the sample anisotropy, respectively. (After Safron, N.S. et al., *J. Appl. Phys.* 103, 033507, 2008. With permission.)

for mapping current densities in thin-film conductors and semiconductor devices [7–10]. The tool is noninvasive, operates under ambient conditions, and is capable of spatial resolution better than 100 nm. The MTJ sensor developed for this tool allows engineers to measure currents as small as a microampere in deeply embedded microstructures. The technique has been applied to fault isolation and failure analysis in ICs. Figure 12.22

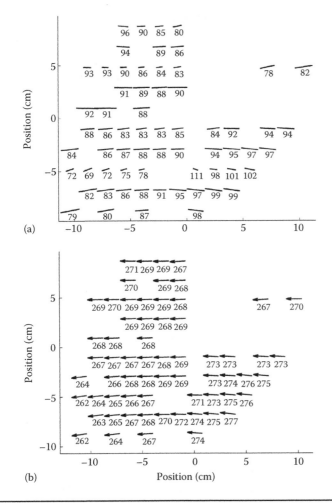

FIGURE 12.21 (a) Plot of the positional distribution of junction anisotropy angle and magnitude for an MTJ wafer. The numbers below each line show the anisotropy angle in degrees, while the length of each line is proportional to the anisotropy strength. (b) Distribution of the extracted pinned layer direction for the same sample as a function of wafer position. (After Safron, N.S. et al., *J. Appl. Phys.* 103, 033507, 2008. With permission.)

displays a current density map in a failed IC device using a scanning MTJ microscope.

The embedded structures in ICs emit a microscopic magnetic field. Imaging this field can reveal the spatial variation of current density and the physical layout of a circuit. The whole IC structure is transparent to the magnetic field. The need for such a microscope is compelling and immediate in the semiconductor industry. The IC engineers can use the microscope to survey the operation of tiny components and to pinpoint defects down to the smallest units, such as transistors and copper interconnects.

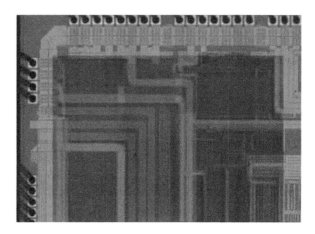

FIGURE 12.22   Map of current density distribution in a failed IC. The map is obtained using a scanning MTJ magnetic microscope. The current density image is automatically overlaid over a high-resolution optical image of the device.

There are several competing technologies available to image the field today. Magnetic force microscopy (MFM) [76], scanning electron microscopy with polarization analysis (SEMPA) [77], and magneto-optical microscopy [78] are capable of imaging magnetic structures. They are not, however, suitable for imaging fields created by electrical currents and, in particular, those from high-frequency currents. In some cases, the sample preparation is destructive, resulting in sample loss for further analysis. The scanning SQUID microscope [79] is poor in resolution, with a lower limit of about 5 μm, and requires cryogenic cooling. A Hall probe can operate under ambient conditions, but its sensitivity is low. Ultrasensitive MTJ sensors are superior to other existing magnetic sensors for magnetic current density microscopy [7–10]. This imaging technology is quite similar to medical x-ray imaging. In a computed tomography (CT) scan, x-rays gather information about a body and inverse-calculate the body's internal structure. The current density microscope gathers the magnetic field information in a circuit and inverse-calculates its current distribution.

The scanning MTJ microscope is also a sensitive magnetic microscope that can directly image the local magnetic field on an absolute scale. The traditional scanning magnetic microscope using magnetic force only senses the local field gradient. The scanning MTJ microscope has also been used in research into "atom chips" used in a Bose–Einstein condensate [80].

12.2.10.2 Spintronics Immunoassay

Scientists have researched a spintronics-based on-chip immunoassay system [81–86]* for the identification of biomolecules (bacteria, viruses, proteins, and DNA). Using MTJ sensor arrays (see Figure 12.23), it is possible to create a powerful immunoassay system that is fast, sensitive, specific, and portable. By using magnetic nanoparticles (NPs) to tag the DNA, viruses, bacteria,

---

* See Chapter 14, Volume 3.

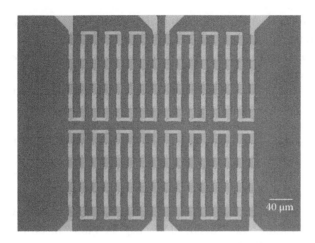

FIGURE 12.23   MTJ sensor array for spintronics immunoassay application.

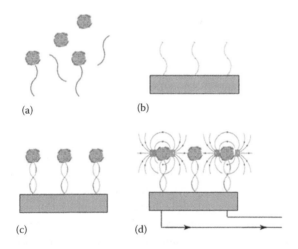

FIGURE 12.24   Using MTJ sensors to detect magnetic field of NPs bound to specific biomolecules (a through d).

or cancer cells, it is possible to detect the concentration of these agents in a biological solution using magnetic sensing, as shown in Figure 12.24.

Initially, magnetic NPs that bind only to specific molecules are introduced into the sample to be tested. If the molecules, viruses, DNA, or cells that the NPs are designed for are present, they will tightly bind to the NPs. Then, the biological sample will be introduced to the surface of an array of magnetic sensors coated with "probe" molecules (antigens and complementary DNA). These probe molecules will bind tightly to certain specific agents, and agents that do not bind to the probe molecules will be washed away. The sensors can then detect the presence of these NPs by sensing their magnetic field. The strength of the signal is proportional to the concentration of the agent. Potentially, the biomagnetic sensor is sensitive enough to detect a single biomolecule.

This new immunoassay method has some very important implications. Virtually any biomolecule can potentially be detected using this "lab-on-a-chip" technology. All tests and experiments could potentially be conducted on this single device using microfluidic circuitry, such as valves, tubes, mixers, pumps, and channels, on a very small scale. This could make an impact on medical diagnoses as a noninvasive, quick, portable, and relatively inexpensive way to detect health problems.

Both GMR and MTJ sensors have been used in spintronics immunoassay [81–84]. However, in addition to higher MR and sensitivity than GMR, MTJ-based devices boast a number of features that make them a superior sensor candidate for this application. First of all, the junction resistance (10^{-1}–10^8 Ω) can be engineered to suit a specific application. Second, the junction area can be reduced to nanometer scale while still maintaining large MR. This is particularly beneficial for NP detection. Third, an MTJ requires very little power to operate due to the small sense currents. These advantages may allow the MTJ-based immunoassay system to be integrated into silicon circuitry and coupled with microfluidics to create a lab-on-a-chip analysis system.

12.3 CONCLUSIONS

Thin-film magnetic field sensors have had a significant impact in many different technological areas, ranging from data storage to magnetic microscopy. The development of magnetic sensors has benefited from the emerging technical discipline called *spintronics*, which takes advantage of the dual spin and charge properties of electrons to form a new generation of electronic/magnetic devices. In particular, MTJ devices have become an important class of mass-market spintronics components in the information storage and magnetic sensor markets. The large MR in MgO-based MTJ sensors has been key to the increased field sensitivity. MTJ sensors are suitable for ultrafast sensing, as in read/write heads, and applications requiring low power consumption. Continuing research is needed to reduce the intrinsic magnetic noise by using new magnetic materials and designing innovative device structures. New experimental characterization tools, such as CIPT and circle transfer curves, can benefit research into MTJ sensors.

There are a few current objectives in MTJ sensor development. A significant impact on applications will occur if the field detectability can be enhanced to 1 pT at low frequencies (e.g., 1 Hz). The manufacturing cost can be reduced by improving the fabrication process for making uniform sensors across a large wafer. The potential discovery of even larger MR may always lead to improved sensor performance. More efforts are also needed to develop new applications for which the MTJ sensors are capable. The prospects for MTJ sensors in particular and spintronics in general are bright.

ACKNOWLEDGMENTS

The author acknowledges funding by the National Science Foundation under Grant No. DMR-0907353 and by JHU MRSEC (NSF) under Grant No. DMR-0520491. The author wishes to thank W.F. Shen, B.D. Schrag, M.J. Carter, W. Egelhoff, X.Y. Liu, D. Mazumdar, X.J. Zou, W.Z. Zhang, N.S. Safron, and Hai Sang for collaboration and discussion.

REFERENCES

1. S. A. Wolf, A. V. Chtchelkanova, and D. M. Treger, Spin-tronics: A retrospective and perspective, *IBM J. Res. Dev.* **50**, 101 (2006).
2. S. A. Wolf, Spintronics: A spin-based electronics vision for the future, *Science* **294**, 1488 (2001).
3. Frost & Sullivan, *World Magnetic Sensor Components and Modules/Subsystems Markets*, February 2006.
4. H. Koch, Recent advances in magnetocardiography, *J. Electrocardiol.* **37**, 117 (2004).
5. D. Cohen and E. Halgren, Magnetoencephalography, in *Encyclopedia of Neuroscience*, 3rd edn., G. Adelman and B. Smith (Eds.), Elsevier, New York, 2004.
6. C. Chappert, A. Fert, and F. N. Van Dau, The emergence of spin electronics in data storage, *Nat. Mater.* **6**, 813 (2007).
7. B. D. Schrag, X. Liu, W. Shen, and G. Xiao, Current density mapping and pinhole imaging in magnetic tunnel junctions via scanning magnetic microscopy, *Appl. Phys. Lett.* **84**, 2937 (2004).
8. B. Schrag, X. Liu, M. J. Carter, and G. Xiao, Scanning magnetoresistive microscopy for die-level sub-micron current density mapping, in *Proceedings of the 29th International Symposium for Testing and Failure Analysis (ISTFA)*, November 2–6, Santa Clara, CA, Awarded the Outstanding Paper Award at ISTFA, 2003.
9. B. D. Schrag, M. J. Carter, X. Liu, J. Hoftun, and G. Xiao, Magnetic current imaging with magnetic tunnel junction sensors: Case study and analysis, in *Proceedings of the 2006 International Symposium for Testing and Failure Analysis*, Austin, TX, 2006.
10. B. D. Schrag, X. Liu, J. Hoftun, P. Klinger, T. Levin, and D. Vallett, Quantitative analysis and depth measurement via magnetic field imaging, *Electron. Device Fail. Anal.* **74**, 24 (2005).
11. K. Sohraby, *Wireless Sensor Networks: Technology, Protocols, and Applications*, Wiley-Interscience, Hoboken, NJ, 2007.
12. E. Y. Tsymbal, O. N. Mryasov, and P. R. LeClair, Spin-dependent tunnelling in magnetic tunnel junctions, *J. Phys. Condens. Matter* **15**, R109 (2003).
13. R. Popovic, The future of magnetic sensors, *Sens. Actuators A: Phys.* **56**, 39 (1996).
14. J. Lenz, A review of magnetic sensors, *Proc. IEEE* **78**, 973 (1990).
15. J. Lenz and S. Edelstein, Magnetic sensors and their applications, *IEEE Sens. J.* **6**, 631 (2006).
16. P. Ripka, Sensors based on bulk soft magnetic materials: Advances and challenges, *J. Magn. Magn. Mater.* **320**, 2466 (2008).
17. E. Ramsden, *Hall-Effect Sensors: Theory and Applications*, Elsevier/Newnes, Amsterdam, Boston, 2006.
18. M. Baibich, J. Broto, A. Fert et al., Giant magnetoresistance of (001)Fe/(001)Cr magnetic superlattices, *Phys. Rev. Lett.* **61**, 2472 (1988).
19. G. Binasch, P. Grünberg, F. Saurenbach, and W. Zinn, Enhanced magnetoresistance in layered magnetic structures with antiferromagnetic interlayer exchange, *Phys. Rev. B* **39**, 4828 (1989).
20. M. Kuzminski, K. Nesteruk, and H. Lachowicz, Magnetic field meter based on giant magnetoimpedance effect, *Sens. Actuators A: Phys.* **141**, 68 (2008).
21. The Nobel Prize in Physics 2007, https://www.nobelprize.org/prizes/physics/2007/summary/ (accessed on October 9, 2007).
22. J. Moodera, L. Kinder, T. Wong, and R. Meservey, Large magnetoresistance at room temperature in ferromagnetic thin film tunnel junctions, *Phys. Rev. Lett.* **74**, 3273 (1995).
23. T. Miyazaki and N. Tezuka, Giant magnetic tunneling effect in $Fe/Al_2O_3/Fe$ junction, *J. Magn. Magn. Mater.* **139**, L231 (1995).

24. M. Julliere, Tunneling between ferromagnetic films, *Phys. Lett. A* **54**, 225 (1975).
25. W. Butler, X. Zhang, T. Schulthess, and J. MacLaren, Spin-dependent tunneling conductance of Fe|MgO|Fe sandwiches, *Phys. Rev. B* **63**, 054416 (2001).
26. J. Mathon and A. Umerski, Theory of tunneling magnetoresistance of an epitaxial Fe/MgO/Fe(001) junction, *Phys. Rev. B* **63**, 220403(R) (2001).
27. S. S. P. Parkin, C. Kaiser, A. Panchula et al., Giant tunnelling magnetoresistance at room temperature with MgO (100) tunnel barriers, *Nat. Mater.* **3**, 862 (2004).
28. S. Yuasa, T. Nagahama, A. Fukushima, Y. Suzuki, and K. Ando, Giant room-temperature magnetoresistance in single-crystal Fe/MgO/Fe magnetic tunnel junctions, *Nat. Mater.* **3**, 868 (2004).
29. J. Zhu and C. Park, Magnetic tunnel junctions, *Mater. Today* **9**, 36 (2006).
30. S. Yuasa and D. D. Djayaprawira, Giant tunnel magnetoresistance in magnetic tunnel junctions with a crystalline MgO(001) barrier, *J. Phys. D Appl. Phys.* **40**, R337 (2007).
31. B. D. Schrag and G. Xiao, Submicron electrical current density imaging of embedded microstructures, *Appl. Phys. Lett.* **82**, 3272 (2003).
32. D. Wang, C. Nordman, J. Daughton et al., 70% TMR at room temperature for SDT sandwich junctions with CoFeB as free and reference layers, *IEEE Trans. Magn.* **40**, 2269 (2004).
33. X. Liu and G. Xiao, Thermal annealing effects on low-frequency noise and transfer behavior in magnetic tunnel junction sensors, *J. Appl. Phys.* **94**, 6218 (2003).
34. M. Tondra, J. M. Daughton, D. Wang, R. S. Beech, A. Fink, and J. A. Taylor, Picotesla field sensor design using spin-dependent tunneling devices, *J. Appl. Phys.* **83**, 6688 (1998).
35. D. D. Djayaprawira, K. Tsunekawa, M. Nagai et al., 230% room-temperature magnetoresistance in CoFeB/MgO/CoFeB magnetic tunnel junctions, *Appl. Phys. Lett.* **86**, 092502 (2005).
36. S. Ikeda, J. Hayakawa, Y. Ashizawa et al., Tunnel magnetoresistance of 604% at 300 K by suppression of Ta diffusion in CoFeB/MgO/CoFeB pseudo-spin-valves annealed at high temperature, *Appl. Phys. Lett.* **93**, 082508 (2008).
37. L. Jiang, H. Naganuma, M. Oogane, and Y. Ando, Large tunnel magnetoresistance of 1056% at room temperature in MgO based double barrier magnetic tunnel junction, *Appl. Phys. Exp.* **2**, 083002 (2009).
38. W. Egelhoff, Jr., V. Hoink, J. Lau, W. Shen, B. Schrag, and G. Xiao, Magnetic tunnel junctions with large tunneling magnetoresistance and small saturation fields, *J. Appl. Phys.* **107**, 09C705 (2010).
39. R. C. Chaves, P. P. Freitas, B. Ocker, and W. Maass, MgO based picotesla field sensors, *J. Appl. Phys.* **103**, 07E931 (2008).
40. R. C. Chaves, P. P. Freitas, B. Ocker, and W. Maass, Low frequency picotesla field detection using hybrid MgO based tunnel sensors, *Appl. Phys. Lett.* **91**, 102504 (2007).
41. S. Ingvarsson, G. Xiao, R. A. Wanner et al., Electronic noise in magnetic tunnel junctions, *J. Appl. Phys.* **85**, 5270 (1999).
42. S. Ingvarsson, G. Xiao, S. Parkin, W. Gallagher, G. Grinstein, and R. Koch, Low-frequency magnetic noise in micron-scale magnetic tunnel junctions, *Phys. Rev. Lett.* **85**, 3289 (2000).
43. E. R. Nowak, M. B. Weissman, and S. S. P. Parkin, Electrical noise in hysteretic ferromagnet–insulator–ferromagnet tunnel junctions, *Appl. Phys. Lett.* **74**, 600 (1999).
44. C. Ren, X. Liu, B. Schrag, and G. Xiao, Low-frequency magnetic noise in magnetic tunnel junctions, *Phys. Rev. B* **69**, 104405 (2004).
45. S. H. Liou, R. Zhang, S. E. Russek, L. Yuan, S. T. Halloran, and D. P. Pappas, Dependence of noise in magnetic tunnel junction sensors on annealing field and temperature, *J. Appl. Phys.* **103**, 07E920 (2008).

46. D. Mazumdar, X. Liu, B. D. Schrag, W. Shen, M. Carter, and G. Xiao, Thermal stability, sensitivity, and noise characteristics of MgO-based magnetic tunnel junctions (invited), *J. Appl. Phys.* **101**, 09B502 (2007).

47. N. A. Stutzke, S. E. Russek, D. P. Pappas, and M. Tondra, Low-frequency noise measurements on commercial magnetoresistive magnetic field sensors, *J. Appl. Phys.* **97**, 10Q107 (2005).

48. D. Mazumdar, W. Shen, X. Liu, B. D. Schrag, M. Carter, and G. Xiao, Field sensing characteristics of magnetic tunnel junctions with (001) MgO tunnel barrier, *J. Appl. Phys.* **103**, 113911 (2008).

49. R. Guerrero, M. Pannetier-Lecoeur, C. Fermon, S. Cardoso, R. Ferreira, and P. P. Freitas, Low frequency noise in arrays of magnetic tunnel junctions connected in series and parallel, *J. Appl. Phys.* **105**, 113922 (2009).

50. K. Klaassen, X. Xing, and J. van Peppen, Signal and noise aspects of magnetic tunnel junction sensors for data storage, *IEEE Trans. Magn.* **40**, 195 (2004).

51. Y. Tserkovnyak and A. Brataas, Shot noise in ferromagnet–normal metal systems, *Phys. Rev. B* **64**, 214402 (2001).

52. R. Guerrero, F. Aliev, Y. Tserkovnyak, T. Santos, and J. Moodera, Shot noise in magnetic tunnel junctions: Evidence for sequential tunneling, *Phys. Rev. Lett.* **97**, 266602 (2006).

53. N. Smith and P. Arnett, White-noise magnetization fluctuations in magnetoresistive heads, *Appl. Phys. Lett.* **78**, 1448 (2001).

54. N. Smith, Modeling of thermal magnetization fluctuations in thin-film magnetic devices, *J. Appl. Phys.* **90**, 5768 (2001).

55. V. D. Tsiantos, T. Schrefl, W. Scholz, and J. Fidler, Thermal magnetization noise in submicrometer spin valve sensors, *J. Appl. Phys.* **93**, 8576 (2003).

56. W. Egelhoff Jr., P. Pong, J. Unguris et al., Critical challenges for picoTesla magnetic-tunnel-junction sensors, *Sens. Actuators A Phys.* **155**, 217 (2009).

57. G. M. Jaramillo, M. Chan, and D. A. Horsley, Integration of MEMS actuators with magnetic tunnel junction sensors, *2007 IEEE Sensors*, Atlanta, GA, pp. 1380–1383, 2007.

58. A. S. Edelstein, G. A. Fischer, M. Pedersen, E. R. Nowak, S. F. Cheng, and C. A. Nordman, Progress toward a thousandfold reduction in $1/f$ noise in magnetic sensors using an ac microelectromechanical system flux concentrator (invited), *J. Appl. Phys.* **99**, 08B317 (2006).

59. A. Guedes, S. B. Patil, S. Cardoso, V. Chu, J. P. Conde, and P. P. Freitas, Hybrid magnetoresistive/microelectromechanical devices for static field modulation and sensor $1/f$ noise cancellation, *J. Appl. Phys.* **103**, 07E924 (2008).

60. Y. Lu, R. A. Altman, A. Marley et al., Shape-anisotropy-controlled magnetoresistive response in magnetic tunnel junctions, *Appl. Phys. Lett.* **70**, 2610 (1997).

61. X. Liu, C. Ren, and G. Xiao, Magnetic tunnel junction field sensors with hard-axis bias field, *J. Appl. Phys.* **92**, 4722 (2002).

62. P. W. T. Pong, B. Schrag, A. J. Shapiro, R. D. McMichael, and W. F. Egelhoff, Hysteresis loop collapse for linear response in magnetic-tunnel-junction sensors, *J. Appl. Phys.* **105**, 07E723 (2009).

63. Y. Jang, C. Nam, J. Y. Kim, B. K. Cho, Y. J. Cho, and T. W. Kim, Magnetic field sensing scheme using CoFeB/MgO/CoFeB tunneling junction with superparamagnetic CoFeB layer, *Appl. Phys. Lett.* **89**, 163119 (2006).

64. W. Shen, B. Schrag, A. Girdhar, M. Carter, H. Sang, and G. Xiao, Effects of superparamagnetism in MgO based magnetic tunnel junctions, *Phys. Rev. B* **79**, 014418 (2009).

65. A. Anguelouch, B. D. Schrag, G. Xiao et al., Two-dimensional magnetic switching of micron-size films in magnetic tunnel junctions, *Appl. Phys. Lett.* **76**, 622 (2000).

66. B. D. Schrag, A. Anguelouch, S. Ingvarsson et al., Néel "orange-peel" coupling in magnetic tunneling junction devices, *Appl. Phys. Lett.* **77**, 2373 (2000).

67. B. D. Schrag, A. Anguelouch, G. Xiao et al., Magnetization reversal and interlayer coupling in magnetic tunneling junctions, *J. Appl. Phys.* **87**, 4682 (2000).

68. L. Li, X. Liu, and G. Xiao, Microstructures of magnetic tunneling junctions, *J. Appl. Phys.* **93**, 467 (2003).

69. W. Shen, D. Mazumdar, X. Zou, X. Liu, B. D. Schrag, and G. Xiao, Effect of film roughness in MgO-based magnetic tunnel junctions, *Appl. Phys. Lett.* **88**, 182508 (2006).

70. X. Liu, D. Mazumdar, W. Shen, B. D. Schrag, and G. Xiao, Thermal stability of magnetic tunneling junctions with MgO barriers for high temperature spintronics, *Appl. Phys. Lett.* **89**, 023504 (2006).

71. G. Landry, Y. Dong, J. Du, X. Xiang, and J. Q. Xiao, Interfacial capacitance effects in magnetic tunneling junctions, *Appl. Phys. Lett.* **78**, 501 (2001).

72. C. Zhang, X. Zhang, P. Krstić, H. Cheng, W. Butler, and J. MacLaren, Electronic structure and spin-dependent tunneling conductance under a finite bias, *Phys. Rev. B* **69**, 134406 (2004).

73. P. Padhan, P. LeClair, A. Gupta, K. Tsunekawa, and D. D. Djayaprawira, Frequency-dependent magnetoresistance and magnetocapacitance properties of magnetic tunnel junctions with MgO tunnel barrier, *Appl. Phys. Lett.* **90**, 142105 (2007).

74. D. C. Worledge and P. L. Trouilloud, Magnetoresistance measurement of unpatterned magnetic tunnel junction wafers by current-in-plane tunneling, *Appl. Phys. Lett.* **83**, 84 (2003).

75. N. S. Safron, B. D. Schrag, X. Liu et al., Magnetic characterization of magnetic tunnel junction devices using circle transfer curves, *J. Appl. Phys.* **103**, 033507 (2008).

76. D. Bonnell, *Scanning Probe Microscopy and Spectroscopy: Theory, Techniques, and Applications*, Wiley-VCH, New York, 2001.

77. M. R. Scheinfein, J. Unguris, M. H. Kelley, D. T. Pierce, and R. J. Celotta, Scanning electron microscopy with polarization analysis (SEMPA), *Rev. Sci. Instrum.* **61**, 2501 (1990).

78. M. R. Freeman, Advances in magnetic microscopy, *Science* **294**, 1484 (2001).

79. S. Chatraphorn, E. F. Fleet, F. C. Wellstood, L. A. Knauss, and T. M. Eiles, Scanning SQUID microscopy of integrated circuits, *Appl. Phys. Lett.* **76**, 2304 (2000).

80. M. Volk, S. Whitlock, C. H. Wolff, B. V. Hall, and A. I. Sidorov, Scanning magnetoresistance microscopy of atom chips, *Rev. Sci. Instrum.* **79**, 023702 (2008).

81. W. Shen, X. Liu, D. Mazumdar, and G. Xiao, In situ detection of single micronsized magnetic beads using magnetic tunnel junction sensors, *Appl. Phys. Lett.* **86**, 253901 (2005).

82. W. Shen, B. D. Schrag, M. J. Carter et al., Detection of DNA labeled with magnetic nanoparticles using MgO-based magnetic tunnel junction sensors, *J. Appl. Phys.* **103**, 07A306 (2008).

83. W. Shen, B. D. Schrag, M. J. Carter, and G. Xiao, Quantitative detection of DNA labeled with magnetic nanoparticles using arrays of MgO-based magnetic tunnel junction sensors, *Appl. Phys. Lett.* **93**, 033903 (2008).

84. P. P. Freitas, R. Ferreira, S. Cardoso, and F. Cardoso, Magnetoresistive sensors, *J. Phys. Condens. Matter* **19**, 165221 (2007).

85. D. Baselt, A biosensor based on magnetoresistance technology, *Biosens. Bioelectron.* **13**, 731 (1998).

86. L. Ejsing, M. F. Hansen, A. K. Menon, H. A. Ferreira, D. L. Graham, and P. P. Freitas, Planar Hall effect sensor for magnetic micro- and nanobead detection, *Appl. Phys. Lett.* **84**, 4729 (2004).

Magnetoresistive Random Access Memory

Johan Åkerman

13.1 INTRODUCTION

13.1.1 BACKGROUND

While long-term information storage has been dominated by magnetic technologies such as hard drives and tape drives for the last 50 years, the time when magnetic core memories dominated random access memories (RAM) is long gone (for reviews on early memory technologies see, e.g., [1, 2]). When static RAM and dynamic RAM, both based on thin film semiconductor technology, were introduced in the 1970s, they rapidly overtook magnetic core memory thanks to their superiority in immediate miniaturization, continued scalability, and ever-decreasing production cost per memory bit (see, e.g., [3]). Attempts at developing alternative magnetic memory technologies to compete with semiconductor memories, such as thin film magnetic bubble memory, did not succeed commercially.

In the mid-1990s, with the looming success of giant magnetoresistance (GMR)–based spin valves as read heads in hard drives, Stuart Wolf at the Defense Advanced Research Projects Agency (DARPA) initiated a program targeting the development of magnetoresistive RAM (MRAM). At about the same time, significant tunneling magnetoresistance (TMR) was demonstrated at room temperature in magnetic tunnel junctions (MTJs) [4, 5]. Together, the available funding and the breakthroughs in spin valve and MTJ technologies spurred an intense research effort, which led to the birth of the entire new field of spintronics—short for spin electronics.

On July 10, 2006, a little over 10 years after the initial DARPA program, the first commercial MRAM was introduced into the market by Freescale Semiconductor. Freescale then span off its MRAM business into a separate company, Everspin Technologies, which diversified the introductory 4 MB MRAM into similar products having different storage capacities and form factors.

The commercial use of MRAM is steadily increasing, and next-generation MRAM technologies are being actively researched and developed in many laboratories and companies around the world. In this chapter, I have chosen to focus on so-called *toggle MRAM*, developed by Motorola, Freescale Semiconductor, and Everspin Technologies, (for a description of MRAM researched and developed by IBM, see, e.g., [6]). Toggle MRAM has recently been followed by spin transfer torque MRAM (STT-MRAM), which offers much higher storage density and greater scalability. The potential and challenges of STT-MRAM are very briefly discussed at the end of this chapter.

13.1.2 MRAM PROPERTIES

MRAM combines a set of properties not simultaneously available in any other memory technology: non-volatility, low-voltage operation, unlimited read and write endurance, high-speed read and write operation, excellent reliability, and radiation hardness (for earlier reviews on MRAM, see, e.g., [6–9]). For example, static RAM (SRAM) has excellent read and write speeds and is readily integrated into embedded applications but lacks non-volatility. Dynamic RAM (DRAM) combines high read and write speeds with a minimal cell size. However, its use of a leaky storage capacitor requires continuous refresh cycles, which leads to high power consumption, lack of non-volatility, and increased process complexity in embedded applications. Flash memory is non-volatile and can be fabricated with very high storage capacity. However, in many applications where high capacity is not a target, its slow write mode and limited write endurance put it at a disadvantage. While ferroelectric RAM (FeRAM) is also non-volatile and offers better write performance than Flash, it is based on materials that dramatically increase the complexity of process integration. As a consequence of these different limitations, most systems combine a number of memory types to achieve optimal operation. MRAM, on the other hand, combines the most critical properties, and thanks to this universal nature, it can eliminate or minimize the need for multiple memories in many applications where maximized storage capacity is not the main driver. For MRAM to also address applications where high storage capacity is needed, it must shrink its cell size significantly, and it is now generally believed that this can only be achieved to a sufficient degree by replacing toggle MRAM with STT-MRAM, where bit cell programming is no longer carried out by magnetic fields in proximity to the MTJ but by spin polarized current through the device.

13.1.3 DEVELOPING A MEMORY TECHNOLOGY

For the last 20 years, the solid state memory market has been entirely dominated by SRAM, DRAM, and Flash. While the core technologies of these dominant memory types have remained largely unchanged, they have benefited from systematic evolutionary innovation toward denser, faster, and cheaper memories. To introduce an entirely new and competing memory technology is quite a daunting task, since there is less established processing and test experience to draw from. While it is quite possible to demonstrate a novel memory technology on a laboratory scale, with a focus on single bit operation at optimum conditions, it is a completely different challenge to develop a fully functioning memory with reliable operation for tens of years over a wide range of operating conditions.

Figure 13.1 shows a schematic of the different aspects of MRAM that have to be developed to deliver a fully operational memory into the marketplace. It is divided into two important and distinctly different parts: I) the fundamentals of single bit operation and II) the realities of array operation. One may very well define and design an operating scheme that works in a limited parameter space, but with the many unknowns and imperfections of the real world, it does not take long before unexpected failure mechanisms dominate over functional operation. The devil is in the distributions: distributions of fundamental intrinsic properties, distributions of extrinsic properties, and distributions of usage conditions. While statistical analysis of accelerated tests of the fundamental properties can be used to extrapolate to actual operating conditions, only a fully tested memory array can teach us about the extrinsic failure mechanisms that will dominate the behavior in the field.

13.2 SINGLE BIT FUNDAMENTALS

As one embarks on the road to a new memory technology, one first has to find or design some sort of hysteresis, that is, some material or system property with memory. In SRAM, one typically uses six transistors in a flip-flop architecture, while DRAM and Flash rely on stored electric charge. In MRAM, the obvious choice is the magnetic hysteresis associated with the spontaneous magnetization of ferromagnetically ordered materials (see, e.g., [10]). Magnetic anisotropy, the property that makes the energy of a magnetic system depend on the angle of the magnetization, can be used to create two well-defined energy minima for the magnetization, which can then be interpreted as two digital states: 0 and 1. In MRAM, one of the ferromagnetic

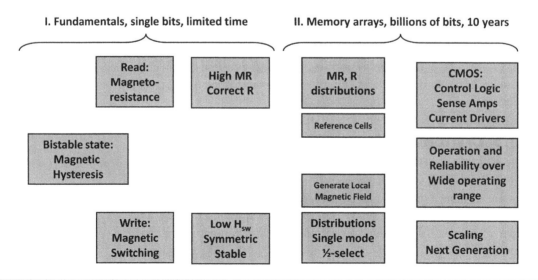

FIGURE 13.1   Different aspects of MRAM that have to be developed for a fully functional memory. While all fundamental aspects were addressed early on, and often by academic research groups, significant challenges arise when full arrays and operations over 10 years must be addressed for an actual product.

layers—the so-called *free layer*—has two such well-defined stable states, which represent the 0 and 1 of the MRAM bit cell, whereas the other ferromagnetic layer—the so-called *fixed layer*—is kept in a single well-defined magnetic state and serves as a reference to the free layer magnetization direction.

One then needs both a way to detect which state the free layer magnetization occupies at any given time (read the memory bit) and a way to prepare the free layer magnetization in either of the states (write the memory bit). In MRAM, the resistance of the bit cell MTJ depends strongly on the relative orientation of the free and fixed layer magnetizations, through the TMR effect, and is used to detect the direction of the free layer magnetization with respect to a fixed layer magnetization (see Chapters 9 through 11 for a thorough description of MTJs and TMR). External magnetic fields are then used to switch the free layer orientation between its two stable states.

These are all the prerequisites for single bit MRAM operation. In the following sections, I will describe the fundamental properties governing single bit operation, the additional challenges of full array operation, the long-term reliability of MRAM, and finally, the emerging MRAM applications.

13.2.1 THE 1T-1MTJ MEMORY CELL

In the so-called 1-transistor-1-magnetic-tunnel-junction (1T-1MTJ) MRAM bit cell [11] used in all commercial MRAM, a single MTJ is connected to ground via a single, minimum-sized, n-type field-effect transistor (FET) isolation transistor (Figure 13.2). The bottom electrode of the MTJ is connected to the drain of its transistor, and its top electrode is connected to the "bit line." The gate of the transistor is connected to a "word line," which runs perpendicular to the bit line. During read operation, the word line turns on all transistors along its length, and the bit line applies a bias of about 0.3 V to all MTJs along its length. As a result, current will only flow through the MTJ where the word and bit lines intersect, and by measuring the current, the resistance state of the MTJ can be determined. While many other cells are "half-selected", that is, either their transistor is open or their MTJ top electrode is biased, these bit cells do not contribute to the resulting current fed into the read-out circuitry (sense amplifier).

A similar selection scheme is used for programming (writing) the bit cell. Two sets of write lines run perpendicular to each other, and only where two energized write lines intersect will a bit cell be written by the combined magnetic field pulses from both write lines (Figure 13.3).

However, in contrast to the read-out selection, where half-selected bits do not interfere with the read-out, the bit cells that are half-selected during writing may inadvertently switch (grey bits in Figure 13.3). Until the invention of toggle switching (see following section), this half-select problem dominated write failures in full arrays and initially threatened the success of the entire MRAM program.

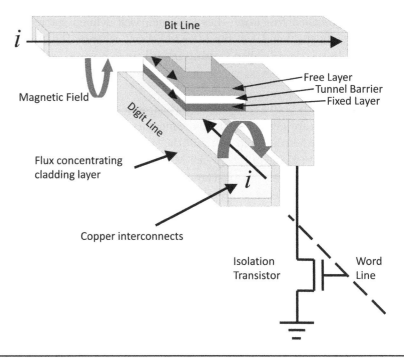

FIGURE 13.2 MRAM memory cell with one MTJ, one read-select transistor, and two write lines (bit line and digit line) for write selectivity. Each write line has flux concentrating cladding on three of the four sides of the Cu interconnect to enhance and focus the magnetic field created during writing.

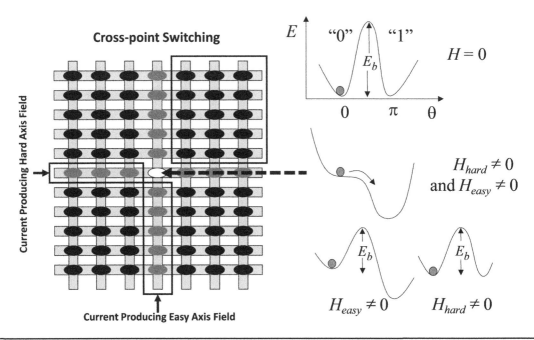

FIGURE 13.3 Conventional MRAM array showing cells in three fundamentally different states during writing: the fully selected bit (white), the half-selected bits (gray), and the unselected bits (black). To the right are shown the corresponding energy barriers.

13.2.2 Two Stable States

In MRAM, digital information is stored in the direction of the free layer magnetization of the MTJ. It is therefore critical that the free layer has two well-defined stable states and that switching between these two states is not only reproducible but identical between MTJs. Magnetic anisotropy in the free layer creates an energy barrier to magnetization reversal, E_b, separating the two states. One typically tries to minimize the material anisotropy, that is, magneto-crystalline and magnetostrictive anisotropies, since their direction and variation are difficult to control and lead to wide switching distributions. Magnetostriction is held at zero, since strain can be quite large, arising from the different processing steps and different materials used, and furthermore, may change substantially with temperature due to different thermal expansion coefficients for different materials.

Shape anisotropy, on the other hand, is primarily used to control the switching properties and critically depends on the quality of the lithographic process used in defining the bit. The elongated elliptical shape of the free layer defines an easy (long) and hard (short) axis. Either of the two stable magnetization states points along the easy axis (Figure 13.3). The barrier to reversal, E_b, is at a maximum when no magnetic field is applied. Regardless of direction, a magnetic field will always reduce the reversal barrier. Any applied field can be decomposed into an easy (H_{easy}) and a hard (H_{hard}) axis field component, where H_{hard} acts to reduce the energy barrier, and Heasy not only reduces the barrier but makes it asymmetric so as to favor one of the two stable states. At small enough dimensions, when the free layer is a single magnetic domain, one may expect magnetization reversal to proceed by coherent rotation. In the case of an ellipsoidal particle, such a reversal can be described by a Stoner–Wohlfarth (SW) switching asteroid [12]. However, in planar geometry, the free layer is only elliptical viewed from the top down but square in cross section and thus, not ellipsoidal, which would be required for true single domain behavior. The actual switching data of elliptical free layers hence deviates somewhat from the ideal SW behavior.

13.3 MEMORY ARRAYS

13.3.1 The 4 MB Array

The first commercially available MRAM, a 4 MB MR2A16A chip from Freescale Semiconductor (today Everspin Technologies), contained 4,194,304 MTJs (not counting redundancy and error correction bits) divided into 16 I/O areas, each containing 262,144 MTJs. Both intrinsic material and lithography variations, and extrinsic defects from, for example, dust and particulates, play major roles in determining the final operating windows for both read and write. When moving from single bits to millions of bits, the focus hence shifts from the intrinsic fundamental properties to the *distributions* of these properties. The final memory design must provide enough read and write *margin* and in addition, sufficient redundancy and error correction bits to ensure error-free operation throughout the specified lifetime of the memory.

13.3.2 RESISTANCE DISTRIBUTIONS

For read-out, a reference generator circuit produces a signal at the midpoint between the high and low states of the bits. Several MTJ bits produce the midpoint value, which hence follows i) resistance variation across the wafer, ii) changes from wafer to wafer, and iii) changes due to temperature effects. Besides variations in operating conditions and the resistance variation across a single wafer and between different wafers, there also exist resistance variations, or distributions, within a given array. Good array uniformity is critical for correct read-out in the mid-point reference scheme. It is important to realize that the resistance distribution within an array is much larger than what one would expect from measured gradients across a wafer. Since memory arrays are only a few millimeters across, a 5% wafer-level resistance uniformity would result in bit resistance variations within an array of only about 0.05%. However, resistance sigma within an array is about 1% at best.

Figure 13.4a illustrates the effect of random resistance variations within an array on MRAM read-out. During the read operation, the circuit compares the resistance of the bit with the reference to determine whether it is in the high or the low state. In the low state, bits with resistance on the high side of the distribution will fall closer to the reference than the average bit. Similarly, the bits on the low side of the distribution in the high resistance state will also be closer to the reference cell. To have working megabyte memories, the circuit must be able to correctly read the state of these bits in the tails of the distributions. The circuit is designed to operate with a signal reduced by 6σ, resulting in a useable resistance change for the circuit ΔR_{use}, as illustrated in Figure 13.4a. A larger ΔR_{use} allows a more aggressive circuit design with faster access time. The intrinsic quality of the MTJ barrier material plays a major role in determining the resistance distributions. Rougher barrier interfaces lead to larger local resistance variations across the MTJ surface and therefore, larger final bit-to-bit variation. Since the resistance

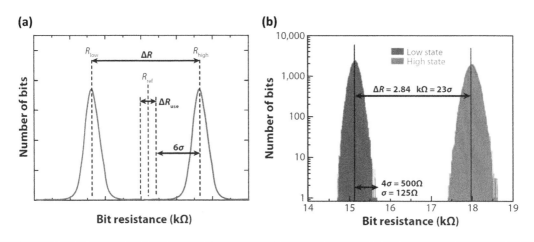

FIGURE 13.4 (a) Schematic representation of the effect of Gaussian MTJ resistance distributions resulting in the concept of an effective read margin when the full distributions are taken into account. (b) Actual MTJ resistance distributions in the high and low resistance state respectively, showing sufficient read margin for operation. (After Engel, B. et al., *IEEE Trans. Magn.* 41, 132, 2005. With permission.)

depends exponentially on barrier thickness, even small changes in roughness can cause large variations in resistance. For large roughness, conduction is largely dominated by a limited number of so-called *hot-spots*, where most of the current is flowing [13–15]. Any variation in bit area, due to lithography or etch variations, will also cause variations in bit resistance. Figure 13.4b shows measured distributions for an array made of MTJ material with an optimized array uniformity showing a 20σ margin. Here, improvement in the MTJ material led to large separation and improved uniformity.

13.3.3 SWITCHING DISTRIBUTIONS AND THE HALF-SELECT PROBLEM

As discussed earlier, material and shape anisotropy are the dominant factors governing magnetization reversal. They also determine the magnitude of the (easy axis) switching field (H_{sw}) for a given applied hard axis field. Consequently, any variation in material quality or bit shape will inevitably lead to bit-to-bit variations of the switching astroid. While both the easy and the hard axis properties are affected, one typically quantifies the bit-to-bit variation as a distribution, ΔH_{sw}, of easy axis switching fields at a constant hard axis field. In this statistical description, H_{sw} denotes the average switching field of the ensemble of bits. While the bit-to-bit switching behavior may appear highly reproducible in a small laboratory experiment, the situation is quite different in an array that can consist of over four million bits. A broad ΔH_{sw} can severely limit the operating window in realistic devices. In addition, as the bit size is reduced, the relative switching distribution $\sigma sw = \Delta H_{sw}/H_{sw}$ increases, reflecting the increasing difficulty in maintaining a consistent bit shape at smaller dimensions.

The switching distribution has two important consequences. First, to program all the bits in a 4 Mb array with the same switching field throughout the chip, the applied field needs to be larger than the bit with the highest switching field. With over two million bits, and assuming a Gaussian distribution of switching fields, this is equivalent to a minimum field of $H_{sw} + 6\Delta H_{sw}$. However, as a consequence of the grid layout of the write lines, for every bit being programmed (the fully selected bit) with a combination of a hard axis field and an easy axis field of $H_{sw} + 6\Delta H_{sw}$, more than a thousand other bits (so-called *half-selected bits*) along the two write lines experience either the applied hard axis field alone or $H_{sw} + 6\Delta H_{sw}$ without a hard axis field. Consequently, the easy axis switching field used throughout the array must also be kept below a certain *maximum* value related to the bit with the *lowest* switching field when no hard axis field is applied. If this distribution is ignored, half-selected bits may inadvertently switch and lead to writing errors, so-called *half-select disturbs*. For error-free programming, conventional MRAM (in contrast to toggle MRAM described in the next section) must keep the switching distribution and the half-select disturb distribution well separated. Between these two distributions, there is a resulting operating window where all bits can be programmed without errors or disturbs. A schematic of the operating window superposed on the switching astroid is shown in Figure 13.5a.

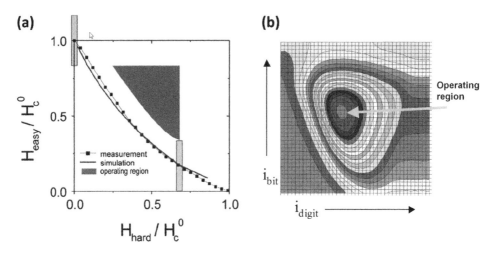

FIGURE 13.5 (a) Constructed write operating region of conventional MRAM. The solid points are quasi-static switching measurements and the solid line is the expected switching curve from micromagnetic simulation. The rectangles represent the switching field distribution within an entire array, which effectively limits the operating region (triangular area). (After Engel, B. et al., *IEEE Trans. Magn.* 41, 132, 2005. With permission.) (b) Actual measurement of the write operating region of a conventional MRAM array showing insufficient write margin for error-free operation.

Already in the early period of MRAM development, it became apparent that the half-select problem amounted to a formidable challenge. In addition to the distributions described earlier, the problem is exacerbated by other sources of variations such as complementary metal-oxide-semiconductor (CMOS) transistor variations (i.e., variations in the actual switching current provided by each write line driver), variations in the write line resistances, and variations in the magnetic cladding properties along the write line. The final blow to conventional MRAM switching is dealt when the true nature of the probabilistic single bit switching is analyzed, and extrapolations to 10 years of operation are considered. Due to thermal energy, the half-select disturb field for a single bit has an intrinsic temperature-dependent distribution. The more disturb attempts each bit is given, the more one probes the tail of this intrinsic distribution, and the further away from the average disturb field one has to operate to ensure zero disturbs.

An example of a measured switching astroid of a 256 kbit conventional MRAM demonstrator is shown in Figure 13.5b. While an optimum region does exist, this region is still not error free. A fully functioning operating region could not be strictly defined for this demonstrator.

13.3.4 SAVTCHENKO SWITCHING AND TOGGLE MRAM

At the turn of the millennium, the half-select challenge increasingly appeared as an insurmountable brick wall. While the continued material and lithography improvements had indeed led to important demonstrations of functional MRAM arrays, even the best distributions were nowhere near the required separations. Predictions of continued incremental improvement did not look promising, and the chances that MRAM would become feasible looked quite slim. As we now know, commercial conventional MRAM never saw the light of day.

The half-select problem was instead solved through an ingenious invention by the late Motorola staff scientist Leonid Savtchenko. The invention includes a novel free layer structure and bit orientation, along with the use of a particular set of timed current pulses, which, in a *coup de grâce*, completely removes the limitations that plagued conventional switching [16]. The mechanism proposed by Savtchenko completely removes half-select disturbs. The single free layer is replaced by a synthetic antiferromagnetic (SAF) free layer; that is, a trilayer structure where two magnetically soft ferromagnetic layers couple antiferromagnetically through a non-magnetic spacer, for example, Ru. This is shown schematically in Figure 13.6a.

The response of the SAF to external fields bears little resemblance to the traditional switching astroid. Instead of the magnetic layers in the SAF aligning with the applied field, the generic response of the SAF is a 90 degree orientation of both magnetic moments with respect to the applied field, a so-called *spin-flop* orientation (Figure 13.6b). If each layer has a finite anisotropy field, a certain critical spin-flop field H_{sw} must first be overcome. In this spin-flopped state, each moment will cant slightly in the direction of the applied field; that is, the SAF layers will scissor an angle δ away from perfect anti-alignment. Through this scissoring mechanism, the SAF can lower its total magnetic energy, and the scissoring angle depends on the competition between the Zeeman energy due to the applied field and the antiferromagnetic exchange energy through the spacer layer [17].

To derive H_{sw}, we first analyze the case where neither of the magnetic layers has any anisotropy field, H_k (Figure 13.7a). In an applied field, H, the angular dependent energy of the SAF can be expressed as

$$E(\theta,\delta) = -K_S \cos(2\delta) - 2M_S H \sin(\theta)\sin(\delta), \qquad (13.1)$$

where

M_S is the saturation magnetization of the magnetic layers
K_S is the antiferromagnetic exchange anisotropy between the layers

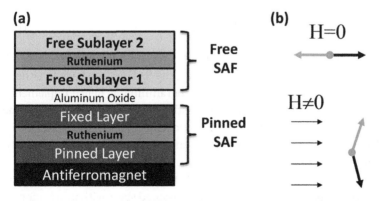

FIGURE 13.6   (a) Schematic of a Toggle-MRAM MTJ with a SAF free layer. (b) The spin-flop response of a SAF free layer when an external field is applied.

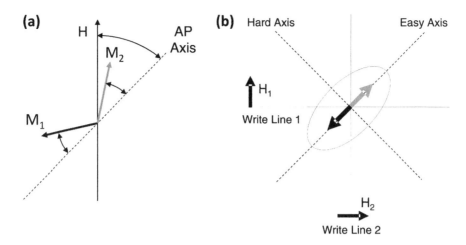

FIGURE 13.7 (a) Definitions of the angles used in the derivation of the spin-flop field. (b) Schematic Toggle-MRAM bit having its easy-axis oriented at 45 degrees with respect to write lines.

Minimizing with respect to δ, one finds the energy as a function of field angle

$$E(\theta) = -\frac{M_S^2 H^2}{2K_S}\sin^2(\theta) + C, \tag{13.2}$$

where C is a constant independent of θ. The applied field hence acts as an anisotropy *orthogonal* to its own direction. If we now add H_k of each layer and analyze the case when H is applied along the same axis as H_k, the angular dependence of the energy changes to

$$E(\theta) = M_S H_k \left(1 - \frac{H^2}{H_k H_{sat}}\right)\sin^2(\theta), \tag{13.3}$$

where we have defined the saturation field of the SAF as $H_{sat} = 2K_S/M_S$. For a spin-flop reorientation to occur, the expression in the parenthesis of Equation 13.3 must be negative; that is, the critical flop field is the arithmetic mean of the anisotropy field and the saturation field:

$$H_{sw} = \sqrt{H_k H_{sat}}. \tag{13.4}$$

However, overcoming the flop field is not sufficient for well-defined switching. If the field is again reduced below H_{sw}, the SAF may go back to either of its two stable zero-field states. Actual programming is hence implemented using a bit orientation at 45 degrees with respect to the two write lines (Figure 13.7b) in combination with tailored field pulses.

The programming pulse sequence and the resulting behavior of the moment of two magnetic layers, represented by arrows, are depicted in Figure 13.8. When write line 1 turns on (t_1), the magnetic field from this single line will prepare the SAF in an initial scissored state, with its net moment trying to align with the applied field. With write line 1 still on, the second

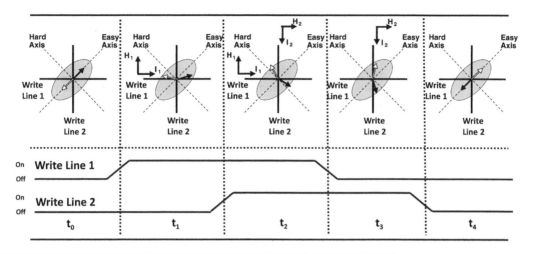

FIGURE 13.8 Programming of a Toggle-MRAM cell. Write pulses are applied sequentially to rotate the free layer SAF 180 degrees regardless of its initial state.

write line turns on (t_2), and the resulting fields add up vectorially to overcome the critical flop field, H_{sw}. While the SAF is now in its fully spin-flopped state, no actual switching has yet taken place. To finally switch the bit, the first write line turns off (t_3), resulting in a final rotation of the SAF toward a scissored state with the net SAF moment largely parallel with the applied field from write line 2. In the final step, the switching is completed by turning off write line 2 (t_4). During this field pulse sequence (t_0 to t_4), the two individual layers have performed a 180 degree rotation. Due to symmetry, if the process is repeated with the same polarity pulses, the two layers of the SAF will again rotate 180 degrees in the same manner. The programming scheme is hence of a toggling type; that is, it does not switch into a predetermined state, but instead, it toggles from one state to the other regardless of the initial state. This toggling process is a result of the carefully balanced nature of the SAF, and any significant imbalance of the two moments will render it ineffective. The toggling scheme also has three direct consequences for the programming circuit: i) it removes the need for bipolar current drivers for the easy axis field, which then simplifies the design and allows the use of smaller transistors; ii) it requires a decision write scheme, where the bit state is first read and only toggled if the new data differs from the existing data; and iii) it effectively reduces the average programming power consumption by 50%, since for half of the programming events, the decision circuitry will simply not energize the write lines.

However, the dominating advantage of toggle MRAM is its greatly enhanced selectivity. If only a single write line current is applied (half-selected bits) as in t_1, the 45 field angle cannot switch the state. In fact, the single-line field *raises* the switching energy barrier of those bits so that they are stabilized against reversal during the field pulse. This is in stark contrast to the conventional approach, where all of the half-selected bits have their switching energy reduced and are therefore more susceptible to thermally excited switching events. The phase relationship of the combined pulses is therefore required for switching and results in an improvement of several orders of magnitude of the bit selectivity in a full memory array.

Similarly to the switching astroid of a conventional MRAM bit, the switching response curve for a toggle MRAM bit can be constructed as a function of the two write lines, as shown in Figure 13.9a. The upper right and lower left quadrants are for the same polarity field pulses and will allow toggling. The white regions are either insufficient field magnitudes or improper field directions for toggling. The green regions are above the toggling transition for all bits and result in 100% toggling. The orange regions represent the bit-to-bit distribution of switching thresholds that must be overcome for reliable switching of all bits in an array. Figure 13.9b is a switching versus current map for an entire 4 Mb memory. The test pattern was a checkerboard/inverse checkerboard of alternating "1" and "0" that was written at each current and read back to verify. In the region below the switching threshold, no bits changed state, and hence, there were no disturbs from half selects. A large operating region is observed above the threshold consistently with the single bit characteristics. The contours in the transition region just at the threshold are a measure of the bit-to-bit switching distribution.

13.4 RELIABILITY

One of the key properties of MRAM is its very high reliability. The promise that a programmed "1" or "0" is indeed stored as a "1" or "0," and that this stored information remains stable for over 10 or 20 years under all use conditions, is of paramount importance. Additionally, the information must be read back correctly at all times, and no wear-out may occur as a function

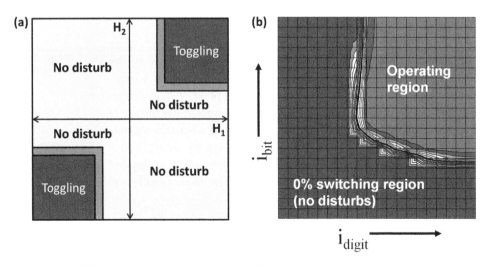

FIGURE 13.9 Ideal switching map of Toggle-MRAM showing two square toggling regions when both pulses have the same polarity and both pulses have sufficient magnitude. b) An actual switching map (quadrant 1) of an entire 4Mb array showing a large operating region for 10 ns write pulses. (After Engel, B. et al., *IEEE Trans. Magn.* 41, 132, 2005. With permission.)

of the number of reads or writes. While these may sound like quite natural demands on a memory, reliability issues are often overlooked during the early stages of development of a new technology. All well-established memory technologies have a long history of detailed reliability studies, and all failure mechanisms are analyzed and modeled to ensure zero bit failures at usage conditions. If a new memory technology cannot demonstrate that all possible failure mechanisms have been studied, analyzed, and modeled, it is unlikely that it will gain any market acceptance. Reliability is hence part of the main function of a memory, and the technology must design for good reliability at as early a stage as possible.

In MRAM, reliability can be divided into well-known CMOS reliability issues and new failure mechanisms related to the MRAM module itself. The possible MRAM failure mechanisms can in turn be divided into MTJ reliability, write line reliability (since the magnetic cladding is an entirely new part of the device), and long-term effects of temperature on bit stability and write currents.

13.4.1 MTJ Reliability

The two main types of MTJ failures are time dependent dielectric breakdown (TDDB) and time dependent resistance drift (TDRD), both induced by voltage stress to the junction. At constant voltage stress, dielectric breakdown is detected as an abrupt increase of the MTJ current due to a short forming through the tunneling barrier [18]. Since the metallic conduction through the short is mostly spin independent [19]), the increase in current is accompanied by a loss of essentially all the MR. From the memory's point of view, a shorted MTJ results in a stuck-at-zero bit failure.

Dielectric breakdown can be successfully studied in single MTJs under ramped voltage stress, yields a rapid assessment of the average quality of the MTJ barrier, and provides rapid feedback for material and process improvements [20, 21]. However, for accurate modeling of TDDB and for extrapolations to usage conditions and long life-times, it is necessary to study much larger arrays of actual memory bits to get enough statistics and to make sure that the correct failure mechanisms are indeed studied. For this purpose, custom designed 1 kB memory reliability circuits, with the same MTJ as in the full 4 MB array but with additional voltage read-out points, were designed to allow accurate measurements of the applied bias stress over the bit as well as the bit resistance during stress [7]. Figure 13.10a shows typical TDDB data obtained from this array stressed at three different temperatures (125 °C, 150 °C, and 175 °C) and five different nominal bias levels (1.1 V, 1.15 V, 1.2 V, 1.25 V, and 1.3 V). The actual bias level during stress is continuously measured, and the exact values are used in the TDDB analysis. Each stress cell had a total of 32 bits with all bits driven to failure for optimal statistics. As expected for dielectric breakdown, the data can be well fitted (straight lines) by a model based on Weibull life-time distributions, linear bias voltage acceleration (linear-E model), and an Arrhenius-type thermal activation.

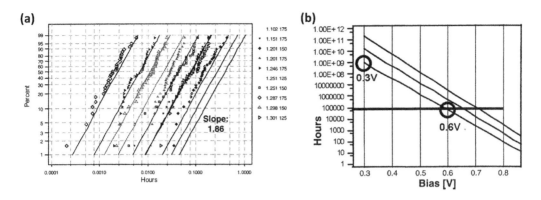

FIGURE 13.10   (a) Time-dependent dielectric breakdown data from a dedicated 1 kb MRAM array. In the legend to the right of the plot, the first number indicates the applied voltage stress and the second number indicates the temperature. (b) Extrapolation to 100ppm failures with 95% single sided confidence. The horizontal line indicates 10 years of operation and determines the upper limit of MTJ bias (0.6V) for reliable operation. (After Åkerman, J. et al., *IEEE Trans. Dev. Mat. Rel.* 4, 428, 2006. With permission.)

In contrast to other solid state memories, the MRAM bit experiences no applied bias stress during data retention, since the information is stored magnetically at zero voltage. As a consequence, only 16 memory bits are stressed simultaneously at any given time during the lifetime of the memory, and the typical stress duty cycle of each bit is *extremely small*. While the exact customer usage is unknown, all possible usage falls somewhere between perfectly uniform memory usage and reading/writing the same 16 bit word for 10 years. As shown in [7]), the rather high value of β (the slope of the straight lines in Figure 13.10a) makes reading the same word over and over again about two orders of magnitude more stressful than the case of uniform usage. Extrapolations to 100 ppm fallout at 95% single sided confidence are shown in Figure 13.10b. The strong bias dependence and the low operating bias of 0.3 V make the intrinsic reliability about four orders of magnitude better than required for 10 years' operation. Alternatively, one can argue that there is a substantial design margin up to 0.6 V should one choose to increase the signal for faster read operation.

The MTJ is also subject to TDRD, which in contrast to TDDB, shows highly deterministic and uniform behavior over time, in that the resistance of all bits drifts about the same amount for the same stress conditions. In Figure 13.11a, we show resistance drift at 120 °C for bias levels ranging from 0.6 to 1.2 V. Since increasing the bias stress only shifts the curves along the log t axis, the time axis can be re-scaled ($t \Rightarrow t^*$) to make all curves overlap (i.e., plot R vs. the scaled t^* instead of the original t). The single master curve, shown in Figure 13.11b, indicates that bias stress only accelerates the rate of resistance drift without introducing new sources of drift. In the inset of Figure 13.11b, it is shown that t^* versus bias voltage follows a linear-E model (the activation energy is reduced linearly with the electric field E) and that an extrapolation to operating bias (0.25 V in this case) yields a scaling factor of 3×10^6 for 10 years of stress at 0.25 V, which is equivalent to 1 minute and 45 seconds of stress at

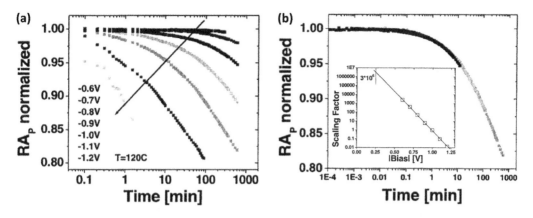

FIGURE 13.11   (a) Normalized time-dependent resistance drift for different value of the applied bias at an ambient temperature of 120 C. (b) The same data after a scaling procedure allowing for extrapolation to lower bias and longer times as well as a master curve for resistance drift. (After Åkerman, J. et al., *IEEE Trans. Dev. Mat. Rel. 4*, 428, 2006. With permission.)

1.0 V, and the corresponding resistance drift can be directly read out from the graph. For example, 10 years' normal operation at 120 °C and 100% (25%) duty cycle would result in R drift of –2.2% (–1.0%). Just as for TDDB, the worst case for resistance drift is single word usage. At continuous operation, the actual bias stress is only applied for about 50% of the cycle time. As long as the circuit can handle roughly 2% resistance drift, no read-out failures are expected due to this mechanism for 10 years at 120 °C.

13.4.2 WRITE LINE RELIABILITY

Sufficient metal interconnect reliability is a standard requirement for any CMOS process, and the required MRAM write currents and operating conditions can be designed within the known reliability performance. However, the addition of magnetic cladding onto three out of four surfaces of the metal interconnects is a significant unknown parameter and must be evaluated for reliability. Figure 13.12a shows a Weibull plot of the electromigration failure of standard Cu interconnects and Cu interconnects with cladding. As can be seen, the magnetic cladding actually improves the median of the fail time by as much as 40 times, which is similar to other reports showing that a hard film cladding can improve the electromigration of Cu significantly [22, 23].

13.4.3 TEMPERATURE EFFECTS

Temperature has two important effects on MRAM performance: i) instantaneous changes, which are typically reversible when the temperature is reduced, and ii) long-term irreversible changes after exposure to high temperature for extended times. Among the first type, we find changes in read signal and write current distributions due to the temperature dependence of the intrinsic magnetic properties of the magnetic layers, and the resistance and magnetoresistance of the MTJ [24]. The second, and irreversible, type is

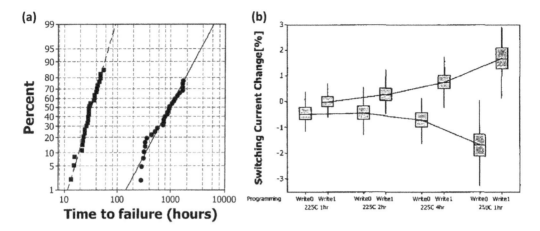

FIGURE 13.12 (a) Distribution of write line failure times due to electromigration. The left data set shows the result from standard Cu interconnects and the right data set shows an improvement when the magnetic cladding is added to the write lines. (After Åkerman, J. et al., *IEEE Trans. Dev. Mat. Rel.* 4, 428, 2006. With permission.) (b) Change in the switching current after heating the patterned Toggle-MRAM bit to different temperatures and for different times. (After Åkerman, J. et al., *IEEE Proceedings 43rd International Reliability Physics Symposium*, 163, 2005. With permission.)

typically due to interdiffusion within the full MTJ stack and can become a reliability issue if not modeled and designed for.

For example, the toggling scheme described earlier relies on both the detailed moment balance between the two NiFe layers in the SAF free layer and the exchange coupling mediated by the Ru layer. Therefore, factors determining the reliability of the write currents are the thermal endurance of the antiferromagnetic exchange coupling and the thermal endurance of the relative moment balance between the two constituent magnetic layers. This can be studied both in continuous films and in patterned structures to distinguish between intrinsic diffusive effects in the stack and increased diffusion from the sides of the patterned structures. Figure 13.12b shows the percentage change in write currents versus several high-temperature anneals in 150 4 MB arrays. While the average of the program current does not change, consistently with no expected change in the antiferromagnetic exchange coupling, there is a significant drift in the individual program currents. This asymmetry in write-0 and write-1 current is a sign of an increased imbalance in the free layer SAF. Since bulk wafers show no change in the free layer moment balance at these temperatures, the observed changes are related to the bit processing. The rate of change is approximately five times greater at 250 °C than at 225 °C, which suggests less than 1% change in switching current after 10 years' operation at 110 °C, which is well within the design limits.

13.5 SPIN TRANSFER TORQUE MRAM

The greatest weakness of MRAM is its large cell size. The first 4 MB toggle MRAM, discussed earlier, used 180 nm CMOS technology and had a cell size of 1.25 μm^2. Although scaling toward 90 nm CMOS was successfully

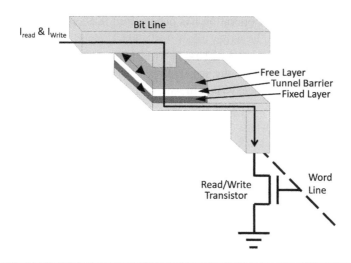

FIGURE 13.13   STT-MRAM memory cell with one MTJ and one read/write-select transistor.

demonstrated early on [25] and reduced the possible cell size to 0.29 μm^2, the actual cell size in later products, such as 16°MB toggle MRAM, has not decreased, since a smaller MTJ results in wider write distributions.

By replacing the required external magnetic field to write the MRAM bit with the direct internal effects of the spin polarized current passing through the bit, an STT-MRAM bit cell can be designed (Figure 13.13) without any need for write lines (see Chapters 7, 8, and 13 for a detailed description of STT). The resulting bit cell is both smaller and less complex than the toggle MRAM cell. The expected cell size using 65 nm CMOS is about 0.04 μm^2, which is similar to Flash at the 65 nm node and is expected to scale as favorably as Flash to smaller dimensions. STT-MRAM hence has the potential to become a true universal memory in which the good functional properties present in toggle MRAM are combined with high storage density and good scalability.

While the first demonstrations of STT-MRAM used magnetic layers similar to those in toggle MRAM, it was soon realized that in-plane STT-MRAM would never scale well due to issues with long-term data retention [26]. The main drawback for in-plane magnetic materials is that STT switching requires the free layer magnetization to overcome the very large out-of-plane demagnetizing field, while data retention is only governed by the in-plane energy barrier. The most promising solution is instead to change to magnetic layers with large perpendicular anisotropy, so-called *perpendicular STT-MRAM*, where both the free and fixed layer magnetizations are oriented perpendicularly to the film plane in their remanent state (in other words, aligned with the direction of the current through the device). This geometry allows both thermal stability and programming currents as low as 9 μA, which is well below the estimated maximum sustainable CMOS transistor current of 20 μA, at the dimensions needed for Gbit STT-MRAM [27].

Once the single bit operation has been solved in STT-MRAM, all possible distributions, and how they affect read and write performance, must be addressed and understood. In STT-MRAM, one must control, or manage, *three* independent distributions, which all are related to the current through the device: read resistance, write current, and MTJ breakdown voltage distributions. For example, a particular challenge is to minimize the write current while also making sure that one does not accidentally write a bit during reading. Such "read disturbs" might turn out to be as detrimental to STT-MRAM as half-select disturbs were to conventional MRAM. Several fully functional STT-MRAM chips have now been demonstrated, all based on perpendicular materials [28, 29]. In August 2014, Everspin announced the first 64 MB STT-MRAM, and 2 years later, in August 2016, they shipped the first commercial samples of 256 MB STT-MRAM, claiming to have 1 GB versions under development.

13.6 APPLICATIONS

With its unique combination of key memory properties, MRAM has particular potential in applications where traditional memory technology cannot provide reliable and secure solutions. The combination of non-volatility with high write speeds and very high reliability makes it attractive for cache memories in Redundant Array of Independent Disks (RAID) storage, where critical cache information can be updated at the same speed as with SRAM but with the added benefit of being retained during power failures. Similar continuous back-up applications are applicable in a wide range of industrial applications, such as communication, transportation, military, and avionic systems, where critical information needs to be retained at all times and with the highest demand on reliability. The very wide operating range at both low and high temperatures also makes MRAM attractive for highly reliable usage under demanding conditions, such as in automotives and in space. In 2016, Everspin claimed it had shipped over 60 million discrete and embedded products to a total of over 500 customers.

As early toggle MRAM product examples, Siemens's Simatic Multipanel MP 277 and MP 377 human machine interface (HMI) for industrial automation systems were able to retain software-programmable logic controller process data without battery back-up. As a clear demonstration of the very high reliability of toggle MRAM, Siemens had, over the course of 2 years and over 100,000 shipped multipanels, not recorded a single memory failure. Toggle MRAM was also employed in Emerson Network Power's single-board computers used in industrial, medical, military, aerospace, and telecom platforms. In these applications, MRAM allows the capture of system data logs that are not impacted by memory wear-out or power failure. In particular, the high write speed and high reliability also allow each unit to receive frequent new programs and data updates with guaranteed data retention over 20 years even at very high temperatures. Finally, Airbus also used toggle MRAM to replace SRAM and Flash in the flight control computer on their Airbus A350 XWB civil aircraft.

The most recent STT-MRAM application comes from Aupera Technologies, who integrate Everspin's 64 and 256 MB STT-MRAM into their All Flash Array system for video data storage, processing, and analytic application.

13.7 CONCLUSIONS

The technology behind commercially available toggle MRAM has been described, as well as the challenges and solutions associated with its development over the last 15 years. The transition from toggle MRAM to STT-MRAM has been briefly described as well. The use of both toggle MRAM and STT-MRAM is currently spreading into many different application spaces, in particular where its combination of non-volatility, fast read/write, unlimited endurance, and high reliability is used to replace a combination of other memory technologies.

REFERENCES

1. J. A. Rajchman, Computer memories—A survey of the state-of-the-art, *Proc. IRE* **49**, 104 (1961).
2. L. A. Russel, R. M. Whalen, and H. O. Leilich, Ferrite memory systems, *IEEE Trans. Magn.* **4**, 134 (1968).
3. W. M. Regitz and J. A. Karp, Three-transistor-cell 1024-bit 500-ns MOS RAM, *IEEE J. Sol. State Circuits* **SC-5**(5), 42 (1970).
4. T. Miyazaki and N. Tezuka, Giant magnetic tunneling effect in $Fe/Al_2O_3/Fe$ junction, *J. Magn. Magn. Mater.* **139**, L231 (1995).
5. J. S. Moodera, L. R. Kinder, T. M. Wong, and R. Meservey, Large magnetoresistance at room temperature in ferromagnetic thin film tunnel junctions, *Phys. Rev. Lett.* **74**, 3273 (1995).
6. W. J. Gallagher and S. S. P. Parkin, Development of the magnetic tunnel junction MRAM at IBM: From first junctions to a 16-Mb demonstrator chip, *IBM J. Res. Dev.* **50**, 5 (2006).
7. J. Åkerman, M. DeHerrera, M. Durlam et al., Magnetic tunnel junction based magnetoresistive random access memor, in *Magnetoelectronics*, M. Johnson (Ed.), Elsevier, pp. 231–272, 2004.
8. J. Daughton, Magnetoresistive random access memories, in *Magnetoelectronics*, M. Johnson (Ed.), Elsevier, pp. 231–272, 2004.
9. J.-G. Zhu, Magnetoresistive random access memory: The path to competitiveness and scalability, *Proc. IEEE* **96**, 1786 (2008).
10. R. O'Handley, *Modern Magnetic Materials: Principles and Applications*, Wiley-Interscience, 1999.
11. P. K. Naji, M. Durlam, S. Tehrani, J. Calder, and M. F. DeHerrera, A 256 kb 3.0V 1T1MTJ nonvolatile magnetoresistive RAM, *ISSCC Digest of Technical Papers*, February 2001, pp. 122–123.
12. E. C. Stoner and E. P. Wohlfarth, A mechanism of magnetic hysteresis in heterogeneous alloys, *Phil. Trans. Roy. Soc. A* **240**, 599 (1948).
13. V. Da Costa, C. Tiusan, T. Dimopoulos, and K. Ounadjela, Tunneling phenomena as a probe to investigate atomic scale fluctuations in metal/oxide/metal magnetic tunnel junctions, *Phys. Rev. Lett.* **85**, 876 (2000).
14. C. W. Miller, Z. Li, J. Åkerman, and I. K. Schuller, Impact of interfacial roughness on tunneling conductance and extracted barrier parameters, *Appl. Phys. Lett.* **90**, 043513 (2007).
15. D. Rabson, B. J. Jönsson-Åkerman et al., Pinholes may mimic tunneling, *J. Appl. Phys.* **89**, 2786 (2001).

16. B. Engel, J. Åkerman, B. Butcher et al., A 4Mb toggle MRAM based on a novel bit and switching method, *IEEE Trans. Magn.* **41**, 132 (2005).

17. J.-G. Zhu, Spin valve and dual spin valve head with synthetic antiferromagnets, *IEEE Trans. Magn.* **35**, 655 (1999).

18. W. Oepts, H. J. Verhagen, W. J. M. de Jonge, and R. Coehoorn, Dielectric breakdown of ferromagnetic tunnel junctions, *J. Appl. Phys.* **73**, 2363 (1998).

19. J. Åkerman, R. Escudero, C. Leighton et al., Criteria for ferromagnetic–insulator–ferromagnetic tunneling, *J. Magn. Magn. Mat.* **240**, 86 (2002).

20. J. Åkerman, M. DeHerrera, J. M. Slaughter et al., Intrinsic reliability of AlOx-based magnetic tunnel junctions, *IEEE Trans. Magn.* **42**, 2661 (2006).

21. J. Åkerman, P. Brown, M. DeHerrera et al., Demonstrated reliability of 4-Mb MRAM, *IEEE Trans. Dev. Mat. Rel.* **4**, 428 (2006).

22. D. A. Gajewski, B. Feil, T. Meixner, M. Lien, and J. Walls, Electromigration performance enhancement of Cu interconnects with PVD Ta cap, *IEEE International Reliability Physics Symposium Proceedings*, pp. 627–628, 2004.

23. C.-K. Hu, L. Gignac, E. Liniger et al., Comparison of Cu electromigration lifetime in Cu interconnects coated with various caps, *Appl. Phys. Lett.* **83**, 869 (2003).

24. J. Åkerman, I. V. Roshchin, J. M. Slaughter, R. W. Dave, and I. K. Schuller, Origin of temperature dependence in tunneling magnetoresistance, *Europhys. Lett.* **63**, 104 (2003).

25. M. Durlam, T. Andre, P. Brown et al. 90 nm toggle MRAM array with 0.29 μm^2 cells, *2005 Symposium on VLSI Technology*, 10B–2, 2005.

26. D. Thibaut, Scalability of magnetic random access memories based on an in-plane magnetized free layer, *Appl. Phys. Expr.* **4**, 093001 (2011).

27. H. Yoda, T. Kishi, T. Nagase et al., High efficient spin torque transfer writing on perpendicular magnetic tunnel junctions for high density MRAMs, *Curr. Appl. Phys.* **10**(Supplement 1), e87 (2010).

28. G. Jan, L. Thomas, L. Son et al., Demonstration of fully functional 8Mb perpendicular STT-MRAM chips with sub-5 ns writing for non-volatile embedded memories, *2014 Symposium on VLSI technology: Digest of Technical Papers*, 2014.

29. N. D. Rizzo, D. Houssameddine, J. Janesky et al., A fully functional 64 Mb DDR3 ST-MRAM built on 90 nm CMOS technology, *IEEE Trans. Magn.* **49**, 4441 (2013).

Emerging Spintronic Memories

Stuart Parkin, Masamitsu Hayashi, Luc Thomas, Xin Jiang, Rai Moriya, and William Gallagher

14.1 INTRODUCTION

The electronics industry is composed of three major components: computing (logic and memory), data storage, and communications. One of the most important limitations of computing devices is the ever increasing power consumption required by higher processing speeds. One means of reducing the power needed is the development of multi-core processors. Such processors, however, demand much larger memories, so that the proportion of a silicon chip dedicated to memory will increase from typical values today of ~60–70% to perhaps 90% in a few years' time [1]. The size and consequent cost of such chips could be significantly reduced by the development of a very high-density, high-performance memory, if this can done at a sufficiently low cost.

Traditional computing systems use a complex hierarchy of volatile and non-volatile (NV) memory technologies (largely to manipulate data) and data-storage technologies (largely to store data) to achieve a reasonable balance of performance, cost, and endurance. This leads to considerable complexity in operating systems and software applications, which adds cost and decreases reliability. Magnetic disk drives are the cheapest form of non-archival data storage with a cost about 10–100 times lower per bit than solid state memory technologies. Disk drives are capable of inexpensively storing huge amounts of data: typical capacities exceed several hundred gigabytes today. However, hard disk drives (HDDs) have a limited mean time to failure. A hard drive includes a fixed read/write head and a moving medium on which data is written. Devices with moving parts always are intrinsically unreliable with a possibility of mechanical failure and complete loss of stored data.

The cheapest solid state memory is FLASH, which has begun to displace disk drives for consumer applications; for example, for more reliable, lighter, and compact laptops and, perhaps paradoxically, for high-performance storage systems, where large numbers of input-output requests are called for [2]. FLASH memory relies on a thin layer of polysilicon that is disposed in oxide on a transistor's on-off control gate. This layer of polysilicon is a floating gate isolated by the silicon from the control gate and the transistor channel. FLASH memory is by far the cheapest form of solid state memory today, because it is a very dense memory; that is, the cell is relatively simple in structure, and each memory cell uses only a single transistor and therefore occupies a very small area of the silicon wafer. In some FLASH memories, more than 1 bit, perhaps 2 or 4, may be stored per transistor, although this usually leads to degraded performance and lifetime. Since the polysilicon slightly deteriorates after each write cycle due to the very high voltages and associated electric fields required, FLASH memory cells begin to lose data after approximately 100,000 to one million write cycles for single bit memories, and as little as 10,000 write cycles for multibit memories. While this is adequate for many applications, particularly where the number of write operations is limited, FLASH memory cells would begin to fail rapidly if used constantly to write new data, such as in a computer's main memory.

Furthermore, the data access time for FLASH memory is much too long for most computationally intensive applications.

Spintronic memories are emerging as one of the next generation storage technologies to replace HDDs and current solid state memories [3]. In this section, the concept, the underlying physics, and the state of the technology of these emerging spintronic memories are briefly reviewed.

14.2 MRAM

Magnetic random access memory (MRAM) stores data as the direction of magnetization of a ferromagnetic material [5]. One approach to MRAM uses a magnetic tunneling junction (MTJ) as the memory cell. The basic structure of the MTJ is a sandwich of two layers of ferromagnetic or ferrimagnetic material separated by a thin insulating material (see Chapters 11–15, Volume 1). In a typical MTJ, the magnetization direction of one magnetic layer (the "free layer") responds to a small magnetic field, whereas the other magnetic layer (the "reference layer") has a fixed magnetization direction. The resistance of the MTJ depends on the relative magnetization directions of the free and reference layer moments and is typically a minimum (maximum) when the two layers are parallel (antiparallel) to each other. MTJs for MRAM applications are engineered so that the magnetization of the free and reference layers lies along a single direction—the easy axis of magnetization—which is usually determined by the magnetic shape anisotropy of the MTJ nano-device, although the intrinsic magnetic crystalline anisotropy of the magnetic materials themselves can also be used. The latter is obviously critical for MTJ elements that are perpendicularly magnetized. In this way, data in the form of "ones" and "zeros" can be stored within the direction of magnetization of the free layer of the MTJ.

The concept of MRAM, based on MTJs, was first proposed by IBM in 1995 [4]. More than a decade later, in 2006, the first commercially available MRAM chips came onto the market from EverSpin, an independent company spun out from Freescale Semiconductor (itself originally part of the Motorola Corporation). The first IBM and the EverSpin MRAM chips [6], just 4 MB in size, can be considered as "first generation" MRAMs, in which the state of the MTJ memory cells is written by local magnetic fields. Localized magnetic fields at a given MTJ element are used to switch the direction of the magnetization of the free layer. A cross bar array of metal wires is used to generate local magnetic fields. In Figure 14.1, an array of MTJ memory elements is shown connected in parallel to a lower set of "word" lines and an upper set of "bit" lines. The "word" lines are oriented orthogonal to the bit lines. Currents are passed along the metal bit and word lines to generate magnetic fields. The strength of these magnetic fields is directly proportional to the current along the wires, and the field is generated uniformly along the wire. Thus, all the MTJs connected to one word line (or similarly, one bit line) will be subjected to the same magnetic field. The field (current) must be chosen so that, although it will disturb the magnetic state of each MTJ along that word or bit line, it is not large enough to change the

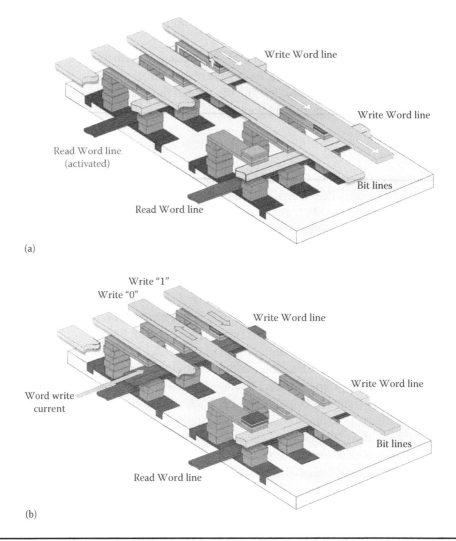

FIGURE 14.1 Diagram of a magnetic field programmable MRAM. (a) Reading a bit. Voltage is applied to the desired Read Word line (to enable the transistors in that word line) and voltage at desired bit line is measured. (b) Writing bits. Current is passed through the desired Write Word line and appropriate bit lines. Superposition of the fields generated by two currents orients the moment of the free layer in the MTJ in the desired direction. The polarity of the moment is determined by the direction of the current in the Bit line; current flowing one direction will flip it one way, and current in the opposite direction will flip it the other way. (After Parkin, S.S.P. et al., *Proc. IEEE* 91, 661, 2003. With permission.)

magnetic state of the MTJ. After the current is removed, the magnetic state of the MTJ will return to its previous state. To write one MTJ element, a current is passed along both the bit line and the write line between which that MTJ is situated. The vector combination of the orthogonal magnetic fields from the bit and word lines is designed to exceed the threshold switching field for that element, so that its state can be set to either a "one" or a "zero." Typically, the polarity of either the bit line or word line current is changed to set the MTJ element to one of its two possible states.

Figure 14.2 shows a scanning electron microscope cross-section of the first ever MRAM chip, which was fabricated by IBM [7]. This chip, which

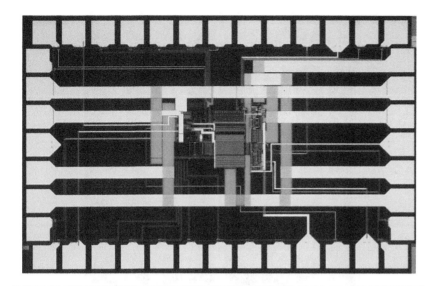

FIGURE 14.2 Top down optical micrograph of a 1 Kb CMOS MRAM array, the first MRAM chip (using field writing) fabricated at IBM in 2000.

consists of a 1 kB complementary metal-oxide-semiconductor (CMOS) MRAM array, was fabricated using IBM's 6SF-CMOS technology, which is an advanced process technology with a front-end-of-line (FEOL) minimum feature size of 0.25 μm and a 0.4 μm minimum feature size for the second level of metallization and above. This first chip used aluminum wiring. At the time this chip was fabricated in around 1999, 200 mm wafer diameter MRAM processing equipment was not available within IBM, so following the fabrication of metal vias above the metal 2 layer, the wafers were diced into 1 inch square pieces. The MTJ and the final wiring layer were then fabricated using research-scale deposition and patterning systems. Using these devices, ~2.7 ns read and write access times were demonstrated [7]. This excellent performance resulted from the tunability of the resistance of MTJs over many orders of magnitude, so allowing optimization of their resistance with regard to that of the CMOS circuits [4]. Note that, as shown in Figure 14.1, there is no connection between the word line and the MTJ device. Rather, an additional metal lateral contact line is used to connect the lower electrode of the MTJ to the output of a transistor, whose state (whether on or off) is controlled by a read word line connected to the gate of the transistor. Thus, each MTJ device is connected to a transistor for the purposes of eliminating current that would otherwise leak through the MTJs in the cross-point array during reading [3].

Encouraged by these results, IBM undertook the development of full 200 mm wafer scale MRAM technology, initially in a joint development alliance with Infineon based on 180 nm CMOS technology [5]. Figure 14.3 shows an image of a 16 Mbit MRAM product demonstrator chip that was developed as an outgrowth of that effort, while Figure 14.4 shows a scanning electron microscopy cross-section from within the cell array [8]. The chip was designed as a 1Mb ×16 SRAM (i.e., 16 memory blocks each 1 Mb in size) with an asynchronous interface and a 1.42 μm^2 cell size and had a read and write performance of 30 ns. For more details, see DeBrosse et al. [8] and Maffitt et al. [9].

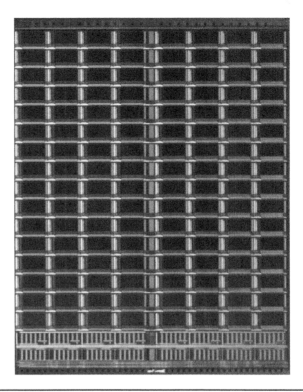

FIGURE 14.3   Image of a 16 Mbit conventional MRAM (field writing) product demonstrator chip developed using a full 200 mm-wafer scale 180 nm CMOS technology (2004).

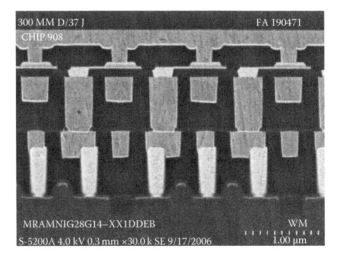

FIGURE 14.4   TEM (transmission electron microscopy) cross section image from within the cell array of a 16 Mbit conventional MRAM product demonstrator chip.

While the concept of field writing of MTJs seems simple and straightforward, it has several drawbacks. Most importantly, this method of writing scales very poorly, as the MTJ devices are shrunk in size to allow for higher-capacity MRAM chips. In particular, the strength of the magnetic fields generated by the word and bit line currents shrink in proportion with the radius of the write and bit lines for the same current density, so that ever larger current densities must be used to generate the same field strength, eventually resulting in problems such as electro-migration. Moreover, due to the superparamagnetic effect, thermal fluctuations can cause the magnetic moment of the free layer to overcome the magnetic anisotropy energy barrier and so reverse the free layer's magnetization direction unless the energy barrier is large enough. As the MTJ element shrinks in size, the switching field will necessarily increase so that sufficiently large magnetic shape anisotropy energy barriers are maintained to meet the typical 10 year memory lifetimes required for a NV memory chip. The larger switching fields correspond to higher currents needed along the word and bit lines. Such high currents not only consume more power but require larger transistors to generate these currents.

Another drawback of field writing of the MTJ elements is that the elements that are half-selected (i.e., those elements that see only one field from either the write or the bit lines) are more susceptible to thermal fluctuations and may be disturbed during the writing process of the selected cell. Finally, the magnetic state of the half-selected cells may be subject to the phenomenon of creep, whereby their magnetic state may be slightly altered each time they are half-selected, and after an accumulation of half-select processes, they may change their magnetic state [10].

To overcome some of the drawbacks of field writing (Stoner–Wohlfarth writing), a "toggle" mode writing scheme was proposed [11], which uses a free layer composed of an artificial antiferromagnet (AAF) formed from two magnetic layers coupled antiferromagnetically through a thin layer of ruthenium [3, 12]. This scheme is described in detail in Chapter 13, Volume 3. The main advantage of this scheme is not a lowering of the writing currents or fields needed but rather, that there is a much larger range of combinations of bit line and write line fields in which the MTJ element can be written without upsetting the half-selected bits. Indeed, paradoxically, the half-selected bits actually become more stable to thermal upsets during the toggle-writing process by using an AAF free layer [11]; that is, the energy barrier to thermally excited switching of the half-selected bits increases during the toggle mode writing of the selected bits on the same word or bit lines.

Due to the high fields and associated high word and bit line currents needed for both Stoner–Wohlfarth and toggle mode writing MRAM, these schemes do not support scaling of MRAM to densities that could make MRAM more than just a niche memory technology. However, an alternative writing method, which is based on recent developments in MTJ materials and on a new spintronics phenomenon, is potentially highly attractive. This scheme, spin transfer torque (STT) writing, uses a current that is passed directly through the MTJ device to "write" data bits. A current that is passed between the pinned layer and the free layer is spin polarized [13] and consequently carries spin angular momentum [14, 15]. When the spin angular momentum is transferred from the tunneling electrons to the magnetic moments of the free layer atoms, then these moments are subject to a torque. This spin transfer–induced torque results in excitation, rotation, and eventual reversal of the free layer magnetization direction if the torque is large enough and sufficient spin angular momentum is transferred. This will occur when a large enough current flows for a sufficiently long time (see Chapter 8, Volume 1). Depending on the direction of the flow of spin-polarized electrons, the free layer magnetization can be rotated toward or away from the direction of magnetization of the pinned layer. Thus, by passing a sufficiently large current through the MTJ element in one direction or the other, the element can be written to a "zero" or a "one." Spin transfer switching of the magnetization has been experimentally observed in a wide variety of MTJ nanopillars since its first observation in manganite-based tunnel junctions [16]. STT writing is the dominant writing mechanism when the lateral dimensions of the MTJ nano-pillars are below $\sim100 \times 100$ nm^2 (since the field from the current increases with the radius of the device [for constant current density], these fields may influence the writing of the bit and can dominate the spin torque for larger-area devices). Since the threshold current for STT writing increases approximately with the magnetic moment of the free layer, reducing the size of the MTJ element (for the same thickness) reduces the threshold writing current accordingly. Thus, the current needed for STT writing scales with the area of the MTJ nanoelement, which is very attractive.

There are several important issues associated with STT writing. Perhaps of paramount importance is that the current for reading must be well below that needed for writing, so that inadvertent upsets of the state of the MTJ elements is avoided. Since there will inevitably be a distribution of writing threshold currents for different devices due to various factors such as variations in (i) the size and shape of the nanoelement, (ii) the thickness of the magnetic layers within the free layer, (iii) the spin polarization of the tunneling current due to barrier thickness inhomogeneities (the polarization of the current typically increases with thickness for thinner barriers due to improvements in the insulating characteristics of the barrier or due to increased spin filtering but typically decreases for much thicker barriers due to impurity-mediated tunneling), (iv) stray fields from neighboring elements, and so on, this requirement is quite stringent. Narrowing the distribution of threshold switching currents is very important. Lowering the writing current is important for several reasons. First, lowering the power used to operate the MRAM chip requires the lowest possible writing current. Second, to build the densest possible MRAM chip, the writing current must be provided by a minimum-area transistor so that the area of a single memory cell can be at its smallest possible size (for a single bit cell) of $\sim4f^2$, where f is the minimum length possible in that CMOS technology node. When f $\sim$ 90 nm, this means that the writing current must be below ~100 μA. Currents well below this level have been demonstrated in perpendicularly magnetized MTJ elements [17]. Third, the writing current must be such that the electric field across the MTJ tunnel barrier is well below the breakdown electric field of the dielectric material from which the barrier is formed. Since dielectric breakdown can occur gradually over many write operations even when the applied voltage is well below the breakdown voltage for a single write operation, this is a very stringent requirement.

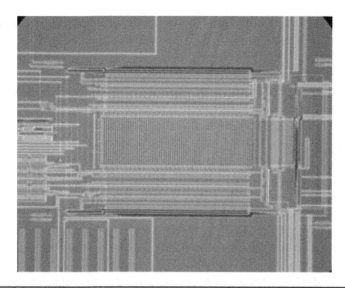

FIGURE 14.5   Image of a 4 kbit experimental array of STT-MRAM chip fabricated in a 200 mm, 180 nm ground rule CMOS technology (2007).

Figure 14.5 shows an image of a 4 kbit experimental MTJ array fabricated in a 200 mm, 180 nm ground rule CMOS technology as described in Assefa et al. [18]. The current focus of the development of STT-MRAM at IBM is to simultaneously satisfy tight distribution requirements for write threshold control while maintaining high breakdown voltages and low disturbs from read operations [19].

14.3 RACETRACK MEMORY

14.3.1 CONCEPT

Racetrack memory is a storage-class memory technology proposed by IBM in 2002 [20, 21]. Like magnetic HDDs, racetrack memory stores data within a persistent pattern of magnetic domains that are accessed dynamically for writing, reading, or modification (i.e., changing the state of a domain wall [DW] bit), but unlike mechanical HDDs, it uses no moving mechanical parts. The data pattern is stored as a series of magnetic DWs in a microscopic ferromagnetic wire (the racetrack) and accessed by using current pulses to move the entire pattern, intact, along the wire to reading and writing elements integrated within the device. Each storage element of the device will comprise a writing element, a magnetic racetrack, and a reading element. As the data is stored as magnetic domains, it has the same NV nature as data stored in a magnetic disk drive.

A very important feature of racetrack memory is that the series of magnetic DWs in the racetrack are moved around the racetrack not by a magnetic field but rather, by using the spintronic concept of *spin momentum transfer* [14, 15]. A spin-polarized current carries spin angular momentum. When this spin angular momentum is delivered to a magnetic DW, the DW can be moved [22–27]. In distinct contrast to their motion under the influence of a magnetic field, neighboring DWs in a magnetic nanowire are moved by current in the same direction along the nanowire, defined by the direction of current (or rather, the flow of spin-polarized electrons). The magnetic field will drive neighboring DWs in opposite directions and so will cause the DWs to annihilate one another. Thus, using current, a sequence of DWs can be moved along a magnetic nanowire in one direction or the other depending on the direction of the current flow, one of the key concepts of racetrack memory.

Most conventional solid state memories require one or more transistors or switches per data bit, so that the density and thereby the cost of such memories are determined, to a large extent, by the size of a single transistor, the number of transistors needed per memory cell, and the cell layout and architecture. Moreover, since CMOS is by far the most well-established microelectronic technology and is clearly the cheapest of today's semiconductor technologies due to the massive scale of silicon foundries, any competitive solid state technology will need to take advantage of and be based on CMOS. The concept of the racetrack memory is that many bits, as many as 10 to 100 or more, can be stored *per transistor*, thereby dramatically increasing the storage density.

The greatest improvement in density is achieved in the racetrack memory concept shown in Figure 14.6, in which the magnetic racetrack is arranged perpendicular to the surface of a silicon wafer (in which CMOS circuits are deployed). In this "vertical" or three-dimensional (3D) racetrack memory (VRM), the magnetic wire extends vertically above the read

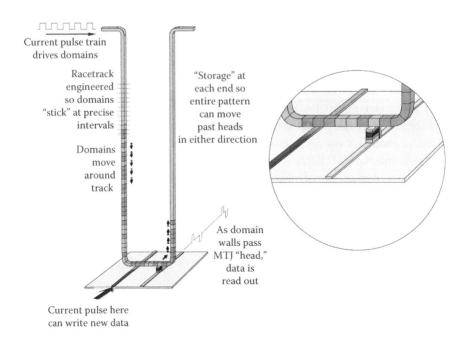

Current pulse train drives domains

Racetrack engineered so domains "stick" at precise intervals

"Storage" at each end so entire pattern can move past heads in either direction

Domains move around track

As domain walls pass MTJ "head," data is read out

Current pulse here can write new data

FIGURE 14.6 Vertical configuration magnetic Racetrack Memory element. A magnetic tunnel junction (MTJ) device is used for reading data in the racetrack.

and write elements in a "U"-shaped racetrack. The DWs in which the data is encoded are written and read in the bottom of the U of the racetrack in Figure 14.6, where read and write elements are located. In this 3D design, a very high data storage density is achieved per unit area of the silicon wafer by storing large numbers of bits in the vertical dimension of the device.

As illustrated in Figure 14.6, the VRM may include storage zones (or reservoirs) in the magnetic racetrack, so the intact DW pattern may be driven back and forth repeatedly as needed. Thus, to move any particular DW in the series of DWs in a single racetrack to these elements, the racetrack needs to be about two times longer than the sequence of DWs. One can consider the racetrack to be composed of a "storage" region and a "reservoir" region at either end of the racetrack. These regions will dynamically occupy different portions of the racetrack as the DWs are moved back and forth during the operation of the memory. Other unidirectional designs are possible, in which data is written at one end of the wire and read at the other, by temporarily storing the data in CMOS memory elements.

While the VRM has the potential to create a solid state storage device with the capacity of a magnetic HDD, that is, about 100 times greater than that of a solid state memory in the same area of silicon, and at about the same cost per bit as an HDD, there are considerable technical challenges in building the vertical racetrack. An alternative racetrack memory device is shown in Figure 14.7. In this horizontal or two-dimensional version of the racetrack memory (HRM), data is stored in a long straight racetrack element parallel to the surface of the silicon wafer. The device, as illustrated, has centrally located read and write elements, and the data pattern may be shifted back and forth across them as necessary by varying the polarity of the current pulses. The horizontal racetrack is not as dense as the vertical racetrack but nevertheless, has the potential to be as dense as or denser than FLASH memories but with much greater reading and writing performance. Moreover, even the HRM has the potential to be much denser than STT-MRAM.

Racetrack encompasses the most recent advances in the field of metal spintronics. The spin-valve read head enabled a thousand-fold increase in the storage capacity of the HDD in the decade from 1997 to 2007; since about 2007, the MTJ has supplanted the spin-valve in nearly all HDD recording heads because of its higher signal [3]. MTJs also form the basis of modern MRAM, in which the magnetic moment of one electrode is used to store a data bit, as discussed in Section 14.2. Whereas MRAM uses a single MTJ element to store and read one bit, and HDDs use a single spin-valve or MTJ sensing element to read the approximately 100 GB of data in a modern drive, the racetrack uses one sensing device to read 10–100 bits.

The actual performance of the HRM depends, in detail, on the characteristics of the magnetic DWs in the racetrack, how closely the DWs can be spaced along a racetrack, and the velocity with which the DWs can be driven around the racetrack with current. Overcoming these challenges is the central focus of current research and is briefly discussed in the following sections.

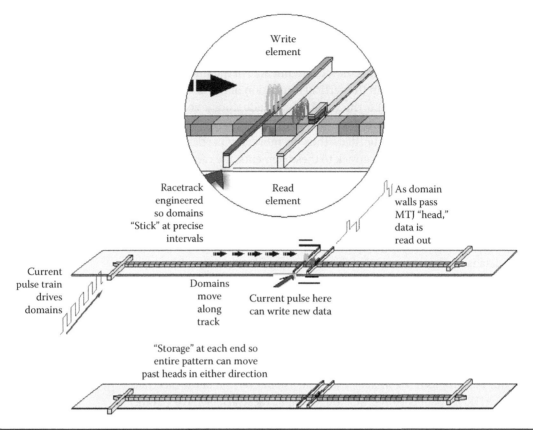

FIGURE 14.7 Horizontal configuration magnetic Racetrack Memory element.

14.3.2 DOMAIN WALLS IN MAGNETIC NANOWIRES

The physical width of DWs in the racetrack memory depends on both the magnetic properties of the material used and the racetrack dimensions. Two very different situations occur depending on whether the racetrack is made of magnetically "soft" or "hard" materials. These two classes of materials can be characterized by the ratio of the magnetic anisotropy energy, determined, for example, by the crystal structure or surface effects, to the magnetostatic energy, which depends on the shape of the racetrack and its magnetization. This ratio is often called the *quality factor* $Q = K/2\pi M_S^2$, where K is the anisotropy constant and M_S the saturation magnetization of the material [28].

Soft materials, defined by $Q \ll 1$, are typically made of alloys of Fe, Co, and Ni. The most commonly used alloy is $Fe_{19}Ni_{81}$ (permalloy). For such materials, the shape anisotropy dominates, and the magnetization is preferably aligned along the racetrack. Domains magnetized in opposite directions point either toward one another or away from one another, and they are separated by head-to-head and tail-to-tail DWs, respectively. Both the internal structure of the DWs (that is, the detailed magnetization profile) and their width depend on the size and shape of the track [29, 30]. Figure 14.8a, b show representative DW structures observed in permalloy nanowires with widths

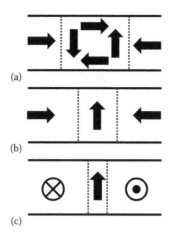

FIGURE 14.8 Schematic illustration of three possible domain wall (DW) structures for racetrack memory (a) Head to Head (HH) vortex, (b) HH transverse, and (c) Bloch domain wall.

of several hundred nanometers and thicknesses of a few tens of nanometers. In these wires, two different structures can be stabilized, the so-called *transverse* and *vortex* DWs. The most stable DW structure depends on the product of the track width and its thickness. Transverse walls are favored for thin and/or narrow tracks, whereas vortex walls have the lowest energy for wider and/or thicker tracks. However, both types of walls can be observed for a fairly wide range of wire widths and thicknesses, and in addition, the wall structure can be strongly influenced by the details of the mechanism by which a DW is injected into the nanowire [31–33].

Vortex and transverse walls have different widths, and their current-driven dynamics is also quite different, as will be discussed later. For both DW structures, the physical width w scales with the track width f, and for tracks made of permalloy, it can be approximated by the expressions: for transverse walls, $w_{TW} \sim f$, and for vortex walls, $w_{VW} \sim 3\pi/4f \sim 2.35f$ [30, 34]. These length scales define the upper bound of the linear density of the racetrack memory, that is, the maximum number of DWs per unit length, in the case of soft materials. The actual minimum spacing between DWs in racetracks without pinning sites will be determined, rather, by the magnetostatic interactions between neighboring DWs.

In the case of hard materials, with $Q \gg 1$, magnetic domains are aligned along the anisotropy axis. The most promising class of hard materials has strong perpendicular magnetic anisotropy (PMA). For this class of materials, for example Co/Pt [35] or Co/Ni [36] multilayers with ultrathin magnetic layers, the magnetization of the domains points out of the plane of the racetrack. The properties of the DWs separating domains (Bloch DW; Figure 14.8c) are essentially determined by the properties of the material (exchange and anisotropy energies) and are much less strongly influenced by the racetrack dimensions. The physical DW width can be extremely small, for example, ~15 nm in Co/Pt multilayers, which makes these DWs potentially very attractive for dense racetrack memories.

14.3.3 THERMAL STABILITY OF PINNED DOMAIN WALLS

In racetrack memory, the data pattern is stored as a series of magnetic DWs. Such a pattern of DWs may be realized, for instance, by introducing uniformly spaced pinning sites along the racetrack. These pinning sites will also act to prevent the DWs from drifting out of position, perhaps propelled by small stray currents or magnetic fields or because the controlling current pulses are not exactly of the correct intensity and duration.

To estimate the required pinning strength, it is convenient to introduce the concept of thermal fluctuations. In the electronics industry, it is common to define a number in units of $k_B T$, where k_B is the Boltzmann constant and T is room temperature, to represent the stability of the data bit. For example, $60\,k_B T$ indicates that the data bit will hold its value for more than 10 years at room temperature. Thermal fluctuations and the superparamagnetic limit are key limitations to the areal density of HDDs, for which bits are stored in many weakly coupled magnetic grains. These are also critical issues for STT-MRAM, because the storage layer must be extremely thin (a few nanometers) to achieve current-induced switching for small current densities. The scaling of the racetrack memory is much more favorable, because DWs can be much larger and therefore are less susceptible to thermal fluctuations. The energy barrier, ΔE, of a DW localized at a pinning site can be readily derived from micromagnetic simulations. ΔE depends in a rather complex way on the DW structure, as well as the detailed shape and depth of the pinning site, but can be described by $\Delta E = K V_{DW}$, where V_{DW} is the volume of the DW and K is an effective anisotropy constant. For soft magnetic materials such as permalloy, K (typically of the order of ~5×10^5 to 5×10^6 erg/cm^3) is determined by the shape anisotropy of the racetrack. Since the DW width is proportional to the width of the racetrack f, the volume of the DW is simply given by $V_{DW} = \kappa t f^2$, where t is the racetrack thickness and κ is a numerical factor that depends on the DW structure ($\kappa = 1$ and 2.35 for transverse and vortex DWs, respectively [30, 34]). The energy barrier gives rise to a pinning field H_p, which is simply given by $H_p = K/M_S$ (assuming a parabolic pinning potential well). Thus, the thermal stability target $\Delta E = n k_B T$ at room temperature is simply given by the track dimensions and magnetization and the pinning field: $\Delta E = n k_B T = \kappa t f^2 M_S H_p$. The pinning required for a typical target value of n = 60 can then be readily calculated for different DW structures as a function of the track thickness.

14.3.4 CURRENT NEEDED TO MOVE DOMAIN WALLS

The critical current density for DW motion, J_C, is key to the performance of racetrack memory, since it is the main source of power dissipation during the DW shift operation. Experimentally, J_C is ~10^8 A/cm^2 for racetracks made of permalloy or closely related materials. There are reports of lower values for multilayered structures (spin-valves) and nanowires with perpendicular anisotropy (Co/Pt, Co/Ni), but few data are available, and more research is needed to explore the wide range of potential ferromagnetic materials.

It is crucial that a solid understanding of the mechanisms of current-driven DW motion be available so as to guide experiments toward reducing the critical current density. Let us first compare DW motion driven by current and magnetic field. The field analog of the critical current density J_c is the propagation (or coercive) field H_p, below which no motion occurs. The origin of this propagation field is extrinsic: for an ideal racetrack, without any defects or roughness, DWs would move for any non-zero applied field, albeit very slowly. Thus, non-zero values of the propagation field are directly related to defects, which provide local pinning sites for the DW. In fact, H_p is proportional to the pinning strength, which can be tuned, for example, by fabricating notches with variable depths.

The relationship between critical current and pinning strength for vortex DWs is shown in Figure 14.9 [21]. These data suggest the existence of two different regimes. For relatively weak pinning (below ~15 Oe), the critical current density scales linearly with the pinning field. For the lowest pinning strength (~5 Oe), the critical current is of the order of 10^8 A/cm². For stronger pinning (>15 Oe), the critical current appears to saturate and becomes independent of pinning strength. In this regime, however, DW motion requires very high current densities, resulting in severe Joule heating, and experimental data are thus difficult to interpret.

According to microscopic theoretical models of the interaction between current and the DW, there are two different contributions by which the current interacts with a DW. The first contribution is the same as that which gives rise to switching of magnetic moments in nanopillars made from giant magneto-resistive spin valves or MTJs as used in STT-MRAM [37, 38]. This contribution is usually referred to as the *adiabatic spin torque*. The efficiency of the adiabatic spin torque is proportional to the rate of spin angular momentum transfer u (unit of velocity, m/s), which depends on the current density J, the spin polarization of the current P, and the saturation magnetization of the racetrack M_S as follows: $u = \mu_B J P / e M_S$ (with e the electron

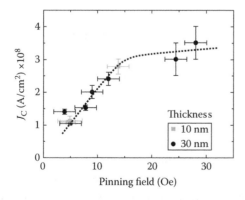

FIGURE 14.9 Threshold current density (JC) needed to move a vortex domain wall plotted as a function of the pinning strength, represented by the pinning field. Wire widths and thicknesses are varied. (After Parkin, S. et al., *Science* 320, 190, 2008. With permission.)

charge and μ_B the Bohr magneton). Quantitatively, for tracks made of permalloy ($M_s = 800$ emu/cm^3) and assuming a spin polarization $P = 0.5$, $u = 100$ m/s corresponds to a current density $J \sim 2.8 \times 10^8$ A/cm^2. Note that the quantity u is equal to the maximum DW velocity when driven by adiabatic spin torque alone.

The second contribution from the current is often called the *non-adiabatic spin torque* or β-*term* and acts like a magnetic field localized at the DW [23–27]. The magnitude of this spin torque is proportional to βu. It is assumed that the dimensionless constant β is proportional to the degree of spin-dependent scattering that takes place at the DW; that is, the DW resistance. Typically, the DW resistance increases when the DW width is reduced, thereby increasing the magnitude of β and making the DW move with less current. Note that the DW velocity is of the order of $(\beta/\alpha)u$ for a non-zero β-term, where α is the Gilbert damping that represents magnetization relaxation.

From the theoretical model, it has been predicted that the existence of the non-adiabatic torque has dramatic consequences for the current-driven DW dynamics; that is, the critical current becomes extrinsic and scales with the pinning strength. Interestingly, the critical current shown in Figure 14.9 seems to scale with the pinning strength in the low pinning regime, suggesting a purely extrinsic origin (a similar dependence was observed in multilayers with perpendicular anisotropy [39]). However, we cannot rule out the existence of a non-zero critical current for very low pinning (<5 Oe). Further work is needed to achieve such an extremely low pinning.

To date, the critical current to move a transverse DW in zero field has not yet been determined. It appears, as is consistent with theoretical models, that the threshold current for the motion of a transverse DW is higher than that for a vortex DW. It could be that in nanowires of appropriate dimensions (e.g., with squarer cross-sections), transverse DWs could be moved.

Finally, the critical currents needed to move DWs in magnetic nanowires are listed in Table 14.1 for various materials. The pinning strength, typically the coercivity of the wire, is also shown in Table 14.1.

14.3.5 READING AND WRITING ELEMENTS

One reading device consists of a magnetic element on top of the track and separated from it by a tunnel barrier, thereby forming an MTJ (see Figure 14.10). Here, within the reading device, the free layer is formed from the magnetic domain of the racetrack. The tunnel barrier and the reference layer are deposited on top of the racetrack. The magnetization of the reference layer is fixed by exchange coupling to an adjacent antiferromagnetic film such as IrMn. In the reading cycle, the bit to be read is shifted so that it is underneath the reading device. Since the resistance of the MTJ depends on the magnetization direction of the bit with respect to that of the reading element, the bit can be read out by measuring the MTJ resistance. (Note that the reading element could also be formed under the racetrack or to one side of the racetrack).

TABLE 14.1
Critical currents and propagation fields to move domain walls in nanowires

Material	$J_c \times 10^8$ (A/cm^2)	H_p (Oe)	Notes
NiFe [21, 40]	~0.6–1×10^8	~5–13	Weak pinning
NiFe [21]	~3×10^8	~30	Strong pinning
CoFeB [41]	~1×10^8	~17	
CoFeB [42]	~4×10^6	~20?	Co/Cu/CoFeB multilayers
Co/Ni [39]	~5×10^6	~150	Multilayers, spin valve (PMA)
Co/Ni [43, 44]	~5×10^7	~100	Multilayers (PMA)
CoCrPt [45]	~1×10^8	~500	(PMA)
GaMnAs [46]	~8×10^4	~50	Temperature <80 K (PMA)
SrRuO$_3$ [47]	~1×10^5–1×10^6	~50–500	Temperature <140 K (PMA)
Co [48]	<9.8×10^6	A few tens of Oersteds?[a]	Pt|0.3 nm Co|Pt (PMA)
Co [49, 50]	~1–1.5×10^8	>500 Oe?[a]	Pt|0.6 nm Co|AlO$_x$ (PMA)
TbFeCo [51]	~4–6×10^6	1000–2200	Tb$_{30}$Fe$_{58}$Co$_{12}$ (PMA)

[a] Pinning field of the wire not specified in these reports.

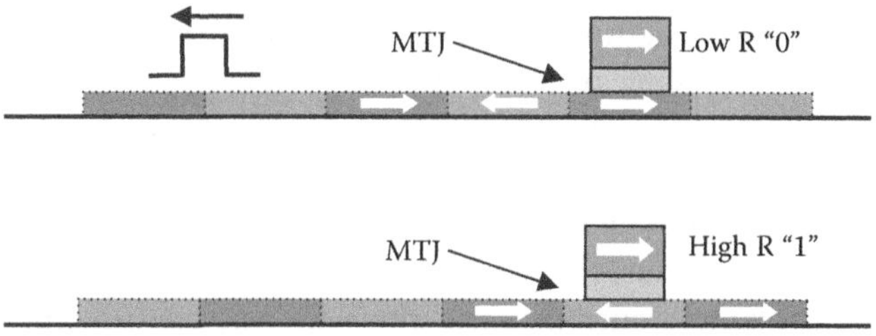

FIGURE 14.10 Reading domain wall bits using a magnetic tunnel junction sensing device.

One method for bit writing is by generating localized magnetic fields in the vicinity of the racetrack to reverse the magnetization direction in a small length of the racetrack and thereby write a magnetic DW. The magnetic field needs to be localized so that it does not perturb the neighboring bits. The simplest method to generate localized magnetic fields is to pass a current pulse along a conducting wire positioned perpendicular to and close to the racetrack. The width of the conducting wire and the distance between it and the racetrack determine how far along the racetrack the Oersted magnetic field will spread. The size of the magnetic field depends on the geometry of the conducting wire; for a given current density, the field magnitude increases with the size of the conducting wire; however, the power needed to generate the current scales with the size as well.

A simple model can be used to estimate the magnitude of the magnetic field needed to nucleate a DW; that is, to write a bit. In this model, the magnetic moments under the conducting line are assumed to rotate coherently from one direction to the other via a configuration where the magnetization points perpendicular to the nanowire's long axis. To cause magnetization rotation, the Zeeman energy from the magnetic field must overcome the magnetostatic energy barrier when the magnetization lies perpendicular to the nanowire's long direction. The size of the nucleation field, therefore, depends on the nanowire's dimension as well as its saturation magnetization. Typically, a few nanoseconds (>1–2 ns) is long enough to cause DW nucleation when the local field is larger than the threshold field for nucleation.

14.3.6 SHIFT REGISTER DEMONSTRATION

There have been several studies recently that have reported the demonstration of a magnetic shift register using the STT-induced motion of DWs in magnetic nanowires. The first such demonstration is shown in Figure 14.11a–c, in which a 3 bit DW shift register memory was demonstrated in a permalloy nanowire [52]. The nanowire is made of a 10 nm thick permalloy film, patterned into the form of a 200 nm wide, 6 μm long nanowire using electron beam lithography. The shift register is a 3 bit unidirectional serial-in, serial-out memory in which the data is coded as the magnetization direction of individual domains (Figure 14.11a). A left (right) pointing domain represents a 0 (1). The data is written into the left-most domain in Section A–B, is shifted by two domains, and is then read from the state of the right-most domain. Here, instead of the state of the right-most domain being read directly, the magnetization configuration of each of the three domains is inferred from the nanowire's resistance. Due to the relatively large anisotropic magnetoresistance of permalloy, the wire resistance provides information on the number of DWs stored in the nanowire and thus, the magnetic configuration. The writing is carried out by applying a current pulse along the left contact line (vertical line in Figure 14.11a), which generates a localized magnetic field and thus, can inject a DW into the nanowire. Shifting of the data bits (that is, the domains and the DWs) is carried out by injecting a nanosecond-long current pulse ($\sim 1.5 \times 10^8$ A/cm^2) into the nanowire. The electron flow direction associated with the current pulse points to the right and so moves the DWs also to the right via STT.

An example of the shift register operation is shown in Figure 14.11b. Here, a data sequence of 010111 is written into the register. The top panel of Figure 14.11b shows the evolution of the nanowire resistance, and Figure 14.11c shows the corresponding magnetization configuration. In Figure 14.11b, the shaded regions correspond to a shift operation, whereas the light regions correspond to a write operation. As the writing/shifting operation is carried out, the resistance of the nanowire changes according to what is expected from the change in the number of DWs inserted in the nanowire, as shown in Figure 14.11c. The bottom panel of Figure 14.11b shows the logic table of each bit. From the logic table, one can find how the data sequence (i.e., 010111) is transferred along the nanowire after each write/shift operation.

The input sequence is accurately transferred to the output after two write/ shift operations. In this example, the cycle time to write and shift 1 bit is ~30 nanoseconds. This is determined by the write time (here ~3–4 nanoseconds) and the time to shift the series of DWs by one domain length. This time is

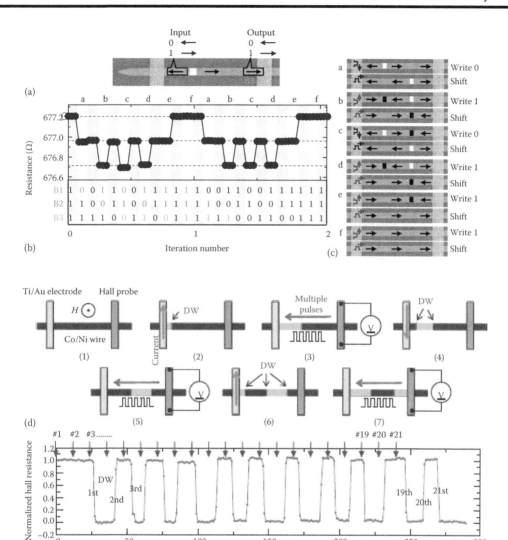

FIGURE 14.11 (a–c) A three bit unidirectional magnetic domain wall shift register using permalloy nanowires. (a) Data are encoded by the magnetization direction of three domains in the nanowire. (After Hayashi, M. et al., *Science* 320, 209, 2008. With permission.) (b) Nanowire resistance variation when a current pulse sequence is used to write and shift domain walls along the register. The sequence, "010111" is written and shifted two times in succession. The table shows the corresponding evolution of the states of the three bits during these operations. (c) Schematic illustration of the magnetic configuration of the nanowire during the shift register operation. (d, e) A unidirectional magnetic domain wall shift register using Co/Ni multilayered perpendicularly magnetized nanowires. (d) (1)–(7) Schematic illustrations of measurement procedure to create and shift multiple domain walls. (e) Normalized Hall resistance as a function of number of pulses when multiple domain walls are introduced into the wire and shifted toward the same direction. The arrows indicate the timing of the domain wall creation. (After Chiba, D. et al., *Appl. Phys. Express* 3, 073004, 2010. With permission.)

determined by the velocity of the DWs, which here is ~150 m/s (current density is ~1.5 × 10⁸ A/cm²) [31].

In this first report, up to three DWs were moved with current in 6 µm long nanowires (subsequently, six walls were reported to move using a 20 µm long nanowires). However, this report also illustrated the importance of pinning for closely packing the DWs for high areal density memories. That is, due to the magnetostatic interaction among the DWs, the spacing between them is limited to a minimum distance of a few microns when the pinning field is of the order of 5–10 Oe. As discussed earlier, stronger pinning leads to higher threshold current densities.

These first experiments, which demonstrated the motion of a series of DWs at high speed using nanosecond current pulses, proved the viability of the shift register racetrack memory concept.

An example of a more recent study demonstrating a current-driven DW shift register operation is shown in Figure 14.11d and e [53]. In this study, a perpendicularly magnetized material is used. As discussed in Chapter 8, Volume 1, theoretical models suggest that the threshold current for the motion of DWs in perpendicularly magnetized nanowires should be lower than in in-plane magnetized nanowires. Here, 500 nm wide, 16 μm long nanowires, formed from multilayered $[0.3 \text{ Co}|0.9 \text{ Ni}]_4|0.3$ Co films (thickness in nanometers) deposited on undoped Si substrates, were used to control the position of multiple DWs with current pulses. A tantalum Hall bar was attached perpendicular to the wire to detect the domain wall passing under the Hall bar after each shifting and writing operation. The shifting operation was carried out with current pulses whose current density was ~1.3×10^8 A/cm^2. Multiple current pulses were used to conduct one shifting operation, and the total length of the pulse was ~132 ns. The wall velocity at this current density is reported to be ~38 m/s. Figure 14.11d shows the magnetization configuration during the writing/shifting operation, and Figure 14.11e shows the corresponding resistance change of the Hall bar. From the Hall bar resistance change, it can be inferred that a total of 21 DWs have been injected and shifted using the current pulses. Note that the pinning field here is reported to be of the order of ~100 Oe. At this pinning strength, dense packing of the DWs can be expected. However, due to the spread of the localized field during the writing operation, the separation distance between neighboring DWs was limited to only a few microns in these experiments as well.

Similar experiments have also been conducted on other perpendicularly magnetized material; for example, nanowires made from ultrathin Co films (5 Ta |2.5 Pt|0.3 Co|1.5 Pt films [thickness in nanometers]) [48]. The shifting of the DWs here resulted from a thermally activated process of DW creep. In this process, lower current density is required to move the DWs compared with the abovementioned cases. However, since the wall motion takes place via a thermally activated hopping mechanism, the velocity is very low and thus requires longer current pulses. Shifting of multiple DWs was observed using a magneto optical Kerr effect (MOKE) microscope. Successful operation of the shifting was observed when 50 ms long current pulses with a current density of ~9.8×10^6 A/cm^2 were used.

In these experiments, the current densities needed to move DWs along the racetracks are similar for fixed DW velocity. Even though the minimum current densities needed to move DWs may have been lower in the case of perpendicularly magnetized nanowires compared with the first experiments with in-plane magnetized permalloy [21, 52] and related alloys [54], the observed current-driven DW velocities were also much lower [48, 53]. It is thus important to distinguish the threshold current above which DWs can be initially moved from the current density needed to achieve a fixed DW velocity. The latter appears to be approximately the same for perpendicularly magnetized as for in-plane magnetized nanowires, but the threshold current appears to be lower for the former.

One way to reduce the current density needed to manipulate DWs is to use a domain wall resonance excitation effect [15, 55, 56]. A DW trapped in a pinning potential can be considered as a classical particle trapped in a parabolic potential, which is analogous to a mass in a harmonic oscillator. When a current pulse whose length is matched to the resonance period of the parabolic potential is applied, the motion of the wall within the potential can be amplified, thus allowing sub-threshold depinning (the threshold here corresponds to the threshold current density when long pulses or direct current are used) [57, 58]. Reduction of the threshold current density by more than a factor of five has been demonstrated by using matched bipolar current pulses [34]. Engineering the pinning potential profile is needed to take full advantage of this effect.

14.3.7 SINGLE DOMAIN WALL RACETRACK MEMORY

NEC Corp. has developed what one might describe as a single DW racetrack memory (or otherwise, a cross-point DW-motion MRAM using current-induced motion of a single DW per bit) [59]. (Figure 14.12). Each memory cell is composed of one MTJ and two transistors. The "free layer" is an elongated ferromagnetic layer, and a tunnel barrier/ferromagnetic layer element is placed on top of the free layer to probe the magnetization direction of the free layer via tunneling resistance measurements. The ends of the free layer's magnetization are pinned by adjacent "pinned layers." The magnetization directions of the pinned layers are opposite to each other, thus forming one

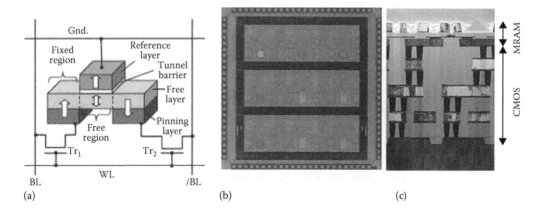

FIGURE 14.12   (a) Cell structure of a DW-motion MRAM. The cell has two transistors and one magnetic tunnel junction. (b) Photograph of a 4k-array and (c) cross-sectional TEM image of the cells. (After Fukami, S. et al., in *Symposium on VLSI Technology Digest of Technical Papers*, June 16–18, 2009, Honolulu, HI, pp. 230–231, 2009. With permission.)

DW in the free layer. A perpendicularly magnetized Co/Ni laminated film was used for the free layer in the demonstration reported.

The target market of the DW-motion MRAM is high-speed memory macros embedded in a next generation System on a Chip (SoC). Compared with conventional MRAM, including STT-MRAMs, the main advantage of DW-motion MRAM is the separation of the writing and read-out circuits and thus, the larger margin for the read-out levels. Due to its reading and writing method, DW-motion MRAM does not suffer from disturbing the stored information during its reading operation and can achieve little cross talk among the cells during its writing operation. It has been reported that DW-motion MRAM can operate with a write current lower than 0.2 mA and a 2 ns writing time with sufficient thermal stability [59].

14.3.8 RACETRACK MEMORY V2.0 AND BEYOND

Recently, a new class of DWs, often referred to as *chiral domain walls* [60], have emerged in heterostructures with broken structure inversion symmetry and strong spin orbit coupling. It has been recently demonstrated [61, 62] that Neel walls with fixed chirality, either right handed or left handed, can be created if the Dzyaloshinskii–Moriya (DM) antisymmetric exchange interaction [63–65] develops at the interface of the ultrathin magnetic layer. Strong interface DM interaction [61, 62, 66–71] has been found at the interface of a magnetic (FM) layer and a heavy metal (HM) layer, typically one of the 5d transition metals. As the DM interaction will cancel out if the top and bottom interfaces of the FM layer are identical, systems with broken structural inversion symmetry are required to observe its effect [61]. Typical heterostructures studied thus far consist of a thin FM layer sandwiched by two different HM layers (i.e., HM1/FM/HM2) or by an HM layer and an insulating oxide layer (e.g., HM/FM/oxide).

Interestingly, trains of chiral DWs can be driven by current in sync via the spin Hall effect of the HM layer [72, 73]. The spin Hall effect [74–76] generates spin current that diffuses into the neighboring FM layer. Such diffusive spin current will exert STT [77–79] on the magnetic moments at the interface between the FM layer and the HM layer. The magnetic moments of Neel walls, with fixed handedness, rotate according to the STT, which leads to the motion of DWs. The DWs, regardless of their type ($\uparrow\downarrow$ and $\downarrow\uparrow$ walls), move in the same direction. When the handedness is changed, the direction to which the DWs move is reversed. For Bloch DWs, the diffusive spin current does not exert torque on the DW magnetic moment, resulting in negligible DW motion.

The velocity of the current-driven motion of chiral DWs can be increased significantly when Neel walls coupled via interlayer exchange coupling [80] are used. Current-driven velocities exceeding 750 m/s have been reported in exchanged coupled Co/Ni multilayers (Ru was used as the spacer layer) [81]. The increase in the DW velocity is due to the decrease in the effective saturation magnetization of the system. Engineering the film stacking thus allows significant improvement in the racetrack performance [82].

14.3.9 OUTLOOK

The development of a high-performance, NV MRAM began with the proposal in 1995, and the subsequent first demonstration in 1999, by IBM of a fully functional 1 kbit MTJ array [4, 7]. Since this early demonstration that even MTJs with only 20% tunnel magnetoresistance (TMR) and a Stoner–Wohlfarth field writing mechanism would allow a high-speed memory (~3 ns), it is now generally accepted that to build a useful MRAM chip, MTJ memory elements, rather than elements based on metallic spin-valve devices, are needed due to both the very high tunneling magnetoresistance of MTJs as well as their much higher resistance as compared with spin-valve devices. The magnitude of the TMR signal from an MTJ has evolved enormously from 1975, with the initial discovery of the TMR phenomenon at low temperature and very low bias voltages [13], to the observation of modest TMR effects at room temperature in 1995 [83, 84] in devices using conventional alumina tunnel barriers, to giant TMR effects in MTJs with MgO tunnel barriers sputter deposited on amorphous substrates in 2004 [85], to TMR values of more than 600% at room temperature in these same MgO tunnel barrier–based MTJs [86]. Although the only MRAM chips that are available today for commercial applications are based on conventional magnetic field-writing, several prototype chips have been developed that use STT writing. This method overcomes the innate problems of scaling field writing to small dimensions, but to date, the current densities needed for high-speed STT writing using practical MTJ devices are still too high to enable high-density and therefore, price-competitive MRAM chips. Moreover, the MTJ to MTJ distributions of the read currents within a sub-array on an MRAM chip must be sufficiently narrow that the highest current to read any device must not overlap with the lowest current needed to write any device within the same array. Similarly, the highest write current must not exceed the breakdown voltage of any device in the same sub-array. Furthermore, these requirements hold not just for a single read or write operation but for as many of these operations as will take place in the lifetime of the MRAM chip (typically 10 years). The magnitude of the reading and writing currents must be carefully designed so that these processes are truly independent of one another.

The racetrack memory is conceptually very different from MRAM. It overcomes the challenge of reading and writing an MTJ memory element without dielectric breakdown of the MTJ tunnel barrier by using separate devices for reading and writing. This is possible in racetrack memory even while racetrack is innately a much denser memory than MRAM. This is because racetrack, unlike MRAM and many other resistive random access memories, stores many bits per memory cell in the form of a series of DWs along magnetic nanowires. The possibility of moving such a series of DWs to and fro along nanowires by using STT in the form of nanosecond-long current pulses has been demonstrated [52]. Thus, the fundamental concept underlying racetrack memory has been validated.

Clearly, racetrack memory is in a nascent stage of development compared with MRAM, but it has much greater potential. To achieve its potential, racetrack memory faces many exciting and interesting challenges. These include (i) the reliable generation of DWs with sufficiently similar mobilities, for example, by creating DWs with the same structure, (ii) the repeated motion of these DWs back and forth along a racetrack without their transformation into different DW structures, (iii) the manipulation of a series of DWs without their annihilation unless planned, (iv) the engineering and control of the magnetostatic fringing fields between DWs within and between racetracks so that these do not impede the motion of the DWs, and (v) sufficiently low current densities to move DWs along the nanowire so that the nanowires are not much heated, DW structures are not transformed, and the shift energy is minimized. Other challenges include the development of novel schemes for writing DWs at sufficiently low energy per operation and various integration challenges, including the integration of writing and especially reading devices that do not impede the motion of the DWs along the racetracks, the development of processes for creating uniform DW pinning sites, and the development of processes for building vertical racetracks.

Notwithstanding these challenges, racetrack memory is likely to be only one of many new memory, storage, and even logic technologies that go beyond the constraints embodied in conventional two-dimensional memory and logic devices. The cost and performance of most conventional memory devices, including MRAM, are controlled by those of the access device, namely, the transistor, to which each device is tethered. Racetrack memory fundamentally overcomes such limitations by using the latest developments in spintronics materials and phenomena to ultimately allow the storage of data bits in the third dimension above the two-dimensional array of transistors that will be used to create, manipulate, and read this data.

REFERENCES

1. See, for example, the International Technology Roadmap for Semiconductors: www.itrs2.net.
2. G. Zhang, L. Chiu, C. Dickey, L. Liu, P. Muench, and S. Seshadri, in *26th IEEE Symposium on Massive Storage Systems and Technologies (MSST 2010)*, Incline Village, Nevada, 2010.
3. S. S. P. Parkin, X. Jiang, C. Kaiser, A. Panchula, K. Roche, and M. Samant, Magnetically engineered spintronic sensors and memory, *Proc. IEEE* **91**, 661 (2003).
4. S. S. P. Parkin, K. P. Roche, M. G. Samant et al., Exchange-biased magnetic tunnel junctions and application to non-volatile magnetic random access memory, *J. Appl. Phys.* **85**, 5828 (1999).
5. W. J. Gallagher and S. S. P. Parkin, Development of the magnetic tunnel junction MRAM at IBM: From first junctions to a 16-Mb MRAM demonstrator chip, *IBM J. Res. & Dev.* **50**, 5 (2006).
6. J. M. Slaughter, Materials for magnetoresistive random access memory, *Ann. Rev. Mater. Res.* **39**, 277 (2009).
7. R. Scheuerlein, W. Gallagher, S. Parkin et al., A sub-10ns read and write time non-volatile memory array using a magnetic tunnel junction and FET switch in each cell, in 2000 *IEEE International Solid State Circuits Conference* Digest Tech. Papers, 218, 2000.

8. J. DeBrosse, D. Gogl, A. Bette et al., A high-speed 128-kb MRAM core for future universal memory applications, *IEEE J. Solid State Circuits* **39**, 678 (2004).

9. T. M. Maffitt, J. K. DeBrosse, J. A. Gabric et al., Design considerations for MRAM, *IBM J. Res. & Dev.* **50**, 25 (2006).

10. S. Gider, B.-U. Runge, A. C. Marley, and S. S. P. Parkin, The magnetic stability of spin-dependent tunneling devices, *Science* **281**, 797 (1998).

11. B. N. Engel, J. Åkerman, B. Butcher et al., A 4-Mbit toggle MRAM based on a novel bit and switching method, *IEEE Trans. Magn.* **41**, 132 (2005).

12. S. S. P. Parkin and D. Mauri, Spin-engineering: Direct determination of the RKKY far field range function in Ruthenium, *Phys. Rev. B* **44**, 7131 (1991).

13. M. Julliere, Tunneling between ferromagnetic films, *Phys. Lett.* **54A**, 225 (1975).

14. J. Slonczewski, Current driven excitation of magnetic multilayers, *J. Magn. Magn. Mat.* **159**, L1 (1996).

15. L. Berger, Emission of spin waves by a magnetic multilayer traversed by a current, *Phys. Rev. B* **54**, 9353 (1996).

16. J. Z. Sun, Current-driven magnetic switching in manganite trilayer junctions, *J. Magn. Magn. Mat.* **202**, 157 (1999).

17. T. Kishi, H. Yoda, T. Kai et al., in *2008 IEEE International Electron Devices Meeting*, San Francisco, p. 12.6, 2008.

18. S. Assefa, J. Nowak, J. Z. Sun et al., Fabrication and characterization of MgO-based magnetic tunnel junctions for spin momentum transfer switching, *J. Appl. Phys.* **102**, 063901 (2007).

19. R. Beach, T. Min, C. Horng et al., in *Electron Devices Meeting, 2008. IEDM 2008.* IEEE International, 2008.

20. S. S. P. Parkin, in *United States Patent and Trademark Office*, United States Patent and Trademark Office (Eds.), IBM, USA, 2004.

21. S. S. P. Parkin, M. Hayashi, and L. Thomas, Magnetic domain-wall racetrack memory, *Science* **320**, 190 (2008).

22. L. Berger, Exchange interaction between ferromagnetic domain wall and electric current in very thin metallic films, *J. Appl. Phys.* **55**, 1954 (1984).

23. G. Tatara and H. Kohno, Theory of current-driven domain wall motion: Spin transfer versus momentum transfer, *Phys. Rev. Lett.* **92**, 086601 (2004).

24. S. Zhang and Z. Li, Roles of nonequilibrium conduction electrons on the magnetization dynamics of ferromagnets, *Phys. Rev. Lett.* **93**, 127204 (2004).

25. A. Thiaville, Y. Nakatani, J. Miltat, and Y. Suzuki, Micromagnetic understanding of current-driven domain wall motion in patterned nanowires, *Europhys. Lett.* **69**, 990 (2005).

26. S. E. Barnes and S. Maekawa, Current-spin coupling for ferromagnetic domain walls in fine wires, *Phys. Rev. Lett.* **95**, 107204 (2005).

27. M. D. Stiles, W. M. Saslow, M. J. Donahue, and A. Zangwill, Adiabatic domain wall motion and Landau-Lifshitz damping, *Phys. Rev. B* **75**, 214423 (2007).

28. A. Hubert and R. Schäfer, *Magnetic Domains: The Analysis of Magnetic Microstructures*, Springer, Berlin, 1998.

29. R. D. McMichael and M. J. Donahue, Head to head domain wall structures in thin magnetic strips, *IEEE Trans. Magn.* **33**, 4167 (1997).

30. Y. Nakatani, A. Thiaville, and J. Miltat, Head-to-head domain walls in soft nano-strips: A refined phase diagram, *J. Magn. Magn. Mat.* **290–291**, 750 (2005).

31. M. Hayashi, L. Thomas, C. Rettner, R. Moriya, Y. B. Bazaliy, and S. S. P. Parkin, Current driven domain wall velocities exceeding the spin angular momentum transfer rate in permalloy nanowires, *Phys. Rev. Lett.* **98**, 037204 (2007).

32. M. Hayashi, L. Thomas, C. Rettner, R. Moriya, and S. S. P. Parkin, Direct observation of the coherent precession of magnetic domain walls propagating along permalloy nanowires, *Nat. Phys.* **3**, 21 (2007).

33. M. Klaui, C. A. F. Vaz, J. A. C. Bland et al., Head-to-head domain-wall phase diagram in mesoscopic ring magnets, *Appl. Phys. Lett.* **85**, 5637 (2004).

34. L. Thomas and S. S. P. Parkin, Current induced domain-wall motion in magnetic nanowires, in *Handbook of Magnetism and Advanced Magnetic Materials*, H. Kronmüller and S. S. P. Parkin (Eds.), John Wiley & Sons Ltd., London, Vol. 2, p. 942, 2007.

35. D. Ravelosona, D. Lacour, J. A. Katine, B. D. Terris, and C. Chappert, Nanometer scale observation of high efficiency thermally assisted current-driven domain wall depinning, *Phys. Rev. Lett.* **95**, 117203 (2005).

36. S. Mangin, D. Ravelosona, J. A. Katine, M. J. Carey, B. D. Terris, and E. E. Fullerton, Current-induced magnetization reversal in nanopillars with perpendicular anisotropy, *Nat. Mater.* **5**, 210 (2006).

37. J. C. Sankey, Y.-T. Cui, R. A. Buhrman, D. C. Ralph, J. Z. Sun, and J. C. Slonczewski, Measurement of the spin-transfer-torque vector in magnetic tunnel junctions, *Nat. Phys.* **4**, 67 (2008).

38. H. Kubota, A. Fukushima, K. Yakushiji et al., Quantitative measurement of voltage dependence of spin-transfer torque in MgO-based magnetic tunnel junctions, *Nat. Phys.* **4**, 37 (2008).

39. D. Ravelosona, S. Mangin, J. A. Katine, E. E. Fullerton, and B. D. Terris, Threshold currents to move domain walls in films with perpendicular anisotropy, *Appl. Phys. Lett.* **90**, 072508 (2007).

40. A. Yamaguchi, T. Ono, S. Nasu, K. Miyake, K. Mibu, and T. Shinjo, Real-space observation of current-driven domain wall motion in submicron magnetic wires, *Phys. Rev. Lett.* **92**, 077205 (2004).

41. L. Heyne, M. Klaui, D. Backes et al., Relationship between nonadiabaticity and damping in permalloy studied by current induced spin structure transformations, *Phys. Rev. Lett.* **100**, 066603 (2008).

42. S. Laribi, V. Cros, M. Muñoz et al., Reversible and irreversible current induced domain wall motion in CoFeB based spin valves stripes, *Appl. Phys. Lett.* **90**, 232505 (2007).

43. H. Tanigawa, T. Koyama, G. Yamada et al., Domain wall motion induced by electric current in a perpendicularly magnetized Co/Ni nano-wire, *Appl. Phys. Express* **2**, 053002 (2009).

44. S. Fukami, Y. Nakatani, T. Suzuki, K. Nagahara, N. Ohshima, and N. Ishiwata, Relation between critical current of domain wall motion and wire dimension in perpendicularly magnetized Co/Ni nanowires, *Appl. Phys. Lett.* **95**, 232504 (2009).

45. H. Tanigawa, K. Kondou, T. Koyama et al., Current-driven domain wall motion in CoCrPt wires with perpendicular magnetic anisotropy, *Appl. Phys. Express* **1**, 011301 (2008).

46. M. Yamanouchi, D. Chiba, F. Matsukura, and H. Ohno, Current-induced domain-wall switching in a ferromagnetic semiconductor structure, *Nature* **428**, 539 (2004).

47. M. Feigenson, J. W. Reiner, and L. Klein, Efficient current-induced domain-wall displacement in $SrRuO_3$, *Phys. Rev. Lett.* **98**, 247204 (2007).

48. K.-J. Kim, J.-C. Lee, S.-J. Yun et al., Electric control of multiple domain walls in Pt/Co/Pt nanotrack with perpendicular magnetic anisotropy, *Appl. Phys. Express* **3**, 083001 (2010).

49. T. A. Moore, I. M. Miron, G. Gaudin et al., High domain wall velocities induced by current in ultrathin Pt/Co/AlOx wires with perpendicular magnetic anisotropy, *Appl. Phys. Lett.* **93**, 262504 (2008).

50. I. M. Miron, P. J. Zermatten, G. Gaudin, S. Auffret, B. Rodmacq, and A. Schuhl, Domain wall spin torquemeter, *Phys. Rev. Lett.* **102**, 137202 (2009).

51. S. Li, H. Nakamura, T. Kanazawa, X. Liu, and A. Morisako, Current-induced domain wall motion in TbFeCo wires with perpendicular magnetic anisotropy, *IEEE Trans. Magn.* **46**, 1695 (2010).

52. M. Hayashi, L. Thomas, R. Moriya, C. Rettner, and S. S. P. Parkin, Current-controlled magnetic domain-wall nanowire shift register, *Science* **320**, 209 (2008).

53. D. Chiba, G. Yamada, T. Koyama et al., Control of multiple magnetic domain walls by current in a Co/Ni nano-wire, *Appl. Phys. Express* **3**, 073004 (2010).

54. R. Moriya, M. Hayashi, L. Thomas, C. Rettner, and S. S. P. Parkin, Dependence of field driven domain wall velocity on cross-sectional area in $Ni_{65}Fe_{20}Co_{15}$ nanowires, *Appl. Phys. Lett.* **97**, 142506 (2010).

55. E. Saitoh, H. Miyajima, T. Yamaoka, and G. Tatara, Current-induced resonance and mass determination of a single magnetic domain wall, *Nature* **432**, 203 (2004).

56. R. Moriya, L. Thomas, M. Hayashi, Y. B. Bazaliy, C. Rettner, and S. S. P. Parkin, Probing vortex-core dynamics using current-induced resonant excitation of a trapped domain wall, *Nat. Phys.* **4**, 368 (2008).

57. L. Thomas, M. Hayashi, X. Jiang, R. Moriya, C. Rettner, and S. S. P. Parkin, Oscillatory dependence of current-driven magnetic domain wall motion on current pulse length, *Nature* **443**, 197 (2006).

58. T. Nozaki, H. Maekawa, M. Mizuguchi et al., Substantial reduction in the depinning field of vortex domain walls triggered by spin-transfer induced resonance, *Appl. Phys. Lett.* **91**, 082502 (2007).

59. S. Fukami, T. Suzuki, K. Nagahara et al., in *Symposium on VLSI Technology Digest of Technical Papers*, p. 230, 2009.

60. M. Heide, G. Bihlmayer, and S. Blugel, Dzyaloshinskii-Moriya interaction accounting for the orientation of magnetic domains in ultrathin films: Fe/W(110), *Phys. Rev. B* **78**, 140403 (2008).

61. K.-S. Ryu, L. Thomas, S.-H. Yang, and S. Parkin, Chiral spin torque at magnetic domain walls, *Nat. Nanotechnol.* **8**, 527 (2013).

62. S. Emori, U. Bauer, S.-M. Ahn, E. Martinez, and G. S. D. Beach, Current-driven dynamics of chiral ferromagnetic domain walls, *Nat. Mater.* **12**, 611 (2013).

63. I. E. Dzyaloshinskii, Thermodynamic theory of weak ferromagnetism in antiferromagnetic substances, *Sov. Phys. JETP* **5**, 1259 (1957).

64. T. Moriya, Anisotropic superexchange interaction and weak ferromagnetism, *Phys. Rev.* **120**, 91 (1960).

65. A. Fert, Magnetic and transport properties of metallic multilayers, *Mater. Sci. Forum* **59&60**, 439 (1990).

66. M. Bode, M. Heide, K. von Bergmann et al., Chiral magnetic order at surfaces driven by inversion asymmetry, *Nature* **447**, 190 (2007).

67. S. Heinze, K. von Bergmann, M. Menzel et al., Spontaneous atomic-scale magnetic skyrmion lattice in two dimensions, *Nat. Phys.* **7**, 713 (2011).

68. G. Chen, T. Ma, A. T. N'Diaye et al., Tailoring the chirality of magnetic domain walls by interface engineering, *Nat. Commun* **4**, 2671 (2013).

69. J. Torrejon, J. Kim, J. Sinha et al., Interface control of the magnetic chirality in CoFeB/MgO heterostructures with heavy-metal underlayers, *Nat. Commun.* **5**, 4655 (2014).

70. K. S. Ryu, S. H. Yang, L. Thomas, and S. S. P. Parkin, Chiral spin torque arising from proximity-induced magnetization, *Nat. Commun.* **5**, 3910 (2014).

71. C. Moreau Luchaire, C. Moutafis, N. Reyren et al., Additive interfacial chiral interaction in multilayers for stabilization of small individual skyrmions at room temperature, *Nat. Nanotechnol.* **11**, 444 (2016).

72. A. Thiaville, S. Rohart, E. Jue, V. Cros, and A. Fert, Dynamics of Dzyaloshinskii domain walls in ultrathin magnetic films, *Europhys. Lett.* **100**, 57002 (2012).

73. A. V. Khvalkovskiy, V. Cros, D. Apalkov et al., Matching domain-wall configuration and spin-orbit torques for efficient domain-wall motion, *Phys. Rev. B* **87**, 020402 (2013).

74. M. I. Dyakonov and V. I. Perel, Current-induced spin orientation of electrons in semiconductors, *Phys. Lett. A* **35**, 459 (1971).

75. A. Hoffmann, Spin Hall effects in metals, *IEEE Trans. Magn.* **49**, 5172 (2013).

76. J. Sinova, S. O. Valenzuela, J. Wunderlich, C. H. Back, and T. Jungwirth, Spin Hall effects, *Rev. Mod. Phys.* **87**, 1213 (2015).

77. J. C. Slonczewski, Current driven excitation of magnetic multilayers, *J. Magn. Magn. Mat.* **159**, L1 (1996).

78. S. Zhang, P. M. Levy, and A. Fert, Mechanisms of spin-polarized current-driven magnetization switching, *Phys. Rev. Lett.* **88**, 236601 (2002).

79. M. D. Stiles and A. Zangwill, Anatomy of spin-transfer torque, *Phys. Rev. B* **66**, 014407 (2002).

80. S. S. P. Parkin, N. More, and K. P. Roche, Oscillations in exchange coupling and magnetoresistance in metallic superlattice structures—Co/Ru, Co/Cr, and Fe/Cr, *Phys. Rev. Lett.* **64**, 2304 (1990).

81. S.-H. Yang, K.-S. Ryu, and S. Parkin, Domain-wall velocities of up to 750 m s^{-1} driven by exchange-coupling torque in synthetic antiferromagnets, *Nat. Nanotechnol.* **10**, 221 (2015).

82. S. Parkin and S.-H. Yang, Memory on the racetrack, *Nat. Nanotechnol.* **10**, 195 (2015).

83. T. Miyazaki and N. Tezuka, Giant magnetic tunneling effect in Fe/Al$_2$O$_3$/Fe junction, *J. Magn. Magn. Mat.* **139**, L231 (1995).

84. J. S. Moodera, L. R. Kinder, T. M. Wong, and R. Meservey, Large magnetoresistance at room temperature in ferromagnetic thin film tunnel junctions, *Phys. Rev. Lett.* **74**, 3273 (1995).

85. S. S. P. Parkin, C. Kaiser, A. Panchula et al., Giant tunneling magnetoresistance at room temperature with MgO (100) tunnel barriers, *Nat. Mater.* **3**, 862 (2004).

86. S. Ikeda, J. Hayakawa, Y. Ashizawa et al., Tunnel magnetoresistance of 604% at 300 K by suppression of Ta diffusion in CoFeB/MgO/CoFeB pseudo-spin-valves annealed at high temperature, *Appl. Phys. Lett.* **93**, 082508 (2008).

GMR Spin-Valve Biosensors

Jung-Rok Lee, Richard S. Gaster, Drew A. Hall, and Shan X. Wang

15.1 OVERVIEW OF BIOSENSING

Magnetic biosensors, particularly giant magnetoresistive (GMR) biosensors based on spin-valve sensors, used to detect surface binding reactions of biological molecules labeled with magnetic particles possess the potential to compete with fluorescent-based biosensors, which currently dominate the application space in both research and clinical settings. Magnetic biosensing offers several key advantages over conventional optical techniques and other competing sensing modalities. First, the samples (blood, urine, serum, etc.) naturally lack any detectable magnetic content, providing a sensing platform with a very low background. Additionally, magnetic tags do not suffer from problems that have plagued fluorescent labels such as label-bleaching and autofluorescence. Second, the sensors can be arrayed and multiplexed to perform complex protein or nucleic acid analysis in a single assay without resorting to optical scanning. Finally, the sensors are compatible with standard silicon integrated circuit (IC) technology, allowing them to be manufactured cheaply with integrated electronic readout in mass quantities and to be deployed in a single-use, disposable format. The technology is scalable and capable of being integrated, making it appealing for point-of-care (POC) testing applications.

Recent work in our group further shows that magnetic biosensors are able to simultaneously measure multiple binding reactions in a wash-free assay in real time. Furthermore, the sensors and the magnetic tags can be manufactured on the nanoscale, making the detector comparable in size to the biomolecules being sensed. This unique set of properties makes them practical and useful in both clinical and research settings.

15.2 HISTORY OF MAGNETIC BIOSENSING

To our knowledge, the first use of magnetic nanoparticles as labels in immunoassays reported in literature was in 1997 by a group of German researchers [1]. The measurement used a superconducting quantum interference device (SQUID) to detect binding events of antibodies labeled with magnetic tags. While it was successful, the operating conditions required liquid helium cooling and a magnetically shielded room, limiting the practicality of SQUID-based biosensors. In 1998, Baselt et al. first demonstrated the detection of magnetic nanoparticles using GMR sensors with GMR multilayers [2]. GMR sensors have the advantage of room-temperature operation and simpler instrumentation, making them more attractive, particularly for portable applications. However, proper design of GMR biosensors, magnetic tags, biochemistry, and electronics is necessary for high-sensitivity bioassays, and this proves to be much more sophisticated than initially perceived. This chapter will highlight some of these advances.

All spintronic sensors, such as anisotropic magnetoresistive (AMR), GMR multilayers, GMR spin-valves, and magnetic tunnel junctions (MTJs), share a common principle of operation whereby a free magnetic layer (or layers) responds to a change in the local magnetic field and causes a change in the electrical resistance of the sensor. Inductive sensors, on the other hand, are only capable of monitoring the rate of temporal change in the magnetic field. The rapid development of magnetic biosensing in this decade was the result of leveraging the existing sensing technology used in hard disk drives. In the 1990s, hard drive manufacturers replaced the inductive pick up coil in the read head with an AMR sensor, which in turn was replaced with a spin-valve (and subsequently an MTJ) for higher sensitivity and scaling requirements in magnetic recording [3]. Since the initial conception of magnetic biosensors in the late 1990s, several groups have continued the work on research and development of magnetic biosensing technology [4–11], and applications of magnetic sensors to biological molecule detection with various magnetic particles have been reviewed elsewhere [12]. Table 15.1, while not exhaustive, summarizes some of the research groups and commercial corporations currently investigating magnetoresistive biosensors. It is evident from the diversity in the table that the field is still progressing and has yet to converge on an optimal design. Furthermore, there are several research groups currently investigating using MTJs as biosensors [13, 14]. Presently, the main issue limiting the adoption of MTJs is that with the large area of the devices, a single pin-hole defect renders the device unusable. Whether MTJs will take over biological sensing applications, as they did with hard disk drives, remains to be seen.

15.3 BIOLOGY AND CHEMISTRY

An important innovation of using GMR sensors as biosensors is the effective, reliable, and reproducible integration of the biology and chemistry with the inorganic sensor surface. By immobilizing recognition molecules complementary to biological molecules, such as proteins or nucleic acids, it is possible to detect binding events of proteins or nucleic acids to the recognition molecules in real time. In this chapter, we will discuss protein detection; however, nucleic acid–based detection is very similar [11]. Historically, antibodies (typically immunoglobulin G [IgG]) have been the recognition molecule of choice for protein-based detection assays, as they are highly specific to one unique region of one protein, known as an *epitope* of the protein. This antibody–protein interaction is typically referred to as a *lock-and-key* system due to the high specificity of the interaction. It is this specificity that allows researchers to selectively capture one specific type of protein in a vast background of proteins.

TABLE 15.1
Comparison of GMR/MTJ Biosensors from Various Groups

Institution and Site	Principal Investigators	Magnetic Particles	Sensor Technology
Naval Research Laboratory, Washington	Whitman, LJ	Dynal M280 2.8 μm	GMR, Multi-segment
NVE, Eden Prairie	Tondra, M		1.6×8000 μm, 42 kΩ
IST	Ferreira, HA	Nanomag-D 250 nm	SV, Single-segment
Lisbon, Portugal	Freitas, PP		2.5×100 μm, 1 kΩ
University of Bielefeld, Germany	Reiss, G	Bangs CM01N 350 nm	GMR Spiral
	Brueckl, H		1×1800 μm, 12 kΩ
Stanford University, Stanford	Wang, SX	Miltenyi MACS 40 nm	SV, Multi-segment
	Pourmand, N		1.5×2800 μm, 45 kΩ
University of Minnesota, Minneapolis	Wang, J-P	Miltenyi MACS 40 nm	SV, Multi-segment
			0.75×120 μm, 3 kΩ est.
Brown University, Providence	Xiao, G	Dynal M280 2.8 μm	MTJ, Ellipse Patch
			2×6 μm, 142 Ω
Phillips Research, Netherlands	Prins, M	Ademtech 300 nm	GMR, Gradiometer
			3×100 μm, 250 Ω est.

Source: Dill, K., Liu, R.H., and Grodzinski, P., 2009. *Microarrays: Preparation, Microfluidics, Detection Methods, and Biological Applications, Integrated Analytical Systems,* Springer, New York, 2009.

The overall structure of IgG antibodies is remarkably similar, whether it is reactive to a cancer tumor marker or an infectious pathogen (Figure 15.1). Every IgG antibody contains two long heavy chains and two short light chains held together via inter-chain disulfide bonds and electrostatic interactions. Both the light chains and the heavy chains can be divided into two distinct regions: a constant region at the base of each chain and a variable region at the tip. The constant region at the base of the heavy chains is typically referred to as the *Fc* region (which will be mentioned later). At the tip

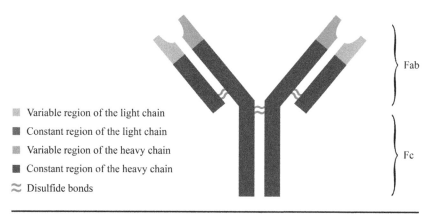

Variable region of the light chain
Constant region of the light chain
Variable region of the heavy chain
Constant region of the heavy chain
≈ Disulfide bonds

Fab

Fc

FIGURE 15.1 Structure of an IgG antibody.

of each chain is the highly variable region known as the *Fab* region, which contains the antigen-binding sites. This highly variable region is where the specificity of the protein–antibody interaction takes place and the lock-and-key interface occurs. By choosing an antibody that reacts with a specific biomarker of interest, one can selectively capture that biomarker of choice over a sensor surface to facilitate protein detection.

15.3.1 SANDWICH ASSAY

The most effective method of detecting proteins in a solution is via a "sandwich assay." Typically known for its use in the enzyme-linked immunosorbent assay (ELISA), the sandwich assay involves the formation of a three-layered structure, in which two antibodies (aptamers, diabodies, Fab fragments, etc. can also be used; however, antibodies are by far the most common recognition molecule used for protein detection) form a sandwich around the protein (also called the *analyte*) of interest (Figure 15.2a). One of the antibodies, typically referred to as the *capture antibody*, is directly immobilized on the sensor surface. To make the sandwich assay highly specific, a monoclonal capture antibody is traditionally used. A solution of monoclonal antibodies means that every antibody in the solution has the exact same Fab region and therefore, will bind to only one epitope on one protein. The capture antibody makes up the foundation of the sandwich assay and acts to selectively capture specific proteins of interest directly over the sensor surface.

The second antibody, known as the *detection antibody*, is delivered in solution and binds to a second epitope on the captured protein of interest. The detection antibody is typically polyclonal and pre-modified with a reactive chemistry, enabling facile attachment of the detection antibody to the tag of interest. A polyclonal antibody solution is one in which all the antibodies react with the same protein; however, they may bind to different epitopes on that protein. Therefore, the Fab region is not identical across all the antibodies in a polyclonal solution. Typically, the antibody is modified with biotin and the tag is modified with streptavidin, since the biotin–streptavidin interaction is one of the strongest non-covalent receptor–ligand interactions

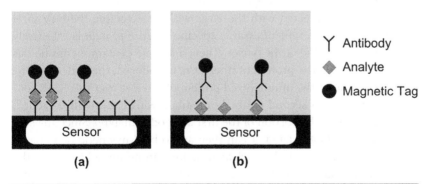

FIGURE 15.2   Schematic of (a) a sandwich assay and (b) a reverse phase assay for protein detection or for antibody detection.

in biochemistry (dissociation constant $K_D \approx 10$ fM). In the ELISA, the tag of interest is typically fluorescent or enzymatic. However, when using GMR biosensors, the tag of interest is magnetic. Therefore, the more protein that is present in the system, the more detection antibody binds, and the more magnetic tag binds. As the number of magnetic tags increases over the sensor, the magnetoresistance in the underlying GMR sensor changes proportionally, producing larger signals [16]. In this way, quantitative protein detection is possible with this assay.

15.3.2 REVERSE PHASE ASSAY

A *reverse phase* assay can be used to detect proteins of interest in many patients' blood samples simultaneously by reorganizing the traditional sandwich assay. In the reverse phase assay, instead of functionalizing a capture antibody onto the sensor surface, patient samples containing proteins of interest such as cell lysates are immobilized onto the GMR sensor array (at least one sensor per sample). Then, a solution containing antibodies complementary to the protein of interest is introduced and will bind to the immobilized protein of interest over the GMR sensor. Subsequently, a second antibody modified with biotin will be added, which typically recognizes the Fc portion of the first antibody. Since the second antibody is biotinylated, it can then bind to magnetic tags coated with streptavidin in a similar fashion to the detection antibody and magnetic tag interaction in the traditional sandwich assay (Figure 15.2b). Note that the scheme in Figure 15.2b can also be accomplished with one biotinylated detection antibody if it directly binds to the protein of interest.

In addition, if one separates the protein of interest from the sample prior to sensor immobilization, it is possible to increase capture protein density.

It is also noteworthy that the schematic in Figure 15.2b can be further modified to detect antibodies in the patient's blood that react against a specific protein of interest. These antibodies can be either auto-reactive antibodies (antibodies against self-proteins typically found in autoimmune disorders) or antibodies conferring immunity (antibodies against infectious agents). If a physician detects antibodies against the HIV capsid protein in a patient sample, for example, then the physician can confirm that the patient has been infected with the virus without detecting the HIV viral protein itself. In this type of assay, a specific *capture protein* is selectively immobilized onto the GMR sensor. Then, a sample containing antibodies complementary to the protein of interest is introduced; the patient's antibodies will bind to the immobilized protein of interest over the GMR sensor. Subsequently, a second antibody modified with biotin will be added, which binds to the Fc portion of the first antibody (analyte). Since the second antibody is biotinylated, it can then act to bind magnetic tags coated with streptavidin. Similarly, nuclear proteins can be immobilized on the GMR sensors and used to detect auto-reactive antibodies in serum samples from individuals with systemic lupus erythematosus (SLE), as previously demonstrated [17].

15.3.3 Capture Antibody or Capture Protein Immobilization

The capture antibody or capture protein can be mounted onto the sensor surface via a variety of different methods. Ideally, the antibody or protein of interest will be attached to the sensor surface while retaining the native conformation and activity. The immobilization can be performed by non-specific physical absorption, in which electrostatic forces between the antibody and a thin film polymer on the substrate dictate the binding. For example, a highly cationic polymer can be used to non-specifically attract negatively charged proteins or antibodies to the sensor surface. Polymers such as polyethyleneimine have been shown to provide an effective and simple means of physical absorption–based antibody immobilization [9].

Alternatively, a covalent binding chemistry can be employed, in which new molecular bonds are formed, physically connecting the recognition molecule (e.g., antibody, nucleic acid, or protein in the case of the reverse phase assay) to the sensor surface. This method of antibody immobilization is typically more effective and reproducible than direct physical absorption; however, it is more challenging to implement and more time consuming to run. Fancy tricks such as orienting antibodies in a specific direction via selectively binding to the Fc fragment of the capture antibody or using protein A or protein G immobilization are also possible. For the detection of antibodies to peptides, target peptides including negative controls can be biotinylated to be anchored on the surface of streptavidin-coated sensors. In this case, to reduce non-specific binding of magnetic tags to the surface, magnetic particles with protein A or G can be used instead of streptavidin-coated magnetic particles [18]. Due to space limitations, only a few of the most common methods of covalent antibody immobilization will be discussed here; however, numerous effective methodologies exist. For more details on bioconjugation of sensor surfaces, refer to *Bioconjugate Techniques* [19].

Antibodies are made out of four strings of amino acids joined via disulfide bonds to form the Y-shaped structure presented earlier. Each amino acid in the string has a specific functional group dangling. Many of these functional groups have reactive moieties that can be used to form covalent bonds for surface immobilization. Silanes, polymers, thiols, cross-linkers, and so on have all been shown to be effective methods of covalent antibody immobilization. Accordingly, choosing which chemical modification scheme to use may depend on the amino acid sequence of the antibody. All antibodies, however, consistently have an amino-terminal end, a carboxy-terminal end, and disulfide bonds connecting the strings of amino acids in the antibody. Accordingly, for every receptor–ligand combination, there are a number of effective ways one can fasten the recognition molecule of choice to the sensor surface.

Perhaps the most common method of antibody immobilization is via the formation of covalent bonds from thiol groups on antibodies to gold surfaces. Since all antibodies contain inter-chain disulfide bonds, it is possible to use a selective reducing agent such as 2-mercaptoethylamine (2-MEA) or dithiothreitol (DTT) to reduce only the disulfide bonds connecting the heavy chains of IgG. This will reproducibly create "half" antibodies with exposed sulfhydryl groups at the hinge region of the antibody for gold–thiol conjugation. Alternatively, if one wants to avoid reducing the antibody into half antibodies, Traut's Reagent (2-iminothiolane) can be used to thiolate primary amines on antibodies.

Another common method of antibody immobilization is via the use of primary amines on antibodies. Amine-reactive chemical reactions are commonly used, because they are rapid and typically provide stable amide bonds between the antibody and the polymer on the sensor surface. In addition, virtually all antibodies are equipped with primary amines, making the chemistry generalizable. Therefore, any antibody can be conjugated to surface immobilized polymers, silanes, or self-assembled monolayers (SAMs).

Arguably, the most critical aspect to all surface functionalization is the surface uniformity. Since GMR sensors, if coated with ultrathin passivation layers, are proximity-based biosensors [9], it is essential to ensure that the tag in the sandwich assay remains at a reproducible distance from the sensor surface across all the sensors within a chip and from chip to chip. Therefore, every component of the sandwich assay must remain at a reproducible distance from the sensor. If there are surface non-uniformities such as microbubbles or particulate debris, then the sensor-to-sensor reproducibility will be severely compromised. In addition, the closer the particle is to the sensor surface, the higher the signal will be. So, to obtain maximum sensitivity in the device, the polymer layers should be made as thin as possible without sacrificing the ability to bind the recognition molecule of interest.

15.3.4 BLOCKING OF SENSOR SURFACE TO REDUCE NON-SPECIFIC BINDING

The main goal in antibody surface functionalization is to make the sensor surface as reactive as possible, but only to one unique capture antibody or capture protein. This will facilitate the maximum possible amount of capture agent binding. Once the capture antibody or capture protein is immobilized on the sensor surface, however, the highly reactive surface must be neutralized or blocked to prevent any future non-specific binding of protein, detection antibody, or tag. In all biosensors, such antifouling chemistry is essential to reproducible and reliable protein detection. The most common method of blocking a sensor surface is to saturate the surface with a common, high-abundance protein, such as albumin or casein. These high-concentration proteins will bind to the remaining reactive groups on the thin film and therefore prevent any future proteins or antibodies from binding to the sensor surface in the absence of the capture antibody or capture protein of interest. Therefore, on addition of sample in the subsequent step of the sandwich assay, only proteins complementary to the selected capture antibodies of interest will bind, and all the non-complementary proteins will be washed away.

15.4 MAGNETIC LABELS

In magnetic biosensors, the magnetic label is a very important component of the system. Unfortunately, there are several factors that constrain the choice of label, such as the size and the functionalization. Larger ferromagnetic particles are initially attractive, since they exhibit a large magnetic moment and will thus be easier to detect. Using large, micron-sized magnetic tags, Li et al. have demonstrated detection down to a single particle [7]. Although exceptionally sensitive protein detection and magnetic manipulation are possible with large magnetic particles, more reproducible protein detection, with minimal non-specific binding and higher stability, is possible when using smaller tags. Furthermore, large micron-sized labels cannot be used for basic science applications in which the kinetics of the binding reactions is studied, because the labels significantly degrade the mobility of the antibodies. For these reasons, the focus in our group has shifted toward smaller, nanometer-sized magnetic particles. With the use of magnetic tags on the order of 50 nm (46 ± 13 nm), the particles are more stable in solution (they do not settle over the sensor) and diffuse more rapidly to the detection antibody. At this size scale, the particles can be superparamagnetic, which helps to reduce aggregation and chaining. Companies such as Miltenyi Biotec offer clusters of 10 nm Fe_2O_3 superparamagnetic particles embedded in a dextran shell, which are pre-functionalized with streptavidin or other reactive groups. Custom-made clusters of superparamagnetic nanoparticles and monodisperse iron oxide nanoparticles [20] are also promising for magnetic labels.

To overcome the difficulties associated with increasing the size of the superparamagnetic particles (aggregation, size distribution, and coercivity) for larger moment, there has been active research investigating magnetic nanoparticles with tunable magnetic properties [21, 22]. Nanoparticles created by nanoimprint lithography containing a synthetic antiferromagnetic (SAF) film stack have large saturation magnetic moments, are monodisperse (at a dimension of 100 nm or less), and do not aggregate (in the absence of applied field). These particles decouple the size restriction associated with superparamagnetic particles, and as a result, the magnetic moment can be substantially larger under externally applied fields, leading to higher signals. Furthermore, the properties of these particles can be tuned by adjusting the thickness of the spacer layer or magnetic layers, making them highly attractive for biological sensing, separation applications, and imaging.

15.5 GMR SENSORS

To quantify the amount of biomolecules present, the GMR sensors transduce the stray field from the magnetic tags into an electrical signal. The stray field, however, is extremely small, because the moment of the labels is also small. This requires the sensor to be extremely sensitive to minute magnetic field changes, making GMR spin-valve sensors preferable to AMR sensors or GMR multilayer sensors. GMR multilayer sensors, consisting of antiferromagnetically coupled magnetic layers, tend to have lower field sensitivity than spin-valves because of higher saturation fields. Spin-valves are well suited for biosensing application due to their high sensitivity, high linearity, small hysteresis, and compact feature size. The design of spin-valve biosensors was previously reviewed [16]. Figure 15.3 shows an illustration of spin-valves used as biosensors. The sensors are passivated with an ultrathin oxide to prevent the biological and chemical solutions used to functionalize the surface from causing corrosion. A thicker passivation layer is used to create a reference sensor that will be unable to sense magnetic tags and therefore allow us to monitor systematic fluctuations in the electronics (Figure 15.3c). In addition, a control sensor is typically coated with proteins, such as bovine serum albumin (BSA) or non-complementary antibodies to monitor non-specific binding in the assay (Figure 15.3b).

Unlike memory applications where the desired output of the spin-valve is binary, a linear analog signal is desired for quantitative detection [23]. The sensors are typically biased such that the pinned magnetization (M_p) is fixed in the transverse direction and the free magnetization (M_F) can rotate freely in the plane. In this configuration, the electrical resistance (R) can be expressed as shown in Equation 15.1 (this will be derived in the next section), where ΔR_{max} is the magnetoresistance and θ_f is the angle of the free layer with respect to the x-axis (in the longitudinal direction of the spin-valve).

$$R = R_0 + \frac{1}{2}\Delta R_{max} \sin\left(\theta_f\right). \tag{15.1}$$

For small θ_f, the resistance versus field transfer curve has a linear dependence. To establish this bias condition, an external magnetic field (H_b) is applied to align the free layer along the x-axis. Alternatively, the sensor can be designed such that the shape anisotropy is sufficient to bias the sensor without the need for an external magnetic field.

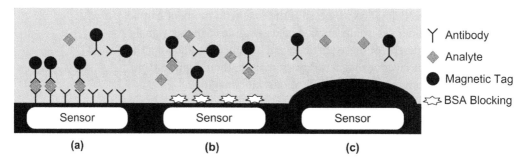

FIGURE 15.3 A sandwich assay with capture antibody, antigen, detection antibody, and magnetic particle. BSA blocking is shown on the control sensor. (a) A biologically active sensor, (b) an unfunctionalized control sensor, and (c) a passivated reference sensor.

15.5.1 SIGNAL PER MAGNETIC LABEL

Before designing the electronics for a spin-valve biosensor, we need to calculate the signal range (resistance change per label). The Stoner–Wohlfarth (SW) model, originally proposed to analyze the magnetic hysteresis of a single-domain particle, was extended to model spin-valves, because the thin ferromagnetic layers in spin-valves are often in a single-domain state [24]. In general, all of the ferromagnetic layers in a spin-valve need to be considered for the total energy expression in the SW model. However, in cases where the strong exchange bias fields effectively pin the magnetization of the pinned layers under small external fields, the free layer alone can be used for the energy expression. As a rough approximation, the average magnetic field caused by a magnetic label positioned in the center of the sensor can be approximated as a dipole, given in Equation 15.2, where the vector $\hat{r}$ points from the center of the tag to the free layer, which has a thickness t [25] (Figure 15.4). This analysis better represents the transduction mechanism when micron-sized magnetic particles are used, with sizes much bigger than the thickness of the sensor.

$$\widehat{H} = \frac{1}{lwt} \int\limits_{-l/2}^{l/2} \int\limits_{-w/2}^{w/2} \int\limits_{-t/2}^{t/2} \left[\frac{3(\hat{m} \cdot \hat{r})\hat{r}}{r^5} - \frac{\hat{m}}{r^3} \right] dx\,dy\,dz. \tag{15.2}$$

The moment of the magnetic particle with a bias (H_b, applied along the x-axis) and tickling field (H_t, applied along the y-axis) is shown in Equation 15.3.

$$\hat{m} = \frac{4}{3}\pi r_b^3 \chi \left(H_b \hat{x} + H_t \hat{y} \right). \tag{15.3}$$

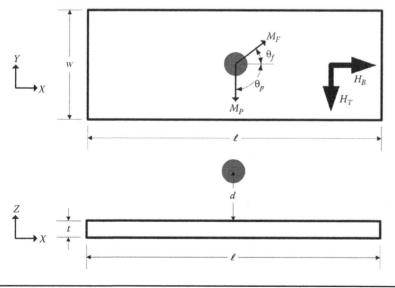

FIGURE 15.4   (Top) Top-down view and (bottom) side view of spin-valve sensor and magnetic tag. Usually $\theta_p = -90°$.

Since the thickness of the free layer is typically significantly smaller than the other dimensions, the field induced in the spin-valve in the x and y directions can be approximated by Equation 15.4.

$$\overline{H_x} = -\frac{8m_x}{\left(l^2 + 4d^2\right)\sqrt{l^2 + w^2 + 4d^2}}$$

$$\overline{H_y} = -\frac{8m_y}{\left(w^2 + 4d^2\right)\sqrt{l^2 + w^2 + 4d^2}}. \tag{15.4}$$

The total energy of the free layer is taken as the sum of energies arising from anisotropy and the external fields, including the bias, tickling, and dipole fields. The angle θ_f represents the rotation of the free layer magnetization. An additional term, H_{other}, is included to account for transverse fields due to other sources, including interlayer coupling as well as fringing fields from the sense current and pinned layers. For simplicity, these can be considered to be negligible in most practical cases.

$$E = \frac{1}{2}H_K M_F \sin^2\theta_f - \left(H_b + \overline{H_x}\right)M_F\cos\theta_f$$
$$-\left(H_t + H_{\text{other}} + \overline{H_y}\right)M_F\sin\theta_f. \tag{15.5}$$

The total anisotropy field (H_K) is the sum of the uniaxial anisotropy field (H_U) and the demagnetizing field (H_D) at the center of the free layer, where the dipole field strength is at its maximum.

$$H_D = \left(N_y - N_x\right)M_F, \quad H_K = H_U + H_D. \tag{15.6}$$

The demagnetizing factors used to calculate H_D are approximated in the expressions given by Li et al. [26].

$$N_x = \frac{8tw}{l\sqrt{l^2 + w^2}}, \quad N_y = \frac{8tl}{w\sqrt{l^2 + w^2}}. \tag{15.7}$$

The first derivative of the energy expression is set to zero to minimize the total energy and solved for θ_f.

$$\frac{dE}{d\theta_f} = M_F\left[H_K\sin\theta_f\cos\theta_f + \left(H_b + \overline{H_x}\right)\sin\theta_f\right.$$
$$\left. -\left(H_t + H_{\text{other}} + \overline{H_y}\right)\cos\theta_f\right] = 0. \tag{15.8}$$

Solving Equation 15.8 and approximating it with the Taylor expansion (valid for $\theta \le 25°$), we arrive at the relatively simple expression Equation 15.9.

$$\sin\theta_f = \frac{M_Y}{M_F} \approx \tanh\left(\frac{H_t + H_{\text{other}} + \overline{H_y}}{H_K + H_b + \overline{H_x}}\right). \tag{15.9}$$

In all practical settings, the magnetization of the pinned layer is fixed in the transverse direction such that $\theta_p = -\pi/2$; thus, the resistance of the spin-valve is given as previously stated (Equation 15.1).

The derivation presented is sufficient for a rough approximation; however, several assumptions were made along the way to make the equations tractable. More accurate results can be obtained by using a finite element model simulator, such as ANSYS Maxwell simulator, or a micromagnetic simulator, such as the open-source object-oriented micromagnetic simulation framework (OOMMF). By using the simulators, additional parameters can be studied, such as the effect of particle location or particle-to-particle interactions.

For nanoparticles whose size is comparable to the thickness of the spin-valve sensor, the particles are also magnetized by the local magnetic field from the free layer. Therefore, a more thorough consideration is required whereby magnetization in the free layer is taken into account [27]. Under an applied tickling field, the free layer is magnetized in the transverse direction, such that the local magnetic field near the edge of the sensor (or between segments in multi-segment spin-valve sensors) is further enhanced, while that on top of the sensor is reduced (Figure 15.5a). Combined with the external tickling field, this local effective field induces different interactions between the free layer and magnetic nanoparticles located at different positions around the sensor. By using this spatially heterogeneous local effective magnetic field, the magnetic field from a particle at different positions along the surface geometry was analyzed using Equation 15.3 normalized by the volume of the nanoparticle (Figure 15.5b). The result shows that a nanoparticle next to the sensor can produce a signal an order of magnitude higher than in other locations.

15.5.2 CHARACTERIZATION OF GMR SPIN-VALVE SENSORS

To convey the design process for using a spin-valve as a biosensor, we will start with a representative spin-valve and design the interface electronics.

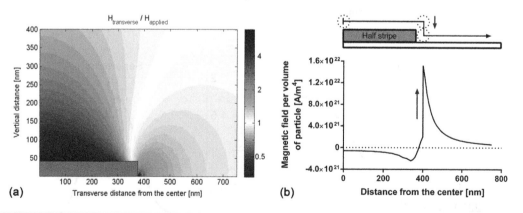

FIGURE 15.5 (a) The effective transverse magnetic field around the half of sensor stripe. The ratio of local magnetic field to the applied field was plotted in log scale. (b) The magnetic field from a nanoparticle at different locations on the surface normalized by the volume of the nanoparticle. (After Lee, J.R. et al., *Sci. Rep.* 6, 18692, 2016. With permission.)

Figure 15.6a shows the transfer curve (resistance as a function of magnetic field) of a spin-valve by sweeping the field strength along the hard axis. The sensor has a minimum resistance of 2190 Ω in the parallel state and a maximum resistance of 2465 Ω in the antiparallel state, corresponding to a MR ratio of 12%. The sensitivity of the sensor, calculated by differentiating the transfer curve, is shown in Figure 15.6b and is maximum when no field is applied, tapering off as the field strength is increased.

For the sensor to be realistic, it will also exhibit realistic noise characteristics. At low frequencies, the noise is dominated by the flicker noise: noise power that has an inverse proportionality to frequency (1/f) and is typically believed to originate from slow defect motions and magnetic domain rotations [28]. The corner frequencies (where the 1/f noise intersects the thermal noise floor) are 900 Hz, 2.25 kHz, and 10 kHz for currents of 100, 250, and 500 μA, respectively. These values are consistent with typical values found in literature and actual devices [29]. The sensor dimensions are such that sufficient shape anisotropy exists not to require an external magnetic bias field.

The flicker noise of the sensor makes it prohibitive to operate at direct current (dc) for sensitive detection. To escape the additional noise penalty incurred when operating at low frequencies, the signal is modulated up to a higher frequency with the use of an alternating external magnetic field, referred to as the *tickling field*, applied in the transverse direction (along the *y*-axis). The tickling field can also serve to magnetize the magnetic particles if they are superparamagnetic or SAF. Modulating the signal to a higher frequency is a common technique used in sensing, often called *chopping* in optical systems. Many research groups [4, 6] have opted to integrate the ability to generate the tickling field on the same die as the sensors. While there are many advantages to integrating the coil, namely, precise alignment capabilities (due to photolithography), compactness, and the ability to actively attract the particles to the sensor surface, they often suffer from poor field uniformity [4].

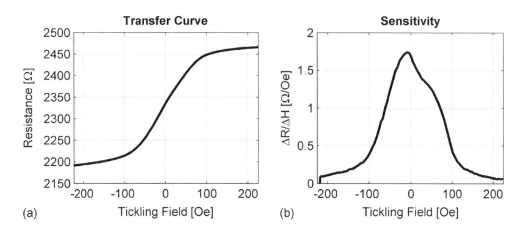

FIGURE 15.6 (a) Transfer curve of a representative spin-valve sensor. (b) Sensitivity of spin-valve in (a).

15.6 SENSOR INTERFACE ELECTRONICS

For a sensor interface, the circuit should be transparent to the sensor, following the adage "Do No Harm" [30]. The sensor parameters for our spin-valve sensor are summarized in Table 15.2. As a case study, two front-ends will be designed to meet different design objectives: a small sensor array (eight sensors) meant for portable applications and a larger array (64 sensors) for benchtop applications.

TABLE 15.2
Summary of Representative Spin-Valve Sensor Parameters

Sensor Property	Value
Nominal Resistance	2.5 kΩ
Magnetoresistance	12%
Breakdown Voltage	>0.5 V
Optimum Tickling Field with Particles	30 Oe
Thermal Noise	6.4 nV/√Hz
Flicker Noise Corner Frequency	900 Hz @ 100 µA
	2.25 kHz @ 250 µA
	10 kHz @ 500 µA

Before we dive into the two designs, there are system-level considerations that can be made to further improve the signal-to-noise ratio (SNR) in both designs. The resistance change per particle is very small (often micro-ohms to milliohms) and can be easily swamped out in the presence of temperature variations and noise. It was previously stated that by modulating the magnetic field, it is possible to modulate the signal to a frequency beyond the flicker noise of the sensor. However, there is a subtle problem with only modulating the field: the sensors have leads, and from Faraday's law, a current will be induced in these leads by the time-varying electromagnetic field (EMF). This parasitic coupling occurs at the same frequency as the signal from the particles. However, if the excitation current (or voltage) applied to the sensor is also modulated at a different frequency, the output of the sensor will appear like a mixer [4, 31]. In this scheme, both the resistance of the device and the current through the device are a function of time (Equation 15.10).

$$R(t) = R_0 + \Delta R_{max} \cos(2\pi f_f t)$$

$$I(t) = I_0 \cos(2\pi f_c t)$$

$$V_{GMR}(t) = R(t) \cdot I(t)$$

$$V_{GMR}(t) = I_0 R_0 \cos(2\pi f_c t) + \frac{I_0 \Delta R_{max}}{2} \cos\big(2\pi(f_c - f_f)t\big)$$
$$+ \frac{I_0 \Delta R_{max}}{2} \cos\big(2\pi(f_c + f_f)t\big). \tag{15.10}$$

The spectrum at the output of the sensor (Figure 15.7) will have four spectral components: a parasitic tone from the EMF (at f_f, the coil frequency, omitted in the derivation of Equation 15.10 for simplicity), a carrier tone (at f_c, the frequency of the modulated excitation), and two side tones at $f_c \pm f_f$. This modulation scheme separates the resistive and the magnetoresistive components of the spin-valve, while only the magnetoresistive component changes under the influence of the magnetic labels. The change in amplitude of the side tone is linearly proportional to the number of magnetic labels, from which the concentration of protein or nucleic acid can be inferred. This

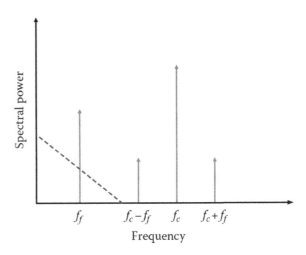

FIGURE 15.7   Spectrum at the output of the sensor using double modulation scheme. The dashed line represents flicker noise.

double modulation scheme improves the SNR of the side tone. In practical situations, the overall SNR of the biosensing system should be dominated by the non-specific binding and not the electronic noise.

15.6.1 Eight-Channel Front-End Design

For the small sensor array case study, the target application is a portable biosensor [32, 33]. The design objectives are to minimize the power consumption and component count without substantially sacrificing the sensitivity. This design will have a minimal analog-to-digital converter (ADC) back-end and very little computational power, necessitating that the demodulation be done in the analog domain. Furthermore, to minimize the form factor of the device, the excitation coil is integrated with the sensors implemented as metal traces on the same die. As such, the coil frequency can operate at a frequency much beyond the flicker noise corner frequency of the sensor. A popular method for high-precision resistance measurement is the Wheatstone bridge. It is constructed of two spin-valve sensors and two programmable resistors. One of the sensors is biologically active, and the other is a reference sensor (shared among the other eight sensors) that is magnetically isolated (from magnetic nanotags, typically with a thick passivation layer) but physically near the active sensor, so that it experiences the same sensing environment (such as temperature changes, magnetic bias, tickling field gradients, etc.). The layout in Figure 15.8 shows how the reference sensor can be placed to maximize the matching between all of the biologically active sensors to minimize process variations. Furthermore, dummy sensors can be added around the outside of the sensors to further improve matching if necessary.

To reduce the component count and hence, the form factor, the bridge is time domain multiplexed (TDM), so that only one sensor is read at any given time. The differential output of the bridge cancels the common mode signal experienced by both the active sensor and the reference sensor (such as

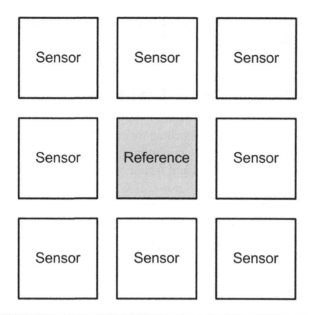

FIGURE 15.8 Sensor layout with reference sensor for matching purposes.

temperature swings). To measure the small resistance change of a magnetic label, a high-gain instrumentation amplifier is used to amplify the small differential output of the bridge. A high-pass (HP) resistor–capacitor (RC) filter blocks the dc component of the signal and adjusts the common mode of the input signal to an appropriate voltage for the instrumentation amplifier. When considering the choice of amplifier, the input-referred noise and the flicker noise corner frequency are important parameters and should be chosen such that the sensor dominates the noise budget. The output of the amplifier is demodulated with a mixer to minimize the ADC requirements. A simple RC low-pass filter removes all but the lower side tone from the spectrum. Our circuit architecture is shown in Figure 15.9. The Wheatstone bridge and high-pass RC filter were used by several research groups [2, 34].

The architecture presented in the preceding text is simple to design yet very effective for small arrays and applications, where the read-out time does not need to be fast. Balancing the bridge requires tuning the two programmable resistors to compensate for mismatch in the sensors due to process variations. If the bridge is not balanced, the offset can cause the high-gain

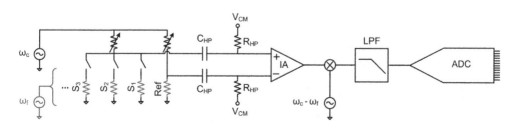

FIGURE 15.9 Circuit architecture for portable biosensor application using time domain multiplexing and double modulation.

differential amplifier to saturate. For small arrays, this tuning step is not a problem; however, it becomes cumbersome for larger arrays, limiting the utility of this architecture in some applications.

15.6.2 64-CHANNEL FRONT-END DESIGN

The larger array of 64 sensors is meant to be used in a research laboratory or a clinical setting, where it is a benchtop instrument [35–38]. In this setting, power and component count are not limiting factors. Instead, achieving the best sensitivity, reproducibility, and fastest readout is the design objective. Furthermore, the computational power is not limited, since it will be performed on a modern computer rather than a microprocessor as for the portable biosensor. Due to the additional computational power and the desire to read out the array as quickly as possible, this design will employ frequency domain multiplexing (FDM) in addition to TDM. By the use of FDM, multiple sensors can be read out simultaneously, reducing the time needed to read out the entire array. The FDM is implemented by exciting sensors with different carrier frequencies ($\omega_1, \omega_2, ..., \omega_n$) [31]. To implement the multiplexing scheme, this design (Figure 15.10) will use a transimpedance amplifier (TIA). The input of the amplifier is a virtual ground providing an easy node to sum the currents from each of the sensors. The demodulation of the signals will be performed in the digital domain after the signals have been digitized.

Unlike the previous example, which used a bridge interface, this interface does not cancel the carrier signal. To remedy this, a replica path is added for carrier suppression, whereby the sensors are replaced with resistors. The output of this path is just the carrier tones without the side tones (the signals of interest). The carrier tones of the sensors are suppressed or canceled completely by subtracting the output of the replica path. This

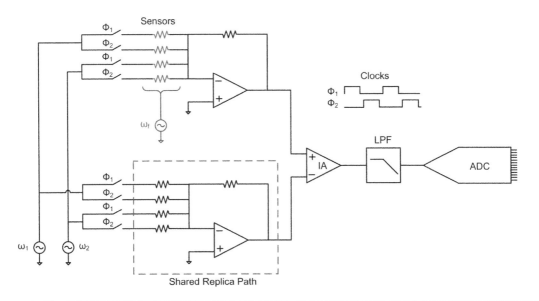

FIGURE 15.10 Circuit architecture using TIA with time domain multiplexing, frequency domain multiplexing, and double modulation.

carrier suppression technique helps to reduce the dynamic range requirement of the back-end ADC.

Since the reference sensor and the biologically active sensors are acquired independently (the difference was acquired in the previous example), this interface provides an additional degree of flexibility, since the reference sensor can be chosen after the experiment is completed. There is also an opportunity to use correction algorithms and digital signal processing to improve many of the non-ideal characteristics of the sensors. For example, the temperature dependence of the sensors can be corrected for in software [15, 36] without requiring heater elements to precisely regulate the temperature [39, 40]. Additionally, correction algorithms can be designed to enhance the reproducibility of the platform, filter out common mode signals, and correct for process variations.

In addition, an imager-like architecture was developed to read out an array of 256 GMR sensors individually [41]. The acquisition system contains 16 readout columns to individually access the sensors using FDM and TDM. This multi-domain modulation scheme reduces the flicker noise and the read-out time. Each readout column consists of four digital-to-analog converters (DAC) for carrier tone suppression, a pseudo-differential TIA, an ADC driver, and a second order, highly oversampled $\Sigma\Delta$ modulator.

15.7 REAL-TIME PROTEIN ASSAYS

One of the key issues in using GMR sensors as biosensors is effectively integrating the biology with the electronics. If recognition molecules complementary to the analytes of interest are immobilized directly over individually addressable GMR sensors, biological molecules of interest can be captured over the sensor surface and detected with the use of magnetic tags. By using magnetic tags and GMR biosensors, real-time protein detection is possible with a broad linear dynamic range and with high multiplex capability.

The ability to monitor protein binding events in real time provides several advantages over end-point measurements used in the ELISA. On incubation of the magnetic nanoparticle tags, it is possible to monitor the change in magnetic content over the sensor surface over time. This provides the user with the ability to distinguish proper nanoparticle binding kinetics, where the signal follows a smooth and continuous absorption pattern (Figure 15.11a), from irregular events due to either non-specific binding or errors in the electronics and sensor (Figure 15.11b–d). If a sensor corrodes during the experiment or there is atypical non-specific binding, for example, the binding signals will look abnormal, and the user will be able to recognize that an error in the system has occurred and in many cases, what went wrong (Figure 15.11). In the ELISA end-point measurements, however, an abnormally high or low data point cannot be rationalized in this manner, as there is no information as to what took place during the reaction. Therefore, eliminating that abnormal data point from the test set may be controversial.

In addition, some reactions take longer than others to reach saturation. In an end-point measurement of an unknown sample, one must blindly

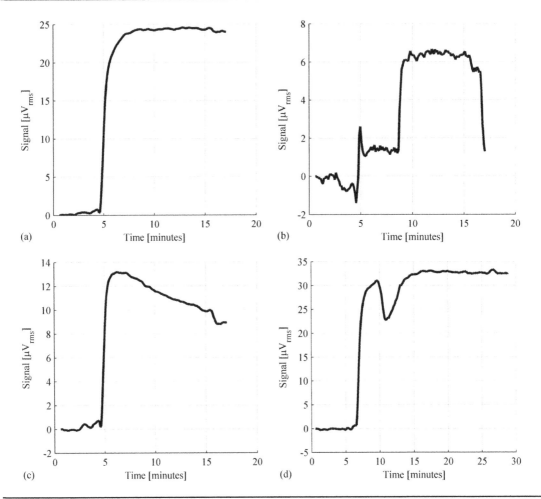

FIGURE 15.11 (a) Normal binding curve in comparison with (b–d) abnormal binding curves.

choose when to take the final point measurement and hope that the reaction is complete or at least reproducible across all experiments. By monitoring the sensors in real time, however, it is possible to observe saturation in the signal (Figure 15.12). For low protein concentration experiments, saturation may be achieved in a minute or two, whereas in the ELISA, the technician must wait the full hour or two (depending on the protocol) prior to any analysis.

To translate the final saturation signal from real-time measurements into a quantifiable protein concentration, a calibration curve must first be generated (Figure 15.13). The calibration curve is created by spiking known concentrations of protein into the medium of interest (phosphate buffered saline [PBS], serum, urine, saliva, etc.) at a range of different concentrations. Therefore, by comparing the signal generated from an unknown sample with the predetermined signals on the calibration curve, it is possible to approximate the concentration of the unknown sample based on the trend observed with the known samples. In addition, the limit of detection, the

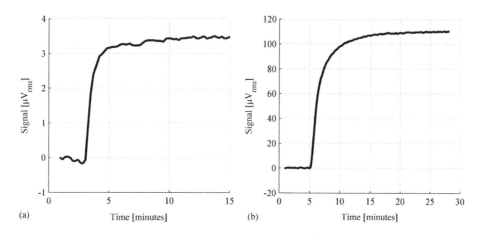

FIGURE 15.12   Example of a binding curve with (a) low concentration and (b) high concentration.

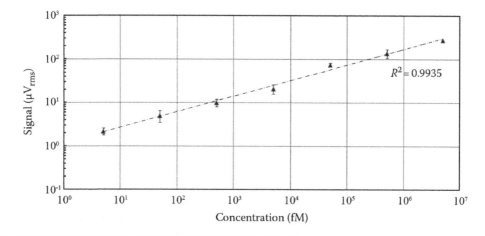

FIGURE 15.13   Calibration curve to translate signal into concentration for a protein.

linear dynamic range of analyte concentrations, and the range of linearity can be determined by plotting the calibration curve. Since each antibody–antigen interaction is unique, every new combination of antibodies and antigens warrants a new calibration curve.

According to the calibration curve in Figure 15.13, GMR biosensors are capable of detection down to femtomolar concentrations without biological amplification or sample preconcentration. Since the calibration curve is linear (on a log–log plot) down to this concentration regime, highly sensitive and quantitative protein detection is possible when using GMR biosensors. In addition, it is important to note that the background signal (where BSA or a non-complementary antibody is immobilized over the sensor surface instead of a capture antibody) remains flat throughout the entirety of the experiment (Figure 15.14), proving negligible magnetic tag settling and minimal non-specific binding. It is customary to define the lower limit of detection to be the average signal on the control sensors (functionalized with non-complementary antibody) plus three standard deviations.

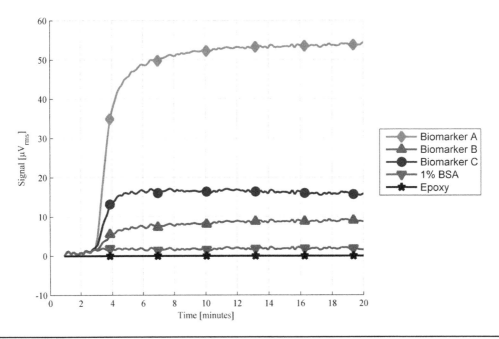

FIGURE 15.14 Triplex experiment illustrating real-time binding curves of three different biomarkers at different concentrations and two negative controls.

In addition, since GMR sensors can be developed into sensor arrays, multiplex protein detection is also possible in a single assay. Accordingly, arrays of up to 64 unique sensors have been demonstrated [9] where an array of sensors is simultaneously monitored with several unique capture antibodies in real time. In Figure 15.14, a triplex protein detection experiment is presented in which three different capture antibodies were functionalized over different sensors in the array. In this assay, Biomarker A was present at the highest concentration and accordingly provided the largest signal, while Biomarker B and Biomarker C were present at successively lower concentrations, revealing lower signals. In addition, the BSA control and epoxy-coated control sensors provide consistently flat signals.

Recently, more applications of GMR sensors have been reported. With sandwich assays, it has been demonstrated that GMR sensors are capable of detecting cancer biomarkers [35], radiation biomarkers [37], cardiovascular biomarkers [42], and allergens in food [38]. In addition, by monitoring signals in real time, GMR sensors have been used for kinetic studies in protein–protein interactions [17, 43] and a platform to check the cross-reactivity of antibodies [44]. Combined with microfluidic chips, wash-free assays with various particles were performed for C-reactive protein [45], and cross-reactivity free measurement was demonstrated for cancer biomarkers by a compartmentalization approach [46]. Furthermore, the signal acquisition system and magnetic coil were miniaturized for portable applications in POC settings. The miniaturized system can be combined with a smartphone [47] or operated as a stand-alone with minimal display [32].

15.8 DESIGN TRADE-OFFS AND OPTIMIZATION

In adapting GMR spin-valves for biosensing applications, there are many design parameters with complex trade-offs and interdependencies. It is not surprising that a single optimal design does not exist for all applications. The following will briefly discuss a few of the design issues, such as breakdown voltage, magnetic field strengths, resistance, noise, and sensor area, as well as the dependencies among them.

Since the magnetic field at the sensor surface from a magnetic label is roughly proportional to the distance cubed, it is highly beneficial to minimize the separation between the label and the sensor. Some of the distance is fixed by the biological material and the surface functionalization, but the distance is typically dominated by the passivation layer on top of the sensor. Reducing the thickness of the passivation layer comes at the cost of reduced breakdown voltage and increased capacitive coupling. Furthermore, beyond a certain thickness, there are diminishing returns. A reduced breakdown voltage may limit the sense current and ultimately reduce the sensitivity of the device, negating any gain obtained from the reduced thickness.

The second set of design trade-offs concerns the geometry, resistance, and noise of the sensor. The resistance is determined by the geometry, the length (L), and the width (W) of the sensor, along with the sheet resistivity (ρ_s) of the film stack. From an electronics perspective, the resistance of the sensor should not be too low, because it would be difficult to design a front-end circuit with lower noise. Similarly, the resistance should not be too large for noise reasons. Sensors with narrow widths are more sensitive, particularly when used in conjunction with small labels. To quantify the relationship between the sensor resistance and the noise, Equation 15.11 shows the thermal (Johnson) noise power present in all conductors and its relationship to the geometry of the sensor:

$$\overline{V_{th}^2} = 4k_B T R \Delta f = 4k_B T \frac{L}{W}\rho_s \Delta f. \tag{15.11}$$

The flicker noise power can be approximated from Hooge's empirical equation (Equation 15.12) where α is a dimensionless constant and N is the number of charge carriers in the conductive layers with a thickness t_c and a charge carrier density of D_c [48].

$$\overline{V_{1/f}^2} = \frac{\alpha}{Nf}(IR)^2 \Delta f = \frac{\alpha}{LWt_c D_c f}\left(I\frac{L}{W}\rho_s\right)^2 \Delta f. \tag{15.12}$$

The geometrical trade-offs can be combined by examining the SNR. There are two possible regions: when the noise is dominated by flicker noise (Equation 15.13) and when the noise is primarily thermal noise. At modest field modulation frequencies (~200 Hz), the former case typically holds. The resistance change per particle is inversely proportional to W^β, where β is an empirically fitted parameter (2.536 in Li's measurements) and independent of the length [16]. When the sensor is operating in the flicker noise region, reducing the sensor area generally increases SNR.

$$\text{SNR} = \frac{I \Delta R_{pp}}{\sqrt{(\alpha/Nf)(IR)^2 \Delta f}} \propto \frac{I(1/W^\beta)}{\sqrt{(\alpha/LWt_c D_c f)I^2(\rho_s L/W)^2 \Delta f}} \quad (15.13)$$

$$\propto \frac{1}{L^{1/2}W^{\beta-3/2}} \propto \frac{1}{L^{1/2}W^{1.036}}.$$

However, from a kinetics perspective, a large sensor coverage area is beneficial to reduce the time required to capture detectable amounts of analytes, attenuate stochastic noise in low-concentration measurements, and improve reproducibility. Unfortunately, for a single stripe spin-valve, the coverage area is directly proportional to the resistance. Shapes other than stripes are also possible; however, curved or round shapes may hinder coherent domain wall rotations and reduce the sensor linearity. Furthermore, the shape of the sensor can also be used to favor a specific magnetization by providing shape anisotropy. With sufficient shape anisotropy, the biasing field may not even be necessary. The geometry of the sensor is a complex trade-off between the noise, sensitivity, and coverage area.

Finally, the amplitude of the tickling field is a parameter that can be used to optimize the system performance. As the amplitude of the tickling field (typically significantly lower than the saturation field) increases, the moment of a superparamagnetic or SAF label also increases. Counteracting this, the sensitivity of the sensor decreases as the field is increased (Figure 15.6b). By multiplying these two curves and differentiating, we can find the maximum signal per particle of the entire system for a given sensor and label pair. For this particular combination of sensor and magnetic nanoparticle, we calculate that each particle induces a 4.6 $\mu\Omega$ change in the resistance. Detection down to a few thousand 50 nm particles has been previously demonstrated [49].

15.9 SUMMARY

In conclusion, by properly designing GMR spin-valve biosensors, labeling detection antibodies with appropriate and colloidally stable magnetic nanoparticle tags, and integrating effective and reliable biochemistry and electronics, we have shown that GMR biosensors are capable of exceptionally sensitive and selective multiplex protein detection either in a single reaction well or in combination with microfluidic channels. The combination of small sample volumes, high sensitivity, real-time signal acquisition, and multiplex capability establishes GMR biosensor arrays as one of the most compelling biosensor platforms for both basic science research and clinical medicine applications. We anticipate that there are many other exciting applications for GMR sensors beyond magnetic recording.

ACKNOWLEDGMENTS

This work would not have been possible without the contributions from the members and collaborators of the Wang Group. In particular, the authors wish to thank Sebastian J. Osterfeld, Heng Yu, Guanxiong Li, Shu-Jen Han, Liang Xu, Robert L. White, Nader Pourmand, Ronald Davis, Robert J. Wilson, Boris Murmann, and Sanjiv S. Gambhir for their underlying work and valuable discussions. This work was supported in part by National Cancer Institute Grants 1 U54CA119367 and N44CM-2009-00011, National Science Foundation (NSF) Grant ECCS-0801385-000, Defense Threat Reduction Agency Grant HDTRA1-07-1-0030-P00005, the Defense Advanced Research Projects Agency/Navy Grant N00014-02-1-0807, and the National Semiconductor Corporation. R.S.G. acknowledges financial support from Stanford Medical School MSTP program and an NSF graduate research fellowship.

REFERENCES

1. R. Kotitz, H. Matz, L. Trahms et al., SQUID based remanence measurements for immunoassays, *IEEE Trans. Appl. Supercond.* **7**, 3678 (1997).
2. D. R. Baselt, G. U. Lee, M. Natesan, S. W. Metzger, P. E. Sheehan, and R. J. Colton, A biosensor based on magnetoresistance technology, *Biosens. Bioelectron.* **13**, 731 (1998).
3. S. X. Wang and A. M. Taratorin, *Magnetic Information Storage Technology, Electromagnetism*, Academic Press, San Diego, CA, 1999.
4. B. M. de Boer, J. A. Kahlman, T. P. Jansen, H. Duric, and J. Veen, An integrated and sensitive detection platform for magneto-resistive biosensors, *Biosens. Bioelectron.* **22**, 2366 (2007).
5. D. L. Graham, H. A. Ferreira, and P. P. Freitas, Magnetoresistive-based biosensors and biochips, *Trends Biotechnol.* **22**, 455 (2004).
6. M. Koets, T. van der Wijk, J. T. van Eemeren, A. van Amerongen, and M. W. Prins, Rapid DNA multi-analyte immunoassay on a magneto-resistance biosensor, *Biosens. Bioelectron.* **24**, 1893 (2009).
7. G. X. Li, V. Joshi, R. L. White et al., Detection of single micron-sized magnetic bead and magnetic nanoparticles using spin valve sensors for biological applications, *J. Appl. Phys.* **93**, 7557 (2003).
8. S. P. Mulvaney, K. M. Myers, P. E. Sheehan, and L. J. Whitman, Attomolar protein detection in complex sample matrices with semi-homogeneous fluidic force discrimination assays, *Biosens. Bioelectron.* **24**, 1109 (2009).
9. S. J. Osterfeld, H. Yu, R. S. Gaster et al., Multiplex protein assays based on real-time magnetic nanotag sensing, *Proc. Natl. Acad. Sci. U S A* **105**, 20637 (2008).
10. J. Schotter, P. B. Kamp, A. Becker, A. Puhler, G. Reiss, and H. Bruckl, Comparison of a prototype magnetoresistive biosensor to standard fluorescent DNA detection, *Biosens. Bioelectron.* **19**, 1149 (2004).
11. L. Xu, H. Yu, M. S. Akhras et al., Giant magnetoresistive biochip for DNA detection and HPV genotyping, *Biosens. Bioelectron.* **24**, 99 (2008).
12. J. R. Lee, D. M. Magee, R. S. Gaster, J. LaBaer, and S. X. Wang, Emerging protein array technologies for proteomics, *Expert Rev. Proteomics* **10**, 65 (2013).
13. T. M. Almeida, M. S. Piedade, J. Germano et al., Measurements and modelling of a magnetoresistive biosensor, in *Biomedical Circuits and Systems Conferences 2006 (BioCAS 2006)*, London, UK, 2006.

14. W. F. Shen, X. Y. Liu, D. Mazumdar, and G. Xiao, In situ detection of single micron-sized magnetic beads using magnetic tunnel junction sensors, *Appl. Phys. Lett.* **86**, 25309 (2005).

15. K. Dill, R. H. Liu, and P. Grodzinski, *Microarrays: Preparation, Microfluidics, Detection Methods, and Biological Applications, Integrated Analytical Systems*, Springer, New York, 2009.

16. S. X. Wang and G. Li, Advances in giant magnetoresistance biosensors with magnetic nanoparticle tags: Review and outlook, *IEEE Trans. Magn.* **44**, 1687 (2008).

17. J. R. Lee, D. J. Bechstein, C. C. Ooi et al., Magneto-nanosensor platform for probing low-affinity protein-protein interactions and identification of a low-affinity PD-L1/PD-L2 interaction, *Nat. Commun.* **7**, 12220 (2016).

18. J. R. Lee, D. J. Haddon, H. E. Wand et al., Multiplex giant magnetoresistive biosensor microarrays identify interferon-associated autoantibodies in systemic lupus erythematosus, *Sci. Rep.* **6**, 27623 (2016).

19. G. T. Hermanson, *Bioconjugate Techniques*, Second Edition, Academic Press, San Diego, CA, 2013.

20. S. Sun, H. Zeng, D. B. Robinson et al., Monodisperse MFe_2O_4 (M = Fe, Co, Mn) nanoparticles, *J. Am. Chem. Soc.* **126**, 273 (2004).

21. W. Hu, C. R. J. Wilson, A. Koh et al., High-moment antiferromagnetic nanoparticles with tunable magnetic properties, *Adv. Mater.* **20**, 1479 (2008).

22. W. Hu, R. J. Wilson, C. M. Earhart, A. L. Koh, R. Sinclair, and S. X. Wang, Synthetic antiferromagnetic nanoparticles with tunable susceptibilities, *J. Appl. Phys.* **105**, 7B508 (2009).

23. G. Li, S. Sun, R. J. Wilson, R. L. White, N. Pourmand, and S. X. Wang, Spin valve sensors for ultrasensitive detection of superparamagnetic nanoparticles for biological applications, *Sens. Actuators A Phys.* **126**, 98 (2006).

24. M. Labrune, J. C. S. Kools, and A. Thiaville, Magnetization rotation in spin-valve multilayers, *J. Magn. Magn. Mater.* **171**, 1 (1997).

25. G. X. Li and S. X. Wang, Analytical and micromagnetic modeling for detection of a single magnetic microbead or nanobead by spin valve sensors, *IEEE Trans. Magn.* **39**, 3313 (2003).

26. G. X. Li, S. X. Wang, and S. H. Sun, Model and experiment of detecting multiple magnetic nanoparticles as biomolecular labels by spin valve sensors, *IEEE Trans. Magn.* **40**, 3000 (2004).

27. J. R. Lee, N. Sato, D. J. Bechstein et al., Experimental and theoretical investigation of the precise transduction mechanism in giant magnetoresistive biosensors, *Sci. Rep.* **6**, 18692 (2016).

28. H. Wan, M. M. Bohlinger, M. Jenson, and A. Hurst, Comparison of flicker noise in single layer, AMR and GMR sandwich magnetic film devices, *IEEE Trans. Magn.* **33**, 3409 (1997).

29. R. J. M. van de Veerdonk, P. J. L. Belien, K. M. Schep et al., 1/f noise in anisotropic and giant magnetoresistive elements, *J. Appl. Phys.* **82**, 6152 (1997).

30. K. A. A. Makinwa, M. A. P. Pertijs, J. C. van der Meer, and J. H. Huijsing, Smart sensor design: The art of compensation and cancellation, in *Esscirc 2007: Proceedings of the 33rd European Solid-State Circuits Conference*, 76–82, 2007.

31. S. J. Han, L. Xu, H. Yu et al., CMOS integrated DNA microarray based on GMR sensors, in *2006 International Electron Devices Meeting*, Vols 1 and 2, 451–454, 2006.

32. R. S. Gaster, D. A. Hall, and S. X. Wang, NanoLAB: An ultraportable, handheld diagnostic laboratory for global health, *Lab Chip* **11**, 950 (2011).

33. D. A. Hall, S. X. Wang, B. Murmann, and R. S. Gaster, Portable biomarker detection with magnetic nanotags, in *Proceedings of 2010 IEEE International Symposium on Circuits and Systems*, p. 1779, 2010.

34. C. R. Tamanaha, S. P. Mulvaney, J. C. Rife, and L. J. Whitman, Magnetic labeling, detection, and system integration, *Biosens. Bioelectron.* **24**, 1 (2008).

35. R. S. Gaster, D. A. Hall, C. H. Nielsen et al., Matrix-insensitive protein assays push the limits of biosensors in medicine, *Nat. Med.* **15**, 1327 (2009).

36. D. A. Hall, R. S. Gaster, S. J. Osterfeld, B. Murmann, and S. X. Wang, GMR biosensor arrays: Correction techniques for reproducibility and enhanced sensitivity, *Biosens. Bioelectron.* **25**, 2177 (2010).

37. D. Kim, F. Marchetti, Z. Chen et al., Nanosensor dosimetry of mouse blood proteins after exposure to ionizing radiation, *Sci. Rep.* **3**, 2234 (2013).

38. E. Ng, K. C. Nadeau, and S. X. Wang, Giant magnetoresistive sensor array for sensitive and specific multiplexed food allergen detection, *Biosens. Bioelectron.* **80**, 359 (2016).

39. S.-J. Han, H. Yu, B. Murmann, N. Pourmand, and S. X. Wang, A high-density magnetoresistive biosensor array with drift-compensation mechanism, in *IEEE International Solid-State Circuits Conference 2007 (ISSCC 2007)*, San Francisco, CA, 2007.

40. W. Hua, C. Yan, A. Hassibi, A. Scherer, and A. Hajimiri, A frequency-shift CMOS magnetic biosensor array with single-bead sensitivity and no external magnet, in *IEEE International Solid-State Circuits Conference (ISSCC 2009)*, San Francisco, CA, 2009.

41. D. A. Hall, R. S. Gaster, K. A. A. Makinwa, S. X. Wang, and B. Murmann, A 256 pixel magnetoresistive biosensor microarray in 0.18 μm CMOS, *IEEE J. Solid-State Circuits* **48**, 1290 (2013).

42. Y. Wang, W. Wang, L. Yu et al., Giant magnetoresistive-based biosensing probe station system for multiplex protein assays, *Biosens. Bioelectron.* **70**, 61 (2015).

43. R. S. Gaster, L. Xu, S. J. Han et al., Quantification of protein interactions and solution transport using high-density GMR sensor arrays, *Nat. Nanotechnol.* **6**, 314 (2011).

44. R. S. Gaster, D. A. Hall, and S. X. Wang, Autoassembly protein arrays for analyzing antibody cross-reactivity, *Nano Lett.* **11**, 2579 (2011).

45. D. J. B. Bechstein, J. R. Lee, C. C. Ooi et al., High performance wash-free magnetic bioassays through microfluidically enhanced particle specificity, *Sci. Rep.* **5**, 11693 (2015).

46. D. J. Bechstein, E. Ng, J. R. Lee et al., Microfluidic multiplexed partitioning enables flexible and effective utilization of magnetic sensor arrays, *Lab Chip* **15**, 4273 (2015).

47. J. Choi, A. W. Gani, D. J. Bechstein, J. R. Lee, P. J. Utz, and S. X. Wang, Portable, one-step, and rapid GMR biosensor platform with smartphone interface, *Biosens. Bioelectron.* **85**, 1 (2016).

48. F. N. Hooge, 1/F Noise Is No Surface Effect, *Phys. Lett. A* **29**, 139 (1969).

49. D. A. Hall, R. S. Gaster, T. Lin et al., GMR biosensor arrays: A system perspective, *Biosens. Bioelectron.* **25**, 2051 (2010).

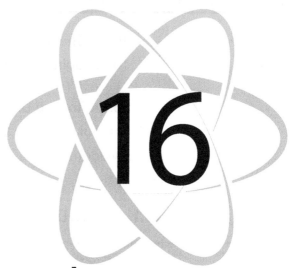

Semiconductor
Spin-Lasers

**Igor Žutić, Jeongsu Lee, Christian Gøthgen,
Paulo E. Faria Junior, Gaofeng Xu,
Guilherme M. Sipahi, and Nils C. Gerhardt**

16.1 INTRODUCTION

Introducing spin-polarized carriers in semiconductor lasers offers an alternative path to realize spintronic applications, beyond the usually employed magnetoresistive effects. Through carrier recombination, the angular momentum of the spin-polarized carriers is transferred to photons, thus leading to the circularly polarized emitted light. Such spin-lasers provide an opportunity to extend the functionality of spintronic devices, as well to exceed the performance of conventional (spin-unpolarized) lasers, from reducing the lasing threshold to improving their dynamical performance and digital operation. Following the introduction to conventional lasers, this chapter describes various elements and operating regimes of spin-lasers. Anticipated future directions and breakthroughs in spin-lasers are closely connected to the advances in more traditional areas of spintronics.

The word *laser* is an abbreviation for *light amplification by stimulated emission of radiation*. However, considering its modern applications, it is perhaps better viewed as a light-generating device rather than a light amplifier [1]. The operating principles of lasers can be traced back to the prediction of stimulated emission by Albert Einstein in 1917 [2]. One of the key laser characteristics is its dependence of the emitted light on pumping or injection. We will distinguish conventional lasers and spin-lasers depending on whether this pumping/injection introduces spin-unpolarized or spin-polarized carriers, respectively. Typically, in conventional lasers, two regimes are distinguished. For low pumping, there is no stimulated emission, and the laser operates as an ordinary light source: the emitted photons are incoherent. However, the higher pumping regime with stimulated emission and coherent light makes the laser a unique light source. The intensity of pumping/injection above which a phase-coherent emission of light occurs is called the *lasing threshold*. Important advances in semiconductor lasers can be associated with the reduction of threshold current density for the onset of lasing. It can be shown that such a reduction would not only decrease the power consumption of lasers but could also significantly enhance their dynamic performance [3]. A schematic illustration in Figure 16.1 shows the historical evolution of threshold reduction in semiconductor lasers, realized by gradually replacing a bulk-like active region (known also as the *gain medium*, where the lasing action takes place) with structures of reduced dimensionality: quantum wells (QWs), quantum wires, and quantum dots (QDs).

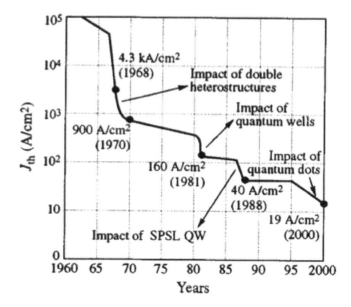

FIGURE 16.1   The evolution of threshold current for semiconductor lasers. (After Alferov, Z.I., *Rev. Mod. Phys.* 73, 767, 2001. With permission.)

There is also an alternative approach to reduce the lasing threshold using spin-polarized carriers generated by circularly polarized light or electrical spin injection. The principles of optical spin injection have been extensively studied for over 50 years [5–8]. Their application to QWs, which are typically used as the active region of spin-lasers, is discussed in Chapter 2, Volume 2. The angular momentum of absorbed circularly polarized light is transferred to the semiconductor. Electron orbital momenta are directly oriented by light and through spin–orbit coupling electron spins become polarized. While initially holes are also polarized, their spin polarization is lost almost instantaneously due to much stronger spin–orbit coupling in the valence band. The spin–orbit coupling, as the leading mechanism for the spin relaxation, is discussed in Chapter 1, Volume 2. For practical applications, it would even be more desirable to use electrical spin injection and simplify the integration with conventional electronics. We omit any detailed discussion of the electrical spin injection, which has been extensively addressed, for example, in Chapters 5, Volume 1, Chapters 2, 4–6, Volume 2, and Chapters 4, 5, 7–9, 17, Volume 3. Figure 16.2 illustrates the theoretical prediction for the threshold reduction for completely spin-polarized injection. The key parameter to assess such a reduction is the spin relaxation time, the timescale it takes for the spin imbalance to be lost. This time strongly depends on the specific material and the geometry used for the active region of a laser. For a very short spin relaxation time, spin-lasers have no threshold reduction as compared with their conventional (spin-unpolarized) counterparts.

Both conventional lasers and spin-lasers share three main elements: the active (gain) region, the resonant cavity (resonator), and the pump, which injects (optically or electrically) energy/carriers. A typical vertical geometry for so-called *vertical cavity surface emitting lasers* (VCSELs) [10] used in spin-lasers is illustrated in Figure 16.3. It is similar to the Faraday geometry of spin light emitting diodes (spin-LEDs) [7], which leads to well-defined

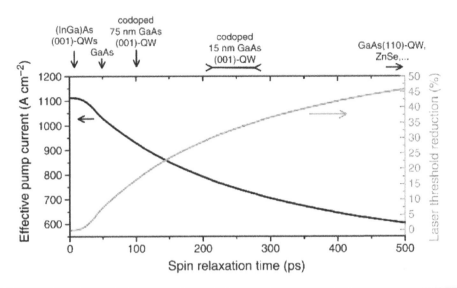

FIGURE 16.2   The laser threshold at room temperature (left axis) and the reduction of the laser threshold (right axis) as a function of spin relaxation time. The spin relaxation time of several semiconductors is depicted on the top axis. (After Oestreich, M. et al., *Superlattices Microstruct.* 37, 306, 2005. With permission.)

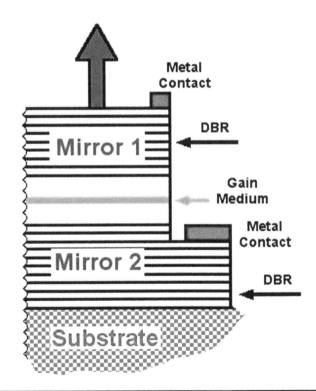

FIGURE 16.3   VCSEL. The resonant cavity is formed by a pair of parallel highly reflective mirrors made of distributed Bragg reflectors (DBRs), a layered structure with varying refractive index, and the gain medium (active region), typically consisting of QWs or QDs. Electrical injection is realized using two metal contacts, and the direction of the emitted light is perpendicular to the top surface and the gain medium.

optical selection rules, further discussed in Chapter 2, Volume 2. Some of the spin-lasers are readily available, since they are based on the commercially available semiconductor lasers, to which subsequently a source of circularly polarized light is added. For electrical spin injection, one could employ ferromagnetic metals at the position of the bottom metallic contact indicated in Figure 16.3. The spin-lasers offer practical paths to room-temperature spin-controlled devices that are not limited to widely employed magnetoresistive effects (another possibility is discussed in Chapter 17, Volume 3). Some of the applications include reconfigurable interconnects, high-speed modulators, and secure communications. In steady-state operation, spin-lasers have demonstrated threshold reduction as compared with their conventional counterparts [11–14]. It was predicted theoretically [15] that this threshold reduction would lead directly to improved dynamic operation of spin-lasers, including an enhanced bandwidth and better switching properties. The spin polarization acts as a new "knob" for controlling the laser's operation. The change in spin polarization not only modifies the lasing threshold but can also be effectively used to turn the laser *on* and *off*, even at the fixed injection [11, 15]. To facilitate the operation of spin-lasers, it is important to consider materials and geometries in which (1) the net spin polarization can be sustained over a relatively long time and (2) there is a large net (and preferably externally controllable) spin polarization of injected carriers at room temperature.

In this chapter, we first provide a description of spin-unpolarized lasers that can be readily generalized to spin-lasers, which are subsequently analyzed in both their steady-state and dynamical operations. While most of our discussions pertain to QW spin-lasers, which have also been predominantly used experimentally, we also provide an outlook for the emerging QD spin-lasers.

16.2 CONVENTIONAL SEMICONDUCTOR LASERS

16.2.1 BUCKET MODEL

A simple description of conventional (spin-unpolarized) lasers can be considered as the special limit of our model of spin-lasers. This simplification allows us to develop a transparent approach and analytical results for laser operation as well as to systematically include the effects of spin polarization and spin relaxation.

One of the most useful analogies to illustrate the laser operation is provided by the bucket model, previously only considered for conventional lasers [16]; we generalize it for spin-lasers in Section 16.3.2 [17]. This model, depicted in Figure 16.4, allows us to establish an intuitive connection to a standard description of a conventional laser operation using rate equations (REs) [4, 18, 19]. For example, the water level represents electron and hole densities, which will be denoted by n and p. The faucet represents the injection of charge carriers (electrons and holes) into the lasers, while the outgoing water corresponds to the photon density, S. We will only be considering light, represented by S, which is supported by the resonant cavity. The recombination and the gain function, which corresponds to the stimulated emission in the active region, will be denoted by $R(n,p)$ and $g(n,p,S)$.

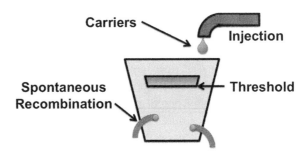

FIGURE 16.4 Bucket model of a conventional laser. The faucet represents the injection or pumping. The small leaks in the bucket represent spontaneous recombination processes, and the large opening near the top of the bucket represents the lasing threshold. At low injection, in Regime I, the water level is low, and the some of this water will trickle out of the leaks. A modest increase in the injection will raise the water level within the bucket, but the amount of escaping water will remain small. At high injection, Regime II, the water will reach the large opening near the top, and from there it will gush out. Additional water will lead only to a small change in the water level, but the output will increase rapidly as compared with Regime I.

The bucket model can be effectively used to illustrate the two characteristic regimes of the laser operation. At the low injection, which we call Regime I, there is only a negligible output of light. The operation of a laser is similar to that of an LED, discussed also in Chapter 2, Volume 2. The spontaneous recombination is responsible for the emitted light. At higher injection, when the water starts to gush out of the large slit in Figure 16.4, the lasing threshold is reached. At the threshold injection, J_T, stimulated emission starts, and the emitted light intensity increases significantly. Regime II, for $J > J_T$ corresponds to fully lasing operation, in which the stimulated recombination is the dominant mechanism of light emission.

As illustrated in Figure 16.3, the gain medium does not occupy the entire resonant cavity; the ratio of gain medium volume to the total cavity volume is given by the dimensionless constant Γ. The fraction of the spontaneous recombination, which produces light that couples to the cavity resonance, is modeled by a constant β. In typical VCSELs, small $\beta \propto 10^{-4}-10^{-3}$ implies that there would be a negligible intensity of light below the lasing threshold. The optical losses from different sources, except from the stimulated absorption, are modeled by the photon lifetime τ_{ph}.

16.2.2 CHARACTERISTIC PROCESSES AND RATE EQUATIONS

From the description of the bucket model in Figure 16.4, it is possible to establish a connection with the REs given in terms of carrier and photon densities:

$$\frac{d}{dt}\alpha = J_\alpha - g(n, p, S) - R_{sp}(n, p), \tag{16.1}$$

$$\frac{d}{dt}S = \Gamma g(n, p, S) + \Gamma\beta R_{sp}(n, p) - \frac{S}{\tau_{ph}}, \tag{16.2}$$

where $\alpha = n$, p, and we consider the densities that are uniform in the gain medium. The expressions for gain and recombination are generally

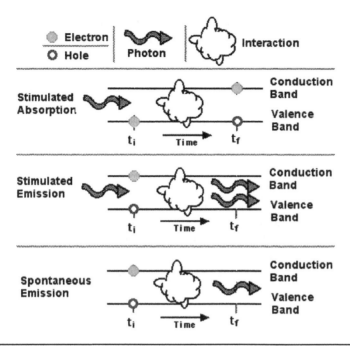

FIGURE 16.5 Schematic description of emission and absorption processes. Stimulated absorption: an incident photon strikes an electron in the valence band, promoting the electron to the conduction band, leaving a hole in the valence band, and annihilating the photon. Stimulated emission: an incident photon triggers a conduction band electron to relax into an empty state, a hole, in the valence band, releasing an additional photon that has the same energy, phase, and momentum as the incoming photon. Spontaneous emission: a conduction band electron relaxes into an empty valence band state, a hole, releasing a photon in the process.

nonlinear functions of the corresponding densities. In Equations 16.1 and 16.2, the contributions for gain and recombination have opposite sign for carrier and photon densities; the processes that reduce the total number of carriers also increase the number of photons. The prefactor Γ, introduced in Section 16.2.1, signifies that the changes of carriers and photons take place in different effective volumes (the gain medium occupies a small fraction of the resonant cavity). To further specify the REs (Equations 16.1 and 16.2), we depict the relevant processes in Figure 16.5. For laser operation, characteristic carrier densities exceed the corresponding equilibrium values, which, when combined with the charge neutrality ($n = p$), allows further simplifications. In the spontaneous recombination (the bottom of Figure 16.5), we distinguish two simple forms: quadratic recombination (QR), $R_{sp} = Bnp/2$, where B is a temperature-dependent constant [20], and linear recombination (LR), $R_{sp} = n/\tau_r$, where τ_r is the recombination time. LR is a particularly good approximation at high n. QR and LR can be viewed as different limits of the general expression for $R_{sp}(n, p)$. For the moderate injection considered, the Auger recombination, which involves three carriers, is typically negligible. However, for higher densities, this process (expected to be $\propto n^3$) becomes quadratic [21], just like the QR considered here.

The gain function can be interpreted as the difference between stimulated emission and absorption, illustrated in Figure 16.5. A specific carrier density at which this difference vanishes is known as the *transparency density*, n_{tran}. It is customary [16] to linearize the gain function around n_{tran}:

$$g(n, S) \approx g_0(n - n_{tran})S, \qquad (16.3)$$

where g_0 is the gain coefficient. Since, experimentally, S does not increase indefinitely with increasing injection, the gain saturation is usually incorporated as

$$g(n, S) \equiv g_0 \frac{n - n_{tran}}{1 + \varepsilon S} S, \qquad (16.4)$$

where ε is the phenomenological gain compression (saturation) factor that models the influence of spectral hole burning and carrier heating, relevant at high injection levels ($J \gg J_T$) [22]. Taken together with the charge neutrality ($n = p$ and $J_n = J_p \equiv J$), the holes can be decoupled from the REs:

$$\frac{d}{dt} n = J - g_0 \frac{n - n_{tran}}{1 + \varepsilon S} S - R_{sp}(n), \qquad (16.5)$$

$$\frac{d}{dt} S = \Gamma g_0 \frac{n - n_{tran}}{1 + \varepsilon S} S + \Gamma \beta R_{sp}(n) - \frac{S}{\tau_{ph}}. \qquad (16.6)$$

These REs are widely used for conventional lasers [3]. Their generalization is typically the starting point to describe spin-lasers.

16.3 SPIN-LASERS: STEADY STATE

16.3.1 EXPERIMENTAL IMPLEMENTATION

Most of the spin-lasers are realized as VCSELs (recall Figure 16.3) with the active region consisting of III-V QWs. Optically or electrically injected/pumped spin-lasers rely on the principles of spin injection. Advances in increasing the efficiency of spin injection, pursued as an effort to improve various spintronic devices, therefore directly aid the enhanced operation of spin-lasers. For example, using an MgO tunnel barrier (discussed further in Chapters 11, 12, Volume 3), which dramatically increases both the magneto-resistance in spin-valves and spin injection in semiconductors [23–25], could also be a promising direction for the next generation of spin-lasers.

The first demonstration of the threshold reduction, shown in Figure 16.6, used optical spin injection in an (In,Ga)As QW-based spin-laser. Slow changes in the injected polarization at the fixed input power clearly lead to the change in the intensity of the emitted light (Figure 16.6b). These results suggest a new path to modulate the laser operation with a fast polarization modulation, explored further in Section 16.4. While this early work was limited to 6 K operation temperature, subsequent efforts have also shown room-temperature optically injected spin-lasers [12].

Alternatively, electrical spin injection employing ferromagnetic contacts has been used to realize the threshold reduction, albeit not yet at room temperature [13, 26, 27]. Figure 16.7 shows the emitted light intensity in an electrically pumped spin-laser using Fe/AlGaAs Schottky tunnel barrier as the spin injector and (In,Ga)As QW as the gain medium.

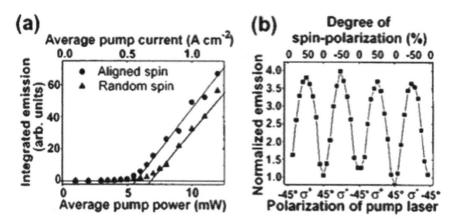

FIGURE 16.6 Threshold reduction in a spin-laser using optical spin injection. (a) The time-integrated intensity after pulsed optical injection is plotted for the 50% spin-aligned electrons (circles) and randomly oriented spins (triangles). The device yields a clear reduction of the laser threshold for spin-polarized injection. (b) Change in the VCSEL emission power by continuously changed pump polarization under 6.5 mW constant pump power. The emission maxima coincide with circular pump polarization and thus with the maximum spin polarization of the excited electrons. Linearly polarized pumping creates random spin orientation, causing a drop below threshold, which demonstrates that the laser emission depends not only on the pump power but also on the spin orientation. (After Rudolph, J. et al., *Appl. Phys. Lett.* 82, 4516, 2003. With permission.)

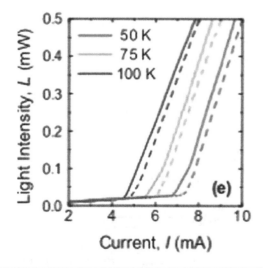

FIGURE 16.7 The intensity of emitted light as a function of injection proves the threshold reduction in a spin-laser using electrical spin injection. The solid (dashed) lines correspond to the injection of spin-polarized (spin-unpolarized) electrons. For every temperature, the onset of lasing is consistently lower for spin injection, which is realized with the aid of an applied magnetic field to ensure an out-of-plane component of magnetization in Fe injector (having an in-plane easy axis). (After Holub, M. et al., *Phys. Rev. Lett.* 98, 146603, 2007. With permission.)

These impressive experimental advances offer new paradigms in spintronic devices that would not be limited to the magnetoresistive effects. However, the theoretical understanding of spin-lasers has been far from complete, even in steady-state operation. This has been somewhat surprising, since the experimental results were supplemented with seemingly a well-established RE description, known already at the textbook level in the spin-unpolarized case and considered in Section 16.2.2. However, numerical results using such REs contain a large number of material parameters, making it difficult to systematically elucidate how the spin-dependent effects influence the device operation. As we shall see, even what is the maximum threshold reduction was not correctly understood, warranting a careful development of spin-polarized REs in this section, followed by systematic study of their analytical solutions in the different operating regimes of spin-lasers.

16.3.2 SPIN-POLARIZED BUCKET MODEL

To gain an insight into the operation of spin-lasers, we first develop a simple description by generalizing the bucket model to the spin-polarized case [17]. We introduce various spin-resolved quantities to model different projections of the spin or helicities of light. The total electron/hole density can be written as the sum of the spin-up (+) and the spin-down (−) electron/hole densities, $n = n_+ + n_-$ and $p = p_+ + p_-$. Analogously, we write the total photon density as the sum of the positive (+) and negative (−) helicities, $S = S^+ + S^-$. Spin injection (optical or electrical) $J = J_+ + J_-$ implies an unequal decomposition $J_+ \neq J_-$ for the two spin projections. As shown in Figure 16.8, they can be represented by two separate water faucets. For example, we can think of the hot- and cold-water analogy, corresponding to spin-up and spin-down carriers. The two halves of the bucket are only coupled through a narrow tube, implying that their content is not completely separated. This situation corresponds to the process of spin relaxation, which mixes spin-up and spin-down populations.

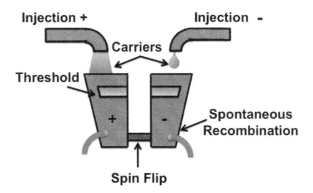

FIGURE 16.8 Spin-polarized bucket model. The two halves of the bucket, representing two spin populations, are separately filled and only connected through a narrow tube to model the spin relaxation. The small holes represent carrier loss through spontaneous recombination, and the large opening near the top denotes the lasing threshold. (After Gøthgen, C., Steady-state analysis of semiconductor spin-lasers, PhD thesis, 2010. With permission.)

It is instructive to consider two limits for the coupling of the two halves of the bucket. If the connecting tube has a negligibly small diameter, the two populations will not mix, corresponding to an infinite spin relaxation time. In the opposite limit, if the connecting tube is very wide, the two unequal populations will equilibrate immediately. This leads to negligible carrier polarization, as expected in the limit of vanishing spin relaxation time, for which spin-lasers and conventional lasers behave in the same way.

Based on this very simple model, we can develop an intuitive understanding for the operation of spin-lasers. With the two unequal faucets, we expect two different lasing thresholds. The half of the bucket that is more effectively filled will overflow sooner, signaling that the lasing will first occur due to (predominantly) one type of (majority spin) carrier. Only later, when the other half becomes sufficiently filled, is lasing expected for the other (minority) carriers. We can then identify three different operating regimes for spin-lasers: (1) spin-LED regime: in the limit of low injection, below the lasing of majority carriers, the laser resembles spin LED for both polarizations of light. There is no stimulated emission, and the emitted light intensity is low; (2) mixed or spin-filtering regime: above the threshold for majority carriers and below the threshold of minority carriers, only one spin population is responsible for the lasing, while the other still operates as a spin LED; (3) fully lasing regime: for injection: above the value of the minority lasing threshold, both spin populations will contribute to lasing.

Qualitatively, this picture of three operating regimes in a spin-laser can be sketched as shown in Figure 16.9. In the approximation of $\beta = 0$ (we examine its accuracy in Figure 16.10), Regime I, that of spin LED, yields negligible

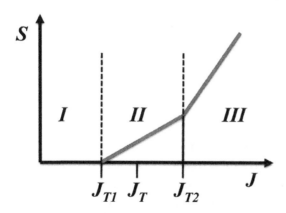

FIGURE 16.9 Total photon density, S, from a spin-polarized laser as a function of injection, J, according to the spin-polarized bucket model. At low pumping, Regime I, both spin-up and spin-down carriers are in the spin LED regime and thus, with negligible emission. In Regime II, the majority spin is lasing, while the minority spin is still in the LED regime; thus, all of the light is from recombination of carriers with majority spin. In Regime III, both spin-up and spin-down are lasing. J_{T1} is the injection at which the majority spin begins to lase, and J_{T2} indicates the injection at which the minority spin begins to lase. The kink of the total photon density S at J_{T2} is from the minority light that begins to contribute to the total photon density. J_T is the spin-unpolarized lasing threshold.

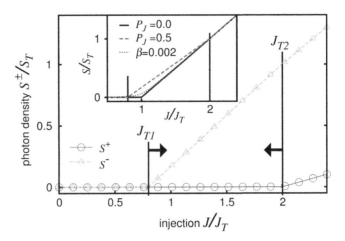

FIGURE 16.10   Photon densities of the left-(S^-) and right-(S^+) circularly polarized light as a function of electron current J with polarization $P_J=0.5$, infinite electron spin relaxation time, and LR. J is normalized to the unpolarized threshold current J_T and $S^\pm$ to $S_T = J_T \Gamma \tau_{ph} = S(2J_T)$. The vertical lines indicate J_{T1} and J_{T2} thresholds; the arrows show their change when P_J is reduced. Inset: total photon density ($S = S^+ + S^-$) for an unpolarized laser ($P_J=0$) and two spontaneous-emission coupling coefficients ($\beta = 0, 0.002$), as well as a spin-laser with $P_J=0.5$, $\beta=0$. (After Gøthgen, C. et al., *Appl. Phys. Lett.* 93, 042513, 2008. With permission.)

emission. In Regime II, we expect what can be called a *spin-filtering effect*, even for a modest polarization of the injected carriers (the two slightly unequal intensities of the faucets); there would be completely polarized light (of only one helicity, since the minority spin carriers do not lase). In Regime III, there is a kink in the emitted light showing the onset of lasing due to the minority carriers. From a closer look at Figure 16.7, we could infer a similar kink at the light intensity of ~0.1 mW, more pronounced at $T = 50$ and 75 K. This kink is completely absent in the absence of spin injection (dashed lines), further corroborating trends that can be inferred from the spin-polarized bucket model, which provides useful guidance for experiments on spin-lasers [11–13].

However, it is also important to elucidate some of the inherent limitations in using this spin-polarized bucket model. While we are not aware of any prior work on the spin-polarized bucket model, a similarly sophisticated description of the spin-laser was used by other researchers to infer what is intuitively expected. In particular, since one-half of the bucket holds only half as much of the water as the whole bucket, the limit of completely spin-polarized injection (only one faucet contributes) should give the maximum possible threshold reduction—half of the spin-unpolarized injection; recall Figure 16.8. Our studies [28] reveal that this maximum can actually be exceeded [29]. The reason for the discrepancy is also apparent from the consideration of the spin-polarized bucket model—complete symmetry between electrons and holes is assumed, extending also to their spin-polarized properties. This questionable approximation can be already traced back to an influential work on spin-effects in lasers [30]. Within this approximation we consider spin-up and spin-down carriers without distinguishing whether they are electrons or holes. Rather than attempting to formulate a more complex bucket model, we next develop the appropriate REs that will give us a systematic description of spin-lasers.

16.3.3 SPIN-POLARIZED RATE EQUATIONS

When generalizing REs (Equations 16.5 and 16.6), for spin-lasers it is crucial to consistently include the effect of holes. Typically, the spin relaxation time of holes is much shorter than for electrons [7], $\tau_s^p \ll \tau_s^n$, implying that the holes can be considered unpolarized with $p_\pm = p/2$. In previous work [11, 13], this effect was introduced only in the recombination term. In the following, we show its importance for the gain term by revealing that the electron densities $n_\pm$ are not separately clamped (pinned).

We generalize the gain (recall Equation 16.4) in the spin-polarized case as $g(n,S) \to g_\pm(n_\pm,p_\pm,S) = g_0(n_\pm + p_\pm - n_{\text{tran}})S^\mp/(1 + \varepsilon S)$. This form, even for $\varepsilon = 0$, differs from the previously employed gain expressions [11, 13], as it explicitly contains the hole density, but it coincides with the rigorously derived gain expression from semiconductor Bloch equations [20]. The gain compression term is an approximate description of, mainly, spectral hole burning and carrier heating [31]. For simplicity, we also use εS for spin-lasers.

The charge neutrality, $p_\pm = p/2 = n/2$, allows us to recover the spin-unpolarized limit for the gain expression as well as to decouple the REs for electrons from those for holes. The spin-polarized REs for electrons thus become

$$\frac{dn_\pm}{dt} = J_\pm - g_\pm(n_\pm,S)S^\mp - \frac{n_\pm - n_\mp}{\tau_s^n} - R_{sp}^\pm, \tag{16.7}$$

$$\frac{dS^\pm}{dt} = \Gamma g_\mp(n_\mp,S)S^\pm + \beta \Gamma R_{sp}^\mp - \frac{S^\pm}{\tau_{ph}}, \tag{16.8}$$

where the notation is extended from Equations 16.5 and 16.6 for conventional lasers. An extra term represents the (electron) spin relaxation $(n_\pm - n_\mp)/\tau_s^n$. Here, the recombination $R_{sp}^\pm$ generalizes the previous QR and LR forms [32] $R_{sp}^\pm = Bn_\pm p_\pm = Bn_\pm n/2$ and $R_{sp}^\pm = n_\pm/\tau_r$.

To describe the solutions of Equations 16.7 and 16.8, we introduce polarizations of injected electron current

$$P_J = \frac{J_+ - J_-}{J}, \tag{16.9}$$

as well as of electron and photon densities $P_n = (n_+ - n_-)/n$, $P_s = (S^+ - S^-)/S$. Electrical injection in intrinsic QWs, using Fe or FeCo, allows [7, 25, 33–35] $|P_n| \sim 0.3$–0.7 with similar values for $|P_J|$, while $|P_n| \to 1$ is attainable optically at room temperature [7].* In the unpolarized limit, $J_+ = J_-$, $n_+ = n_-$, $S^+ = S^-$, and $P_J = P_n = P_S \equiv 0$. As in recent experiments [13], we consider a continuous-wave

---

* We mostly consider $P_J \geq 0$, while the results for $P_J < 0$ can be deduced easily.

TABLE 16.1
Parameters Used to Model Spin-Lasers

Quantity		Value
$S^{\pm}$	Photon density	$\sim 10^{13}$ cm^{-3}
$(n, p)_{\pm}$	Carrier density	$\sim 10^{17}$ cm^{-3}
n_{tran}	Transparency density	4×10^{17} cm^{-3}
$J_{\pm}$	Carrier injection rate	$\sim 10^{27}$ cm^{-3} s^{-1}
P_J	Degree of injection polarization	0, 0.5, 0.9
g_0	Gain coefficient	$\sim 10^{-4}$ cm^3 s^{-1}
ε	Gain compression factor	$0-2 \times 10^{-18}$ cm^3
Γ	Optical confinement factor	0.029
β	Spontaneous emission factor	$0, 10^{-4}, 2 \times 10^{-3}$
τ_r	Recombination time	200 ps
τ_{ph}	Photon lifetime	1 ps
τ_s^n	Electron spin relaxation time	200 ps$- \infty$
$w = \tau_r / \tau_s^n$	Normalized spin relaxation rate	0, 1, 5

From: Lee, J. *et al.*, Appl. Phys. Lett., 97, 041116, 2010. With Permission.

operation and look for steady-state solutions of Equations 16.7 and 16.8. Guided by the experimental range [11, 13] of $\beta \sim 10^{-5}-10^{-3}$, we mostly focus on the limit $\beta = 0$, for which all the operating regimes of the spin-lasers can be simply described. Additionally, we consider $\varepsilon = 0$, relevant for moderate pumping intensities and parameters, from Holub et al. [13]. While most of our results are analytical and readily adapted for different laser materials and geometries, in Table 16.1 we summarize the characteristic parameter values we used, guided by the experiments [11–13].

16.3.4 Light-Injection Characteristics

We first show analytical results for an *unpolarized* laser and LR in the inset of Figure 16.10. Injection current density is normalized to the unpolarized threshold value, $J_T = N_T/\tau_r$, with N_T denoting the total electron density at (and above) the threshold, $N_T = n(J \geq J_T) = (\Gamma g_0 \tau_{ph})^{-1} + n_{\text{tran}}$, while photon density is normalized to $S_T = J_T \Gamma \tau_{ph}$. A small difference in $S(J)$ between the vanishing (solid line) and finite β (dotted line, $\beta = 0.002$ overestimates the experimental values [11, 13]) shows the accuracy of $\beta = 0$ approximation. As inferred from the bucket model, the unpolarized case has two regimes: LED-like, with negligible stimulated emission, for $J < J_T$ and a fully lasing regime for $J > J_T$.

With finite P_J and $\tau_s^n \to \infty$, as expected from the spin-polarized bucket model (recall Figures 16.8 and 16.9), we reveal a more complicated behavior shown in Figure 16.10, main panel. The two threshold currents, J_{T1} and J_{T2} ($J_{T1} \leq J_T \leq J_{T2}$; the equalities hold only when $P_J = 0$), delimit three regimes of a spin-laser. (i) For $J < J_{T1}$, it operates as a spin-LED [7, 33]. (ii) For $J_{T1} \leq J_T \leq J_{T2}$, there is mixed operation: lasing only with left-circularly polarized light S^- (we assume $J_+ > J_-$), which can be deduced from the spin-dependent gain in

Equations 16.7 and 16.8, while S^+ is still in a spin-LED regime ($S^+ \to 0$ for $\beta \to 0$). (iii) For $J \geq J_{T2}$, it is fully lasing with both $S^\pm > 0$.

RE description of spin-unpolarized lasers for $\varepsilon = 0$ reveals that the carrier densities are clamped above J_T; this effect corresponds to pinning of the water level in the bucket (Figure 16.4). We find a related effect in the steady-state spin-lasers when $\varepsilon = \beta = 0$. For $J > J_{T1}$, Equation 16.8 for S^- can be divided by S^-, showing that the quantity $n_+ + p_+ = n_+ + n/2$ (we recall the charge neutrality condition and $\tau_s^p \to 0$) is clamped at N_T. If $J_{T1} < J < J_{T2}$, then neither n_+ nor n_- is separately clamped. If $J > J_{T2}$, then $n_- + n/2 = N_T$ must hold in addition to the previous condition $n_+ + n/2 = N_T$. These two conditions yield $n_\pm = N_T/2$, independently of P_J. Thus, above J_{T2}, the sum of Equation 16.8 for S^- and S^+ reduces to the usual unpolarized equation, and S is independent of P_J (inset of Figure 16.10). However, if $P_J \neq 0$, then we still find $S^- \neq S^+$ (Figure 16.10, $J > J_{T2}$).

Most of the results discussed earlier do not change qualitatively for a finite τ_s^n or QR. The general features, such as the existence of three operating regimes, remain the same. Gain compression will change J_{T2}, but even for $\varepsilon = 2 \times 10^{-16}$ cm^3 (~4× larger than in Ref. [13]), the relative increase will be only ~1%. However, the photon and carrier densities (for a given J) as well as the threshold J_{T1} depend quantitatively on the spin-flip rate and the recombination form. We investigate this dependence in Figure 16.11, which shows the evolution of photon and carrier polarizations with the injection current. We consider both LR and QR forms and express our results using the ratio of the radiative recombination and spin relaxation times: $w = \tau_r/\tau_s^n$. For QR, the unpolarized threshold is $J_T = BN_T^2/2$, and we define $\tau_r = N_T/J_T$ by analogy with the LR case.

When $0 < J < J_{T1}$, then $P_n(J)$ depends on the recombination form; P_n is constant for LR, while for QR, it grows monotonically [6, 36]. Only for a very long spin relaxation time, $w \to 0$, $P_n(J < J_{T1}) \to P_J$. When $\beta = 0$, there is no amplified spontaneous emission, so that $P_S \equiv 0$. When $J_{T1} < J < J_{T2}$, $P_S = -1$ for any w; recall Figure 16.10. With only a partially polarized electron current, the spin-laser emits fully circularly polarized light, analogous to the spin-filtering effect in magnetic materials. In the same regime, P_n decreases with J, because while n_- grows, n_+ must drop to maintain $n_+ + n/2 = (3n_+ + n_-)/2 = N_T$.

For $P_J \neq 0$, we can define the (normalized) spin-filtering interval

$$d = \frac{J_{T2} - J_{T1}}{J_T}, \tag{16.10}$$

which is widest for $\tau_s^n \to \infty$, when we obtain

$$\text{LR:} \quad d = \frac{3|P_J|}{2 - |P_J| - |P_J|^2}, \tag{16.11}$$

$$\text{QR:} \quad d = \frac{|P_J|(8 + |P_J|)}{4 - 3|P_J|^2 - |P_J|^3}, \tag{16.12}$$

showing that d diverges for $P_J = 1$ and vanishes if the injection is spin-unpolarized. When τ_s^n decreases, the lower threshold J_{T1} grows toward its

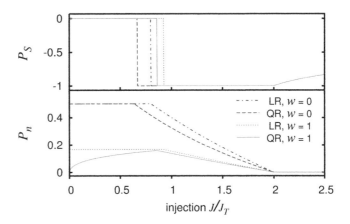

FIGURE 16.11   Electron (P_n) and photon (P_S) density polarizations as a function of injection current J, with polarization $P_J = 0.5$ and different ratios of recombination and electron spin relaxation time, w. LR and QR recombination are shown. J is normalized to the unpolarized threshold J_T. (After Gøthgen, C. et al., *Appl. Phys. Lett.* 93, 042513, 2008. With permission.)

unpolarized counterpart J_T and d contracts, as seen in Figure 16.11. We find that $J_{T2} = J_T/(1 - P_J)$, independent of τ_s^n. This should hold up to very small values of τ_s^n ($w \sim 10^2$). In the extreme case of even smaller τ_s^n, for $J_{T1} < J < J_{T2}$, the majority and minority spin densities become almost equal, so that $\beta \sim 10^{-4}$ can effectively drive the device to the fully lasing regime. For typical values [11, 13] of $w \sim 5$, setting $\beta = 0$ is an accurate assumption. At J_{T2}, we recall $n_+ = n_- (= N_T/2)$, so that the spin-relaxation term in Equation 16.7 for n_- is zero. The equation reduces to $J_- \equiv (1 - P_J)J/2 = R_{sp}^-$, explaining why J_{T2} depends only on P_J but not on w. For the same reason, P_S and P_n are independent of w when $J \geq J_{T2}$. The quantity P_S increases ($|P_S|$ decreases) with J as $P_S = -P_J J/(J - J_T)$ to the asymptotic value $P_S = -P_J$. Thus, P_J can be inferred from P_S only at sufficiently high injection. In all three regimes, defined by Figure 16.10, we find that $0 \leq P_n < P_J$ (only for $\tau_s^n \to \infty$ and $J < J_{T1}$, $P_n \to P_J$). For $J > J_{T2}$, we find $P_n = 0$. Therefore, P_n is not enhanced in the spin-laser's active region.

16.3.5 THRESHOLD REDUCTION

Experiments in Figures 16.6 and 16.7 have demonstrated that injecting spin-polarized carriers reduces the threshold current in a laser; that is, $J_{T1}(P_J \neq 0) < J_T$. We quantify this threshold-current reduction [11, 13] as

$$r = \frac{J_T - J_{T1}}{J_T}, \tag{16.13}$$

which can be obtained analytically from REs (Equations 16.7 and 16.8) to yield

$$\text{LR:} \quad r = \frac{|P_J|}{2 + |P_J| + 4w}, \tag{16.14}$$

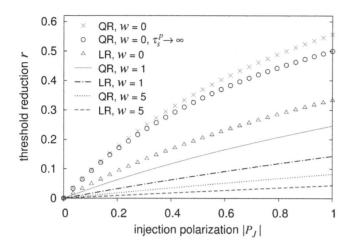

FIGURE 16.12   Threshold current reduction r as a function of injection polariza-tion P_J. Shown are both LR and QR for various w, the ratio of recombination time and electron spin relaxation time. The hole spin relaxation time $\tau_s^p = 0$, except for the circles, which show the previously assumed "ideal" case, in which $\tau_s^p \to \infty$. (After Gøthgen, C. et al., *Appl. Phys. Lett.* 93, 042513, 2008. With permission.)

$$
\text{QR:} \quad r = 1 - \frac{\left[1 - 2w + \sqrt{(1+2w)^2 + 4|P_J|w}\right]^2}{\left(2 + |P_J|\right)^2}. \tag{16.15}
$$

We find $r > 0$, for any $P_J \neq 0$, $\tau_s^n > 0$, and both LR and QR. As expected, $r \to 0$ for $\tau_s^n \to 0$, independently of the recombination form, since we have assumed $\tau_s^p = 0$. We illustrate $r(P_J)$ for both LR and QR in Figure 16.12 and choose $w = 1,5$, close to the typical values $\tau_r \sim 1$ ns and $\tau_s \sim 100$ ps for QWs at room temperature [11]. From Figure 16.12 or Equations 16.14 and 16.15, one can see that when $\tau_s^p = 0$, $r(P_J, w)$ for LR is always smaller than for QR. For $|P_J| = 1$ and $\tau_s^n \to \infty$ (but still $\tau_s^p = 0$), we obtain the maximum threshold reductions

$$
\text{LR:} \quad r = \frac{1}{3}, \quad \text{QR:} \quad r = \frac{5}{9}. \tag{16.16}
$$

Remarkably, for QR, the maximum r is larger than the previously assumed possible $r = 1/2$ [11, 13] (recall also Figure 16.2), even though holes are com-pletely unpolarized ($\tau_s^p = 0$). This can be explained as follows. First, we cal-culate n from Equation 16.8 for S^-. The threshold density is reached when $n_+ + n/2 = N_T$, which gives $n = (2/3)N_T$ for $P_J = 1$; that is, for $n_+ = n$. Thus, the threshold *electron density* is reduced by only 1/3. Assuming $S^- = 0$ at the threshold, the threshold current from Equation 16.7 is given by $J_+ = Bn_+n/2$. If $P_J = 1$, then $J = J_+ = Bn^2/2 = (2/3)^2 J_T$, which yields the 5/9 *current* reduction, a direct consequence of the recombination $\propto n^2$ (note that for LR, both the N_T and J_T are reduced only by 1/3).

For comparison, we calculate r for QR in the *a priori* "ideal" case, where $n_\pm = p_\pm$ for any injection J, and *both* $\tau_s^n, \tau_s^p \to \infty$. Therefore, the gain and QR terms in Equation 16.7 for n_+ (n_-) and in Equation 16.8 for S^- (S^+) must be modified by replacing $n/2$ with n_+ (n_-). Surprisingly, we obtain $r(|P_J| = 1) = 1/2$, a smaller reduction than in the case of unpolarized holes. A simple calculation yields $n = N_T/2$; that is, the reduction of the threshold *electron density* is larger than for $\tau_s^p = 0$. However, the interband recombination R_{sp}^+ is more efficient, because none of the holes undergo spin-flip. Thus, we obtain $J_+ = J_{T1} = B n_+^2 = B N_T^2/4 = J_T/2$. For $\tau_s^p = 0$ and $P_J = 1$, half of the holes are inaccessible for recombination with the fully spin-polarized electrons. This lowers the injection necessary to overcome the recombination losses, and the interplay of stimulated and spontaneous recombination leads to a smaller r in the "ideal" case.

16.4 SPIN-LASERS: DYNAMIC OPERATION

The most attractive properties of conventional lasers usually lie in their dynamical performance [3]. Here, we explore intriguing opportunities offered by spin-polarized modulation and study them with the REs' description. Each of the key quantities, X (such as, J, S, n, and P_J), can be decomposed into a steady-state X_0 and a modulated part $\delta X(t)$, $X = X_0 + \delta X(t)$. Important properties for the modulation response of lasers can be deduced from their analogy with the driven damped harmonic oscillator [37].

We focus on the amplitude and polarization modulation (AM, PM). AM for a steady-state polarization implies $J_+ \neq J_-$ (unless $P_J = 0$):

$$AM: \quad J = J_0 + \delta J \cos(\omega t), \quad P_J = P_{J0}, \tag{16.17}$$

where ω is the angular modulation frequency. AM is illustrated in Figure 16.10.

Similarly to the steady-state analysis in Section 16.3, $P_J \neq 0$ leads to unequal threshold currents J_{T1} and J_{T2} (for $S^\mp$, majority and minority photons). For the injection $J_{T1} < J_T < J_{T2}$, we expect a modulation of fully polarized light, even for a partially polarized injection. Such a modulation can be contrasted with PM, which also has $J_+ \neq J_-$, but J remains constant.*

$$PM: \quad J = J_0, \quad P_J = P_{J0} + \delta P_J \cos(\omega t). \tag{16.18}$$

Recent progress in electrically tunable P_J [38] suggests that fast PM could be realized in future spin-lasers. Currently, optically injected lasers with controllable circular polarization are more promising for implementing PM [11, 12, 39–41]. Slow PM using a Soleil–Babinet polarization retarder at a fixed $J (J_{T1} < J < J_{T2})$ leads to 400% modulation of laser emission [11]. Fast ($\omega/2\pi \sim 40$ GHz) PM can be implemented using, for example, a coherent electron spin precession in a transverse magnetic field for a pulsed laser emission with

---

* This is not essential but simplifies our analytical results.

alternating circular polarization [42] or a mode conversion in a ridge wave-guide electro-optic modulator [43]. As we show, the advantages of spin-lasers are also expected for $J \neq J_0$; that is, for *AM*. We present our results for LR, while the advantages for QR can be even larger, as can be inferred from Figure 16.12.

16.4.1 QUASISTATIC REGIME

In the quasistatic regime ($\omega \ll 1/\tau_{ph}, 1/\tau_r$), we can use the steady-state results from Section 16.3 to obtain *AM* (*PM*) with $J_0(P_{J0})$ substituted by $J(t)$ ($P_J(t)$), and the injection $J_\pm$ will be in phase with the response $n_\pm$ and $S^\mp$. For typical parameters in Table 16.1, we confirmed numerically that this regime is valid up to $\omega/2\pi \sim 10$ MHz. The steady-state results ($\varepsilon = \beta = 0$) for the two threshold currents (see Figures 16.10 and 16.13)

$$J_{T1} = \frac{J_T}{1 + |P_J|/2(1 + 2w)}, \quad J_{T2} = \frac{J_T}{1 - |P_J|}, \tag{16.19}$$

remain directly applicable for *AM* and *PM*. To reduce J_{T1} and improve dynamic performance, a small ratio of the recombination and spin relaxation times $w \sim 0.1$ ($\tau_s^n > 0.4$ ns, $\tau_r \sim 40$ ps at 300 K) can be realized in (110) GaAs-based QW spin-lasers [44–46], while $w \sim 1$ in (100) GaAs QW [7, 12, 45]. As in the steady state, J_{T1} and J_{T2} in Equation 16.19 delimit the three regimes of a spin-laser. For large *AM* and *PM*, these regimes determine $S^\pm(t)$, normalized to S_T, as shown in Figure 16.14. *AM* in the upper panel reveals $S^\pm$ near $\omega t = 0, 2\pi$, which corresponds to $J > J_{T2}$ (a fully lasing regime) and a constant $n_+ = n_-$. With time evolution, the laser enters the mixed regime $J_{T1} < J < J_{T2}$ (only lasing with S^-, Figure 16.13). If $P_J \geq 0$, for both *AM* and *PM*, the photon densities for $w = 0$ in Figure 16.14, as well as for an arbitrary w, can be expressed as

$$\frac{S^-}{S_T} = \frac{2 + 4w}{3 + 4w}\left[\frac{J}{J_T}\left(1 + \frac{P_J}{2 + 4w}\right) - 1\right], \quad \frac{S^+}{S_T} = 0, \tag{16.20}$$

where J and P_J are given by either Equation 16.17 or 16.18. For $P_J \leq 0$, the expressions for S^+ and S^- in Equation 16.20 are interchanged. Finally, near $\omega t = \pi, 3\pi$, no emitted $S^\pm$ implies $J < J_{T1}$ (the spin-LED regime) for *AM*. In the *PM* case (right panels), the injection is chosen such that $J > J_{T1}$ (S^- is present). The fully lasing regime, near $\omega t = \pi, 3\pi$, corresponds to constant $n^+ = n^-$, even as $P_J(t)$ varies.

16.4.2 SMALL SIGNAL ANALYSIS

The quasi-static approach allowed us to consider analytically large-signal modulation (for both *AM* and *PM*), usually only studied numerically. We next turn to the complementary approach for laser dynamics; that is, small signal analysis (SSA), limited to a small modulation ($|\delta J/J_0| \ll 1$ for

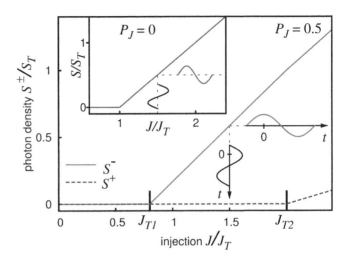

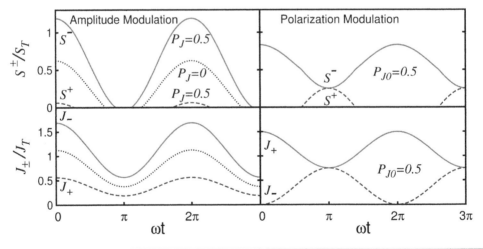

FIGURE 16.13 *AM.* Circularly polarized photon densities, $S^\pm$, as a function of injection J with polarization $P_J \equiv P_{J0} = 0.5$, for infinite electron spin relaxation time. Vertical and horizontal harmonic curves show the modulation of (input) current and the resulting modulation of the (output) light. *AM* of a partially polarized $J < J_{T2}$ leads to the modulation of the fully polarized output light (no S^+ component). Inset: For *AM* in a spin-unpolarized laser [3], S^+ and S^- undergo identical modulations. (After Lee, J. et al., *Appl. Phys. Lett.* 97, 041116, 2010. With permission.)

FIGURE 16.14 Time-dependence of the photon densities and injection in the quasi-static regime for large *AM* and *PM*. The results show coupling of $J_\pm$ with $S_\mp$ (Equations 16.7 and 16.8) given at $J_0 = 1.5 J_T$, $w = 0$ $(\tau_s^n \to \infty)$, and $\beta = \varepsilon = 0$. Left *AM*: $\delta J/J_0 = 0.5$ and $P_{J0} = 0.5$ (Equation 16.17), solid and broken lines. Reference results are given by dotted lines for $P_J = 0$: $S/2 = S^- = S^+$ (upper panel) and $J/2 = J_+ = J_-$ (lower panel). For $J > J_{T2}$, S^+ is in phase with S^-. Right *PM*: $\delta P_J = 0.5$ and $P_{J0} = 0.5$ (Equation 16.18). For $J > J_{T2}$, S^+ has an opposite phase to S^-. (After Lee, J. et al., *Appl. Phys. Lett.* 97, 041116, 2010. With permission.)

AM and $|\delta P_J| \ll 1$, $|P_{J0} \pm \delta P_J| \leq 1$ for *PM* [47])* but valid for all frequencies. We confirmed that the two approaches coincide in the common region of validity. Our SSA for spin-lasers proceeds as in conventional lasers [3, 48]. The decomposition $X = X_0 + \delta X(t)$ for $J_\pm$, $n_\pm$, and $S^\pm$ is substituted in Equations 16.7 and 16.8. The modulation terms can be written as $\delta X(t) = Re[\delta X(\omega) e^{-i\omega t}]$. We then analytically calculate $\delta n_\pm(\omega)$, $\delta S^\pm(\omega)$, and the appropriate generalized frequency response functions $R_\pm(\omega) = |\delta S^\mp(\omega)/\delta J_\pm(\omega)|$, which reduce to the conventional form [3] $R(\omega) = |\delta S(\omega)/\delta J(\omega)|$, in the $P_J = 0$ limit. From SSA, we obtain the resonance in the modulation response $R_\pm(\omega)$, also known as *relaxation oscillation frequency*, $\omega_R/2\pi$. For $P_{J0} = 0$ and $J > J_{T2}$, we find

$$\omega_R^{AM} \approx \sqrt{2}\omega_R^{PM} \approx \left[\Gamma g_0 \left(\frac{N_T}{\tau_r} \right) \left(\frac{J_0}{J_T} - 1 \right) \right]^{1/2}$$

$$= \left(\frac{g_0 S_0}{\tau_{ph}} \right)^{1/2},$$ (16.21)

where ω_R^{AM} recovers the standard result [3]. For $P_{J0} \neq 0$ and $J_{T1} < J < J_{T2}$

$$\omega_R^{AM} = \omega_R^{PM} \approx \left[\Gamma g_0 \left(\frac{N_T}{\tau_r} \right) \left[\left(1 + \frac{|P_{J0}|}{2} \right) \frac{J_0}{J_T} - 1 \right] \right]^{1/2}$$

$$= \left(\frac{3 g_0 S_0^-}{2 \tau_{ph}} \right)^{1/2}.$$ (16.22)

Such an ω_R corresponds to a peak in the frequency response and can be used to estimate its bandwidth [3], a frequency where the normalized response $R(\omega/2\pi)/R(0)$ decreases to −3 dB.

We can now look for possible advantages in the dynamic operation of spin-lasers as compared with their conventional counterparts. From Equation 16.22, we infer that the spin-polarized injection increases ω_R and thus, increases the laser bandwidth—an important figure of merit [3].

In Figure 16.15, these trends are visible in the normalized frequency response. Results for $P_{J0} = 0$ (using finite ε and β [12, 13]) show that our analytical approximations for $\varepsilon = \beta = 0$ are an accurate description of *AM* and *PM* at moderate pumping power. The increase of ω_R and the bandwidth with P_{J0}, for *AM* and *PM*, can be understood as the dynamic manifestation of threshold reduction with increasing P_{J0}. With $\omega_R \propto (S_0^-)^{1/2}$ (Equation 16.22), the situation is analogous to conventional lasers: ω_R and the bandwidth both increase with the square root of the output power [3, 48] ($S_0^+ = 0$ for $J_{T1} < J < J_{T2}$). An important advantage of spin-lasers is that the increase in S_0^- can be achieved even at *constant* input power (i.e., $J - J_T$) simply by increasing P_{J0}. In the limit of slow modulation, this behavior can already be seen

---

* These constraints could be relaxed with the generation of pure spin currents implying $P_J > 1$, demonstrated optically and electrically, as discussed in Žutić et al. [7].

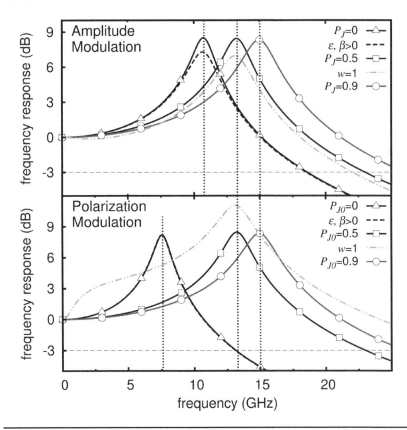

FIGURE 16.15 Small signal analysis of AM ($\delta J/J_0 = 0.01$) and PM ($\delta P_J = 0.01$), at $J_0 = 1.9 J_T$. The frequency response function $|\delta S^-(\omega) / \delta J_+(\omega)|$ is normalized to the corresponding $\omega = 0$ value. Results are shown in the limit of $w = 0$, and $\beta = \epsilon = 0$, except broken lines for $w = 1$, $\beta = 10^{-4}$, and $\epsilon = 2 \times 10^{-18}$ cm^3. The frequency response at -3 dB value gives the bandwidth of the laser [3]. The vertical lines denote approximate peak positions evaluated at $w = \epsilon = \beta = 0$ from Equation 16.22, except for PM at $P_{J0} = 0$ evaluated from Equation 16.21. When $P_{J0} = 0$ there is a small difference between $\epsilon = \beta = 0$ and finite ϵ, β results for AM, the difference is nearly invisible for PM. (After Lee, J. et al., *Appl. Phys. Lett.* 97, 041116, 2010. With permission.)

from Figure 16.6b. Additionally, a larger P_{J0} allows a larger J_0 (maintaining $J_{T1} < J_0 < J_{T2}$), which can further enhance the bandwidth, as seen in Equation 16.22. For $P_{J0} = 0.9$, J_0 can be up to $10 J_T$ from Equation 16.19.

We next examine the effects of finite τ_s^n, shown for $w = 1$ and $P_{J0} = 0.5$. AM results follow a plausible trend: ω_R and the bandwidth monotonically decrease and eventually attain "conventional" values for $\tau_s^n \to 0 (w \to \infty)$. The situation is rather different for PM: seemingly detrimental spin relaxation *enhances* the bandwidth and the peak in the frequency response, as compared to the long τ_s^n limit ($w = 0$). A shorter τ_s^n will reduce P_J and thus the amplitude of modulated light. Since $\delta S^-(0)$ decreases faster with w than $\delta S^-(\omega > 0)$, we find an increase in the normalized response function, shown in Figure 16.15. The increase in the bandwidth comes at the cost of a reduced modulation signal.

The above considered trends allow us to infer some other possible advantages of *PM* at fixed injection. In the quasi-static approximation, for

$J > J_{T2}$, constant $n_+ = n_- = N_T/2$ implies that PM would be feasible at a reduced chirp (α-factor), since $\delta n(t)$, which is a chirp source in AM, is eliminated [49]. These findings pertinent to the limit of low-frequency and $\epsilon = \beta = 0$, can be combined with SSA in Figure 16.15, revealing substantially smaller ϵ, β effects for PM than AM, to suggest that the reduced chirp and therefore desirable switching properties of spin-lasers can be expected for a broad range of parameters [49].

16.4.3 LARGE-SIGNAL ANALYSIS

We next turn to the large-signal analysis [50] for abrupt changes between *off* and *on* injection, J_{off}, J_{on}, important for high modulation frequency, short optical pulse generation, and high bit-rate in telecommunication [51, 52]. As the choices of gain regions increase, beyond the usually employed GaAs- or InAs-based QWs with negligible spin relaxation times for holes, τ_s^p, it is helpful to generalize our description of spin-lasers and explicitly consider spin polarization of holes in a separate RE analogous to Equation 16.7. $\tau_s^p \neq 0$ could be important in gain regions based on QDs, GaN [53, 54] or transition metal dichalcogenides [50, 55].

In conventional lasers a step increase in injection leads to a turn-on delay time, t_d, required for carrier density to build up to the threshold value before light is emitted, generally detrimental for the high-speed communication. Such delay can be obtained for $\beta = 0$ (see Equation 16.8) and $J_{off} < J_T < J_{on}$ by [52],

$$t_d = \tau_r \ln\left[\left(J_{on} - J_{off}\right) / \left(J_{on} - J_T\right)\right], \tag{16.23}$$

for the linear recombination. For the quadratic recombination [53]

$$t_d = \tau_r \sqrt{\frac{J_T}{J_{on}}} \left[\tanh^{-1}\sqrt{J_T / J_{on}} - \tanh^{-1}\sqrt{J_{off} / J_{on}}\right], \tag{16.24}$$

where $\tau_r = 1/(Bn_T)$. Another turn-on characteristic of the large-signal modulation is an overshoot of the emitted light and its damped ringing, often undesirable and implying distortion for a pulsed laser operation [51, 52].

Could these turn-on characteristics be improved in spin-lasers, and what is the corresponding effect of spin relaxation time? We first examine the maximum asymmetry between the $\tau_s^n > 0$ and $\tau_s^p = 0$, for optimal threshold reduction r (recall Figure 16.12). Our results in Figure 16.16 provide a direct comparison between large-signal modulation in conventional and spin-lasers having the same *on*-state photon density after the end of turn-on transients. From Equations 16.23 or 16.24 we can also infer trends for spin-lasers: an increase in τ_s^n leads to the threshold reduction and thus the monotonic decrease of t_d. These equations also apply for spin-lasers by replacing J_T with J_{T1}, as long as $J_{off} < J_{T1} < J_{on}$. For a conventional laser in Figure 16.16a, $t_d = 0.64$ ns from Equation 16.24 ($\beta = 0$), or $t_d = 0.66$ ns ($\beta = 1.8 \times 10^{-4}$)

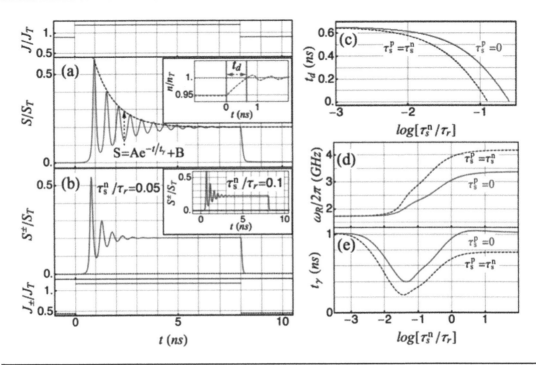

FIGURE 16.16 Large signal modulation. Time-dependence of *AM* of (a) injection, emitted light, and (inset) carrier density for a conventional laser, all normalized to $P_J=0$ threshold values. Delay and decay times, t_d and t_γ. (b) *PM* of injection, emitted light, for a spin-laser with $\tau_s^n/\tau_r = 0.05$, and (inset) emitted light for $\tau_s^n/\tau_r = 0.1$, in both cases $P_J^n = 1, \tau_s^p = 0$. (c) t_d, (d) ω_R, and (e) t_γ shown as a function of τ_s^n. (After Lee, J. et al., *Appl. Phys. Lett.* 105, 042411, 2014. With permission.)

obtained numerically. The delay for a spin-laser is reduced in Figure 16.16b to $t_d=0.53$ ns, even for a short spin relaxation time, $\tau_s^n/\tau_r = 0.05$, and to $t_d=0.36$ ns for $\tau_s^n/\tau_r = 0.1$ (inset). In Figure 16.16c we show $t_d(\tau_s^n)$, for both $\tau_s^p = 0\,(P_J^n = 1,\ P_J^p = 0)$ and $\tau_s^p = \tau_s^n\,(P_J^n = P_J^p = 1)$. For $\tau_s^n/\tau_r \ll 1$, t_d obtained from Figure 16.16a is recovered.

The step-function injection causes an abrupt increase and overshoot in the photon and carrier densities. Through damped oscillations, their densities relax to the steady-state values. From ringing patterns in Figures 16.16a and (b) we can estimate: (i) a modulation bandwidth from the corresponding relaxation oscillation frequency, $\omega_R \propto 1$/period of damped oscillations [52], and (ii) the decay time of the damped oscillations, t_γ. The ringing pattern confirms another advantage of spin-lasers: enhanced ω_R, as compared to the conventional counterparts (recovered for $\tau_s^n, \tau_s^p \to 0$). Such ω_R can be identified with the SSA relaxation oscillation frequency which determines the modulation bandwidth, enhanced in spin-lasers with the threshold reduction [15], see Section 16.4.2 for linear recombination. In Figure 16.16d we see a monotonic increase in ω_R with τ_s^n, exceeding 100% for $\tau_s^n = \tau_s^p$. Using SSA with $P_J=1$ and $\tau_s^n \to \infty$ when $J_{T1} < J_T < J_{T2}$, we find for quadratic recombination [50],

$$\omega_R^2 = \Gamma g_0 \left[\left(\frac{3}{2}\right)J - \left(\frac{2}{3}\right)J_T\right], \quad \omega_R^2 = \Gamma g_0 \left[2J - J_T\right], \qquad (16.25)$$

for $\tau_s^p = 0$ and $\tau_s^p \to \infty$, respectively. These results match well numerically extracted ω_R from a ringing pattern and confirm a large (>100%) bandwidth enhancement in spin-lasers, shown in Figure 16.16d. For a bandwidth enhancement, comparable τ_s^n and τ_s^p are desirable, unlike when maximizing r, where their asymmetry is required [28, 50].

The damping in the ringing pattern, previously only considered in conventional lasers where, near J_{Th} it is dominated by the recombination processes and photon decay. A parametrization of this damping by t_γ (Figure 16.16a) reveals a peculiar behavior. Let us focus on the $\tau_s^p = 0$ case. In a spin-laser, the damping is enhanced by the spin relaxation due to $n_+ \neq n_-$: a shorter τ_s^n would provide a stronger damping. However, spin-lasers reduce to conventional lasers for $\tau_s^n \to 0$; the injected spin polarization is immediately lost and the damping mechanism from spin relaxation is suppressed. Comparing then Figure 16.16a (viewed as the spin-laser with $\tau_s^n \to 0$) with Figure 16.16b and its inset confirms the *nonmonotonic* damping of ringing shown in Figure 16.16e.

How can we understand this nonmonotonic dependence of $t_\gamma(\tau_s^n)$? It is instructive to view the effective decay as an interplay of the spin-independent and spin-dependent part: $1/t_\gamma = 1/t_\gamma^0 + 1/t_\gamma^{sn}$. This can be viewed as Matthiessen's rule for different scattering mechanisms [50]. Similarly, it is customary to express [51] the carrier lifetime t_c as a combination of radiative (R) and nonradiative (NR) contributions: $1/t_c = 1/t_c^R + 1/t_c^{NR}$. In the limit of $\tau_s^n \to 0$, the spin contribution vanishes $1/t_\gamma^{sn} \to 0$ and $t_\gamma = t_\gamma^0 \approx 1\text{ns}$ (Figure 16.16e), as in the conventional lasers. For the maximum decay $\left[\log(\tau_s^n/\tau_r) \approx -1.5\right]$ the two contributions should be comparable: $1/t_\gamma^{\text{MIN}} = 2/t_\gamma^0$, that is, $t_\gamma^{\text{MIN}} \approx 0.5\text{ns}$, similar to the actual value of $t_\gamma^{\text{MIN}} \approx 0.4\text{ns}$, for $\tau_s^p = 0$. Turning now to the case of $\tau_s^n = \tau_s^p$, we infer an additional hole-spin contribution $1/t_\gamma^{sp}$, which leads to $t_\gamma^{\text{MIN}} = t_\gamma^0/3 \approx 0.3\text{ns}$. We see in Figure 16.16e that $t_\gamma^{\text{MIN}} = 0.2\text{ns}$ for $\tau_s^p = \tau_s^n$ is lower than the $\tau_s^p = 0$ case, as predicted above.

Our analysis of spin-lasers with the explicit inclusion of spin relaxation times for both electrons and holes reveals that optimizing the performance in spintronic devices is much more complex than simply requiring suppressed spin relaxation. Considering that the spin relaxation times for a given material could be readily changed by an applied magnetic field (and even electric field) [7], it is possible to test some of our predicted trends in already fabricated spin-lasers. For applications of spin-lasers relying on their threshold reduction, it could be desirable to have a large asymmetry between τ_s^n and τ_s^p. While such a situation is usually associated with bulk III-V semiconductors as the active regions, another interesting possibility is presented in MoS_2, where τ_s^n is actually much longer [55]. Advances in transition metal dichalcogenides (see Chapter 5, Volume 3) and their very large binding energy of excitons, could make them promising candidates for room temperature spin-lasers.

16.5 MICROSCOPIC DESCRIPTION OF SPIN-LASERS

To better understand the operation of spin-lasers and how their performance could exceed that of best conventional lasers it is important to complement their RE description with a microscopic picture. We focus on the gain region which usually includes III-V QWs or QDs. At the level of REs, effective mapping between QW- and QD-based lasers can be formulated [37, 56].

16.5.1 OPTICAL GAIN

The key effect of the gain region is to produce a stimulated emission and coherent light that makes the laser such a unique light source. The corresponding optical gain that describes stimulated emission, under sufficiently strong pumping/injection of carriers, can be illustrated in Figures 16.17a and b for both conventional and spin-lasers, respectively. Neglecting any losses in the resonant cavity, such a gain provides an exponential growth rate with the distance across a small segment of gain material [51]. Since both static and dynamic operations of spin-lasers crucially depend on their corresponding optical gain, we consider its microscopic description derived from an accurate electronic structure of an active region.

16.5.2 THEORETICAL FRAMEWORK

We focus here on the QW implementation also found in most of the commercial VCSELs [57]. To obtain an accurate electronic structure in the active region, needed to calculate optical gain, we use the 8×8 $k \cdot p$ method [58, 59]. Previous calculations were performed using 6×6 $k \cdot p$ method in which the electronic structure for conduction and valence band is assumed decoupled [60]. The total Hamiltonian of the QW system, with the growth axis along the z direction, can be written as,

$$H_{\mathrm{QW}}(z) = H_{\mathrm{kp}}(z) + H_{\mathrm{st}}(z) + H_{\mathrm{o}}(z), \tag{16.26}$$

where $H_{\mathrm{kp}}(z)$ denotes the $k \cdot p$ term, $H_{\mathrm{st}}(z)$ describes the strain term, and $H_{\mathrm{o}}(z)$ includes the band-offset at the interface that generates the QW energy profile.

Considering that common nonmagnetic semiconductors are well characterized by the vacuum permeability, μ_0, a complex dielectric function

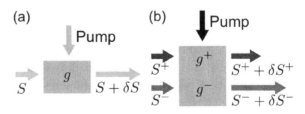

FIGURE 16.17 Schematic of the optical gain, g, for (a) conventional and (b) spin laser. With pumping/injection, a photon density S increases by δS as it passes across the gain region. In the spin laser this increase depends on the positive (+)/negative(−) helicity of the light. (After Faria Junior, P.E. et al., *Phys. Rev. B* 95, 115301, 2017. With permission.)

$\varepsilon(\omega) = \varepsilon_r(\omega) + \varepsilon_i(\omega)$, where ω is the photon (angular) frequency, can be used to simply express the dispersion and absorption of electromagnetic waves. The absorption coefficient describing gain or loss of the amplitude of an electromagnetic wave propagating in such a medium is the negative value of the gain coefficient (or gain spectrum) [3]. This gain coefficient corresponds to the ratio of the number of photons emitted per second per unit volume and the number of injected photons per second per unit area, therefore having a dimension of 1/length, it can be microscopically expressed as,

$$g^a(\omega) = -\frac{\omega}{cn_r}\epsilon_i^a(\omega), \tag{16.27}$$

where c is the speed of light, n_r is the dominant real part of the refractive index of the material [61] and $\varepsilon_i^a(\omega)$ is the imaginary part of the dielectric function which generally depends on the polarization of light, a, given by

$$\varepsilon_i^a(\omega) = C_0 \sum_{c,v,\vec{k}} \left| p_{cv\vec{k}}^a \right|^2 \left(f_{v\vec{k}} - f_{c\vec{k}} \right) \delta\left[\hbar\omega_{cv\vec{k}} - \hbar\omega \right], \tag{16.28}$$

where:

the indices c (not to be confused with the speed of light) and v label the conduction and valence subbands, respectively,

$\vec{k}$ is the wave vector,

$p_{cv\vec{k}}^a$ is the interband dipole transition amplitude which is directly calculated using $k \cdot p$ wave functions,

$f_{c(v)\vec{k}}$ is the Fermi-Dirac distribution for the electron occupancy in the conduction (valence) subbands,

$\hbar$ is the Planck's constant,

$\omega_{cv\vec{k}}$ is the interband transition frequency, and

δ is the Dirac delta-function, which is often replaced to include broadening effects for finite lifetimes [3, 20].

The constant C_0 is $C_0 = 4\pi^2 e^2 / (\varepsilon_0 m_0^2 \omega^2 \Omega)$, where e is the electron charge, m_0 is the free electron mass, and Ω is the QW volume.

Using the dipole selection rules for the spin-conserving interband transitions, the gain spectrum,

$$g^a(\omega) = g_+^a(\omega) + g_-^a(\omega), \tag{16.29}$$

can be expressed in terms of the contributions of spin-up and -down carriers. To obtain $g_{+(-)}^a(\omega)$, the summation of conduction and valence subbands is restricted to only one spin: $\Sigma_c \to \Sigma_{c+(-)}$ and $\Sigma_v \to \Sigma_{v+(-)}$ in Equation 16.28.

To see how spin-polarized carriers could influence the gain, we show chemical potentials, $\mu_{C(V)}$, for a simplified conduction (valence) band in Figure 16.18a. In order take into account the spin imbalance in the active region, we introduce $\mu_{C\pm}$,

$$n_\pm = \int_0^\infty dE\rho_{C\pm}(E) / \left[1 + \exp\left(E - \mu_{C\pm}\right) / k_B T \right], \tag{16.30}$$

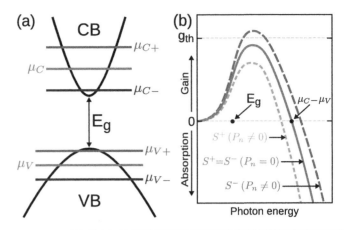

FIGURE 16.18   (a) Energy band diagram with a bandgap E_g and chemical potentials in conduction (valence) bands, μ_C (μ_V) that in the presence of spin-polarized carriers become spin-dependent: $\mu_{C\,(V)+} \neq \mu_{C\,(V)-}$, unlike the rest of our analysis, here holes are spin-polarized. (b) Gain spectrum for unpolarized (solid) and spin-polarized electrons (dashed curves). Positive gain corresponds to the emission and negative gain to the absorption of photons. The gain threshold g_{th}, required for lasing operation, is attained only for S^- helicity of light. (After Faria Junior, P.E. et al., *Phys. Rev. B* 92, 075311, 2015. With permission.)

where:

   E is the energy,

   $\rho_{C\pm}(E)$ is density of states in the conduction band for the respective spin,

   k_B is the Boltzmann constant, and

   T the absolute temperature.

Unlike the conventional chemical potential, $\mu_{C\pm}$ cannot be determined independently because each band does not contain pure spin states, but mixed states. However, when the spin mixing is not significant, we could assume that $\mu_{C\pm}$ are almost independent. Similarly, one can introduce $\mu_{V\pm}$, but we mostly consider that holes are not spin-polarized, $\mu_{V+} = \mu_{V-} = \mu_V$.

Such different chemical potentials lead to the dependence of gain on the polarization of light, described in Figure 16.18b. Without spin-polarized carriers, the gain is the same for S^+ and S^- helicity of light. In an ideal semiconductor laser, $g > 0$ requires a population inversion for photon energies, $E_g < \hbar\omega < (\mu_C - \mu_V)$, where E_g is the bandgap. However, a gain broadening is inherent to lasers and, as depicted in Figure 16.18b, $g > 0$ even below the bandgap, $\hbar\omega < E_g$. If we assume $P_n \neq 0$ and $P_p = 0$ (accurately satisfied, as spins of holes relax much faster than electrons), we see different gain curves for S^+ and S^-. The crossover from emission to absorption is now in the range of $(\mu_{C-} - \mu_{V-})$ and $(\mu_{C+} - \mu_{V+})$.

For our microscopic description of spin-lasers we focus on a (Al,Ga)As/GaAs-based active region, a choice similar to many commercial VCSELs. We consider an $Al_{0.3}Ga_{0.7}$ As barrier and a single 8 nm thick GaAs QW. The corresponding electronic structure of both band dispersions and the density

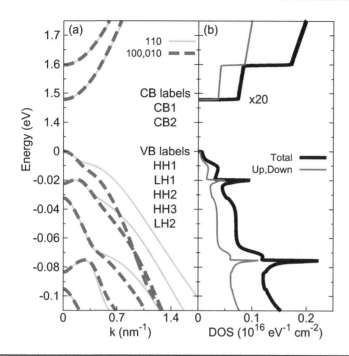

FIGURE 16.19   (a) Band structure for the $Al_{0.3}Ga_{0.7}As$/GaAs QW for different k-directions along [100], [010], and [110]. (b) DOS calculated from (a). The conduction band DOS is multiplied by a factor of 20 to match the valence band scale. The bandgap is $E_g = 1.479$ eV (CB1-HH1 energy difference). (After Faria Junior, P.E. et al., *Phys. Rev. B* 92, 075311, 2015. With permission.)

of states (DOS) is shown in Figure 16.19. Our calculations yield two confined CB subbands: CB1, CB2, and five VB subbands, labeled in Figure 16.19a by the dominant component of the total envelope function at $\vec{k} = 0$. The larger number of confined VB subbands stems from larger effective masses for holes than electrons. These differences in the effective masses also appear in the DOS shown in Figure 16.19b.

As we seek to describe the gain spectrum in the active region, once we have the electronic structure, it is important to understand the effects associated with carrier occupancies though injection/pumping. The carrier population [51] is given using the product of the Fermi-Dirac distribution and the DOS for CB and VB for both spin projections.

16.5.3 SPIN-DEPENDENT OPTICAL GAIN AND BIREFRINGENCE

From the conservation of angular momentum and polarization-dependent optical transitions we can understand that even in conventional lasers carrier spin plays a role in determining the gain. However, in the absence of spin-polarized carriers (without a magnetic region or an applied magnetic field), the gain is identical for the two helicities: $g^+ = g^-$, and we recover a simple description (spin- and polarization-independent) from Figure 16.17b. In our notation, $g_{\pm}^{\pm}$, the upper (lower) index refers to the circular polarization (carrier spin) (recall Equation 16.29).

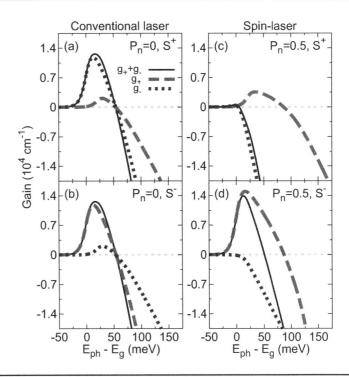

FIGURE 16.20 Gain spectra shown as a function of photon energy measured with respect to the energy bandgap. Conventional laser, $P_n = 0$ for (a) S^+ and (b) S^- light polarization. Spin-lasers, $P_n = 0.5$ for (c) S^+ and (d) S^- light polarizations. The carrier density $n = p = 3 \times 10^{12}$ cm^{-2} and $T = 300$ K. (After Faria Junior, P.E. et al., *Phys. Rev. B* 92, 075311, 2015. With permission.)

This behavior can be further understood from the gain spectrum in Figures 16.20a and b, where we recognize that $g^+ = g^-$ requires: (i) $g_+^+ = g_+^-$ and $g_+^+ = g_-^-$, dominated by CB1-HH1 (1.479 eV $= E_g$) and CB1-LH1 (1.501 eV) transitions, respectively (recall Figure 16.19). No spin imbalance implies spin-independent μ_C and μ_V (Figure 16.18a) and thus $g^\pm$, $g_+^\pm$ and $g_-^\pm$, all vanish at the photon energy $E_{ph} = \hbar\omega = \mu_C - \mu_V$. Throughout our calculations we choose a suitable cosh^{-1} broadening [20] with FWHM of 19.75 meV, which accurately recovers the gain of conventional (Al,Ga)As/GaAs QW systems.

We next turn to the gain spectrum of spin-lasers. Why is their output different for S^+ and S^- light, as depicted in Figure 16.17b? Changing only $P_n = 0.5$ from Figures 16.20a and b, we see very different results for S^+ and S^- light in Figures 16.20c and d. $P_n > 0$ implies that $\mu_{C+} > \mu_{C-}$, leading to a larger recombination between the spin-up carriers ($n_+ p_+ > n_- p_-$) and thus to a larger g_+ for S^+ and S^- (long dashed line) than g_- (short dashed line). The combined effect of having spin-polarized carriers and different polarization-dependent optical transitions for spin-up and -down recombination is then responsible for $g^+ \neq g^-$, given by solid lines in Figures 16.20c and d. For this case, the emitted light S^- exceeds that with the opposite helicity, S^+, there is a gain asymmetry (also known as the gain anisotropy) [13, 27, 40] another consequence of the polarization-dependent gain. The gain asymmetry is one of the key figures of merit for spin-lasers and can be viewed as crucial for their spin-selective properties, including robust spin-filtering or spin-ampli-fication, in which even a small P_n (few percent) in the active region leads to an almost complete polarization of the emitted light (of just one helicity) [62]. The zero gain is attained at $\mu_{C+}- \mu_V$ for spin-up (long dashed curves) and $\mu_{C-}-\mu_V$ for spin-down transitions (short dashed curves). The total gain, including both of these contributions, reaches zero at an intermediate value. Without any changes to the band structure, a simple reversal of the carrier spin-polarization, $P_n \rightarrow -P_n$, reverses the role of preferential light polarization.

An important implication of an anisotropic dielectric function is the phenomenon of birefringence in which the refractive index, and thus the phase velocity of light, depends on the polarization of light [51]. Due to phase anisotropies in the laser cavity and the polarization- dependence of the refractive index, the emitted frequencies of linearly polarized light in the x- and y-directions (S^x and S^y) are usually different. Such birefringence is often undesired in conventional lasers since it is the origin for the typical complex polarization dynamics and chaotic polarization switching behavior in VCSELs [30, 63–66]. While strong values of birefringence are usually con-sidered to be an obstacle for the polarization control in spin-polarized lasers [40, 67–69], the combination of a spin-induced gain asymmetry with bire-fringence in spin-VCSELs allows to generate fast and controllable oscillations between S^+ and S^- polarizations [70, 71]. The frequency of these polarization oscillations are determined by the linear birefringence in the VCSEL cavity and can be much higher than ω_R. This may open the path towards ultrahigh bandwidth operation for optical communications [50, 70, 72].

In order to investigate birefringence effects in the active region of a con-ventional laser, we consider uniaxial strain by extending the lattice constant in x-direction. For simplicity, we assume the barrier to have the same lattice constant as GaAs, 5.6533 Å, in y-direction. Therefore, both barrier and well regions will have the same extension in x-direction. For $a_x = 5.6544$ Å we have the corresponding element of the strain tensor $\varepsilon_{xx} \sim 0.019\%$, while $a_x = 5.6566$ Å gives $\varepsilon_{xx} \sim 0.058\%$. Besides the differences induced in the band structure, the uniaxial strain also induces a change in the dipole selection rules between S^x and S^y light polarizations, which can be seen in the gain spectra $g^x \neq g^y$.

To calculate the birefringence coefficient in the active region, we use the definition [73],

$$\gamma_p(\omega) = -\frac{\omega}{2n_e n_g} \delta\varepsilon_r(\omega), \qquad (16.31)$$

where:

   ω is the frequency of the longitudinal mode in the cavity,
   n_e the effective index of refraction of the cavity, and
   n_g the group refractive index.

For simplicity, we assume $n_e = n_g$. The real part of the dielectric func-tion can be obtained from the imaginary part using the Kramers-Kronig relations [3]. We consider three resonant cavity positions: c_1, c_2, c_3 (vertical

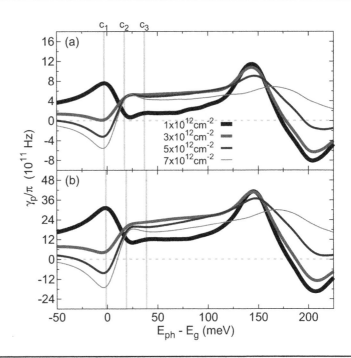

FIGURE 16.21 Birefringence coefficient as a function of photon energy considering (a) $\varepsilon_{xx} \sim 0.019\%$ and (b) $\varepsilon_{xx} \sim 0.058\%$. Just an increase of 0.0022 Å in a_x increases γ_p by approximately 3 times. The two peaks, around $E_{ph}-E_g \sim 0$ meV and $E_{ph}-E_g \sim$ 150 meV are related to transitions from CB1 and CB2. Transitions related to CB2 are in the absorption regime. (After Faria Junior, P.E. et al., *Phys. Rev. B* 92, 075311, 2015. With permission.)

lines in Figure 16.21), defining the corresponding energy of emitted photons $c_1 = 1.48$ eV–1.479 eV (CB1-HH1 transition), $c_2 = 1.5$ eV–1.501 eV (CB1-LH1 transition) and $c_3 = 1.52$ eV (at the high energy side of the gain spectrum).

We present the birefringence coefficient in Figure 16.21a and b for $\varepsilon_{xx} \sim 0.019\%$ and $\varepsilon_{xx} \sim 0.058\%$, respectively. This strain in the active region, responsible for modest changes in the gain spectra, produces birefringence values of the order of 10^{11-12} Hz which may be exploited to generate fast polarization oscillations. Increasing the strain amount by ~0.04% from case (a) to case (b), the value of γ_p increases by approximately 3 times. We also included in our calculations spin-polarized electrons and notice that they have only a small influence in the birefringence coefficient. Although they slightly change $|g^x|$ and $|g^y|$, the asymmetry is not affected at all for small P_n of 10%–20%, relevant values in real devices. Comparing different cavity designs we observe that for c_1, the value of γ_p strongly decreases and also changes sign with the carrier density, n. In contrast, for c_2 and c_3, γ_p is always positive. For large anisotropies in the DBR, the birefringence coefficient is on the order of 10^{10} Hz, consistent with the measurements given in Ref. [63]. Therefore, for the investigated strain conditions, the main contribution to γ_p comes from the active region and it is a very versatile parameter that can be fine-tuned using both carrier density and cavity designs, possibly even changing its sign and reaching carrier density-independent regions [59].

16.5.4 TOWARD ULTRAFAST SPIN-LASERS

The potential to exceed the operation frequency of conventional lasers has already been hinted with the first demonstration of spin-lasers [42, 74]. The modulation of optically injected (In,Ga)As-based QW VCSEL at 10 K was realized by the Larmor precession of the carrier spins in applied magnetic field, reaching 120 GHz at 11 T. Even though such a large magnetic field is impractical, it showed one of the paths to enhance the dynamical operation of lasers by employing spin-polarized carriers. The prospect of ultrafast spin-lasers is not limited just to their higher frequency or a larger dynamical bandwidth. They could also reduce the undesired frequency modulation (chirp), provide faster switching and improved eye diagrams with lower bit-error rates which describe the quality of a digital signal [49, 50, 75].

In addition to the faster dynamical operation that could be attributed to the reduced threshold of spin-lasers, it is useful to expand our discussion from Section 16.5.3 and explore how a birefringence-induced PM can overcome the usual AM frequency limitation in best conventional lasers of $\omega_R/2\pi < 40$ GHz [60]. The proof of concept can be seen in Figures 16.22a and b from unoptimized commercial lasers in which conventional electrical spin injection is supplemented with optical injection to create spin-polarized carriers at 300 K [70, 76]. With PM, $\delta P_J(t)$, there is a much faster response in P_S (circular polarization degree) than in the response of the emitted light intensity, S, after AM, $\delta J(t)$. Interpreting Figures 16.22a and b relies on a very influential spin-flip model (SFM) [30, 77] developed for dynamical operation and linear polarization switching in conventional VCSELs and later also used in spin-lasers [70, 76, 78]. Unlike the intensities $S^{\pm}$ used in Equations 16.7 and 16.8, the SFM contains helicity amplitudes of the light field $E^{\pm}$ [30, 77],

$$\frac{dE^{\pm}}{dt} = \left(\frac{1}{2\tau_{ph}}\right)(1 + i\alpha_0)(N \pm m_z - 1)E^{\pm} - (\gamma_a + i\gamma_p)E^{\mp}, \qquad (16.32)$$

where N and m_z are the normalized carrier and spin density, α_0 is the linewidth enhancement factor and τ_{ph} given in Equation 16.2. The evolution and the coupling of E^+, E^- depends also on the cavity anisotropies: in the refraction index-birefringence (linear), γ_p, and in the absorption of $E^{\pm}$–dichroism, γ_a [79]. The other SFM equations for N and m_z resemble Equations 16.7 and 16.8, containing photon intensities $|E^{\pm}|^2$. However, this SFM has serious limitations: $\tau_s^n = \tau_s^p$, no gain saturation, and no spectral information. Since $\tau_s^n \neq \tau_s^p$ already alters the steady-state response, the predictive power of the SFM is uncertain, but the SFM may provide some guidance for an observed dynamical operation in Figure 16.22 and complement our microscopic gain calculations predicting a large strain-induced birefringence [59], discussed in Section 16.5.3.

Due to the cavity anisotropy, VCSELs operate in one of two orthogonal linearly-polarized modes [30], with mode splitting γ_p: $\omega_{1,2} \approx \omega_0 \pm \gamma_p$. Recalling Equation 16.31, under uniaxial strain, $g^x \neq g^y$ and $S^x \neq S^y$ [59],

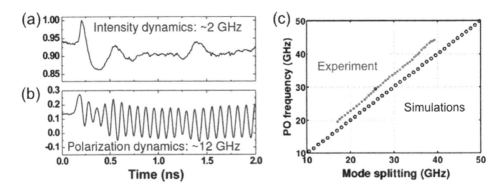

FIGURE 16.22 Oscillations of the emitted light for (a) intensity and (b) photon density polarization (P_S) under hybrid excitation: continuous electrical pumping and pulsed (3 ps) optical pumping with circularly polarized light. (After Gerhardt, N.C. et al., *Appl. Phys. Lett.* 99, 151107, 2011. With permission.) (c) A comparison between the birefringence-induced mode splitting and the oscillation frequency of the right circularly polarized laser mode. (After Lindemann, M. et al., *Appl. Phys. Lett.* 108, 042404, 2016. With permission.)

the resulting birefringence induces beating between S^+ and S^- leading to the P_S oscillations. This is confirmed experimentally by mechanically deforming VCSELs in Figure 16.22b: a higher deformation increases the frequency of P_S oscillations. Subsequent experiments for an even higher measured mode splitting from strain-induced birefringence demonstrate an unprecedented frequency range of >200 GHz and ultrafast dynamical operation for directly modulated spin-lasers [80]. Surprisingly the resulting nearly an order of magnitude increase in the bandwidth, as compared to the state-of-the-art conventional lasers, is also supporting an ultralow power consumption in high-birefringence lasers [80].

16.6 OUTLOOK AND CONCLUSIONS

The study of spin-lasers is still an emerging area of spintronics currently conducted by a modest number of research groups. However, impressive accomplishments have already been realized. This pertains not only to demonstrating performance of lasers potentially superior to their conventional counterparts, but also to elucidating novel concepts and operation principles in spintronics. For example, simultaneous spin polarization of electrons and holes, a coupling between spin, carrier, and photon dynamics, amplification, and strong nonlinear response, offer many unexplored opportunities. The modulation of spin polarization in lasers has motivated introducing the concept of spin interconnects [81, 82] while using the spin-polarized carriers can also enable coherent emission of sound [83].

For practical applications it would be important to go beyond the commonly employed optically-pumped lasers and realize their room-temperature electrical operation. Some of the related challenges can be seen from an early demonstration of electrical spin injection in a spin VCSEL [13], depicted in Figure 16.23a. With the distance of several μm between the spin injector (Fe) and the gain region [(In,Ga)As QW], through spin relaxation the spin polarization of injected carriers is substantially reduced when it reaches the gain

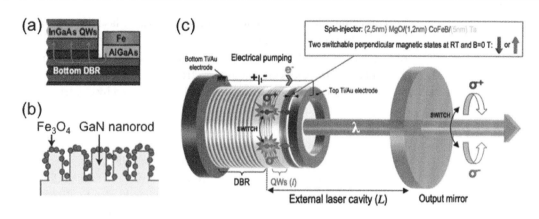

FIGURE 16.23 (a) Electrical spin injection in VCSEL from Fe spin injector to the gain region, InGaAs QWs. (After Holub, M. et al., *Phys. Rev. Lett.* 98, 146603, 2007. With permission.) (b) Integrating Fe_3O_4 nanomagnets with the gain region, GaN nanorods. (After Cheng, J.-Y. et al., *Nat. Nanotechnol.* 9, 845, 2014. With permission.) (c) A proposal for electrically injected spin-laser with an external cavity. With a perpendicular magnetization, in the absence of applied magnetic field it would be possible to switch between injected spin-up/down electrons and emitt light of different helicity. (After Frougier, J. et al., *Opt. Express* 23, 9573, 2015. With permission.)

region [84], limiting the electrical operation to −100 K [13]. Replacing the QW by QD gain region can enhance the operation temperature up to −230 K using MnAs injector [85]. A strong confinement in QDs can suppress the effect of spin–orbit coupling, a leading mechanism for the spin relaxation of carriers and thus enhance the spin relaxation time [7]. Experiments showing room temperature spin injection in MnAs-based spin-LEDs [86] are encouraging that MnAs could enable a room temperature injection in spin-lasers.

While almost all spin-lasers have been based on zinc-blende semiconductors, such as GaAs or InAs, a spin-laser with an gain region made of a wurtzite semiconductor (GaN-based) has so far been the only case of an electrically manipulated spin-laser at room temperature [53]. That operation was realized through spin-filtering by integrating nanomagnets with the active region of a VCSEL as illustrated in Figure 16.23b. The nanomagnets were too small to support ferromagnetism and a modest applied (0.35 T) was needed to magnetically align them. An alternative path to electrical operation of spin-lasers at room temperature is sought with vertical external cavity surface emitting lasers (VECSELs) [67], shown in Figure 16.23c. They could enable depositing a thin-film ferromagnet just 100–200 nm away from the active region, sufficiently close to attain a considerable spin polarization of carriers in the gain region at room temperature. With their external cavity, there is a flexibility to accurately design the gain anisotropy [87] and influence the polarization dynamics and operation of VECSELs which are known to have a desirable noise performance, beneficial to microwave photonic applications. In contrast to VCSELs [54, 76, 80], currently the effort in VECSELs was focused on reducing the birefringence [69, 87, 88]. However, modeling VECSELs and their sources of anisotropy, which directly influence polarization dynamics, could also give useful insights for the operation of other spin-lasers [87].

Following our suggestion for atomically-thin spin-lasers based on a gain region made of transition metal dichalogenides [50], we anticipate that their room-temperature electrical operation could be realized through magnetic proximity effects. A leaking magnetism from a nearby magnet would create proximity-induced spin-splitting that could be changed by electric gating or altering the direction of magnetization [89–93]. Moreover, magnetic proximity effects in transition metal dichalcogenides could be used to transform the optical selection rules and tightly-bound electron-hole pairs [93], providing an intriguing opportunity for tunable spin-lasers [89].

An independent progress in spintronics to store and sense information using magnets with a perpendicular anisotropy could also be directly beneficial for spin-lasers. Electrical spin injection usually relies on magnetic thin films with in-plane anisotropy requiring a large applied magnetic field to achieve an out-of-plane magnetization and the projection of injected spin compatible with the carrier recombination of circularly polarized light in a VCSEL geometry. However, a perpendicular anisotropy could provide an elegant spin injection in remanence [94–96], avoiding the technologically undesirable applied magnetic field. The progress in fast magnetization reversal [97, 98] (see also Chapter 10, Volume 1) and spin-LEDs showing electrical helicity switching [38, 99], could stimulate implementing all-electrical schemes for spin modulation in lasers that were shown to yield an enhanced bandwidth in lasers [15, 37, 50, 70, 71, 80, 100].

As can already be seen from the topics of spin injection and spin relaxation, the future studies of spin-lasers, and the prospect for their commercial viability, will increasingly depend on many developments in spin transport and magnetism, discussed throughout this book.

ACKNOWLEDGMENTS

This work has been supported by the NSF ECCS-1810266, NSF ECCS-1508873, NSF CAREER ECCS-054782, US ONR N000141310754, N000141712793, DOE-BES DE-SC0004890, DFG GE1231/2-2, and DFG MI607/9-2. We thank S. Bearden, G. Boeris, H. Dery, M. R. Hofmann, M. Lindemann, M. Ostereich, R. Michalzik, R. Oszwałdowski, A. Petrou, T. Pusch, and E. Wasner for valuable discussions.

REFERENCES

1. V. M. Ustinov, A. E. Zhukov, A. Yu. Egorov, and N. A. Maleev, *Quantum Dot Lasers*, Oxford University, New York, 2003.
2. A. M. Prochorov, Nobel Lecture, http://nobelprize.org/nobel_prizes/physics/laureates/1964/prokhorov-lecture.html (accessed on March 17, 2011).
3. S. L. Chuang, *Physics of Optoelectronic Devices*, 2nd edn., Wiley, New York, 2009.
4. Z. I. Alferov, Nobel lecture: The double heterostructure-concept and its applications in physics, electronics, and technology, *Rev. Mod. Phys.* **73**, 767 (2001).
5. G. Lampel, Nuclear dynamic polarization by optical electronic saturation and optical pumping in semiconductors, *Phys. Rev. Lett.* **20**, 491 (1967).

6. F. Meier and B. P. Zakharchenya (Eds.), *Optical Orientation*, North-Holland, New York, 1984.

7. I. Žutić, J. Fabian, and S. Das Sarma, Spintronics: Fundamentals and applications, *Rev. Mod. Phys.* **76**, 323 (2004).

8. J. Fabian, A. Mathos-Abiague, C. Ertler, P. Stano, and I. Žutić, Semiconductor spintronics, *Acta Phys. Slov.* **57**, 565 (2007).

9. M. Oestreich, J. Rudolph, R. Winkler, and D. Hägele, Design considerations for semiconductor spin lasers, *Superlattices Microstruct.* **37**, 306 (2005).

10. S. F. Yu, *Analysis and Design of Vertical Cavity Surface Emitting Lasers*, Wiley, New York, 2003.

11. J. Rudolph, D. Hägele, H. M. Gibbs, G. Khitrova, and M. Oestreich, Laser threshold reduction in a spintronic device, *Appl. Phys. Lett.* **82**, 4516 (2003).

12. J. Rudolph, S. Döhrmann, D. Hägele, M. Oestreich, and W. Stolz, Room-temperature threshold reduction in vertical-cavity surface-emitting lasers by injection of spin-polarized carriers, *Appl. Phys. Lett.* **87**, 241117 (2005).

13. M. Holub, J. Shin, and P. Bhattacharya, Electrical spin injection and threshold reduction in a semiconductor laser, *Phys. Rev. Lett.* **98**, 146603 (2007).

14. M. Holub and P. Bhattacharya, Spin-polarized light-emitting diodes and lasers, *J. Phys. D Appl. Phys.* **40**, R179 (2007).

15. J. Lee, W. Falls, R. Oszwałdowski, and I. Žutić, Spin modulation in semiconductor lasers, *Appl. Phys. Lett.* **97**, 041116 (2010).

16. M. A. Parker, *Physics of Optoelectronics*, CRC Press, New York, 2005.

17. C. Gøthgen, Steady-state analysis of semiconductor spin-lasers, PhD thesis, University at Buffalo, New York, 2010.

18. H. Dery and G. Eisenstein, Self-consistent rate equations of self-assembly quantum wire lasers, *IEEE J. Quantum Electron.* **40**, 1398 (2004).

19. H. Dery and G. Eisenstein, The impact of energy band diagram and inhomogeneous broadening on the optical differential gain in nanostructure lasers, *IEEE J. Quantum Electron.* **41**, 26 (2005).

20. W. W. Chow and S. W. Koch, *Semiconductor-Laser Fundamentals: Physics of the Gain Materials*, Springer, New York, 1999.

21. J. Hader, J. V. Moloney, and S. W. Koch, Beyond the ABC: Carrier recombination in semiconductor lasers, *Proc. SPIE* **6115**, 61151T (2006).

22. H.-H. Lee and W. Casperson, Gain and saturation in semiconductor lasers, *Opt. Quant. Electron.* **25**, 369 (1993).

23. S. S. P. Parkin, C. Kaiser, A. Panchula, P. M. Rice, B. Hughes, M. Samant, and S.-H. Yang, Room temperature tunneling magnetoresistance in magnetic tunnel junctions with MgO(100) tunnel barriers, *Nat. Mater.* **3**, 862 (2004).

24. S. Yuasa, T. Nagahama, A. Fukushira, Y. Suzuki, and K. Ando, Giant room-temperature magnetoresistance in single-crystal Fe/MgO/Fe magnetic tunnel junctions, *Nat. Mater.* **3**, 868 (2004).

25. G. Salis, R. Wang, X. Jiang, R. M. Shelby, S. S. P. Parkin, S. R. Bank, and J. S. Harris, Temperature independence of the spin-injection efficiency of a MgO-based tunnel spin injector, *Appl. Phys. Lett.* **87**, 262503 (2005).

26. D. Basu, D. Saha, C. C. Wu, M. Holub, Z. Mi, and P. Bhattacharya, Electrically injected InAs/GaAs quantum dot spin laser operating at 200 K, *Appl. Phys. Lett.* **102**, 091119 (2008).

27. D. Basu, D. Saha, and P. Bhattacharya, Optical polarization modulation and gain anisotropy in an electrically injected spin laser, *Phys. Rev. Lett.* **102**, 093904 (2009).

28. C. Gøthgen, R. Oszwałdowski, A. Petrou, and I. Žutić, Analytical model of spin-polarized semiconductor lasers, *Appl. Phys. Lett.* **93**, 042513 (2008).

29. I. Vurgaftman, M. Holub, B. T. Jonker, and J. R. Mayer, Estimating threshold reduction for spin-injected semiconductor lasers, *Appl. Phys. Lett.* **93**, 031102 (2008).

30. M. San Miguel, Q. Feng, and J. V. Moloney, Light-polarization dynamics in surface-emitting semiconductor lasers, *Phys. Rev. A* **52**, 1728 (1995).

31. G. P. Agrawal, Spectral hole-burning and gain saturation in semiconductor lasers: Strong-signal theory, *J. Appl. Phys.* **63**, 1232 (1988).
32. I. Žutić, J. Fabian, and S. C. Erwin, Spin injection and detection in silicon, *Phys. Rev. Lett.* **97**, 026602 (2006).
33. A. T. Hanbicki, B. T. Jonker, G. Itskos, G. Kioseoglou, and A. Petrou, Detection of spin-polarized electrons injected into two-dimensional electron gas, *Appl. Phys. Lett.* **81**, 2131 (2002).
34. A. T. Hanbicki, O. M. van't Erve, R. Magno, G. Kioseoglou, C. H. Li, B. T. Jonker, G. Itskos, R. Mallory, and A. Petrou, Analysis of the transport process providing spin injection through an Fe/AlGaAs Schottky barrier, *Appl. Phys. Lett.* **82**, 4092 (2003).
35. T. J. Zega, A. T. Hanbicki, S. C. Erwin, I. Žutić, G. Kioseoglou, C. H. Li, B. T. Jonker, and R. M. Stroud, Determination of interface atomic structure and its impact on spin transport using Z-contrast microscopy and density-functional theory, *Phys. Rev. Lett.* **96**, 196101 (2006).
36. I. Žutić, J. Fabian, and S. Das Sarma, Spin injection through the depletion layer: A theory of spin-polarized p-n junctions and solar cells, *Phys. Rev. B* **64**, 121201(R) (2001).
37. J. Lee, R. Oszwaldowski, C. Gothgen, and I. Žutić, Mapping between quantum dot and quantum well lasers: From conventional to spin lasers, *Phys. Rev. B* **85**, 045314 (2012).
38. N. Nishizawa, K. Nishibayashi, and H. Munekata, Pure circular polarization electroluminescence at room temperature with spin-polarized light-emitting diodes, *Proc. Nat. Acad. Sci.* **114**, 1783 (2017).
39. S. Hövel, N. C. Gerhardt, C. Brenner, M. R. Hofmann, F.-Y. Lo, D. Reuter, A. D. Wieck, E. Schuster, and W. Keune, Spin-controlled LEDs and VCSELs, *Phys. Status Solidi (a)* **204**, 500 (2007).
40. A. Bischoff, N. C. Gerhardt, M. R. Hofmann, T. Ackemann, A. Kroner, and R. Michalzik, Optical spin manipulation of electrically pumped vertical-cavity surface-emitting lasers, *Appl. Phys. Lett.* **92**, 041118 (2008).
41. H. Ando, T. Sogawa, and H. Gotoh, Photon-spin controlled lasing oscillations in surface-emitting lasers, *Appl. Phys. Lett.* **73**, 566 (1998).
42. S. Hallstein, J. D. Berger, M. Hilpert, H. C. Schneider, W. W. Rühle, F. Jahnke, S. W. Koch, H. M. Gibbs, G. Khitrova, and M. Oestreich, Manifestation of coherent spin precession in stimulated semiconductor emission dynamics, *Phys. Rev. B* **56**, R7076 (1997).
43. J. D. Bull, N. A. F. Jaeger, H. Kato, M. Fairburn, A. Reid, and P. Ghanipour, 40 GHz electro-optic polarization modulator for fiber optic communication systems, in *Photonic North 2004: Optical Components and Devices*, J. C. Armitage, S. Fafard, R. A. Lessard, and G. A. Lampropoulos (Eds.), *Proc. SPIE*, Bellingham, WA, **5577**, 133–143.
44. Y. Ohno, R. Terauchi, T. Adachi, F. Matsukara, and H. Ohno, Spin relaxation in GaAs (110) quantum wells, *Phys. Rev. Lett.* **83**, 4196 (1999).
45. H. Fujino, S. Koh, S. Iba, T. Fujimoto, and H. Kawaguchi, Circularly polarized lasing in a (110)-oriented quantum well vertical-cavity surface-emitting laser under optical spin injection, *Appl. Phys. Lett.* **94**, 131108 (2009).
46. K. Ikeda, T. Fujimoto, H. Fujino, T. Katayama, S. Koh, and H. Kawaguchi, Switching of lasing circular polarizations in (110)-VCSEL, *IEEE Photonic Tech. L.* **21**, 1350 (2009).
47. M. J. Stevens, A. L. Smirl, R. D. R. Bhat, A. Najmaie, J. E. Sipe, and H. M. van Driel, Quantum interference control of ballistic pure spin currents in semiconductors, *Phys. Rev. Lett.* **90**, 136603 (2003).
48. A. Yariv, *Optical Electronics in Modern Communications*, 5th edn., Oxford University, New York, 1997.
49. G. Boeris, J. Lee, K. Vyborny, and I. Žutić, Tailoring chirp in spin-lasers, *Appl. Phys. Lett.* **100**, 121111 (2012).

50. J. Lee, S. Bearden, E. Wasner, and I. Žutić, Spin-lasers: From threshold reduction to large-signal analysis, *Appl. Phys. Lett.* **105**, 042411 (2014).

51. L. A. Coldren, S. W. Corzine, and M. L. Mašović, *Diode Lasers and Photonic Integrated Circuits*, 2nd edn., Wiley, Hoboken, 2012.

52. K. Petermann, *Laser Diode Modulation and Noise*, Kluwer Academic, Dordrecht, 1988.

53. J.-Y. Cheng, T.-M. Wond, C.-W. Chang, C.-Y. Dong, and Y.-F Chen, Self-polarized spin-nanolasers, *Nat. Nanotechnol.* **9**, 845 (2014).

54. P. E. Faria Junior, G. Xu, Y.-F. Chen, G. M. Sipahi, and I. Žutić, Wurtzite spin lasers, *Phys. Rev. B* **95**, 115301 (2017).

55. K. F. Mak, K. He, J. Shan, and T. F. Heinz, Control of valley polarization in monolayer MoS$_2$ by optical helicity, *Nat. Nanotechnol.* **7**, 494 (2012).

56. R. Oszwałdowski, C. Gøthgen, and I. Žutić, Theory of quantum dot spin lasers, *Phys. Rev. B* **82**, 085316 (2010).

57. R. Michalzik, *VCSELs Fundamentals, Technology and Applications of Vertical-Cavity Surface-Emitting Lasers*, Springer, Berlin, 2013.

58. P. Yu and M. Cardona, *Fundamentals of Semiconductors*, 2nd edn., Springer, New York, 2010.

59. P. E. Faria Junior, G. Xu, J. Lee, N. C. Gerhardt, G. M. Sipahi, and I. Žutić, Towards high-frequency operation of spin-lasers, *Phys. Rev. B* **92**, 075311 (2015).

60. M. Holub and B. T. Jonker, Threshold current reduction in spin-polarized lasers: Role of strain and valence-band mixing, *Phys. Rev. B* **83**, 125309 (2011).

61. H. Haug and S. W. Koch, *Quantum Theory of Optical and Electronic Properties of Semiconductors*, 4th edn., World Scientific Publishing, Singapore, 2004.

62. S. Iba, S. Koh, K. Ikeda, and H. Kawaguchi, Room temperature circularly polarized lasing in an optically spin injected vertical-cavity surface-emitting laser with (110) GaAs quantum wells, *Appl. Phys. Lett.* **98**, 081113 (2011).

63. M. P. van Exter, A. K. Jansen van Doorn, and J. P. Woerdman, Electro-optic effect and birefringence in semiconductor vertical-cavity lasers, *Phys. Rev. A* **56**, 845 (1997).

64. M. Sondermann, M. Weinkath, and T. Ackemann, Polarization switching to the gain disfavored mode in vertical-cavity surface-emitting lasers, *IEEE J. Quantum Electron.* **40**, 97 (2004).

65. M. Virte, K. Panajotov, H. Thienpont, and M. Sciamanna, Deterministic polarization chaos from a laser diode, *Nat. Photonics* **7**, 60 (2012).

66. R. Al-Seyab, D. Alexandropoulos, I. D. Henning, and M. J. Adams, Instabilities in spin-polarized vertical-cavity surface-emitting lasers, *IEEE Photon. J.* **3**, 799 (2011).

67. J. Frougier, G. Baili, I. Sagnes, D. Dolfi, J.-M. George, and M. Alouini, Accurate measurement of the residual birefringence in VECSEL: Towards understanding of the polarization behavior under spin-polarized pumping, *Opt. Expr.* **23**, 9573 (2015).

68. N. Yokota, R. Takeuchi, H. Yasaka, and K. Ikeda, Lasing polarization characteristics in 1.55-μm spin-injected VCSELs, *IEEE Photon. Tech. Lett.* **29**, 711 (2017).

69. T. Fordos, H. Jaffres, K. Postava et al., Eigenmodes of spin vertical-cavity surface-emitting lasers with local linear birefringence and gain dichroism, *Phys. Rev. A* **96**, 043828 (2017).

70. N. C. Gerhardt, M. Y. Li, H. Jahme, H. Höpfner, T. Ackemann, and M. R. Hofmann, Ultrafast spin-induced polarization oscillations with tunable lifetime in vertical-cavity surface-emitting lasers, *Appl. Phys. Lett.* **99**, 151107 (2011).

71. H. Höpfner, M. Lindemann, N. C. Gerhardt, and M. R. Hofmann, Controlled switching of ultrafast circular polarization oscillations in spin-polarized vertical-cavity surface-emitting lasers, *Appl. Phys. Lett.* **104**, 022409 (2014).

72. N. C. Gerhardt and M. R. Hofmann, Spin-controlled vertical-cavity surface-emitting lasers, *Adv. Opt. Techn.* **2012**, 268949 (2012).

73. J. Mulet and S. Balle, Spatio-temporal modeling of the optical properties of VCSELs in the presence of polarization effects, *IEEE J. Quantum Electron.* **38**, 291 (2002).

74. M. Oestreich, J. Hübner, D. Hägele, M. Bender, N. Gerhardt, M. Hofmann, W. W. Rühle, H. Kalt, T. Hartmann, P. Klar, W. Heimbrodt, and W. Stolz, Spintronics: Spin electronics and optoelectronics in semiconductors, in *Advances in Solid State Physics*, B. Kramer (Ed.), Vol. 41, Springer, Berlin, pp. 173–186, 2001.

75. E. Wasner, S. Bearden, J. Lee, and I. Žutić, Digital operation and eye diagrams in spin-lasers, *Appl. Phys. Lett.* **107**, 082406 (2015).

76. M. Lindemann, T. Pusch, R. Michalzik, N. C Gerhardt, and M. R. Hofmann, Frequency tuning of polarization oscillations: Toward high-speed spin-lasers, *Appl. Phys. Lett.* **108**, 042404 (2016).

77. J. Martin-Regalado, F. Prati, M. San Miguel, and N. B. Abraham, Polarization properties of vertical-cavity surface-emitting lasers, *IEEE J. Quantum Electron.* **33**, 765 (1997).

78. S. S. Alharthi, A. Hurtado, V.-M. Korpijarvi, M. Guina, I. D. Henning, and M. J. Adams, Circular polarization switching and bistability in an optically injected 1300 nm spin-vertical cavity surface emitting laser, *Appl. Phys. Lett.* **106**, 021117 (2015).

79. Y. B. Band, *Light and Matter: Electromagnetism, Optics, Spectroscopy, and Lasers*, Wiley, New York, 2007.

80. M. Lindemann, G. Xu, T. Pusch, R. Michalzik, M. R. Hofmann, I. Žutić, and N. C. Gerhardt, Ultrafast spin-lasers, *Nature* **568**, 212 (2019).

81. H. Dery, Y. Song, P. Li, and I. Žutić, Silicon spin communication, *Appl. Phys. Lett.* **99**, 082502 (2011).

82. I. Žutić and H. Dery, Taming spin currents, *Nat. Mater.* **10**, 647 (2011).

83. A. Khaetskii, V. N. Golovach, X. Hu, and I. Žutić, Proposal for a phonon laser utilizing quantum-dot spin states, *Phys. Rev. Lett.* **111**, 186601 (2013).

84. H. Soldat, M. Y. Li, N. C. Gerhardt, M. R. Hofmann, A. Ludwig, A. Ebbing, D. Reuter, A. D. Wieck, F. Stromberg, W. Keune, and H. Wende, Room temperature spin relaxation length in spin light-emitting diodes, *Appl. Phys. Lett.* **99**, 051102 (2011).

85. D. Saha, D. Basu, and P. Bhattacharya, High-frequency dynamics of spin-polarized carriers and photons in a laser, *Phys. Rev. B* **82**, 205309 (2010).

86. E. D. Fraser, S. Hegde, L. Schweidenback, A. H. Russ, A. Petrou, H. Luo, and G. Kioseoglou, Efficient electron spin injection in MnAs-based spin-light-emitting-diodes up to room temperature, *Appl. Phys. Lett.* **97**, 041103 (2010).

87. M. Alouini, J. Frougier, A. Joly, G. Baili, D. Dolfi, and J.-M. George, VSPIN: A new model relying on the vectorial description of the laser field for predicting the polarization dynamics of spin-injected V(e)CSELs, *Opt. Express* **26**, 6739 (2018).

88. A. Joly, G. Baili, M. Alouini, J.-M. George, I. Sagnes, G. Pillet, and D. Dolfi, Compensation of the residual linear anisotropy of phase in vertical-external-cavity-surface-emitting laser for spin injection, *Opt. Lett.* **42**, 651 (2017).

89. I. Žutić, A. Matos-Abiague, B. Scharf, H. Dery, and K. Belashchenko, Proximitized materials, *Mater. Today*, **22**, 85 (2019).

90. P. Lazić, K. D. Belashchenko, and I. Žutić, Effective gating and tunable magnetic proximity effects in two-dimensional heterostructures, *Phys. Rev. B* **93**, 241401(R) (2016).

91. J. Xu, S. Singh, J. Katoch, G. Wu, T. Zhu, I. Žutić, and R. K. Kawakami, Spin inversion in graphene spin valves by gate-tunable magnetic proximity effect at one-dimensional contacts, *Nat. Commun.* **9**, 2869 (2018).

92. C. Zhao, T. Norden, P. Zhao, Y. Cheng, P. Zhang, F. Sun, P. Taheri, J. Wang, Y. Yang, T. Scrace, K. Kang, S. Yang, G. Miao, R. Sabirianov, G. Kioseoglou, A. Petrou, and H. Zeng, Enhanced valley splitting in monolayer WSe$_2$ due to magnetic exchange field, *Nat. Nanotechnol.* **12**, 757 (2017).

93. B. Scharf, G. Xu, A. Matos-Abiague, and I. Žutić, Magnetic proximity effects in transition-metal dichalcogenides: Converting excitons, *Phys. Rev. Lett.* **119**, 127403 (2017).

94. A. Sinsarp, T. Manago, F. Takano, and H. Akinaga, Electrical spin injection from out-of-plane magnetized FePt/MgO tunneling junction into GaAs at room temperature, *Jpn. J. Appl. Phys.* **46**, L4 (2007).

95. S. Hövel, N. C. Gerhardt, M. R. Hofmann, F.-Y. Lo, A. Ludwig, D. Reuter, A. D. Wieck, E. Schuster, H. Wende, W. Keune, O. Petracic, and K. Westerholt, Room temperature electrical spin injection in remanence, *Appl. Phys. Lett.* **93**, 021117 (2008).

96. J. Zarpellon, H. Jaffrès, J. Frougier, C. Deranlot, J. M. George, D. H. Mosca, A. Lemaitre, F. Freimuth, Q. H. Duong, P. Renucci, and X. Marie, Spin injection at remanence into III-V spin light-emitting diodes using (Co/Pt) ferromagnetic injectors, *Phys. Rev. B* **86**, 205314 (2012).

97. S. Garzon, L. Ye, R. A. Webb, T. M. Crawford, M. Covington, and S. Kaka, Coherent control of nanomagnet via ultrafast spin torque pulses, *Phys. Rev. B* **78**, 180401(R) (2008).

98. A. V. Kimel, A. Kirilyuk, P. A. Usachev, R. V. Pisarev, A. M. Balbashov, and Th. Rasing, Ultrafast non-thermal control of magnetization by instantaneous photomagnetic pulses, *Nature* **435**, 655 (2005).

99. N. Nishizawa, K. Nishibayashi, and H. Munekata, A spin light emitting diode incorporating ability of electrical helicity switching, *Appl. Phys. Lett.* **104**, 111102 (2014).

100. D. Banerjee, R. Adari, M. Murthy, P. Suggisetti, S. Ganguly, and D. Saha, Modulation bandwidth of a spin laser, *J. Appl. Phys.* **109**, 07C317 (2011).

17

Spin Logic Devices

Hanan Dery

17.1 INTRODUCTION

Designs using thin metallic layers, based on either giant magnetoresistive effect [1, 2] or tunneling magnetoresistance [3], have already found application in hard drive read heads, in magnetic random access memories, and in magnetic field sensors (see Chapter 5, Volume 1, and Chapter 14, Volume 3). These designs employ a magnetoresistive spin-valve effect: The current passing through them depends on the magnetization configuration of two terminals. Over the last two decades, the advances in magnetic storage have enabled a 1000-fold increase in the capacity of computer hard drives using metal-based spin valves. Despite this remarkable success, there are only a few seminal attempts to propose logic gates in all-metallic magnetic systems. Cowburn and Welland [4] have implemented a room temperature magnetic quantum cellular automata network of submicrometer magnetic dots interacting via magnetostatic interactions. To trigger a logic operation, an applied oscillating magnetic field generates a magnetic soliton that carries information through the network. The logic output is then encoded in the resulting magnetic configuration. Similar proposals have used either domain walls to propagate information or shape anisotropy [5, 6]. A different all-metallic magneto-logic paradigm relies on magnetic tunneling junctions as building blocks [7–9]. The output of a logic operation is decoded from the current amplitude flowing through a combination of these junctions. Similarly to the case of magnetic random access memories, the logic operands are encoded using multiple bit (current) lines. This in turn changes the resistance of the junction, and the added effect from various junctions denotes an output of a logic operation.

The main advantage of magneto-logic designs lies in their scaling potential. For comparison, using the 40 nm technology node in a complementary metal oxide semiconductor, the lateral size of a typical two input NAND gate with a moderate fan-out capability is about 0.5 μm^2. In this view, magneto-logic gate implementations can provide an improvement of about two orders of magnitude [4, 10]. Another important benefit of prospective magneto-logic gates is their intrinsic run time reprogrammability; the logic functionality of a magneto-logic gate can be changed at the switching speed of the operands at its input [8, 10]. Conventional implementations, on the other hand, rely on field programmable gate arrays at which the circuit is programmed via a flash memory or an electrically erasable programmable read-only memory. The speed gap between the logic circuitry and these memories is such that these applications are practically reprogrammed only during bootup time. Moreover, the limited writing endurance of these memories questions their feasibility in runtime reprogrammable applications. Currently, the main drawback of proposed magneto-logic schemes is that the output of a logic operation cannot be carried into the input of another logic operation fast enough without having to go through a conversion step. We are still lacking efficient propagation or conversion schemes to cascade logic operations.

The research in the emerging field of semiconductor spintronics has brought theoretical proposals of various devices exploiting the spin degree of freedom [11–20]. Remarkable progress has been made in understanding the basic physics of semiconductor spintronics on account of the advances in diluted ferromagnetic semiconductors [21–23], in discoveries of new spin phenomena [24, 25], and in spin injection into semiconductors from ferromagnetic metals [22, 26–37]. Unlike metals, semiconductors have a relatively low carrier density that can be drastically changed either by sample preparation using inhomogeneous doping or by applied stimuli using gate voltage or light-irradiation. This enables their use in digital logic for information processing. Yet, only a few viable approaches have been proposed to use semiconductor spintronics [38–40], hampering the unification with traditional electronics (e.g., the integration of magnetic memory into a semiconductor chip). In this chapter, we review paths to break this impasse by using spin accumulation in nonlocal geometries [41–43], discussed also in Chapter 5, Volume 1, and Chapters 4, 5, 7–9, Volume 3, as the basis for spin-based logic operations in semiconductors [10, 44].

In the following section, we briefly review spin-dependent transport through a Schottky barrier and in a bulk semiconductor channel (see Chapters 2, 3, and 6, Volume 3, for experimental efforts). Then, we analyze the steady-state spin accumulation profile in two-terminal devices and in lateral multi-terminal devices. We then incorporate a time-dependent analysis in which a technique to generate digital signals is introduced. In addition, the working principles of semiconductor magneto-logic reprogrammable gates are presented, and we use these devices to illustrate potential applications. In the final section, we also study semiconductor devices with magnetic contacts, whose magnetization configuration is non-collinear. A complementary path to spin logic, employing spin waves, is discussed in Chapter 19, Volume 3.

17.2 SPIN TRANSPORT IN SEMICONDUCTORS

Theoretical analysis of spin injection from metals into semiconductors shows that junctions with large resistance are necessary for the current to be polarized [45–47]. More precisely, since the spin-depth conductance of the semiconductor $G_{sc} = \sigma/L$ (with conductivity σ and spin diffusion length L) is much smaller than its metal counterpart $G_m = \sigma_m/L_m$, for spin injection to occur, the junction conductance G has to fulfill $G \leq G_{sc}$. On the other hand, with an ohmic contact ($G \gg G_{sc}$), the spin accumulation in the semiconductor channel vanishes, since the ferromagnetic contact acts as a powerful spin sink. This "conductivity mismatch" effect was first analyzed by Johnson and Silsbee [48] and discussed in Chapter 5, Volume 1. The spin injection constraint, $G \leq G_{sc}$, is easily achieved by an insulator barrier or by the naturally formed Schottky barrier at the interface between a metal and a semiconductor [49, 50]. The electrochemical potential is discontinuous at the junction in the case of a thin tunneling barrier. This discontinuity alleviates the conductivity mismatch between the semiconductor and the ferromagnetic metal. The spin polarization of the current across the junction is then determined by the spin selectivity of the barrier, $\Delta G = G_+ - G_-$, where G_s

are the conductances for spin $s = \pm$ with the spin quantization axis directing along the magnetization axis of the ferromagnet. Other than non-negligible spin selectivity, an efficient spin injection occurs when $G \sim G_{sc}$, for which the ratio between the spin accumulation density and the total electron density reaches its maximum. On the other hand, an over-resistive barrier such that $G \ll G_{sc}$ results in a small spin accumulation [45, 51, 52]. In this view, a Schottky barrier is more suitable than a thin insulator spacer due to its reduced barrier height.

A Schottky barrier between a metal and a semiconductor is created by redistribution of charges in the space charge layer [50]. The barrier width is denoted by d, and the barrier height measured from the Fermi level of the metal is denoted by ϕ_B. For a uniformly n-doped semiconductor, the barrier shape is approximately parabolic:

$$\phi(x) = \frac{e^2 n_{sb} x^2}{2\varepsilon_{sc}}, \tag{17.1}$$

where:

     n_{sb} is the doping concentration in the Schottky barrier region

     e is the elementary charge

     ε_{sc} is the static permittivity of the semiconductor (typically about 10^{-12} F/cm)

The barrier is defined by the depleted region between the interface ($\phi(d) = \phi_B$) and the edge of the semiconductor flat-band region ($\phi(0) = 0$). In a typical metal/semiconductor interface where $\phi_B = 0.7$ eV and $n_{sb} \sim 10^{17}$ cm^{-3}, the barrier width is $d \sim 100$ nm. At lower concentrations, the tunneling processes are completely irrelevant, and the current is due to purely classical thermionic emission, which depends solely on the barrier height, not on its width or shape. Even if this current is spin polarized, its total density is too small to create an appreciable spin accumulation. For greater concentrations, the tunneling dominates the transport through the junction, where one usually achieves an appreciable current density already at low bias conditions when thin barriers with $d \leq 10$ nm are employed ($n_{sb} \geq 10^{19}$ cm^{-3}). To achieve such thin barriers yet with the bulk of the semiconductor having a much lower carrier density, a strongly inhomogeneous doping profile has to be used near the interface [31, 53]. The need for a much lower carrier density in the bulk region, $n_0 \ll n_{sb}$, stems from the previously mentioned condition of optimal spin accumulation density when $G \sim \sigma/L$. Since the barrier conductance, G, is rather low even when $d \sim 10$ nm, this "spin-impedance" matching is met in bulk regions which are moderately doped and have a relatively long spin diffusion length. This can be achieved when $n_0 < 10^{17}$ cm^{-3}, at which the bulk semiconductor conductivity is relatively small, and the spin diffusion length is relatively large. The latter is valid, since the magnitudes of both the Elliot–Yafet and Dyakonov–Perel mechanisms, which dominate the spin relaxation in n-type semiconductors, are reduced when the doping levels are non-degenerate [54]. As a result of the doping inhomogeneity between the bulk and the barrier regions, a potential well is likely to be created between these regions [55–57]. Even with a careful doping design in which there is no well in equilibrium, at forward bias, when fewer electrons need to be depleted from the semiconductor, the well creation is inevitable. The spin-related effects of this potential well may contribute to the spin accumulation in the bulk semiconductor region [58, 59]. Finally, the donor identity in the depletion region plays an important role in the electron spin of multivalley semiconductors such as silicon or germanium. The spin relaxation in highly doped regions of these elemental semiconductors is governed by the spin–orbit coupling of the impurity [60–67]. It is therefore beneficial to choose light-element donors such as P in these semiconductors.

17.2.1 Spin Injection

An important quantity in the description of spin transport is a spin-dependent electrochemical potential $\mu_s(x)$. It is defined as

$$\mu_s(x) = \mu_s^c(x) - e\phi(x), \tag{17.2}$$

where μ_s^c is the chemical potential for spin s, ϕ is the electrostatic potential, and $e > 0$ is the elementary charge. The spin splitting of the electrochemical potential, $\Delta\mu = \mu_+ - \mu_-$, corresponds to the presence of non-equilibrium spin density (spin accumulation). In a non-magnetic material, $\Delta\mu \neq 0$ means $\Delta n = n_+ - n_- \neq 0$, where n_s is the density of electrons with spin s. In ferromagnet/semiconductor systems, the large difference of conductivities, $\sigma_{fm} \gg \sigma_{sc}$, allows us to disregard both spatial and spin dependence of μ_s in the ferromagnet and to use a single value of chemical potential μ_{fm}. In what follows, we will simplify our analysis by assuming quasi-neutrality in the bulk semiconductor region ($n_+ + n_- = n_0$) and by assuming a linear proportionality between μ_s and the spin accumulation density, $\delta n_s = n_s - n_0/2$. The linear proportionality is valid when $|\delta n_s| \ll n_0$, and quasi-neutrality holds if the dielectric response time of the semiconductor, $\sigma_{sc}/\varepsilon_{sc}$, is much shorter than the spin relaxation time and the transit time in the semiconductor channel [68, 69]. Using these assumptions, we can define the bias voltage, V, applied across a ferromagnet/semiconductor junction as the difference between μ_{fm} and the average electrochemical potential in the semiconductor flat-band region, $\mu_{fm} - \mu_0 = -eV$, where $\mu_0 = (\mu_+ + \mu_-)/2$. This follows the conventional notation in which electrons flow from (into) the semiconductor in forward (reverse) bias, $V > 0$ ($V < 0$).

The current density due to tunneling of free electrons is calculated by the Landauer–Büttiker formalism assuming coherent transport across the barrier regions [49]:

$$j_{s,t} = \frac{em_{sc}}{4\pi^2\hbar^3} \int_0^\infty dE \left[f_{fm}(E) - f_{sc,s}(E) \right] \int_0^\infty dE_t T_s(E - E_t, eV), \tag{17.3}$$

where:

f_{fm} is the Fermi–Dirac distribution at the ferromagnetic metal (with our assumption of a single μ_{fm})

$f_{sc,s}$ is the Fermi–Dirac distribution at the bulk semiconductor region

Its spin dependence is introduced via μ_s. The zero reference energy is the flat conduction band in the bulk semiconductor region. E and E_t are, respectively, the total electron's energy and the electron's energy due to transverse motion. m_{sc} is the electron effective mass in the semiconductor region. The spin-dependent transmission coefficient, T_s, is a function of the kinetic energy of the impinging electron, $E{-}E_t$, and of the potential drop across the barrier, eV. The transmission magnitude is governed by the exact shape of the barrier. To quantify the spin-dependent transmission coefficient, a simple effective mass band structure is invoked. This model is sufficient for simulating the spin-polarization of the tunneling current across the Schottky barrier, and it is also capable of reproducing the spin polarization seen in Fe/GaAs experiments [27, 28, 37]. In this view, we assume that the spin-based tunneling stems from the spin-dependent kinetic energies in the ferromagnetic side of the junction. and we neglect the electronic structure of the interfacial atomic layer [36, 70–73]. To readily calculate the second integral in Equation 17.3, specular transmission is assumed, which conserves the transverse wave vectors due to a motion in parallel to the junction's interface, $k_{\parallel}^{fm,s} = k_{\parallel}^{sc}$. Neglecting small corrections due to spin–orbit coupling and using the effective mass approximation, the electron's velocity and wave vector are related via $v = \hbar k/m$, and the transmission is calculated from particle and current conservation; continuity of the wave function, and its first derivative divided by the effective mass. In a rectangular barrier with width d, the transmission is analytical and is given by

$$T_s\left(E-E_t, eV\right) \simeq \frac{16 v_{fm,s} v_{sc} e^{-2\kappa_b d}}{v_b^2 + v_{fm,s}^2 + v_{sc}^2 + v_{fm,s}^2 v_{sc}^2/v_b^2}. \tag{17.4}$$

The validity of this expression is restricted to the small transmission ($T \ll 1$). All velocities are along the normal of the interface, where the spin selectivity of the transmission is solely due to the spin-dependent velocities in the ferromagnetic contact, $v_{fm,s}$. The velocities in the semiconductor, v_{sc}, and in the barrier, v_b, are spin independent. The latter is an *effective tunneling velocity* given by

$$v_b = \frac{\hbar \kappa_b}{m_{sc}} = \sqrt{\frac{2}{m_{sc}}\left(\phi_B - eV - E + E_t\right)}. \tag{17.5}$$

In semiconductors with small electron mass, $m_{sc} \ll m_0$, when the bias is either reverse or moderately forward, then the effective barrier velocity, v_b, dominates the denominator of Equation 17.4, and the ratio between the spin-up and spin-down transmissions is nearly identical to the ratio of their Fermi velocities in the ferromagnetic contact. In iron, this ratio is about $v_+/v_- \sim 2.5$ [74].

This picture is changed for electrons that are photoexcited in the barrier region, so that their tunneling energy is high above the quasi-Fermi levels of both sides of the junction. In this case, the Fermi velocities in the ferromagnetic contact become more dominant in the denominator of Equation 17.4, and the spin polarization can switch its sign [75]. In triangular or parabolic barriers, the transmission is mainly affected by the change of κ_b and v_b [17]. In addition to the fact that spin-dependent properties of the interfacial atomic layers were neglected, the validity of this calculation is restricted to a finite bias range around equilibrium. The reverse bias range is limited by the transport across a wider depletion region with enhanced electric field [76, 77] and by the enhanced spin relaxation of injected hot electrons [78]. In forward bias, the restriction to a small bias range is twofold. First, electrons tunnel to states above the Fermi level in the ferromagnetic side, where new bands may exist. Second, the barrier conductance increases exponentially with the forward bias, and therefore, the restriction for $G \leq G_{sc}$ is eventually violated.

The description of transport is complete when we relate the spin-dependent electrochemical potentials in the semiconductor channel with their boundary conditions at the semiconductor/ferromagnetic junctions. To simplify the analysis, non-degenerate semiconductors and low electric fields are assumed. The electrochemical potentials are given by [79]

$$\mu_s = k_B T \ln\left(\frac{n_0 + 2\delta n_s}{n_0}\right) - e\phi \approx 2k_B T \frac{\delta n_s}{n_0} - e\phi, \qquad (17.6)$$

where the second expression is the linear approximation valid for small spin accumulation density compared with the total electron density. The spatial form of the electrochemical potential in a one-dimensional semiconductor channel follows the spin diffusion equation [45–47]

$$\nabla^2 \mu_s = \frac{\partial \mu_s}{\partial t} + \frac{\mu_s - \mu_{-s}}{2L^2}, \qquad (17.7)$$

where the spin diffusion length, $L = \sqrt{D\tau_s}$, is defined in terms of the diffusion constant D and the spin relaxation time, τ_s. In this formalism, we employ an adiabatic approximation with respect to sub-picosecond time scale processes: the momentum scattering and dielectric relaxation [68]. The time-dependent analysis will be discussed in Section 17.4.1. In steady state conditions ($\partial/\partial t = 0$), the general solution in one dimension is given by

$$\mu_s(x) = A + Bx \pm \left(Ce^{x/L} + Me^{-x/L}\right), \qquad (17.8)$$

where A, B, C, and M are constants to be determined from the boundary conditions. The spin-s current is related to the slope of the electrochemical potential by

$$j_s = \frac{\sigma_s}{e}\nabla\mu_s = \sigma_s E + eD\nabla n_s, \qquad (17.9)$$

where σ_s is the conductivity for spin s. In the linear regime of small spin accumulation in non-degenerate semiconductors, the conductivity and diffusion constants are all spin independent to the first-order approximation

$$\sigma_s = \frac{\sigma}{2} = ev\frac{n_0}{2}, \quad D = v\frac{k_B T}{e},$$
(17.10)

where v is the spin-independent mobility. Therefore, the only way for the semiconductor to support spin accumulation is by creating a net spin density, $n_+ \neq n_-$. As a result, the linear and exponential terms in Equation 17.8 are related, respectively, to the drift and diffusion terms in Equation 17.9. To keep a compact analysis, we match the spin-dependent bulk current (Equation 17.9) at its boundary with the spin-dependent tunneling current (Equation 17.3) by linearizing the latter around equilibrium:

$$j_{s,t} \approx \frac{G_s}{e}\left(\mu_{fm} - \mu_s\right),$$
(17.11)

G_s is the barrier conductance at low bias (slope of the J-V curve near zero bias). Its spin dependence is governed by the integrated transmission coefficient. As discussed earlier, the ratio G_+/G_- is equal to the ratio of the velocities of carriers with different spins in the ferromagnetic contact.

17.3 TWO-TERMINAL DEVICES

We are now in a position to discuss the magnetoresistance (MR) effect in a biased ferromagnet/semiconductor/ferromagnet system. Figure 17.1a sketches a one-dimensional spin valve where the semiconductor channel is sandwiched between two ferromagnetic contacts. The width of the channel is w, and it is assumed to be shorter than the spin diffusion length, so that we can linearize the exponential terms in Equation 17.8. Due to the small bias between the ferromagnetic contacts, the total barrier conductances are the same on both sides of the channel. For parallel (P) magnetization configuration, the spin-dependent conductances are the same in the right and left barriers, $G_+^R = G_+^L$ and $G_-^R = G_-^L$, whereas in the antiparallel (AP) configuration, we have $G_+^R = G_-^L$ and $G_-^R = G_+^L$. Using these conditions and the former assumption of $G < \sigma/L$, Equations 17.8 through 17.11 provide a solution for the electrochemical potential in the semiconductor channel:

$$\frac{2\mu_\pm^P(x)}{eV} \approx 1 + \frac{Gx}{\sigma} \pm \frac{F}{2}\frac{Gx}{\sigma},$$
(17.12)

$$\frac{2\mu_\pm^{AP}(x)}{eV} \approx 1 + \frac{Gx(1-r)}{\sigma} \mp \frac{F}{2}\frac{Gw}{\sigma}\left(\frac{2L^2}{w^2} + \frac{x^2}{w^2}\right),$$
(17.13)

where:

    V is the applied voltage
    $F = (G_+ - G_-)/(G_+ + G_-)$ is the finesse (spin selectivity) of the barrier
    r denotes the MR effect of this device

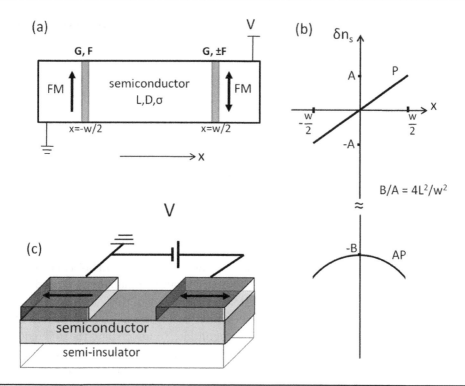

FIGURE 17.1 (a) A one-dimensional biased spin valve. (b) The spin accumulation profile in the semiconductor channel in parallel and antiparallel configurations. (c) A realistic lateral spin valve geometry. The signal is improved when the semiconductor channel is etched away at the edges.

The MR is defined by the relative change of the total current across the device between the parallel and antiparallel configurations:

$$r \equiv \frac{I_P - I_{AP}}{I_P} \simeq F^2 \left(1 + \frac{\sigma w}{2L^2 G} \right)^{-1}. \tag{17.14}$$

This expression is valid when both w/L and LG/σ are small fractions (say $\leq 1/3$). One can notice that the MR effect is larger with improved matching between the barrier conductance, G, and the spin-depth factor of the semiconductor, σ/L. Nonetheless, the magnetoresistive effect is rather small when applying typical parameters. For example, consider a $w = 200$ nm of 10^{16} cm^{-3} n-type GaAs region in between two iron contacts at room temperature. This doping corresponds to $\sigma = 10$ $(\Omega \cdot \text{cm})^{-1}$ and $L = 1$ μm [54]. The barrier parameters are taken from a typical Fe/GaAs spin light emitting diode [29]. The total barrier conductance is $G = 1000$ Ωcm^{-2}, and its finesse, measured from the emitted circular polarization, can reach $F \sim 1/3$ [29, 30].

This value is also comparable to our estimated value of $(v_{fm,+} - v_{fm,-})/(v_{fm,+} + v_{fm,-})$ in iron. Plugging these parameters results in $\sim 1\%$ magnetoresistive effect. Contrary to such a small effect, the spin accumulation profile of the two magnetization configurations differs substantially.

The separation between the spin-dependent electrochemical potentials is given by

$$\Delta\mu^{P}(x) \approx F \cdot eV \cdot \frac{Gx}{2\sigma}, \tag{17.15}$$

$$\Delta\mu^{AP}(x) \approx F \cdot eV \cdot \frac{Gw}{4\sigma}\left(\frac{4L^2}{w^2} + \frac{2x^2}{w^2}\right), \tag{17.16}$$

where $-w/2 \le x \le w/2 < L$. One can see that the ratio between the spin accumulation densities can easily take values of 100 or more:

$$\frac{\Delta\mu^{AP}}{\Delta\mu^{P}} = \frac{\Delta n^{AP}}{\Delta n^{P}} \ge \left(\frac{2L}{w}\right)^2. \tag{17.17}$$

Figure 17.1b shows the spin accumulation profile in both magnetization configurations. The large difference in spin accumulation profiles does not affect the spin-dependent currents due to the fact that only the base levels of the spin accumulation profiles differ by a factor of $4L^2/w^2$, whereas their slopes are of the same order of magnitude. To understand the difference between the spin accumulation profiles, we can picture the following. In a parallel configuration, the reversed bias contact injects electrons whose net spin-polarization is identical to the preferred extracted spin by the forward biased contact. Since there is no bottleneck, the spin accumulation density in the channel is very small. Figure 17.1b also shows that in the parallel configuration, the spin accumulation changes its sign across the semiconductor channel. The reason is again simple; next to the injector, there is an accumulation of spin-up electrons, which matches the spin of the majority population in the ferromagnetic contact. Next to the extractor, there is a depletion of spin-up electrons, since they can easily exit the channel, whereas the spin-down electrons find it "harder" to tunnel into the ferromagnetic contact. This scenario changes in the antiparallel configuration, where the injector preferentially injects spin-up electrons, whereas the extractor preferentially extracts spin-down electrons. This bottleneck results in a spin accumulation profile at which the spin polarization in the channel keeps its sign, and its magnitude increases dramatically.

17.3.1 LATERAL GEOMETRY

Similarly to the structure of conventional field effect transistors, micro- and nano-lithography allow us to fabricate lateral spin valve devices [27, 35, 42]. In this view, we need to adjust the previous analysis, in which we have used a simplified one-dimensional geometry to describe the general solution of the spin diffusion equation (Equation 17.7). To analyze the spin diffusive transport in realistic lateral devices, as shown in Figure 17.1c, we consider an effective one-dimensional transport equation derived from the three-dimensional spin diffusion equation. This effective one-dimensional diffusion equation can accurately describe the spin diffusion in a layer of material of thickness h smaller than the spin diffusion length L and covered by contacts with junction conductances G smaller than the conductance σ/h of the underlying semiconductor layer. In a structure like the one shown in Figure 17.1c, we calculate the spin diffusion by introducing the layer-averaged electrochemical potential $\xi_s(x)$ in the semiconductor channel:

$$\xi_s = \frac{1}{h}\int_0^h dy\, \mu_s(x,y).$$

(17.18)

In regions of the semiconductor channel that are not covered by ferromagnetic contacts, this layer-averaged electrochemical potential is identical to $\mu_s(x)$, and we can proceed with the analysis after Equation 17.8. However, under the contacts when we integrate out the y dependence from Equation 17.7, we obtain the approximate equation [51]

$$\frac{\partial^2 \xi_s}{\partial x^2} = \frac{\xi_s - \xi_{-s}}{2L^2} + \frac{2G_s}{\sigma h}\left(\xi_s - \mu_{fm}\right),$$

(17.19)

where μ_{fm} is the electrochemical potential in the ferromagnetic contact, and the second term was derived by incorporating the boundary condition from Equation 17.11. As before, this equation is valid when assuming small electric fields and small spin accumulations (so that $\sigma_s \approx \sigma/2$). The general solution of this equation is given by

$$\xi_\pm(x) = -eV + (1\pm\lambda)\left[pe^{\lambda_s x}qe^{-\lambda_s x}\right]$$
$$+ (\lambda\mp 1)\left[re^{\lambda_c x} + se^{-\lambda_c x}\right],$$

(17.20)

where:

    $p, q, r,$ and s are constants to be determined by boundary conditions
    $-eV$ is the potential level in the ferromagnetic contact

The other parameters are

$$\lambda = \cot\left[\frac{1}{2}\tan^{-1}\beta\right],$$

(17.21)

$$\lambda_{(s,c)}^2 = \left[\alpha + 1 \pm \sqrt{1+\beta^2}\right]\Big/\left(2L^2\right)$$
$$(\alpha,\beta) = 2L^2\left(G_+ \pm G_-\right)\big/(\sigma h),$$

(17.22)

where the first of each pair of symbols (s, c) or (α, β) takes the upper sign. Consider the case that spin selectivity is robust, so that α and β are comparable. If $\alpha \ll 1$, then $\lambda_c \ll 1/L$, and $\lambda_s \sim 1/L$. The s-mode is limited by the spin diffusion constant, and it corresponds to spin accumulation ($\lambda \gg 1$ in this case). If $\alpha \gg 1$, then both eigenvalues are nearly independent of L, and neither of the eigenvectors is a pure spin mode $\lambda \simeq 1$; the inhomogeneity of injection dominates the spatial dependence. To get a unique solution of the problem, the electrochemical potentials and their slopes are matched between covered and uncovered sections of the semiconductor channel (Equations 17.8 and 17.20). At the outermost edges of the device, the spin currents are zero ($\partial \xi_{\pm}/dx = 0$). For Fe/GaAs structures with $G_s \sim 10^3\ \Omega^{-1}\mathrm{cm}^{-2}$, this approximate formalism gives results indistinguishable from exact numerical calculations of a two-dimensional spin diffusion equation [51]. The magnetoresistance of a lateral spin valve device whose resistance is dominated by the tunneling contacts is slightly changed compared with Equation 17.14:

$$r \simeq F^2\left(1 + \frac{\upsilon}{A} \cdot \frac{\sigma}{2L^2 G}\right)^{-1},\tag{17.23}$$

where:
  υ is the volume of the channel
  A is the area of the contacts

which are assumed to have the same total conductance and area for simplicity.

The simple two-terminal geometry of a spin valve captures the essence of a logic exclusive OR (XOR) operation if we consider the spin accumulation magnitude as the output and the magnetization configuration as the input. If we define binary "0" and "1" as the two possible magnetization directions of the easy axis of the contacts, then the possible parallel magnetization configuration inputs of a spin-valve are "00" or "11," and the possible antiparallel magnetization configurations are "01" or "10." In the next section, we elaborate on logic and memory device schemes whose outputs display clearly distinct current signals. This is achieved by using the spin accumulation profiles rather than their slopes. We will show a universal, reprogrammable magneto-logic gate that relies on the addition of XOR operations.

17.4 MULTI-TERMINAL DEVICES

To achieve a more efficient electrical expression of spin accumulation, one has to move beyond a passive two-terminal spin valve device and consider a multi-terminal system in which additional external stimuli can control the magnetoresistive effects. In the following, we review several proposals of devices consisting of more than two ferromagnetic terminals connected to a semiconductor channel. Similarly to the all-metallic nonlocal spin valve, the common feature of these devices is the use of a ferromagnetic contact to sense the spin accumulation in the non-magnetic channel [41–43]. Figure 17.3a shows a device that consists of three ferromagnetic tunneling contacts deposited on top of a paramagnetic semiconductor channel. The outer edges of the semiconductor channel are removed to confine the spin accumulation under and between the contacts [51]. In steady state, the current passes only between the left (L) and middle (M) contacts, driven by the voltage V_L. The right (R) contact is connected to a capacitor (C), which in steady state enforces zero charge current across this terminal. In Section 17.4.4, we will consider general non-collinear cases. Here, we focus on the collinear case, in which all of the magnetization directions are either parallel or antiparallel to the $+z$ direction ($\theta_i = 0$ or π). The steady state value of the (single) electrochemical potential level inside the right contact is thus self-adjusted according to

$$J_R = 0 = G_{R,+}\left(\mu_{R0} - \xi_{R,+}\right) + G_{R,-}\left(\mu_{R0} - \xi_{R,-}\right), \qquad (17.24)$$

and thus,

$$\mu_{R0} = \xi_{R,0} + \frac{F_R}{2}\Delta\xi_R, \qquad (17.25)$$

where $\xi_{R,0} = (\xi_{R,+} + \xi_{R,-})/2$, $\Delta\xi_R = \xi_{R,+} - \xi_{R,-}$ and $F_R = (G_{R,+} - G_{R,-})/(G_{R,+} + G_{R,-})$. $\xi_{R,\pm}$ denote the averaged spin-dependent electrochemical potentials in the semiconductor channel beneath the right contact. The positions of the energy levels of $\xi_{R,+}$ and $\xi_{R,-}$ are not affected by the parameters of the right contact (as long as it is a tunneling contact). Due to the zero steady state current through the right contact, the creation of spin accumulation under this contact ($\Delta\xi_R$) is driven solely by the diffusion of spin-polarized electrons from the left part of the device, where charge current flows between the spin-selective middle and left contacts. Specifically, we can get only two possible sets of values for $\xi_{R,+}$ and $\xi_{R,-}$, which correspond to the spin accumulation pattern of a parallel or an antiparallel magnetization configuration of the left and middle contacts. In the parallel configuration, the spacing, $\Delta\xi_R^P$, is considerably smaller than in the antiparallel configuration, $\Delta\xi_R^{AP}$. In contrast to $\xi_{R,\pm}$, the value of the self-adjusting electrochemical potential inside the right contact, μ_{R0}, depends on the magnetization direction of the right contact. As can be understood from Equation 17.25, the value of μ_{R0} is closer to $\xi_{R,+}$ if $F_R > 0$, which is the case when the magnetization of the right contact is parallel to the magnetization of the middle contact. Similarly, μ_{R0} is closer to $\xi_{R,-}$ if $F_R < 0$, which is the case when the magnetization of the right contact is antiparallel to the magnetization of the middle contact. Figure 17.3b shows an energy diagram of the electrochemical potential levels for the possible magnetization collinear configurations with a pinned middle contact. Diagrams I and II (III and IV) denote antiparallel (parallel) magnetization configurations of the middle and left contacts. The solid (green) lines denote the energy levels of $\xi_{R,\pm}$, and the dashed lines (red) denote the possible energy levels of μ_{R0}. Given that the middle contact is pinned, there are four possible magnetization configurations in total (denoted by three arrows above each diagram). The energy scales in this graph correspond to a semiconductor channel with a spin diffusion length of $L = 1$ µm and conductivity $\sigma = 3$ ($\Omega \cdot$cm)$^{-1}$. The three barriers are identical, $G = 10^4\ \Omega^{-1}$cm^{-2} and $F = \pm 1/3$, where the sign depends on whether the magnetization direction is "1" or "0." The bias is $V_L = 0.1$ V, and the lateral dimensions are $w_L = w_M = w_R = 400$ nm, $w_d = 200$ nm, $h = 100$ nm. Finally, it is mentioned that the spin-dependent currents $j_{R,+}$ and $j_{R,-}$ are constantly flowing across the right junction even in steady state conditions. Their sum (charge current) is zero, and they flow in opposite directions, since μ_{R0} lies between $\xi_{R,+}$ and $\xi_{R,-}$: electrons with a spin direction that matches the accumulated spin in the channel flow into the ferromagnetic contact, and the same number of electrons with opposite spin direction flow into the semiconductor. Inside the ferromagnetic material, these spin currents decay to zero within a few nanometers due to the short spin diffusion length of these materials [80]. That is, spin currents do not (and are not required to) reach the external capacitor in Figure 17.3a.

Recently, Wen et al. experimentally demonstrated a three-terminal magneto-logic gate that functions as XOR in a graphene spintronic device at room temperature [44], see also Chapter 5, Volume 3. By carefully tuning the bias current between the two left input electrodes, shown in Figure 17.2a, and adding an offset voltage in the detection loop, a clear non-zero output current (logic "1") is observed when the two inputs are antiparallel, with an absolute zero output current (logic "0") when the two inputs are parallel (Figure 17.2b). These results provide the proof-of-concept demonstration for a class of magneto-logic devices based on spin accumulation and establish

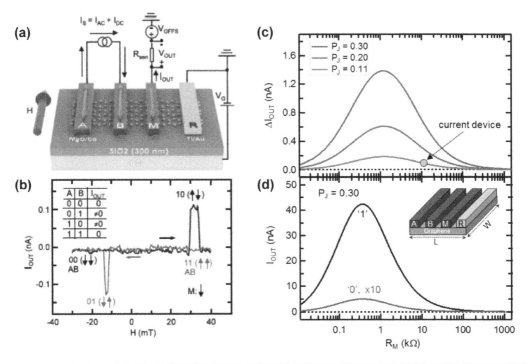

FIGURE 17.2   Experimental demonstration of graphene XOR magnetologic gate. (a) Cartoon of experimental device structure and measurement setup. A, B, and M are MgO/Co electrodes. Spin channel is a single layer of graphene. R is a Ti/Au nonmagnetic reference electrode used as ground point. I_{OUT} and V_{OUT} are the measured current and voltage signal, respectively. R_{sen} is a variable resistor. V_{OFFS} is an alternating current (ac) voltage source. An external magnetic field H is applied to the easy axis of the electrodes. Center-to-center distances of the electrodes are: $L_{AB}=1.6$ μm, $L_{BM}=1.8$ μm, $L_{MR}=7.85$ μm. Graphene width along H direction is ~4.3 μm. (b) I_{OUT} measured as a function of H. Black (red) curve indicates H sweeps upwards (downwards). Vertical arrows indicate the magnetization states of A and B. Top left inset: truth table of XOR logic operation. (c) Optimizing the output current signal. Signal difference between "1" and "0" logic output $\Delta I_{OUT}(=|I_{OUT}('1')-I_{OUT}('0')|)$ as a function of R_M for different spin polarization of contacts, assuming $P_A=P_B=P_M=P_J$, and $P_R=0$. Grey dot represents our current device parameters. (d) Output signal I_{OUT} ("1" and "0") as a function of RM for an optimized device geometry. Signal for "0" is magnified 10 times. Inset: optimized device structure. There is no graphene beyond electrodes A and R. The whole device length is $L=350$ nm. Graphene width is $W=500$ nm. Each electrode (A, B, M, and R) has a width of 50 nm, and the center-to-center distance between adjacent electrodes is 100 nm. Spin polarization P_J is 0.3. (After Wen, H. et al., *Phys. Rev. Appl.* 5, 044003, 2016. With permission.)

the feasibility of the universal five-terminal magneto-logic gate (MLG) that will be discussed in Section 17.4.2. Furthermore, the signal size of the logic "1" output can be significantly enhanced by reducing the device size according to numerical simulation, shown in Figure 17.2c and d, making it promising for future spintronic device applications.

17.4.1 TIME-DEPENDENT ANALYSIS

The device shown in Figure 17.3a can also be used for the dynamical readout of magnetization alignment or to electrically measure the magnetization dynamics [12, 81]. Consider, for example, a 2π rotation of the right magnet in

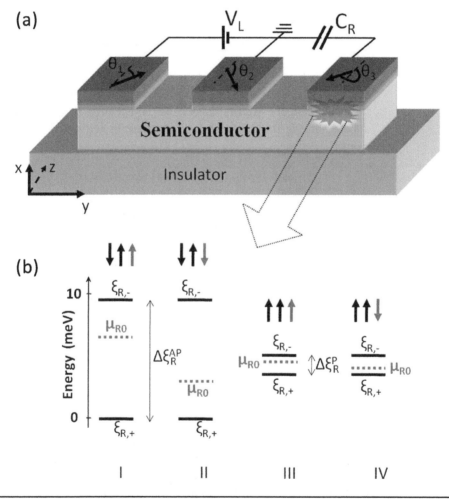

FIGURE 17.3 (a) A three-terminal non-local spin valve attached to a capacitor which guarantees a zero steady state current across the right terminal. The magnetization directions are denoted by the angles θ_i (from the z-axis in the yz plane). The left and middle contacts are biased, and their configuration sets the information. The floating contact (attached to the capacitor) is for reading the stored information (see text). (b) Positions of the electrochemical potentials for various magnetization configurations when $\theta_i = 0$ or π. The solid lines denotes the levels in the channel beneath the R contact. The dashed lines correspond to the level inside the R contact. There are four possible magnetization configurations if the middle contacts are pinned in the up direction.

the *yz* plane (starting from the +*z* direction), where the magnetization direction of the middle and left contacts is fixed at parallel or antiparallel configurations ($\theta_1 = (0 \text{ or } \pi), \theta_2 = 0$). After half of the rotation cycle, the spin-dependent conductances, $G_{R,+}$ and $G_{R,-}$, exchange roles, and by the end of the cycle (2π), they recover their original values. As will be later explained, this rotation is described by $F_R(t) = F_0 \cos(\omega_r t)$. To explain the resulting transient current across the right magnet during the magnetization dynamics, we will first consider only the external capacitor and ignore the intrinsic Schottky barrier capacitance. If the external capacitor charging time is fast enough with respect to ω_r^{-1}, then the electrochemical potential of the right contact is instantaneously being readjusted during the rotation time. Its value guarantees that at steady state, $J_R = 0$, and therefore, μ_{R0} follows Equations 17.24 and 17.25. Since μ_{R0} adiabatically follows $F_R(t)$, the peak current value is proportional to

$$J_p \propto C\omega_r F_0\left(\Delta\xi_R\right) \ll GF_0\left(\Delta\xi_R\right),$$

where C is the external capacitance per unit area, and the inequality corresponds to the fast charging time of a small capacitor $C/G \ll 1/\omega_r$. If, on the other hand, the external capacitance is very large, then μ_{R0} is only slightly affected during the dynamics, and the current across the right contact is simply

$$J_R\left(t\right) = G_{R,+}\left(t\right)\left(\mu_{R0} - \xi_{R,+}\right) + G_{R,-}\left(t\right)\left(\mu_{R0} - \xi_{R,-}\right),$$

where $G_{R,\pm}(t) = G_R(F_R(t) \pm 1)/2$. In this case, the peak of the transient current across the right contact is $GF_0(\Delta\xi_R)$ provided that initially ($t \leq 0$), the current is zero. Thus, if $C\omega_r/G \gg 1$, then the peak current is independent of C. In prospective integrated circuit applications, the size of C is limited by the available circuit area, and in what follows, we will work in the case that $C\omega_r/G \sim 1$. The figure of merit of this operation is that the amplitude of the resulting current oscillation, $J_R(t)$, is much smaller when the magnetization configuration of the left and middle contacts is parallel compared with the antiparallel configuration $\left(\Delta\xi_R^P \ll \Delta\xi_R^{AP}\right)$.

Using the small capacitance analysis, in the antiparallel configuration, this perturbation refers to a swing of μ_{R0} between its two possible values, as shown by the dashed lines of Diagrams I and II in Figure 17.3b. In the parallel configuration, the swing is between the dashed lines of Diagrams III and IV. If we rotate the magnetization of the left rather than of the right ferromagnetic contact, then the measurement of the accompanying transient current ($J_R(t)$) also allows electrical readout of magnetization dynamics at the left contact. Here, the potential perturbation refers to the swing of $\xi_{R,\pm}$ between its two possible values as shown by the solid lines of Diagrams I or II and those of Diagrams III or IV.

To change the magnetization of any of the ferromagnetic contacts, the planar structure should be augmented by a set of current-carrying lines known from magnetic random access memories [82]. In these devices, there are wide metallic strips (so-called *bit* and *word lines*) running above and below the magnets that are to be addressed. The upper and lower wires are at a 90° angle, so that every magnet is (when looking from above) located at an intersection of two wires. To switch its magnetization, the current pulses are passed through the appropriate bit and word lines. These currents generate transient local magnetic fields (through Ampere's law), which can switch the magnetization of the addressed terminal. The two lines are necessary for addressing purposes in a matrix consisting of lateral multi-terminal devices. Only the magnet located at the intersection of two lines is switched. The other magnets over (under) which the activated current-carrying line passes are unaffected, as the current magnitudes are such that a magnetic field from a single line is unable to switch the magnetization. Furthermore, using the two lines (giving two perpendicular magnetic fields that add up) is advantageous, since the presence of magnetic field components non-collinear with the magnet's easy axis aids fast switching [83].

To model the transient behavior, we add the time dependence to the formalism of lateral spin diffusion. The relevant characteristic time scales are the magnetization rotation time and the capacitor charging/discharging time, C/G, where we assume that the contacts are the most resistive elements in the circuit. As discussed before, C/G is chosen to be of the order of the magnetization dynamics whose fastest time scale is of the order of 0.1 ns [84, 85]. For example, for $G \sim 10^4\, \Omega^{-1}\mathrm{cm}^{-2}$ and a capacitor area of 1 μm², the external capacitance is $C = 40$ fF. This conductance value maximizes the spin accumulation density $(G \sim \sigma/L)$ in a semiconductor channel whose conductivity is $1 \sim \Omega^{-1}\mathrm{cm}^{-1}$ and whose spin diffusion length is about 1 μm (e.g., a non-degenerate n-type GaAs at room temperature). We mention that if we were to use all-metallic versions of this device, then the typical spin accumulation in a paramagnetic metal corresponds to $\Delta\mu$ of less than 1 μV [41, 42]. With such a small voltage swing on the capacitor, for the transient current, $C(\partial\Delta\mu/\partial t)$, to be measurable, one has to use a nearly macroscopic capacitor, which rules out application in integrated circuits. Again, the small carrier density in a semiconductor (allowing $\Delta\mu \sim 10$ mV) is indispensable.

The time-dependent transport in the semiconductor channel beneath a ferromagnetic contact is described via the layer-averaged electrochemical potentials, $\xi_{\pm}$. By repeating the steady state analysis that led to Equation 17.19 from Equation 17.7, we can write the time-dependent diffusion equation for the spin splitting of the electrochemical potentials $\Delta\xi = \xi_+ - \xi_-$:

$$\frac{\partial \Delta\xi}{\partial t} = D \frac{\partial^2 \Delta\xi}{\partial x^2} + \frac{\beta(t)}{\tau_{sr}}\left(\mu^m - \xi\right) - \frac{\alpha}{2\tau_{sr}}\Delta\xi - \frac{\Delta\xi}{\tau_s}, \qquad (17.26)$$

α and β are dimensionless parameters given by Equation 17.22. The dynamics of magnetization is parameterized by $\beta(t) \sim \Delta G(t)$, which characterizes the contact polarization only along the z-axis. If we deal with a coherent precession of magnetization, then this is an approximation. In general, one should take into account the non-collinearity of spins and the magnets during the rotation time. However, for tunneling barriers, the non-trivial effects of this non-collinearity are expected to be small, and the only thing that matters is the average polarization along the z direction [86, 87]. Thus, we can model the influence of the contact with magnetization making an angle $\theta(t)$ with the z-axis by assuming that $\beta(t) \sim \Delta G \cos\theta(t)$. On the other hand, if the magnetization reversal is incoherent (e.g., proceeding by nucleation of domains with opposite magnetization; [88], the parameter $\beta(t)$ describes an area average of spin-selectivity of magnetically inhomogeneous contact, and it is naturally proportional to the z component of the contact's magnetization.

The dielectric relaxation time, $\tau_d = \sigma/\varepsilon_{sc}$, is much faster than the time scale of magnetization dynamics and spin diffusion (about 100 fs for a non-degenerate semiconductor with $n_0 = 10^{16}$ cm^{-3}). Thus, quasi-neutrality in the channel is assumed at all times; $\delta n_+(t) + \delta n_-(t) = 0$. In the linear regime under consideration (when $\Delta\xi < k_B T$), the average electrochemical potential $\xi = (\xi_+ + \xi_-)/2$ is equal to $-e\phi$. At every moment of time, ξ fulfills the Laplace equation with boundary conditions given by currents at the interfaces. In the time-dependent case, these currents include also a displacement current connected with the charging of the barrier capacitance C_B. A Schottky barrier is a dipole layer, and its capacitance can have a strong effect on the dynamics of currents on the time scales of interest here. Taking this displacement current into account, the boundary conditions for the spin currents are

$$j_s = \frac{G_s}{e}\left(\mu_{fm}(t) - \xi_s(t)\right) + \frac{c_B}{2e}\frac{\partial}{\partial t}\left(\mu_{fm}(t) - \xi(t)\right), \qquad (17.27)$$

where c_B is the barrier capacitance per unit area, and $\xi = (\xi_+ + \xi_-)/2$. The first term on the right hand side is the time-dependent tunneling current (Equation 17.3), and the second term represents the carriers that flow toward the barrier but do not tunnel through it. Instead, they stay in the semiconductor close to the barrier, making the depletion region slightly thinner or wider. The charge involved in this process is negligible compared with the charge already swept out from the semiconductor, so we can keep c_B constant. For small spin splitting (so that the conductivities $\sigma_+ \simeq \sigma_-$), the same number of carriers of each spin are going to be brought from the channel into the barrier, and the displacement current is the same for each spin in Equation 17.27. For layer-averaged ξ, we get then

$$\frac{\partial^2\xi}{\partial x^2} = -\frac{\alpha}{2L^2}\left(\mu_{fm} - \xi\right) + \frac{\beta(t)}{4L^2}\Delta\xi - \frac{c_B}{\sigma h}\frac{\partial}{\partial t}\left(\mu_{fm} - \xi\right). \qquad (17.28)$$

The description of $\Delta\xi$ and ξ in semiconductor regions that are not covered by ferromagnetic contacts follows Equations 17.26 and 17.28 with $\alpha = \beta = 0$. The total solution is achieved by the conservation of carriers and currents throughout the semiconductor channel. Mathematically, this requires a continuity of $\Delta\xi$ and ξ and their first x-derivatives at all times at the boundaries between covered and uncovered regions. The magnetization dynamics of the ith contact translates into time dependence of β_i, driving the spin diffusion in Equation 17.26 and the semiconductor electric potential in Equation 17.28. From ξ_s, we calculate the current $I_R(t)$ charging the capacitor C. The electrochemical potential of the R terminal $\mu_R = -eV_R$ changes according to $dV_R/dt = I_R/C$. Examples of calculations for two possible modes of operation (sensing the L dynamics and reading out the L/M alignment) are shown in Figure 17.4.

17.4.2 REPROGRAMMABLE MAGNETO-LOGIC GATE

The same physical principle of operation can be harnessed to achieve a more complicated functionality. In Figure 17.5a, we present a scheme of a five-terminal system [10] in which the electric sensing of spin accumulation is used to perform a logic operation; that is, two bits of input are converted into a binary output signal. This is a reprogrammable magneto-logic gate. As mentioned in the introduction, spintronic logic gates have been proposed in purely metallic systems, but this is the first proposal that employs semiconductors as active elements of the system.

The system presented in Figure 17.5a works in the following way. The charge currents are flowing between two pairs of terminals (X and A, Y and B), between which the bias V_{dd} is applied. Depending on the alignment of these pairs of magnets, different patterns of spin accumulation are created in the channel: if both X/A and Y/B are AP, the spin accumulation underneath M is large; if only one pair of contacts is AP, the spin accumulation is approximately two times smaller; and if both pairs are P, there is very small $\Delta\xi$ beneath the M terminal. The M contact is used to directly express the differences in the average spin accumulation beneath it.

The logic inputs are encoded by the magnetization directions of A, B, X, and Y terminals. We will concentrate on the case in which A and B magnetization directions are preset, defining the logic function of the gate. This reprogrammability is an important feature of magnetization-based logic [8, 10]. X and Y are then the logic operands, and the output is generated when the M magnet is rotated by 2π, triggering a transient $I_M(t)$ current of amplitude proportional to the spin accumulation in the middle of the channel. Let us focus on the example of the NAND gate, as any other

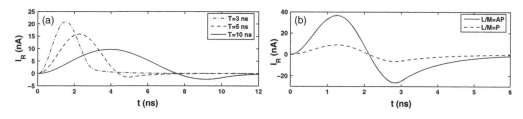

FIGURE 17.4 (a) R current signal for reversal of L magnetization occurring on a time scale of 3, 5, and 10 ns starting from AP alignment of L relative to M magnet. (b) R current signal for 2π rotation of R magnet for P and AP alignments of L and M magnets. The period of rotation is 3 ns. The conductance of the barriers $G = 10^4 \, \Omega^1 cm^{-2}$. The area of a junction is 1 μm^2, and the barrier thickness is taken to be 10 nm, resulting in junction resistance $R_B = 10 \, k\Omega$ and capacitance $C_B = 10$ fF. The external capacitance is $C = 40$ fF. The channel is GaAs at room temperature, with carrier density $n_0 = 10^{16} \, cm^{-3}$.

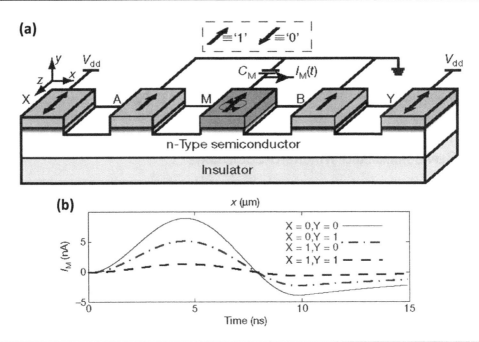

FIGURE 17.5 (a) A five-terminal magneto-logic gate (MLG). The logic inputs "0" and "1" are encoded by the magnetization direction of the A, B, X, and Y terminals (see text for details). As shown here, the gate is set to work as a NAND operation between X and Y (A and B are fixed to "1" values). In the read-out phase, the magnetization of the middle (M) terminal is rotated by 2π, or the back-gate voltage V_G is increased. (b) The $I_M(t)$ transient current triggered by rotation of the M magnetization by 2π. The solid line corresponds to the logic output "1". (After Dery, H. et al., *Nature* 447, 573, 2007. With permission.)

logic function can be realized by using a finite number of such gates. For the NAND operation, A and B magnets are set parallel to each other in a direction defining the logic "1." The amplitude of the $I_M(t)$ oscillation is two times larger for $X =$ "0" and $Y =$ "0" compared with the case when one of them is "0" and the other is "1," and for the "11" case, the current is negligible. This is shown in Figure 17.5b. The transient current can be captured by an external electronic circuit and then used to control a suitable write operation applied to a magnetic contact of another gate. For the technical details of this electrical conversion circuit, see the supplementary material of [10]. In the next section, we will consider this conversion circuitry as a black box and will present a common logic application which may benefit by using these devices beyond their trivial non-volatility merit.

17.4.3 CONTENT ADDRESSABLE MEMORIES

There is a significant gap between the number of devices (and thus parallelism) available in each new generation of technology and our ability to express parallelism in programs. Despite improvements in the software and hardware support, parallel programming remains challenging for the computing community at large. Clearly, in addition to traditional general-purpose parallel programming, we need other mechanisms to help bridge the gap. One important alternative mechanism is a content addressable memory

(CAM) [89]. A CAM is an example of an associative memory in which the search data (the key) is simultaneously compared with an array of data stored in memory cells to find matches. For example, in a CAM, we would ask for the list of addresses at which the word "John" is stored. On the other hand, in the standard computer memory (random access memory) we wish to know what is the information that is stored at a specified address (one piece of data at a time). Of these two cases, only the associative search provides a very significant amount of *data parallelism*, and as such, CAMs have found many uses in areas such as routing, searching, and network security and in general-purpose microprocessors (e.g., table lookaside buffers or load-store queues). In practical circuit designs, however, broadcasting the key presents technical challenges. Long buses with high capacitive loads are needed to simultaneously deliver the key to a large number of memory cells. This results in long delays and high power consumption and partly contributes to the size limitation in many applications of CAMs. An example of the importance of CAMs and their limitations is evident in the *Load-Store Queue*, which is a crucial microarchitectural structure in modern high-end microprocessors. It allows the processor to speculatively execute instructions out of original, program-specified order for high performance and yet guarantee their sequential semantics. The larger the queue, the more instruction-level parallelism the processor can exploit. The state-of-the-art designs use two queues to track loads and stores separately and yet can only allow about 20–30 entries in each queue [90, 91]. Compared with the size of other microarchitectural structures (e.g., register files), these are small sizes, and even these are sometimes obtained via non-trivial architectural speculations so as not to violate timing budget [92]. Scaling up these queues without affecting cycle time is a significant challenge in the design of microprocessors.

Using the magneto-logic gates, spin-based circuits have a unique advantage to support a large fan-out and make large CAM structures straightforward to implement and scale. The key factor is that transmitted information is determined by the direction of the current that magnetizes the ferromagnetic contacts. The amount of current needed for the magnetization is smaller than that needed to charge the distributed intrinsic capacitance of many memory cells along very long buses (together with parasitic capacitances). Figure 17.6 shows the bit line circuitry of a spin-based CAM. The current direction in the bit line encodes the search information bit in one of the operand contacts of each of the magneto-logic gates under (or above) the line. The adjacent contact holds the encoded bit information of the gate. If the magnetization directions are dissimilar/similar, then we have a mismatch/match, and the spin accumulation in the semiconductor channel is high/low. As such, an XOR operation between the search and information bits can tell the matching result, and this can be read by perturbing the magnetization of an additional contact next to the operand contacts [10]. In addition, the half-selection problem of magnetic random access memories [82], which inhibits their down-scaling, is irrelevant in the magnetic CAM, since the search is made simultaneously in all the memory cells (full selection of all bits).

The availability of energy-efficient *scalable* CAM designs will have a tremendous impact on general-purpose as well as application-specific

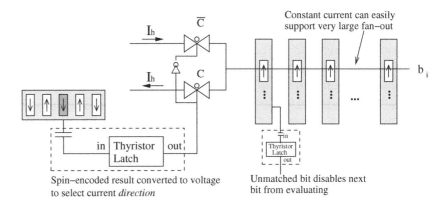

FIGURE 17.6 Schematics of an associative search in a proposed spin-based content addressable memory. The thyristor latch translates the current signal out of a five-terminal magneto-logic gate into a digital voltage level (left side). This in turn controls the direction of the current in the search bit-line (denoted by b_i) according to the pass-gates and the NOT gate in between. The search is performed simultaneously in all the other five-terminal magneto-logic gates under/above the bit line without the need to charge/discharge the intrinsic capacitance of any of these gates. Once a match is identified between the searched bit and the encoded information of a given gate (XOR operation), then the accompanied thyristor latch will enable the search in the magneto-logic gate at the next bit line (a similar structure to the b_i line). This procedure allows the search to be pipelined and conserves power (the search is disabled at the $i+1$ bit if the search of the i-bit failed in a given word). The details of the thyristor latch are explained in (Dery, H. et al., *Nature* 447, 573, 2007) and its supplementary material.

processors. Possibly the most immediate and far-reaching impact is the capability to build on-chip caches with extremely high associativity. Highly associative caches can significantly cut down cache miss rate and thus reduce the need for slow, energy-intensive off-chip memory accesses.

17.4.4 NON-COLLINEAR CONFIGURATIONS

Up to this point, we have treated the spin polarization in the channel, $p = (n_+ - n_-)/n_0$, as a scalar that implicitly relies on the assumption that the net spin has a fixed direction throughout the semiconductor channel. This description is valid in collinear systems at which the magnetization directions in all of the ferromagnetic elements share a common (easy) axis. In a more general, non-collinear configuration, the boundary conditions impose a change in the direction of the spin polarization during the transport in the channel [86]. For a general coordinate system in spin space, the spin-dependent electron density is described by a 2×2 matrix:

$$\hat{\mathbf{n}}(\mathbf{r},t) = \frac{n_0}{2}\left(\hat{I} + \mathbf{p}(\mathbf{r},t)\cdot\hat{\boldsymbol{\sigma}}\right), \tag{17.29}$$

where:

$\hat{\boldsymbol{\sigma}}$ is the Pauli matrix vector

$\mathbf{p}$ has the magnitude p along the + spin direction

Using this notation and repeating the analysis, one can derive the components of the spin diffusion equation [69]

$$\frac{1}{D}\frac{\partial p_i}{\partial t} + \frac{p_i}{\ell_{sf}^2} = \sum_j \left(\frac{\partial^2 p_i}{\partial x_j^2} + \frac{E_j}{V_T}\frac{\partial p_i}{\partial x_j} \right), \tag{17.30}$$

where i (j) enumerate the x, y, and z coordinates in spin (real) space. The components of the charge current density (vector) and of the spin current density (second-rank tensor) are

$$J_j = \sigma_0 E_j, \tag{17.31}$$

$$\mathcal{J}_{i,j} = \sigma_0 \left(V_T \frac{\partial p_i}{\partial x_j} + E_j p_i \right). \tag{17.32}$$

The description of transport is complete when the spin-polarized currents across the SC/FM junctions are expressed in terms of the spin polarization vector at the semiconductor side of the junction, $\mathbf{p}(\mathbf{r}, t)$. We follow the notation by Brataas et al. [86, 87] and write the population distribution matrices on both sides of the junction:

$$\hat{f}_{sc}(\varepsilon) = \frac{1}{2} e^{(\mu_0 - \varepsilon)/k_B T} \left(\hat{I} + \mathbf{p} \cdot \hat{\boldsymbol{\sigma}} \right), \tag{17.33}$$

$$\hat{f}_{fm} = \frac{1}{2} \left(1 + e^{(\varepsilon - \mu_0 + qV)/k_B T} \right)^{-1} \hat{I}, \tag{17.34}$$

where ε denotes the energy. As mentioned before, due to the conductivity mismatch, the ferromagnetic side is a reservoir with a constant chemical potential, $\mu_0 - qV$, where $\mu_0 = \mu_c - q\phi(\mathbf{r}, t)$ is evaluated at the semiconductor side of the junction, and V is the voltage drop across the semiconductor/ferromagnet (SC/FM) junction. The applied bias voltage is related to the electrical field via $\mathbf{E} = -\nabla\phi$. The population distribution matrices assume a non-degenerate semiconductor for which the spin-dependent electrochemical potentials, $\mu_\pm$, have the following relation with the net spin polarization $p = |\mathbf{p}|$:

$$\frac{\mu_\pm(\mathbf{r}, t)}{k_B T} = \frac{\mu_0}{k_B T} + \ln\left(1 \pm p(\mathbf{r}, t) \right). \tag{17.35}$$

For compact notation, the boundary conditions of each SC/FM junction are written in a spin coordinate system in which the z-axis is collinear with the magnetization direction of the corresponding ferromagnetic contact. With this simplification, the reflection matrices are diagonal:

$$\hat{r}_{sc}(\varepsilon_\perp, V) = \begin{pmatrix} r_\uparrow & 0 \\ 0 & r_\downarrow \end{pmatrix}, \quad \hat{r}_{fm}(\varepsilon_\perp, V) = \begin{pmatrix} \tilde{r}_\uparrow & 0 \\ 0 & \tilde{r}_\downarrow \end{pmatrix}. \tag{17.36}$$

The reflection coefficients in the left (right) matrix are of electrons from the semiconductor (ferromagnetic) side of the junction. For a given material system, these coefficients vary with the voltage drop and with the longitudinal energy, $\varepsilon_\perp$, which denotes the impinging energy of electrons due to their motion toward the SC/FM interface. The up and down arrows denote, respectively, the majority and minority spin directions in the ferromagnetic contact, where we have set the $+z$ direction parallel to the majority direction. Using the Landauer–Büttiker formalism, the tunneling current across the SC/FM junction is given by [75, 86, 87]

$$\hat{J}(V) = \int\limits_0^\infty d\varepsilon \hat{j}(\varepsilon) = \frac{q}{\hbar} \int\limits_0^\infty d\varepsilon \int\limits_0^{k_\varepsilon} \frac{d^2 k_\parallel}{(2\pi)^2}$$

$$\left\{ \left[\hat{f}_{fm} - \hat{r}_{fm} \hat{f}_{fm} \hat{r}_{fm}^\dagger \right] - \left[\hat{f}_{sc} - \hat{r}_{sc} \hat{f}_{sc} \hat{r}_{sc}^\dagger \right] \right\}. \tag{17.37}$$

The first (second) term in square brackets is related to the transmitted current from the ferromagnet (semiconductor) due to electrons whose total and longitudinal energies are ε and $\varepsilon_\perp$, respectively. The zero energy refers to the bottom of the semiconductor conduction band. The inner integration is carried over transverse wave vectors due to a motion in parallel to the SC/FM interface, and its upper integration limit, k_ε, denotes the wave vector amplitude of an electron with energy ε. In the chosen spin coordinate system, the transmitted spin current from the ferromagnetic side is nonzero only along the z direction. Its spin-up and spin-down components are proportional, respectively, to $(1-|r_\uparrow|^2)$ and $(1-|r_\downarrow|^2)$, where we have rendered the fact that $\left| r_{\uparrow(\downarrow)} \right|^2 = \left| \tilde{r}_{\uparrow(\downarrow)} \right|^2$. The transmitted current from the semiconductor side, on the other hand, includes off-diagonal mixed terms that are proportional to $r_\uparrow r_\downarrow^*$. This is the case when $\mathbf{p}$ and $\mathbf{z}$ are neither parallel nor antiparallel due to the flow of electrons from/into ferromagnetic contacts that have non-collinear magnetization directions and are located within about a spin diffusion length.

The boundary conditions across an SC/FM junction were derived using the assumption that the spin-z-axis is collinear with the majority spin direction in the FM. However, to consider all (non-collinear) ferromagnetic terminals and the semiconductor channel as one system, one should use a single spin reference coordinate system. Specifically, we transform the general expression in Equation 17.37 into this new "contact-independent" coordinate system. We use $\tilde{x}$, $\tilde{y}$, and $\tilde{z}$ to represent the contact-dependent coordinates (for which the reflection matrices are diagonal), and we reserve x, y, and z for coordinates in the contact-independent system. The angle of the majority spin direction in the ith contact with respect to $+z$ is denoted by θ_i (in the yz plane). The relation between the spin current densities in these two frames is given by

$$\begin{pmatrix} J_{x,\alpha} \\ J_{y,\alpha} \\ J_{z,\alpha} \end{pmatrix} = \begin{pmatrix} 1 & 0 & 0 \\ 0 & \cos\theta_i & \sin\theta_i \\ 0 & -\sin\theta_i & \cos\theta_i \end{pmatrix} \begin{pmatrix} J_{\tilde{x},\alpha} \\ J_{\tilde{y},\alpha} \\ J_{\tilde{z},\alpha} \end{pmatrix}. \tag{17.38}$$

The charge current density does not depend on the spin space coordinate. Similarly, the spin polarization vector transformation follows:

$$
\begin{pmatrix} p_{\bar{x}} \\ p_{\bar{y}} \\ p_{\bar{z}} \end{pmatrix} = \begin{pmatrix} 1 & 0 & 0 \\ 0 & \cos\theta_i & -\sin\theta_i \\ 0 & \sin\theta_i & \cos\theta_i \end{pmatrix} \begin{pmatrix} p_x \\ p_y \\ p_z \end{pmatrix}.
\tag{17.39}
$$

Figure 17.7 shows the transient currents through the right contact of Figure 17.3a during its magnetization dynamics (2π rotation within 3 ns). The four curves denote various fixed magnetization directions of the left contact, $\theta_1 = (0, \pi/2, \pi, \text{ or } 3\pi/2)$, whereas the middle contact is pinned ($\theta_2 = 0$). We have included interface capacitance $c_b = 1$ μF cm^{-2} for the total charge current. These results also include the escape current mechanism due to localization of electrons at the interface region (the details of this effect are explained in [69]). The external capacitor is $C = 4$ fF, and the applied bias is 0.1 V. The Schottky barrier is assumed to be highly doped (2×10^{19} cm^{-3}), so that the total barrier conductance is about $G \approx 2 \times 10^4$ Ω^{-1} cm^{-2}. By detecting the signal shape, one can use such non-volatile magnetic transistors to hold more information than in binary devices.

17.5 SUMMARY

We have reviewed the physics of spin injection and studied the presence of spin accumulation in the semiconductor at small bias conditions. We have described how spin-dependent properties could be used to electrically sense

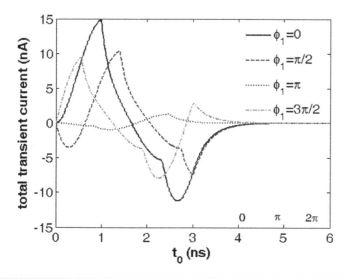

FIGURE 17.7 Transient currents of the right semi-floating contact by rotating its magnetization a single clockwise rotation from +z within 3 ns. The curves denote different fixed magnetization directions of the left contact, whereas the magnetization direction of the middle contact is pinned ($\theta_2 = 0$). (After Song, Y. et al., *Phys. Rev. B* 81, 045321, 2010. With permission.)

the spin accumulation in a semiconductor. The multi-terminal logic and memory devices that we have described rely only on spin selectivity of the junctions and the presence of spin accumulation. As recently experimentally demonstrated in graphene-based systems, these devices can work at room temperature and may improve common logic circuitry such as CAM [93]. In addition, by exploiting non-collinear magnetization configuration schemes, we have shown the increased storage capabilities of these devices.

ACKNOWLEDGMENTS

I would like to thank Łukasz Cywiński, Michael Huang, and Yang Song for their active contribution in writing this chapter. I would also like to thank Lu Sham, Parin Dalal, Berry Jonker, Ian Appelbaum, and Paul Crowell for many helpful discussions. This work was supported by the Air Force Office of Scientific Research (AFOSR) Discovery Challenge Thrusts (DCT) program and by National Science Foundation (NSF) contract No. ECCS-0824075.

REFERENCES

1. M. N. Baibich, J. M. Broto, A. Fert et al., Giant magnetoresistance of (001)Fe/(001)Cr magnetic superlattices, *Phys. Rev. Lett.* **61**, 2472 (1988).
2. G. Binasch, P. Grünberg, F. Saurenbach, and W. Zinn, Enhanced magneto-resistance in layered magnetic structures with antiferromagnetic interlayer exchange, *Phys. Rev. B* **39**, R4828 (1989).
3. J. S. Moodera, L. R. Kinder, T. M. Wong, and R. Meservey, Large magnetoresistance at room temperature in ferromagnetic thin film tunnel junctions, *Phys. Rev. Lett.* **74**, 3273 (1995).
4. R. P. Cowburn and M. E. Welland, Room temperature magnetic quantum cellular automata, *Science* **287**, 1466 (2000).
5. D. A. Allwood, G. Xiong, C. C. Faulkner, D. Atkinson, D. Petit, and R. P. Cowburn, Magnetic domain-wall logic, *Science* **309**, 1688 (2005).
6. A. Imre, G. Csaba, L. Ji, A. Orlov, G. H. Bernstein, and W. Porod, Majority logic gate for magnetic quantum-dot cellular automata, *Science* **311**, 205 (2006).
7. A. T. Hanbicki, R. Magno, S.-F. Cheng, Y. D. Park, A. S. Bracker, and B. T. Jonker, Nonvolatile reprogrammable logic elements using hybrid resonant tunneling diode-giant magnetoresistance circuits, *Appl. Phys. Lett.* **79**, 1190 (2001).
8. A. Ney, C. Pampuch, R. Koch, and K. H. Ploog, Programmable computing with a single magnetoresistive element, *Nature* **425**, 485 (2003).
9. R. Richter, L. Bär, J. Wecker, and G. Reiss, Nonvolatile field programmable spinlogic for reconfigurable computing, *Appl. Phys. Lett.* **80**, 1291 (2002).
10. H. Dery, P. Dalal, L. Cywiński, and L. J. Sham, Spin-based logic in semiconductors for reconfigurable large-scale circuits, *Nature* **447**, 573 (2007).
11. C. Ciuti, J. P. McGuire, and L. J. Sham, Spin-dependent properties of a twodimensional electron gas with ferromagnetic gates, *Appl. Phys. Lett.* **81**, 4781 (2002).
12. L. Cywiński, H. Dery, and L. J. Sham, Electric readout of magnetization dynamics in a ferromagnet-semiconductor system, *Appl. Phys. Lett.* **89**, 042105 (2006).
13. S. Datta and B. Das, Electronic analog of the electro-optic modulator, *Appl. Phys. Lett.* **56**, 665 (1990).
14. J. Fabian and I. Žutić, Spin-polarized current amplification and spin injection in magnetic bipolar transistors, *Phys. Rev. B* **69**, 115314 (2004).

15. J. Fabian, I. Žutić, and S. Das Sarma, Magnetic bipolar transistor, *Appl. Phys. Lett.* **84**, 85 (2004).

16. J. P. McGuire, C. Ciuti, and L. J. Sham, Theory of spin transport induced by ferromagnetic proximity on a two-dimensional electron gas, *Phys. Rev. B* **69**, 115339 (2004).

17. V. V. Osipov and A. M. Bratkovsky, Efficient nonlinear room-temperature spin injection from ferromagnets into semiconductors through a modified Schottky barrier, *Phys. Rev. B* **70**, 205312 (2004).

18. J. Schliemann, J. C. Egues, and D. Loss, Nonballistic spin-field-effect transistor, *Phys. Rev. Lett.* **90**, 146801 (2003).

19. I. Žutić, J. Fabian, and S. Das Sarma, Spintronics: Fundamentals and applications, *Rev. Mod. Phys.* **76**, 323 (2004).

20. I. Žutić, J. Fabian, and S. C. Erwin, Bipolar spintronics: From spin injection to spin-controlled logic, *J. Phys. Condens. Matter* **19**, 165219 (2007).

21. T. Dietl, H. Ohno, F. Matsukura, J. Cibert, and D. Ferrand, Zener model description of ferromagnetism in zinc-blende magnetic semiconductors, *Science* **287**, 1019 (2000).

22. Y. Ohno, D. K. Yound, B. Beschoten, F. Matsukura, H. Ohno, and D. D. Awschalom, Electrical spin injection in a ferromagnetic semiconductor heterostructure, *Nature* **402**, 790 (1999).

23. S. J. Potashnik, K. C. Ku, S. H. Chun, J. J. Berry, N. Samarth, and P. Schiffer, Effects of annealing time on defect-controlled ferromagnetism in $Ga_{1-x}Mn_xAs$, *Appl. Phys. Lett.* **79**, 1495 (2002).

24. Y. K. Kato, R. C. Myers, A. C. Gossard, and D. D. Awschalom, Observation of the spin Hall effect in semiconductors, *Science* **306**, 1910 (2004).

25. J. Wunderlich, B. Kaestner, J. Sinova, and T. Jungwirth, Experimental observation of the spin-Hall effect in a two-dimensional spin-orbit coupled semiconductor system, *Phys. Rev. Lett.* **94**, 047204 (2005).

26. I. Appelbaum, B. Huang, and D. J. Monsma, Electronic measurement and control of spin transport in silicon, *Nature* **447**, 295 (2007).

27. S. A. Crooker, M. Furis, X. Lou et al., Imaging spin transport in lateral ferromagnet/semiconductor structures, *Science* **309**, 2191 (2005).

28. A. T. Hanbicki, B. T. Jonker, G. Itskos, G. Kioseoglou, and A. Petrou, Efficient electrical spin injection from a magnetic metal/tunnel barrier contact into a semiconductor, *Appl. Phys. Lett.* **80**, 1240 (2002).

29. A. T. Hanbicki, O. M. J. van 't Erve, R. Magno et al., Analysis of the transport process providing spin injection through an Fe/AlGaAs Schottky barrier, *Appl. Phys. Lett.* **82**, 4092 (2003).

30. X. Jiang, R. Wang, R. M. Shelby et al., Highly spin-polarized room-temperature tunnel injector for semiconductor spintronics using MgO(100), *Phys. Rev. Lett.* **94**, 056601 (2005).

31. B. T. Jonker, Progress toward electrical injection of spin-polarized electrons into semiconductors, *Proc. IEEE* **91**, 727 (2003).

32. B. T. Jonker, G. Kioseoglou, A. T. Hanbicki, C. H. Li, and P. E. Thompson, Electrical spin-injection into silicon from a ferromagnetic metal/tunnel barrier contact, *Nat. Phys.* **3**, 542 (2007).

33. X. Lou, C. Adelmann, S. A. Crooker et al., Electrical detection of spin transport in lateral ferromagnet-semiconductor devices, *Nat. Phys.* **3**, 197 (2007).

34. B.-C. Min, K. Motohashi, C. Lodder, and R. Jansen, Tunable spin-tunnel contacts to silicon using low-work-function ferromagnets, *Nat. Mater.* **5**, 817 (2006).

35. D. Saha, M. Holub, and P. Bhattacharya, Amplification of spin-current polarization, *Appl. Phys. Lett.* **91**, 072513 (2007).

36. T. J. Zega, A. T. Hanbicki, S. C. Erwin et al., Determination of interface atomic structure and its impact on spin transport using Z-contrast microscopy and density functional theory, *Phys. Rev. Lett.* **96**, 196101 (2006).

37. H. J. Zhu, M. Ramsteiner, H. Kostial, M. Wassermeier, H.-P. Schönherr, and K. H. Ploog, Room-temperature spin injection from Fe into GaAs, *Phys. Rev. Lett.* **87**, 016601 (2001).

38. J. S. Friedman, N. Rangaraju, Y. I. Ismail, and B. W. Wessels, A spin-diode logic family, *IEEE Trans. Nanotechnol.* **11**, 1026 (2012).

39. J. S. Friedman and A. V. Sahakian, Complementary magnetic tunnel junction logic, *IEEE Trans. Electron. Devices* **61**, 1207 (2014).

40. J. S. Friedman, B. W. Wessels, G. Memik, and A. V. Sahakian, Emitter-coupled spintTransistor logic: Cascaded spintronic computing beyond 10 GHz, *IEEE J. Emerg. Sel. Top. Circuits Syst.* **5**, 17 (2015).

41. F. J. Jedema, A. T. Filip, and B. J. van Wees, Electrical spin injection and accumulation at room temperature in an all-metal mesoscopic spin valve, *Nature* **410**, 345 (2001).

42. Y. Ji, A. Hoffman, J. E. Pearson, and S. D. Bader, Enhanced spin injection polarization in Co/Cu/Co nonlocal lateral spin valves, *Appl. Phys. Lett.* **88**, 052509 (2006).

43. M. Johnson, Bipolar spin switch, *Science* **260**, 320 (1993).

44. H. Wen, H. Dery, W. Amamou et al., Experimental demonstration of XOR operation in graphene magnetologic gates at room temperature, *Phys. Rev. Appl.* **5**, 044003 (2016).

45. A. Fert and A. Jaffrès, Conditions for efficient spin injection from a ferromagnetic metal into a semiconductor, *Phys. Rev. B* **64**, 184420 (2001).

46. E. I. Rashba, Theory of electrical spin injection: Tunnel contacts as a solution of the conductivity mismatch problem, *Phys. Rev. B* **62**, R16267 (2000).

47. G. Schmidt, D. Ferrand, L. W. Molenkamp, A. T. Filip, and B. J. van Wees, Fundamental obstacle for electrical spin injection from a ferromagnetic metal into a diffusive semiconductor, *Phys. Rev. B* **62**, R4790 (2000).

48. M. Johnson and R. H. Silsbee, Thermodynamic analysis of interfacial transport and of the thermomagnetoelectric system, *Phys. Rev. B* **35**, 4959 (1987).

49. R. Stratton, Tunneling in schottky barrier rectifiers, in *Tunneling Phenomena in Solids*, E. Burstein and S. Lundqvist (Eds.), Plenum Press, New York, p. 105, 1969.

50. S. M. Sze, *Physics of Semiconductor Devices*, John Wiley, New York, 1981.

51. H. Dery, L. Cywiński, and L. J. Sham, Lateral diffusive spin transport in layered structures, *Phys. Rev. B* **73**, 041306 (2006).

52. A. Fert, J.-M. George, H. Jaffrès, and R. Mattana, Semiconductors between spin-polarized sources and drains, *IEEE Trans. Electron. Dev.* **54**, 921 (2007).

53. C. Adelmann, J. Q. Xie, C. J. Palmstrøm, J. Strand, J. Wang, and P. A. Crowell, Effects of doping profile and post-growth annealing on spin injection from Fe into (Al,Ga)As heterostructures, *J. Vac. Sci. Technol. B* **23**, 1747 (2005).

54. G. E. Pikus and A. N. Titkov, Spin relaxation, in *Optical Orientation*, F. Meier and B. P. Zakharchenya (Eds.), North-Holland, New-York, p. 73, 1984.

55. J. M. Geraldo, W. N. Rodrigues, G. Medeiros-Ribeiro, and A. G. de Oliveira, The effect of the planar doping on the electrical transport properties at the Al:n-GaAs(100) interface: Ultrahigh effective doping, *J. Appl. Phys.* **73**, 820 (1993).

56. V. I. Shashkin, A. V. Murel, V. M. Daniltsev, and O. I. Khrykin, Control of charge transport mode in the Schottky barrier by d-doping: Calculation and experiment for Al/GaAs, *Semiconductors* **36**, 505 (2002).

57. M. Zachau, F. Koch, K. H. Ploog, P. Roentgen, and H. Beneking, Schottky-barrier tunneling spectroscopy for the electronic subbands of a δ-doping layer, *Solid State Commun.* **59**, 591 (1986).

58. H. Dery and L. J. Sham, Spin extraction theory and its relevance to spintronics, *Phys. Rev. Lett.* **98**, 046602 (2007).

59. P. Li and H. Dery, Tunable spin junction, *Appl. Phys. Lett.* **94**, 192108 (2009).

60. O. Chalaev, Y. Song, and H. Dery, Suppressing the spin relaxation of electrons in silicon, *Phys. Rev. B* **95**, 035204 (2017).

61. S. Dushenko, M. Koike, Y. Ando, T. Shinjo, M. Myronov, and S. Shiraishi, Experimental demonstration of room-temperature spin transport in n-type germanium epilayers, *Phys. Rev. Lett.* **114**, 196602 (2015).
62. Y. Fujita, M. Yamada, S. Yamada, T. Kanashima, K. Sawano, and K. Hamaya, Temperature-independent spin relaxation in heavily doped n-type germanium, *Phys. Rev. B* **94**, 245302 (2016).
63. M. Ishikawa, T. Oka, Y. Fujita, H. Sugiyama, Y. Saito, and K. Hamaya, Spin relaxation through lateral spin transport in heavily doped n-type silicon, *Phys. Rev. B* **95**, 115302 (2017).
64. Y. Song, O. Chalaev, and H. Dery, Donor-driven spin relaxation in multivalley semiconductors, *Phys. Rev. Lett.* **113**, 167201 (2014).
65. Y. Song and S. Das Sarma, Impurity-driven two-dimensional spin relaxation induced by intervalley spin-flip scattering in silicon, *Phys. Rev. Appl.* **7**, 014003 (2017).
66. V. Sverdlov and S. Selberherr, Silicon spintronics: Progress and challenges, *Phys. Rep.* **585**, 1 (2015).
67. M. Yamada, Y. Fujita, M. Tsukahara, S. Yamada, K. Sawano, and K. Hamaya, Large impact of impurity concentration on spin transport in degenerate n-Ge, *Phys. Rev. B* **95**, 161304(R) (2017).
68. R. A. Smith, *Semiconductors*, Cambridge University Press, Cambridge, 1978.
69. Y. Song and H. Dery, Spin transport theory in ferromagnet/semiconductor systems with noncollinear magnetization configurations, *Phys. Rev. B* **81**, 045321 (2010).
70. W. H. Butler, X.-G. Zhang, X. Wang, J. van Ek, and J. M. MacLaren, Electronic structure of FM–semiconductor–FM spin tunneling structures, *J. Appl. Phys.* **81**, 5518 (1997).
71. A. N. Chantis, K. D. Belashchenko, D. L. Smith, E. Y. Tsymbal, M. van Schilfgaarde, and R. C. Albers, Reversal of spin polarization in Fe/GaAs (001) driven by resonant surface states: First-principles calculations, *Phys. Rev. Lett.* **99**, 196603 (2007).
72. O. Wunnicke, Ph. Mavropoulos, R. Zeller, P. H. Dederichs, and D. Grundler, Ballistic spin injection from Fe(001) into ZnSe and GaAs, *Phys. Rev. B* **65**, 241306(R) (2002).
73. M. Zwierzycki, K. Xia, P. J. Kelly, G. E. W. Bauer, and I. Turek, Spin injection through an Fe/InAs interface, *Phys. Rev. B* **67**, 092401 (2003).
74. J. C. Slonczewski, Conductance and exchange coupling of two ferromagnets separated by a tunneling barrier, *Phys. Rev. B* **39**, 6995 (1989).
75. C. Ciuti, J. P. McGuire, and L. J. Sham, Spin polarization of semiconductor carriers by reflection off a ferromagnet, *Phys. Rev. Lett.* **89**, 156601 (2002).
76. J. D. Albrecht and D. L. Smith, Spin-polarized electron transport at ferromagnet semiconductor Schottky contacts, *Phys. Rev. B* **68**, 035340 (2002).
77. S. Saikin, A drift-diffusion model for spin-polarized transport in a two-dimensional non-degenerate electron gas controlled by spin–orbit interaction, *J. Phys. Condens. Matter* **16**, 5071 (2004).
78. S. Saikin, M. Shen, and Cheng, M.-C., Spin dynamics in a compound semiconductor spintronic structure with a Schottky barrier, *J. Phys. Condens. Matter* **18**, 1535 (2006).
79. Z. G. Yu, and M. E. Flatté, Spin diffusion and injection in semiconductor structures: Electric field effects, *Phys. Rev. B* **66**, 235302 (2002).
80. B. A. Gurney, V. S. Speriosu, J. P. Nozieres, H. Lefakis, D. R. Wilhoit, and O. U. Need, Direct measurement of spin-dependent conduction-electron mean free paths in ferromagnetic metals, *Phys. Rev. Lett.* **71**, 4023 (1993).
81. L. Cywiński, H. Dery, P. Dalal, and L. J. Sham, Electrical expression of spin accumulation in ferromagnet/semiconductor structures, *Mod. Phys. Lett. B* **21**, 1509 (2007).
82. S. Tehrani, B. Engel, J. M. Slaughter et al., Recent developments in magnetic tunnel junction MRAM, *IEEE Trans. Magn.* **36**, 2752 (2000).

83. B. C. Choi and M. R. Freeman, Nonequilibrium spin dynamics in laterally defined magnetic structures, in *Ultrathin Magnetic Structures III*, B. Heinrich and J. A. C. Bland (Eds.), Springer-Verlag, Berlin, p. 211, 2004.

84. Th. Gerrits, H. A. M. van den Berg, J. Hohlfeld, L. Bär, and Th. Rasing, Ultrafast precessional magnetization reversal by picosecond magnetic field pulse shaping, *Nature* **418**, 509 (2002).

85. W. Schumacher, C. Chappert, P. Crozat, R. C. Sousa, P. P. Freitas, and M. Bauer, Coherent suppression of magnetic ringing in microscopic spin valve elements, *Appl. Phys. Lett.* **80**, 3781 (2002).

86. A. Brataas, Y. V. Nazarov, and G. E. W. Bauer, Finite-element theory of transport in ferromagnet-normal metal systems, *Phys. Rev. Lett.* **84**, 2481 (2000).

87. A. Brataas, Y. V. Nazarov, and G. E. W. Bauer, Spin-transport in multi-terminal normal metal-ferromagnet systems with non-collinear magnetizations, *Eur. Phys. J. B* **22**, 99 (2001).

88. B. C. Choi, M. Belov, W. K. Hiebert, G. E. Ballentine, and M. R. Freeman, Ultrafast magnetization reversal dynamics investigated by time domain imaging, *Phys. Rev. Lett.* **86**, 728 (2001).

89. T. Kohonen, *Content-Addressable Memories*, Springer-Verlag, Heidelberg, 1980.

90. Compaq Computer Corporation, *Alpha 21264/EV6 Microprocessor Hardware Reference Manual*, September 2000. Order number: DS-0027B-TE.

91. J. Tendler, J. Dodson, J. Fields, H. Le, and B. Sinharoy, POWER4 system microarchitecture, *IBM J Res. Develop.* **46**, 5 (2002).

92. D. Boggs, A. Baktha, J. Hawkins et al., The Microarchitecture of the Intel Pentium 4 Processor on 90 nm Technology, *Intel Technol. J.* **8**, 1 (2004).

93. H. Dery, H. Wu, B. Ciftcioglu et al., Spintronic manoelectronics based on magneto-logic gates, *IEEE Trans. Electron. Dev.* **59**, 259 (2012).

18
Spin Wave
Logic Devices

Alexander Khitun and Ilya Krivorotov

18.1 INTRODUCTION

Spintronics is an emerging approach toward novel computing devices, which takes advantage of the spin degree of freedom in addition to the electric charge of electrons [1]. The variety of spin logic devices described in Chapter 17 illustrates the intriguing possibility of building scalable energy-efficient devices using the spin of electrons. However, efficient spin injection and long spin diffusion length in the channel of a spin-field-effect-transistor (FET) are the two major challenges inherent in all spin-FETs. When injected into a semiconductor channel, the spins of conduction electrons are subject to different relaxation mechanisms (e.g., Elliott [2]; D'yakonov and Perel' [3]; Bir et al. [4]), which reduce the spin polarization (see Chapter 1, Volume 2, of this book for a detailed discussion of these relaxation mechanisms). All scattering mechanisms tend to equalize the number of spin up and spin down electrons in a non-magnetic semiconductor channel. In turn, the variation of the spin polarization among the ensemble of conducting electrons reduces the on/off ratio. In the best case scenario, materials with high mobility and low scattering (e.g., graphene) show electron spin diffusion length on the order of several micrometers at room temperature [5].

The problems associated with the limited spin diffusion length can be resolved by using collective spin phenomena, where the interactions among a large number of spins make the system more immune to spin scattering. A spin wave is a collective classical oscillation of spins around the equilibrium direction of magnetization with well-defined wavelength, phase, and direction of propagation. Spin waves appear in magnetically ordered materials, and a quantum of a spin wave is referred to as a *magnon* (see Chapter 6, Volume 1, for basics of spin waves). The collective nature of spin wave phenomena manifests itself in relatively long coherence length, which reaches tens of micrometers in conducting ferromagnetic materials (e.g., $Ni_{81}Fe_{19}$ [6]) and can exceed millimeters in ferrite insulators (e.g. $Y_3Fe_2(FeO_4)_3$ [7]) at room temperature. The first operational spin wave-based logic device was experimentally demonstrated by Kostylev et al. in 2005 [8]. The authors constructed a Mach–Zehnder-type spin wave interferometer to demonstrate the output voltage modulation as a result of spin wave interference. The schematics of this spin wave device operating as an XOR gate are shown in Figure 18.1. The phase difference among the spin waves propagating in the arms of the interferometer is controlled by an Oersted magnetic field produced by the electric gate currents I applied to the two arms of the interferometer, resulting in either constructive or destructive interference of spin waves propagating in the two arms. Subsequently, exclusive-not-OR and not-AND gates have been experimentally demonstrated in a similar Mach–Zehnder-type structure [9]. This first working prototype stimulated a great deal of interest in spin wave logic devices [8–14]. In this chapter, we present an overview of the spin wave logic, including devices for Boolean and non-Boolean logic circuits. Fundamentals of magnon-based computing are discussed in Chapter 6, Volume 1, of this book.

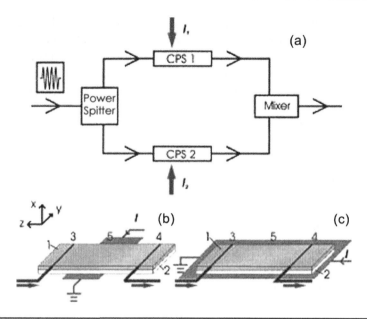

FIGURE 18.1   (a) Schematic diagram of Mach–Zehnder interferometer. (b) Controlled phase shifter (CPS) based on backward volume magnetostatic spin wave (BVMSW) propagation. (c) CPS based on magnetostatic surface spin wave (MSSW) propagation. 1: ferromagnetic film, 2: nonmagnetic substrate, 3: input strip-line microwave transducer, 4: output microwave transducer, 5: control-current stripe conductor. (Panel (a) after Kostylev, M.P. et al., *Appl. Phys. Lett.* 87, 153501, 2005. With permission.)

18.2 BOOLEAN-TYPE LOGIC GATES

Boolean spin wave circuits are aimed at providing the same basic set of logic gates (AND, OR, NOT) for general-type Boolean computing as provided by the conventional transistor-based circuit. The advantage of using waves (i.e., spin waves) is the ability to exploit the waveguides as passive logic elements for controlling the phase of the propagating wave. Waveguides of the same length but with different width or composition introduce different phase change to the propagating spin waves. The latter offers an additional degree of freedom for logic circuit construction. In addition, the use of spin wave interference is efficient for building high fan-in devices, which is a significant advantage over the transistor-based circuits [15]. It is possible to build multifunctional logic gates (e.g., majority [MAJ] or MOD), which enables the construction of all other types of Boolean logic gates with fewer components [16].

18.2.1 GENERAL SCHEMATICS OF SPIN WAVE DEVICES

The schematics of the spin wave MAJ+NOT gate logic circuit are shown in Figure 18.2 [17]. The circuit consists of the following elements: (i) magneto-electric (ME) cells, (ii) magnetic waveguides serving as spin wave buses, and (iii) a phase shifter. The ME cell provides inter-conversion between the applied electric voltage and the spin wave signal. It can be used to both excite the spin wave via the application of a voltage pulse and to electrically read out the output spin wave signal. The operation of the ME cell is based on the magneto-electric coupling effect (such as that present in multiferroics), which enables magnetization control by an electric field and vice versa. The waveguides are simply strips of ferromagnetic material (e.g., NiFe or yttrium iron garnet [YIG]), whose purpose is to transmit the spin wave signal in a particular direction with minimal loss of amplitude. The phase shifter is a passive element providing a π-phase shift to the propagating spin waves. Examples of such a phase shifter include a section of a waveguide with different width and magnetic domain wall.

The principle of operation of the logic gate in Figure 18.2 is described as follows. The input information is received in the form of voltage pulses. Binary inputs 0 and 1 are encoded in the polarity of the voltage applied to the input ME cells (e.g., +10 mV corresponds to logic state 0, and –10 mV corresponds to logic state 1). The polarity of the applied voltage defines the initial

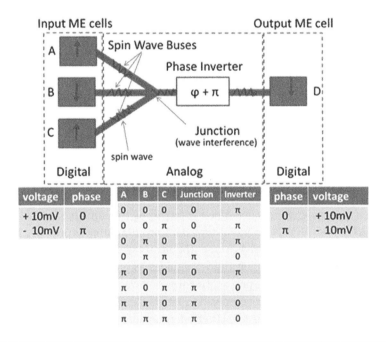

voltage	phase	A	B	C	Junction	Inverter	phase	voltage
+ 10mV	0	0	0	0	0	π	0	+ 10mV
- 10mV	π	0	0	π	0	π	π	- 10mV
		0	π	0	0	π		
		0	π	π	π	0		
		π	0	0	0	π		
		π	0	π	π	0		
		π	π	0	π	0		
		π	π	π	π	0		

FIGURE 18.2 Schematic view of a majority-type spin wave logic circuit. The gate has three inputs (A, B, C) and a single output D. The inputs and the output are the ME cells connected via the ferromagnetic waveguides serving as spin wave buses. Voltage pulses of opposite polarity applied to the input cells generate spin waves of the same amplitude but different initial phase 0 or π, corresponding to logic 0 and 1, respectively. The waves propagate through the waveguides and interfere at the point of junction. The phase of the wave transmitted through the junction corresponds to the majority of the interfering input wave phases. The phase of the transmitted wave is inverted (via, e.g., passing through a domain wall). The truth table illustrates the phase data processing. The phase of the transmitted wave defines the final magnetization of the output ME cell D. The circuit can operate as NAND or NOR gate for inputs A and B depending on the third input C (NOR if C = 1, NAND if C = 0). (After Khitun, A., *J. Appl. Phys.* 111, 054307, 2012. With permission.)

phase of the spin wave signal. The mechanism of transduction of the voltage polarity into the spin wave phase can be understood as follows. The ME coupling converts the applied voltage into an effective magnetic field pulse that rotates the equilibrium direction of magnetization. As a result, the magnetization starts to precess around the new equilibrium direction and thereby generates a propagating disturbance in the form of a spin wave. Since the voltage-induced torque applied to the magnetization has opposite sign for opposite voltage polarities, the phase of the generated spin wave will also differ by π for opposite polarities of the applied voltage. Thus, the input information is translated into the phase of the excited wave (e.g., initial phase 0 corresponds to logic state 0, and initial phase π corresponds to logic state 1). Then, the waves propagate through the magnetic waveguides and interfere at the point of waveguide junction. For any junction with an odd number of interfering waves, there is a transmitted wave with non-zero amplitude. The phase of the wave exiting the junction always corresponds to the majority of the phases of the interfering waves (for example, the transmitted wave will have phase 0, if there are two or three input waves with 0 phase). The wave transmitted through the phase shifter accumulates an additional π-phase shift (i.e., phase $0 \rightarrow \pi$, and phase $\pi \rightarrow 0$). Finally, the spin wave signal reaches the output ME cell. The output cell has two stable magnetization states. At the moment of spin wave arrival, the output cell is placed into a metastable state by a readout initialization voltage pulse (magnetization is along the hard axis perpendicular to the two stable states). The *phase* of the incoming spin wave defines the direction of the magnetization relaxation in the output cell [16, 18]. The process of magnetization switching in the output ME cell is associated with the change of electrical polarization in the multiferroic material and can be recognized by the induced voltage across the ME cell [19] (e.g., +10 mV corresponds to logic state 0, and –10 mV corresponds to logic state 1). The truth table in Figure 18.2 shows the input/output phase correlation. The waveguide junction works as a MAJ logic gate. The amplitude of the transmitted wave depends on the number of the in-phase waves, while the phase of the transmitted wave always corresponds to the majority of the phase inputs. The π-phase shifter works as an inverter in the phase space. The three-input one-output gate in Figure 18.2 can operate as either a NAND or a NOR gate for inputs A and B depending on the third input C (NOR if C = 1, NAND if C = 0). Such a gate can be a universal building block for any Boolean logic gate construction. In general, wave-based logic circuits may process a number of bits in parallel by exploiting wave superposition. For instance, the multifunctional logic gate shown in Figure 18.2 can be modified for multi-frequency operation. The same circuit consisting of waveguides, junctions, and phase shifters can operate in the range of frequencies $(f_1, f_2, \ldots f_n)$, which is defined by the waveguide's bandwidth and the size and composition of the junctions and phase shifters. Each of the frequencies $(f_1, f_2, \ldots f_n)$ is considered as an independent information channel, where logic 0 and 1 are encoded into the phase of the propagating spin wave. An example of a multi-frequency circuit is described in Khitun [17].

In addition to the phase-based spin wave logic, devices based on the amplitude of spin wave as a state variable have been proposed and realized. A notable example is the magnon transistor, which is based on the nonlinear interaction of magnons [20]. The operation of such a three-terminal device based on a YIG waveguide resembles that of the conventional charge-based FET. In the absence of magnons injected via the gate terminal, magnons can propagate in the YIG waveguide between the source and drain terminals with negligible scattering. Therefore, the magnon transistor is in the "on" state. When a sufficiently high density of magnons is injected through the gate terminal, spin waves propagating from the source are scattered on the magnons injected through the gate terminal and never reach the drain. In this case, the magnon transistor is in the "off" state. Digital logic circuits based on such magnon transistors can have architectures very similar to those made from conventional FETs. However, spin wave amplifiers are needed to ensure proper signal fan-out.

18.2.2 Spin Wave Components and Prototypes

During the past decade, three- and four-terminal spin wave devices exploiting spin wave interference have been experimentally demonstrated [15, 21]. Figure 18.3a shows the top view of a four-terminal spin wave device. The sample consists of a silicon substrate, a 20 nm thick layer of $Ni_{81}Fe_{19}$, a 300 nm layer of silicon dioxide, and a set of five conducting wires on top. The distance between the wires is 2 μm. Three of the five wires are used as input ports, and the two other wires are connected in an inductive pickup loop to detect the inductive voltage produced by the output spin wave signal. The direction of the current in the wires defines the phase difference between the excited spin waves (i.e., the waves have the same initial phase if the direction of the current is the same, and there is a π-phase difference if the current flows in the opposite direction). The plot in Figure 18.3b shows the output inductive voltage for different combinations of spin wave phases. The blue and the red curves correspond to the case when all three waves are excited in phase, and the black and the green curves correspond to the case when one of the waves has a π-phase difference with the others. The experimental data show the change of the phase of the output voltage as a function of the input combination (i.e., 000, π00, 0π0, 00π corresponds to the output phase 0, and πππ, 0ππ, π0π, ππ0 results in the change of the output phase voltage on π). This logic device operates as a MAJ gate for binary phase variables [22]. The data are taken at room temperature in a 95 Oe bias magnetic field (perpendicular to the spin wave propagation) and 3 GHz frequency.

The use of electric-current wires for spin wave excitation is not energy efficient (i.e., more than picojoules per spin wave), as only a small fraction of electromagnetic energy is transferred to the spin wave. An energy-efficient method for the excitation and detection of spin waves is multiferroic or ME energy conversion between the electric and magnetic domains [23–26]. Recently, spin wave excitation and detection by synthetic multiferroic elements has been experimentally demonstrated [27]. The schematics

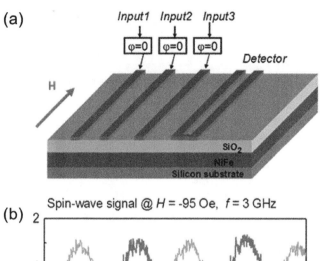

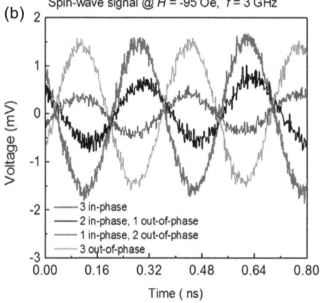

FIGURE 18.3   (a) Schematics of a four-terminal magnonic device. The device structure comprises a silicon substrate, a 20 nm thick layer of permalloy, a layer of silicon dioxide, and a set of five conducting wires on top (three wires to excite three spin waves, and the other two wires connected in a loop to detect the inductive voltage). The phase of the excited spin wave (0 or π) is controlled by the phase of the excitation microwave current. (b) Experimental data showing the inductive voltage as a function of time. The curves of different color correspond to the different combinations of the phases of the interfering spin waves. (After Khitun, A. et al., *Annual Report of Western Institute of Nanoelectronics* Abstract 2.1, 2009. With permission.)

of the experiment and the experimental data are shown in Figure 18.4. Two synthetic multiferroic elements were used to excite and detect spin waves propagating in a permalloy waveguide (the distance between the excitation and the detection elements is 40 μm). The multiferroic element consists of a layer of piezoelectric (PZT) and a layer of magnetostrictive material (Ni). An electric field applied across the piezoelectric produces stress, which, in turn, affects the magnetization of the magnetostrictive material. Thus, the application of alternating voltage to the multiferroic element results in

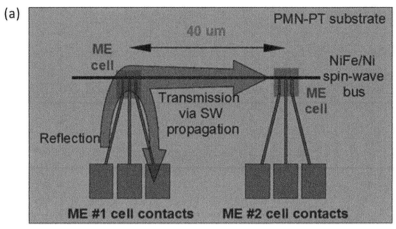

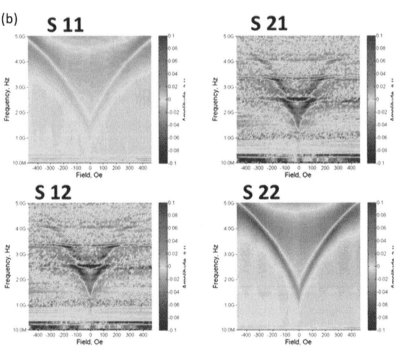

FIGURE 18.4   (a) Schematics of the experiment on spin wave excitation and detection by multiferroic elements (ME cells). (b) Collection of the experimental data (S11, S12, S21, and S22 parameters) obtained at different frequencies and bias magnetic field.

magnetization oscillation, which can propagate as a spin wave in the permalloy waveguide. Vice versa, the oscillation of magnetization in the magnetostrictive layer results in an alternating voltage signal generated by the ME cell due to the stress produced by the oscillation. The experimental data in Figure 18.4b show the excitation, propagation, and detection of spin waves in an ME-based spin wave gate at different operation frequencies and bias magnetic fields. The use of multiferroics has resulted in the energy being reduced to approximately 10 fJ per spin wave cycle [27].

18.2.3 COMPARISON WITH COMPLEMENTARY METAL-OXIDE-SEMICONDUCTOR (CMOS)

The principle of operation of spin wave devices is different from the conventional CMOS technology, and the design rules adopted in CMOS are not applicable to the spin wave circuits using spin wave phase as the state variable. To compare spin wave devices with CMOS, we present the estimates on the circuit level, including area, time delay, energy per operation, and functional throughput. The area of the spin wave logic circuit is defined by several parameters: the size of the ME cell ($F \times F$); the number of ME cells per circuit, N_{ME}; and the length L_{swb} and the width W_{swb} of the spin wave buses. These parameters are related to each other via the same physical quantity—the wavelength of the spin wave. Theoretically, the feature size F of the ME cell can be much smaller than the wavelength λ of the information-carrying spin waves. On the other hand, the length of the ME cells should be similar to the wavelength $F \sim \lambda$ for efficient spin wave excitation via the ME coupling. The width of the spin wave bus W_{swb} is also related to the wavelength λ via the dispersion law. However, the width of the spin wave bus can be much smaller than the wavelength. In our estimates, we assume the feature size of the ME cell F to be equal to the wavelength λ ($F \approx \lambda$), $W_{swb} << L_{swb}$, and L_{swb} to be one or one and a half times the wavelength depending on the particular logic circuit (e.g., $L_{swb} = \lambda$ for a buffer gate, $L_{swb} = \lambda/2$ for an inverter). The number of ME cells per circuit varies depending on the circuit functionality. At present, there is no empirical rule to estimate the size of the magnonic logic circuits based on the number of ME cells. In the following, we present estimates for the area A of some logic circuits [16].

$$A = F \times (2F + \lambda) \approx 2\lambda^2\text{-Buffer}$$

$$A = F \times (2F + \lambda/2) \approx 2.5\lambda^2\text{-Inverter}$$

$$A = F \times (3F + \lambda + \lambda) \approx 3\lambda^2\text{-AND gate}$$

$$A = (3F + 2\lambda) \times (2F + \lambda) \approx 15\lambda^2\text{-MAJ gate/MOD2 gate}$$

$$A = (3F + 2\lambda) \times (3F + 2\lambda) \approx 25\lambda^2\text{-Full Adder Circuit.}$$

Time delay per circuit is a sum of (i) the time required to excite spin waves by the input ME cells t_{ext}, (ii) propagation time for spin waves from the input to the output cells t_{prop}, and (iii) the time of magnetization relaxation in the output ME cells t_{relax}:

$$t_{delay} = t_{ext} + t_{prop} + t_{relax}.$$

The minimum time delay for spin wave excitation is limited by the RC delay of the electric part, where R is the resistance of metallic interconnects, and C is the capacitance of the ME cell. Typically [15], the RC delay is much shorter than the time required for the spin wave to propagate between the excitation and the detection ports. The propagation time can be estimated by dividing the length of the spin wave bus connecting the most distant input and the output cells by the spin wave group velocity v_g, $t_{prop} = L_{swb}/v_g$. The group velocity depends on the material and geometry of the bus as well as the specific spin wave mode. The typical group velocity of backward volume magnetostatic spin waves propagating in conducting ferromagnetic materials (e.g., NiFe) is approximately 10^6 cm/s [30].

We want to emphasize the difference between the volatile and non-volatile magnonic circuits in terms of the operation speed. The speed of operation of the volatile magnonic circuits is limited only by the spin wave excitation time, the length of the circuit, and the spin wave group velocity, while the non-volatile circuits require an additional time for the output bi-stable ME cell switching. The relaxation time of the output ME cell depends on the material properties of the magnetostrictive material (e.g., damping parameter α). Realistically, the minimum time delay required for magnetization reversal t_{relax} of a bi-stable nano-magnet (thermal stability > 40) is about 100 ps, which may be much longer than the propagation time (e.g., 100 nm/10^6 cm/s = 10 ps).

The energy per operation in the magnonic logic circuits depends on the number of ME cells and the energy required for magnetization rotation in each cell. It is important to note that the electric field required for magnetization rotation in Ni/PZT synthetic multiferroic is about 1.2 MV/m [29]. The latter promises a very low, in the order of attojoules, energy per switch achievable in nanometer-scale ME cells (e.g., 24 aJ for 100 nm × 100 nm ME cell with 0.8 μm PZT). Thus, the maximum power dissipation density per 1 μm² area circuit operating at 1 GHz frequency can be estimated as 7.2 W/cm² (three input cells per one frequency). The addition of an extra operating frequency would linearly increase the power dissipation in the circuit. The upper limit for the acceptable power dissipation in magnonic circuits may be higher than that of their silicon counterparts, as the metallic waveguides can be placed on the non-magnetic metallic base (e.g., Cu) to enhance the thermal transport.

The comparison between the magnonic and CMOS-based logic devices should be made at the circuit level by comparing the overall circuit parameters, such the number of functions per area per time, time delay per operation, and energy required for a logic function. In Table 18.1, we summarized the estimates for a magnonic full adder circuit and compare them with the parameters of the CMOS-based circuit. The data for the full adder circuit

TABLE 18.1
Comparison between the Magnonic and the Conventional Full Adder Circuits

	45 nm	32 nm	$\lambda = 45$ nm	$\lambda = 32$ nm
Area	6.4 μm²	3.2 μm²	0.05 μm²	0.026 μm²
Time delay	12 ps	10 ps	13.5 ps/0.1 ns	9.6 ps/0.1 ns
Functional throughput	1.3 × 10⁹ Ops/(ns cm²)	3.1 × 10⁹ Ops/(ns cm²)	1.48 × 10¹¹ Ops/(ns cm²)	4.0 × 10¹¹ Ops/(ns cm²)
Energy per operation	12 fJ	10 fJ	24 aJ	15 aJ
Static power	>70 nW	>70 nW	–	–

made on 45 nm and 32 nm CMOS technology is based on the ITRS projections [30] and available data on current technology [31]. The data for the magnonic circuits is based on the design described in Khitun and Wang [16] and the estimates made earlier. Magnonic circuits are predicted to have a significant ~100× advantage in minimizing circuit area due to the lower number of elements required per circuit (e.g., five ME cells versus 25–30 CMOSs). At the same time, magnonic logic circuits would be slower than their CMOS counterparts. In Table 18.1, we have shown two numbers for time delay corresponding to volatile and non-volatile circuits. The delay time of the volatile circuit is mainly defined by the spin wave group velocity, while the delay time of the non-volatile circuit is restricted by the relaxation time of the output ME cell. The most prominent (~1000×) advantage over CMOS circuitry is expected to be in minimizing power consumption. Besides the great reduction of active power, there is no static power consumption in magnonic logic circuits based on non-volatile magnetic cells. The overall functional throughput is about 100 times higher for magnonic logic circuits due to the smaller circuit area.

We also estimate the additional functional throughput enhancement due to the use of multiple frequencies. In Figure 18.5, we present the estimates on the functional throughput in operations per nanosecond per centimeter squared for the full adder circuit built of scaled CMOS (blue markers) and the magnonic multi-frequency circuit (red markers). The estimates for the CMOS full adder circuit are based on the data for the 32 nm

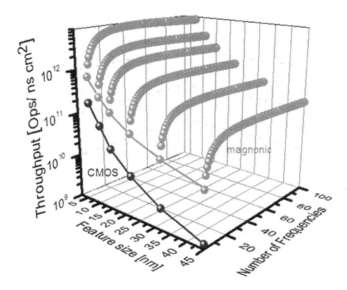

FIGURE 18.5 Numerical estimates on the functional throughput for the full adder circuit built of CMOS (blue markers) and magnonic multi-frequency circuit (red markers). The estimates for the CMOS circuit are based on the 32 nm CMOS technology. The estimates for smaller feature size are extrapolated: the area per circuit scales as ×0.5 per generation, and the time delay scales as ×0.7 per generation. (After Khitun, A., *J. Appl. Phys.* 111, 054307, 2012. With permission.)

CMOS technology (area = 3.2 μm^2, time delay = 10 ps from Chen [31].) The estimates for further generations are extrapolated by using the following empirical rule: the area per circuit scales as ×0.5 per generation, and the time delay scales as ×0.7 per generation. The estimates for the magnonic circuit are based on the multi-frequency model presented in Ref. [17]. The plot in Figure 18.5 shows the relative functional throughput enhancement as a function of the number of frequencies N (independent information channels). It should be noted that the introduction of an additional frequency channel is associated with the additional area and time delay introduced by each new input/output port. There is some optimum number of the operational channels providing maximum functional throughput, which varies for different logic circuits.

In conclusion, Boolean-type spin wave logic devices provide an alternative approach to logic circuit construction by exploiting the intrinsic non-linearity of magnetic materials. In contrast to spin-FETs, most magnonic devices do not require highly efficient spin injection, and room temperature–operating prototypes have been demonstrated. Non-volatility is the main advantage inherent in all spin wave device schemes, which may potentially eliminate the need for static power consumption. The future success of spin wave logic is mainly tied to the realization of the most energy-efficient mechanism for local magnetization control. The development and use of multiferroic materials is a promising approach capable of scaling down energy per magnetization reversal to the attojoule level. At the same time, the use of nanomagnets for information storage implies certain limits on the operation speed, which is restricted by the time required for magnetization reversal. It is unlikely that devices based on ferromagnets will be able to compete with CMOS in speed. Nevertheless, magnetic logic circuits can provide higher functional throughput or perform the same data processing tasks with lower energy consumption than CMOS by implementing complex logic functions and/or reconfigurability.

18.3 NON-BOOLEAN SPIN WAVE LOGIC DEVICES

In contrast to the Boolean logic gates for general-type data processing, non-Boolean circuits are designed for one or several specific logic operations. Data search and image processing are examples of widely used tasks that require a significant amount of resources from a general-type processor. In this section, we consider examples of non-Boolean data processing with magnonic holographic devices [32].

18.3.1 MAGNONIC HOLOGRAPHIC MEMORY

Magnonic holographic memory (MHM) is a type of holographic device similar to the holographic devices developed in optics but using spin waves instead of light [33]. The schematics of the MHM are shown in Figure 18.6a. The core of the MHM structure is a magnetic matrix consisting of a grid of magnetic waveguides with nano-magnets placed on top of the waveguide

(a)

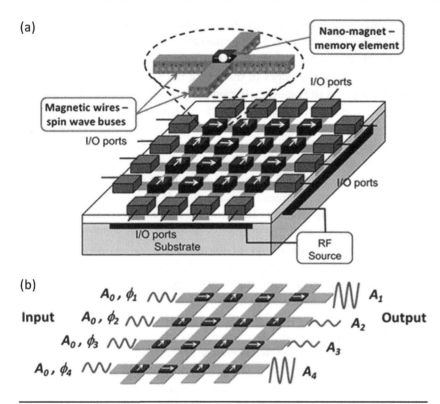

(b)

FIGURE 18.6   (a) Schematics of MHM consisting of a 4×4 magnetic matrix and an array of spin wave generating/detecting elements. For simplicity, the matrix is depicted as a two-dimensional grid of magnetic wires with just four elements on each side. These wires serve as a medium for spin wave propagation. The nano-magnets on the top of the junctions are memory elements, where information is encoded into the magnetization state. The spins of the nano-magnet are coupled to the spins of the magnetic wires via the dipole–dipole or exchange interaction. (b) Illustration of the principle of operation. Spin waves are excited by the elements on one or several sides of the matrix (e.g., left side), propagate through the matrix, and are detected on the other side (e.g., right side) of the structure. All input waves are of the same amplitude and frequency. The initial phases of the input waves are controlled by the generating elements. The output waves are the results of the spin wave interference within the matrix. The amplitude of the output wave depends on the initial phases and the magnetic states of the junctions.

junctions. The waveguides serve as a medium for spin wave propagation (spin wave buses). The buses can be made of a magnetic material such as $Y_3Fe_2(FeO_4)_3$ (YIG) or permalloy ($Ni_{81}Fe_{19}$), ensuring maximum possible group velocity and minimum attenuation for the propagating spin waves at room temperature. The nano-magnets placed on top of the waveguide junctions are the memory elements holding information encoded in the magnetization state. The nano-magnets can be designed to have two or several thermally stable states for magnetization, where the number of states defines the number of logic bits stored in each junction. The spins of the nano-magnet are coupled to the spins of the junction magnetic wires via the exchange and/or dipole–dipole coupling affecting the phase of the propagation of spin waves. The phase change received by the spin wave depends on the strength and the direction of the magnetic field produced by the nano-magnet. At the same time, the spins of the nano-magnet are affected by the local magnetization change caused by the propagating spin waves. The input/output ports are located at the edges of the waveguides. These ME elements are designed to convert the input electric signals into spin waves, and vice versa, convert the output spin waves into electric signals. The details of the read-in and read-out processes are presented in Khitun [33]. The incident spin wave beam is produced by the number of spin wave generating elements (e.g., by the elements on the left side of the matrix as illustrated in Figure 18.6b). All the elements are biased by the same radiofrequency (RF) generator exciting spin waves of the same frequency f and amplitude A_0, while the phase of the generated waves is controlled by the direct current voltages applied individually to each element. Thus, the elements constitute a phased array allowing one to artificially change the angle of illumination by providing a phase shift between the input waves. Propagating though the junction, spin waves accumulate an additional phase shift $\Delta\phi$, which depends on the strength and the direction of the local magnetic field provided by the nano-magnet, H_m. The output signal is a result of superposition of all the excited spin waves, which travelled along the different paths through the matrix. The amplitude of the output spin wave is detected by the voltage generated in the output element (e.g., either the inductive or the ME voltage produced by the spin waves at the output junction). The amplitude of the output voltage is within the ultimate values corresponding to the maximum when all the waves are coming in-phase (constructive interference), and the minimum when the waves cancel each other (destructive interference). The output voltage at each port depends on the magnetic states of the nano-magnets within the matrix and the initial phases of the input spin waves. To recognize the internal state of the magnonic memory, the initial phases are varied (e.g., from 0 to π). The ensemble of the output values obtained at the different phase combinations constitutes a hologram, which uniquely corresponds to the internal structure of the matrix.

The schematics of the experimental setup and the first experimental demonstration of a two-bit MHM prototype are shown in Figure 18.7 [34]. The magnetic matrix is a double-cross structure made of YIG epitaxially grown on gadolinium gallium garnet ($Gd_3Ga_5O_{12}$) substrate with (111) crystallographic orientation. The YIG film has a ferromagnetic resonance (FMR) line width of $2\Delta H \approx 0.5$ Oe, saturation magnetization $4\pi M_s = 1750$ G, and thickness $d = 3.6$ μm. This material is chosen due to its long spin wave coherence length and relatively low damping [7], which makes it the best candidate for room-temperature spin wave device prototyping. The length of the whole structure is 3 mm; the width of the arm is 360 μm. There are two micro-magnets on the top of the cross junctions. These magnets act as the memory elements, where logic bits are encoded into the two possible directions of their magnetization. There are six micro-antennas fabricated on the top of the YIG waveguides. These antennas are used to excite spin waves in YIG material and to detect the inductive voltage produced by the propagating spin waves. The input and the output micro-antennas are connected to a

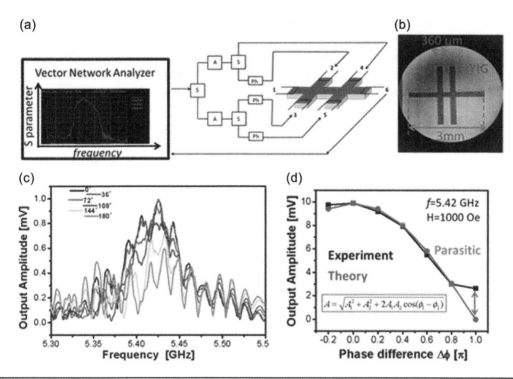

FIGURE 18.7 (a) The schematics of the experimental setup for MHM characterization. The device under study is a double-cross YIG structure with six micro-antennas fabricated on the edges. The input and the output micro-antennas are connected to the Hewlett-Packard 8720A Vector Network Analyzer (VNA). The VNA generates an input RF signal in the range from 5.3 to 5.6 GHz and measures the S parameters showing the amplitude and the phases of the transmitted and reflected signals. (b) Optical image of the YIG double-cross structure. The length of the structure is 3 mm, and the arm width is 360 μm. (c) Transmitted signal S_{12} spectra for the structure without micro-magnets. Two input signals are generated by the micro-antennas 2 and 3. The curves of different color show the output inductive voltage obtained for several values of the phase difference between the two interfering spin waves. (d) The slice of the data taken at the fixed frequency of 5.42 GHz (black curve). The red curve shows the theoretical values obtained by the classical equation for the two interfering waves.

Hewlett-Packard 8720A Vector Network Analyzer (VNA). The VNA generates an input RF signal and measures the S-parameters showing the amplitudes and phases of the transmitted and reflected signals. The prototype is placed inside an electro-magnet allowing variation in the bias magnetic field from −1000 to +1000 Oe. The input from the VNA is split between the four inputs via two power splitters, where the amplitudes of the signals are equalized by attenuators (step ±1 dB). The phases of the signal provided to Ports 3 and 4 are controlled by the two phase shifters (±2°). A photo of the YIG structure is shown in Figure 18.7b.

The graph in Figure 18.7c shows the raw data collected for the structure with just two micro-antennas. First, test experiments are carried out for the structure without magnets placed on the junctions. The graph shows the amplitude of the output inductive voltage versus the excitation frequency in the range from 5.30 to 5.55 GHz. The curves of different color depict the output signal amplitude measured for several values of the phase difference Δϕ between the inputs 2 and 3. This data shows that the output signal amplitude oscillates as a function of frequency and the phase difference between the two generated spin waves. The frequency dependence of the output is attributed to the effect of spin wave confinement within the structure, while the phase-dependent oscillations reveal the interference nature of the output signal. In Figure 18.7d, we show the slice of the data taken at the fixed frequency of 5.42 GHz. The experimental data is fitted well by the classical equation for two interfering waves. The only notable discrepancy is observed at $\Delta\phi = \pi$, where the experimental value is non-zero. This non-zero value can be explained by the presence of parasitic signals arising from

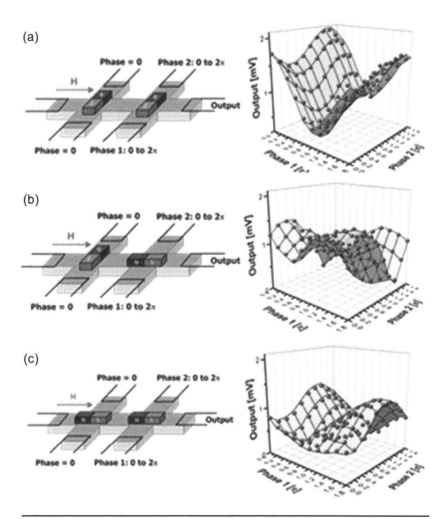

FIGURE 18.8 A set of three holograms obtained for the three configurations of the top micro-magnets as illustrated by the schematics in the left column: (a) both micro-magnets aligned perpendicular to the long axis; (b) the magnets orthogonal; and (c) both magnets parallel to the long axis. The red markers show the measured output voltage amplitude as a function of two input signal phases, while two other input signal phases are fixed. The cyan surface is a computer-reconstructed 3-D plot. The excitation frequency is 5.4 GHz; the bias magnetic field is 1000 Oe. All experiments are done at room temperature.

reflecting waves, direct coupling between the input/output ports, structure imperfections, and so on.

Figure 18.8 shows the set of three holograms obtained for the first two-bit MHM prototype. There are two magnets made of a cobalt magnetic thin film with a length of 1.1 mm, a width of 360 μm, and a coercivity of 200–500 Oe. Figure 18.8 shows the output voltage obtained for different magnet configurations: (a) two micro-magnets aligned in the same direction perpendicular to the long axis of the structure; (b) the magnets oriented in orthogonal directions; and (c) both magnets directed along the long axis. The red markers show the measured data (inductive voltage in millivolts) as a function of the phases of two input signals, while the phases of the other two input signals are kept fixed. The cyan surface is a computer-reconstructed three-dimensional (3-D) plot. The excitation frequency is 5.40 GHz; the bias magnetic field is 1000 Oe. All measurements are made at room temperature.

As one can see from Figure 18.8, the state of the micro-magnet significantly changes the output signal of MHM. The three holograms clearly demonstrate the unique signature defined by the magnetic state of the micro-magnet. The internal state of the holographic memory can be reconstructed by the difference in amplitude as well as the phase-dependent distribution of the output. The obtained experimental data do not show any significant signatures of thermal noise. The latter can be explained by taking into account that the flicker noise level in ferrite structures usually does not exceed −130 dBm [35]. At the same time, the direct coupling between the input/output antennas in our case is on the order of −70 to −80 dBm. The main challenge with further magnonic hologram development is associated with downscaling of the operation wavelength. The experimental data presented here are obtained for magnetostatic spin waves with a wavelength of hundreds of micrometers. Scaling down the wavelength to values below 100 nm makes exchange coupling the dominant mechanism (so-called *exchange spin wave*), which significantly changes the spin wave dispersion [36]. As of today, experimentally realized spin wave devices rely on magnetostatic spin waves (e.g., Bailleul et al. [37]; Covington et al. [38]) while devices based on propagation and interference of exchange spin waves remain largely unexplored.

18.3.2 Data Processing with Magnonic Holographic Devices

The correlation between the phases of the input spin waves and the output inductive voltage can be also exploited for building special-task data processing units. For instance, pattern recognition is a widely used procedure, which has multiple applications in text classification, speech recognition, radar processing, and biology [39–41]. Antivirus software installed on a computer system checks the incoming data string to see whether it matches one of the data strings (i.e., a computer virus) stored in memory. This task requires the exact matching of the input and stored data strings and can be performed by the general-type processor within an acceptable time frame. The same task can be accomplished more efficiently by using holographic devices [42]. Wave interference is the key mechanism allowing us to reconstruct/recognize a certain pattern, even if some part of the data is missing. In contrast to general-type logic processors, the increased number of missing bits does not result in computational overload for holographic-type devices. However, to date, several technological challenges are currently delaying the practical implementation of optical holographic devices [43]. One of the key obstacles is on-chip compatibility with conventional electronic integrated circuits; it may appear more practically feasible to use some other types of waves (e.g., spin waves) for building on-chip holographic co-processors.

18.3.2.1 Pattern Recognition

The photo and the schematics of the eight-terminal MH device for pattern recognition are shown in Figure 18.9. The core of the structure is a magnetic matrix comprising a 2×2 grid of magnetic waveguides with magnets placed on top of the waveguide junctions. The waveguides are made of single crystal 3.6 μm thick YIG film epitaxially grown on top of a gadolinium gallium garnet substrate using the liquid-phase transition process. After the films were grown, micro-patterning was performed by laser ablation using a pulsed infrared laser ($\lambda \approx 1.03$ μm) with a pulse duration of ~256 ns. The YIG matrix has the following dimensions: the length of each waveguide

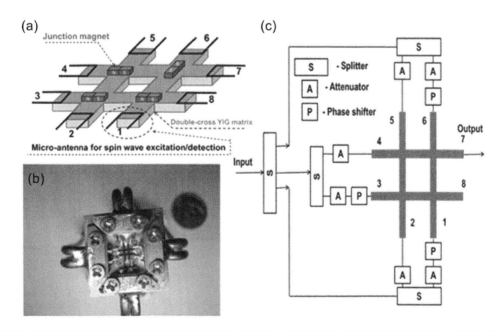

FIGURE 18.9 (a) Schematics of the eight-terminal MHM prototype made of YIG with four micro-magnets placed on the top of the cross junctions. The core of the structure is a magnetic matrix comprising a 2×2 grid of magnetic waveguides with magnets placed on top of the waveguide junctions. There are eight micro-antennas fabricated on the edges of each waveguide. (b) Optical image of the prototype device packaged. The YIG matrix has the following dimensions: the length of each waveguide is 3 mm; the width is 360 μm; and the YIG film thickness is 3.6 μm. The length of each magnet is 1.1 mm, the width is 360 μm. and the coercivity of each magnet is in the 200–500 Oe range. (c) Microwave measurement circuit schematics. The antennas are connected to a Hewlett-Packard 8720A Vector Network Analyzer (VNA) via a number of splitters (S), attenuators (A), and phase shifters (P).

is 3 mm; the width is 360 μm; and the YIG film thickness is 3.6 μm. The length of each magnet is 1.1 mm, the width is 360 μm, and the coercivity is in the 200–500 Oe range. There are eight micro-antennas fabricated at the edges of each waveguide. The antennas were fabricated from a gold wire and mechanically placed directly on top of the YIG film. Spin waves were excited by the magnetic field generated by the alternating (ac) electric current flowing through the antenna(s). The detection of the transmitted spin waves is via inductive voltage measurements, as described in Covington et al. [6]. The antennas are connected to a Hewlett-Packard 8720A VNA via a number of splitters/combiners with phase shifters and attenuators included in the system. The connection schematics are shown in Figure 18.9c. The VNA allowed the S-parameters of the system to be measured, showing both the amplitude of the signals as well as the phase of both the transmitted and reflected signals. Samples were tested inside a GMW 3472-70 electro-magnet system, which allowed the biasing magnetic field to be varied from –1000 to

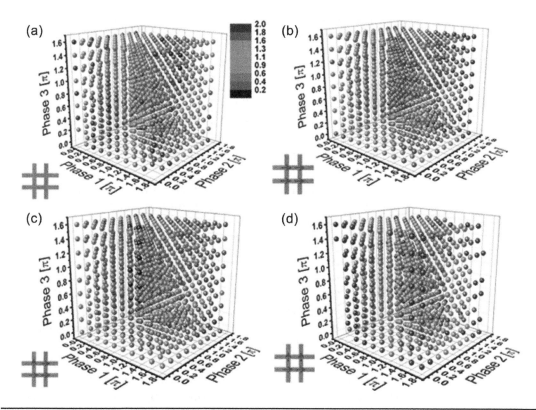

FIGURE 18.10 Experimental data obtained for four magnonic configurations of the system illustrated on the inset schematics: (a) YIG matrix without magnets; (b) matrix with all four magnets oriented parallel to each other; (c) one of the magnets rotated by 90°; (d) two of the four magnets rotated by 90°. The x-, y-, and z-axes in each plot represent the phases of the spin waves generated at input ports 1, 3, and 5, respectively. The level of the output voltage in millivolts is depicted by the color of the markers, with the black color representing lower-voltage outputs and blue color representing higher-voltage outputs. In all experiments, bias magnetic field of 1000 Oe was applied along the line connecting Ports 2 and 5. All experiments are performed at room temperature.

+1000 Oe. The input power is 12.5 µW per input port. All measurements are made at room temperature.

Figure 18.10 shows experimental data obtained for four magnonic matrix configurations with different orientations of the junction magnets [44]. The x-, y-, and z-axes in each plot show the phases of the input spin waves generated at the input ports 1, 3, and 5, respectively. The other three input antennas (labeled 2, 4, and 6) generate spin waves of identical constant phase. Hereafter, we take this phase to act as the reference (0) and define the relative phase change at Ports 1, 3, and 5 with respect to the reference one. The amplitudes of the spin waves generated by the six input antennas are adjusted by the attenuators attached to the inputs (see Figure 18.1) to equalize the transmitted signals at the output port. The inductive output voltage is detected at Port 7. Each plot in Figure 18.10 is a collection of data obtained by varying the three input phases, which appears as a cube in the 3-D phase space. The level of the output voltage is depicted by the color of the markers, with the red color representing lower-voltage outputs and blue color representing higher-voltage outputs. This is the inductive voltage generated in the micro-antenna at Port 7 by the time-varying magnetic flux caused by the propagating and interfering spin waves. The direct coupling between the input and output micro-antennas has been extracted by the standard procedure as described in REF. Covington et al. [6]; Chen et al. [45]; Counil et al. [46]. There are four plots in Figure 18.10. The first plot (Figure 18.10a) shows the output for the structure without magnets. Figures 18.10b–d show the output for specific configurations of micro-magnets. As shown in Figure 18.10, each of the magnet configurations produces a unique correlation between the input and the output. The positions of the maxima and minima (red and blue color markers) depend on the orientation of the junction magnets.

This correlation between the input spin wave phases and the output voltage amplitude is the basis of the pattern recognition procedure. We consider input information encoded into the phases (Phase 1, Phase 2, and Phase 3) of spin waves generated at Ports 1, 3, and 5, respectively. We employ 10 values (i.e., 0π, 0.2π, 0.4π, 0.6π, 0.8π, 1.0π, 1.2π, 1.4π, 1.6π, and 1.8π) of the input phases, which can represent the decimal numbers (0, 1, 2, 3, 4, 5, 6, 7, 8, 9). Thus, every input phase combination can be treated as a three-digit decimal number from 000 to 999, where the first digit corresponds to Phase 1, the second digit to Phase 2, and the third digit to Phase 3. For example, the phase combination of $(0\pi,0\pi,0\pi)$ corresponds to number 0, the phase combination of $(0\pi,0.2\pi,0\pi)$ corresponds to number 10, and the phase combination of $(0.4\pi,0.6\pi,0.8\pi)$ corresponds to number 234. We classify all possible input patterns as "recognized" or "non-recognized" by the level of the inductive voltage produced in the selected output ports (e.g. Port 7 in our experiments). For example, the pattern is "recognized" (i.e., stored in MHM) if $V_{ind}<V_r$, where V_r is the reference voltage (e.g., 0.6 mV). All patterns providing output voltage higher than the reference one are considered as "non-recognized" (i.e., not stored in MHM). This final step of pattern recognition requires an analog-to-digital output conversion, which can be accomplished by conventional electronic circuits and is a typical step for light-based techniques [47].

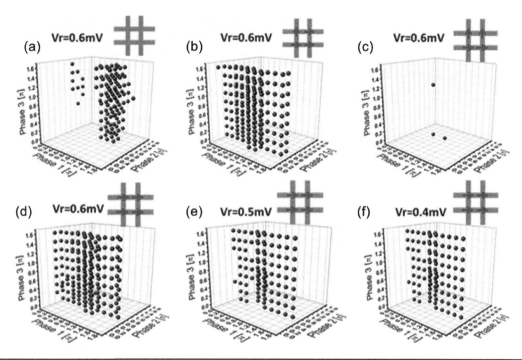

FIGURE 18.11 Experimental data converted to digital output. The black markers depict input phase combinations with output voltages lower than the reference. (a–d) show the digital output for the different magnet configurations obtained at the same reference voltage vr = 0.4 mV. (d–f) show the output for the same magnet configuration plotted for different reference voltages vr: (d) 0.4 mV, (e) 0.5 mV, and (f) 0.6 mV, respectively. The phase image expands for higher reference voltage and shrinks for a lower reference voltage.

To illustrate this procedure, we converted the color plots from Figure 18.10 into black and white plots shown in Figure 18.11. Figure 18.11a–d show the "recognized output" for the four different magnet configurations, where black markers correspond to the output voltages lower than 0.6 mV. The magnet configuration is schematically depicted on the top of each graph. Comparing the four graphs, one can observe the difference in the number of "recognized" states for each configuration. For instance, there are about 20 recognized states for the magnet configurations shown in Figure 18.11a, b, and d, while there are only three states in Figure 18.11c. Translating the phase combinations into numbers, one can determine the set of numbers kept in memory for each configuration (e.g., 31, 32, 35, 36, 37, 38, 39, … 938, 939 for Figure 18.11b; 97, 365, 565 for Figure 18.11c). There are other possible ways of coding information into the phase combinations. For instance, the three-digit numbers used in our example can be converted into a binary code where 0 and 1 correspond to the white and black pixels of some image. In this scenario, the MHM device can determine whether or not the input image is stored in the memory. It is important to note that the proposed technique allows us to recognize similar patterns (e.g., images that differ for some number of pixels) by decreasing/increasing the reference voltage. Figure 18.11e and f show the same data as in Figure 18.11d but taken with different reference voltages: vr = 0.4 mV, vr = 0.5 mV, and vr = 0.6 mV, respectively. The phase image expands for a higher reference voltage and shrinks for a lower reference voltage. In general, the change of the reference voltage is a powerful tool allowing us to identify similar patterns within a certain Hamming distance [48].

18.3.2.2 Prime Factorization

The use of the output signal phase in addition to its amplitude provides additional information that can be used for solving complex problems such as prime factorization. Prime factorization is the process of finding the set of prime numbers that multiply together to give the original integer N. The most naive approach to this problem is sequentially checking all possible products of all numbers from 2 to $\sqrt{N}$, where the number of operations increases exponentially with the increase of the input state N. Even the most efficient digital prime factoring algorithms (e.g., Fermat's factoring algorithm [49]) take an enormous number of operations to find the primes of sufficiently large N. This explains the extensive use of prime encoding in information security [50]. Quantum computing has emerged as a promising computational paradigm in part due to its ability to solve prime factorization problem more efficiently than with conventional digital computers. P. Shor has developed a polynomial-time quantum algorithm for the factoring problem, and its computational complexity has been proved to be $O((\log N)^2$ $(\log \log N)(\log \log \log N))$, which provides a fundamental advantage over any type of digital-type computing [51]. Although the fundamental advantage of quantum computing is undisputable, its practical realization is associated with multiple technological challenges required for quantum entanglement implementation. The prime factorization problem can be also solved by using classical wave interference. This approach has been intensively studied in optics [52]. Here, we present an example of solving the prime factorization problem by using a magnonic holographic device.

The factoring procedure includes three major steps. First, the general-type computer calculates the sequence of numbers $m^k \mathrm{mod}(N)$, where m is a randomly chosen positive integer, $k = 0,1,2,3,4,5,6....$ Second, the obtained numbers are converted into the spin waves in MHM. The waves are excited sequentially, whereby each new number in the modular sequence adds a spin wave to the interference pattern. The output detects the phase and the amplitude of the signal produced by the spin wave interference. In this scenario, the output of MHM oscillates with the period r, where r is the period of the calculated sequence of numbers. Finally, the primes are found by the general-type computer based on the obtained period r: $\gcd(m^{r/2} + 1, N)$ and $\gcd(m^{r/2} - 1, N)$. This procedure is illustrated in Figure 18.12.

For example, the procedure of finding the primes of the number 15 is the following. We choose $m = 7$ and calculate $7^k \mathrm{mod}(15)$ for $k = 0,1,2,3,...,20$. The calculations of the mod function are done by the general-type computer. The first calculated number is 7. It is converted into a spin wave with amplitude $A_1 = A_0$ and phase ϕ_1. The second calculated number is 4. It is a new number in the sequence, which is converted into a spin wave with amplitude $A_2 = A_0$ and phase $\phi_2 \neq \phi_1$. As the next two numbers (i.e., 13 and 1) are

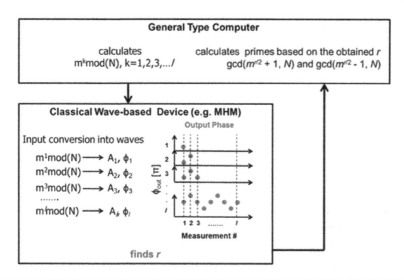

FIGURE 18.12 Illustration of the prime factorization procedure. The general-type computer calculates the sequence of numbers $m^k \bmod(N)$, where m is a randomly chosen positive integer, $k(0,1,2,3,4,5,6\ldots)$. Next, the obtained numbers are converted into the spin waves in the MHM device. The waves are excited sequentially, whereby each new number in the modular sequence adds a spin wave to the interference pattern. The output of MHM oscillates with the period r, where r is the period of the calculated sequence of numbers. Finally, the primes are found by the general-type computer based on the obtained period r: $\gcd(m^{r/2}+1, N)$ and $\gcd(m^{r/2}-1, N)$.

calculated, two more waves are added to the input. As in the previous cases, these waves have the same amplitude A_0, and phases ϕ_3 and ϕ_4, respectively. Thus, each distinct number in the modular function is converted into a spin wave with a distinct phase (i.e., $\phi_1 \neq \phi_2 \neq \phi_3 \neq \phi_4$). The fifth computation gives number 7, which has been already related to the spin wave of phase ϕ_1. In this case, the amplitude of the spin wave at Port 1 is doubled: $A_1 = \eta A_0$, where η is some positive number; and so on; every new number in the sequence adds a new spin wave signal, while the amplitude of the signal is multiplied by η every time the same number is found. The phase of the output oscillates with the same period as the period of the modular function. For instance, the phase of the output is the same for Step 4 (all antennas generate spin waves of amplitude A_0), Step 8 (all antennas generate spin waves of amplitude ηA_0), Step 12, and so on with period $r = 4$. That is the key idea of the proposed MHM approach for finding the period of the modular function by exploiting spin wave interference. Finally, the primes are found by the general-type computer based on the obtained period r: $\gcd(m^{r/2}+1, N)$ and $\gcd(m^{r/2}-1, N)$.

A photo of the MHM device and the schematics of the experimental setup are shown in Figure 18.13. Figure 18.13a shows the device packaged. The device consists of a double-cross structure (as shown in Figure 18.13b) made of YIG epitaxially grown on gadolinium gallium garnet substrate with crystallographic orientation (111). This material is chosen due to its long spin wave coherence length and relatively low damping [7], which make it the best candidate for room-temperature spin wave device prototyping. YIG film has

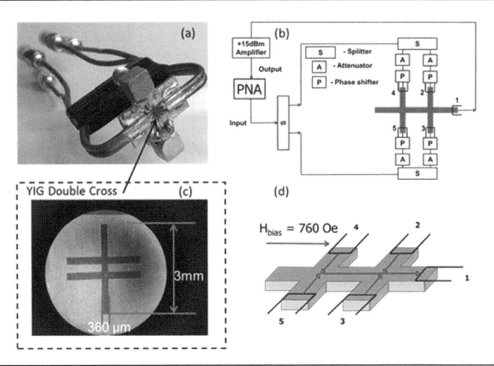

FIGURE 18.13   (a) Photo of the packaged device. (b) Photo of the double-cross $Y_3Fe_2(FeO_4)_3$ structure. The length of the structure is 3 mm, the width of the arm is 360 μm, and the thickness of the YIG film is 3.6 μm. (c) The schematics of the experimental setup. There are five micro-antennas fabricated on top of the YIG structure. The antennas are connected to the programmable network analyzer (PNA) via the set of splitters (depicted as S). Four antennas (#2, 3, 4, and 5) are to excite spin waves, and one antenna (#1) is to detect the inductive voltage produced by the spin wave interference. There is a system of phase shifters (depicted as P) and attenuators (depicted as A) to control the phase and the amplitude of the spin wave signals. (d) Schematics of the YIG structure with five micro-antennas. The device is placed inside an electro-magnet producing bias magnetic field of 760 Oe.

FMR linewidth $2\Delta H \approx 0.5$ Oe, saturation magnetization $4\pi M_s = 1750$ G, and thickness = 3.6μm. The length of the whole structure is 3 mm; the width of the arm is 360 μm. There are six micro-antennas fabricated on the top of the YIG waveguides. These antennas are aimed to excite spin waves and to detect the inductive voltage produced by the interfering spin waves. The excitation of the spin waves is accomplished by passing ac electric current through the antenna contour, which in turn, generates an ac magnetic field and may excite propagating spin waves in the waveguide under the resonance conditions. The inductive voltage is a result of the spin wave interference, where the propagating spin waves alter the magnetic flux from the surface. The inductive voltage has maxima when spin waves are coming in-phase and shows minima when spin waves are coming out-of-phase. The set of equations connecting the amplitudes and the phases of multiple spin waves and the resultant inductive voltage can be found in Gertz [53]. Both the phase and the amplitude of the inductive voltage are measured at the output. The details of the inductive voltage measurements using micro-antennas can be found in International Technology Roadmap for Semiconductors [30].

In this experiment, just one antenna was used for the output voltage detection (i.e., Antenna #1 in Figure 18.13c), and four antennas for spin wave excitation (i.e., antennas numbered 2, 3, 4, and 5 in Figure 18.13c). The input and the output micro-antennas are connected to a Keysight Technologies programmable network analyzer (PNA) N5221A-217. The PNA generates an input RF signal and measures the S_{12} parameters showing the amplitude and the phase of the transmitted signal. The input from the PNA is split among the four inputs via the system of splitters, phase shifters, and attenuators as shown in Figure 18.2c. The output voltage from Antenna #1 is amplified +15 dBm. The YIG device is placed inside an electro-magnet allowing variation in the bias magnetic field from −1000 to +1000 Oe. The external magnetic field is directed in-plane as shown in Figure 18.12b. Before the experiment, the region in the frequency-bias magnetic field space where both backward volume magnetostatic spin waves (BVMSW) and magnetostatic surface spin waves (MSSW) can propagate was found. The latter is critically important for the operation of the cross-shape magnonic devices, as the input spin waves initially propagate perpendicular to the bias field (MSSW), while they reach the output propagating along the external magnetic field (BVMSW). The most prominent overlap takes place at the frequency 4.165 GHz and 760 Oe bias magnetic field. All experimental data were collected at this particular combination.

Spin waves were sequentially excited by the four antennas and detected the phase and the amplitude of the output inductive voltage at Port 1. Prior to the experiment, we used the system of attenuators as depicted in Figure 18.13b to equalize the output inductive voltage produced by the independently working antennas (i.e., Ports #2, 3, 4, and 5). We set up the following phase difference between the input ports: 0π (Port #2), $\pi/3$ (Port #3), $2\pi/3$

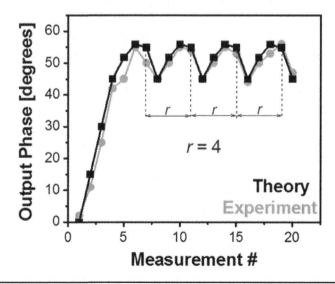

FIGURE 18.14 Experimental and theoretical data showing the phase of the output signal for the 20 sequential measurements. The red markers and red curve show the experimental data. The black markers and the curve show the theoretical data for the ideal case with zero input phase/amplitude variations.

(Port #4), and $\pi/2$ (Port #5). Then, we carried out 20 sequential measurements. During this experiment, we changed the amplitude of the spin wave generating antenna. For instance, the amplitude of the spin wave signal at Port #2 was reduced approximately by a factor two ($\eta = 0.5$) at the measurement number 5. In Figure 18.14, we plotted the phase of the output for the 20 sequential measurements (red markers and red curve). As expected, the phase of the output voltage oscillates with a period $r = 4$, which is the period of the calculated modular function 7,4,13,1,7,4,13,1.... The latter allows us to find the primes of number 15: 5 and 3 (i.e., greatest common divisor [gcd] ($7^{4/2} + 1$, 15) and gcd($7^{4/2} - 1$, 15)). More details on the experiment can be found in Khivintsev et al. [54]. It should be clear that the classical wave-based logic circuits cannot compete with quantum computers in efficiency. However, the use of these devices may complement conventional digital logic circuits in special-type data processing (i.e., prime factorization).

18.4 SUMMARY

In this chapter, we considered different ways of using spin waves for building logic gates. Boolean-type spin wave devices hold the potential for building scalable and ultra-low power logic circuits. In this approach, spin waves are used for information transfer between the input and the output multiferroic cells. The switching of the output cells depends on the phase of the input spin wave. The latter enables data processing during spin wave propagation by implementing phase shifters and by using spin wave interference (e.g., MOD and MAJ gates). In turn, it may be possible to build some types of logic circuits with a smaller number of components than required for transistor-based designs [55]. The latter may lead to a functional throughput advantage over the scaled CMOS [56]. However, it is difficult to expect that magnetic logic circuits will ever replace electronic digital logic circuits. It is more feasible to combine the advantages of both approaches by building hybrid logic architectures with both magnetic and electronics elements. Non-volatile magnetic components may be of great practical value for implementation in field programmable gate arrays (FPGA) and other types of architectures with built-in memory.

Non-Boolean magnonic devices for special-task data processing promise a significant functional throughput enhancement over the scaled CMOS. According to the estimates [33], the functional throughput of MHM may exceed 10^{18} bits/s/cm^2, which is orders of magnitude higher than for any general single-core processor. Holographic logic units can be used for solving the NP class of problems (i.e., prime factorization). The efficiency of holographic computing with classical waves is intermediate between digital logic and quantum computing, allowing us to solve a certain class of problems fundamentally more efficiently without the need for quantum entanglement. Image recognition and processing are among the other potential applications for magnonic holographic devices able to process a larger number of bits/pixels in parallel within a single core.

REFERENCES

1. International Technology Roadmap for Semiconductors, Emerging Research Devices, www.itrs.net/Links/2013ITRS/2013Chapters/2013ERD.pdf, 2013.
2. R. J. Elliott, Theory of the effect of spin-orbit coupling on magnetic resonance in some semiconductors, *Phys. Rev.* **96**, 266 (1954).
3. M. I. D'yakonov and V. I. Perel, Spin relaxation of conduction electrons in noncentrosymmetric semiconductors, *Sov. Phys. Semicond.* **13**, 3581 (1971).
4. G. L. Bir, A. G. Aronov, and G. E. Pikus, Spin relaxation of electrons due to scattering by holes, *Sov. Phys. JETP* **69**, 1382 (1975).
5. P. Zhang and M. W. Wu, Electron spin diffusion and transport in graphene, *Phys. Rev. B* **84**, 045304 (2011).
6. M. Covington, T. M. Crawford, and G. J. Parker, Erratum: Time-resolved measurement of propagating spin waves in ferromagnetic thin films, *Phys. Rev. Lett.* **89**, 237202 (2002).
7. A. A. Serga, A. V. Chumak, and B. Hillebrands, YIG magnonics, *J. Phys. D Appl. Phys.* **43**, 264002 (2010).
8. M. P. Kostylev, A. A. Serga, T. Schneider, B. Leven, and B. Hillebrands, Spin-wave logical gates, *Appl. Phys. Lett.* **87**, 153501 (2005).
9. T. Schneider, A. A. Serga, B. Leven, and B. Hillebrands, Realization of spin-wave logic gates, *Appl. Phys. Lett.* **92**, 022505 (2008).
10. K.-S. Lee and S.-K. Kim, Conceptual design of spin wave logic gates based on a Mach-Zehnder-type spin wave interferometer for universal logic functions, *J. Appl. Phys.* **104**, 053909 (2008).
11. A. Khitun and K. Wang, Nano scale computational architectures with Spin Wave Bus, *Superlattices Microstruct.* **38**, 184 (2005).
12. A. V. Chumak, V. I. Vasyuchka, A. A. Serga, and B. Hillebrands, Magnon spintronics, *Nat. Phys.* **11**, 453 (2015).
13. V. V. Kruglyak, S. O. Demokritov, and D. Grundler, Magnonics, *J. Phys. D Appl. Phys.* **43**, 264001 (2010).
14. B. Lenk, H. Ulrichs, F. Garbs, and M. Munzenberg, The building blocks of magnonics, *Phys. Rep.* **507**, 107 (2011).
15. P. Shabadi, A. Khitun, P. Narayanan et al., Towards logic functions as the device, in *Proceedings of the Nanoscale Architectures (NANOARCH), 2010 IEEE/ACM International Symposium*, pp. 11–16, 2010.
16. A. Khitun and K. L. Wang, Non-volatile magnonic logic circuits engineering, *J. Appl. Phys.* **110**, 034306 (2011).
17. A. Khitun, Multi-frequency magnonic logic circuits for parallel data processing, *J. Appl. Phys.* **111**, 054307 (2012).
18. A. Khitun, M. Bao, and K. L. Wang, Spin wave magnetic nanofabric: A new approach to spin-based logic circuitry, *IEEE Trans. Magn.* **44**, 2141 (2008).
19. S. Dutta, S.-C. Chang, N. Kani et al., Non-volatile clocked spin wave interconnect for beyond-CMOS nanomagnet pipelines, *Sci. Rep.* **5**, 9861 (2015).
20. A. V., Chumak, A. A. Serga, and B. Hillebrands, Magnon transistor for all-magnon data processing, *Nat. Commun.* **5**, 4700 (2014).
21. Y. Wu, M. Bao, A. Khitun, J.-Y. Kim, A. Hong, and K. L. Wang, A three-terminal spin-wave device for logic applications, *J. Nanoelectron. Optoelectron.* **4**, 394 (2009).
22. S. Klingler, P. Pirro, T. Brächer, B. Leven, B. Hillebrands, and A. V. Chuma, Design of a spin-wave majority gate employing mode selection, *Appl. Phys. Lett.* **105**, 152410 (2014).
23. A. Khitun, D. E. Nikonov, and K. L. Wang, Magnetoelectric spin wave amplifier for spin wave logic circuits, *J. Appl. Phys.* **106**, 123909 (2009).
24. J. Zhu, J. A. Katine, G. E. Rowlands et al., Voltage-induced ferromagnetic resonance in magnetic tunnel junctions, *Phys. Rev. Lett.* **108**, 197203 (2012).
25. T. Nozaki, Y. Shiota, S. Miwa et al., Electric-field-induced ferromagnetic resonance excitation in an ultrathin ferromagnetic metal layer, *Nat. Phys.* **8**, 491 (2012).

26. Y-J. Chen, H. K. Lee, R. Verba et al., Parametric resonance of magnetization excited by electric field, *Nano Lett.* **17**, 572 (2016).
27. S. Cherepov, P. Khalili Amiri, J. G. Alzate et al., Electric-field-induced spin wave generation using multiferroic magnetoelectric cells, in *Proceedings of the 56th Conference on Magnetism and Magnetic Materials (MMM 2011), DB-03*, Scottsdale, Arizona, 2011.
28. A. Khitun et al., Demonstration of Majority Logic Gate with Spin-waves, *Annual Report of Western Institute of Nanoelectronics*, Abstract 2.1 (2009).
29. T. K. Chung, S. Keller, and G. P. Carman, Electric-field-induced reversible magnetic single-domain evolution in a magnetoelectric thin film, *Appl. Phys. Lett.* **94**, 132501 (2009).
30. International Technology Roadmap for Semiconductors. Semiconductor Industry Association, 2007.
31. A. Chen, Private communication, 2010.
32. P. Ambs, Optical computing: A 60-year adventure, *Adv. Opt. Technol.* **12**, 1 (2010).
33. A. Khitun, Magnonic holographic devices for special type data processing, *J. Appl. Phys.*, **113**, 164503 (2013).
34. F. Gertz, A. Kozhevnikov, Y. Filimonov, and A. Khitun, Magnonic holographic memory, *IEEE Trans. Magn.* **51**, 4002905 (2015).
35. E. Rubiola, Y. Gruson, and V. Giordano, On the flicker noise of ferrite circulators for ultra-stable oscillators, *IEEE Trans. Ultrason. Ferroelectr. Freq. Control* **51**, 957 (2004).
36. B. A. Kalinikos and A. N. Slavin, Theory of dipole-exchange spin-wave spectrum for ferromagnetic-films with mixed exchange boundary-conditions, *J. Phys. C Solid State Phys.* **19**, 7013 (1986).
37. M. Bailleul, D. Olligs, C. Fermon, and S. Demokritov, Spin waves propagation and confinement in conducting films at the micrometer scale, *Europhys. Lett.* **56**, 741 (2001).
38. M. Covington, T. M. Crawford, and G. J. Parker, Time-resolved measurement of propagating spin waves in ferromagnetic thin films, *Phys. Rev. Lett.* **89**, 237202 (2002).
39. S. Tong and D. Koller, Support vector machine active learning with applications to text classification, *J. Mach. Learn. Res.* **2**, 45 (2002).
40. A. K. Jain, R. P. W. Duin, and J. C. Mao, Statistical pattern recognition: A review, *IEEE Trans. Pattern Anal. Mach. Intell.* **22**, 4 (2000).
41. W. Al-Nuaimy, Y. Huang, M. Nakhkash, M. Fang, V. Nguyen, and A. Eriksen, Automatic detection of buried utilities and solid objects with GPR using neural networks and pattern recognition, *J. Appl. Geophys.* **43**, 157 (2000).
42. B. M. Watrasie, Character recognition by holography, *Nature* **216**, 302 (1967).
43. D. A. B. Miller, Are optical transistors the logical next step?, *Nat. Photonics* **4**, 3 (2010).
44. A. Kozhevnikov, F. Gertz, G. Dudko, Y. Filimonov, and A. Khitun, Pattern recognition with magnonic holographic memory device, *Appl. Phys. Lett.* **106**, 142409 (2015).
45. Y.-C. Chen, D.-S. Hung, Y.-D. Yao, S.-F. Lee, H.-P. Ji, and C. Yu, Ferromagnetic resonance study of thickness-dependent magnetization precession in $Ni_{80}Fe_{20}$ films, *J. Appl. Phys.* **101**, 09C104 (2007).
46. G. Counila, J.-V. Kim, T. Devolder, P. Crozat, and C. Chappert, Magnetic anisotropy of epitaxial MgO/Fe/MgO films studied by network analyzer ferromagnetic resonance, *J. Appl. Phys.* **98,** 023901 (2005).
47. O. Golani, L. Mauri, F. Pasinato et al., A photonic analog-to-digital converter using phase modulation and self-coherent detection with spatial oversampling, *Opt. Expr.* **22**, 12273 (2014).

48. T. Koide, H. R. Mattausch, Y. Yano, T. Gyohten, Y. Soda, A nearest-Hamming-distance search memory with fully parallel mixed digital-analog match circuitry, *ASP-DAC 203: Proceedings of the Asia and South Pacific Design Automation Conference*, 591–592, 2003.

49. R. S., Lehman, Factoring large integers, *Math. Comput.* **28**, 637 (1974).

50. Q. Zheng, H. Chen, and G. Chen, The principle and implementation of a public-key cryptosystem and the RSA scheme, *Mini-Micro Syst.* **9**, 16 (1988).

51. P. Shor, Polynomial-time algorithms for prime factorization and discrete logarithm problems, *SIAM J. Comput.* **26**, 1484 (1997).

52. A. A. Rangelov, Factorizing numbers with classical interference: Several implementations in optics, *J. Phys. B Atom. Mol.* **42**, 021002 (2009).

53. F. Gertz, A. V. Kozhevnikov, Y. A. Filimonov, D. E. Nikonov, and A. Khitun, Magnonic holographic memory: From proposal to device, *IEEE J. Explor. Solid-State Comput. Devices Circuits* **1**, 67 (2015).

54. Y. Khivintsev, M. Ranjbar, D. Gutierrez et al., Prime factorization using magnonic holographic devices, *J. Appl. Phys.* **120**, 123901 (2016).

55. A. Khitun and K. L. Wang, Non-volatile magnonic logic circuits engineering, *J. Appl. Phys.* **110**, 034306 (2011).

56. A. Sarkar, D. E. Nikonov, I. A. Young, B. Behin-Aein, and S. Datta, Charge-resistance approach to benchmarking performance of beyond-CMOS information processing devices, *IEEE Trans. Nanotechnol.* **13**, 143 (2014).

Index

PI Group (UK) Ltd, Croydon, CR0 4YY

/10/2024

95-0015